journey into light

JOURNEY INTO LIGHT

life and science of C.V. Raman

g. venkataraman

published by
INDIAN ACADEMY OF SCIENCES
in co-operation with
INDIAN NATIONAL SCIENCE ACADEMY

Indian Academy of Sciences
P.B. No. 8005, Sadashivanagar
Bangalore 560 080, India

Telephone: 342546   Telex: 0845-2178 ACAD IN
Telefax: 0812 340492

Distributed by
Oxford University Press
YMCA Library Building, Jai Singh Road, New Delhi 110001
Bombay Calcutta Madras
Oxford New York Toronto
Melbourne Tokyo Hong Kong

© 1988 Indian Academy of Sciences
First published 1988

ISBN 81-85324-00-X

Typeset and printed at Thomson Press (India) Ltd
Faridabad, India

In fond memory of Suresh

Preface

I do not remember when I first heard about Raman, but it certainly must have been when I was quite young. Not only was Raman a household name during that period in all South Indian middle class families, but fond relatives often insisted on calling me G. V. Raman, and predicting for me (unjustifiably!) a brilliant future. In addition, a distant uncle was a Professor in the Indian Institute of Science when Raman was its Director, and the India Meteorological Department where my father worked had many students of Raman, several of whom were our family friends. Prominent among them was Raman's brother C. Ramaswamy whom I got to know quite well when he and my father served together in Karachi during the War years.

In college, Raman was no longer just a name or even the winner of the Nobel Prize. He was the teacher of one of my teachers (Dr P. S. Srinivasan), he was the product of a rival college, and he had theories of the violin and of specific heats which were possible examination questions. Slowly his personality began to captivate me. A couple of years later when I spent a few fleeting months at the Indian Institute of Science, Bangalore, I often used to pass by the Raman Research Institute but was intimidated by the warning sign asking visitors not to disturb. The early phase of my research involved a study of phonons by neutron scattering, and inevitably I became interested in Raman's theory of lattice vibrations. I wrote for and received a whole collection of reprints (which I still have). Raman's theory was controversial, and this triggered even more my curiosity about the man. Over the years I have talked to many of his associates and students, and collecting Raman stories became a favourite pastime.

In spite of the deep fascination he had for me, I somehow never created any opportunity for meeting him face to face. Looking back I realize how easily I could have achieved this through the good offices of several people who knew Raman. It is a matter of eternal regret for me that I let slip several such opportunities. The only time I saw him was at a lecture he delivered at the Presidency College, Madras, in 1957. That was the year of the Centenary of the Madras University, and I was in Madras managing a science exhibition put up by the Department of Atomic Energy. A classmate of mine who was then a Demonstrator in Physics in the Presidency College invited me to the lecture. I sat there spellbound. As we were coming out after the lecture, I overheard someone saying that the lecture was uninteresting. I was stunned, for I had just heard a masterly exposition on the long-standing problem of superconductivity. (It would be just a short while before Bardeen, Cooper and Schriffer cracked the problem.) My friend then explained to me that many members of the audience were drawn from the humanities departments and they usually came because Raman was well known for his colourful dismissal of his rivals and their theories! On this occasion he had dealt with a subject far removed from his normal preoccupations, which explained its lack of appeal to the non-scientific section of the audience. For me, it was a revelation of Raman as a teacher *par excellence.*

Initially I had no intention of writing this biography. The thought simply did not cross my mind since there were many others better qualified, who not only knew Raman personally but had even worked with him. Thus, when I was first approached I was not only surprised but also quite hesitant. However, subsequent discussions with several people, in particular Professor Ramaseshan, made me change my mind. I allowed myself to be persuaded by the argument that, not having known Raman personally, I would be free from bias. As I began to collect material in preparation for the writing, I soon found that Raman's life was far from the fairy tale it is often projected to be in popular articles and booklets. Instead I discovered to my utter astonishment that it was an epitome of the trials and tribulations of Indian science itself, and all of a sudden his life began to acquire a deeper meaning and significance. It also became clear to me that there are many lessons to be learnt from Raman's life even as we are groping today for excellence in our science. As the reader would observe, these thoughts have influenced my treatment a great deal. In a sense, the story of Raman is the story of science in modern India.

When I actually got down to the task of writing the biography, there was the question of whom it should be addressed to. Considering that the book had been commissioned by our two Academies, it clearly had to be a scientific biography with appeal to the entire community of scientists. Accordingly I have endeavoured to treat Raman's discoveries somewhat at the level of *Scientific American.* At the same time, keeping in mind the fact that Raman was a physicist, I have occasionally included some technicalities that would interest physicists. Such sections, marked with a star ☆ (or portions between two stars), may be skipped by non-specialists without detriment. In his day, Raman was an important public figure who played a seminal role in the development of science in India. For this reason, I have devoted a considerable amount of attention to the life of Raman as well. One hopes that this would be of special interest to the younger generation. Old-timers may consider some of the background material included in the earlier chapters as rather excessive. My own view is that this would permit a better appreciation of those times as well as of the man. A whole generation has grown up which scarcely realizes that India *was* a colony and that life was *very* different (at least in some respects) in those days. As suggested by a friendly critic, I have included notes wherever appropriate, especially where historical events or personalities are involved. These should also be useful to the foreign readers (of whom one hopes there would be many, considering the wide use of the Raman effect).

It is obvious that the task attempted here could not have been accomplished without generous help and assistance from several individuals as well as institutions. Since the list is long, I make the acknowledgements separately. Here I would just like to say one final word to the reader before he or she begins a perusal of the book. In many respects Raman was like a child and yet he was also a complex personality. He lived and worked in a country which was in bondage for the better part of his life, and half a world away from where the action was. And there were no means of rapid

Preface

communication then as we have now. Raman had many faults but it is not my job to judge him; that is the business of history. I have attempted only a portrait, a task which enabled me to live briefly in those exciting but turbulent days. That in itself has been a rich and satisfying experience. My satisfaction would be complete if my portrayal is fair.

> *Turning the accomplishment of many years*
> *Into an hour-glass: for the which supply,*
> *Admit me Chorus to this history;*
> *Who, prologue-like, your humble patience pray,*
> *Gently to hear, kindly to judge, our play.*

(Henry V, Act I)

April 1988
Hyderabad

G. VENKATARAMAN

ACKNOWLEDGEMENTS

This is my third book, but it has been an altogether different experience. Unlike on earlier occasions, I owe much to many, and it is a pleasure to record my gratitude. Naturally I must start with expression of thanks to the Indian National Science Academy and the Indian Academy of Sciences for inviting me to write this biography. C. N. R. Rao, the moving spirit behind the invitations, has subsequently provided much encouragement in his capacity as the Editor of Publications of the Indian Academy of Sciences.

As declared in the Preface, the invitations notwithstanding, it was S. Ramaseshan who helped me overcome my initial diffidence. Actually, Ramaseshan has been a pillar of strength throughout this venture and, adapting the words of Raman, it would require more than a mere acknowledgement to express the diverse ways in which I am indebted to him. All I can say to you, Professor, is a huge THANK YOU! Several people (besides Ramaseshan of course) have been kind enough to share their experiences with and their perception of Raman, principal among them being S. Bhagavantam, S. Chandrasekhar (of Chicago), A. Jayaraman, C. Ramaswamy and B. V. Thosar.

When I started on the literature survey, I was initially thrilled by the fact that Raman had published his first paper even while in college but when I began to hunt for that paper my thrill turned into dismay; digging as far back as the turn of the century is no easy task! My problem was eventually solved by A. Ratnakar who not only produced this and all the other reprints I subsequently wanted, but also anticipated my needs and drew my attention to many I was not even aware of.

Raman's papers are exciting to read but not always easy to understand, and where I floundered, I have been helped out by M.C. Valsakumar and to some extent by V.C. Sahni. The manuscript was prepared while I was at the Indira Gandhi Centre for Atomic Research, and I am grateful to that organization for the support extended and the facilities provided. Particular thanks are due to C. V. Sundaram who greatly stimulated my thoughts during numerous lunch-hour discussions.

The typescript owes much to the diligence of P. Subba Rao and T.D. Sundarakshan, particularly the latter, who literally worked with demonic energy. The graphics are substantially the handiwork of P. Harikumar, while the computer graphics were provided by G. Athithan and Sunder Kingsley. Naga Nirmala and Anand Parthasarathy simplified index preparation by skilfully harnessing the computer. Artistic embellishments are due to Kamalesh, working with whom has been a delight and an education. To my good friend R. P. Riesz, I am not only indebted for the beautiful colour reproductions of some of Raman's experiments, but even more so for his stubborn insistence that the cover should not feature a "Nature fake". He also read through portions of the book, particularly the sections dealing with optics.

A large number of publishers and institutions have been kind enough to permit quotation and reproduction of source material, acknowledgements for which are being made separately. However, special mention must be made of the kind efforts of R. W. Cahn in obtaining for me access to the Born–Rutherford correspondence. The photographs, both black and white and colour, are part of a collection made over several years at the Raman Research Institute. Most of them appear in the book *C. V. Raman in Pictures* by S. Ramaseshan, in which the credits for individual photographs are given. They appear in this book thanks to the generosity of the Raman Research Institute, P. Ramachandra Rao providing useful assistance in organizing the compilation. The delightful cartoon on the back cover is the creation of the incomparable R. K. Laxman whom it is a pleasure to thank.

Having worked closely with publishing staff for the first time, I realize what a heavy load the publication staff of the Indian Academy of Sciences carries. I am therefore most grateful to them for the help rendered to me. Messrs Thomson Press, the printers, have throughout been most obliging and co-operative.

The task of preparing the manuscript and seeing it through print took over twenty months, during which period I have received staunch support from all members of my family. (In a sense, Raman also became one of us.) My wife Saraswati, in particular, has provided invaluable encouragement and understanding, especially during the last few critical months when my concentration has tended to falter.

The title is derived from an unusual exhibition of paintings (inspired by the Raman effect!) held by the well-known artist M. F. Hussain.

October 1988
Hyderabad

G. VENKATARAMAN

SOURCE CREDITS

Acknowledgement is made for material taken from sources for which the following publishers/organizations hold copyright.

Academic Press Inc. (London) Ltd and M.V. Berry, for figures 9.7, 9.8, 9.12, 9.13 and 9.14, and some of the mathematical aspects of the Raman–Nath theory in Chapter 9, from *The Diffraction of Light by Ultrasound* by M.V. Berry (1966). Addison-Wesley Publishing Co., Reading, Mass., USA, for quotations on pages 79 and 490, from *Feynman's Lectures on Physics* (1964). American Institute of Physics, New York, for quotations on pages 65, 70, 71–72 and 75–76, and figures 4.3, 4.8, 4.52, 4.54, 4.81, 4.82, 8.36 (G. Shirane and Y. Yamada, 1969), 10.10 (J.L. Warren *etal.*, 1966) and 10.11 (S.A. Solin and A.K. Ramdas, 1970) from *Physical Review*; figures 9.10 and 9.11 (A. Alippi *etal.*, 1975) from *Applied Physics Letters*; quotations on pages 205, 206, 211 and 214–215 (I.L. Fabelinskii, 1978) from *Soviet Physics Uspekhi*.

Acknowledgements

Andhra Pradesh Academy of Sciences, Hyderabad, and S. Bhagavantam, for quotations on pages 351, 418, 461, 464, 470 and 471 from *Professor Chandrasekhara Venkata Raman* by S. Bhagavantam (1972). Bharatiya Vidya Bhavan, Bombay, for quotations on pages 468–469, 470–471 and 517–518 from *Bhavan's Journal* (December 1970) Calcutta University Press, for quotations on pages 102–104 from *Sir Asutosh Mookerjee Silver Jubilee Volume*, vol. 2 (1922); quotation on pages 192–193 from *Molecular Diffraction of Light* by C.V. Raman (1922). Cambridge University Press, for quotations on pages 68, 126–127 and 139 from *Scientific Papers of Lord Rayleigh* (1900–1920). Chicago University Press, for quotations on pages 139–140 and 173 from *Astrophysical Journal*. Clarendon Press, Oxford, for figure 10.9 from *Dynamical Theory of Crystal Lattices* by M. Born and K. Huang (1954); Clarendon Press and Abraham Pais for quotations on pages 193, 513 and 520 from '*Subtle is the Lord...*' *The Science and the Life of Albert Einstein* by Abraham Pais (1982). Current Science Association, Bangalore, for quotations on pages 87, 128, 283–284, 306, 330, 340, 349, 408–409, 420–421, 456, 508 and 509, and figure 12.27 (S. Ramaseshan and R. Nityananda, 1986) from *Current Science*. Dover Publications Inc., New York, for quotations on pages 79 and 99 from *On the Sensations of Tone* by H. Helmholtz, transl. by A. Ellis (1954); figure 4.25 from *The Theory of Sound* by Lord Rayleigh (reprint 1945). Dowden, Hutchinson and Ross, Philadelphia, for quotation on page 110 and figure 4.25 from *Musical Acoustics, Part I*, edited by C.M. Hutchins (1975). Heyden, London, for quotations on pages 215–216, 251, 344 and 389 from *Proceedings of the Sixth International Conference on Raman Spectroscopy*, edited by E.D. Schmid et al. (1978). *The Hindu*, Madras, for quotation on pages 514–515. Indian Association for the Cultivation of Science, Calcutta, for quotations on pages 30, 31, 32, and 33 from *A Century* (1975); quotations on pages 49 and 50 from the Annual Reports of the IACS; quotations on pages 84–85 and 88, and figures 4.16, 4.17, 4.18, 4.20 and 4.37 from *Bulletin of the IACS*; quotations on pages 94–95, 108, 151, 152 and 179–180, and figures 4.21, 4.22, 4.23, 4.24, 4.33, 4.34, 4.62, 4.85, 8.2 and 8.3 from *Proceedings of the IACS*; quotations on pages 234 and 237–238, and figures 4.26 and 4.63 from *Indian Journal of Physics*. Indian Institute of Science, Bangalore, for quotations on pages 11, 26–27, 38, 182, 208, 215, 265, 465 and 471 from *C.V. Raman Memorial Lecture 1978* by S. Ramaseshan. Indian Science Congress Association, for quotations on pages 58 and 196. Macmillan Magazines Ltd, London, for quotations on pages 37, 65, 70, 156, 171, 173, 187, 194, 236, 240, 331 and 332–333, and the full papers (letters) of C.V. Raman on pages 206–208 and 213–214 from *Nature*. McGraw-Hill Book Co., New York, and D.A. Long, for figures 7.1, 7.3, 7.4c and 7.7 from *Raman Spectroscopy* by D.A. Long (1977). National Research Council of Canada, Ottawa, for fig. 9.6 (F.H. Sanders, 1936) from *Canadian Journal of Research*. Nehru Memorial Museum and Library, New Delhi, for quotations on pages 457, 458, 461, 462 and 463–464 from *Jawaharlal Nehru on Science-Speeches delivered at the Annual Sessions of the Indian Science Congress* by Baldev Singh (1986). Nobel Foundation, Stockholm, for quotation on page 196 from the Nobel lecture of C.V. Raman (1930). Noordhoff, Leiden, for figure 8.39 (Steigmeier et al.) from *Anharmonic Lattices, Structural Transitions and Melting*, edited by T. Riste (1974). Optical Society of America, Washington DC, for quotation on page 177 and figure 4.83 from *Journal of the Optical Society of America*. Orient Longman, Bombay, for quotation on pages 457–458 from *Collected Works of Meghnad Saha*, edited by S. Chatterjee (1987). Oxford University Press, New Delhi, for quotation on page 518 from *Jawaharlal Nehru–A Biography* by Sarvepalli Gopal. Philosophical Library, New York, for extracts on pages 474–494 from Raman's radio talks published in book form as *The New Physics: Talks on Aspects of Science* (1951). Physical Society, London, for quotations on pages 50, 147–149 and 169, and figure 4.59 from *Proceedings of the Physical Society*. Publications Division, Govt. of India, New Delhi, and P.R. Pisharoty, for extracts from Krishnan's diary on pages 196–198 from *C.V. Raman* by P.R. Pisharoty (1982). The Royal Institution, London, for quotation on page 354 from *Commemoration Lecture* by Sir John Cockcroft (1967) and quotations on pages 357 and 358 from *Commemoration Lecture* by M.G.K. Menon (1967). The Royal Society, London, for quotations on pages 27 and 35 from *Biographical Memoirs of the Royal Society*, vol. 17 (1971), and quotations on pages 354, 355 and 358 from *Biographical Memoirs*, vol. 13 (1967); quotations on pages 102, 127, 169, 186, 188, 189, 190 and 215, and figures 4.27a, 4.28 and 4.67 from *Proceedings of the Royal Society*. Springer–Verlag, Berlin, for quotation on page 111 from *Handbuch der Physik*, vol. VIII (1927) and part of figure 12.22 from *Handbuch*, vol. XXV, part 1 (1961). Syndics of the Cambridge University Library, for extracts on pages 271–274 and 410 from the correspondence between Max Born and Lord Rutherford, Tata Institute of Fundamental Research, Bombay, for quotation on pages 356–357 from *Collected Scientific Papers of Homi Jehangir Bhabha*, edited by B.V. Sreekantan, Virendra Singh and B.M. Udgaonkar (1985). Taylor and Francis Ltd, London, for quotations on pages 263, 264, 395 and 520 from *My Life–Recollections of a Nobel Laureate* by M. Born (1978); quotations on pages 70, 72, 77, 78, 157, 160, 169 and 170, and figures 2.2, 2.3, 2.4, 2.5, 2.6, 4.14, 4.19, 4.48, 4.49, 4.51, 4.55, 4.64, 4.68, and 4.69 from *Philosophical Magazine*. University Grants Commission, New Delhi, for quotations on pages 39, 53, 507 and 510 from *Indian Journal of Physics Education*. John Wiley and Sons, London, for quotation on pages 54–55 (G. Torkar, 1986) from *Journal of Raman Spectroscopy*. Publishers of *Advances in Physics* for quotation on page 402; *Reviews of Modern Physics* for quotation on pages 394–395; *Journal of the Acoustical Society of America* for figure 4.38 (B.S. Ramakrishna and M.M. Sondhi, 1954); *Naturwissenschaften* for the paper of G. Landsberg and S. Mandel'shtam on pages 209–210.

CONTENTS

Preface   vii
Acknowledgements   xi

1 The Background   1

2 The Early Years   9

    2.1 Tanjore   9
    2.2 Ancestors and Parentage   11
    2.3 Naming System   11
    2.4 Raman at School   12
    2.5 Off to College   13
    2.6 The First Papers   15
    2.7 Books that Influenced Raman   23
    2.8 Career and Marriage   25

3 Oh Calcutta!   29

    3.1 The IACS – Its Origin   29
    3.2 Enter Raman   34
    3.3 Palit Professorship   37
    3.4 Calcutta School of Physics   39
    3.5 Other Giants   42
    3.6 The Principal Lieutenants   46
    3.7 The Second Decade   48
    3.8 The Exit   55

4 Glimpses from the Golden Era   61

    4.1 Studies on Vibration   61
    4.2 Raman and Musical Instruments   79
    ☆4.3 A Mathematical Interlude   112
    4.4 Whispering Gallery   126
    4.5 Studies in Optics   127
    4.6 Studies on Impact   173
    4.7 X-rays   178
    4.8 An Appraisal   183

5	**Elementary, My Dear Watson!**	**185**
	5.1 The Beginning	185
	5.2 On Board the SS *Narkunda*	186
	5.3 Why is the Sea Blue?	187
	5.4 The Einstein–Smoluchowski Formula	188
	5.5 Back to the Blue Seas	189
	5.6 A Dash up the Mountains	191
	5.7 Some Reflections	192
	5.8 On with the Charge	194
	5.9 Feeble Fluorescence	194
	5.10 The Discovery	196
	5.11 The Mechanism	199
	5.12 The Follow-up	204
	5.13 Who Got There First?	204
	5.14 A Mild Confrontation	213
	5.15 The Honours	215
☆**6**	**I say, What is this Raman Effect?**	**219**
	6.1 Optical Inhomogeneity and Light Scattering	219
	6.2 Light Scattering and the Dipole Approximation	221
	6.3 Rayleigh Scattering	222
	6.4 Polarization Effects in Rayleigh Scattering	223
	6.5 Scattering of Light by Liquids	224
	6.6 Raman Effect in the Polarizability Picture	226
	6.7 Time-dependent Perturbation Theory and Raman Scattering	227
	6.8 Raman Scattering by Molecules	228
	6.9 Intensity Formulae	229
	6.10 Raman Scattering by Crystals	232
	6.11 Quantum Theory of Light Scattering	232
	6.12 Spin of the Photon	234
	6.13 Anti-Stokes Scattering	239
7	**Brighter than a Thousand Suns**	**241**
8	**On to Bangalore**	**255**
	8.1 The IISc – Its Origin	255
	8.2 The Institute Before Raman	258
	8.3 Enter Raman	261
	8.4 The Irvine Report	265
	8.5 The Inside Story	271
	8.6 The Resignation	275
	8.7 Some Reflections	281
	8.8 Science in Bangalore	283
	8.9 The Franklin Medal	349

Contents xvii

8.10	The Academy	350
8.11	Visit of Gandhiji	350
8.12	Fiftieth Birthday	351
8.13	Bhabha and Raman	353
8.14	Into Retirement	360

9 Son et Lumiere   365

9.1	The Problem	365
9.2	Raman's Ideas	367
9.3	Experimental Verification	371
9.4	Applications	373
☆9.5	The Theoretical Aspects	377

10 The Born–Raman Controversy   383

10.1	The Beginnings of Crystal Dynamics	383
10.2	The Debye Theory	384
10.3	The Born Theory	384
10.4	Periodic Boundary Conditions	385
10.5	Raman's Interest in Crystal Dynamics	388
10.6	Raman Spectrum of Diamond	389
10.7	Raman's Theory	391
10.8	The Dispute	393
10.9	History's Verdict	396
☆10.10	The Theoretical Aspects	402

11 The Academy   407

12 The Final Years   417

12.1	The Raman Research Institute	418
12.2	Science at the Raman Institute	422
12.3	Raman and Pancharatnam	430
☆12.4	Pancharatnam and Berry's Phase Angle	454
12.5	The Recluse	456
12.6	The End	465

13 Sharing the Pleasures   473

14 Looking Back   495

14.1	Raman the Scientist	496
14.2	Raman the Man	497
14.3	A Comparison	499
14.4	Hazards in the Indian Scene	501
14.5	An Introspection	502

Notes to Chapters	**505**
References	**523**
Appendices	
1 Honours	531
2 Bibliography	533
Name Index	**551**
Subject Index	**559**

1 The Background

The impact of Western culture on India was the impact of a dynamic society, of a 'modern' consciousness, on a static society wedded to medieval habits of thought.... And yet, curiously enough the agents of this historic process were not only wholly unconscious of their mission in India but, as a class, actually represented no such process.

— JAWAHARLAL NEHRU,
in **The Discovery of India**

Amongst all the scientific discoveries of modern times, the Raman effect is perhaps the best known to the Indian public, and for understandable reasons. Chandrasekhara Venkata Raman, the discoverer of this effect, was a scientist of great merit. In fact, in India he was a rare exception in his passionate commitment to science and in his personal involvement with scientific research till the very end of his long life. Nevertheless, one cannot describe his life simply in terms of the science he did. This may sound strange, especially if one thinks of Helmholtz or of Lord Rayleigh with whose works Raman's own had much in common. But the life of a scientist in India, particularly in those days, was not the life of a gentleman of leisure. It was a perpetual struggle, often bitter, against various odds – some due to the backwardness of the country, some due to geographic isolation, and others arising from diverse hostile forces natural in a cramped and foreign-dominated environment. And yet Raman, and also many contemporaries of his, did manage to spark the growth of science (physics in particular) in India. As we look back one hundred years after his birth, what we perceive in him is not merely a great scientist but a microcosm of India with all her problems.

The nineteenth century saw classical physics in all its glory. What a galaxy of masters – Faraday, Boltzmann, Maxwell, Helmholtz, Rayleigh, Kelvin,...! Such were their triumphs that some even felt that there was little else to be discovered. Of course this belief was soon shattered but the new prophets really took off where the old ones left, with some like Lorentz, Thomson and Sommerfeld, for example, bridging the gap. Most of the action was in Europe but America had its quota of heroes like Gibbs and Michelson. By contrast, in India there was an almost complete vacuum.

India is a land of ancient civilizations, and through the ages has produced great savants, scholars and religious preceptors. Literature, art and architecture have always flourished, and from the days of Alexander and of the Roman Empire, there have been contacts with the outside world. Inevitably the sciences also took root, and until the Renaissance, Indian science was about as good as any elsewhere [1.1].

Europe went through a dark age and came out of it with a bang. Where science was concerned, Galileo and Newton marked a turning point. In India there were neither major ups nor downs, and when Europe began to gallop after the Renaissance, India simply slipped behind. But science was not yet the fountainhead of technology, and manufacture still depended on basic human skills and craftsmanship. Thus, for a while, Indian products continued to be comparable to those made in Europe and were often superior, like, for example, the Dacca muslin. There are records of ships having been built in India for the British. The Industrial Revolution struck a sharp blow and changed even that. India was now left quite behind, a position from which she is still struggling to emerge.

The political complexion of the country during that period made its own contribution to the slide into backwardness. Coming as traders in the sixteenth century, the European powers quickly established bridgeheads, and, taking advantage of internal dissensions, rapidly expanded their respective spheres of influence. Capturing territories was the next step, which inevitably led to rivalry amongst the powers themselves. In the resulting struggle the British East India Company emerged the victor, directly controlling large parts of the country and influencing the rest through the princes who ruled them. Following the popular uprising in 1857 (referred to by the British as the Sepoy Mutiny), the British Government stepped in to take over the possessions of the Company, and India became a Crown Colony. Life continued but the spirit was crushed. It would be close to a hundred years before the foreign yoke was thrown off.

The British had colonies throughout the world but India was in a different class altogether. The colony was neither new (and thus open to large-scale settling by the White man) nor a land of primitive tribes. On the contrary it was a land of great civilizations, traditions and culture. The British fully understood this, and came to equilibrium with this unique situation. They were content to protect their interests – military, political and economic – leaving the natives largely to themselves as long as British suzerainty was not threatened. Of course things became different after the introduction of Western education, for it brought about, among other things, an awakening.

Western science came to India before Western education[1]. This is not as paradoxical as it sounds, for initially, Western science stayed exclusively with the Europeans. Roughly three phases can be identified in the introduction of Western science into India [1.1]: (i) use of science by the trading companies for understanding the natural resources of the country, the flora, the fauna, the mineral wealth, the geology, etc., particularly in so far as such knowledge would secure military and economic advantages, (ii) establishment of scientific services beneficial to the ruling power, and (iii) exposure of the native population to modern science via Western education.

The first phase involved mainly individual effort to which the Jesuit missionaries (who came with the Portuguese) contributed much. Subbarayappa [1.1], for example, records, "...in the field of geography of India, the latitude and longitude measurements which were determined scientifically by some of the Jesuit missionaries contributed not a little to the geographical knowledge of India."

The missionaries were not the only curious ones. There were many in the service of the trading companies who had an equally insatiable thirst for knowledge. Thus there is the example of the Dutch Governor Hendrik van Reede tot Drakenstein who studied the medicinal plants of Kerala and published a 12-volume series in Amsterdam. The British East India Company also promoted studies of the flora, fauna, etc. with, of course, its own interest in view. It is estimated that at the end of the nineteenth century, there were as many as 457 persons interested in botanical explorations, among whom 104 were surgeons or physicians and 111 were administrators. There were others engaged similarly in survey, astronomical observations, and so on.

As the political ambitions of the East India Company grew, it found it necessary to organize specialized scientific services instead of relying on individuals. Survey being of strategic importance, a Trigonometric Survey of India was created. Several other agencies and provincial organizations were created which were later to become nuclei of all-India Departments like the India Meteorological Department etc.

By the middle of the nineteenth century there was a fair amount of scientific activity in India in the Western tradition, but controlled almost entirely by Europeans and heavily slanted towards the interests of the colonial power. However, much good also came out of this effort, especially in the shape of valuable monographs like *Birds of India*, *Fishes of India*, and *Handbook of Cyclonic Storms in the Bay of Bengal*. There were also many journals like the *Transactions of the Medical and Physical Society*, *Indian Journal of Medical Science*, *Madras Monthly Journal of Medical Science*, and so on.

There are some interesting snippets from that period. For instance, in 1864, the Surveyor-General, Walker, applied for permission to the Secretary of State for India to undertake a series of pendulum experiments in connection with the trigonometric survey of the country. The Royal Society lent not only its support but also some equipment like an astronomical clock and two invariable pendulums. The Imperial Academy of Sciences in St Petersburg (later to become Leningrad) in Russia lent two convertible pendulums (which had already been used on the Russian arc) in order to establish a connection between the Indian and Russian experiments.

The creation of the Solar Observatory at Kodaikanal is another example. A commission of enquiry set up to ascertain the causes of the Madras famine of 1876-77 brought to the notice of the Government a correlation between the seasonal distribution of rain in India and sunspot periodicity and recommended that necessary steps be taken for solar observations. The Kodaikanal observatory was a direct outcome of this recommendation.

The first major scientific society to be founded in India was the Asiatick Society (later renamed the Asiatic Society of Bengal). The idea for such a society took shape during a meeting of thirty European intellectuals of Calcutta on January 15, 1784. The moving spirit was Sir William Jones, a puisne judge of the Supreme Court of Judicature at Fort William in Bengal. Jones was a scholar of repute, and already a Fellow of the Royal Society when he came to India in 1783.

In 1808 it was recommended to the Society that two Committees be formed – one for "Natural History, Physics, Medicine, Improvement of the Arts, and whatever is comprehended in the general term of Physics", and the other for Literature. From 1818 onwards, the Physical Committee did valuable work. In 1832 the Society started publishing a journal of its own – the *Journal of the Asiatic Society*. Initially it carried mainly papers of a literary character but pretty soon it became a periodical for important scientific communications. In fact, Sir J. C. Bose[2] published his first paper in this journal in May 1895.

The formation of the Asiatic Society is a landmark. Though initially it was for the Europeans and by the Europeans, it brought home to Indians the culture and the ethos of modern science. And in its own way, the Asiatic Society inspired the formation of several other scientific bodies like the Agricultural Society of India, the Bombay Natural History Society, and, much later, other specialist organizations like the Mining and Geological Institute of India, the Indian Mathematical Club, etc.

As already remarked, initially (i.e., in the late eighteenth and early nineteenth centuries) the pursuit of science was an all-European affair. Indians came in only peripherally, like the native doctors of Malabar who helped the Dutchman Drakenstein in the collection of medicinal plants. There was little else Indians could do at that stage, not having been trained in modern scientific methods. It was only after the introduction of Western education that the first trickle of Indian scientists appeared on the scene.

The story of Western education in India is quite fascinating in itself. In the early part of the nineteenth century, there were hardly any organized schools. Learning was the prerogative of a privileged few. Imparted in the traditional style, it largely revolved around religious training – our version of seminaries, one might call it. An exception to this pattern was the missionary school. But such schools were few in number, and concentrated mostly around Calcutta, Madras and Bombay where British residents were in maximum numbers.

In 1811 Lord Minto observed:

> It is a common remark that science and literature are in a progressive state of decay among the natives of India.... The number of the learned is not only diminished, but the circle of learning, even among those who still devote themselves to it, appears to be considerably contracted. The abstract sciences are abandoned, polite literature neglected, and no branch of learning cultivated but what is connected with the peculiar religious doctrines....

Lord Minto traced this decay to the lack of encouragement "which was formerly afforded by princes, chieftains and opulent individuals under the native government". He was unhappy that Britain had done little to arrest this decline. Perhaps as a result of this, at the time of the renewal of its Charter in 1813, the East India Company was asked to assume legal responsibility for educating Indians. Soon after, the great social reformer Raja Ram Mohan Roy started an English medium school for Hindu boys. Later it became the Hindu College, and the nucleus of the present Presidency College in Calcutta. Raja Ram Mohan Roy also petitioned the Governor-General to encourage instruction in scientific subjects, and not "to lead the minds of youth with grammatical niceties and metaphysical distinctions of little or no practical use to the possessors or to society".

In 1834 Lord Macaulay came to India and recommended the use of English as a medium of instruction, and the promotion of Western learning. Nevertheless, Western education did not spread widely. The year 1854 marked a turning point, for in that year, Charles Wood (later Lord Halifax) prepared an Educational Despatch in which he recommended "the diffusion of the improved arts, science, philosophy and the literature of Europe, in short, European knowledge". Soon after, a strong machinery for large-scale education with a Director of Public Instruction for each province[3] was created. Science education was a part of this new plan. In addition, three universities were set up in 1857, in Calcutta, Madras and Bombay, on the pattern of the London University. Towards the close of the nineteenth century there were about 200 colleges, mostly located in the big cities and towns. Nevertheless, they were sufficiently scattered throughout the country, and young Indians finally had access to modern education, at least up to a certain level.

It took several decades for the Indian public to adjust to this new scheme of things, and it is only after that incubation period that one began to hear of Indians entering professional fields like law, medicine and engineering. Slowly but surely, one also began to hear the names of some outstanding Indian scientists, like Sir J. C. Bose and Sir P. C. Ray[4], for example. But they were a tiny handful, and Indian science was still a set of scattered individuals. There were no great laboratories, nor centres of excellence – at best just a few fledgling scientific institutions, often a one-man show. While Europe had built up a base and a glorious tradition, all that India could manage by the turn of the (nineteenth) century was a faint beginning in physics. But the early pioneers gave a valuable start. There was some semblance of scientific activity which originated from the Indian mind, and this was a great morale booster. It is into such a setting that Raman wandered in 1907. And by dint of his personal genius and dynamism, he built up a thriving school of physics where none existed before.

There is a curious sociological fact relating to British influence which is worth noting before we conclude this background survey. Three communities in different parts of India came into contact with Western education, thought and culture more or less simultaneously. But their responses were altogether different.

The Parsis are a tiny community concentrated mainly in and around Bombay and Surat on the west coast. They came to India from Iran in the eighth century, fleeing the proselytising Islamic wave. Adopting the local language as well as many of the customs, they have had a harmonious existence for centuries. Trade and commerce are their particular forte. Used as they were to adapting, they not only took to English but to Western culture itself when the opportunity came. An important fall-out was that it gave ready access to British circles, especially for those in the upper bracket. In turn, British patronage was duly utilized for promoting the community's interests. Several Parsis served in influential Government bodies, including the most important of them all, the Viceroy's Council[5]. The Parsis are extraordinarily versatile, having produced men of distinction in various fields, including music[6], in numbers far out of proportion to the size of their community.

British presence in the South dates back to the landing in Madras in the early part of the seventeenth century. In spite of this, however, Western culture made hardly any impact on any community, although the English language did. Society was quite feudal and customs were rigid. Nevertheless, when Western education became available many readily took to it, more as a means of employment rather than for gaining entry into British circles. The British were always in need of subordinate administrative staff, and serving the ruler was an accepted tradition in a feudal society. Only, in this case one had to learn a new language. Soon, getting a Government job became the in-thing, not only because it provided security but also as it conferred a certain status as well as some privileges. In a sense, even a petty clerk represented the might of the British Empire.

The Brahmin community in South India took to English education with particular alacrity. Having been exempted by the caste system from manual labour, trade and military service, the Brahmin had few means of earning a livelihood. Some were fortunate enough to be landlords, especially those whose ancestors had received gifts of land from olden-day princes and kings. For the majority, priesthood was the only available profession but it was oversubscribed. A few took up jobs as cooks but even that offered only limited scope since most Brahmins would not serve in non-Brahmin households. And the number of well-to-do Brahmins who could afford cooks was not large.

English education opened a magic door, and all of a sudden one could not only become a clerk or a petty official like a Tahsildar but also a teacher, a lawyer, a doctor, an engineer or, on occasions, even a judge or a senior civil servant. By diligence and hard work, several rose to positions of eminence, and the roster of men of distinction, particularly lawyers and judges, is long.

The South Indian Brahmin differed in one important respect from the Parsi. He accepted only the English language but not the custom or the culture, preferring his own. Orthodoxy was rigid, and the only concession a sartorial one. He rejected the trousers in favour of the *dhoti* to which he was accustomed, but accepted the shirt,

The Background

the coat and even the tie. Above the neck it was back to his own style – a prominent caste mark on the forehead, a tuft concealed by a white turban and, at times, a pair of ruby or diamond ear-rings! In later years trousers became more common but *dhoti*-clad lawyers and judges wearing black coats and turbans were not an uncommon sight until recently.

The British came to Bengal shortly after the landing in Madras but their impact on that part of the country was quite different. As in South India, young students drank deep from Shakespeare, Milton and other poets and writers of great distinction. But they also avidly read Voltaire, Rousseau and Karl Marx. Socialism and Marxism had a natural intellectual appeal, and the revolutionary spark was lit in many. It would however be an oversimplification to say that Western education merely created an interest in leftist ideologies. It was much more than that. There was in fact a general awakening with respect to art, literature, culture, philosophy and science – in short, there was an urge for self-discovery. Commenting on this period, the noted chemist Acharya Prafulla Chandra Ray observed: "There was ferment all round. A new world had been opened out; new aspirations were awakened. Roused from a period of stupor and stagnation, Young Bengal began to realize that there were immense possibilities in the Hindu nation." No one symbolized this resurgent spirit better than Swami Vivekananda[7]. But there were a host of others too – Bankim Chandra Chatterjee[8], Rabindranath Tagore[9], Aurobindo[10], and so on, each a beacon to the nation in his own special way.

To the Englishmen of those days Calcutta was not only the capital of British India but, with its clubs and upper class society, was also somewhat like the London of the East. To the people of Bengal however, Calcutta, bursting with an *avant-garde* spirit, was more like the Paris of the East. There was nothing like Calcutta not only in the whole of India but in fact east of Suez. Dickens wrote *A Tale of Two Cities*; Calcutta was two cities rolled into one!

Each of the three communities mentioned above produced a distinguished physicist who plays an important role in the present narrative – Bhabha, Raman and Saha, respectively. As was only natural, each was shaped by the cultural influence of the community he belonged to. The lives of these three men overlapped but, more important, their paths crossed. Given their distinctive personalities, there were sharp differences and at times even conflicts, all of which were to have an impact on the growth of physics in India. In this book Raman is the central character, but inevitably, the others also make brief appearances.

The lives of these three make fascinating reading not only for the science they did, but also for the struggle they waged and the odds they had to overcome. How much more could Raman and Saha in particular have contributed had they lived in Europe in the mainstream of physics? Or how much more could they have achieved in India itself had the country been free and not in thralldom? Interesting as they are, these are but wistful questions, lying beyond the edge of history. Our own concern is largely with events as they actually happened.

2 The Early Years

2.1 Tanjore

In the Hindu tradition, the Kaveri is one of the sacred rivers of India. Rising near Mercara in the Western Ghats, it has a rather short run before fanning out and flowing into the Bay of Bengal. Gemini astronauts have taken a beautiful picture of peninsular India showing clearly the entire Kaveri river including the delta region.

The Kaveri delta is noted not merely as the granary of the South but more so as the seat of ancient culture whose various manifestations are visible even today. Temples built during the reigns of many kings who ruled in that region dot the landscape in large numbers, and the most famous of them all, the *Brihadeeswarar* temple in Tanjore (now Thanjavur)[1], is not only majestic but unique in many ways. Tanjore was the capital of the famous Chola kingdom, which vied with others in the South for the promotion of literature and the various fine arts.

In the late seventeenth century, the well-known Maratha prince Sivaji made a southern foray, capturing large chunks of territories. Tanjore was one of these, where he installed his chieftain. The Marathas contributed their own inputs to the cultural milieu, especially by injecting Sanskrit influence into a region predominantly Tamil. One of the rulers of Tanjore, named Tulaja, wrote a treatise on music entitled *Sangitasaramrita*. Notable among Sivaji's descendants was Prince Sarabhoji, a great patron of music and the founder of the famous *Saraswati Mahal*, a depository library for works of art, literature and music. Sarabhoji was a good and gentle prince but quite naive. It is said that when the Marquis of Wellesley[2] went on a rampage annexing territories for the East India Company, Sarabhoji thought he could keep the Company soldiers away by sprinkling holy water from the Ganges on the palace walls! Fortunately, Wellesley allowed Sarabhoji to continue his rule but extracted an annual subsidy. The reign of Sarabhoji coincided with the appearance of the great trinity of Carnatic music[3], namely, Tyagaraja, Muthuswamy Dikshitar and Syama Sastri. These great composers left behind an incredible heritage which forms the backbone of modern Carnatic music. Music was not an abstract art but intimately related to religion. Since religion dominated society, music was everywhere all the time, and nobody escaped its magic.

Unlike in the Western world, music evolved largely around individual, vocal singing – in praise of the Almighty of course. But there were accompanying instruments of various kinds, including percussion instruments which provided the *tala* or temporal rhythm. It is interesting that the violin, an instrument of European origin, not only gained ready acceptance amongst the musicians of that period but was soon elevated to a position of importance[4]. While accepting the instrument, a certain change was also introduced in the playing posture. To suit the convenience of the performers, who always squatted on the floor, the violin was held, not horizontally but in an inclined manner, with the lower end resting on an outstretched foot.

The sway of music over the people of the Tanjore region becomes visible every year in January when the birth anniversary of Tyagaraja is celebrated. Musicians flock from all over to Tiruvaiyaru (where Tyagaraja lived and died) to pay homage to the saint-composer. The *Aradhana*, as the festival is called, is a great event widely covered on the radio and television. It is almost axiomatic that all who hail from that region have a strong interest in music.

In Sarabhoji's days, the delta region was largely a network of villages. Later a few small towns, like Kumbakonam, Mayavaram (now Mayiladudurai) and Trichinopoly (now Tiruchirapalli), appeared. There was no town or village without a branch of the Kaveri flowing through it, and every place had a presiding deity together with some folklore attached to it. Life was tranquil and pastoral, with people's lives revolving around agriculture and religion. It was also highly structured. The Brahmins lived in an area called the *Agraharam*, close, naturally, to the local temple. The joint-family system was very much in vogue; all the brothers usually lived together along with their families in the ancestral home, the father playing a patriarchal role. The Brahmin household had three distinct, weakly interacting sets of people. First, there were the menfolk, mostly concerned with supervising agricultural operations. The women and the older girls lived in a world of their own – the kitchen! Then there were the boys, who were subjected to a strong ritualistic drill, but life was cheerful nonetheless – school, games in the evening on the river bank, and of course a stiff dose of prayers and worship. Excitement came mainly through a never-ending succession of religious festivals of various kinds. After school there was college – at least for some – and marriage, strictly arranged of course, usually came even while one was studying. The well-known writer R. K.

Narayan has beautifully captured life in those days in his various novels set in the mythical town of Malgudi.

2.2 Ancestors and Parentage

Raman's ancestors were agriculturists owning land near the villages of Porasakudi and Mangudi in Tanjore district. His paternal grandfather was one Ramanathan, a pious Brahmin and a good Tamil scholar. Raman's father Chandrasekaran was born in 1866 as the second son of Ramanathan and Sitalakshmi Ammal. Their first son died in infancy. Young Chandrasekaran studied in a school in Kumbakonam and passed the Matriculation examination in 1881. College education (which came after Matric) involved first the Intermediate, followed by the BA and the MA degrees, each requiring two years of study. Chandrasekaran did the Intermediate in the SPG (Society for the Promotion of the Gospel) College, Trichinopoly, and then enrolled in the BA class in the Madras Christian College. For some reason he could not complete studies, and he worked as a school teacher for many years. But eventually he went back to the SPG College, obtained the BA degree in physics in 1891, and became a lecturer in the same college.

Soon after passing the Matriculation examination, Chandrasekaran got married to Parvati Ammal, the daughter of Saptarshi Sastri of Tiruvanaikkaval, a village near Trichinopoly on the banks of the Kaveri. The name Tiruvanaikkaval is derived from a legend that an elephant once worshipped there to gain the *darshan* (divine vision) of Lord Siva. Saptarshi Sastri was a great Sanskrit scholar and in his younger days had travelled on foot to distant Bengal (over 2000 km away) to learn *navya nyaya* (modern logic).

Chandrasekaran and Parvati Ammal had eight children: five sons and three daughters. As Ramaseshan [2.1] once put it humorously: "Apparently Chandrasekara Iyer did not believe in family planning. Even if he did, and had stuck to the statutory limit, Raman would still have made it!"

Raman was born on November 7, 1888, in his maternal grandfather's house in Tiruvanaikkaval. In those days the custom was that the expectant mother always went to the home of her parents for the delivery. Raman was the second son; before him came Subramanian, born in 1886.

2.3 Naming System

The naming system in South India must undoubtedly appear quite confusing and chaotic to most readers! However, there *is* a method, though at variance with what is practised elsewhere. The first thing to note is that there are no family names! Every child is given a name, invariably that of a deity, and that is the only name it has. Next, the father's name is added in front which, in abbreviation, serves as an initial[5]. (Occasionally, the name of the hometown is added before that of the father.) Thus the name of Raman's father would be R. Chandrasekaran or Ramanathan Chandrasekaran in full. The name of the sect or subsect one belonged to was often added at the end; with this addition, the above name would be R. Chandrasekara Iyer. In the same manner, the name of Raman's elder brother would be written C. Subramanian or C. Subramania Iyer (he was actually better known as C. S. Iyer). It then was and still is a common practice to name at least one son (usually the eldest) after the paternal grandfather. In keeping with this practice, both Subramanian and Raman named their first sons Chandrasekhar[6]. Of these, Subrahmanyan Chandrasekhar is known to the world as an outstanding astrophysicist, as the Morton D. Hull Distinguished Service Professor in the University of Chicago, and also as a Nobel Laureate[7].

Women also are known by their given names. Thus Raman's mother was S. Parvati before marriage and C. Parvati after, the new initial derived from the husband's name. The suffix Ammal, frequently added, does not denote any caste name but is just an honorific.

Raman was actually given the name Venkataraman at birth, after the Lord of the Seven Hills in the sacred shrine of Tirupati. In school, Venkataraman was first split to Venkata Raman, following which C. Venkata Raman became C. V. Raman, a name that remained. Later, when a knighthood was conferred on Raman, it created problems! In England, William Bragg would have been addressed as Mr Bragg before knighthood and Sir William thereafter. Adopting this practice, *Nature* once referred to Raman as Sir Venkata. In India people tried every combination – Sir Chandrasekhara, Sir Raman and Sir C. V., the last mentioned being the most popular one. The story is told that a young American who once called on Raman in Bangalore decided to get friendly and asked Raman what his first name was. Raman cooly replied, "Sir." The visitor thought for a while and dropped the idea of getting on a first-name basis[8]!

2.4 Raman at School

When Raman was four years old, his father moved to Vizagapatam (now Visakhapatnam) to take up a lecturer's job in the Mrs A. V. Narasimha Rao College. There he taught physics, mathematics and physical geography. Mr Chandrasekara Iyer was also a good athlete, and took active part in sports and physical culture. He was much liked, and there is a portrait of his in the college even today.

On the cultural side, Mr Iyer loved music (naturally), and played well on the violin. He was also an avid reader and collected many good books. As Raman once said, "a good home and a good school may be judged by the kind of books they put in the way of the growing young person for him to feed his mind and emotions upon" [2.2]. In Raman's case, there was nothing wanting on this score.

Unlike his father, young Raman was not strong. But what he lacked in physical abilities, he more than made up in intellectual ones. He really excelled in studies, giving early signs of unusual talent by winning numerous prizes and scholarships.

Apparently, Raman developed an interest in physics even while in school. It is said that he once built a dynamo all by himself. There is also a story that on one occasion when he was bedridden, he was so enamoured of the Leyden jar that he would not go to sleep until his father brought the apparatus from college and demonstrated it.

2.5 Off to College

At the incredibly young age of eleven, Raman passed the Matriculation examination, standing first of course. College was the next obvious step and he joined the AVN College to study for the Intermediate. Again, he won laurels in the university examination, and in 1903 he was off to Madras with a scholarship to study for the BA degree in the Presidency College. Apparently, they had never seen anyone so young before in the College. As Raman himself wrote later:

> Indeed in the first English class I attended, Prof. E. H. Elliot addressing me asked if I really belonged to the Junior BA class, and I had to answer him in the affirmative. He then proceeded to inquire how old I was. [2.3]

The Presidency College was the premier college in the South, to which all ambitious young men aspired to go. Run by the Government, it was essentially a liberal arts college, catering to the entire Madras Presidency, which included all of today's Tamilnadu, a good chunk of Andhra Pradesh, the Malabar region of Kerala, and parts of coastal Karnataka as well[9]. In Raman's days, almost all the professors were Europeans. Writing about them later, Raman said:

> Some of my pleasantest recollections of the four years I spent at college in Madras are of the extraordinary kindness and consideration which I received from the European members of the staff who were then the Heads of Departments of study. Their attitude seems all the more surprising when I look at the undistinguished and diminutive figure of myself thirty-five years ago as it appears in the college photographs of those days. [2.3]

In college, Raman's interests became focused. Besides a flair for physics (about which more will be said shortly), he also developed a great liking for English, stimulated in part by the English classes. Writing about this Raman observed:

> The English classes were conducted by Professors Bilderbeck and Elliot. They held their classes usually in the big lecture hall overlooking the sea, and the seats were so arranged that if the students did not like the lecture, they could instead gaze at the far horizon of the blue sea or count the glittering waves as they crashed down on the beach. Did ever students of the English language have a more marvellous panorama

the contemplation of the beauty of which could lighten their labours? I am almost tempted to compare it with the glorious theatre built by the ancient Greeks on the height of Taormina from which you could see the waves of the Ionian Sea washing the coast of Sicily, or, turning your eyes up, you could see the glittering heights of Mount Etna. It must be said to the credit of the teachers I have mentioned that they often did hold our attention in spite of the lure of the swirling waters of the ocean breaking upon the shore, or was it because of the same fascinating vision of the sea that our minds were better attuned to the complicated beauties of the English language? I have vivid memories of the spirit with which Professor Bilderbeck conducted his classes and sought to infuse into us a due appreciation of the great English writers.

Raman passed the BA examination in 1904 obtaining the first rank in the university and winning gold medals in English and physics. Some of the certificates his teachers gave him are interesting: "The best student I have had in thirty years", "Possessing great alertness of mind and a strong intellectual grasp", "Exhibited an unusual appreciation of English literature and a facility in idiomatic expression", "A young man of independence and strength of character".

When Raman passed the BA examination, his teachers suggested that he should go to England for further studies. But the Civil Surgeon of Madras ruled it out saying that the young and frail Raman would not be able to stand up to the rigours of the English climate – not an unreasonable opinion when one recalls the difficulties experienced by the mathematical genius Ramanujan during his stay in England (from 1914 to 1919). Raman later said of the Civil Surgeon, "I shall ever be grateful to this man!" As it turned out, Raman did not go abroad till he was thirty-three.

Unable to proceed to England for higher studies, Raman enrolled in the MA class in the Presidency College to study physics. The Professor of Physics at that time was one R. Llewellyn Jones, a kindly gentleman but otherwise undistinguished. Of him Raman wrote:

> Professor Jones believed in letting those who were capable of looking after themselves to do so, with the result that during the four years I was at the Presidency College, I enjoyed a measure of academic freedom which seems almost incredible. To mention only one detail, during the whole of my two years' work for the MA degree, I remember attending only one lecture, and that was on the Fabry–Perot by Professor Jones himself.

The Professor's photograph is still to be seen as one enters the lecture hall in the Physics Department (along with those of Raman and Chandrasekhar as well). In 1939 the Indian Academy of Sciences (see Chapter 11) brought out a commemorative volume on the occasion of Raman's fiftieth birthday. Reviewing it, *Nature* complained that the volume "unfortunately omits to mention how much the future Nobel prize winner owed to the then head of the department of physics, the late R. L. Jones". *Nature*'s sympathy for the Englishman is understandable but not justified. As we shall presently see, Prof. Jones contributed little, other than letting Raman have a free hand.

The Early Years

2.6 The First Papers

Raman made good use of the freedom given to him by the kindly Prof. Jones, tinkering around the lab and trying out various things. There was not much available by way of scientific equipment except the usual assortment of lenses, prisms, gratings, sonometer, etc. needed for class work. But for Raman this was ample. Guided by irrepressible curiosity, he posed himself many questions, to settle which he performed suitable experiments. Often the questions were of a type the answers to which were not found in the published literature. Instinctively he had discovered the essence of research, and it was this spirit of enquiry that was to constantly drive him to do experiments throughout his life.

One problem that Raman examined at college was unsymmetrical diffraction. The diffraction of light by a rectangular slit was standard textbook material, of which Raman was well aware. Let us briefly remind ourselves of this conventional wisdom. (Later, in Chapter 4, we will go further into the principles of optics.)

Figure 2.1 shows a schematic of a rectangular slit illuminated by a parallel beam of light. Ray optics would lead us to expect an image of similar shape on a distant

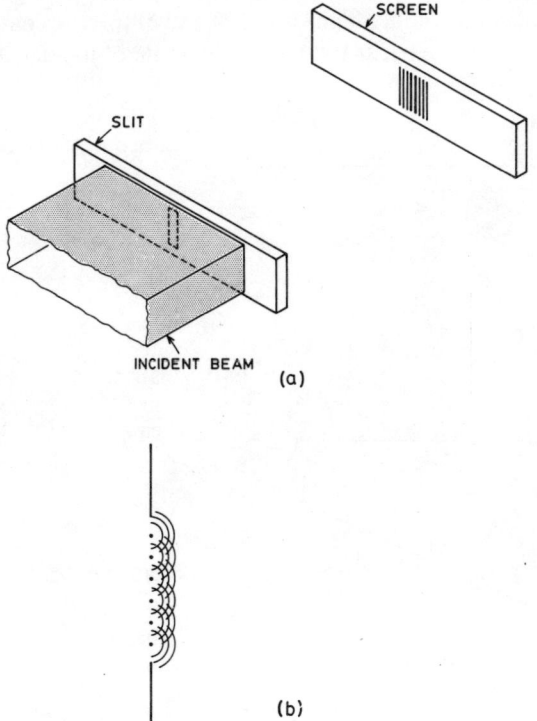

Figure 2.1 (a) Geometry of the standard arrangement to demonstrate diffraction from a slit; (b) illustrates how the pattern actually arises due to interference from neighbouring waves.

screen. Careful examination would however reveal dark and light bands or fringes on either side of the bright central image. The pattern is symmetric and can be readily understood using the wave theory of light. According to the well-known Huygens principle, when light falls on the slit secondary wavelets are set up rather in the manner in which ripples are produced when pebbles are dropped into a pond. As one would have noticed on such occasions, waves from adjacent disturbances can interfere, reinforcing each other in some directions and cancelling in others. The Huygens secondary waves emerging from the slit interfere in a similar manner leading to diffraction bands.

Raman asked: What if the light is incident obliquely, as in Fig. 2.2a? Would there not be an asymmetry in the bands as there would be obstruction to band formation on one side? To answer this question, he used the arrangement shown in Fig. 2.2b. A 2-cm-wide aperture was cut out in a thin sheet of zinc which was then bent in the shape shown, the side vanes serving to cut off unwanted light. Raman found that when incidence was oblique, the diffraction bands were indeed quite unsymmetrical.

The next step was to make the observations a bit more quantitative, for which purpose Raman used the arrangement shown in Fig. 2.3a. A prism of face-width 4.5 cm was mounted on a college spectrometer and illuminated obliquely; the emergent light on the other side was examined as usual using the viewing telescope. Raman found that when the angle of incidence was 85°, the diffraction pattern seen in the field of view was quite symmetrical but when the angle of incidence became

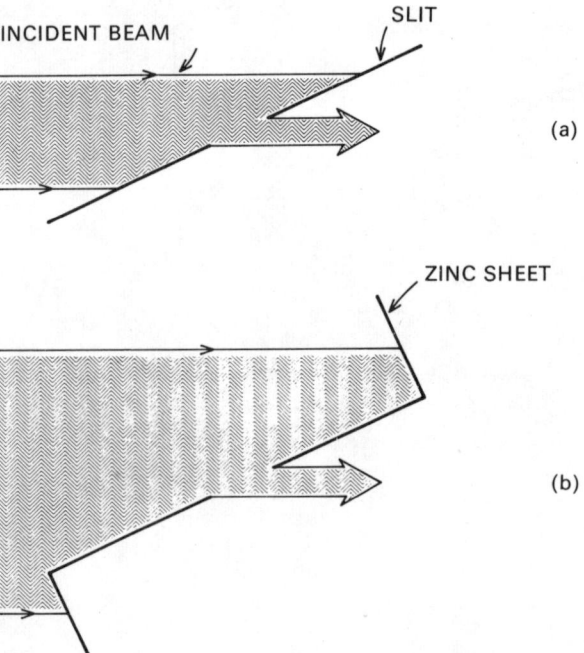

Figure 2.2 (a) Oblique incidence of light on a slit. (b) Arrangement used by Raman to make a preliminary visual survey of the problem of oblique incidence.

The Early Years

greater than 87°, this was no longer true. The observations made by Raman are shown in Table 2.1. Can these be explained? Yes indeed. One proceeds as usual and notes that the condition for interference is that the path difference between rays R_1 and R_2 in Fig. 2.3b must be an integral multiple of the wavelength, i.e.,

$$CB - AD = \pm n\lambda, \quad n \text{ an integer}$$

or

$$a(\cos\theta - \cos\phi) = \pm n\lambda. \tag{2.1}$$

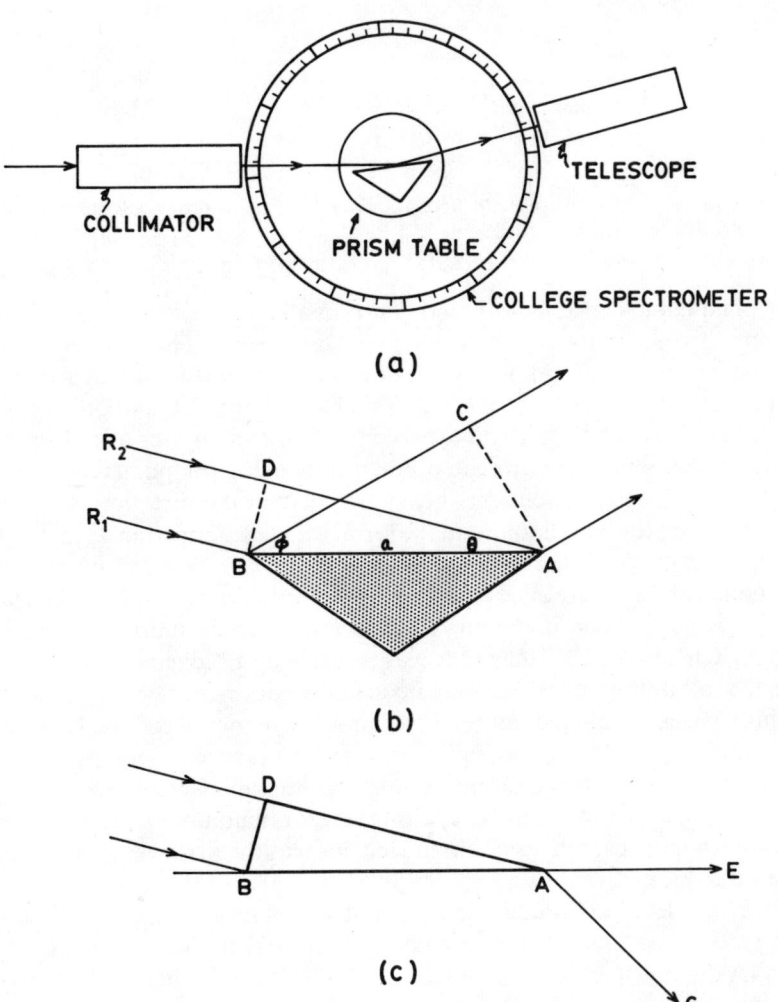

Figure 2.3 (a) Schematic illustration of the arrangement used by Raman to make a quantitative study of asymmetrical diffraction; (b) the geometry of the experiment; (c) shows that *diffracted* rays like AG lying beyond the limiting direction AE are not possible.

Table 2.1. Angular distances between adjacent bands in unsymmetrical diffraction. The conditions of the experiment are: $a = 4.57$ cm, $\lambda = 6500$ Å and $\theta = 1°24'55''$.

Distance from	Observed	Calculated
5th minimum to 4th minimum	110″	108″
4th minimum to 3rd minimum	105″	110″
3rd minimum to 2nd minimum	111″	113″
2nd minimum to 1st minimum	118″	115″
1st minimum on one side to 1st minimum on the other	234″	237″
1st minimum to 2nd minimum	125″	123″
2nd minimum to 3rd minimum	127″	126″
3rd minimum to 4th minimum	132″	130″
4th minimum to 5th minimum	132″	134″

Å (Angstrom) is a unit of wavelength, and is equal to 10^{-8} cm. More popular at present is the SI unit nanometre, which is equal to 10 Å.

For oblique incidence, both θ and ϕ are small as a result of which the diffraction pattern predicted by (2.1) becomes unsymmetrical. The angular displacements computed by Raman taking this feature into account (and treating the direction defined by $\theta = \phi$ as reference) are also shown in Table 2.1, and one sees satisfactory agreement between theory and experiment. One can also see from Fig. 2.3c that the direction BE sets a limit to the diffraction pattern, for points on the surface AB obviously cannot send out secondary wavelets in the direction AG.

Raman wrote up his findings in the form of a manuscript and gave it to Prof. Jones for comments. The latter kept the note with him for several months but offered no opinion. Taking courage, Raman sent the paper on his own to the *Philosophical Magazine* in London under the title "Unsymmetrical diffraction bands due to a rectangular aperture". The paper was published in November 1906 [2.4]. Raman was the sole author and there were no acknowledgements; obviously he could stand on his own legs, though barely eighteen and not yet out of college. Not a great paper, one would agree, but it was sufficiently original and of adequate quality to merit publication by a reputed scientific journal. Raman's achievement is all the more significant because the Presidency College was essentially a teaching college with no tradition whatsoever of research. Indeed, his was the first ever paper to come out of that institution.

One wonders how Raman knew about *Philosophical Magazine*. His college did not receive the journal and my enquiries revealed that the Madras University Library did not start subscribing to it till 1908. It is said that the Connemara Public Library, which was about 5 km away from the college, used to receive many scientific journals on the recommendation of various professors and members of scientific services. Possibly it received *Philosophical Magazine*, and possibly it was here that Raman became aware of this particular journal. We do not know. Whatever it was, Raman certainly took the trouble of being in touch with current literature (at least to the point of knowing about the existence of various journals), a practice rare amongst college students even today.

It would be too facile to dismiss Raman's first paper merely as a commendable display of minor curiosity. The problem of oblique diffraction is actually a bit more involved, and there are many questions not answered in this first paper but to which Raman came back subsequently – questions like: What is the intensity distribution in the bands? Why are the bands of unequal widths? What happens when the width of the slit becomes comparable to the wavelength of light? What happens when the rectangular slit is replaced by a circular slit? How may one measure the intensity of the bands? All these problems were to occupy Raman's attention soon after he went to Calcutta; that story comes later (see Chapter 4). Incidentally, almost immediately following Raman's paper there is one by the famous R. W. Wood of Johns Hopkins University. Much later, Wood was to send a cable to *Nature* hailing the discovery of the Raman effect.

The paper on unsymmetrical diffraction was no flash in the pan, and there was yet another one in *Philosophical Magazine*, also reporting some work done in college [2.5]. This second paper, entitled "The curvature method of determining the surface tension of liquids", was inspired by some remarks on capillarity made by Lord Kelvin in his book *Popular Lectures and Addresses*.

As is well known, surface tension is the property which makes possible liquid drops and soap bubbles. In the case of the latter, there is a very simple result relating T the surface tension to P the excess pressure inside, namely,

$$P = 4T/R,$$

where R is the radius of the bubble. For a spherical liquid drop the corresponding formula is

$$P = 2T/R,$$

since a liquid drop has only one surface compared to a bubble, which has two.

We know that when a liquid is held in a capillary tube, a drop is formed at the lower end. Raman devised a "convenient and fairly accurate" method for measuring T from observations made on the drop. The quantities to be determined from such observations are R and P. Unlike a bubble, the drop at the end of a capillary does not quite have a spherical shape; rather it is a paraboloid of revolution. The section of the drop (see Fig. 2.4) is a parabola, and does not have a unique radius, but in the experiment under consideration, one is interested in a small region of the drop near the very bottom. Writing the equation for AOB as

$$y^2 = 4ax$$

and then approximating the region near the tip to a circle, one easily finds that

$$R = 2a.$$

If one can determine $2a$, then R is known. The problem thus boils down to an accurate study of the profile AOB and the determination of the quantity $2a$ by an analysis of this profile.

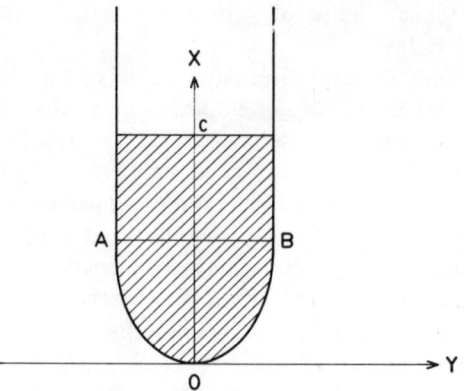

Figure 2.4 Schematic of a liquid drop at the end of a capillary tube. The drop profile is a parabola.

Turning next to the quantity P, one sees that this is clearly equal to the pressure head of the column CO, i.e.,

$$P = g\rho h,$$

where ρ is the density of the liquid and g the acceleration due to gravity. Since both these quantities are known, pressure determination reduces to a measurement of h, the height of the column CO.

It is not easy to hold a drop steady as there may be vibrations. The drop could also fall while measurements are in progress. Raman therefore resorted to a photographic method which permitted a leisurely and accurate analysis of the drop profile. The method employed by him is shown in Fig. 2.5.

> A tube 2 cm diameter is connected up with a tube 6 mm diameter by Caoutchouc tubing. The arrangement shown in the figure is filled up with the liquid. The small tube can easily be adjusted so that the liquid bulges out of both tubes, the concavity being in opposite directions in the two. A plumb-line hanging between the two gives the direction of the vertical. The apparatus is then photographed in the following manner:
>
> A horizontal beam of parallel light is produced by an electric spark placed at the focus of an achromatic lens of long focal length (about 1.5 metres), and is thrown on the apparatus. The shadow cast by the tubes is then photographed on a plate held vertically as near to them as possible. [2.5]

Measurements on the film were made using a microscope. Raman used a biologist's microscope rather than a travelling microscope since the latter permits distance measurements only along one direction whereas he needed to measure both the x and y coordinates of the parabolic profile. But the ordinary microscope, while permitting movements in two perpendicular directions, is not graduated. No problem! Raman placed a "réseau plate" on the photographic film. The "réseau

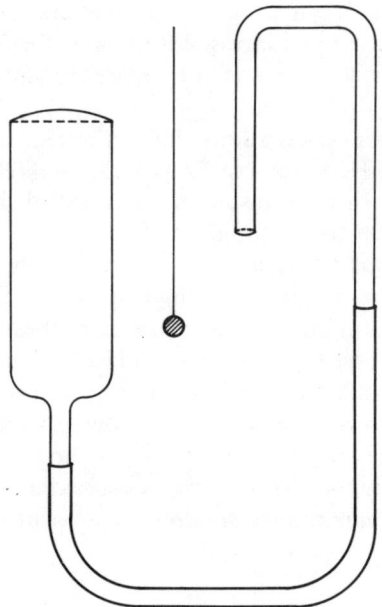

Figure 2.5 Experimental arrangement used by Raman to measure the surface tension of liquids. For explanations, see text.

plate" is a glass plate on which is ruled a set of squares (side $= 0.5027$ cm). It thus served as a graph paper, permitting an accurate determination of the liquid drop profile and hence of R.

In the practical implementation, Raman is careful about many details. He recognizes that the drop does not cast a sharp shadow on the plate but produces a diffraction band, and that the darkest part B of the band is displaced some distance from G the geometric shadow point (see Fig. 2.6). A correction GB is therefore required. Secondly, the curvature of the meniscus in the larger tube needs consideration. If one is not careful, there would be errors in the determination of this curvature, and this was in fact "the weakest point in the whole work". He hoped to eliminate this inaccuracy in future work by using a tube with a diameter greater than 6 cm so that the larger meniscus could be altogether eliminated.

Raman measured T for distilled water, and repeatedly obtained values between 75 and 77 dynes per cm. A previous determination by Hall in 1893 had yielded the value 71.3 dynes per cm at 30°C. Raman is not able to explain the difference but notes that his different trials yielded consistent results. He hoped to investigate later the discrepancy with existing values, but never did!

Once again it is a single-author paper, and once again there are no acknowledgements. If his first paper had a distinguished companion, so did his second one; this time it was a paper by the great Lord Rayleigh, the subject being the sensitiveness of the ear to pitch. Interestingly, Raman himself was to enter this general area later.

Raman's papers attracted Lord Rayleigh's attention, and the two exchanged some correspondence. It is said that Lord Rayleigh addressed Raman as Professor. One cannot blame him for not knowing that Raman was but a mere student still in his teens!

Raman apparently made several other observations while in college but did not publish them for one reason or another. However, one finds occasional references to these in subsequent papers of his. In one instance, he recalled his earlier investigations in the form of a comment to a journal [2.6].

In 1910 a paper entitled "Some curious phenomena observed in connection with Melde's experiment", by one J. S. Stokes, appeared in *Physical Review*. Stokes had connected a linen thread to a vibrating tuning fork so as to excite vibrations in the string. The other end of the string was passed over a pulley and had suitable weights attached to it a few centimetres below the pulley. Upon exciting vibrations of the string, Stokes found that the pulley slowly began to rotate towards the prong of the tuning fork. A drop of kerosene was applied in the groove of the pulley and the direction of rotation reversed. Stokes made various other observations but could not explain them. When the paper appeared, Raman wrote a comment to *Physical Review*, observing:

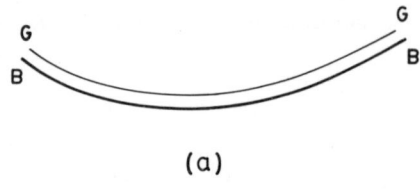

(a)

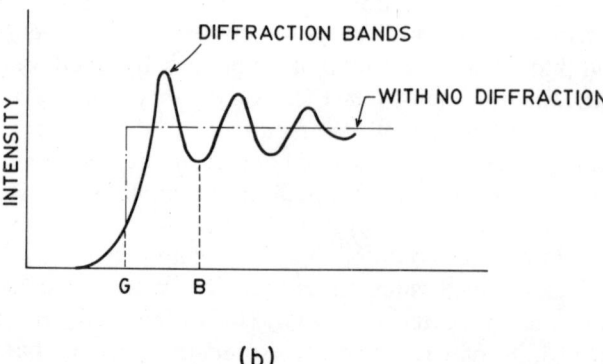

(b)

Figure 2.6 The photograph of the drop does not show a sharp shadow but a series of diffraction bands as in (**a**). The dark edge *BB* which is clearly discernible in the photograph, is displaced from the geometric edge *GG*. Raman therefore applies a correction *GB* to locate the correct geometric shadow. (**b**) A schematic of the intensity variation in the diffraction bands. The intensity pattern in the absence of diffraction is also shown.

> At the conclusion of the paper referred to [i.e., the paper of Stokes], the writer remarks, "As far as the demonstrator has been able to ascertain, these phenomena have not been hitherto observed or described by anyone." I may therefore be permitted to observe that the phenomena described in the paper (with the exception of the effect on the rotation of the pulley of waxing the thread) were observed by me about five years ago, when working in collaboration with Mr V. Apparao at the Presidency College, Madras, and were subsequently shown by us to a large number of others. As, however, other phenomena of interest were then observed which seemed to be of greater importance and which could be explained less readily, I did not seek an opportunity of publishing my observations on the subject.
>
> I may also state that the phenomenon of the rotation of the pulley was independently observed about two years ago by Mr A. W. Porter[10], who published a note on the subject in Knowledge and Scientific News at that time.
>
> My observations furnish a clue for the explanation of an effect which Mr J. S. Stokes says he is unable to account for.... [2.6]

One thing that becomes clear from these early papers is that young Raman was shaping up as an experimentalist. But he was not a mere gadgeteer, though he did like to improvise and innovate. (It is said that to rig up his apparatus, he would sometimes buy odd parts in the Moore Market, famous for its second-hand shops.) Nor was it Raman's aim to just make measurements or collect data. As we shall see in greater detail in Chapter 4, he was driven by a deeper urge – a quest for the aesthetics of natural phenomena. Not surprisingly, he emerged as a superb natural philosopher in the rich tradition of the nineteenth century, with an intuitive approach to natural phenomena. The mathematician Peitgen [2.7] observes that there are two complementary ways of experiencing the natural world – one analytic and the other intuitive. Chandrasekhar and Raman are classic examples of these two approaches. Peitgen also reminds us, "We have become accustomed to seeing them [these two approaches] as opposite poles, yet don't they depend on one another?"

2.7 Books that Influenced Raman

Youth is a period when the personality is moulded. Among the various influences which control this process, books constitute one. What are the books that influenced Raman? He himself provides us the answer.

> I finished my school and college career and my university examinations at the age of eighteen. In this short span of years had been compressed the study of four languages and of a great variety of diverse subjects, in several cases up to the highest university standards. A list of all the volumes I had to study would be of terrifying length. Did these books influence me? Yes, in the narrow sense of making me tolerably familiar with subjects so diverse as Ancient Greek and Roman History, Modern Indian and European History, Formal Logic, Economics, Monetary Theory and Public Finance, the late Sanskrit writers and minor English authors, to say nothing of Physiography, Chemistry and a dozen branches of Pure and Applied Mathematics, and of Experimental and Theoretical Physics. But out of this welter of

> subjects and books, can I pick out anything really to mould my mental and spiritual outlook and determine my chosen path in life? Yes, I can and I shall mention three books.
>
> A purposeful life needs an axis or hinge to which it is firmly fixed and yet around which it can freely revolve. As I see it, this axis or hinge has been, in my own case, strangely enough, not the love of science, not even the love of Nature, but a certain abstract idealism or belief in the value of the human spirit and the virtue of human endeavour and achievement. The nearest point to which I can trace this source of idealism is my recollection of reading Edwin Arnold's great book *The Light of Asia*[11]. I remember being powerfully moved by the story of Siddhartha's great renunciation, of his search for truth and of his final enlightenment. This was at a time when I was young enough to be impressionable, and this reading of the book fixed firmly in my mind the idea that this capacity for renunciation in the pursuit of exalted aims is the very essence of human greatness. This is not an unfamiliar idea to us in India, but it is not always easy to live up to. It has always seemed to me a surprising and regrettable fact that the profound teaching of the Buddha has not left a deeper impression in the life of our country of which he was the greatest son that ever lived. [2.2]

So deep was the impression made on Raman, that in later life he made frequent references to Buddha, a notable occasion being the dinner following the Nobel Prize ceremonies (see Chapter 5).

What about the sciences? Were there not books on science which influenced him? Yes indeed.

> The next set of books that I have to mention is one of the most remarkable works of all time, namely, *The Elements of Euclid*. Familiarity with some parts of Euclid and a certain dislike of its formalism have dethroned this great work from the apparently unassailable position which it occupied in the esteem of the learned world for an almost incredibly long period of time. Indeed, my own early reactions to the compulsory study of Euclid were anything but favourable. The reason for this is, I think, to be found in the excessive emphasis placed on the subject as an intellectual discipline and the undue attention given to details as distinguished from its broader aspects. To put it a little differently, the student of Euclid is invited to took at the trees and to examine their branches and twigs so minutely, that he ceases even to be conscious of the existence of the wood. The real value of Geometry appears when we consider it as a whole, not merely as the properties of straight lines, triangles and circles, but of everything else, curves, figures and solids of all kinds. Thus regarded, Geometry makes a profound appeal both to our senses and to our intellect. Indeed, of all branches of mathematics, it is that which links most closely what we see with the eye with what we perceive by reasoning. The ancient Greeks had a fine sense of the value of intellectual discipline, they had also a fine sense of the beautiful. They loved Geometry just because it had both these appeals. In my early years, it was a great struggle for me to learn to overcome the dislike of the formalism of Euclid and gradually to perceive the fascination and beauty of the subject. Not until many years later, however, did I fully appreciate the central position of Geometry in relation to all natural knowledge. I can illustrate this relationship by a thousand examples but will content myself with remarking that every mineral found in Nature, every crystal made by man, every leaf, flower or fruit that we see growing, every living thing from the smallest to the largest that walks on earth, flies in the air or swims in the water or lives deep down on the ocean floor, speaks aloud of the fundamental role of Geometry in Nature. The pages of Euclid are like the opening bars of the music of the grand opera of Nature's great drama. So to say, they lift the veil and show to our vision a glimpse of a vast world of natural knowledge awaiting study.

The Early Years

This explains Raman's deep interest in crystals and gems, concerning which more will be said later.

Raman spent many years studying vibrations, sound and musical instruments. One can partly attribute this to the deep and abiding love for music which he inherited. But then, why did he not simply relax and enjoy music? Did not Einstein, for example, do that? Why all the studies on tone and its sensations, the physics of musical instruments, and so forth? Was he influenced by something he had read? Yes, the works of the great Helmholtz!

> Of all the great names in the world of learning that have come down to us from the remote past, that of Archimedes, by common consent, occupies the foremost place. Speaking of the modern world, the supremest figure in my judgement is that of Hermann von Helmholtz. In the range and depth of his knowledge, in the clearness and profundity of his scientific vision, he easily transcended all other names I could mention, even including Isaac Newton. Rightly he has been described as the intellectual colossus of the nineteenth century. It was my great good fortune, while I was still a student at college, to have possessed a copy of an English translation of his great work *The Sensations of Tone*. As is well known, this was one of Helmholtz's masterpieces. It treats the subjects of music and musical instruments not only with profound knowledge and insight, but also with extreme clarity of language and expression. I discovered this book for myself and read it with the keenest interest and attention. It can be said without exaggeration that it profoundly influenced my intellectual outlook. For the first time, I understood from its perusal what scientific research really meant and how it could be undertaken. I also gathered from it a variety of problems for research which were later to occupy my attention and keep me busy for many years.

Raman also makes a reference to another great masterpiece of Helmholtz, namely, *The Physiology of Vision.* Unfortunately it was not available to him and he never read it. Interestingly, towards the end of his life, Raman himself wrote a book with the same title.

It should be added that Raman also came under the spell of Lord Rayleigh's works which considerably influenced his thinking on optics. Indeed, Rayleigh's volumes were constant companions to leading scientists of the day, inasmuch as they provided the best introduction to topics in theoretical physics.

2.8 Career and Marriage

Raman passed the MA examination in January 1907, topping as usual and walking away with all the prizes. His heart was clearly in science but opportunities for a research career in India (for Indians!) were nil. Going to England was also ruled out, on health grounds. Government service remained the only other attractive option, having the merit of being safe, secure and prestigious[12]. Well, if it was going to be Government service, then why not in style? The coveted Indian Civil Service (ICS) was the highest echelon in Government service but to enter it one had to study in England and also appear for the examination there, which, in Raman's case, was ruled out on medical grounds. The next best bet was the Financial Civil Service (FCS) of which Raman's brother C. S. Iyer was already a member. The FCS was the forerunner of today's Indian Audit and Accounts Service. Recruitment to it was by an all-India competitive examination, but even to appear for this examination one had to first go through an interview. Raman was duly screened, and, as expected, he stood first in the written examination although he had to study many unfamiliar subjects like history and economics. Raman's brother Mr Ramaswamy told me, "After returning from the screening interview Raman said, 'I took one look at all the candidates who had assembled, and I knew I was going to stand first.'" An early example of the famous Raman ego!

Raman passed the FCS examination in 1907, and while still awaiting a posting he married Lokasundari. Marriages were arranged by parents, a process the starting point of which was a critical analysis of the horoscopes of the boy and the girl to see if they matched. Then followed a visit by the boy and his parents to the house of the girl during which the hapless girl was paraded while her anxious parents duly extolled her capabilities, talents and virtues. A small musical presentation by the girl offered the boy a few fleeting moments to catch a glimpse of the coy girl. The next time he would be seeing her would be on the occasion of the marriage, that is, if the alliance was approved by the boy's party (meaning not only their liking the girl but more so their acceptance of the financial terms offered by her father!).

Raman's marriage, however, followed a daringly different pattern. While studying in the Presidency College, Raman became friendly with one Mr Ramaswamy Sivan, a Freemason, theosophist and a man with progressive views. Raman used to visit Sivan frequently at the latter's home, and on one occasion heard the strains of the *veena*[13] as he was entering. The player was young Lokasundari, Sivan's sister-in-law, from Madurai. Lokasundari was quite talented in the playing of the *veena*, and Raman was attracted to her music. Raman expressed a desire to meet her and upon doing so took an instant liking to her. Lokasundari being of marriageable age (by the standard of those days), her parents and relatives (which included Sivan) were on the look-out for a suitable groom. Impressed by Raman's keen intellect, Sivan cautiously sounded Raman and the latter immediately accepted the idea. The next step was to get the approval of Raman's parents. It was then discovered that, though a Brahmin, Lokasundari belonged to a different subsect. In the language of physics, marriage even among subsects was in those days a forbidden transition! Raman's father was quite liberal-minded and not only accepted the idea of Raman selecting his own bride, but also one from a different subsect. However, many others in the family were not pleased, including Raman's mother. But Raman defied all pressures and insisted on having his own way. What is even more remarkable (indeed even by today's standards), Raman refused dowry!

We cannot leave the subject of Raman's marriage without briefly quoting Ramaseshan:

> The story has it that on the first occasion he saw her, she was playing on the veena the Tyagaraja keertana [composition] '*Rama ni samanam evaro*' [Rama, is there anyone your equal]. We shall never know whether it was by intent or by accident. Anyway, she insists that she still does not know if Raman married her for the extra

The Early Years

allowance of Rs. 150 which the Finance Department gave to its married officers! [2.1]

In later years Lokasundari was better known as Lady Raman. Bhagavantam[14] writes of her:

> Those who have known her ... had often said that her principal interest in life was to enable Professor Raman to carry on his scientific work with efficiency and in an uninterrupted manner.... Seldom did she permit projection in the public of her own personality as distinct from that of her husband. This aspect of hers, besides being in line with the best of Indian traditions, was so noticeable on occasions that she drew the admiration of all concerned. [2.8]

In the middle of 1907, while still in his teens (!), Raman was posted as Assistant Accountant-General in Calcutta (on a salary of Rs. 400 which included the marriage allowance)[15], and thither he went with his young bride. Instead of going West, he went East. Instead of going to the capital of the British Empire, he went to the capital of British India, and instead of entering science, he entered Government service. He was all set to get lost in obscurity and disappear into oblivion, but he did not!

3 Oh Calcutta!

It was fortunate that Calcutta was still the capital of British India when Raman entered Government service, for otherwise he might have been posted somewhere else[1]. There is no question that, irrespective of where he was asked to serve, Raman would definitely have pursued research on the side, as indeed he did during brief spells at Rangoon and Nagpur, places which were wilderness as far as science was concerned. On the other hand, it is equally clear that Calcutta offered to Raman opportunities which no other city in India could have.

It was almost a romantic association between him and that city – while it lasted! In Calcutta Raman was able to give full expression to his scientific creativity and rise to great heights. Calcutta, for its part, acquired the distinction of becoming the premier city for science in the East. A remarkable aspect of Raman's highly successful scientific activities in Calcutta is that he never met the person who made it possible! In fact this great benefactor passed away three years before Raman arrived in that city. That man was Dr Mahendra Lal Sircar.

3.1 The IACS – Its Origin[2]

Mahendra Lal Sircar was born in 1833. He received his early education in the Hare School and the Hindu College (to which a reference has been made in

Chapter 1). Later he joined the Calcutta Medical College and obtained the MD degree in 1863, standing first in the order of merit. Deeply nationalistic in outlook, he allowed his interests to range far outside the medical world. He was particularly concerned that the recently introduced system of education laid greater stress on the arts than on the sciences. Further, there were hardly any opportunities for Indians to pursue advanced science since the few Government centres which existed, with facilities for advanced studies as well as research, were closed to Indians (on the ground that they lacked training).

In 1869 Mahendra Lal Sircar wrote an article in the *Calcutta Journal of Medicine* entitled "On the desirability of a national institution for the cultivation of sciences by the natives of India". After dealing at length with the achievements of the ancient Indians, Sircar argued that the prevailing backwardness of the country was due to backwardness in science. The solution lay, Sircar said, in the vigorous pursuit of the sciences by original research. But how could one go about it, especially when there were no avenues? Mahendra Lal had an answer:

> We want an Institution which will combine the character, the scope and the objectives of the Royal Institution of London and of the British Association for the Advancement of Science. We want an Institution which shall be for the instruction of the masses, where lectures on scientific subjects will be systematically delivered and not only illustrative experiments performed by the lecturers, but the audience would be invited and taught to perform them themselves. And we wish that the Institution be entirely under native management and control.

As it turned out, the institution which finally emerged was, if at all, more akin to the Royal Institution. The counterpart of the British Association for the Advancement of Science had to wait for some more years until the Indian Science Congress Association was born.

The Royal Institution was founded in 1799 "for diffusing knowledge and facilitating the general introduction of useful mechanical inventions and improvements and for teaching by courses of philosophical lectures and experiments the application of science to the common purposes of life". The Institution was made famous by the brilliant researches of many people, Michael Faraday in particular. Years later, in an article entitled "India's debt to Faraday", Raman recalled the inspiration the Royal Institution had provided to Mahendra Lal Sircar [3.1].

To get back to the story of Sircar and his dreams, his article elicited wide support encouraging him to publish a prospectus for the proposed centre in the *Hindoo Patriot* in January 1870. The response was immediate, with several leading persons making contributions in the range of Rs. 1000 to Rs. 2000 (which was really a substantial sum in those days). However, despite these encouraging signs, the project did not get off the ground for several years and had to be kept alive with meetings and discussions! (One sees here a common Indian penchant for putting off decisionmaking often unnecessarily.)

In 1875 Rev. Fr Lafont of the St Xavier's College started a small observatory, which generated a fresh wave of enthusiasm for science. Lafont then joined hands with Sircar, and together they got the Lieutenant-Governor of Bengal, Sir Richard

Temple, interested in Sircar's project. In April of that year, the first meeting of the subscribers took place during which Sircar once again made a fervent plea for science, adding this time also a practical note:

> Now, Gentlemen, for want of such [trained] men here, Government has to bring out men from England whenever any necessity arises for carrying on investigations in any subject, and even for professorships in its educational institutions. Whether, when our Association will be able to furnish such men, Government will accept their services, I cannot venture to say, but then there will be at least no excuse for Government to order out men from England at necessarily heavier expense.

Sircar also firmly believed in total freedom from Government control or interference (a concept Raman was to reiterate later) and observed:

> ... we should endeavour to carry on the work with our own efforts, unaided by Government, or perhaps more properly speaking, without seeking its aid. Now this does not mean that we will not accept any aid from that quarter if it comes to us unasked, and unhampered with conditions and restrictions, excepting the all-important condition of the continuance of the Association. Let me not be misunderstood. I want freedom for the Institution. I want it to be entirely under our own management and control. I want it to be solely native and purely national.

The meeting endorsed the aims and objectives of the proposed science intitute. However, once again nothing much happened for the next several months, except for the happy augmentation of the subscription funds! In November 1875 there was a second meeting of the subscribers but this time there was some progress in that a Provisional Committee was formed to work out the detailed structure of the proposed centre. The Committee soon got to work and noted that while there were several branches of science worth pursuing, there was only Rs. 80,000 in the kitty. Caution was therefore required in making plans, and accordingly the Provisional Committee recommended that a beginning be made only in three areas, namely, physics, chemistry and physiology, since these were "the only branches of science which have received permanent professorships in the Royal Institution". But who would teach these subjects and carry out research? The Committee had an answer. It recommended that for each subject there should be "a head worker, selected from among the graduates of the colleges", who would take charge and do the needful. These head workers had to be paid, for Sircar did not believe in unremunerated workers. "The mind must have leisure to think ..., and this can only be secured by providing for the demands of the stomach" he said.

The institution was formally established at the third meeting of the subscribers, but not before some anxious moments. At that time, there was an organization called the India League which was agitating for the establishment of a technical college on the model of the *technischen hochshulen* of Germany and Switzerland. A vigorous press campaign in support of this idea was launched while Sircar's proposal was ridiculed as pleasure-seeking. Sircar correctly anticipated trouble from this quarter during the (forthcoming) third meeting of the subscribers, and he took the precaution of briefing Sir Richard Temple in advance about his intentions and objectives. This was a wise move, for it was Sir Richard who presided over the fateful third meeting of subscribers.

The meeting began with a presentation by Sircar of the Provisional Committee's report, following which the India Leaguers were invited to voice their views. It was then the turn of the distinguished oriental scholar Dr Rajendra Lal Mitra to take the podium, and he delivered a blistering attack on those who thought that the time was not ripe for India to engage in basic research. He added:

> Time never came to the sluggard, while the active and energetic could always take it by the forelock, and force it to appear.... Science had a higher and nobler claim than the narrow, utilitarian, Benthamite one.... It was the most powerful lever for progress, for the advancement of civilization, for ennobling the mind of man.

The same question of pure versus applied science was to haunt Raman sixty years later.

Thanks to the advance briefing by Sircar, Sir Richard skilfully steered the meeting, lending full support to Sircar's ideas although he was, personally, more inclined towards the technical school. The meeting of the subscribers finally endorsed the recommendations of the Provisional Committee, and a Committee of Management was appointed, with Sir Richard Temple as President and Mahendra Lal Sircar as Secretary.

Six years had passed since the idea was first mooted, and a major milestone had been crossed with the Science Association, as it was then called, formally established. In February 1876 the Government of Bengal offered to the Association for occupation the available premises at 210 Bow Bazar Street. But there were some conditions which the Committee of Management did not quite accept. Instead it requested the Government to make an outright gift of the land and the building. The Government did not agree to this counter-suggestion, but allowed occupation against an investment of Rs. 50,000 in Government securities.

Finally, scientific activity commenced in July 1876 with Rev. Fr Lafont and Mahendra Lal Sircar delivering lectures on topics in physics and Dr Kanai Lal Dey in chemistry. The honorary lecturers were "allowed to charge a fee of eight annas [half a rupee] from all persons other than subscribers and donors of Rupees fifty and upwards". In 1878 a formal constitution was adopted and the organization was named the Indian Association for the Cultivation of Science (IACS), by which name it is still known.

The building obtained from the Government was not quite suitable, besides being old. Laboratory space and a lecture hall were badly needed but no additions or alterations could be made since the Association did not have property rights. In 1880 the Association persuaded the Government to sell the house, and 210 Bow Bazar passed into the ownership of the Association. Construction of the lecture hall was then given priority and the hall was inaugurated in 1884 by Lord Ripon, the Viceroy. This lecture theatre served the Association well, and half a century later, Raman, while referring to his public lectures, observed:

> As the requests to deliver these lectures came from public bodies, they had to be delivered in various parts of the city.... It would have been much more satisfactory if such lectures had been delivered in our own well-equipped Lecture Theatre. [3.2]

While the auditorium did meet some needs of the Association, there were still no laboratories since there was no money left to build them! A timely grant at this stage by the Maharajah of Vizianagaram saved the situation, and a new building with as many as thirteen rooms was built in 1891. Not surprisingly, it was named the Vizianagaram Laboratory.

Although Sircar dreamt of research, teaching remained the main activity of the Association for many years, apart from daily meteorological observations and routine chemical analysis for assorted customers. However, the teaching was at an advanced level, with the speaker often choosing a topic which he himself wanted to learn.

One lecturer of great distinction who participated in this programme was Sir Asutosh Mookerjee, who was later to become Raman's mentor. Sir Asutosh was a luminary of the legal profession, rising eventually to the position of a judge of the Calcutta High Court. He was also a gifted mathematician and a keen member of the Asiatic Society. Sir Asutosh lectured on a wide variety of topics in both physics and mathematics, examples being Stokes' dynamical theory of diffraction, potential theory, electromagnetic theory, hydrodynamics, Clebsch's transformations, applications of Bessel functions and Fourier transformations, and theory of analytic functions. Not only were the lectures widely appreciated, but, more important, many of the topics were current, i.e., they were being actively researched in Europe at that time. For the young science enthusiasts of Calcutta, the lectures of Sir Asutosh were classic models of self-study. Nowhere else in India was such training available. Strangely enough, despite all the invigorating lectures delivered by so many, *research* did not take root in the Association the way Mahendra Lal Sircar wanted. Eminent scientists like Sir J. C. Bose and Sir P. C. Ray (see Sec. 3.5) lectured at the Association but preferred to conduct their researches in their own college.

Mahendra Lal Sircar was keen about research and extolled its value at every annual meeting of the Association. But there was not enough money to appoint full-time research professors like in the Royal Institution, and Sircar yearned in vain for torch-bearers in the tradition of Young, Davy, Faraday and others. Addressing the twenty-fifth annual meeting in 1902, he lamented:

> I do not know how to account for this apathy of our people towards the cultivation of science. And therefore I am forced to confess that I made a mistake in starting the project of founding a Science Association at all, and that I have wasted a life, as I have told you, in attempting to make it a national institution.

In 1903 he was too ill to attend the annual meeting but his message was read. It again revealed his despair.

> ...I am afraid [it] is likely to be my last [address]....I have only now to reiterate my conviction that if our country is to advance at all and take rank and share her responsibilities with the civilized nations of the world, it can only be by means of science or positive knowledge of God's works. To this end I have given the best portion of my life, but I am sorry to leave this world with the impression that my labours have not met with the success which the end aimed at deserves. However, I do not despair of our future. My faith in an over-ruling Providence has not abated an iota on account of my own ill success.... Younger men should come and step in to take my place and work with more energy than I have been able to put forth....

Mahendra Lal Sircar died a disappointed man in February 1904 but his last wish was amply fulfilled when in 1907 there entered the portals of Vizianagaram Laboratory a remarkable person. This young genius was a seeker of the scientific truth, and though he lacked the expert training the Association had been providing to its members, he nevertheless achieved incredible success in research. And he wrought a magic on the Association that even the noble Sircar could never have dreamt of.

3.2 Enter Raman

Raman arrived in Calcutta in June 1907, and immediately rented a house. In those days, hunting for a house was not the nightmare it presently is. By a happy accident, his residence was in Scots Lane, quite close to Bow Bazar Street. Raman was not aware of the IACS then, but pretty soon he spotted the signboard while riding a tram to work. His curiosity was aroused, and on his way back from office he stopped by to investigate. There was an ornamental door at the entrance to the Vizianagaram Laboratory but it was closed. Raman rang to gain admission and after a while it was opened by an attender named Asutosh Dey. It was a historic meeting, for Dey, later more affectionately known as Ashu Babu, was to become Raman's right-hand man for the next quarter of a century or so.

As Raman entered he found the place wore a deserted look, and there was no sign of any activity. Raman was intrigued and asked Dey what the Association was all about. The latter silently led him to the presence of the Honorary Secretary Amrita Lal Sircar, the nephew of the late Mahendra Lal Sircar. Raman repeated his queries to Amrita Lal who in turn wanted to know why Raman was interested. Raman explained that though he was employed in Government service he was keen on pursuing research in his spare time, and by way of credentials cited his research as well as publications while in college. Amrita Lal Sircar could scarcely believe his ears. The late Founder had appealed endlessly, but in vain, for researchers to avail of the facilities of the Association. The magnificent laboratories built through the munificence of the kindly Maharajah of Vizianagaram were lying idle and gathering dust. Here was a young man, already employed and that too not in a scientific service, who, nevertheless, wanted the use of the Association's facilities and, on top of that, was not asking for an honorarium. It was too good to be true! Amrita Lal Sircar shed a silent tear in memory of his departed uncle and, welcoming Raman with open arms, said, "All these years we have been waiting for a person like you." Raman was immediately given full use of the Association's facilities and the freedom to come and go as he liked. Years later Raman acknowledged this, observing,

> My own work at Calcutta commenced in 1907 and was made possible by the special facilities put at my disposal by the Honorary Secretary Mr Amrit Lal Sircar who had the Laboratory kept open at very unusual hours in order that I might carry on research in the intervals of my duties as an officer of the Indian Financial Department. [3.3]

Raman plunged into science right away, leading a hectic double life. We have an account of it from Ramaseshan, who has heard of it from Lokasundari herself:

> Young Lokasundari tells us of the routine – 5.30a.m. Raman goes to the Association. Returns at 9.45 a.m., bathes, gulps his food in haste and leaves for his office, invariably by taxi[3] so that he may not be late. At 5 p.m. Raman goes directly to the Association on his way back from work. Home at 9.30 or 10 p.m. Sundays, whole day at the Association. Truly, not an exciting life for a young bride.

But the young bride was enterprising, and she launched on her own programme of self-education, acquiring eventually an ability to converse and write in English, not a mean achievement considering that she, like most young girls of that period, had received barely any schooling. She had many interesting experiences, and Ramaseshan tells us about one of them,

> There was one, of her going to the beautiful church at the end of Scots Lane when exploring the neighbourhood – but the Brahmin cook (they had one, remember the Rs. 150 married allowance?) left because he did not want to have any truck with people who go to church!

Raman's work at the Association was interrupted for a while when he was posted as the Currency Officer in Rangoon[4]. But this did not keep him from science. He records in a paper published during that period: "This investigation was commenced at the Physical Laboratory of the Indian Association for the Cultivation of Science, Calcutta. My removal from that station necessitated its being completed elsewhere." Raman often worked at home, but this he could afford to do since his experiments required only the proverbial "shoe-string and sealing wax".

While Raman was in Rangoon, his father Chandrasekara Iyer fell ill and Raman came to see him. From some references in his papers, one gathers that during this leave Raman paid a brief visit to the Presidency College to conduct some experiments there. Later Raman's father passed away and after all the associated religious ceremonies were over, Raman rejoined duty. This time he was posted in Nagpur.

In Nagpur, Raman established contact with one Prof. Owen in the College of Science. Later his links with the College continued for years in various forms, and Raman is now remembered in Nagpur through a memorial museum. Lovers of folklore might be interested in the following:

> There is a story, still being recalled by a citizen of Nagpur whose collection of a few hundred-rupee notes was nearly burnt by a blazing fire. The perturbed individual went to the Accountant-General's office and presented the half-burnt bundle, but with little hope of retrieving any. Any other officer would probably have shown him the door, but Raman who was then in the office of the Accountant-General, took the trouble to scrutinize the notes under a magnifying glass, one by one, and instructed the treasurer to give him fresh notes. Raman argued that the numbers on the half-burnt notes were visible and thus it was a genuine case. [3.5]

The outstation duty was fortunately brief, and Raman was soon back in Calcutta. He moved into a house adjoining the Association though the locality was not suitable for residence, and had a door put between his house and the Association so that he could walk into the laboratory any time he wished! Research was resumed with renewed vigour and many topics were studied, concerning which more will be said in the next chapter. A few broad observations are, however, appropriate at this stage.

Firstly, one is struck that Raman worked practically alone during the entire first decade, barring the assistance rendered by Ashu Babu in performing some of the experiments. The kinds of problems Raman studied are interesting. As Ramaseshan observes:

> It is remarkable that every one of them is connected with his direct experience, thus arousing his curiosity. He has heard his father play the violin. He was worked with the sonometer and done Melde's experiment in college. So follow his papers on the bowed string, the struck string, the maintenance of vibrations, resonance, aerial waves generated by impact, the sound of splashes, the singing flame, music from heated metals and many others. [3.4]

The study of vibrations led Raman quite naturally to a study of musical instruments, including those of Indian origin. Unfortunately, Raman's remarkable contributions in this area have been overshadowed by the great discovery he made later. Just to mention a few highlights, he explained why the Indian percussion instruments, the *mridangam* and the *tabla*, produce musical notes in contrast to the Western drum, which is "unmusical and just a noise producer". We are all familiar with the *tambura*. A classical musical concert without it is inconceivable in the Carnatic or Hindustani styles. The *tambura* supplies the *sruti* but have we ever wondered why the drone is so rich? On the other hand, when we pluck a sonometer, its sound is dull and monotonous. Indeed, so vibrant is the sound of the *tambura* that Lionel Fielding used it for starting the signature tune of All India Radio. What is it that gives life to the *tambura*? Raman furnished the answer.

Raman displayed considerable ingenuity in designing his experiments, and my favourite example is the mechanical violin player (see Fig. 4.22). Raman had developed a theory of the violin, and he needed a gadget to test it. As one knows, the violin is played by holding the instrument steady and moving the bow. In Raman's player, the bow is stationary and it is the violin that moves!

Raman firmly believed that scientific research is useless unless the results are promptly published. Thus, after he joined the Association, there was a steady flow of research publications from it. In fact the first ever paper to emanate from the Association was due to Raman, some thirty and odd years after it (i.e., the Association) had come into existence[5]. Under Raman's encouragement, even Ashu Babu once authored a paper on his own for the prestigious *Proceedings of the Royal Society*, a remarkable achievement for one who had never studied in college[6]. Altogether, Raman's first decade at the Association was quite fruitful with as many as twenty-seven publications coming out. In between, Raman picked up the Curzon Research Prize in 1912 and the Woodburn Research Medal in 1913. And he was still a Government officer!

The momentum generated by Raman had many consequences. Firstly, thanks to the increased visibility achieved by the Association, research work in chemistry and biology also took root. Secondly, the Association began to publish its own Bulletin. Of these, Bulletin No. 15, entitled *On the Mechanical Theory of Vibrations of Bowed*

Strings of Musical Instruments of the Violin Family with Experimental Verification of the Results, Part I, is truly a collector's item. Running to 158 pages it is priced at Rs. 2 As. 8 or 3s. 4d. (!). The printing (done at the Baptist Press) is of very good quality despite the presence of a liberal dose of mathematical symbols, and the plates are beautiful.

Raman also lectured extensively, adding further to the image of the Association. Raman's capabilities as a lecturer in popular science are legendary. He held his audience spellbound with his booming voice, superb diction, lively demonstrations and of course his rich humour. As Ramaseshan [3.4] remarks, "He was perhaps the greatest salesman science has ever had in this country."

The reputation of the Association reached even overseas, and in 1916 *Nature* wrote:

> At a time when Indian universities were purely examining bodies so dear to the Philistine soul, when secondary education in India was mainly book-mongery (to call it 'literary' would be a fault to heaven), and literary gentlemen were brought from England to feed raw Indian youths with husks of commentary laboriously ground from the English classics, Dr Mahendra Lal Sircar,... with single-minded devotion... set agoing a society much on the style of the Companies of the Friends of Natural History, the aim of which, to begin with, was, and had to be, generally educative.... By degrees and by the accretion of laboratories for particular studies, the institution, while retaining an educational character, advanced to the differentiated technical stage; and now, beyond its educational purpose, it has become a well-organized and well-equipped institution for original experimental research.

Raman was devoted to science but that did not mean he neglected his normal, official duties. Indeed, he was not merely a conscientious officer but a highly capable one too. The Member for Finance in the Viceroy's Council wrote of him, "We find Venkataraman is most useful in the Finance Department being, in fact, one of our best men." The establishment had high hopes for Raman but a change was soon to occur in his life.

3.3 Palit Professorship

In 1906 Sir Asutosh Mookerjee was appointed the Vice-Chancellor of the Calcutta University, and he promptly seized the opportunity to tone up science education. Undaunted by the absence of Government support, he raised private donations to the tune of several lakhs[7] of rupees and established the University College of Science. Many chairs were also established, especially in the names of the principal donors, namely, Sir Taraknath Palit and Sir Rashbehari Ghosh. The next step was to hire the people. Here too Sir Asutosh displayed keen discrimination, appointing only the very best. Thus Sir P. C. Ray (see Sec. 3.5) was invited to take the Palit Chair in Chemistry, and the four Rashbehari Professorships went to Ganesh Prasad (Applied Mathematics), D. M. Bose (Physics), P. C. Mitra (Chemistry) and S. P. Agharkar (Applied Botany). Three lecturers were also appointed, namely, S. N.

Bose, M. N. Saha and S. K. Mitra, all of whom were to achieve great reputation later. To whom should the Palit Chair for Physics go? For obvious reasons, Sir Asutosh had his eye on Raman but would Raman be willing to give up a well-paying Government job? Were there also not great possibilities for career advancement in that line? Anyway, there was no harm in trying, and an offer was duly made. To Raman, this offer "to an unknown Government official" was an "act of great courage", and he promptly jumped at it. But the Finance Department was not happy. As Ramaseshan [3.4] describes:

> Raman's decision produces consternation in the establishment. There may soon be Indianization, they tell him. As one of the best officers, he may even end up as Member (Finance) in the Viceroy's Council – who knows? But Raman's mind is made up.

The Viceroy's Council was the ultimate, and in that body the position of the Finance Member was the choicest one – yet Raman was prepared to give all that up!

At first it was hoped that Raman could take leave of absence for a while and try out the Professorship but Sir Harcourt Butler, the Member for Education, declined to permit Raman to accept the Palit Chair unless he completely resigned the Government job. Raman was not daunted and decided to sever his connection with the Government, even if it meant a huge pay cut from Rs. 1100 to Rs. 600 per month.

Sir Asutosh was both pleased and touched. Speaking on the occasion of the laying of the foundation stone of the University College of Science he said:

> For the Chair of Physics created by Sir Taraknath Palit, we have been fortunate enough to secure the services of Mr Chandrasekhara Venkata Raman, who has greatly distinguished himself and acquired a European fame by his brilliant researches in the domain of Physical Science, assiduously carried on under the most adverse circumstances amidst the distraction of pressing official duties. I rejoice to think that many of these valuable researches have been carried on in the laboratory of the Indian Association for the Cultivation of Science, founded by our late illustrious colleague, Dr Mahendra Lal Sircar, who devoted a lifetime to the foundation of an institution for the cultivation and advancement of science in this country. I should fail in my duty if I were to restrain myself in my expression of the genuine admiration I feel for the courage and spirit of self-sacrifice with which Mr Raman has decided to exchange a lucrative official appointment with attractive prospect for a University Professorship, which, I regret to say, does not carry even liberal emoluments. This one instance encourages me to entertain the hope that there will be no lack of seekers after truth in the Temple of Knowledge which it is our ambition to erect.

Despite Raman's willingness, the appointment could not be effected immediately as there were some legal problems connected with the Palit endowment. These were eventually overcome but in the meanwhile, yet another problem cropped up. According to one of the stipulations, the occupant of the Palit Chair must have received training in England. Raman, who had thus far never been to England, indignantly refused to go now in order to be "trained". The authorities gracefully yielded, and finally, in 1917, Raman entered the academic world. Years later, the elder statesman Rajaji, paying a tribute to Sir Asutosh Mookerjee, said that but for him, Raman would have retired as a faultless Accountant-General!

3.4 Calcutta School of Physics

The Palit Professor was not required to do any routine teaching to the M A and M Sc classes. Instead his duties were (1) to devote himself to original research in his subject to extend the bounds of knowledge, (2) to stimulate and guide research by students, and (3) to supervise the laboratory in the College of Science. No matter what the formal terms and conditions were, Raman obviously could not be kept away from teaching. Ramdas recalls that

> though under the terms of his appointment as the Palit Professor, he [Raman] was entirely free from any teaching responsibilities, he was equally enthusiastic to take a prominent part in M Sc teaching.... To some of us who had joined the M Sc course at Calcutta, Prof. Raman once made the side remark that the best way for him to master or revise any subject in physics was indeed to lecture on it.
>
> Prof. Raman took 'Electricity and Magnetism' in the year 1920–21 and 'Physical Optics' in 1921–22. Both sets of M Sc students felt that they were indeed listening to a type of inspired teaching to which was brought all the original flavour and excitement of the great giants of the past.... We shared with him much of the excitement and superb thrill that Benjamin Franklin, Oersted, Arago, Gauss, Faraday, Maxwell, Hertz, Lord Kelvin and many others must have felt while they were making their actual discoveries.... Often he used to take the entire forenoon, for 2 and sometimes even 3 hours – such was his tremendous love of teaching.... And after each lecture we used spontaneously to look up original papers and classical treatises like Maxwell's *Electricity and Magnetism*, J. J. Thomson's *Conduction of Electricity*, Faraday's *Experimental Researches*, Lord Rayleigh's and Kelvin's *Collected Papers*, and so forth. [3.6]

Those were the days of the non-cooperation movement started by Mahatma Gandhi.

> Sometimes, bands of students who had given up their studies at the call of Mahatma Gandhi used to squat in front of the Science College... barring entry to professors and students alike. On most of such occasions, by cajolement, entreaty, or otherwise, "Raman Sahib", as he used to be affectionately called in those days, would break the cordon and rush into the class room.

Raman now began operating on two fronts, with the Association becoming (where he was concerned) the research arm of the University. And once again life was hectic. As Ramdas tells us:

> Often he used to work on far into the night and, when exhausted, sleep on a table until the astonished Ashu Babu awakened him the next morning! On most mornings he would come in informal dress to the Association and carry on experiments until 9.30 a.m. when he would remember his lecture engagement at the University College of Science about 4 miles away. A quick return home, a few strategic sweeps of his razor for shaving, a quick bath, a hurried dressing along with the inevitable Madras turban, and he would start at the top of his powerful voice, "Ashu Babu – taxi", which would evoke an equally loud response from Ashu Babu! Snatching his breakfast in a minute or two he would rush into the Association with a sheaf of his lecture notes for the day, jump into the waiting taxi with his laboratory attendant Shivnandan, and after a hectic drive reach the Science College just in time to start his lecture.

Ramdas continues:

> The strain of jointly controlling two laboratories with the high tempo of their activities would have overwhelmed anyone. But Prof. Raman was equal to this gigantic task that he had set for himself. His physical energy was as extraordinary as his mental output. An outstanding characteristic of Raman was that he was seldom relaxed like ordinary persons, but was keyed up to a high pitch of fervour by the continuing thrill or excitement of the ever-expanding avenues of scientific researches that occupied his mind. He was capable of tremendous concentration on intellectual work, as the following incident will exemplify. Once, when he was drafting a rather tough paper for publication in his office at the Association, a Swamiji visited him. He welcomed him very courteously, asked him to take his seat and said that as soon as he finished the work he was busy with, he would be glad to spend some time with the Swamiji. I was present. For nearly a couple of hours he was completely oblivious of the Swamiji. When he had completed the task on hand, he suddenly looked at the Swamiji and told him that as he had only five minutes to spare before attending to another pressing engagement, he would request the Swamiji to explain to him briefly what message he had to give him. The Swamiji was equal to the occasion. He told Raman that deep concentration was what a spiritual person had to practise. And Raman already possessed this great quality as witnessed by the fact that he was not at all aware of anything else while he was bent on his work for two hours. A donation for a charitable cause that the Swamiji requested was generously given by Prof. Raman.

The double-pronged effort led to a symbiotic growth of both the University and the Association, resulting eventually in a thriving school of physics in Calcutta. Even as Raman was playing his role, Sir Asutosh was implementing far-reaching reforms in the University. Reviewing these developments, Raman wrote:

> The only other institution in Calcutta in which research in physics was possible [the reference here is to the Presidency College] was staffed by 'Professors of Physics' drawing extremely high emoluments in the so-called 'Indian Educational Service', but who did no research and were absolutely unknown to the world of science. It is not surprising therefore that for many years things were at a standstill. [3.3]

But what about the famous Sir J. C. Bose? Ah, Raman is careful about that! He writes:

> Naturally in making this statement, I must exclude the name of Dr J. C. Bose, who was an experimenter of distinction. But for some years prior to the date I mention, Dr Bose had turned his attention to biology. The research scholars in physics paid by the Government and attached to Dr Bose were apparently engaged as Laboratory Assistants in his biological work.

Raman adds:

> Thus a decade ago there was at Calcutta a total lack of anything that could be regarded as a real centre of teaching and research in physics. No doubt the subject figured in the curricula of the University, but the higher teaching had latterly been weak, particularly as regards mathematical physics, and research was absolutely at a standstill. A new impetus was obviously required and it was not long in coming.

The new impetus consisted first of a set of New Regulations (introduced by Sir

Asutosh Mookerjee) which strengthened teaching in physics and mathematics. Furthermore,

> lectures were delievered under the auspices of the University by distinguished scientific workers amongst whom may be mentioned Dr A. Schuster and Dr G. T. Walker[8]. These lectures undoubtedly stimulated interest in the study of physics, and brought home to the younger generation of university students the fact that scientific knowledge is essentially a product of the human mind and not simply something to be found in books.

Opportunities were also given to local scientists to show their capabilities. Thus D. N. Mallick delivered lectures on optical theories, which were later published in book form by the Cambridge University Press.

While the regulations applied to the University as a whole (i.e., to all the affiliated colleges), Sir Asutosh tried to provide a cutting edge by creating the College of Science. Commenting on this Raman observed:

> The establishment of the University College of Science has made possible a great advance in the higher teaching of physics in the University. Prior to it, the only institution at Calcutta that was equipped (even in part) for M Sc work in physics was the Presidency College.... Under the new arrangement, the higher teaching of physics has been divided up between the combined staff of the University College of Science and the Presidency College, and the resulting substantial addition to the number of men engaged on the work, and the high qualification of the men attached to the University College, have made a greater degree of specialization and a wider choice of subjects possible.... One respect in which the most substantial advance has been effected is in the teaching of the mathematical aspects of the different branches of physics. This is now possible, because three out of the eight lecturers in physics attached to the University College of Science are men who are first class M Sc's in applied mathematics, and have since made a special study of physics.... It must be obvious to every unprejudiced inquirer that the teaching of physics conducted by men of this stamp cannot possibly be much inferior to that attainable in any country.

Teaching is most effective when the teacher himself possesses an inquiring mind, i.e., is engaged in research. Concerning this Raman remarks:

> The most encouraging sign is the extent to which men who take part in the postgraduate teaching actively interest themselves in research work, and this feature would have been more marked but for the fact that the physical laboratory of the university College is still in its formative stage, and some of the best men in it have their hands full with the administrative detail of laboratory organization. Four out of the eight lecturers in physics attached to the University College, namely, Mr S. K. Banerjee, Mr S. K. Mitra, Mr M. N. Shaha and Mr S. N. Basu, have succeeded in publishing research papers in European journals, and two of the others have investigations in progress which are likely to prove fruitful in the near future[9]. Mr S. K. Banerjee has in particular distinguished himself by his exemplary character and by his remarkable capacity in the fields of mathematical and experimental research. In the course of about three years, he has already published six original papers, and has two more practically ready for publication. In recognition of his work, he has been awarded the Premchand Roychand Scholarship, which of late years has grown to be one of the highest distinctions open to an alumnus of the Calcutta University. Mr Banerjee's papers have attracted attention in Europe, and among those who have expressed their interest may be mentioned Prof. E. H.

Barton, FRS, and Prof. J. H. Vincent, both of whom are well known for their original investigations. Prof. Vincent wrote an account of Mr Banerjee's work specially for the journal 'Knowledge' and Prof. Barton in reviewing Mr Banerjee's work in 'Science Abstracts' suggested that the instrument devised by him and used in his work should be given the name of 'Ballistic Phonometer'. I venture to think that in Mr S. K. Banerjee, the Calcutta University possesses a man who can claim to be regarded as a rising young researcher of the best type. Mr S. K. Mitra has also shown most praiseworthy ability and industry and is a researcher of great promise.

Thus, "an organization for original research of the highest type had been firmly established in the Calcutta University". In fact Calcutta could claim to possess a school of physics

the like of which certainly does not exist in any other Indian university, and which, even now, will not compare very unfavourably with those existing in the best European and American universities.

3.5 Other Giants

We interrupt our narration here to present brief pen portraits of a few other scientific luminaries of that period. This is necessary in order to stress the fact that while Raman was a towering personality, he did not by any means exhaust the Calcutta scientific scene.

Jagadish Chandra Bose

J. C. Bose was perhaps the first Indian to establish an international scientific reputation. Beginning as a physicist, he later changed into a biophysicist (as has already been mentioned).

Bose was born in 1858, and after completing college education in Calcutta, went to England to study medicine. But, coming under the spell of Lord Rayleigh, he shifted to physics to study which he enrolled in the Christ College, Cambridge. Working his way up the educational ladder, he earned the D Sc degree from the London University in 1896, and, upon return to India, he became the first Indian to occupy the post of Professor of Physics at the Presidency College. It is said that the Government paid Bose only a fraction of the salary paid to Englishmen who had served as professors earlier. When Bose discovered this, he protested by refusing to draw salary, though he continued with his professional duties. This *satyagraha* worked and eventually the authorities yielded.

As early as 1884, Bose had published a paper on double refraction in the *Journal of the Asiatic Society of Bengal*. Later he investigated the generation, transmission and reception of electromagnetic waves in the 5 mm to 1 cm region. This is a difficult region to work in, and it is a pity that the pioneering contributions made by Bose are hardly known. In 1895 Bose gave a public demonstration of the transmission of electromagnetic waves through solid walls. This was before Marconi's celebrated experiments, and naturally Bose became famous through this demonstration. In December 1896 he repeated this experiment at the Royal Institution before an audience that included Lord Kelvin.

After the turn of the century Bose shifted his interests to plant physiology and devised ingenious techniques to study the tiny movements of plants in response to stimuli. In 1917 he founded the Bose Institute, again somewhat on the pattern of the Royal Institution. He was knighted in 1917 and elected to the Royal Society in 1920.

Bose had strong literary interests, and was a good friend of the poet Rabindranath Tagore. It is believed that Bose set the physics question paper of the FCS examination for which Raman appeared.

Prafulla Chandra Ray

Acharya Prafulla Chandra Ray was born in 1861. After completing his early college education in the Metropolitan College in Calcutta, he won a scholarship and continued his studies in the University of Edinburgh, finally obtaining the D Sc degree for his researches in organic chemistry. On return to India he got a job as a lecturer in the Presidency College where J. C. Bose was already teaching.

Ray's mentor at Presidency College was Sir Andrew Pedlar, famous for his studies on the chemistry of cobra venom. Ray's first success was the preparation of crystalline mercurous nitrate, earlier regarded as unstable. Later, the study of nitrates and of various mercury salts formed the central theme of Ray's investigations for several years.

Ray soon became Professor in the same college but in 1916 went over to the University College of Science to occupy the Palit Chair in Chemistry. Like J. C. Bose, he too was knighted.

Besides being a pioneer in chemical research, Ray was also instrumental in starting many chemical industries, the most well-known of them being the Bengal Chemicals and Pharmaceutical Works Ltd. Not surprisingly, Ray was sympathetic to applied chemical research (a factor which worked against Raman later in Bangalore).

Swept by the prevailing nationalistic fervour, Ray gave up Western dress, actively promoted ideas of Indian culture, and formed links with literary circles. Later in life, he was always respectfully known as *Acharya*.

Meghnad Saha

Saha plays an important role in this narration on account of his clash with Raman which lasted several years. He was born in 1893 in a village near Dacca, and came from a large family whose traditional business was shopkeeping. Hardship imposed by limited means plus the humiliation faced on account of belonging to a lower caste infused into Saha a bitterness which never left him.

There was no school in Saha's native village but he managed to find a benefactor in a neighbouring town with whose help he worked his way through elementary school. Winning a scholarship, he then went to Dacca to study in a Government high school. The British Governor of Bengal came to Dacca on tour in 1905, and young Saha joined a boycott organized on that occasion. This lost him his scholarship and he then had to struggle through a private school. In 1911 he joined the Presidency College for higher studies. J. C. Bose and P. C. Ray were then on the staff of the College. Saha's contemporaries included S. N. Bose and P. C. Mahalanobis[10]. Saha passed the M Sc examination in 1913 securing the second position, the first going to S. N. Bose.

Saha wanted to sit for the FCS examination but was debarred on account of his earlier political activities. For a while he supported himself by giving private tuitions until Sir Asutosh Mookerjee offered lectureships to him and to S. N. Bose. Both were appointed in the University Department of Mathematics but Saha could not get along with the Professor. Sir Asutosh then transferred both Saha and Bose to the Physics Department where Raman joined them as Palit Professor in 1917. During this period Saha and Bose translated several of Einstein's papers from German into English, an exercise which led Saha to several investigations in electromagnetic theory and relativity.

Saha's greatest contribution was his study of high temperature ionization and the discovery of the famous Saha ionization formula, now a corner-stone of astrophysics.

In 1919 Saha won the Premchand Roychand Scholarship, which enabled him to spend two years in Europe. On return in 1921 he was appointed as Khaira Professor of Physics in the Calcutta University but left it in 1923 to become Professor as well as the Head of the Department of Physics in the Allahabad University. But Saha soon discovered that there was nothing like Calcutta and he yearned to return. (As we shall see, this had an important bearing on the Saha–Raman split.) He eventually returned to Calcutta in 1938 to occupy the Palit Chair vacated by Raman in 1933. Saha retired as Palit Professor in 1953. Meanwhile he was elected the Honorary Secretary of the IACS in 1944 and served as its President between 1944 and 1950. He played an important role in the shifting of the IACS from Bow Bazar to its present campus in Jadavpur. Saha also founded the Institute of Nuclear Physics, now renamed Saha Institute of Nuclear Physics.

Anxious that India should enter the nuclear age, Saha started a cyclotron project in the late thirties. But the project was beset with numerous difficulties and could be completed only much later.

A staunch patriot with a deep concern for the poor and the underprivileged, Saha was always interested in socio-economic matters. He was an ardent socialist, and he admired very much the centralized planning practised in the Soviet Union. It is not surprising therefore that he took active part in the National Planning Commission appointed by the Indian National Congress in 1938 with Jawaharlal Nehru as

Chairman. In 1951 he was elected to Parliament as an independent member.

Saha was quite austere in his personal life, rather harsh in expression, and never bothered to placate others. He was elected FRS in 1927. He also founded the well-known journal *Science and Culture*, to which he made numerous contributions on diverse topics[11].

Satyendra Nath Bose

Satyendra Nath, or Satyen as he was more popularly known, was born on New Year's Day in 1894. Even in school Satyen showed signs of unusual abilities. It is said that once his mathematics teacher awarded him 110 marks out of 100, for Satyen had not only solved all the problems but had also solved many of them in more ways than one. College was no different. The story is told that P. C. Ray used to make Satyen sit by his side on a stool during his lectures to prevent the young lad from asking embarrassing questions!

Satyen Bose always stood first, in the Intermediate, in B Sc and in M Sc (applied mathematics). In 1916 Sir Asutosh Mookerjee appointed him (along with Saha) as a lecturer in the University College. Here Bose had to teach relativity to students of mathematics. It was the first time the subject was being taught in India, and there were hardly any books on the subject. Bose took the opportunity to translate Einstein's paper on general relativity, which, along with a few other important papers, was published as a small booklet by the Calcutta University.

In 1921 Bose left Calcutta to take up a readership in the Dacca University. It was during this period that he wrote the famous paper which introduced Bose statistics. Bose sent the manuscript of his paper to Einstein along with a letter in which he said:

> I have ventured to send you the accompanying article for your perusal and opinion. I am anxious to know what you think of it.... If you think the paper is worth publication, I shall be grateful if you arrange for its publication in *Zeitschrift für Physik*. Though a complete stranger to you, I do not feel any hesitation in making such a request, because we are all your pupils through your writings. I do not know whether you still remember that somebody from Calcutta asked your permission to translate your papers on relativity into English. You acceded to the request. The book has since been published. I was the one who translated your paper on general relativity.

The rest is history. Einstein arranged for the translation and publication of Bose's paper on the statistics of photons, adding the remark that the paper represented substantial progress. Bose statistics is now an integral part of physics.

In 1924 Bose went to Paris on study leave to work in the laboratories of Mme Curie. Berlin was the next stop, where, naturally, Bose met the "Master"[12]. He also came into contact with various other celebrities. Bose returned to Dacca in 1926 and became Professor, and was joined soon by K. S. Krishnan (see the next section) as

Reader. Shortly before Independence, Bose returned to Calcutta to become the Khaira Professor of Physics from which post he retired in 1956. He was elected FRS in 1958[13].

Bose loved science but he loved literature even more. He enjoyed poetry, and translated poems from many languages like French and even Hebrew into Bengali. He firmly believed that science would not spread unless it was communicated in the native language, and worked hard for the teaching of science in Bengali. Bose passed away in 1974.

3.6 The Principal Lieutenants

Our thumb-nail sketches would not be complete without a brief reference to at least some of Raman's own colleagues. Since the list is long, we restrict ourselves to two men who played a key role in the discovery of the Raman effect.

Kalpathi Ramakrishna Ramanathan

K. R. Ramanathan was born in Kalpathi in Palghat district (now a part of Kerala) in 1893. He received his early education in the Victoria College, Palghat, and later obtained the MA degree (in physics) from the Madras University, studying in the Presidency College, Madras. Professor Stephenson of the Maharajah's College of Science, Trivandrum, who was Ramanathan's examiner, was so impressed that he immediately offered the young scholar the post of Demonstrator in his college. Ramanathan worked in Trivandrum for seven years, during which period he also served in the local observatory. This experience enabled him to write a paper on tropical thunderstorms. He returned to meteorology again later.

Attracted by Raman's fame and work, Ramanathan went to Calcutta in 1921 as a University of Madras research scholar. He was one of the earliest to work on the scattering of light (other than Raman, that is), and, in the course of his work, discovered a puzzling phenomenon which he dubbed "feeble fluorescence". Relentless pursuit of this phenomenon was later to culminate in the discovery of the Raman effect. Ramanathan also did some nice work on X-ray diffraction in liquids (see Chapter 4). His publication list shows as many as *ten* papers for the year 1923, in such journals as the *Proceedings of the Royal Society*, *Astrophysical Journal* and *Physical Review*. No wonder he earned the D Sc degree in record time.

Domestic responsibilities forced him to take up a lectureship in Rangoon as soon as he obtained the doctoral degree, but Ramanathan rushed to Calcutta at every conceivable opportunity in order to stay in touch with research. In 1925 he gave up his lectureship to join the India Meteorological Department, where he served till retirement at the age of 55. As a meteorologist, Ramanathan made many significant contributions. He discovered that the lowest temperatures in the stratosphere occurred over the equator and not over the poles as one might normally expect. He won international fame for his studies on ozone in the atmosphere.

After retirement, Ramanathan became the first Director of the Physical Research Laboratory at Ahmedabad, just then founded by Vikram Sarabhai. In 1966 Ramanathan relinquished the directorship of the Laboratory but stayed on as Professor Emeritus. He was quite active in his research till the very end – a touch of Raman, one might say.

It is interesting that Ramanathan went abroad for the *first* time only after he retired from the India Meteorological Department! Nevertheless, international recognition came to him in several forms. He was awarded the International Meteorological Organization Prize, and elected President of the International Association of Meteorologists. He was also elected President of the International Union of Geodesy and Geophysics in 1957, and President of the International Ozone Commission for three terms.

In the country, he was honoured first with the *Padma Bhushan* and later the *Padma Vibhushan* titles. He was the first recipient of the Aryabhata medal of the Indian National Science Academy, and also occupied the prestigious Raman Chair of the Indian Academy of Sciences. He passed away in 1985.

Kariamanickam Srinivasa Krishnan

K. S. Krishnan was born in 1898 in Watrap in Tamil Nadu (then a part of the Madras Presidency). He received his schooling in the neighbouring town of Srivilliputtur, but for college education he went out to Madurai and Madras. From his father, Krishnan inherited an abiding love for religion and philosophy, and also for Tamil and Sanskrit literature. Krishnan was also a voracious reader of Western literature, and his favourites included Plato and Aristotle. He was much influenced by the biographies of famous scientists, and the *Collected Scientific Papers* of Lord Rayleigh were a constant companion throughout his life. A scholar to the core, he made an impact on all he came into contact with. In 1955 he was a guest speaker at the annual banquet of the US National Academy of Sciences where he excelled himself, lecturing on cultural values in technical education. Professor van Vleck (later to win the Nobel Prize) said afterwards, "He quoted extensively from Whitehead, and it was his speech that prompted me to read some of Whitehead's writings."

Krishnan took a master's degree in physics but the only opening then available to him was as Demonstrator in Chemistry in the Madras Christian College. During this period, Krishnan ran an informal but highly successful lunch-hour discussion on diverse topics in physics and chemistry which soon began to attract participants from other colleges as well. One beneficiary later remarked that he had learnt more physics from the lunch-break seminars than from regular class-room lectures.

Attracted by Raman's growing reputation, Krishnan went to Calcutta in 1920. He attended post-graduate lectures in the University College of Science, and Raman, spotting his talent, promptly appointed him as a research assistant in the IACS.

Young Krishnan worked very hard, reporting for work as early as 6 a.m. But in the evenings he would sneak away to watch soccer, and what better place for that game than Calcutta! Like others, he too was swept by the prevailing national fervour and even tried to write his scientific papers in his mother tongue, Tamil! But soon he mellowed (on this count), and gave up such extreme steps. (Later we shall discuss how Krishnan's elevation became the bone of contention between Raman and Saha, leading to a permanent split between the latter two. And in Chapter 5, we shall consider the role he played in the discovery of the Raman effect.)

In 1929 Krishnan went to Dacca as Reader in Physics, and while he was there, commenced his well-known work on the magnetic properties of crystals. He was back in Calcutta in 1933 as the Mahendra Lal Sircar Professor at the IACS. In 1942 he moved over to Allahabad, and, after Independence, became the first Director of the newly-established National Physical Laboratory in New Delhi. Along with Bhabha and Bhatnagar, he served as a member of the Atomic Energy Commission from its inception.

It is striking that though Krishnan played an important role in the discovery of the Raman effect, he did not pursue that subject later. Instead he showed his mettle in other fields like magnetism (as already mentioned), thermal conductivity and thermionics.

Krishnan's work at Dacca and Calcutta attracted attention in far-away England, resulting in invitations from Lord Rutherford and Sir William Bragg. Krishnan was elected to the Royal Society in 1941, and he won several other honours. He passed away in 1961.

3.7 The Second Decade

Resuming the narration, we turn to Raman's second decade at the Association. It was truly epoch-making, with Raman making the greatest discovery of his life. Unlike earlier, Raman was not only able to devote all his time to science, but, thanks to students whom he now began to attract in large numbers, he could also study a richer variety of problems. Old interests still continued but several new ones also developed, like the elastic and optical properties of crystals, the physics of colloids, electric, magnetic and mechanical birefringence, and X-ray diffraction.

In 1919 Amrita Lal Sircar passed away, and Raman was immediately elected the Honorary Secretary. It therefore became his responsibility to prepare and present the annual reports of the Association to the Committee of Management. Instead of being bland, as such documents usually tend to be, Raman's reports are a connoisseur's delight, fully reflecting his vibrant personality. Typically they would begin with the words, "The Honorary Secretary begs to submit to the Committee of

Management the following report on the activities of the Association during the year...", whatever the year was. The report was always in two parts, scientific and administrative, both interesting to the historian.

The scientific parts are a pleasure to read. It was always the intention to give "a general account, in language as free from technicalities as possible, of the scientific results obtained by investigations carried out under the general direction of your Honorary Secretary". The hope was that such an account would be of interest to a wide circle of readers. Here is a sample of Raman's popular exposition:

> The scattering of light by the free surfaces of a great variety of liquids has been extensively investigated by Mr Ramdas.... Whenever we wish to imagine an ideal plane surface we instinctively take the example of the surface of a liquid at rest. This is really a delusion, for, if we were to look at the surface of a liquid magnified, say a billion times, we would see it in a very turbulent state indeed, full of waves nearly as large as those which roll over a sea in a storm. We do not realize this because these waves are beyond the power of detection of our most powerful microscopes.
>
> In a liquid the molecules have lost the great freedom of movement they had in the gaseous state, and, on account of the diminution of the extent of motion, the attraction of neighbouring molecules on each other becomes sensible.... At the surface, the conditions are special. Here the attraction which the molecules have for each other causes them to present a united front to disturbing forces. The surface is, in fact, comparable with a stretched elastic membrane. That is, it requires work to increase its area ever so little. But one should not think that the molecules at the surface are motionless. They also partake in continuous movements and the more vigorous molecules occasionally fly off. Very often molecules of sufficient velocity collide against the surface and occasionally burst through the surface layer or knock out one of the surface molecules into the space above. We call this process evaporation. If the space above is enclosed, a stage is reached when the molecules escaping from the liquid equal in number those re-entering it. This continuous exchange of molecules between the liquid and the vapour is accompanied by the continuous agitation of the surface. We may conceive the surface as full of waves and ripples owing to molecular bombardments, just like the surface of a pond disturbed by throwing stones. We are unable to see these waves and ripples because the lateral bonding of the molecules at the surface into a thin elastic layer prevents the waves from becoming too large. To put it more accurately, we may say that the surface tension forms the controlling force which limits the degree of roughness of a liquid surface. This roughness is of molecular dimensions at ordinary temperatures....
> [3.7]

Raman then goes on to describe the experiments of Ramdas.

The style and clarity of reporting remain the same, whatever the topic – musical instruments, chromatic emulsions, X-ray diffraction, etc. Part I always ended with a list of publications originating from the Association.

Part II dealt with the laboratory, the so-called personal matters (sometimes), the library and the reading room, the workshop, and finally various financial matters. The Association workshop was small but provided valuable service in fabricating parts needed for experiments, like an "oscillating crystal spectrometer" or an "X-ray box", for example. Sometimes, parts needed for the fabrication were picked up from scrap dealers. Describing one such instance of scrap purchase (relating to a magnet needed for studies on diamagnetism), Raman recalled:

> Not possessing the financial resources which seem to lie so close to hand in the case of many American physicists, and perhaps also to some physicists in this country [i.e., England], I wandered about the streets of Calcutta and looked into the old iron shops and discovered a huge dynamo of the Edison type, which had probably figured in some nobleman's house but had been thrown away in favour of some newer model. This dynamo happened to be in perfect order, and we got it for the price of the old copper it contained. The old Edison dynamos are enormous and the marked feature of the electromagnet was its great length of 35 cm, while the gap was very small, only about 1 cm. We found that with a current of about 10 amperes, we had a field of 25,000 gauss. Using the electromagnet we were able to observe magnetic double refraction in practically all the substances we examined. [3.8]

The workshop made a heavy brass stand for the improvised electromagnet derived from the discarded dynamo. The assistance rendered by the workshop was always handsomely acknowledged in the annual reports. Years later, Raman tried to reorganize the workshop in the Indian Institute of Science, Bangalore, to make it equally useful for research there, but in the process stirred up a hornet's nest (see Chapter 8).

No detail was too small. Thus we are informed that

> the water and gas fittings, which were previously taken through the flooring, were laid across the ceiling and a sufficient number of gas and water taps were provided for each room. The almirahs containing valuable apparatus were all rearranged so that different rooms could be allotted for work on X-rays, on light-scattering, on spectroscopy, on magnetism and sound. As a result of the present arrangement, the Association can now provide space for a great number of research students to work conveniently in the various branches of physics and chemistry. [3.7]

The library always received detailed coverage in the reports.

> The very valuable journals and books belonging to the library have been arranged neatly in the hall and long tables, provided with racks for keeping current numbers of the foreign publications, placed along the centre of the hall, for the use of the readers.

There was not enough money to acquire a large number of books but as many as twenty-five leading journals were subscribed to. More important, an even larger number of journals were obtained in exchange for the Association's Bulletin. Among the societies and institutions with which such exchanges were made were the Smithsonian Institution, the Cambridge Philosophical Society, the Akademie der Wissenschaften Leipzig, the Franklin Institute, the Bureau of Standards, the Academy of Sciences Leningrad, the Royal Academy of Sciences Amsterdam, the Accademia Nazionale Rome, the Imperial Academy Tokyo, the Royal Dublin Society, and the Gesellschaft der Wissenschaften Göttingen. Raman always made good use of exchange, and later in Bangalore he not only made the *Proceedings of the Indian Academy* (see Chapter 11) well known by this method, but also secured journals from all over for the library of the Academy. At the Association, he also placed at the disposal of the library the various journals he personally received, including the *Proceedings of the Royal Society* (after he was elected to that body).

It has probably not escaped the notice of the reader that although Raman supervised students working for the doctoral degree, he himself did not possess one.

If the doctorate merely certifies that the holder is capable of original research, then Raman certainly did not need one! In any event, the Calcutta University decided in 1921 to confer on him a doctorate *honoris causa*.

That same year, Raman made his first visit abroad as a delegate to the Universities Congress at Oxford. It was during this trip that he came face to face with famous physicists like J. J. Thomson, Bragg (senior) and Rutherford. But Raman's reputation had preceded him, and his name also was known to them. While in London, he attended a lecture by Rutherford at the Royal Institution. Raman was seated in a back bench, but Rutherford, seeing him, beckoned him to the front row where all the leading physicists were seated. Raman was deeply touched by this gesture[14].

The visit to England was profitable not only in terms of the contacts made, but also in terms of various investigations carried out there. Raman stayed in a room in Putney, London, and, using the facilities extended to him by Prof. A. W. Porter of the University College, carried out studies on conical refraction in biaxial crystals and on smoky quartz. Earlier in Calcutta, Raman had analysed one of Porter's experiments and provided an explanation for some previously unexplained observations (see Chapter 5). This probably provided the necessary link. Prof. Porter also helped Raman to study the whispering gallery phenomenon in St Paul's Cathedral (see Chapter 4). G. L. Mehta (a former Vice-Chairman of the Planning Commission), who was Raman's room-mate in London at that time, recalls how Raman used to return from his experiments in the evening cold and shivering, and, rubbing his hands to keep warm, ask the landlady to serve some hot food.

The return trip to India is historic for it was on this voyage that Raman, attracted by the deep blue of the Mediterranean, became interested in the colour of the sea. In turn this led to a serious study of the scattering of light, culminating in the discovery of the Raman effect (see Chapter 5).

In 1924 Raman was elected Fellow of the Royal Society. He was thirty-six, and the fourth Indian to be so elected. There is a story (unconfirmed but quite in keeping with Raman's personality) that when felicitated and asked what next, he replied, "The Nobel Prize of course!"

An event of great sadness was the passing away of Sir Asutosh Mookerjee. For Raman, it was a deep personal loss. It was indeed a rather unusual bond between the two given their somewhat different tastes, with Sir Asutosh's inclinations being distinctly mathematical and Raman's tendency being to leap over mathematics (see next chapter for further comments on the latter aspect). Clearly Sir Asutosh recognized Raman's genius, not letting differences in style come in the way.

The year 1924 also saw Raman cross the Atlantic for the first time, following an invitation from the British Association for the Advancement of Science to tour Canada. On this tour, Raman opened a discussion meeting in Toronto on the scattering of light. From Canada Raman went to the United States to represent India at the centenary celebrations of the Franklin Institute. Following this, Raman spent four months as Visiting Professor at the Norman Bridge Laboratory of the California Institute of Technology (Caltech), Pasadena, on an invitation from Robert Millikan. It was indeed a most prestigious invitation, for Raman's predecessors were Sommerfeld, Lorentz and Einstein. While at Caltech, Raman not only lectured extensively on various topics, but also tried some experiments on the diamagnetic properties of gases in collaboration with Dumont (later to distinguish himself by the accurate measurement of the fundamental constants). Subsequently, these studies were continued by Hemmer in Pasadena and Vaidyanathan in Calcutta.

Raman visited many centres in the US and met distinguished American physicists like Compton, Langmuir, Coolidge, and others. Compton had recently discovered the Compton effect, and it clearly had messages for light scattering as well (as we shall see in Chapter 5). The visit to America was thus timely.

Soon after their first meeting, Compton invited Raman for breakfast. Raman arrived at Compton's house at the appointed hour and rang the bell. The maid, who had been alerted about the impending visit, opened the door and on seeing Raman exclaimed, "Oh my God! It is a black man." Raman was deeply shaken, and a highly embarrassed Compton rushed to offer apologies and smooth ruffled feelings. But Raman apparently carried memories of that incident for a long time, as indeed of a few other experiences of racial intolerance.

Raman also visited the famous Mount Wilson Observatory, and of this visit he said:

> I came away tremendously impressed by the light-gathering power of the giant 60-inch [$\sim$ 150 cm] and 100-inch [254 cm] reflectors. The great nebula in Orion, for instance, which in ordinary instruments appears as a shapeless area of great luminosity, appeared in the 60-inch reflector as a luminous path of variegated colour, determined by the light emission of the gases of which it is composed.

How happy Raman would have been to know that comparable telescopes are now available in Kavalur, not far from Bangalore!

On his way back from America, Raman made many stopovers in Europe, and among those he met was Prof. Goldschmidt of Oslo. Later, while at the Indian Institute of Science, Bangalore, Raman, acting on a suggestion by Max Born, tried to get Goldschmidt as Visiting Professor. But this and several other similar efforts by Raman were rejected by the Institute management. That story comes later (Chapter 8).

Raman returned from this long trip (perhaps his longest) in March 1925. In August he was off again, this time to the Soviet Union, to participate in the centenary celebrations of the (Leningrad) Academy of Sciences. It is interesting that during this visit, German was the language of communication between Raman and the Soviet physicists. The story goes that in Leningrad, Raman was given a demonstration of a certain experiment. On seeing it, Raman exclaimed: "Das ist sehr schön. Aber ist es nicht Rozhdestvenskii experiment?" (This is very beautiful. But is it not the Rozhdestvenskii experiment?) To this the professor who gave the demonstration replied: "Ja, ich bin Rozhdestvenskii." (Yes, I am Rozhdestvenskii.) [3.9] Raman returned to India in November via Berlin, Paris, Geneva and Rome.

It should not be concluded that Raman had discovered foreign travel! On the contrary, he did not stir out for the next three or four years. He was extremely busy, and it was during this period that the Raman effect was discovered. Ramdas gives us a good glimpse of Raman in this period.

Professor Raman's first two visits abroad and the contacts he made with many scientific men and institutions during these visits brought about a very wholesome change in his habits and methods of work. Previously, he used to work away for very long spells without regard to day or night, oblivious to needs of food, rest, etc. After his foreign experiences, the pattern of his working became more regular and systematic. He took to business-like habits in getting through his very heavy workload of scientific research and direction of his many pupils. Regular hours for meals and short spells of relaxation in the evenings must have been a great relief to Lady Raman.

After he became FRS, the authorities of the Calcutta University were pleased to augment Prof. Raman's salary. About 1924, he acquired a horse and carriage, thus freeing himself from dependence on taxis. The carriage was black while the horse was chestnut brown. There was, of course, a syce to take care of both and to act as a driver. It was a very presentable set-up. Often, while taking his evening drive, if Lady Raman was not accompanying him, he would invite one of his senior scholars to keep him company. I once accompanied him. The evening drive usually led to the Calcutta Maidan, stopping somewhere near the statue of Lord Kitchener or Lord Roberts. Getting down, Prof. Raman would take a brief walk towards Fort William and then take some quick jerks of 'kasrat' and 'dand' for exercise. He would then walk back to his carriage and drive home. [3.6]

About work at the Association, Ramdas has this to say:

The daily activities of the Association usually started by 7 a.m. Having started on an experiment, one would go on working till 1 p.m. After a quick lunch the scholars would be back by 2 p.m. and work on far into the evening and often until 9 or 10 p.m. until the job on hand had reached a satisfactory stage. Working at this furious rate it is no wonder that many of the pupils could work through 5 or 6 major investigations each year. Those who could not cope with such a fast tempo would automatically drop off.

Concerning the student–teacher relationship, Ramdas observes:

He [i.e., Raman] inspired his scholars to use their own initiative and ingenuity to the fullest extent. He would see what was going on and discuss results at intervals. At any given time, however, he would concentrate his attention on the particular scholar who was then entering the most critical phase of his research. Interpretation of results, fruitful suggestions to carry the investigation several steps further and quick discussion of results already obtained for immediate publication resulted from this effective type of collaboration between the Professor and his pupil. Each of his pupils had his opportunity for such exhilarating collaboration at the developing phase of his investigation. All the time, the pupils enjoyed the fullest freedom to think, work and improvise for themselves. Spoon-feeding of any kind was absolutely taboo. A spirit of perfect understanding and goodwill pervaded the entire Association with Ashu Babu, the Assistant Secretary, ever ready to help us with any material or facility that we needed, the scholars themselves helping each other spontaneously.

With the place humming with activity, it was but natural that a number of topics were taken up for investigation. For Raman himself, studies on the scattering of light began to assume importance. True, the colour of the sea had been explained and so had a number of other observations, but there was still something elusive. Loosely dubbed "feeble fluorescence", this phenomenon was first noticed by Ramanathan in 1923 and again by Krishnan in 1925. It came to the fore again in late 1927 during some investigations by Venkateswaran, and Raman now refused to believe that the new and mysterious phenomenon was due to fluorescence. Already he was thinking about an analogue of the Compton effect in light scattering. Was this so-called feeble fluorescence perhaps it? Raman now took direct charge and along with Krishnan launched on an intense series of investigations. All observations were made visually using an arrangement described by Raman as follows:

> As it is very essential to prevent the distraction of the eye by external light and to keep the eye at its maximum sensitiveness to feeble illumination like that due to scattering, the observer located himself in a light-tight wooden cage 4 ft square [~ 1.2 × 1.2 metres] and 8 ft [2.5 metres] high, facetiously known as the 'Black Hole of Calcutta'[15]. [3.10]

Finally, on February 28, 1928, it was conclusively established that the unexplained phenomenon hitherto observed was in fact a new type of scattering mechanism giving rise to what was then called modified radiation. The Raman effect had been discovered, and the world was told about it the following day. The entire story is narrated in detail in Chapter 5.

Excitement now swept the Association, and Raman and his students delighted in displaying the effect to visitors. One such visitor was Satyen Bose, who, after witnessing the demonstration, congratulated Raman and said: "Prof. Raman, you have made a great discovery. I predict that this effect would be called the Raman effect and I also predict that you will get the Nobel Prize for it." Both predictions turned out to be correct!

The summer of 1928 also saw a shy young student-visitor from Madras. That was Chandrasekhar, already developing a strong interest in physics and mathematics. Earlier that year, when Raman stopped over at their house in Madras on his way to deliver the now famous lecture on the Raman effect (see Chapter 5), young Chandra had seen the first-ever Raman spectra which had been recorded[16]. At the Association, Chandra worked alone, but saw a lot of Raman's associates, particularly Krishnan. He spent his time thinking about stars, and his work, entitled "Thermodynamics of the Compton effect with reference to the interior of stars", was later presented as a paper in the Physics Section of the Indian Science Congress when it met in Madras in January 1929. Raman was the General President of that Congress, and S. N. Bose the President of the Physics Section.

Apart from the great discovery, the event of the year (1928) as far as the Association was concerned was the visit of Arnold Sommerfeld. A respected member of the old guard, Sommerfeld was in the hub of the new developments in physics that were sweeping Europe. Already Sommerfeld had met Saha in Germany, and he was attracted to India not merely because of its religion and philosophy but also because

> it was in this ancient land of civilization that, during the last years, strong shoots of modern physics had grown, by which India suddenly emerged in the competition of research as an equal partner with her European and American sisters. [3.11]

Sommerfeld had been invited to America, and he decided to go there via the East. When Raman heard about Sommerfeld's plans, he immediately sent a cable:

Oh Calcutta!

"Calcutta University inviting you, lecture honorarium thousand rupees, kindly wire date arrival India[17]." This was a few weeks before the discovery of the Raman effect.

Sommerfeld made careful plans for his Indian visit, with Saha's help, it is believed. He had many lecturing engagements, one of which was at the Presidency College, Madras. Chandrasekhar was one of the members of that audience, and from Sommerfeld he heard for the first time about the new Fermi–Dirac statistics; the rest is history [3.12].

On October 4, Sommerfeld arrived in Calcutta the "main attraction". At this point, we turn to the diary of Sommerfeld for a glimpse of his stay in that city:

> October 4: Grand reception at Howrah station. Raman, Bose, Krishnan, Senn, Gosh, Mitra... also the German Vice-Consul Eberl at whose place I stay...3 beautiful rooms with bathroom...hibiscus flowers in blossom upto the first floor... then [Sir J. C.] Bose takes me to Raman's institute who shows me papers on diffraction....
> October 6: 8 O'clock in the morning, first special lecture on Kepler problems, discussion until 10 o'clock ... eventually the Raman effect visually: blue filter and compl[ementary] screen, now put in front of the incident light, then in front of the scattered light: difference...
> October 7: Sunday...Wonderful lecture by Raman (also the rotations of the molecules can be seen unresolved as modified rad[iation])....
> October 8–13: Lectures daily from 8–10 with lively discussions[18] ...I saw scattering blue-green on an ice block in the institute, obvious modified scattering. Everything in the institute is very good, but bathroom terrible.... [3.11]

Sommerfeld's interests were not confined to physics alone, and he took time off to visit Benares, "India's Rome", and Agra, the "oriental fairy tale". He is reported to have said that India was a beautiful country, but 50 cm above the ground! On an invitation from the poet Rabindranath Tagore (whom he much admired), Sommerfeld visited Santiniketan, enjoying "the peace of an Indian autumn". On October 26 he sailed from Calcutta: "With two cars to the harbour. Again a garland of Krishnan and flowers from the X-ray man...." Sommerfeld left India with "deepest affection for the highly-gifted, unhappy nation" and "with sincere gratitude for the many acts of friendliness and honourings".

The year had ended well. A major discovery had been made, and honours poured on the discoverer from all over. The most coveted one was of course the Nobel Prize, which came a little later in 1930, twenty-three years after Raman arrived in Calcutta. What a magic he had wrought on the Association and how the Founder must have rejoiced in heaven!

3.8 The Exit

It is said that all good things must come to an end. Certainly Raman's Golden Era in Calcutta did, not in the slow and gradual way one expects from natural ageing processes, but in an abrupt and somewhat unpleasant manner. At the time the painful episode occurred, observers saw only the immediate causes, of which undoubtedly there were many. Some traced it directly to a clash of the strong personalities of Saha and Raman. Indeed, there is merit in all such analyses, but, viewing these events long after they occurred, one is inclined to the view that all the causes earlier suggested were but mere symptoms, the malady actually being something deeper and far beyond the control of the parties involved. For Raman, it meant leaving the Association. The entire sequence of events needs to be discussed dispassionately, as there are lessons in it for us even today.

Success always breeds envy, and Raman's success was no exception. And jealousies become stronger when the stage is crowded and opportunities are few. In the early days when Raman toiled hard and enhanced the reputation of the Association, everyone applauded him. He did it single-handed, working entirely in his spare time. He asked for and received little. But things began to change after he was appointed to the Palit Chair. The impression began to be created that Raman was a has-been physicist working on highly dated problems while a new wave of physics was sweeping Europe. Snide remarks would be made, and there would be sarcastic interruptions during his seminars. Raman smarted under these insults and boiled inside with fury. The Nobel Prize provided some respite but it was short-lived, for doubts began to be raised whether it was really Raman's discovery, and even if it was, whether Raman understood what he had discovered. The unfortunate thing about rumours and innuendos is that they do not need much to feed on.

However, rumours alone are insufficient, and something stronger is needed to foment revolt; it came in the form of grievances. One of the charges against Raman was that he had gathered a South-Indian coterie around himself. Prima facie, this seemed a valid complaint, for, willy-nilly, many students from the South had gathered around Raman after 1920. Before analysing this situation one must digress to discuss some obvious misconceptions.

There is a tendency even today to group people hailing from diverse regions into one category. Much to the annoyance of people from Kerala, Karnataka and Andhra, people south of the Vindhyas are often referred to simply as Madrasis! Likewise, people in the South often fail to perceive the ethnic, language and cultural differences of people from say Bengal, Rajasthan and Maharashtra and tend to categorize them all as simply North Indians. One has learnt to live with such thoughtlessness, occasionally complaining about it in letters to editors, or joking about it. In Raman's case, however, it was no joking matter. Ramanathan, for example, was from Kerala, Bhagavantam from Andhra, and Seshagiri Rao from Mysore; but to Raman's opponents such distinctions were irrelevant. They were all part of the same South-Indian crowd, and they had no business to be in Calcutta and form a sizeable part of the Association. Actually, if one examines the annual reports, one finds that there were students from all parts of India (except, for some unknown reason, from the Bombay Presidency). The Association enjoyed a great reputation, and people came to work there from great distances. We read, for example, that Dhabadghao came from Amaravati to spend a summer recording spectra. R. N. Ghosh of Allahabad came there to study the acoustics of the Stroh violin. There was young Chandrasekhar, a future Nobel Laureate, who came there not to work with anyone in particular but merely to be in a creative atmosphere. One year there was even a visitor from Austria!

All this is qualitative evidence. In terms of numbers, why were there fewer Bengalis than South Indians (assuming one does not object to such an omnibus classification!)? After all, the Association was in the heart of Bengal, and was it not natural to expect more of the locals there? Actually, it is an error to assume that Raman had no Bangali associates, for he had many – B. Banerji, D. Banerjee, K. Banerjee, S. K. Banerji, B. N. Chuckerbutti, P. Das, S. K. Dutta, G. L. Dutta, D. N. Ghosh, M. N. Mitra, S. K. Mitra, B. Roy, S. C. Sircar, N. K. Sur, and so on. True, some of them published on their own, but they all worked on problems identified by Raman. It is unlikely that these people would have enjoyed the Association's facilities unless Raman approved of it, for, remember, he was the Honorary Secretary. The accusation that Raman was anti-Bengali does not appear to be borne out by the facts.

The fact of the matter was that people converged on Calcutta fired by idealism and attracted by Raman's fame. K. R. Ramanathan first came there deputed by the Madras University as a research scholar. Later, when he took up a job in Rangoon, he made a beeline to Calcutta during every vacation so that he could keep in touch with research; and he paid his own way. S. Venkateswaran was already employed in the Test House in Calcutta, and he worked at the Association in his spare time even as Raman himself had done earlier. For Raman, motivation was the supreme factor. If a person came a long distance for no assurances of any kind but stimulated by a passionate desire to do research, it was enough; Raman simply took him. In some instances he even provided food and shelter until the scholars could stand on their own. But this generosity was misunderstood. No credence was given to Raman's lasting affection for Ashu Babu, his eternal regard for Sir Asutosh, and his appreciation for S. K. Banerji. How could all this have been possible if Raman was anti-Bengali?

One finds today many professors of Indian origin in the US, and many of them have had a string of Indian students. And yet we do not hear of faculty upheavals or protests from the locals in America. There is enough room for all, and no one feels threatened. Viewed from this angle, insecurity appears to have been the major sociological factor underlying the crisis Raman faced. Had there been enough faculty and student positions, events might not have reached the proportions they actually did. An added complication was the patriotic fervour sweeping Bengal, and as one knows, extreme nationalism can sometimes breed parochialism.

It is well to remember that sociology plays an important role in a complex country like ours, and, as a result, there are often waves. For instance, shortly after Independence, a very large number of toppers in the Indian Foreign Service (IFS) and the Indian Administrative Service (IAS) examinations came from Madras, but today social dynamics has changed and this is no longer true. Likewise, Bengal appears to be producing more good physicists right now than any other part of the country, while elsewhere, there seems to be an indifference to the basic sciences.

Returning to the unfortunate events under discussion, every conflagration needs a spark, and in this instance also there was one. Around 1932, the prestigious Indian Institute of Science (see Chapter 8) was looking for a new director to replace Sir Martin Forster, FRS, who was about to retire. The Tata family (who as heirs of the founder were represented on the Council of the Institute) informally sounded out Lord Rutherford on possible names for a successor. Rutherford pointed out that it was no longer necessary to look for directors in England when India already had a distinguished candidate in the shape of Raman. The Institute Council then went through a formal process of selection (see Chapter 8), and eventually Raman was offered the directorship. Raman was inclined to try out Bangalore but someone should be in charge in Calcutta. Perhaps it was time that the Association had its own full-time professor as its founder himself had visualized. Also, the finances now permitted it. What better, then, than a Chair named after Mahendra Lal? When Saha, then in Allahabad, heard about this move, he promptly wrote to Raman asking that he (i.e., Saha) be nominated to that Chair. In his reply, Raman, while appreciating Saha's earlier achievements, expressed disappointment that lately Saha had not been much concerned about research; what the Association needed was an active young man on his way up rather than one who had reached a plateau[19]. In Raman's mind, Krishnan was a better candidate. Saha became furious. He had already suffered much in the past, and living in Allahabad was like banishment. An unimaginative registrar (an ex-judge) was placing untold obstacles in his attempts to build a viable department of physics, in short, making his life miserable. Saha naturally longed for his native Bengal and here he was, being kept away by a South-Indian clique. He would not take it lying down. Earlier he had been dissatisfied when he had to serve as a mere lecturer in the University College of Science while Raman with his "antique view of physics" was a prestigious professor. This denial was the last straw. Saha rushed to Calcutta and sought the counsel of Dr Shyama Prasad Mookerjee, the son of Sir Asutosh Mookerjee, and the then Vice-Chancellor, Calcutta University. It was felt that Raman was injurious to the Association as well as to science in Bengal and that he must be curbed. But only the General Body of the Association could restrict his authority, and so far, the General Body had overwhelmingly applauded Raman. It was therefore clearly necessary to induct into the Association people opposed to Raman. This was duly done. The press too was mobilized and editorials appeared, highly critical of Raman's administrative abilities. It must be added that Raman did not make it any easier with his Himalayan ego and his sharp tongue.

On the fateful day of the General Body meeting the battle lines were drawn. The meeting itself was stormy, with Shyama Prasad accusing Raman of betraying Sir Asutosh and Raman retorting that Shyama Prasad did little credit to his departed father. The final outcome was as expected – Raman was voted out of the Honorary Secretaryship, a position he had held with distinction for years. For Raman it was a moment of bitter humiliation. Few tears were shed for him, for he had indeed made many enemies lately by his sharp reactions. Earlier Calcutta rejoiced in his triumphs but now it exulted in his defeat. All that remained for Raman was to make an honourable exit. Fortunately for him, an opportunity for this was available, thanks to the already pending invitation from the Indian Institute of Science, Bangalore. When Raman finally left, it was said[20]:

> Calcutta's loss will be Bangalore's gain. At present Calcutta may be regarded as the centre of scientific research in India; but with the transference to Bangalore of one of her leading investigators she will have to guard her laurels. [3.13]

Oh Calcutta!

At the time Raman joined the Indian Institute of Science, it had no physics department but it was agreed that he could form one. And soon he built up a department which acquired an international reputation. But he had many problems as well, as we shall see in Chapter 8.

One last comment about the Saha–Raman controversy. Much can perhaps be said on both sides, for as well as against. They were both human and had their weaknesses as well as their strengths. Their temperaments differed so strongly that it would have been a miracle if they had not clashed at all. Clashes between strong personalities are not uncommon and they are to be found everywhere and at all times. But in the academic world, one seldom witnesses events such as those which occurred in Calcutta. The basic problem there was that there was not enough room for both the giants. If only the country had been rich enough to afford several faculty and student positions, this conflict might never have gone beyond minor skirmishes. Indeed this is what happens in other countries where people who are strongly opposed are nevertheless able to carry on independently. The Calcutta of those days did not provide enough room for coexistence, and a Darwinian struggle was inevitable. The actual battle itself and the immediate causes for it are largely irrelevant in a historical context.

4 Glimpses from the Golden Era

> Sometimes a thousand twangling instruments
> Will hum about mine ears
>
> **The Tempest** (Act III, sc. II)

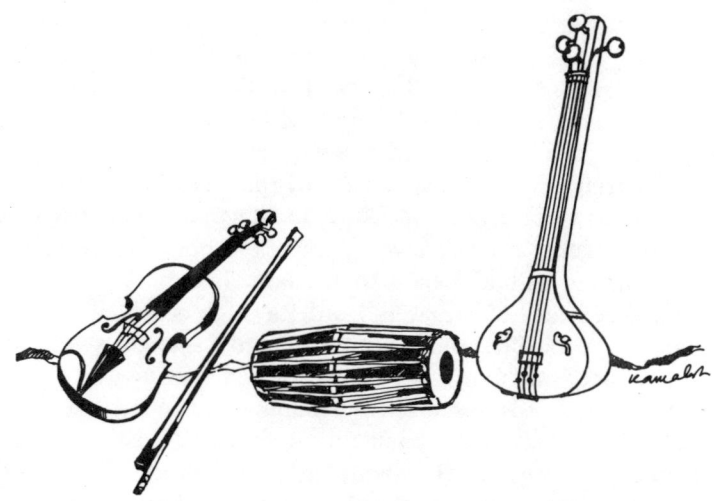

Raman often described his Calcutta days as the Golden Era, and with good reason. Indeed, one would describe this period in such a fashion even without the Raman effect, for it is during this that one sees Raman at his creative best. Starting as a soloist he quickly became a superb conductor of a grand orchestra without at the same time losing any of his individual virtuosity. Throughout, vibrations and optics remained the two principal themes of his research but there were also many other excursions.

4.1 Studies on Vibration

Raman's interest in vibrations may perhaps be traced to the influence which Helmholtz and Lord Rayleigh indirectly had on him.

What a remarkable man Helmholtz was! Writing of his first meeting with him, Lord Kelvin observed:

> I expected to find in him, who is one of the foremost mathematical physicists of Europe, a man somewhat older than myself, and I was not a little astonished when a light blond youth of girlish appearance came towards me.... With respect to analytical acumen, clarity of thought and versatility he surpasses every great scientist I have met; indeed I myself feel a little stupid in his presence.

While Lord Kelvin regarded Helmholtz as the foremost theoretical physicist, the University of Bonn saw him in an altogether different light, appointing him a full Professor of Anatomy and Physiology! And if that is astounding, consider the fact that Helmholtz wrote a masterpiece (also a Raman favourite) which has since become required reading for students of music!!

Unlike Helmholtz, who passed away when Raman was but a mere child, Rayleigh's career overlapped a shade with that of Raman (as briefly seen in Chapter 2). Lord Rayleigh received his early baptism under Routh, the famous "coach" in applied mathematics, as well as under Sir George Stokes. Later, coming under the spell of the *magnum opus* of Helmholtz, he started his investigations in sound, leading also to much correspondence with Maxwell on the subject.

Soon after his marriage, Rayleigh was struck by rheumatic fever. A trip to Egypt was then undertaken as a recuperative measure, and it was on a boat trip up the Nile in 1872 that the famous classic *The Theory of Sound* was born. It is remarkable that the first part was written without access to any large library.

After the premature death of Maxwell, Lord Rayleigh accepted the Cavendish Professorship for five years. One tends to think of Lord Rayleigh as an outstanding theorist which he indeed was but it is seldom realized that he was also a skilled experimenter. As Bruce Lindsay remarks, "He appeared to possess an uncanny power to make the simplest of experiments produce the utmost precision."

Raman had no direct *gurus* but Helmholtz and Rayleigh together provided him with all the inspiration he needed. One is rather reminded of the story of *Ekalavya*[1].

Raman wrote many papers on vibrations. Read in isolation of each other they tend to give the impression of being rather terse, appearing at times like a hasty, impromptu commentary. It is almost as if one is hearing an excited explorer who, landing in a new territory, goes about shouting and exclaiming, enthusiastically describing at random the various sights he sees. However, when Raman's papers are considered in their entirety, one not only sees a progression of ideas but also a touch of genius.

Some Preliminaries

Vibrations and the study of sound are intimately connected since the source of a sound is always a body in a state of to and fro motion, i.e., in vibration. Musical

Glimpses from the Golden Era

sounds are characterized by the periodicity of the vibrations. For describing periodic vibrations we need concepts like period, frequency, amplitude and phase (see Fig. 4.1) but for describing musical tones we need some more, as we shall see later.

Our perception of sound is based on hearing, and the branch of physics that deals with vibrational motions as perceived by the ear is referred to as 'acoustics'. The study of sound involves not only an investigation of the manner in which periodic disturbances are propagated and perceived, but also an enquiry into the generation of the vibrations themselves. Raman's researches covered all these aspects in varying measure.

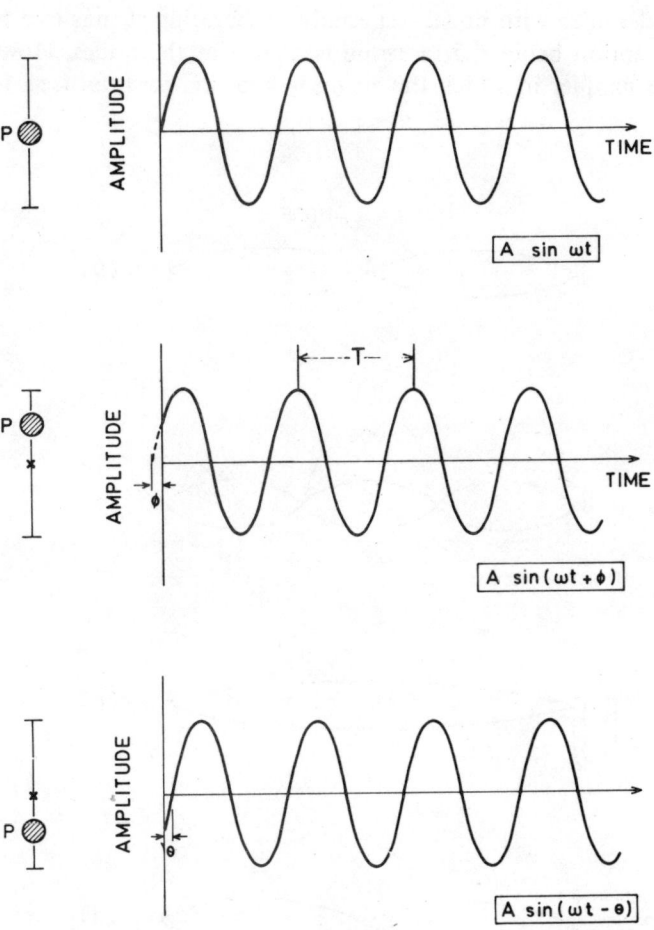

Figure 4.1 Illustration of various concepts relating to the periodic vibrations of a particle P (see left). On the right are sketched the variation of the amplitude with time for various starting configurations. ϕ and θ are phase angles. For simplicity, the curves are taken to be sinusoidal. The frequency is given by $1/T$, where T is the period of vibration.

Much of Raman's work on vibrations was connected with what he called the *maintenance* of vibrations. Sustenance of vibrations is possible only when the vibrating system is connected to an external supply of energy. The question then arises as to which modes of vibrations can be maintained. In turn this led Raman to the study of several musical instruments especially of the stringed variety.

Movement of Nodes

One is quite familiar with nodes, especially of vibrating strings (see Fig. 4.2), the popular conception being that a string is at rest at the nodes. However, if one considers the manner in which the string is kept in vibration (see, for example,

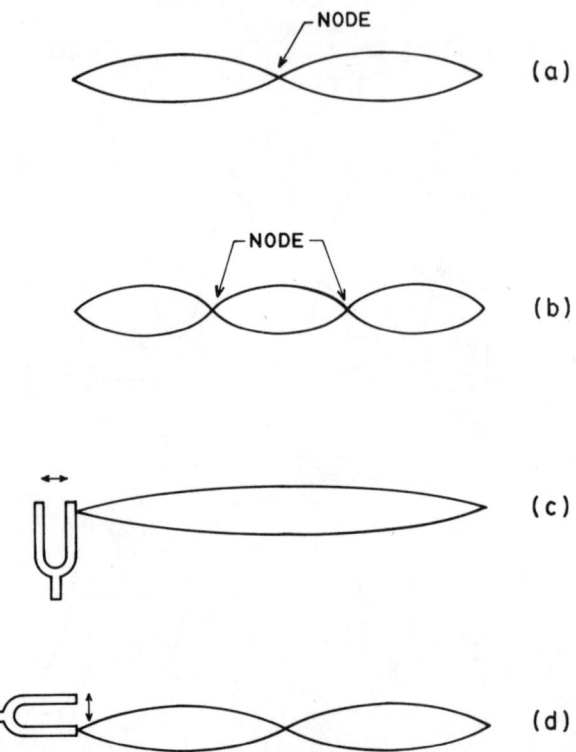

Figure 4.2 Illustration of (a) and (b) the concept of nodes, and (c) and (d) how a string may be set in vibration using a tuning fork. This is the Melde's arrangement. In (c) the excitation is longitudinal while in (d) it is transverse.

Fig. 4.2c), and asks how energy flows in the system, the notion of a node being a point of absolute rest becomes questionable. On this issue Raman wrote:

> It is generally recognized that the nodes of a string which is maintained permanently in oscillation in two or more loops cannot be points of absolute rest, as the energy requisite for the maintenance of the vibrations is transmitted through these points. I have not, however, seen anywhere a discussion or experimental demonstration of some peculiar properties of this small motion. [4.1]

This, Raman proceeded to supply. His set-up is illustrated in Fig. 4.2c, and is known to all students of sound as the Melde's arrangement. And yet from this simple-looking apparatus, Raman squeezed a remarkable variety of results, as we shall presently see. To study the movement of the nodes, Raman illuminated the vibrating string with electrical sparks produced at *twice* the frequency of vibration. The string is therefore

> seen in *two* slowly-moving positions, which represent opposite phases of the actual motion. If the nodes were points of absolute rest, then the two positions seen under the periodic illumination would intersect at fixed points. On account, however, of the small transverse motion at the nodes, the points of intersection or "fictitious nodes" are seen to execute a motion of *large* amplitude parallel to the string.... This motion, best seen under a magnifying glass, is represented in Fig. 1 [our Fig. 4.3], in which nine successive stages at equal intervals of a complete cycle are shown. [4.2]

Raman anticipated such results in 1910, and confirmed the analysis by actual photography (as here described) in 1911.

Localized Change of Phase at the Nodes

In a well-known book on acoustics which dominated the scene before the appearance of the Rayleigh classic, Donkin had suggested that when a string is maintained in vibration by a periodic force applied at some point along its length, the vibrations are in phase with the applied oscillations. Contesting this Raman wrote, "Donkin falls into a curious error in his treatment of the problem of forced oscillations." The error is a lack of *physical* consistency in Donkin's mathematical reasoning which, though it had escaped notice for decades, caught Raman's attention immediately. Not content with merely finding a flaw in someone else's argument, Raman proceeded to devise several methods to study the phase difference between the applied periodic force and the actual vibrations of the string. One of the arrangements used by him is shown in Fig. 4.4 and employs the so-called Lissajous technique.

Lissajous showed that if a point O is subjected to two periodic forces of equal frequency acting at right angles as illustrated in Fig. 4.5, then, depending on the phase difference between the forces, the resultant motion can have a variety of

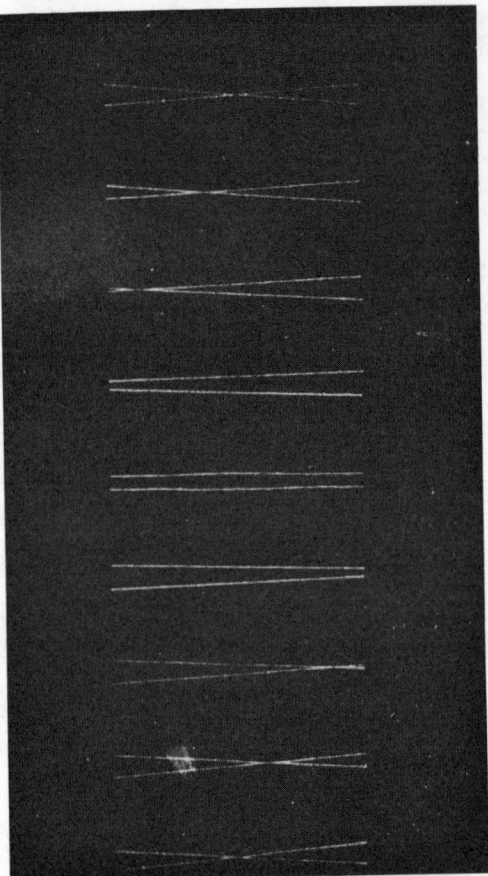

Figure 4.3 Movement of nodes as photographed by Raman. (After ref. [4.2].)

appearances as sketched therein. Raman describes the way he exploited this principle:

> A small mirror M is fixed vertically to the end of a light brass lever L [see Fig. 4.4]. This lever is pivoted upon a vertical axis and is actuated by a stout thread which with one end attached to a prong of the tuning fork T passes under a pulley and then being fixed at the other extremity to the lever, keeps the latter tightly pressed against a spring S. When the tuning fork (which is a massive one) is maintained in oscillation, the mirror M executes oscillations in a plane perpendicular to the excursions of the prongs of the tuning fork, the phases of the former and the latter being identical – this is secured by the inextensibility of the connecting thread. One point A on the string maintained in vibration is brightly illuminated throughout its excursion by means of a cylindrical lens, and the luminous line produced thereby is viewed by reflection, first at the fixed mirror RR and then at the oscillating mirror M. From the Lissajous figure... seen under these circumstances, the phase relation between the vibration of the tuning fork and that of the string... can at once be inferred. [4.3]

Glimpses from the Golden Era

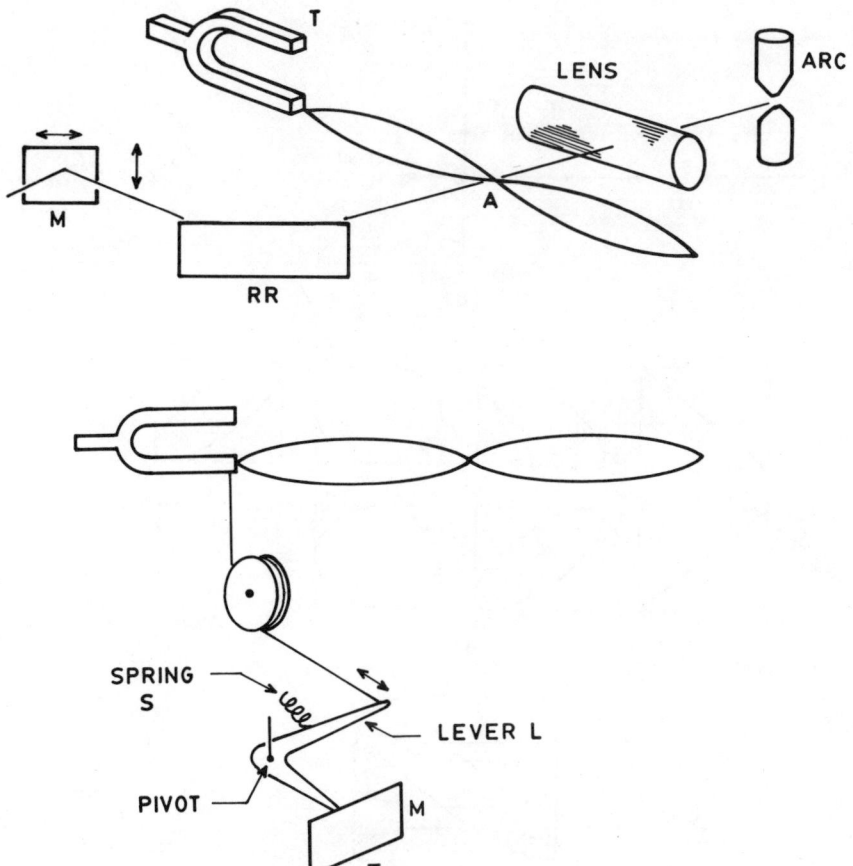

Figure 4.4 Arrangement used by Raman to study phase change at the nodes.

In a nutshell, Raman used a spot of light for playing the role of the point O in Fig. 4.5, the light having previously been bounced from two mirrors vibrating in two perpendicular directions. As he correctly anticipated, there *was* a phase difference to the extent of a quarter of an oscillation at the nodes although elsewhere there was no such difference, precisely as Donkin had argued[2].

Maintenance of Vibrations

Lord Rayleigh was one of the first to address himself to the problem of maintained vibrations (or forced oscillations as we would refer to them today). Focusing on the

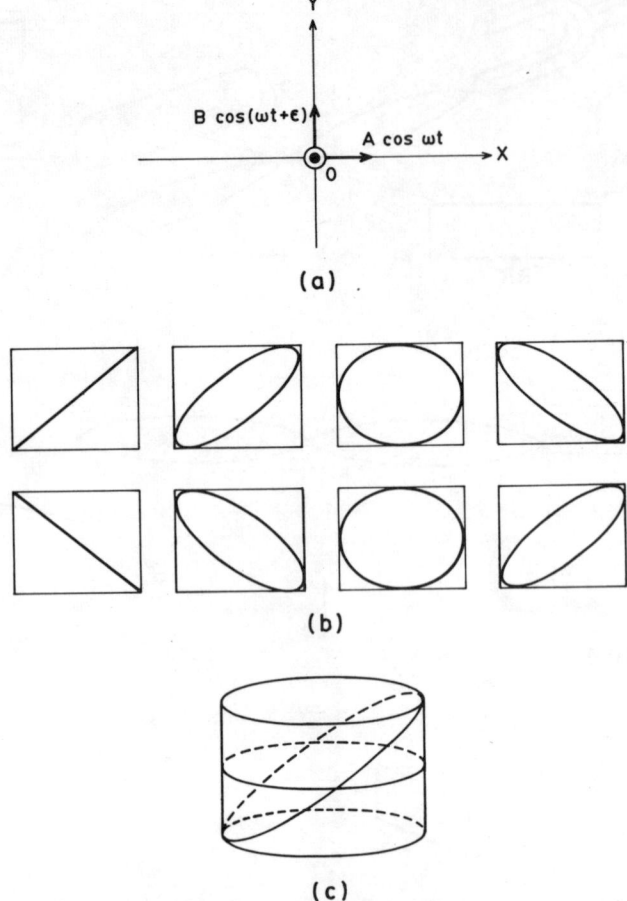

Figure 4.5 A point O is subjected to two forces acting at right angles to each other as in (a). Depending on the phase difference ε, the resultant motion can assume various forms, some of which are illustrated in (b). Lissajous showed that these figures can be obtained by drawing an ellipse on a transparent cylinder as in (c), viewing the cylinder from the side, and rotating the cylinder about its axis.

Melde's experiment since it was the archetype of this kind of problem, he remarked that

> the effect of the motion [of the fork] is to render the tension of the string periodically variable; and at first sight there is nothing to cause the string to depart from its equilibrium condition of straightness. It is known, however, that under these circumstances the equilibrium position may become unstable, and that the string may settle down into a state of permanent and vigorous vibration, whose period is double that of the point of attachment. [4.4]

As we shall soon see, the problem of maintained vibrations is more general than as formulated by Lord Rayleigh. However, in his very first experiments, Raman confined himself to the case studied by Rayleigh who had also himself performed an experiment related to this question. Rayleigh had mounted a tuning fork of frequency 256 upon a resonance-box corresponding to a frequency of 512. When the fork was strongly bowed and let go, the resonance-box was found to emit a firm tone of frequency 512, the primary tone of 256 being partly or at times even entirely suppressed. Raman wondered whether there were other examples of vibrating systems emitting tones of frequency higher than their own, and soon discovered one.

The system Raman first experimented with is shown in Fig. 4.6. A thin board was screwed into a rigid frame leaving the central part capable of vibration. A metallic wire attached perpendicularly to the centre of the board was passed horizontally over a hook and weighted. With the wire in tension, it could either be plucked or bowed. When so disturbed and made to vibrate in one loop, a loud note was emitted by the sounding board which, by using a sonometer, was proven by Raman to have twice the frequency of the wire. Observe that here the sounding board is the analogue of the fork in Melde's experiment. The only difference is that in the latter the fork is bowed to get the string vibrating, whereas in Raman's experiment the string drives the sounding board. Raman also remarks:

> Strange as it may seem, there is in actual use in India, a musical instrument, a rather crude one, it is true, the working arrangements of which are, in essentials, the same [as that in the experiment just described]. It is styled the *Gopijantra* or oftener the *Ectara* [literally meaning single string], and is chiefly used, I find, by vocalists – those of the poorer sort – for striking key notes and marking time. The users of the instrument are apparently totally unaware of its unusual characteristics. [4.3]

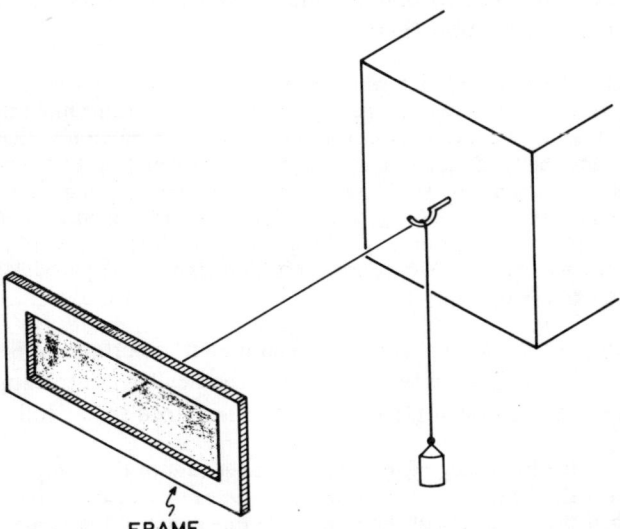

Figure 4.6 A wire in tension is attached to a board stretched on a frame. When the wire is plucked, the board vibrates at twice the frequency of the wire.

What are these unusual characteristics? Raman himself provides the answer:

> In the theory of small vibrations of a stretched string, it is assumed as a working hypothesis that the tension of the string is constant throughout the oscillation. When metallic wires are employed, it is obvious that, since these have a high Young's modulus, the amplitude of oscillation should be very small indeed, if this assumption is to be valid. With larger amplitudes, such as can be obtained by plucking or bowing, we should expect a periodical fluctuation of tension, since, if its extremities are kept fixed, the length of the wire cannot be constant throughout an oscillation.

By an elementary argument relating to the changes in the length of the string, Raman showed that the wire tension varies periodically as cos $2pt$, where p is the frequency of the wire. Observe that the tension varies as *twice* the frequency of the wire. It is this periodic variation of the tension which, when communicated to the sounding board, causes the latter to vibrate at the frequency $2p$. Raman had an *Ectara* specially constructed for his use, and performed many experiments with it.

In 1906 Andrew Stephenson published a paper entitled "On a class of forced oscillations", in which he extended Lord Rayleigh's work. Commenting on this Raman wrote:

> Mr Andrew Stephenson attacks this problem by pushing to a higher degree of approximation the analytical method employed by Hill in discussing certain problems in the Lunar Theory, and by Lord Rayleigh in working out the particular case of double frequency. His analysis... leads to the result that the oscillations of the system may be magnified or maintained under suitable circumstances, if the frequency N of the imposed variation of spring [tension] stands to the frequency N_1 of the oscillation in the relation 2:r where r is any positive integer. [4.5]

Raman is not satisfied with a mere mathematical deduction such as that made by Stephenson. He is of the opinion that

> the beauty and interest of the results obtained by Mr Andrew Stephenson have not been generally realized, otherwise it is nearly certain that something more satisfying in the way of experimental demonstration of these oscillations than mere observation of "instability of equilibrium" in certain cases would have been put in the field. I think an experimentalist would hardly be pleased with anything less than the actual permanent maintenance of oscillations of the type mentioned, i.e., something similar to the experiments of Faraday, Melde and Lord Rayleigh for the case of double frequency, which, as Mr Stephenson points out, is only one particular case of his general theorem. [4.6]

Raman went back to Melde's apparatus and cajoled it to perform as Stephenson's analysis demanded. To start with, by the simple expedient of controlling the string tension he showed that the vibrations could be maintained in the following cases:

> (1) When the frequency of the fork is 2 times that of the string;
> (2) When the frequency of the fork is 2/2 times that of the string;
> (3) When the frequency of the fork is 2/3 times that of the string;
> (4) When the frequency of the fork is 2/4 times that of the string;
> (5) When the frequency of the fork is 2/5 times that of the string;
> And so on. [4.7]

Glimpses from the Golden Era

The first of the above is the well known case of double frequency already discussed by Lord Rayleigh while the second is the more familiar example of resonance. All the others are

> remarkable cases which form apparent exceptions to this law of equality of periods, that is, in which we have marked resonance when the periods of the impressed force and of the system do *not* stand to each other in a relation of approximate equality. [4.8]

So indeed other cases of resonance were possible as Stephenson had predicted. Raman was so fascinated by these resonances that he studied them in various ways. He started by directly photographing them. The results, while quite interesting, were not revealing enough. Therefore he decided to follow the time evolution of the vibrations by the moving-photographic-plate technique which became quite popular with him. To observe the vibrations he used two sources of light (see Fig. 4.7); one was

> a horizontal slit, and the other was a vertical slit placed behind the oscillating string. Both were illuminated by sunlight and had collimating lenses in front of them. The light from the former fell upon a small mirror attached to the prong of the vibrating fork and after reflection fell upon the lens (having an aperture of 4 cm diameter) of a roughly constructed camera. The light from the vertical slit behind the vibrating

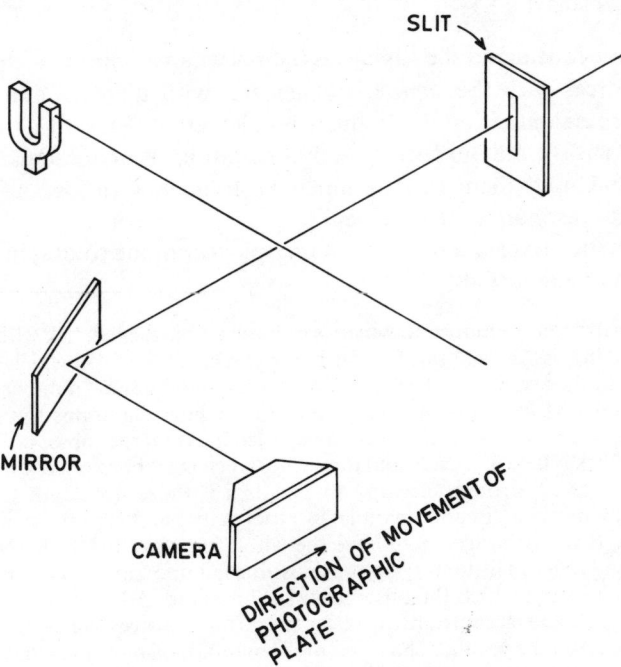

Figure 4.7 Arrangement used by Raman to study vibrations of a string. As described in the text, Raman also monitored the vibrations of the fork to provide a calibration. The reflection system used for the latter purpose is not shown.

string was also reflected into the camera by a fixed mirror. In the focal plane of the camera was placed a brass plate with a vertical slit cut in it. The images of the horizontal and vertical slits fell, one immediately above the other, on the slit in the plate.... The dark slide which held this was moved uniformly by hand in horizontal grooves behind the slit in the focal plane of the camera, while the fork and the string were in oscillation.

Figure 4.8 shows some of Raman's results whose significance may be appreciated by considering, for example, (c). (h) and (i) together. These

represent the type of motion in which the string makes *three* swings for every *two* swings of the fork. But it is evident from Figs. 8 and 8(a) [our Fig. 4.8(h) and (i)] that the successive swings on opposite sides are not all equal in amplitude and the influence of this is also perceptible in Fig. 3 [our Fig. 4.8(c)], having given rise to the appearance of two extra strings, which represent really the turning points of the motion.

The familiar Lissajous technique was also pressed into service. For exploiting it, Raman caused the string to execute a compound motion involving one frequency in one plane and a different one in a perpendicular plane.

Under these circumstances, the motion of a point on the string in a plane transverse to it becomes and remains the appropriate Lissajous figure, and the frequency relation between the component motions is thus rendered evident to inspection in a most striking manner. [4.5]

Earlier we have considered the Lissajous figures when a point O is driven by two perpendicular forces with the *same* frequency but with different amplitudes and phases. In the present situation the frequencies also are different. Figure 4.9 shows the typical patterns we may expect in such a situation. Raman actually observed such patterns and used them to determine the frequency ratios and thence the frequencies of the resonances themselves.

Raman also made clever use of the stroboscopic technique to obtain a picture of the waveform. As he describes:

A Rayleigh synchronous motor on which is mounted a blackened disk with narrow radial slits cut in it is very suitable for this purpose. One of the disks which I use has thirty slits in it, the armature-wheel of the motor having the same number of teeth. The electric current from the self-interrupter fork which maintains the string in vibration also runs the synchronous motor. In making the observations, the stroboscopic disk is held vertically and the string which is set horizontal and parallel to the disk is viewed through the top row of slits, i.e., those which are vertical or nearly so and move in a direction parallel or practically parallel to the string as the disk revolves. It is advantageous to have the whole length of the string brilliantly illuminated and to let as little stray light as possible fall upon the reverse of the disk at some distance from which the observer takes his stand. A brilliant view is then obtained. Under these circumstances, we see the string in successive cycles of phase along its length, and the peculiar character of the maintained motion in these cases is brought out in a very remarkable way. *The string is seen in the form of a vibration-curve* [see Fig. 4.10] which would be identical with those obtained by other methods but for the fact that the amplitude of motion is not the same at all points of the string, being a maximum at the ventral segments and zero at the nodes.

Glimpses from the Golden Era

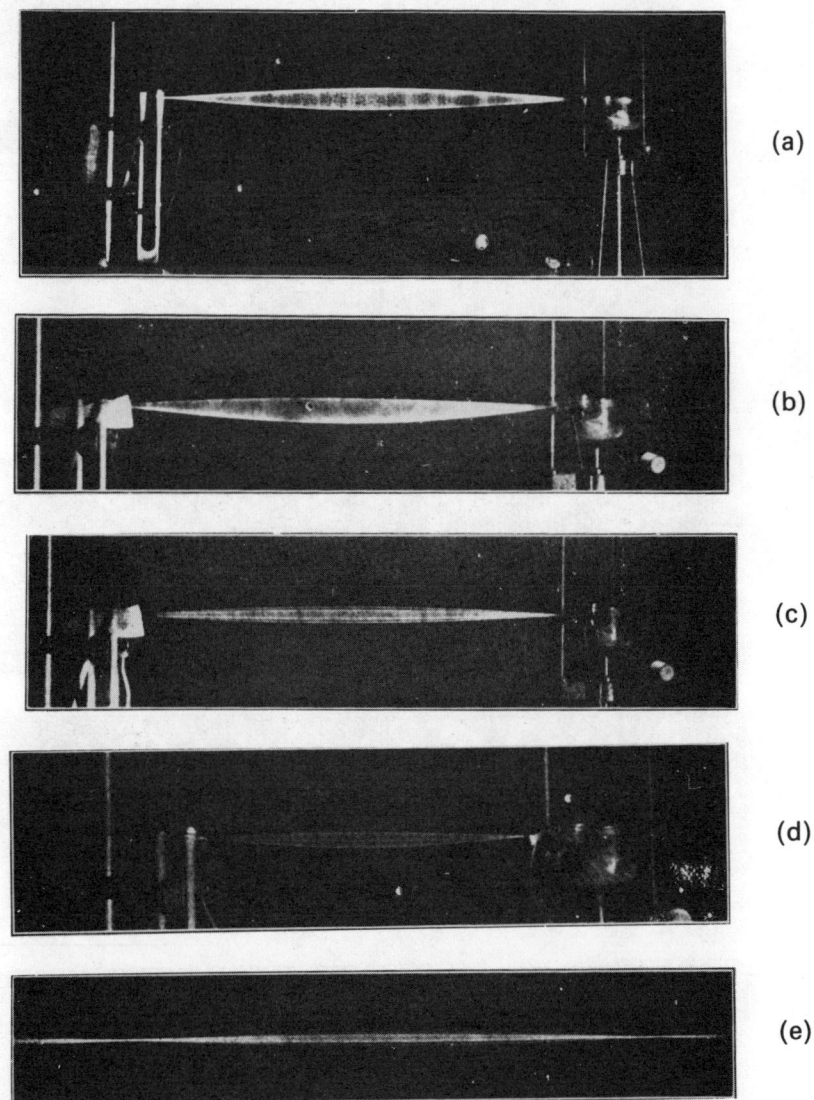

Figure 4.8 Studies on "remarkable resonances". (a)–(e) Photographs of the vibrating string.

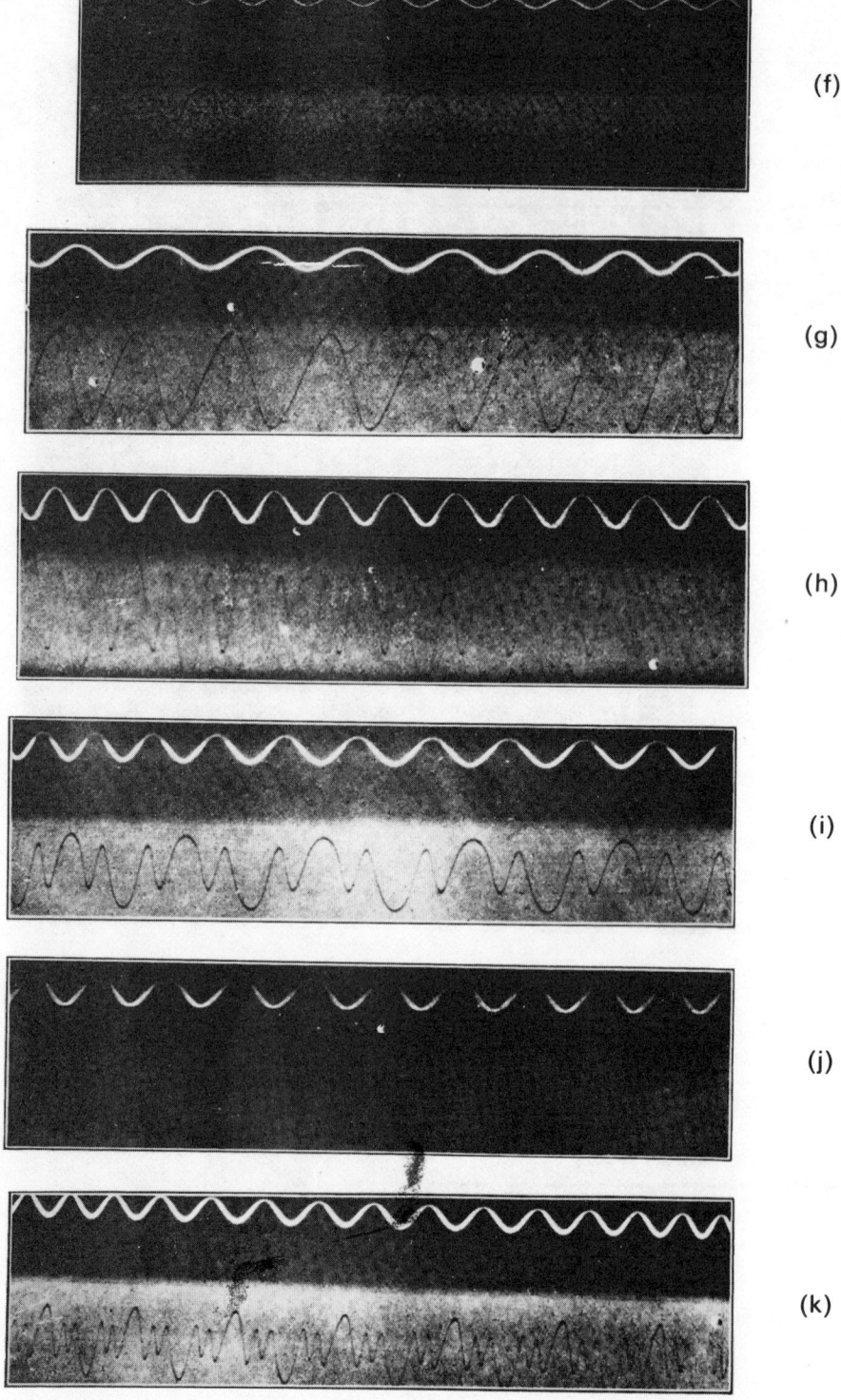

Figure 4.8 (contd.) Studies on "remarkable resonances". **(f)**–**(k)** Variation of amplitude with time of a select point on the string. The calibration curves are also shown. (After ref. [4.7].)

Glimpses from the Golden Era

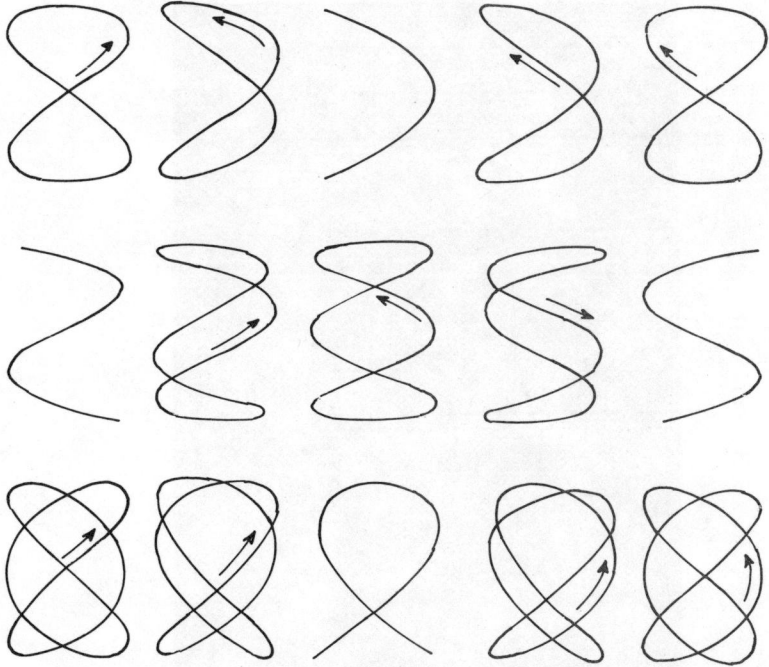

Figure 4.9 A point O is made to execute simultaneously the motions $a\cos(pt - \varepsilon)$ along x and $b\cos qt$ along y. Figure shows the resultant Lissajous patterns which may be expected for representative values of ε for a given (p/q) and for a few different ratios (p/q).

If we use a fork with a frequency of 60 per second, the *free* oscillations of the string should have a frequency of 30 in the case of the first type, 60 in the case of the second, 90 with the third, 120 with the fourth, 150 with the fifth, and so on. With the disk having 30 slits on it we get 60 views per second of any one point on the string, and with the even types of motion, i.e., the second, fourth, etc., the 'vibration-curve' seen through the stroboscopic disk appears single. [4.7]

Raman noted that the stroboscopic method could also be applied to the study of bowed and struck strings but somehow he never appears to have done so himself.

In the classical Melde's experiment, the string is fixed at one end and driven at the other with a fork. What if the string is driven at *both* ends with forks of different frequencies, say N_1 and N_2? One of course expects resonances at frequencies $\frac{1}{2}rN_1$ or $\frac{1}{2}sN_2$, where r and s are positive integers. Raman remarks that in addition to these,

> the observer is surprised and delighted to find, even at a first trial of the experiment, a large number of other cases of vigorous maintenance which have evidently to be ascribed to the joint action of the two forks on the string. Their variety and number is extraordinary, and these, together with the way in which they come rapidly

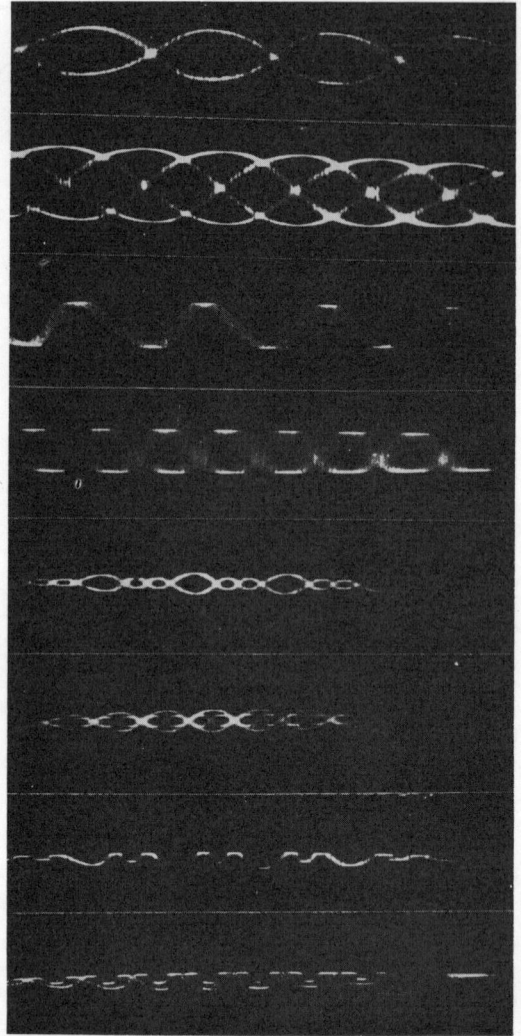

Figure 4.10 Vibrations of a stretched string observed through a stroboscopic disc mounted on a synchronous motor. (After ref. [4.8].)

following one another particularly at the higher frequencies, remind the observer, by a vivid analogy, of the lines in a complicated spectrum-series. It is readily guessed at once that these are cases of 'combinational' resonance in which the frequency of the principal term in the maintained motion is related *jointly* to the frequencies of both the forks.... Under suitable conditions the equilibrium of the system becomes unstable and a vigorous motion is maintained if the frequency of free vibration in any given mode is sufficiently nearly equal to $\frac{1}{2}rN_1 \pm \frac{1}{2}sN_2$.... [4.9]

Raman photographed several of these combinational resonances, although he found that the difference modes, with frequencies $\frac{1}{2}(rN_1 - sN_2)$, were much more difficult to maintain than the summation modes.

The pleasure that Raman unquestionably derived in making various observations should not obscure the fact that he was always equally interested in the underlying physics. While we reserve for later discussion some of the mathematical nuances of the various problems he tackled, it is pertinent to call attention here to an experiment by Raman and Dey where the focus was on the physics of the problem rather than on the aesthetics of the phenomenon [4.10].

It will be recalled that in the Melde's experiment, the *longitudinal* oscillatory force imposed by the fork does not at first disturb the string from its position of equilibrium. However, the imposition of the oscillatory force renders the equilibrium unstable, the string then being thrown into *transverse* oscillations. Raman and Dey investigated instabilities in vibrating systems using an arrangement previously employed by Klinkert in one of his experiments.

A 2-metre-long wire was stretched vertically and its tension was adjusted suitably to make it vibrate in different modes. An electromagnet placed near a selected point on the wire provided the necessary periodic excitation force, the periodicity being obtained by a fork-interrupter of frequency 60 cycles per second. Raman and Dey found that when, after adjusting the tension of the wire to make it vibrate in two or more loops, it was excited by the electromagnet, then initially the wire did vibrate at 60 c/s in two or more segments as adjusted. However, this type of motion soon became unstable giving way to another which was the form finally sustained. As they observe,

> if the wire initially divides up into two segments and vibrates with a frequency of 60, its centre, which at first is a node, gradually acquires a very considerable motion, and the frequency of the vibration alters to 30. Similarly, if the wire initially vibrates in 3 segments, the frequency changes to 20 when the instability sets in; when the initial vibration is in 4 segments, the frequency changes to either 30 or 15 according as the instability does or does not result in a movement of the centre of the wire; and so on. [4.10]

Raman and Dey found that the onset of instability and the transition in the frequency of vibration could take even several minutes. They also showed that it was not necessary to have the wire vibrating first at 60 c/s before transition to a submultiple frequency. After carefully adjusting the tension of the wire so that it could vibrate in two, three, or a larger number of segments as desired, they had the wire stationary and then placed an electromagnet exactly opposite the point where a node of this oscillation would be. According to a well-known principle, such an excitation was not supposed to lead to any kind of forced oscillations; in fact, one should expect the wire to continue to remain stationary. However, Raman and Dey found that when the electromagnet was excited, the state of rest of the wire became unstable and gradually a vibration of a submultiple frequency got built up.

Raman concluded from these studies that non-uniformity of the applied periodic force was the crucial factor in the onset of instability. In recent times, the study of instability has become an important aspect of analytical dynamics. Viewed from that angle, Raman's investigations appear to have been quite ahead of their time.

The maintenance of vibrations is a general problem in mechanics, and the Melde's arrangement merely a convenient test bed for demonstrating the various features. Raman recognized this generality and in fact illustrated it by demonstrating vibrations in a seemingly unlikely system, namely, a synchronous motor. His motor consisted of a wheel of soft iron mounted on an axis with two ball-bearings between the two poles of an electromagnet placed diametrically with respect to the wheel. The wheel had thirty teeth, and when a direct current was passed through the electromagnet, set itself rigidly at rest with a pair of teeth at the ends of a diameter. The equilibrium under such conditions was found to be thoroughly stable. The same was generally true when an intermittent current supplied by a fork-interrupter was used to excite the electromagnet, except in some situations when instability occurred and the wheel began to exhibit steady angular oscillations about its equilibrium position. Once again by resorting to the Lissajous technique Raman was able to establish the frequency ratios for which the maintenance of vibrations occurred.

Encouraged by this success, Raman then tried a cute variation in which the aim was to maintain *rotational* rather than *vibrational* motion. As he describes:

> It is well known that with an intermittent current passing through its electromagnet, the synchronous motor can maintain itself in "uniform" rotation, when for every period of the current, one tooth in the armature-wheel passes each pole of the electromagnet. In other words, the number of teeth passing per second is the same as the frequency of the intermittent current. From a dynamical point of view it is of interest, therefore, to investigate whether the motor could run itself successfully at any speeds other than the 'synchronous' speed. [4.11]

To test this, Raman coupled the synchronous motor to an independent driving system using which arbitrary rotational speeds could be imposed. He found that the driving system

> was very satisfactorily obtained by fixing a small vertical water-wheel to the end of the axis of the motor and directing a jet of water against it. The water-wheel was boxed in to prevent any splashing of water on the observer. By regulating the tap leading up to the jet, the velocity of the latter could be adjusted. The speed of the phonic wheel was ascertained by an optical method....

Raman made observations on the rotational motions using a projection lantern. The motor was placed on the stage of the lantern and the rim of the wheel was focused on a screen[3]. In front of the projection prism was placed a fork-interrupter to provide intermittent illumination. When the fork was vibrated and the motor was set in rotation, a pattern corresponding to the maintained speed became visible on the screen. The experiment consisted in finding the imposed rotational speeds at which the motor "bites". As Raman observes, "When the motor 'bites' the pattern seen becomes stationary and remains so for long intervals of time or even indefinitely." Raman found that rotations could be maintained at 1/2, 3/2, 4/2, 5/2,... times the synchronous speed. He also discovered that uniform rotation could be maintained by exciting the electromagnet of the motor simultaneously by the intermittent currents from two separate fork-interrupters having different frequencies. Uniform rotation was possible at speeds related jointly to the frequencies of the two currents.

Glimpses from the Golden Era

4.2 Raman and Musical Instruments

From vibrations to musical instruments was but a short step for Raman, especially as in instruments like the violin one is concerned with the maintenance of the vibrations – of course with a view to producing musical notes!

In a delightful introduction to the subject of musical tones, Feynman observes:

> Pythagoras is said to have discovered the fact that two similar strings, under the same tension and differing only in length, when sounded together give an effect that is pleasant to the ear *if* the lengths of the strings are in the ratio of two small integers. If the lengths are as one is to two, they correspond to the octave in music. If the lengths are as two is to three, they correspond to the interval between C and G, which is called a fifth. These intervals are generally accepted as 'pleasant' sounding chords.... Pythagoras could only have made his discovery by making an expermental observation.... [An important aspect of this interesting discovery is] it had to do with two notes that *sound pleasant* to the ear. We may question whether we are any better off than Pythagoras in understanding *why* only certain sounds are pleasant to the ear. The general theory of aesthetics is probably no further advanced now than in the time of Pythagoras. In this one discovery of the Greeks there are three aspects: experiment, mathematical relationships, and aesthetics. Physics has made great progress on only the first two parts. [4.12]

Where Raman was concerned, he was deeply involved with all the three aspects.

The player as well as the maker of a musical instrument are both interested in the *quality* of musical tones. As Helmholtz describes, "By quality of a tone we mean that peculiarity which distinguishes the musical tone of a violin from that of a flute or that of a clarinet, or that of the human voice, when all these instruments produce the same note at the same pitch [4.13]." We know that musical notes should be periodic but on what precisely does quality depend? Again Helmholtz gives us the answer: "The quality of the tone depends on the form of the vibrations." Elaborating, Helmholtz observes:

> On exactly and carefully examining the effect produced on the ear by different forms of vibration,... we meet with a strange and unexpected phenomenon.... The ear, when its action has been properly directed to the effect of the vibrations which strike it, does not hear merely that one musical tone whose pitch is determined by the period of the vibrations... but in addition to this becomes aware of a whole series of higher musical tones which we call the *harmonic upper partial tones* and sometimes simply the *upper partials* of the whole musical tone or note, in contradistinction to the *fundamental* or *prime partial tone* or simply the *prime* as it may be called, which is the lowest and generally the loudest of all the partial tones, and by the pitch of which we judge the pitch of the whole *compound musical tone*.... [4.13]

In other words, the *form* of the vibration and its harmonic content become important in the context of musical instruments. This aspect received much attention from Raman.

The process by which instruments produce musical notes is quite complex, particularly in the case of the violin. While the form of the vibration is certainly pertinent, one must also consider various other related factors like the playing technique, the transport of vibrational energy across the system and finally the manner in which sound energy is radiated into air. In other words, one needs not merely a *kinematical* theory which confines its attention to the form of the vibrating string, but also a *dynamical* theory of the entire instrument; Raman investigated both.

Studies on the Violin

Felix Savart, often regarded as the grandfather of violin research, once wrote, "The finest of all instruments is the violin; it has been called king of instruments." Starting with the work of Savart, there has been a long history of ingenious experiments for probing the secrets of this marvellous musical instrument. However, in the nineteenth century there were only a few major islands of activity, firstly due to Savart himself, then Helmholtz and finally Lord Rayleigh. It is only with the work of Raman in the early part of this century that violin research blossomed into a subject of its own, remaining vigorous to the present day.

Raman's involvement with the violin was almost inevitable one might say. As a boy he had heard his father play but when he listened, it was not only with the enthusiasm of a music lover but also with the curiosity of a physicist. There were several questions triggered mainly by his reading of the Helmholtz classic, questions that assumed new significance after his many experiments on the maintenance of vibrations.

Let us start with the instrument itself; it is shown in Fig. 4.11 together with the names of some of the parts. The four strings labelled G, D, A, E correspond (roughly) to the frequencies 196, 294, 440, 659. From the physics point of view there are three major components or subsystems, i.e., the string, the bridge which communicates the vibrational energy of the string to the resonance box, and finally the box itself which takes up the vibrational energy, resonates, and then radiates the energy into air.

The first clear description of the motion of the bowed string was provided by Helmholtz. Though somewhat idealized, it is a good first approximation. When the bow is drawn, the string adheres to the bow and moves at a constant velocity v_B which is the velocity of the bow. However, the string is not carried forward indefinitely – naturally. It slides back for a while, is gripped and carried forward again, slides back, and so on. The string configuration or the displacement pattern then changes in time as in Fig. 4.12, that is, the apex propagates to and fro inside two enveloping parabolae. On the other hand, if one observes any one point on the string as a function of *time* one obtains a characteristic, two-step zigzag curve as shown in Fig. 4.12b. The type of vibration just described is referred to as the *principal* mode. Helmholtz arrived at the form in Fig. 4.12a, essentially from observations such as in Fig. 4.12b.

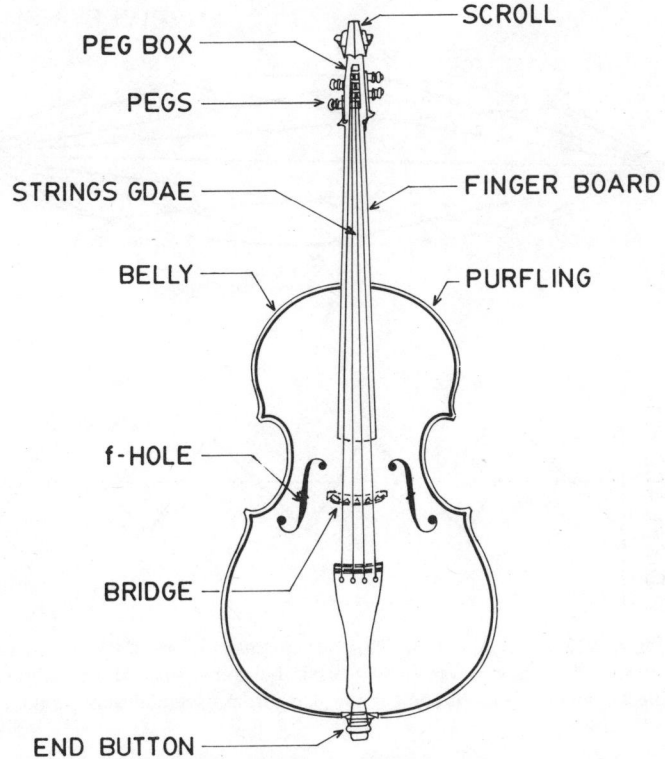

Figure 4.11 The violin and its various parts.

Raman's contributions to the physics of the violin may be summarized as follows:

1) Extension of Helmholtz's ideas to the other modes of vibration via a kinematical analysis
2) Study of the wolf-note
3) Examination of the role of friction between the bow and the string
4) Investigation of the influence of the bridge by a dynamical theory
5) Various experimental studies including measurement of the frequency response of the violin. (Such a response curve is now known as Raman curve.)

Curiously, Raman's first investigations on the subject were not on the violin itself but on an improvised system calculated to reveal certain characteristics of bowed strings [4.14]. Apparently, he had carried out some studies even while in the Presidency College, Madras.

In any vibrating system, the central question relates to the different possible modes of vibration. In the case of the bowed string, a direct mathematical analysis of this problem is both tricky and non-trivial. Raman's approach is ingenious. Guided by an earlier work due to Harnock, he recognized that the key to an understanding

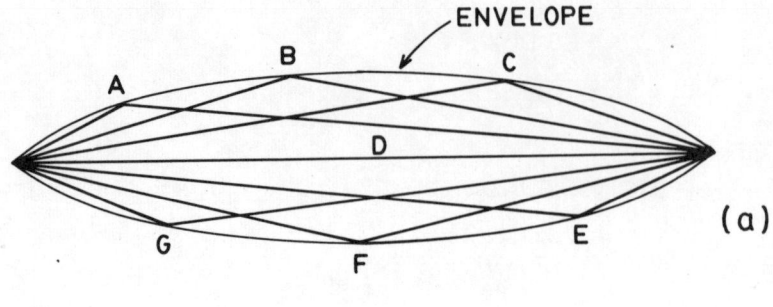

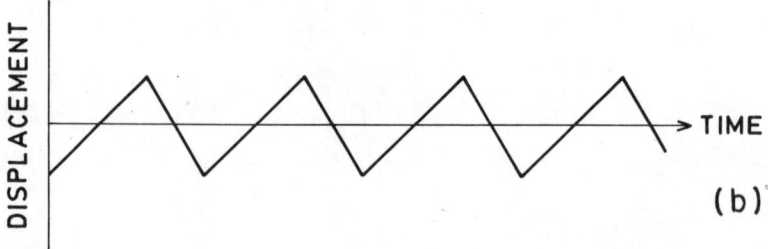

Figure 4.12 A, B,...G in (a) illustrate the dispositions of the bowed string at various instants. The apex moves along the parabolic envelope. (b) The time variation of the displacement of a given point of the string from its equilibrium position.

of the different vibrational modes lay in the study of *discontinuous* velocity waves set up in the string. Accordingly, his first experiments were concerned with the nature of discontinuous wave motion. For a better appreciation of Raman's motivations, it is necessary to digress briefly and consider the nature of the vibrations of a bowed string.

The displacement patterns illustrated in Fig. 4.12 are the result of propagation of two oppositely directed velocity waves which nowadays are called Raman waves (see Fig. 4.13). The propagation velocities of the two Raman waves are numerically equal, and the profile of each wave at every instant describes the particle velocities at all points along the string at that particular instant due to the wave concerned. The net velocity profile obtained by adding together the two Raman waves is shown in Fig. 4.13c. This exercise can be repeated for other instants of time. We will pursue this topic later, and return for the present to the experiments of Raman and Appaswamaiyer who wished to demonstrate that Raman waves on a string lead to displacements exactly as in a bowed string [4.14].

An ordinary simple pendulum was drawn to one side and allowed to swing down. The downward swing was then arrested by a fixed stop or bridge placed between the upper end and the bottom-most position of the bob, roughly three-fourths of the way along the length. The impulse produced by the sudden arrest generates a discontinuous wave motion along the length of the string in the region between the fixed end and the bridge. A small portion of the string was illuminated by an electric

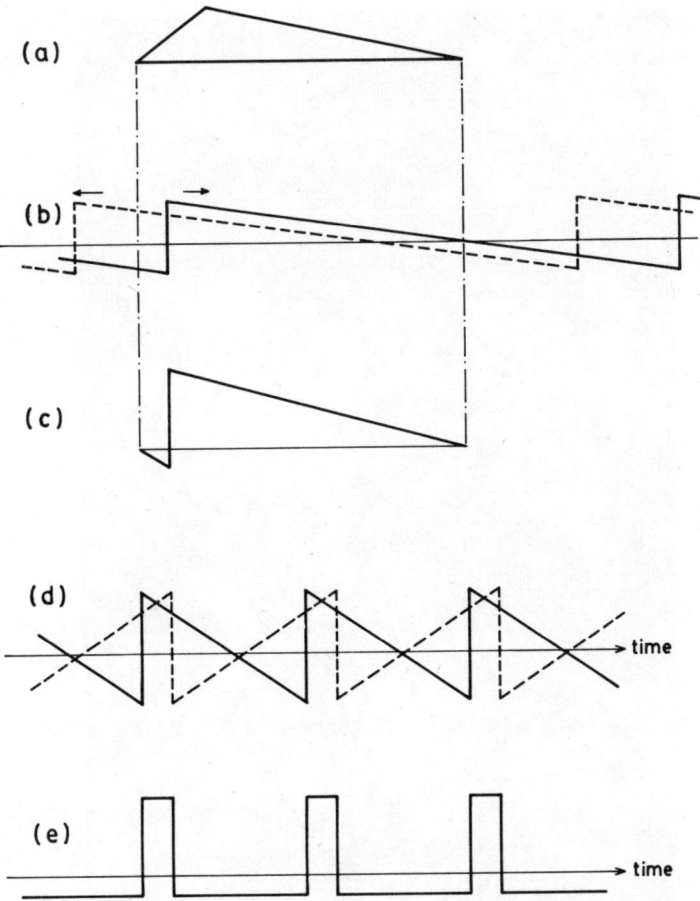

Figure 4.13 (a) Typical displacement pattern. This may be considered to arise due to the two Raman waves in (b). Each describes the velocity of various points along the length. The waves propagate in the directions shown by the arrows. The *combined* effect of the two Raman waves is shown in (c). This gives the net velocity with which every particle along the string moves at the particular instant corresponding to configuration (a). Some points on the string have a positive velocity while others have negative velocity. If one observes how the velocity at the bowing point varies as a function of time, one will obtain the two curves in (d), corresponding to the two Raman waves. The net velocity variation at the bow is shown in (e).

arc and its shadow was captured on a photographic film kept behind and moved uniformly downwards. From typical records, displayed in Fig. 4.14, it can be seen that the form of the vibration "reproduces that of a bowed string in a perfect manner", i.e., every point first moves in one direction with a particular velocity, then reverses and starts moving with a different velocity, and so on, this despite the fact that there was *no* bowing! The conclusion to be drawn is that the *vibration modes of bowed strings are essentially determined by the propagation of discontinuous velocity waves*. Raman

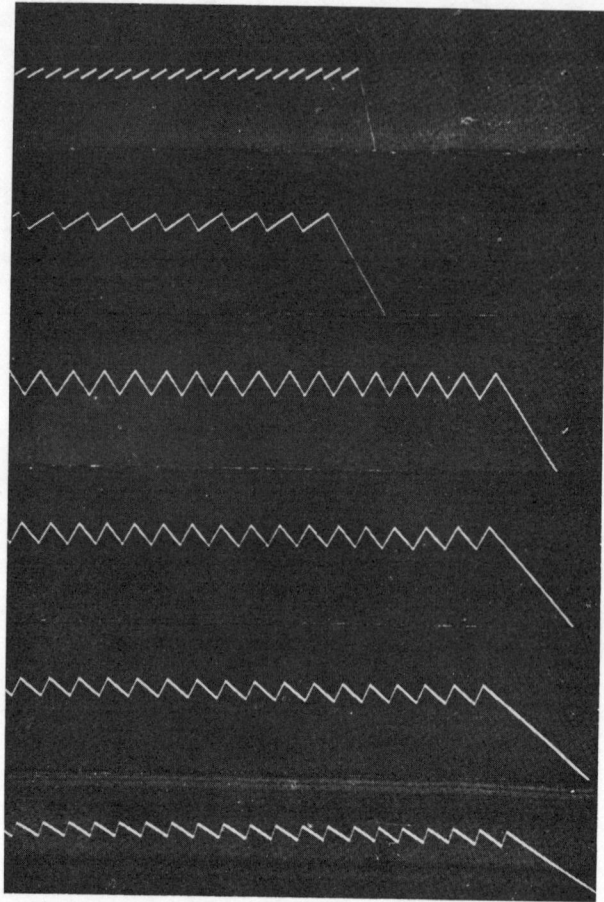

Figure 4.14 Photographs showing the initial movement and the subsequent vibration produced by the imposition of a discontinuous velocity wave. (After ref. [4.14].)

later extended these experiments so that two discontinuities could be imposed on the string instead of just one.

Turning now to Raman's studies on the violin: he was drawn to the subject because not enough was known about the physics of the instrument. No doubt Helmholtz had given a good start but his elucidation was confined to the principal mode of vibration. Subsequently Krigar–Menzel and Raps had carried out extensive experimental work but the information they had generated appeared to remain largely undigested. There was also an analysis of vibrating strings by Andrew Stephenson but it would seem, as Raman puts it,

> that Stephenson was unacquainted with the work of Krigar–Menzel and Raps published in 1891, and that he was, indeed, unaware of many facts which anyone who has experimented with a bow and monochord could readily observe for himself. It is not a matter for surprise therefore that, though Stephenson's paper is

Glimpses from the Golden Era

noteworthy as an attempt to treat the motion of a bowed string as a case of maintained vibration, it takes us little beyond the work of Helmholtz. [4.15]

Further work was clearly needed and Raman had definite ideas of how to go about it. The basic questions to be asked and answered were: What types of Raman waves are induced by bowing, and how are they superposed? Such an analysis is no doubt kinematical in the sense that the motion of the string is considered without any consideration of the forces brought into action by the act of bowing. Nevertheless, kinematics does provide much insight.

A better appreciation of Raman's line of thinking may be had by referring to Fig. 4.15 which offers a perspective plot of the various velocity profiles and displacement patterns associated with the principal mode. At the top are portrayed

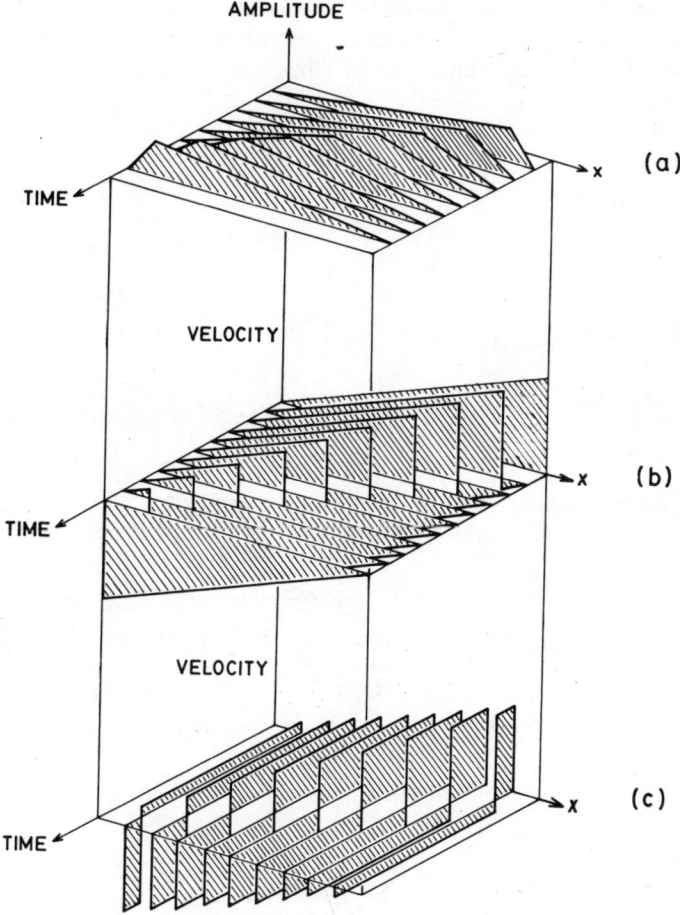

Figure 4.15 (a) Displacement profiles at various instants of time. (b) Velocity profiles at various times. (c) Time variation of velocity at various points along the string.

the displacement patterns at various instants of time. Then come the "snapshots" of the velocity distribution $v(x)$ along the string at different instants of time. From this information one can easily construct graphs of $v(t)$ the time variation of the velocity for various points along the string. A collection of such plots is shown at the bottom, and one observes that every point moves with one of two velocities, namely, $v_\uparrow$ and $v_\downarrow$. At the bowed point, $v_\uparrow = v_B$ the velocity of the bow, but elsewhere this equality does not hold. However, the magnitude of the difference $(v_\uparrow - v_\downarrow)$ remains the same at all points along the string.

Consider next the second type of vibration which arises owing to the presence of *two* discontinuities in the velocity profile (as opposed to one in Fig. 4.15). Since the discontinuities move along the length of the string and get reflected at the ends, they necessarily must cross at some point on the string and much depends on where exactly they do so. Raman showed that if they crossed exactly at the centre, the vibration curve everywhere would be a two-step zigzag. On the other hand if the crossing occurs off-centre, complexities are possible as Fig. 4.16 shows. Even richer detail is encountered in the third type of vibration, a part of which is sketched in

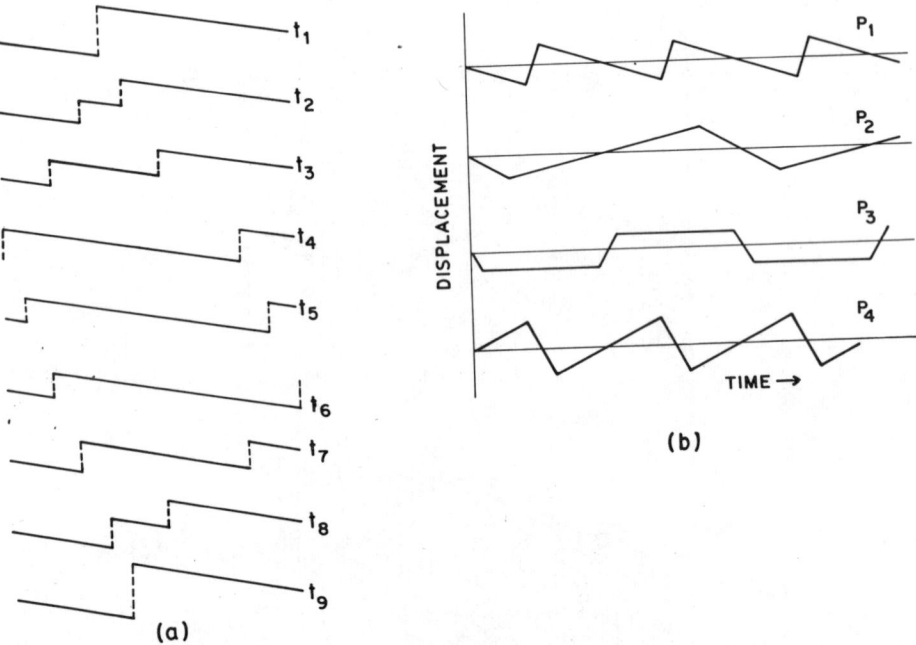

Figure 4.16 (a) Velocity profiles at various instants with two discontinuities present. (b) The displacement versus time curves for some representative points on the string. Observe that both two- and four-step curves are possible.

Glimpses from the Golden Era 87

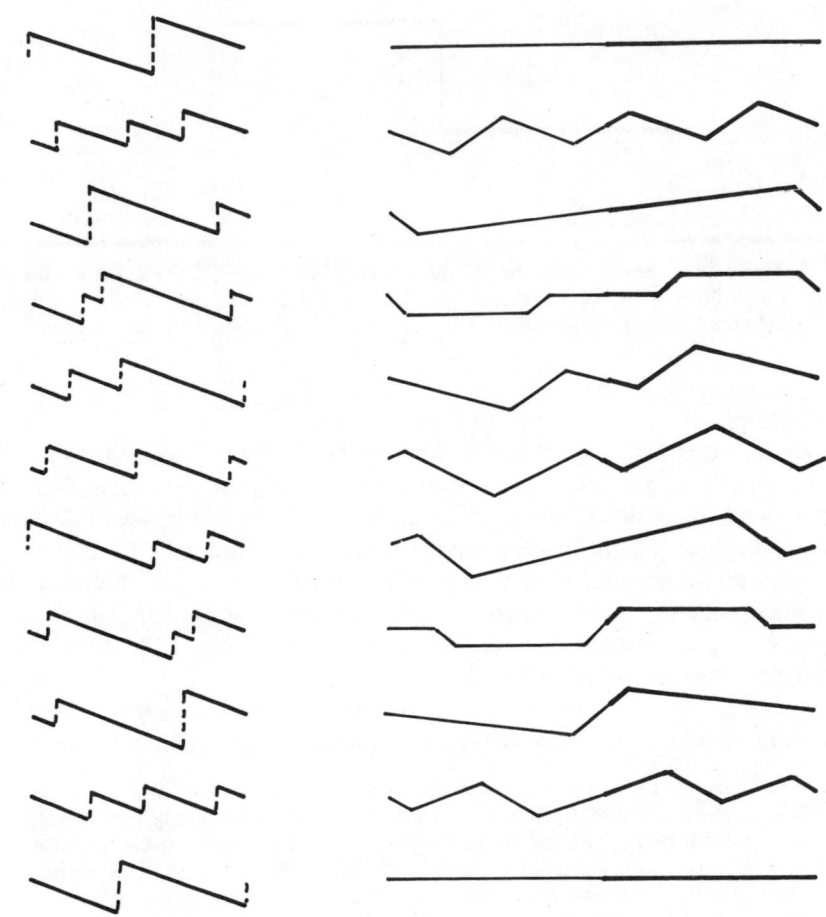

Figure 4.17 On the left are the velocity profiles at various instants corresponding to the third type of vibration. On the right are displacement–time curves for representative points on the string. (After ref. [4.15].)

Fig. 4.17. And when one goes to still higher orders, further ramifications are possible depending on whether the number indicating the order is prime or not! Years later, the mathematician Madhavarao commented:

> A remarkable result of mathematical interest is his analysis of the nature of motion when n, the number of discontinuities, is a prime > 1, or is composite. A careful examination of the manner in which this analysis is accomplished shows that he has not used any sophisticated results of prime number theory, but just the definition of a prime number! [4.16]

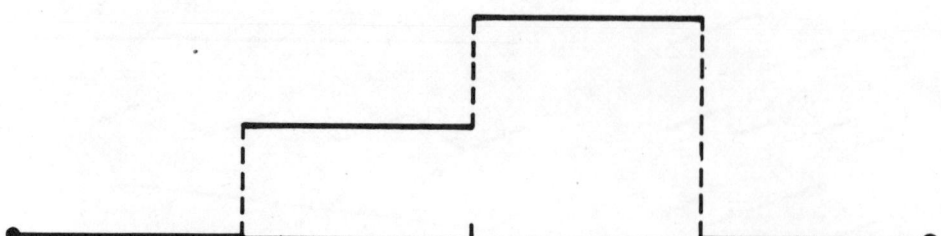

Figure 4.18 Velocity curve for the third type of vibration when the string is bowed at $(5/8)l$. Observe the horizontal steps and contrast with earlier figures of a similar nature. (After ref. [4.15].)

Another example of Raman's intuitive brilliance.

There are also interesting subtleties relating to the position of bowing. As is only to be expected, if the string is bowed at a point corresponding to one of the nodes, then all harmonics having nodes at that point would not be excited. Suppose for instance the third mode is excited, the bowing point being at a distance $(5/8)l$ from one end. Clearly this point corresponds to the 6th node of the 8th harmonic, 11th node of the 16th harmonic, and so on; such harmonics are therefore not excited. The velocity profile under these circumstances would then be as in Fig. 4.18, significantly different from those encountered earlier.

The kinematical analysis presently being sketched is obviously quite oversimplified, sweeping under the rug many real-life complexities. As Raman himself puts it:

> It now remains to consider various subsidiary questions that arise. What are the modifications in the kinematical theory necessary when... the velocity with which the bowed point slips past the hairs of the bow is not exactly constant in each period of vibration? Then again, what is the effect produced by the finiteness of the region with which the bow is in contact, a region which for the purpose of discussion we have so far taken as equivalent to a mathematical point? Does any slipping occur when the string is being carried forward by the bow? Is it possible in practice that by simply removing the bow from a nodal point to another closely contiguous to it, the missing harmonics in any given type of vibration are suddenly restored to their full strength as our kinematical discussion tacitly assumed? Finally we have the all-important question, what are the conditions of excitation, e.g., pressure and velocity of bowing and so on, required for any given type of vibration to be elicited? What part do the instrument on which the string is mounted and the handling and properties of the hairs of the bow play in determining these conditions? What, for instance, is the effect on the motion of the string produced by loading the bridge over which it passes with a mute or otherwise? [4.15]

A most formidable array of questions, as Raman himself remarked!

Raman was preoccupied for nearly three years with finding answers to these questions (in the midst of various other activities and researches of course!). The investigations covered both theoretical and experimental aspects and four major publications emerged. In the process, many questions got clarified. There was, for instance, the problem of the "wolf-note", a troublesome phenomenon manifesting as a jarring note (particularly in the cello), and described variously as an impure and

Glimpses from the Golden Era 89

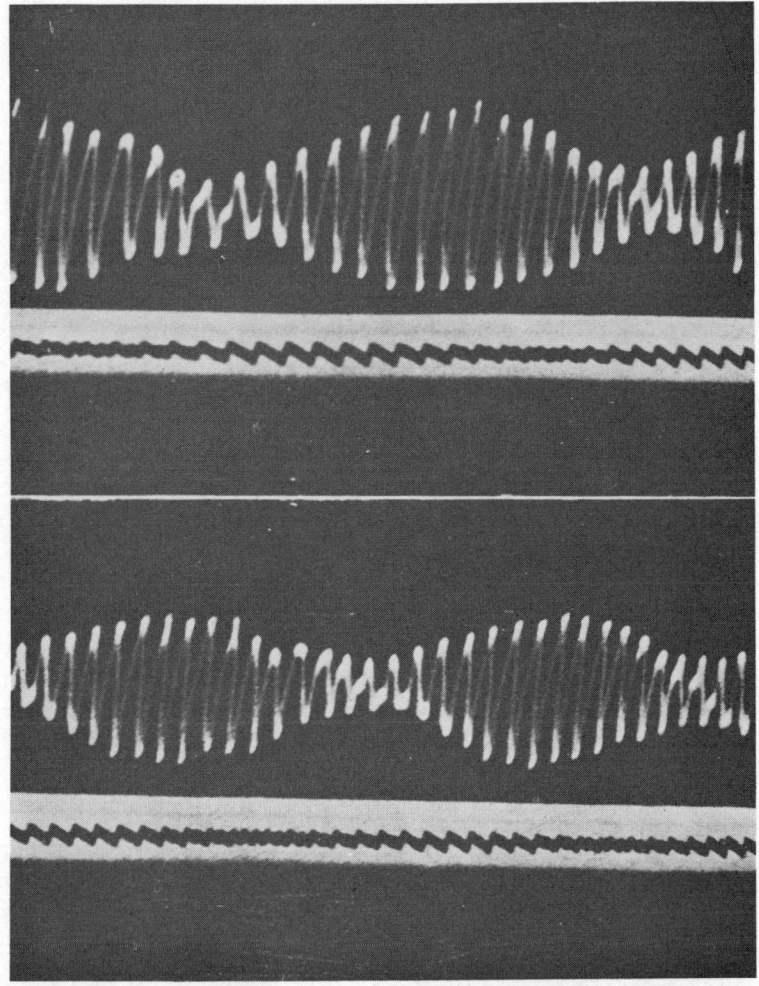

Figure 4.19 Variations associated with the wolf note. (After ref. [4.17].)

wheezy sound, a cyclical stuttering response to the bow, or a cyclic fluctuation of intensity of impure quality. The skilled player of course avoids the "wolf". At times, one also employs a wolf-note eliminator which is attached to the instrument.

From the physics point of view, the interesting question is: Why does a wolf-note occur? G. W. White who studied the movements of the top plate during the excitation of a wolf-note concluded (i) that it occurred at the main resonance of the sounding box, and (ii) that it represented some kind of beat phenomenon, although he did not explain what the two notes were which were beating. Raman disagreed with White, and, based on his own experiments, suggested an alternative scenario [4.17]. Initially when the string is excited the fundamental mode is dominant, and the energy of the string is dissipated through the bridge to the belly. The rate of dissipation of energy increases continually until the demand for energy (by the belly) exceeds the supply (from the string). At this limit the bow cannot maintain the string in its normal mode of vibration in which the fundamental is dominant. The form of the vibration then changes, the second mode taking over since the bow is still moving. As the belly is now no longer being forced to vibrate at its resonance frequency, its vibrations in turn die away and the string regains its original form of vibration. The entire cycle then repeats all over again (see Fig. 4.19). This explanation is now widely accepted as the *physical* interpretation of the wolf-note[4].

As a follow-up to his kinematical analysis, Raman made extensive photographs of the vibrations, some of which are shown in Fig. 4.20. Most of these are taken from his monograph (published as a Bulletin of the Association) to which a reference has already been made (see Chapter 3). The monograph is supposed to be Part I of his report and the table of contents promises 15 sections but, though the text runs to 168 pages, only 12 sections are covered. The monograph concludes with the words "to be continued", but Raman never seems to have found the time to complete all the sections. Indeed, missing Part II seems to have been a fairly common failing with Raman! He was always so busy with his investigations that writing up results sometimes took a lower priority.

Raman's monograph became quite well known in its day, and even today it is frequently cited. However, as one writer comments, the work of Raman, "while included in every bibliography, is mainly read 'by title' in part for the reason that it is out of print and unavailable even in some large libraries". Fortunately, a part of this monograph as well as some of Raman's other papers on the subject have now been included in a collection of benchmark papers on musical acoustics [4.18].

Turning now to the mechanics of playing: there are certain limits to the speed of

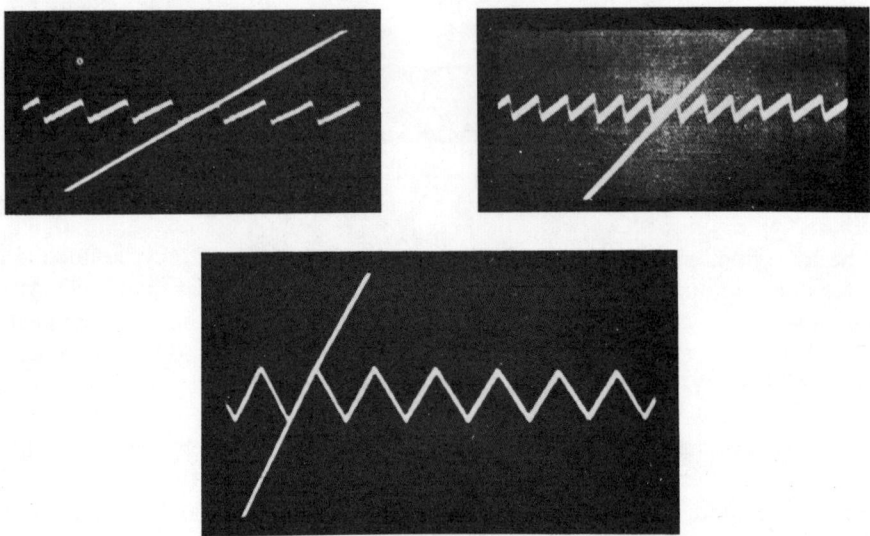

Figure 4.20(a)

Glimpses from the Golden Era

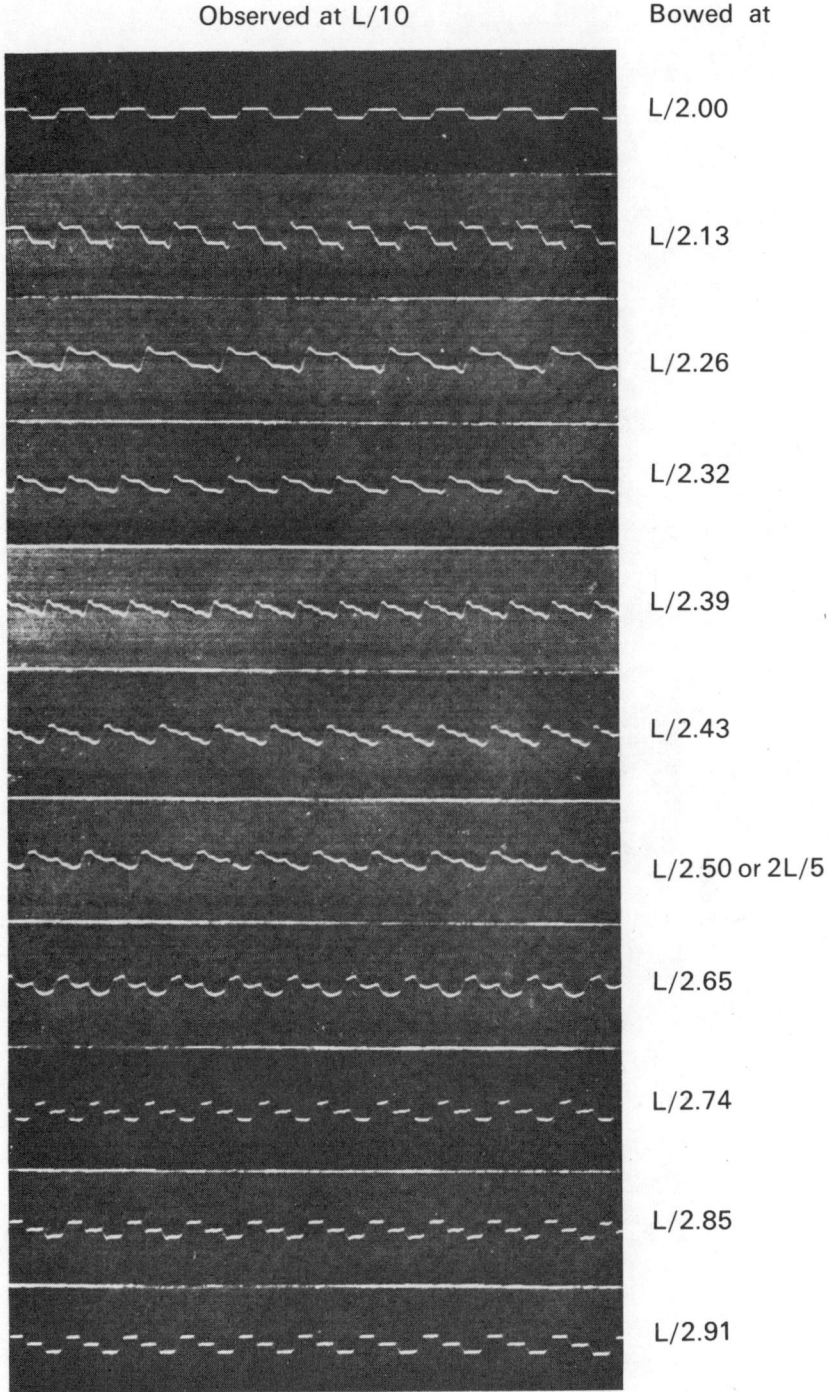

Figure 4.20(b)

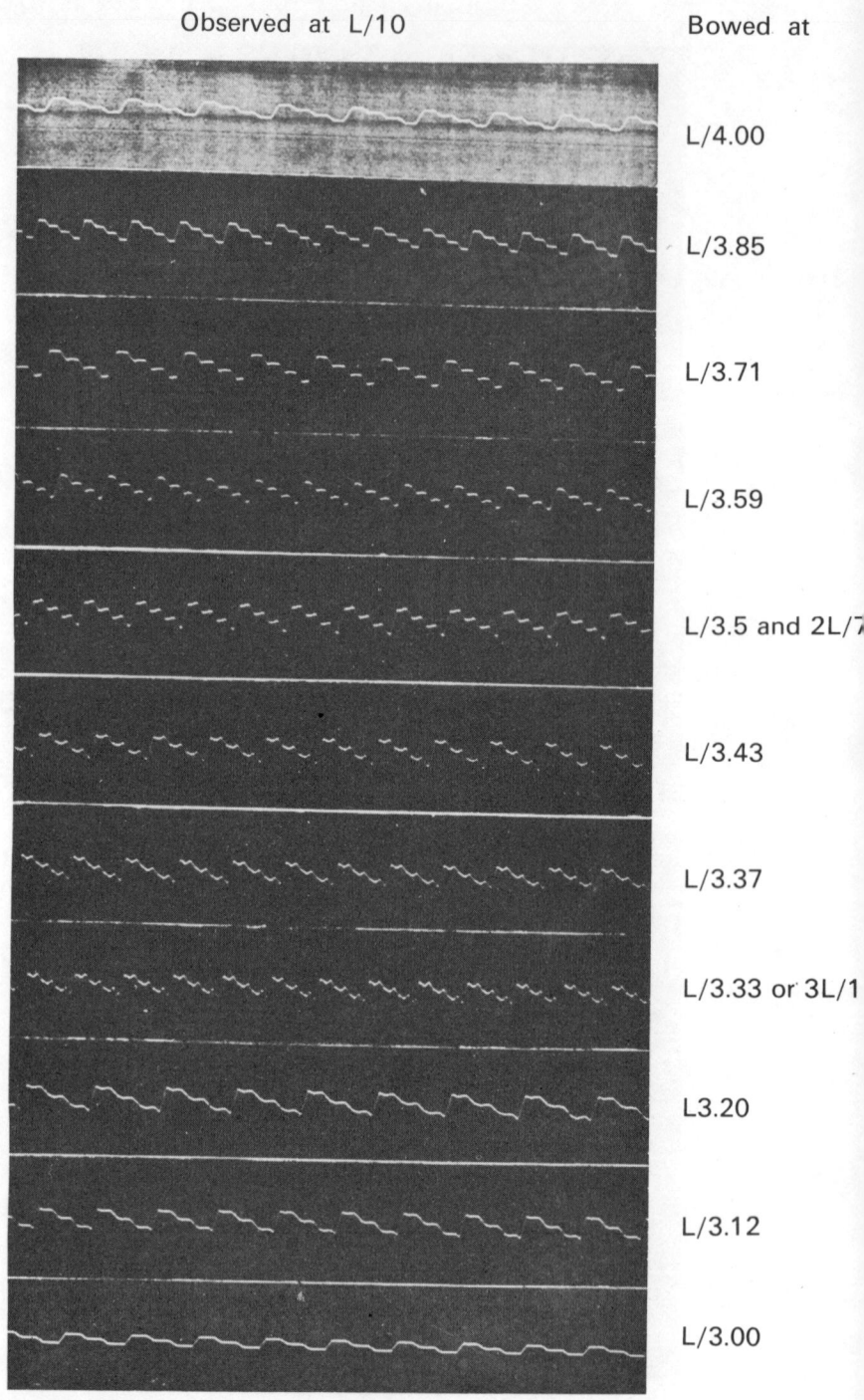

Figure 4.20(c)

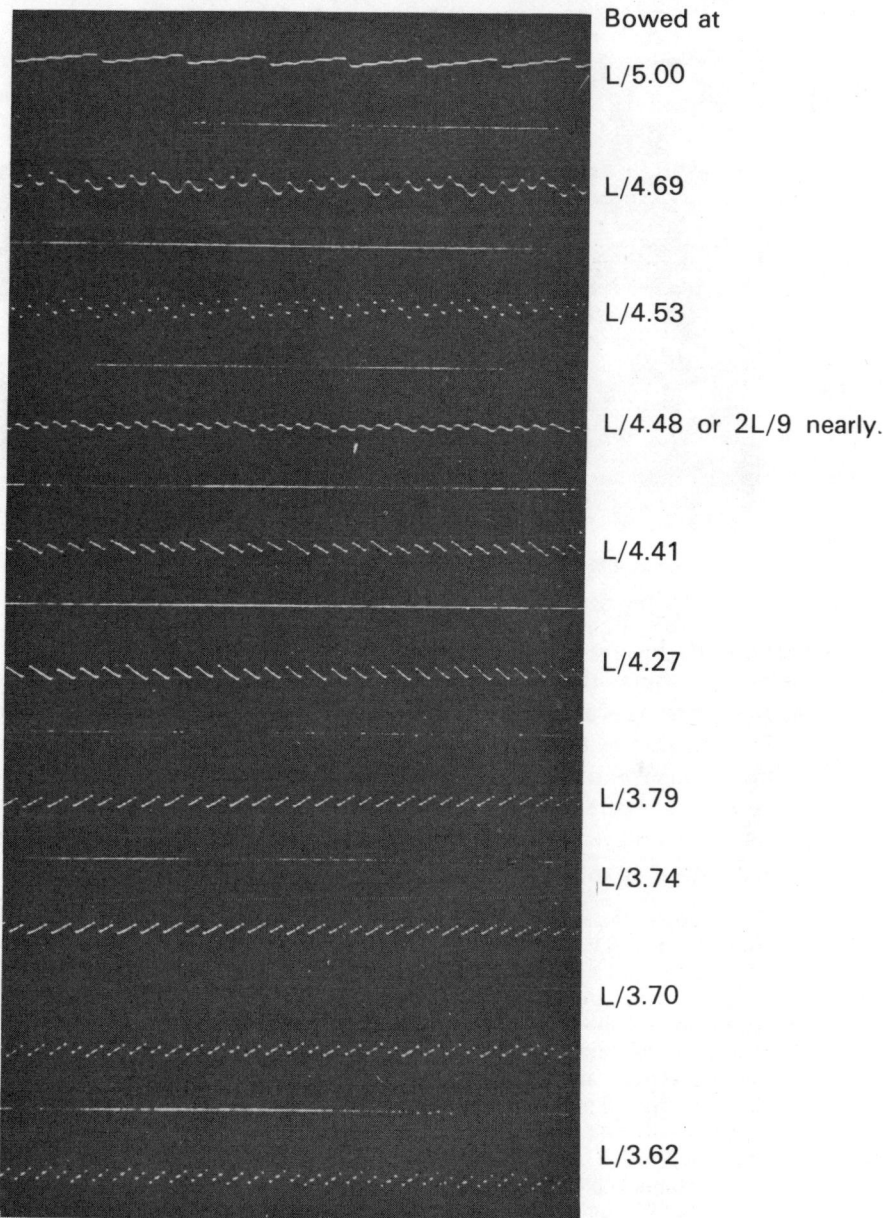

Figure 4.20(d)

Figure 4.20 (a) Record of motion of the bow and the bowed point. (After Bulletin 11 of IACS, 1914.) (b)–(d) Representative results on bowed strings from ref. [4.25]. In each case, the string is bowed at various points and observed at a particular point along its length.

Figure 4.21 The mechanical violin player devised by Raman. (After ref. [4.19].)

the bow, its location on the string, and the force that must be applied. For the violinist of course this is a matter of choosing more or less subconsciously from familiar patterns of experience but a physicist would like to understand the nature of the force exerted by the bow and the limits on this force. Raman gave much thought to this problem, and to test his ideas he devised a mechanical violin player (to which a reference has already been made in Chapter 3) – see Fig. 4.21. An interesting feature of this player is that the bow is kept stationary and it is the violin which is moved[5]! As Raman describes:

> The whole of the apparatus was improvised in the laboratory from such materials as were to hand. The slide and cast-iron track were parts of a disused optical bench. The chain and hubs were spare parts purchased from a cycle dealer. The ball-bearing of the axle of the lever was also part of a cycle. The other fittings were made up in the workshop. The apparatus was driven by a belt running over a conical pulley which in its turn was driven by a belt passing over the pulley of a shunt-wound electric motor controlled by a rheostat.... The mounting of the bow required special attention in order to ensure satisfactory results. As is well known, the violinist in playing his instrument handles the bow in such manner that when it is applied with a light pressure, only a few hairs at the edge touch the string.... [In the mechanical player] the violin-bow is held fixed at the end of a wooden lath, an adjustment being provided so that fewer or more hairs of the bow may be made to touch the string of the violin. The lath itself is balanced after the manner of a steelyard, the axis of the lever being mounted on ball-bearings so as to secure the necessary solidity combined with freedom of movement. The weight of the bow is balanced by a load hung freely near the end of the shorter arm of the lever. The axis of the lever can be raised or lowered to the proper height above the violin such that when the hairs of the bow touch the string, they are perfectly parallel to the cast-iron

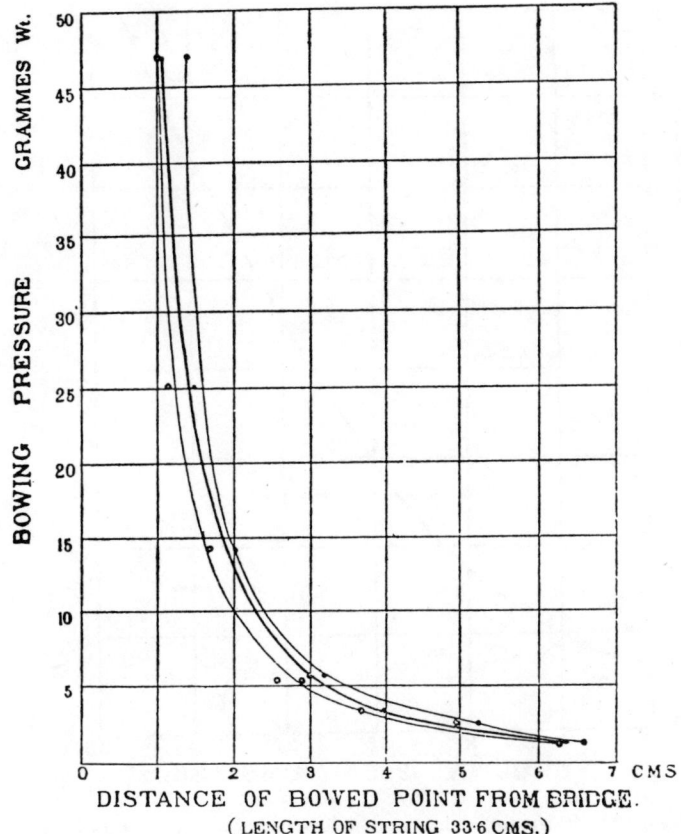

Figure 4.22 Variation of minimum bowing pressure with the point of application of the bow. (After ref. [4.19].)

track along which the violin slides. This is of great importance in order to obtain steady bowing.... [4.19]

In brief, Raman was able to control the bowing speed, the bowing pressure and the distance of the point of bowing from the bridge.

The violin which Raman used was a German copy of Stradivarius[6]. Figure 4.22 shows the results he obtained with the D-string for the *minimum* bowing pressure required to elicit a full steady tone with a pronounced fundamental. For each observation two points are shown, corresponding to the two edges of the bow, reflecting the difficulty in specifying the point of contact since contact is made over a region. The two thin lines show the trend of the experimental results, while the thick line is Raman's prediction.

Violinists often alter the intensity of the tone by varying the speed of the bow. Raman investigated this aspect by bowing the D-string at a fixed point using various speeds. His results, shown in Fig. 4.23, are once again in accord with the expectations of his theory.

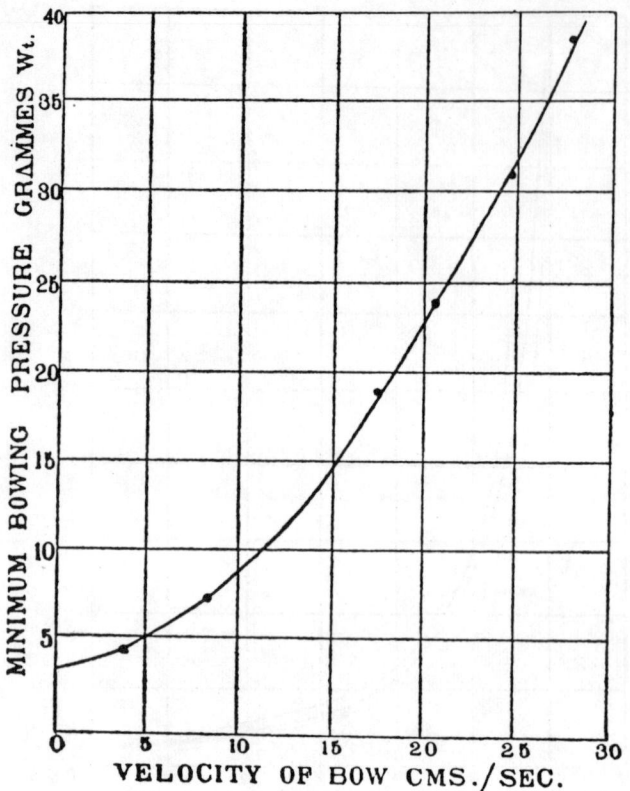

Figure 4.23 Bowing pressure versus velocity of bow. (After ref. [4.19].)

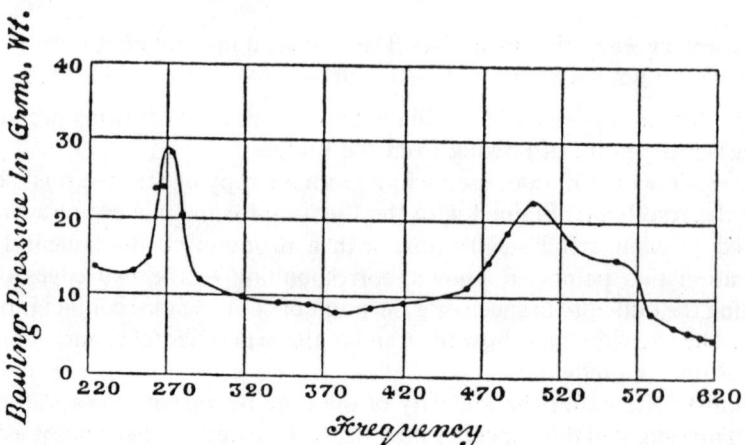

Figure 4.24 Frequency response of the violin. The frequency was varied by stopping down the string. (After ref. [4.19].)

Glimpses from the Golden Era

The pitch of the violin tone depends upon a number of factors like the linear density of the string, its length and its tension. The violinist varies the pitch either by "stopping" down the string on the fingerboard or by passing from one string to another. Raman studied these aspects by experimenting on the G-string, using a light brass clamp for effecting the stopping. When a violinist stops down the string so as to elicit a note of higher pitch, he generally takes the bow up rather near the bridge so as to preserve the relationship between the vibrating length of the string and the distance of the bow from the bridge. Since this was difficult to do in the experiment, Raman kept the distance fixed, arguing that the observed data could be suitably corrected if required. Raman's results (see Fig. 4.24) reveal many interesting features, prominent among which are the resonances. Concerning the latter Raman observes,

It would appear that the gravest resonance of the violin chiefly involves a vigorous oscillation of the air within the belly of the instrument, but not so vigorous an

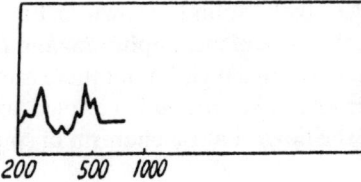

Raman curve

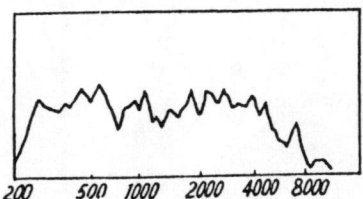

Response curve obtained by Saunders by means of the early method

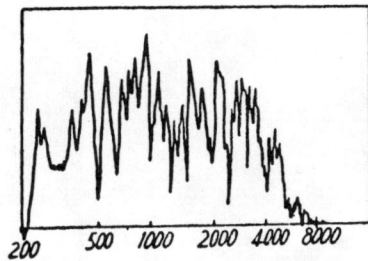

Response curve obtained by Saunders by means of the later method

Figure 4.25 This figure illustrates how the study of Raman curves has improved with time. (After ref. [4.18].)

oscillation of the bridge and belly as in the second and third natural modes of vibration which show the wolf-note phenomenon.

Observe that *more* pressure is required at resonance which is contrary to the usual experience. However, there is a ready explanation which is that at resonance, the energy of the string is most easily drained away by the bridge into the belly. Raman is careful to note that resonance frequencies deduced from bowing pressure are likely to be somewhat different compared to those obtained by more direct means. Raman recommended the frequency response as a good characterization of the violin, and indeed the study of "Raman curves" as they are now called is quite common (see Fig. 4.25)[7].

Raman called attention to many other problems which could be studied using his violin player, but did not bother to pursue these ideas. However, others did, one of them being R. N. Ghosh of Allahabad. Ghosh [4.20] studied not the usual violin but the Stroh violin which, while not musically inferior, is far less complicated in a dynamical sense. The Stroh violin is shown in Fig. 4.26 and consists of five parts– the body, the string, the bridge, the diaphragm and the horn. The upper part is more or less similar to that of the usual violin but there is only one string which passes over a bridge at its lower end. The bridge is coupled to the diaphragm, which in turn communicates with the horn. The mechanism of sound production is simple, and as

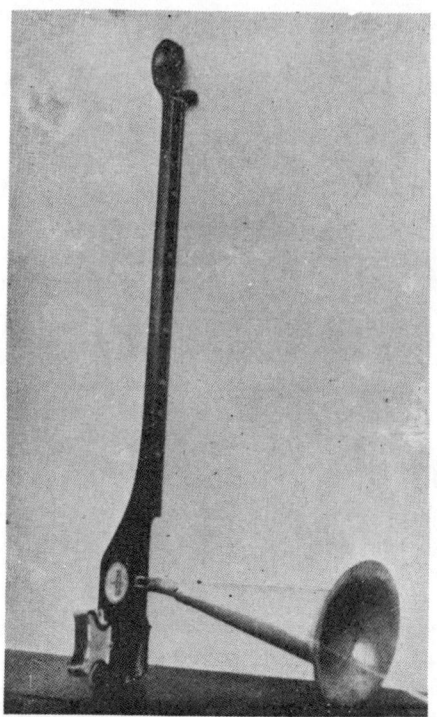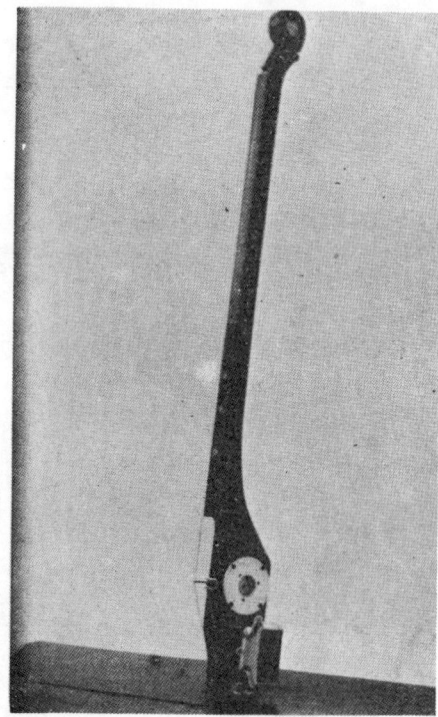

Figure 4.26 Picture of the Stroh violin. (After ref. [4.20].)

in the gramophone, the vibration of the string produces a transverse force on the bridge which, by suitable lever action, communicates a transverse motion to the diaphragm, producing sound waves in the horn. Ghosh verified many aspects of Raman's theory which did not receive attention from the master himself[8].

Piano

Raman's studies on the pianoforte were not as extensive as those on the violin. The essential point from the physics angle is that in the pianoforte the string is excited by striking. Helmholtz regards plucking and striking in the same bracket. There are, however, significant differences which he is careful to recognize and indeed it is these which are responsible for the differences in the musical quality of the tones elicited from plucked instruments like the guitar and those that emanate from the piano. Plucking consists of drawing a string with a finger or a point to one side and then letting it go. In the pianoforte, on the other hand, the string is struck by a hammer-shaped body. Commenting on the striking process, Helmholtz observes:

> If the string is struck with a sharp-edged metallic hammer which rebounds instantly, only the one single point struck is directly set in motion. Immediately after the blow the remainder of the string is at rest. It does not move until a wave of deflection rises, and runs backwards and forwards over the string. This limitation of the original motion to a single point produces the most abrupt discontinuities, and a corresponding long series of upper partial tones, having intensities in most cases equalling or even surpassing that of the prime. When the hammer is soft and elastic, the motion has time to spread before the hammer rebounds. When thus struck the point of the string in contact with such a hammer is not set in motion with a jerk, but increases gradually and continuously in velocity during the contact. The discontinuity of the motion is consequently much less, diminishing as the softness of the hammer increases, and the force of the higher upper partial tones is correspondingly decreased. [4.13]

In slightly more crisp but technical terms, the amplitude of the nth partial varies as $(1/n^2)$ in the case of the plucked string whereas it goes as $(1/n)$ for the struck string. Thus in the pianoforte the upper partials have a stronger say.

When the string of the piano is struck by the hammer, it is given an impulse or rather a step force. As a result, two discontinuous velocity waves are set up on the wire which travel outwards, one on each side. In due course they reach the two ends, are reflected, and return again. Since the duration of contact of the hammer with the wire is finite (why it is so we shall examine presently), the incoming waves are likely to be bounced by the hammer and sent again on an outward journey. This process of bouncing back and forth from the hammer will go on till the contact lasts.

When it strikes the wire, the hammer gets compressed. However, the compression does not attain its maximum value instantaneously but builds up gradually. Meanwhile there are jerks due to the velocity waves periodically bouncing off the hammer, and the net result is that the compression varies in time in a discontinuous manner. After a while, the elastic forces set up inside the hammer become strong enough to oppose the compression and cause a recovery. The mean force then starts decreasing and when the force becomes zero, the wire breaks loose from the hammer.

Raman was interested in knowing (i) how exactly the force due to the hammer varied as a function of time, and (ii) how the duration of contact of the hammer varied as a function of the strike position.

The calculation of the force versus time curve is beset with various mathematical

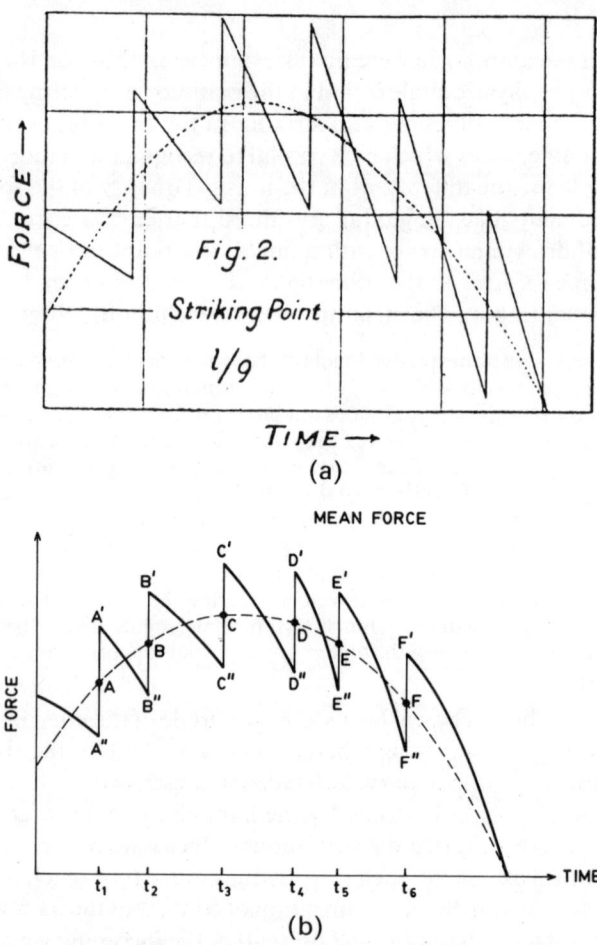

Figure 4.27 (a) Time variation of the striking force in a pianoforte. The solid line is the actual variation while the dashed line shows the mean force. (After ref. [4.21].) (b) The way Raman obtained the results. First he calculated the curve for the mean force using analytical methods. Next he located the times t_1, t_2, etc. when discontinuities occur. The magnitude of the discontinuity was then estimated, and breaks $A'A''$, $B'B''$, $C'C''$, etc. were constructed. The curve was then filled in by free-hand drawing (!), i.e., A' joined to B'', B' to C'', and so on.

Glimpses from the Golden Era

difficulties. Raman overcame these in a characteristically resourceful way [4.21] that would delight physicists but shock mathematicians! The spirit of his approach may be appreciated by referring to Fig. 4.27 which shows a force profile computed by Raman. What one is actually after is the solid curve which, as just noted, is problematical to compute by straightforward mathematical analysis. However, a simple consideration of the propagation of the two discontinuous velocity waves on the wire is enough to reveal when exactly these discontinuities would bounce off the hammer. In other words, one can predict the instants at which discontinuities will occur in the force versus time graph. Further, by an independent analysis, Raman not only knew that the various breaks in the force–time curve were all equal, but also the value of the step. Armed with this knowledge, Raman adopted a short-cut. By rather simple means, he first calculated the time variation of the *mean* force, which in Fig. 4.27a is given by the dashed line. Clearly this is unrealistic. To make it realistic, he identified the times t_1, t_2, etc. at which discontinuities would occur in the force curve. He then added *by hand* the required steps as explained in Fig. 4.27b. The resulting graph then gives access to the duration of contact, which is the time taken for the force to become zero. Figure 4.28a shows how this duration changes when the strike point is varied. To test these findings, Raman and Banerji carried out some experiments. A piano was clearly unsuited since the strike distance had to be varied. Raman and Banerji used instead a rather simple but elegant arrangement.

A steel wire 150 cm long was nickel- and silver-plated and stretched over a sonometer. A small solid brass cylinder was mounted at the end of a light, pivoted

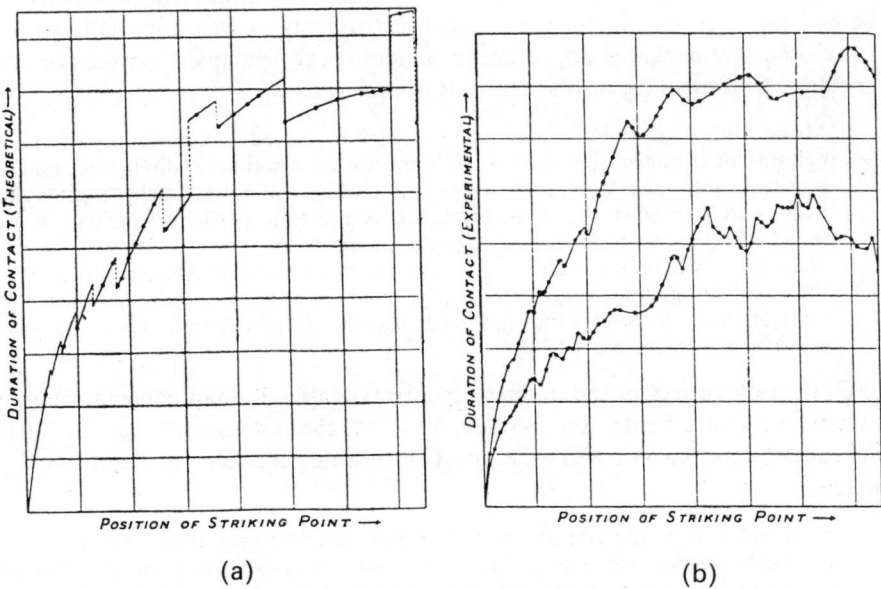

Figure 4.28 (a) Variation of the duration of contact with the striking point. (b) Experimental results for hammers of two masses. (After ref. [4.21].)

shaft, with a provision to impinge transversely on the wire. Immediately on impact, the cylinder and the shaft of the hammer completed an electrical circuit. The duration of the electrical contact (determined using a ballistic galvanometer) then directly gave the duration of the contact of the hammer with the wire. Results obtained in this manner for hammers with two different masses are shown in Fig. 4.28b. Commenting on these, Raman and Banerji observe that

> the general resemblance between the computed and observed curves is obvious. Why there is not a much closer agreement is an open question. It must be remembered that, in many respects, the experimental arrangements do not strictly reproduce the conditions assumed in the theoretical calculations. The finite size of the cylinder and elastic flexure of the shaft, the stiffness of the wire and its yielding at the ends, and the effect of gravity on the motion of the impinging cylinder, are factors which probably influence the results in an appreciable degree. There is no doubt, however, that the experimental results shown in Fig. 4 [our Fig. 4.28b], broadly speaking, confirm the correctness of the theoretical results, and the suitability of the method of calculation set out in the paper. [4.21]

Some time later, Ghosh and Dey [4.22] extended these investigations by actually photographing the impact[9].

Indian Musical Instruments

Given Raman's extensive research on the violin, one should not be surprised that he became interested in Indian musical instruments as well, especially since they "disclosed a remarkable appreciation of acoustical principles" on the part of our ancients. Commenting on the latter, Raman wrote:

> Music, both vocal and instrumental, undoubtedly played an important part in the cultural life of ancient India. Sanskrit literature, both secular and religious, makes numerous references to instruments of various kinds, and it is, I believe, generally held by archaeologists that some of the earliest mentions of such instruments to be found anywhere are those contained in the ancient Sanskrit works. Certain it is that at a very early period in the history of the country, the Hindus were acquainted with the use of stringed instruments excited by plucking or bowing, with the transverse form of the flute, with wind and reed instruments of different types and with percussion instruments. [4.23]

Of the rich variety of Indian instruments available, Raman concentrated on the percussion instruments, the *mridangam* and the *tabla*, and on the stringed instruments, the *tambura* and the *veena*. Concerning percussion instruments, Raman writes:

> As is well known, the vibrations of a circular stretched membrane or drum-head excited by impact are generally of an extremely complex character. Besides the gravest or fundamental tone of the membrane, we have a large retinue of overtones which stand to each other in no sort of musical relation. These overtones are always excited in greater or less degree and produce a discordant effect. All the instruments

Glimpses from the Golden Era

of percussion known to European physicists in which a circular drum-head is employed have therefore to be regarded more as noise producers introduced for marking the rhythm than as musical instruments. This is true even of the kettle-drum which is tuned to a definite pitch and occasionally used in European orchestral music.

Raman continues:

All the instruments of percussion known to European science are thus essentially non-musical and can only be tolerated in open air music or in large orchestras where a little noise more or less makes no difference. Indian musical instruments of percussion however stand in an entirely different category. Times without number we have heard the best singers or performers on the flute or violin accompanied by the well-known indigenous musical drums, and the effect with a good instrument is always excellent. It was this, in fact, that conveyed to me the hint that the Indian instruments of percussion possess interesting acoustic properties, and stimulated the [i.e., Raman's own] research. [4.23]

The *mridangam* (see Fig. 4.29) has

a massive hollow wooden body in the form of two truncated cones put end to end, one of which is longer than the other. Over the two ends of this body are stretched the two drumskins, which are each provided with a tightening ring of leather and are kept in a state of tension by a leather rope which passes through apertures in the rings at 16 equidistant points around the circumference. Eight cylindrical tuning blocks of wood inserted at regular intervals under the tension-rope provide the means for a rough adjustment of tension. The fine adjustment of tension of the smaller drumhead to equality in the 8 octants of the circumference is carried out by

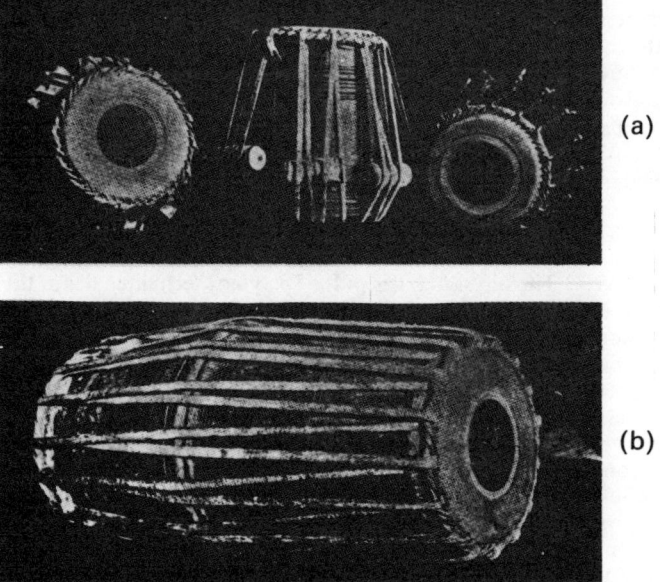

Figure 4.29 The *tabla* (a) and the *mridangam* (b). (After ref. [4.24].)

pulling up or pushing down the tightening ring by stroking it with a small mallet. The large drumhead gives the base note, and its pitch and tone-quality are adjusted by spreading a temporary load of wetted *ata* or wheaten flour over it. The most remarkable feature of the drum is the manner in which the second or smaller drumhead is constructed. This membrane as first put on in the construction of the drumhead is double, the layers being of specially chosen leather of uniform thickness and connected to the tightening ring so as to be in a state of tension. The upper layer is then cut away in the middle exposing a circular area of the lower membrane, and leaving an annular ring of the outer membrane round the margin, of which the width is regulated according to the requirements of the tone-quality. The centre of the exposed circle of the inner membrane thus formed is loaded concentrically in several successive layers of gradually decreasing radii and of graduated thickness with a dark coloured composition which is put on at first in the form of a paste and is then rubbed in till it becomes dry and permanently adherent to the membrane. The composition of this material is finely powdered iron filings, charcoal and starch, and when put on the membrane it is flexible in a noteworthy degree. The putting on of the load is carried out in stages, the sound of the drumhead being continuously tested during the progress of construction. Its final adjustment and regulation of thickness is an art which is handed down from generation to generation as traditional knowledge, and acquired by long training and experience.

As for the *tabla* (see Fig. 4.29), it

consists of two drums played simultaneously with the right hand and left hand respectively. Both consist of wooden or metal shells open at one end only and covered with drumskins. The drumhead of the *tabla* played with the right hand is very similar to that of the *mridangam*. The drum played with the left hand has a firmly adherent composition which is, however, unsymmetrically placed on the membrane. The purpose of such unsymmetrical loading is quite different from that of the symmetrical loading used in the right-hand drum.... The tension arrangements in the *tabla* are similar to those in the *mridangam*, with the difference that the tightening cords simply pass round the closed end of the *tabla*. The number of tightening straps is exactly 16 as in the *mridangam*. [4.24]

A skilled percussionist knows how to exploit the capabilities of the instrument he is playing. Discussing the playing technique, Raman observes:

If the instrument is in itself a noteworthy piece of acoustic workmanship, still more remarkable is the manner in which its acoustic characters are utilized in actual musical practice.... The physical basis of the [playing] technique lies in the manner of striking the drumhead.... The strokes involve the exact regulation of the region of contact, the softness or hardness of the blow, its duration and force, and provide for touching the membrane with some of the fingers either during or after the blow so as to damp out certain harmonics and bring out certain others. Some of the recognized strokes provide for bringing out either the first or the second or the third harmonic practically by itself, or in combination with one or more of the five available tones. The strokes on the drumhead may be either by themselves or may be simultaneous with strokes on the base side of the drum which is tuned to one octave below the pitch of the first drumhead. Over and above this is the fact that the drumming is practically continuous and proceeds on a complex and varied metre and rhythm of its own depending on the accompaniment. All this may serve to give some idea of the extraordinary degree of development which the construction and use of percussion instruments has attained in India. [4.23]

Glimpses from the Golden Era 105

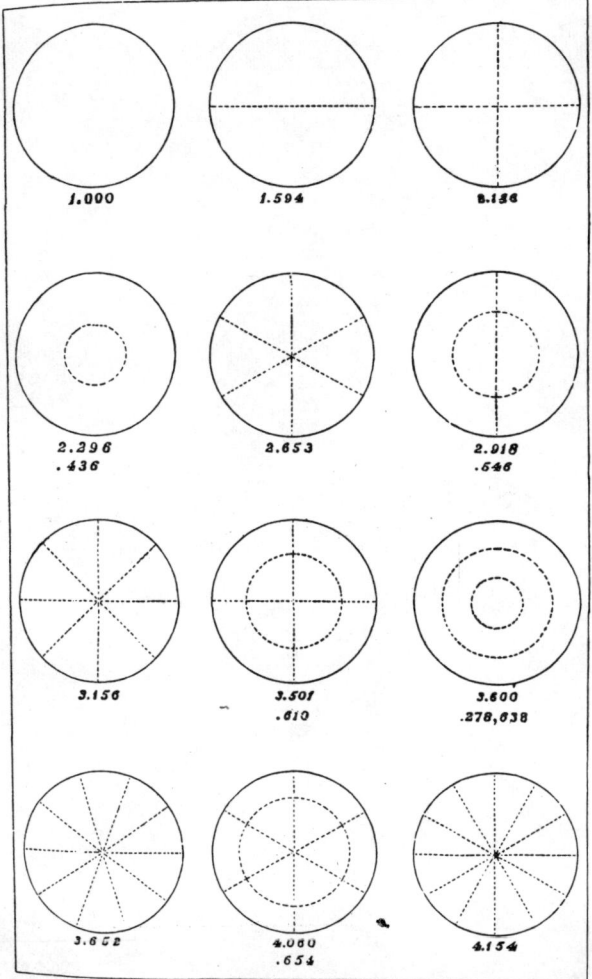

Figure 4.30 The important normal modes of vibration of a circular membrane. The dashed lines show the nodal planes and the nodal circles. The numbers in the upper line give the frequencies, taking the fundamental as unity. The numbers in the lower line give the radii of the nodal circles relative to that of the membrane. (After ref. [4.25].)

To appreciate the physics of the *mridangam* and the *tabla*, one must first consider the vibrations of a *uniform* circular membrane held in tension around its circumference. Lord Rayleigh has discussed this problem in his celebrated book [4.25]. Displayed in Fig. 4.30 are the vibrational patterns corresponding to the various normal modes of the membrane. Their significance may be better understood by consulting Fig. 4.31 which shows the time evolution of one of the vibrational modes[10]. Going back to Fig. 4.30, displayed there are not only the nodal lines and nodal circles, but also the frequencies of the various modes relative to that of the fundamental (which is taken as unity). The unusual feature is that the

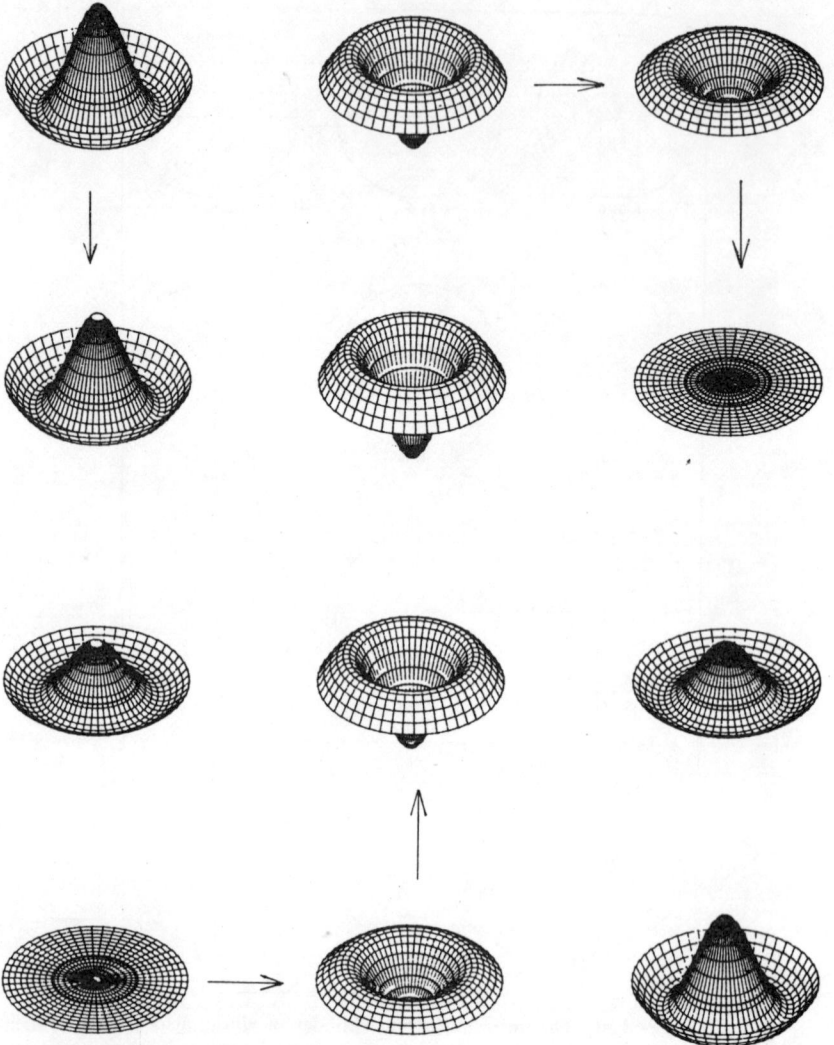

Figure 4.31 Time evolution of the normal mode corresponding to frequency 2.296 in Fig. 4.30. Shown here are the instantaneous dispositions of the membrane at various stages during one complete vibration.

frequencies of the higher modes are not integer multiples of the frequency of the fundamental, as they usually are in all musical instruments. It is not surprising then that Raman describes the Western drum as a mere "noise producer". In the case of the *mridangam* and the *tabla*, the nodal patterns have the same symmetry as before. However, the higher modes have near-harmonic frequencies, contributing to the musical quality of the instrument[11].

Raman investigated the various nodal lines and circles by the simple technique of strewing a little fine sand on the membrane either before or immediately after

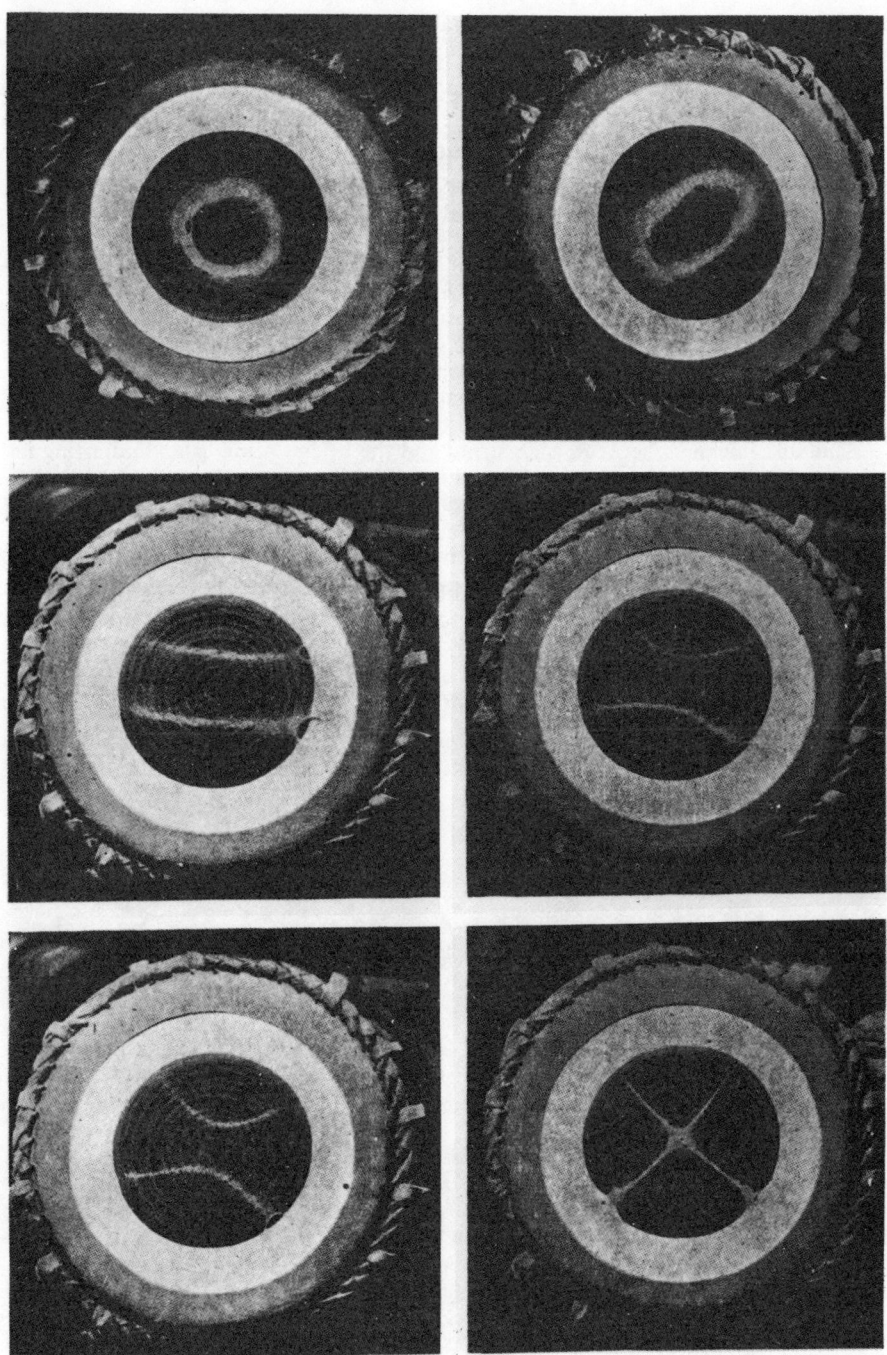

Figure 4.32 Sand patterns on the *mridangam*. (After ref. [4.24].)

striking the drumhead. The sand then gathers along the nodes in a clearcut manner (see Fig. 4.32).

As already remarked in an earlier chapter, the *tambura* (see Fig. 4.33) is used as a drone in accompaniment. It has four metal strings which are stretched over a large resonant body, and can be accurately tuned up to the right pitch by a simple device for the continuous adjustment of tension. In effect, it is a rather simple plucked instrument, yet possessing a rich tonal characteristic. Raman explains how this comes about. He writes:

> The remarkable feature of the *tambura* to which I wish to draw attention is the special form of the bridge fixed to the resonant body over which the strings pass. The strings do not come clear off the edge of a sharp bridge as in European stringed instruments, but pass over a curved wooden surface fixed to the body which forms the bridge [see Fig. 4.34a]. The exact length of the string which actually touches the upper surface of the bridge is adjusted by slipping in a woollen or silken thread of suitable thickness between each string and the bridge below it and adjusting its position by trial. Generally the thread is moved forwards or backwards to such a position that the metal string just grazes the surface of the bridge. [4.26]

The *veena* (see Fig. 4.33) is a fretted instrument and its bridge differs from that of the *tambura* in two respects. Firstly the upper curved surface of the bridge is of metal, and secondly the adjustable thread used for tuning the *tambura* is dispensed with.

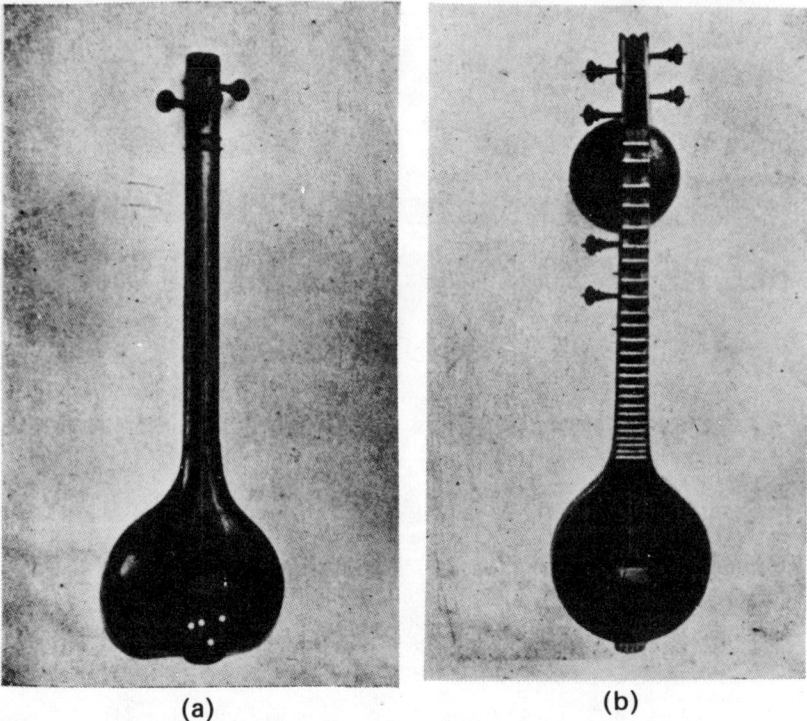

Figure 4.33 The *tambura* (a) and the *veena* (b). (After ref. [4.26].)

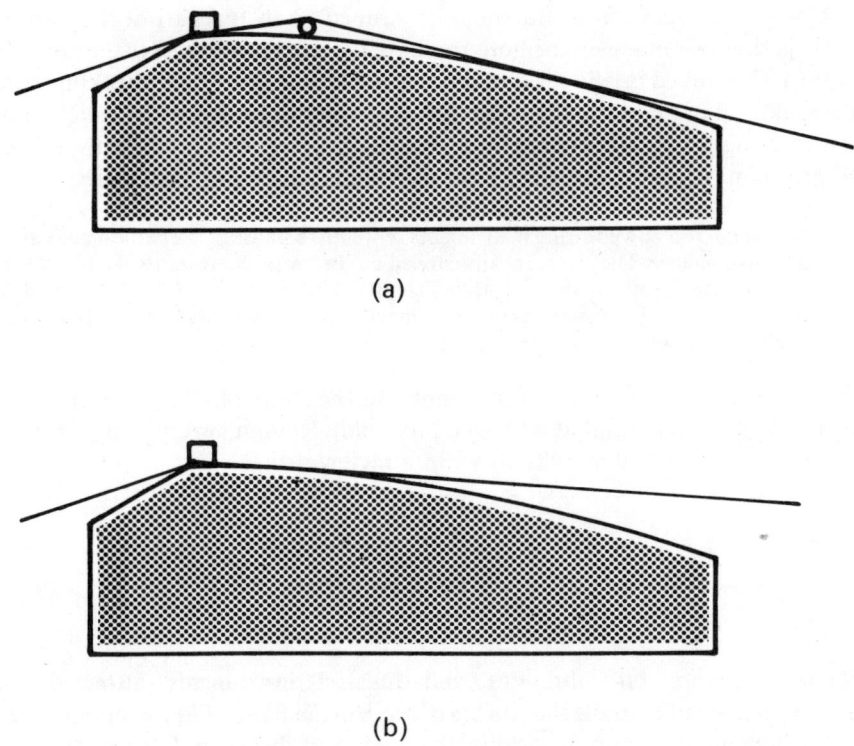

Figure 4.34 Bridges of (a) the *tambura* and (b) the *veena*. (After ref. [4.26].)

Thus the string comes off the curved upper surface at a tangent, as sketched in Fig. 4.34b.

Apparently Raman made detailed studies on the *tambura* and *veena* but published very little. An extensive account which was promised never materialized. Raman's crucial observation is that, unlike in a sonometer, plucking at a node does not suppress the harmonics. As he remarks, "On trying the same experiment [as the sonometer] with the *veena* or the *tambura*, it will be found that the overtone having a node at the plucked point sings out powerfully." Why? Raman hazards a guess:

> The forces exerted by the vibrating string on the bridge must be very different from what they would be for a bridge of ordinary form. It seems probable that by far the greater portion of the communication of energy to the bridge occurs at or near the point of grazing contact. The forces exerted by the string on the bridge near this point are probably in the nature of impulses occurring once in each vibration of the string. This would explain the powerful retinue of overtones including even those absent initially in the vibration of the string. At a slightly later stage, the reaction of the bridge on the string would result in a modification of the vibration form of the latter and bring into existence partials absent initially in it. There would in fact be a continual transformation of the energy of vibration of the fundamental vibration into the overtones.

Under persuasion from Raman, one Ahmed Shah Bukhari of the Government College, Lahore, made some photographs of the vibration curves of the *tambura* while Kar [4.27] studied briefly the dynamics of the bridge. But really speaking, studies on the scale of those performed on the violin were not done. Raman himself was finding it increasingly difficult to spare time for acoustics since a new interest was beginning to grip him. As he wrote years later to the Catgut Acoustical Society[12]:

> My studies on bowed string instruments represent a phase of my earliest activities as a man of science. They were mostly carried out between the years 1914 and 1918. My call to the Professorship at the Calcutta University in July 1917, and the intensification of my interests in *optics* inevitably called a halt to my further studies of the violin family instruments. [4.28]

It was not merely the study of the violin but the study of vibrations and of musical instruments in general that was called to a halt. Raman hardly ever returned to the subject except briefly in 1927 to write a review article.

Handbuch Article

Raman's studies on vibrations and musical instruments attracted sufficient attention abroad to make the editors of the *Handbuch der Physik* series invite him to contribute an article on musical instruments and their tones. It was then the mid-twenties and Raman had already laid aside his work on acoustics, having become deeply engrossed in his studies on the scattering of light. But this invitation was too prestigious to be declined, especially since the editors had made a departure from their normal custom of inviting German authors. Besides, they offered to accept a manuscript in English and take the responsibility of getting it translated into German. Clearly time had to be found for the task.

The physical study of musical instruments as tone generators necessarily involves an investigation of the characteristic features of musical notes. As perceived by the ear, musical notes have three distinctive features, namely, pitch, timbre and loudness. But this does not mean that the perception of tone is purely subjective. There is also an objective part, although from an aesthetic point of view subjective perception assumes importance.

While the question of the relationship between the subjective and the objective aspects of musical notes is a delicate one, it is certainly meaningful to ask what it is from a physics point of view that endows musical instruments of various types with desirable musical qualities. This is the problem that Raman studies in his article [4.29].

He begins by explaining what is meant by pitch, timbre and loudness. The objective features corresponding to the first two are respectively the frequency and the waveform of the vibrations in air. Loudness, on the other hand, is determined by many factors, the chief of them being the amplitude of the vibrations. While the musically trained ear can make accurate estimates of these various attributes, the physicist has his own devices for qualitative and quantitative studies of tone. Raman describes many quaint gadgets for such purposes then in vogue, and one wonders whether some of them can be found even today in museums.

Quantitative analysis of musical notes is very important for the study of musical instruments, for it yields observational material for comparison with the physical theory of such instruments. Moreover, one is able to quantitatively express the differences in timbre in the notes produced by various instruments. Impressive variations in tonal timbre can be achieved by controlling the relative intensities of the partials.

Having laid the foundation, Raman then goes on to discuss the various musical instruments. Naturally the violin takes the pride of place, closely followed by the piano. His treatment of the violin is no doubt crisper than in his monograph but does not adequately reflect his own numerous findings. Then follows a discussion of various aspects of the physics underlying wind instruments, like the flow of air through apertures, energy emission from openings, acoustic impedance, the resonator as a source of sound, vibrations of air columns in tubes, the consequences of the variation of the cross-section of the tube (in many wind instruments, the cross-section increases rapidly towards the open end), blowing of pipes, etc. After these comes the theory of the flute, the oboe, the clarinet, the horn, and various other brass instruments.

The percussion instruments receive brief mention, and Raman uses the occasion to illustrate via a simple model how an inhomogeneous membrane can conceivably produce overtones with harmonic relations.

Next come the church bells and the so-called glass bells. We know it as *jalatarang*, but it is also prevalent in the West. It consists of a series of glass shells filled with water to varying extent so that, when struck with a light wooden rod or mallet, the shells emit notes on a chromatic scale. From the physics point of view, the interest lies in the influence of the liquid on the vibrations of the elastic shell. The underlying hydrodynamic problem was first investigated by Lord Rayleigh. S. K. Banerji made a more thorough study, assuming different shapes for the container – cylindrical, conical and hemispherical. Naturally the theoretical predictions were checked by experiment.

It is interesting that there is no mention of either the *veena* or the *tambura* in the article. One wonders whether they were excluded because of their unfamiliarity to Western readers. Raman concludes his review with the comment:

> Each class of instruments has its own particular timbre which distinguishes it from others and determines its place in the scheme of orchestral music. Thus the string instruments of the violin family – which form the backbone of the orchestra – are distinguished by the fact that they possess an especially expressive timbre which does not tire the ears of the listener and, to a remarkable degree, can be made to match various musical moods. The class of flutes, in contrast, is characterized by a mild, softly flowing timbre. The great multiplicity of sonorous notes of modern composers calls for instruments with even greater variety in timbres.... Among all the instruments, we naturally ascribe the highest degree of individuality to the human voice.

Raman does not deal with this topic but cites instead a companion article in the *Handbuch* series, and also a book by Stumpf where a comparison of the timbre of the human voice with that of musical instruments is presented.

☆ 4.3 A Mathematical Interlude

It is appropriate at this juncture to pause and briefly underscore the mathematical nuances of some of the descriptive material presented earlier. We begin by recalling some familiar facts, starting with the harmonic oscillator equation

$$\ddot{\theta} + q^2\theta = 0. \tag{4.1}$$

As is well known, its solution is given by

$$\theta(t) = A \sin(qt + \varepsilon), \tag{4.2}$$

where the constants A and ε are arbitrary, and are chosen to suit the initial conditions. If damping is present we have

$$\ddot{\theta} + \kappa\dot{\theta} + q^2\theta = 0, \tag{4.3}$$

κ denoting the damping constant. The solution now is

$$\theta(t) = A \exp(-\kappa t) \sin(pt + \varepsilon), \tag{4.4}$$

where

$$p^2 = q^2 - \kappa^2. \tag{4.5}$$

Observe that the frequency of oscillations is modified by damping.

In acoustic systems p and q are usually large and κ is small. Thus one often ignores the difference between p and q while retaining the damping term $\exp(-\kappa t)$ in (4.4).

Solution (4.4) shows that the free oscillations eventually die away. If the oscillations are to be maintained, then clearly there must be continuous energy supply. Suppose now that there is an agency capable of this and that it contributes a periodic force. We then have

$$\ddot{\theta} + \kappa\dot{\theta} + q^2\theta = f \sin nt, \tag{4.6}$$

and the corresponding solution is

$$\theta(t) = A \exp(-\kappa t)\sin(pt + \varepsilon) + \frac{f \sin \delta}{2\kappa n} \sin(nt - \delta), \tag{4.7}$$

where

$$p^2 = q^2 - \kappa^2, \text{ and } \tan \delta = 2\kappa n/(p^2 - n^2). \tag{4.8}$$

As before, A and ε are arbitrary constants to be determined by initial conditions.

Glimpses from the Golden Era

The two terms in (4.7) have distinct physical significance. The first is of the same nature as (4.4), i.e., it is a transient. The second term describes the maintained vibrations which, observe, have the same frequency as that of the applied force. Also the amplitude is completely determined. Our purpose in recalling the forced oscillator is to contrast it with the Melde's problem to which we shall direct attention shortly.

Returning to the free oscillator, (4.3) is inadequate when the amplitude of oscillations is large, in which case one must include nonlinearities. As an illustration consider

$$\ddot{\theta} + q^2\theta + \alpha\theta^2 = 0, \tag{4.9}$$

where, for simplicity, the damping term has been dropped. Using perturbation theory Lord Rayleigh showed [4.25] that to second order,

$$\theta(t) = A\{d + \cos pt + e\cos 2pt + f\cos 3pt + g\cos 4pt\}, \tag{4.10}$$

where A is the arbitrary amplitude,

$$p^2/q^2 = 1 - (5/6)\sigma^2, \quad \sigma^2 = \alpha^2 A^2/q^2, \tag{4.11}$$

and d, e, f, g are constants related to σ^2. Two important features to be noticed are:

(i) The constant term d displaces the equilibrium position to one side, and
(ii) the primary oscillations are accompanied by harmonics. (However, the intensities of the harmonics diminish rapidly as one goes to higher orders.)

Another illustrative case is that of cubic nonlinearity where one has

$$\ddot{\theta} + q^2\theta + \beta\theta^3 = 0.$$

The second harmonic is now absent in the solution, and the higher harmonics commence with the third. Further, there is no shift of the equilibrium position.

Localized Change of Phase

We have already described Raman's experiment relating to this topic. Raman also explained how such a change of phase comes about. Earlier Lord Rayleigh had shown that for a string maintained in vibration by a periodic force $f\cos pt$, the displacements may be described by

$$y(x,t) = \gamma \frac{R_x}{R_l} \cos(pt + \varepsilon_x - \varepsilon_l). \tag{4.12}$$

Here it is assumed that the string is fixed at $x = 0$ and that the periodic force is applied at the other end $x = l$. The quantity γ is defined by

$$y(l,t) = \gamma \cos pt, \tag{4.13}$$

while R_x and ε_x are given by

$$R_x^2 = \sin^2 \alpha x + \frac{\kappa^2 x^2}{4a^2} \cot \alpha x \tag{4.14}$$

and

$$\tan \varepsilon_x = \beta x \cot \alpha x \tag{4.15}$$

respectively. In the above $\alpha = p/a$ and $\beta = -\kappa/2a$, a being the velocity of the vibrational wave. From (4.15) it follows that

$$\varepsilon_x - \varepsilon_l = \tan^{-1}(\beta x \cot \alpha x) - \tan^{-1}(\beta b \cot \alpha b). \tag{4.16}$$

The phase difference $(\varepsilon_x - \varepsilon_l)$ is very small and may be put equal to zero, i.e., every point on the string has the same phase as the exciting force, which is what Donkin claimed. There is, however, an exception missed by Donkin which occurs when $\alpha x = n\pi$, i.e., when x corresponds to a node. Then $\cot \alpha x$ becomes large and correspondingly $(\varepsilon_x - \varepsilon_l) \sim \pi/2$. That such a localized change of phase must occur was first recognized and demonstrated by Raman.

Melde's Experiment and its Extension[13]

Let us start with a vibrating string of length l. Its displacement can be written

$$y = \phi_1 \sin \frac{\pi x}{l} + \phi_2 \sin \frac{2\pi x}{l} + \cdots + \phi_s \sin \frac{s\pi x}{l} + \cdots. \tag{4.17}$$

Here $\phi_1, \phi_2, \ldots$ are the normal coordinates while the sine functions are eigenfunctions. For a freely vibrating (undamped) string, the normal coordinates would satisfy the equation

$$\ddot{\phi}_s + (s\pi a/l)^2 \phi_s = 0, \tag{4.18}$$

where a is the velocity of the wave propagating along the string (for further clarifications regarding the velocity wave, see the subsection on vibrating strings).

The starting point of Lord Rayleigh's analysis of Melde's experiment is the normal mode equation [4.25]

$$\ddot{\theta} + \kappa \dot{\theta} + n^2 \theta = \theta(2\alpha \sin 2pt), \quad \alpha, \kappa \ll 1. \tag{4.19}$$

The distinguishing feature of the above equation is that the forcing term (on the r.h.s.) is *proportional* to θ (contrast with (4.6)). The most general solution to (4.19) may be written in the form

$$\theta(t) = A_1 \sin pt + B_1 \cos pt + A_3 \sin 3pt$$
$$+ B_3 \cos 3pt + \cdots + \{B_0 + A_2 \sin 2pt$$
$$+ B_2 \cos 2pt + A_4 \sin 4pt$$
$$+ B_4 \cos 4pt + \cdots\}. \tag{4.20}$$

In Melde's experiment

$$\theta(t) \approx A_1 \sin pt, \qquad (4.21)$$

and further the even harmonics do not play any role. Recognizing this Lord Rayleigh neglects the terms in the curly bracket in (4.20), thus writing

$$\theta(t) = A_1 \sin pt + B_1 \cos pt + A_3 \sin 3pt + B_3 \cos 3pt + \cdots. \qquad (4.22)$$

The amplitudes of higher orders are of diminishing importance with (A_m, B_m) being $\sim (A_{m-2}, B_{m-2})$. Restricting to first order, one easily finds

$$A_1(n^2 - p^2) - (\kappa p + \alpha)B_1 = 0$$
$$A_1(\kappa p - \alpha) - (n^2 - p^2)B_1 = 0, \qquad (4.23)$$

whence

$$\frac{B_1}{A_1} = \frac{n^2 - p^2}{\kappa p + \alpha} = \frac{\alpha - \kappa p}{n^2 - p^2} = \frac{\sqrt{(\alpha - \kappa p)}}{\sqrt{(\alpha + \kappa p)}}, \qquad (4.24)$$

and

$$n^2 - p^2 = \alpha^2 - \kappa^2 p^2. \qquad (4.25)$$

Thus to a first approximation,

$$\theta(t) \approx P \sin(pt + \varepsilon), \qquad (4.26)$$

where

$$\varepsilon = \tan^{-1}(B_1/A_1). \qquad (4.27)$$

In the above, both p and ε get fixed once α and κ are given. On the other hand the amplitude P is not, which is in contrast to the forced oscillator discussed earlier. Lord Rayleigh notes that this result is true to all orders.

So far not much attention has been paid to the quantity n in Eq. (4.19). One would of course recognize it as the frequency of free and undamped vibrations. Melde's experiment corresponds to the case $n = p$. But it turns out that resonances are also possible when $n = mp$ where m is a positive integer. Correspondingly, the ratio $(v_{string}/v_{fork}) = (m/2)$. Raman studied such resonances, and the analysis of his experiment proceeds as above. The dominant terms are now of order m, the higher order ones being of progressively diminishing importance. However, Raman discovered from the waxing and the waning of the vibrations that the terms of order $(m-2)$ must also be considered. Thus he observes [4.7] that the vibration curve in general is represented by

$$P \sin(mpt + \varepsilon) + Q \sin\overline{(m-2}pt + \varepsilon'). \qquad (4.28)$$

Raman also briefly considered the role of nonlinearities by modifying (4.19) as

$$\ddot{\theta} + \kappa\dot{\theta} + (n^2 - 2\alpha \sin 2pt + \beta\theta^2)\theta = 0. \qquad (4.29)$$

Later this topic was pursued further by Ghosh [4.30].

Vibrating Strings

Consider a string of length l stretched along the x-axis and clamped at its two ends $x = 0$ and $x = l$. The equation of the vibrations is

$$\partial^2 y/\partial t^2 = v^2(\partial^2 y/\partial x^2), \quad (4.30)$$

and its solution may be written

$$y(x, t) = f_1(x - vt) + f_2(x + vt), \quad (4.31)$$

where f_1 and f_2 denote disturbances travelling in the positive and negative x-directions respectively and v denotes the velocity of the disturbances. The Raman waves considered earlier emerge when we take the time derivative of y.

Next one must consider the initial conditions. A popular one corresponds to the

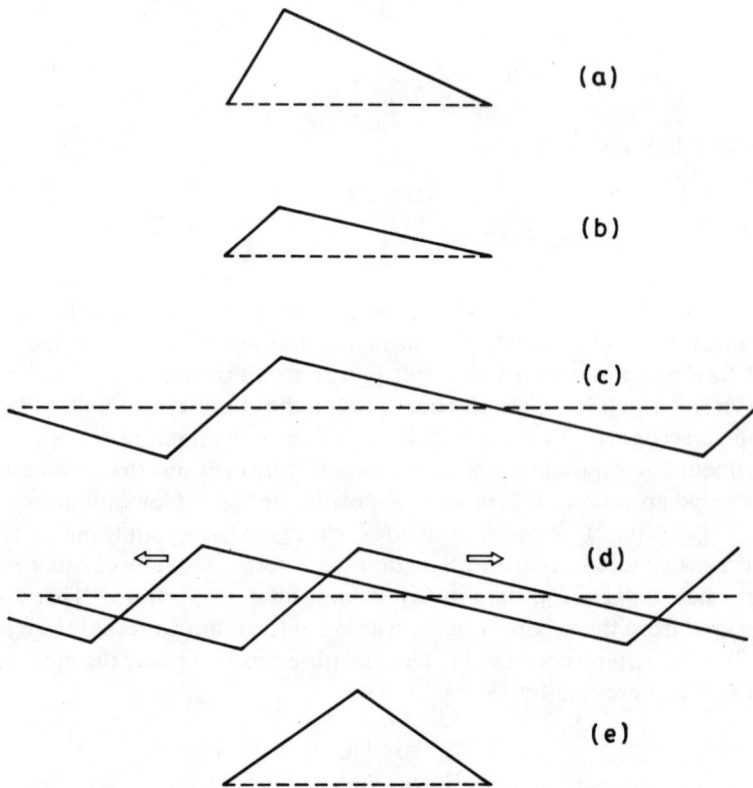

Figure 4.35 (a) Displacement $y(x, 0)$ at time $t = 0$. (b) $u(x) = 1/2\ y(x, 0)$. (c) $u(x)$ extended periodically. (d) Displacements at a later time $t \neq 0$ due to the two travelling waves. (e) Displacement $y(x, t)$ constructed by adding the displacements due to the two waves in (d), over the region 0 to l.

case where the string is given a static displacement, i.e.,
$$y = \phi(x), \quad \dot{y} = 0 \quad \text{for} \quad t = 0. \tag{4.32}$$

We then obtain
$$y(x,t) = 1/2\{\phi(x - vt) + \phi(x + vt)\}$$
$$\equiv f(x - vt) + f(x + vt). \tag{4.33}$$

Using the boundary conditions
$$y = 0 \quad \text{at} \quad x = 0 \quad \text{and} \quad x = l, \tag{4.34}$$

it is easily shown that
$$(i) f(x) = -f(-x),$$
$$(ii) f(x) = f(x + 2l). \tag{4.35}$$

These facts permit a graphical construction of $y(x,t)$ as follows (see Fig. 4.35):

i) First draw $y(x,0)$
ii) Scale it down by a factor of two to obtain $u(x)$.
iii) Extend $u(x)$ periodically.
iv) Treat the curve in (iii) as representing the displacements due to two wave trains travelling in opposite directions and at that instant, i.e., $t = 0$, coincident.
v) Add the two displacements to obtain $y(x,0)$.
vi) Displace the two wave trains by $+vt$ and $-vt$, i.e., shift them to the left and to the right by the amounts indicated. Add; this gives $y(x,t)$.

Repeating step (vi) for various values of t enables one to study the time evolution of the displacement curve. Raman used this technique extensively.

Analytical representation of $y(x,t)$ is obtained through the use of Fourier analysis. Consider, as an example, the case of the plucked string. Let the coordinates of the plucked point be (α, β). Writing the initial configuration as $y = F(x)$ we see that

$$F(x) = \beta x/\alpha \quad \text{for} \quad \alpha > x > 0$$
$$= \beta(l - x)/(l - \alpha) \quad \text{for} \quad l > x > \alpha. \tag{4.36}$$

The most general solution for the displacement may be written as

$$y(x,t) = \sum_{n=1}^{\infty} \sin\frac{n\pi x}{l}(a_n \cos npt + b_n \sin npt). \tag{4.37}$$

It is easily verified that all the b_ns vanish and that

$$a_n = \frac{2\beta l^2}{n^2\pi^2\alpha(l - \alpha)} \sin\frac{n\pi\alpha}{l}. \tag{4.38}$$

Thus

$$y(x,t) = \frac{2\beta l^2}{\pi^2\alpha(l - \alpha)} \sum_n \frac{1}{n^2} \sin\frac{n\pi x}{l} \sin\frac{n\pi\alpha}{l} \cos npt. \tag{4.39}$$

It follows that if the string is plucked at one of the nodes of the nth harmonic, then that harmonic as well as all its overtones are absent.

We consider next the struck string. Suppose that it is struck in the infinitely short region between $x = h$ and $x = h + dx$ and imparted a velocity u. The initial conditions now are:

$y = 0$ for $x = 0$ to $x = l$, except for the interval
$\dot{y} = 0$ for $x = 0$ to $x = l$, between $x = h$ and $x = h + dx$. (4.40)

The general solution may be written

$$y(x,t) = \sum_{n=1}^{\infty} b_n \sin\frac{n\pi x}{l} \sin npt. \tag{4.41}$$

The cosine terms are absent since $y = 0$ at $t = 0$. The coefficients b_n are determined as usual, whereupon one obtains

$$y(x,t) = \frac{c}{\pi l N} \sum_{n=1}^{\infty} \frac{1}{n} \sin\frac{n\pi h}{l} \sin\frac{n\pi x}{l} \sin npt, \tag{4.42}$$

where $N = (p/2x)$ and $c = u\,dx$. Comparing with (4.39) we find (as remarked earlier) that the series for the struck string converges more slowly than for the plucked string.

Reference may now be made to Raman's work on the piano [4.21]. His mathematical analysis is *quite sketchy*, drawing upon various sections of Lord Rayleigh's classic which too skips many details. Upon patiently reconstructing the analysis, Raman's grip on vibrations becomes quite evident. We shall now outline (filling in many steps jumped by Raman!) the way in which Raman arrived at the dashed curve in Fig. 4.27.

Raman supposes that the motion of the string produced by the impact of the hammer can be simulated by the vibrations of a string having a load attached to it at the striking point. The loaded string analogy applies till the hammer breaks contact; thereafter the string executes free periodic vibration of the usual kind.

Consider therefore the problem of the loaded string. Assume that the load M divides the length into two parts a and $(l - a) = b$. For convenience we shall suppose $(l - a) > a$. The vibrations of the string are described by the solutions of

$$\mu\frac{\partial^2 \xi}{\partial t^2} = T\frac{\partial^2 \xi}{\partial x^2} \quad \text{for } x < a \text{ and } x > a. \tag{4.43}$$

Here μ is the mass per unit length of the string and T is the tension. The boundary conditions are

$$\xi(0, t) = \xi(l, t) = 0. \tag{4.44}$$

At $x = 0$ ξ is continuous but $(\partial \xi/\partial x)$ is not. For the moment let us ignore the impulse delivered by the hammer and simply suppose that a periodic force $F \exp(ipt)$ acts at

$x = a$. Assume that the solutions $\xi(x, t)$ may be written

$$\xi(x, t) = u(x) \exp(ipt).$$

Substituting in (4.43) we obtain

$$d^2u/dx^2 + \lambda^2 u = 0, \quad \lambda^2 = \mu p^2/T = p^2/c^2, \tag{4.45}$$

where c is the velocity of the vibrational wave. The solutions of (4.45) in the two regions may be written

$$u_1(x) = A \cos \lambda x + B \sin \lambda x \quad 0 \leqslant x \leqslant a,$$
$$u_2(x) = C \cos \lambda(l - x) + D \sin \lambda(l - x) \quad a \leqslant x \leqslant l. \tag{4.46}$$

Applying the boundary conditions and the continuity condition at $x = a$ one finds

$$u_1(x) = B \sin \lambda x$$

$$u_2(x) = B \frac{\sin \lambda a}{\sin \lambda b} \sin \lambda(l - x). \tag{4.47}$$

Now the force acting at $x = a$ is balanced by the resultant of the tensions acting at that point. The latter quantity is given by

$$T\Delta(\partial \xi/\partial x)_{x=a} = T\{(\partial \xi_2/\partial x)_{x=a} - (\partial \xi_1/\partial x)_{x=a}\}$$

$$= -T\lambda B \frac{\sin \lambda l}{\sin \lambda b} \exp(ipt). \tag{4.48}$$

On the other hand, the force $F \exp(ipt)$ is nothing but the inertial force due to the mass M. If we denote by ξ_a the displacement of the point $x = a$, then the inertial force is given by

$$M \frac{d^2 \xi_a}{dt^2} = -Mp^2 B \sin \lambda a \exp(ipt). \tag{4.49}$$

Equating (4.48) and (4.49) we obtain

$$\mu \sin \lambda l = M\lambda \sin \lambda a \sin \lambda b. \tag{4.50}$$

This equation is the starting point of Raman's analysis. Solving this for the allowed values of λ gives the eigenfrequencies of the loaded string. Note in passing that if we put $M = 0$, we recover the usual result $\lambda_n = (n\pi/l)$ for the unloaded string.

Recognizing that many solutions λ_r are possible from (4.50), we now write the general solution for the string displacement as

$$\xi(x, t) = \sum_r \phi_r \sin c\lambda_r t \begin{cases} \sin \lambda_r x \cdot \sin \lambda_r b, & x < a \\ \sin \lambda_r a \cdot \sin \lambda_r (l - x), & x > a. \end{cases} \tag{4.51}$$

The constants ϕ_r have to be found from the initial conditions. Lord Rayleigh deals with such a problem in article 101 of his book to which Raman now appeals. According to this, if $Z(x)$ is the initial momentum distribution along the string, then

$\xi_a(t)$ may be written

$$\xi_a(t) = \sum_r \sin(c\lambda_r t) \frac{u_r^2(a) \int Z(x)\,dx}{c\lambda_r \int \rho(x) u_r^2(x) x}. \tag{4.52}$$

Here u_r is the displacement amplitude in the rth mode and ρ is the density.

From earlier analysis we know

$$u_r(a) = B \sin \lambda_r a$$

which, for convenience, we rewrite as

$$u_r(a) = \psi \sin \lambda_r a \sin \lambda_r b. \tag{4.53}$$

In our problem,

$$Z(x) = Mv\delta(x - a), \tag{4.54}$$

where v is the initial velocity of M. We note in passing that an impulsive force will have many periodic components. The earlier analysis indicated the effects produced by a typical periodic element.

In evaluating the integral

$$\int \rho u_r^2(x)\,dx,$$

one must not only consider the regions $x < a$ and $x > a$ but also the fact that at $x = a$ there is a mass M. Thus

$$\int \rho(x) u_r^2(x)\,dx = \psi^2 M \sin^2 \lambda_r a \sin^2 \lambda_r b$$

$$+ \psi^2 \mu \left\{ \int_0^a \sin^2 \lambda_r b \sin^2 \lambda_r x\,dx \right.$$

$$\left. + \int_a^l \sin^2 \lambda_r a \sin^2 \lambda_r (l - x)\,dx \right\}.$$

Evaluating the integrals and substituting in (4.52) Raman deduces

$$-M \frac{d^2 \xi_a}{dt^2} = \sum_r \frac{2Mvc\lambda_r \sin(c\lambda_r t)}{(1 + c\mu/M)\{a/\sin^2 \lambda_r a + b/\sin^2 \lambda_r b\}}. \tag{4.55}$$

Expression (4.55) gives the force exerted by the particle on the string as a series in harmonic functions. It is, so to say, something like a Fourier approximation to the discontinuous force exerted by the hammer on the string (while the contact lasts). One knows from mathematics that a Fourier approximation to a discontinuous function converges (under appropriate conditions etc.!) to the mean. In practical terms, Raman is able, by taking a sufficient number of terms, to arrive at the points A, B, C,... in Fig. 4.27b.

Glimpses from the Golden Era

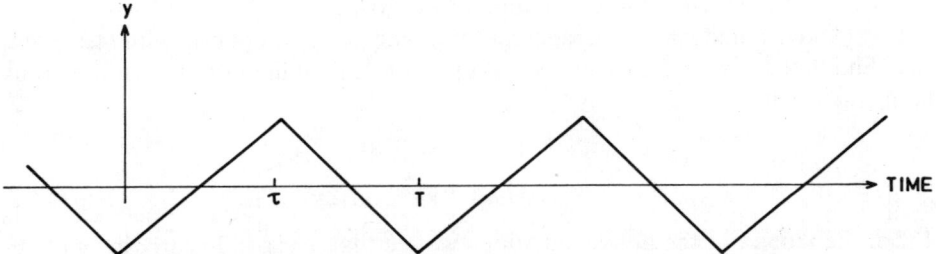

Figure 4.36 Vibration curve for a representative point on the bowed string.

Later, Panchanon Das [4.31] tackled the problem somewhat differently by considering the equation of motion for the string as well as for the hammer. The two equations are coupled.

We next consider the bowed string. Helmholtz, to whom is due the first analytical treatment of the problem, started from his observation that the vibration curve (in the fundamental mode) is a two-step zigzag. Let T be the period of the vibration of the curve and be described at the point x by (see Fig. 4.36)

$$y = ft + h \qquad \text{from } t=0 \quad \text{to } t=\tau, \qquad (4.56)$$
$$y = g(T-t) + h \quad \text{from } t=\tau \quad \text{to } t=T.$$

Thus for $t = \tau$ we have

$$f\tau = g(T-\tau).$$

Now since y is periodic in time, one could expand it in a series as

$$y = A_1 \sin(2\pi t/T) + A_2 \sin(4\pi t/T) + \cdots$$
$$+ B_1 \cos(2\pi t/T) + B_2 \cos(4\pi t/T) + \cdots. \qquad (4.57)$$

Solving for the coefficients, Helmholtz shows that

$$y = \frac{(f+g)T}{\pi^2} \sum_{n=1}^{\infty} \frac{1}{n^2} \sin\frac{n\pi\tau}{T} \sin\frac{2\pi n}{T}\left(t - \frac{\tau}{2}\right). \qquad (4.58)$$

On the other hand, quite independent of Fig. 4.36, one may write

$$y(x,t) = \sum_{n=1}^{\infty} C_n \sin\frac{n\pi x}{l} \sin\frac{2\pi n}{T}\left(t - \frac{\tau}{2}\right)$$
$$+ \sum_{n=1}^{\infty} D_n \sin\frac{n\pi x}{l} \cos\frac{2\pi n}{T}\left(t - \frac{\tau}{2}\right). \qquad (4.59)$$

Comparing the last two equations we see that all the D_ns vanish. After a little algebra one obtains

$$y(x,t) = \frac{8V}{\pi^2} \sum_n \frac{1}{n^2} \sin\frac{n\pi x}{l} \sin\frac{2\pi n}{T}\left(t - \frac{\tau}{2}\right), \qquad (4.60)$$

where V is the amplitude in the middle of the string.

Can (4.60) be made more transparent? Yes indeed, by comparing with (4.39). We then find that at time t, (4.60) represents a pair of straight lines meeting at the point (α, β) such that

$$\beta l^2/[\alpha(l-\alpha)] = \pm 4V,$$
$$\sin(s\pi\alpha/l) = \pm \sin(2\pi st/T). \quad (4.61)$$

From the second of the above equations we see that α varies linearly in time. As Helmholtz further noted, the point of intersection follows a parabolic arc.

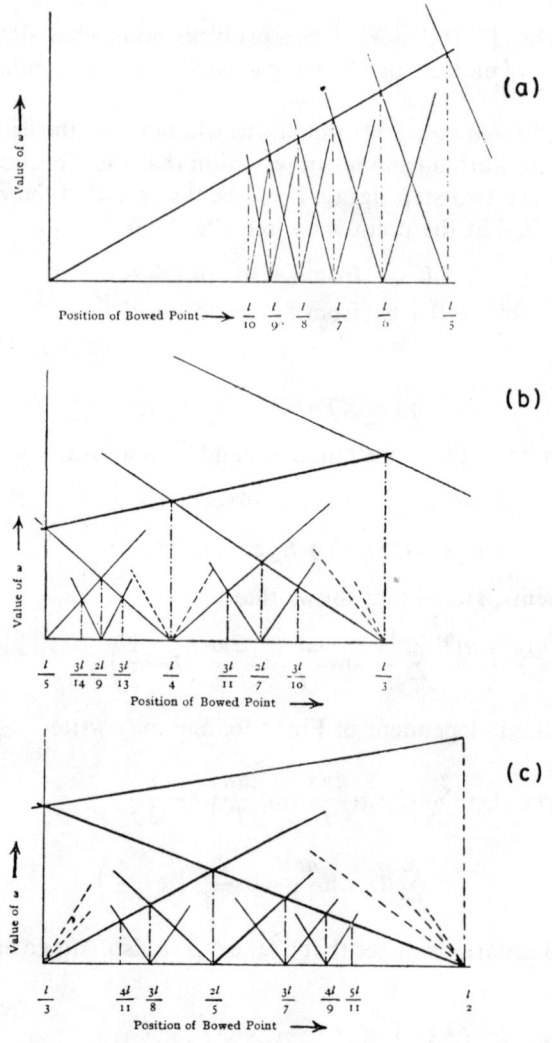

Figure 4.37 Variation of the parameter ω with the position of bowing. For explanations, see text. (After ref. [4.15].)

Raman's kinematical analysis dealt primarily with the higher modes of vibration. The analysis, being quite extensive, can hardly be reproduced here. Raman studied the vibrations that result not only when the bow is applied at a point of irrational division of the string but also those when the bow is applied at a node. Starting with velocity curves with an appropriate number of discontinuities, he painstakingly constructed the displacement profiles and thence the vibration curves of various points on the string. Featuring prominently in his analysis is a parameter ω which is the total fraction of the period of vibration in which a given point moves with the larger of the two velocities (recall that every point moves alternately with one of two velocities). Figure 4.37a shows for example how ω varies for the various modes with the position of bowing. Here the heavy line is the graph of ω for the first type of vibrational mode, and the thin lines meeting in pairs at the points $l/9$, $l/8$, etc. give the values of ω for the 9th, 8th, etc. types respectively. The most noticeable feature in the diagram is the extreme steepness of these lines compared with the slope of the line for the first type. Figures 4.37b and c show further extensions. From an analysis of these curves Raman concluded that while in an ordinary monochord the higher types of vibration may easily be excited by suitable bowing, in the case of the violin with its short and heavily-damped string, only the first type of vibration and its modification are important.

Raman also carried out an extensive study of the frictional effects of the bow, and made numerous plots of how the frictional force varies as a function of time when the string is bowed at various points and in such a manner as to excite vibrations other than the fundamental. In turn this led him to a study of what he calls *transitional forms* of vibrations, "which are intermediate between the irrational type and their rational modifications". Other topics that received attention were the vibrational forms for very small bowing pressures, instability of vibrations under high bowing pressures, effects of yielding of the bridge, and the effect of a mute on the wolf-note.

Physics of the *Mridangam*

Consider first a uniform circular membrane of radius a clamped around its edge. Denoting by ψ the displacement amplitude normal to the plane of the membrane, the differential equation for the vibrations may be written as

$$\frac{\partial^2 \psi}{\partial t^2} = c^2 \left(\frac{\partial^2 \psi}{\partial x^2} + \frac{\partial^2 \psi}{\partial y^2} \right), \qquad (4.62)$$

where c is the velocity of sound. Since the problem has cylindrical symmetry, it is obviously more meaningful to employ polar coordinates. Using them, and assuming ψ has the form $\psi = \Psi \cos pt$, one readily obtains

$$\frac{d^2 \Psi}{dr^2} + \frac{1}{r}\frac{d\Psi}{dr} + \frac{1}{r^2}\frac{d^2 \Psi}{d\theta^2} + \lambda^2 \Psi = 0, \qquad (4.63)$$

where $\lambda = (\omega/c)$.

Lord Rayleigh showed [4.25] that solutions of (4.63) are of the form

$$\Psi(r, \theta) = A_n J_n(\lambda r) \cos n(\theta + \alpha), \tag{4.64}$$

where J_n denotes the spherical Bessel function of order n. The boundary condition requires that

$$J_n(\lambda a) = 0, \tag{4.65}$$

an equation whose roots give the admissible values of λ and therefore of ω.

The most general solution for Ψ is obtained by combining all the particular solutions embodied in (4.64) with all the admissible values of λ and n.

Let us now examine the character of some of the normal modes. If $n = 0$, Ψ is a function of r only, i.e., the solution is symmetrical with respect to the centre of the membrane. The nodes, if any, are concentric circles whose radii are obtained as the roots of the equation

$$J_0(\lambda r) = 0. \tag{4.66}$$

When $n \neq 0$, the equation of the nodal system takes the form

$$J_n(\lambda r) \cos n(\theta - \alpha) = 0. \tag{4.67}$$

There are two types of nodes, circles represented by

$$J_n(\lambda r) = 0 \tag{4.68}$$

and diameters described by

$$\theta = \alpha + (2m + 1)\pi/2n, \tag{4.69}$$

where m is an integer. There are n diameters dividing the circle uniformly. In other respects their position is arbitrary.

Figure 4.30 shows the more important normal modes as described by Lord Rayleigh. The numbers indicate the frequencies of the various modes, taking that of the fundamental as unity. As already remarked, the higher modes do not form a harmonic series.

In his *Handbuch* article, Raman briefly examines a membrane whose density varies as r^{2m-2}, special cases of which were studied by Ghosh [4.32] and by Rao [4.33]. Ramakrishna and Sondhi [4.34] studied the problem again in the fifties as they felt that the power law was basically incapable of explaining some of the observed degeneracies. Instead they suppose that the drumhead is a composite membrane with a radial density profile

$$\begin{aligned}\rho &= \rho_1 \quad 0 \leqslant r \leqslant a, \\ &= \rho_2 \quad a \leqslant r \leqslant b.\end{aligned} \tag{4.70}$$

The membrane is clamped at $r = b$, and is supposed to be under tension T per unit length. Analogous to (4.63), one now has

$$\begin{aligned}\nabla^2 \Psi_1 + \lambda_1^2 \Psi_1 &= 0 \quad 0 \leqslant r \leqslant a, \\ \nabla^2 \Psi_2 + \lambda_2^2 \Psi_2 &= 0 \quad a \leqslant r \leqslant b,\end{aligned} \tag{4.71}$$

Glimpses from the Golden Era 125

where
$$\lambda_1 = \omega/c_1, \quad \lambda_2 = \omega/c_2, \tag{4.72}$$
with
$$c_1^2 = T/\rho_1, \quad c_2^2 = T/\rho_2. \tag{4.73}$$

Ψ_1 and Ψ_2 satisfy the following conditions:

$$\Psi_1 \text{ and } \Psi_2 \text{ are finite over the membrane,} \tag{4.74a}$$

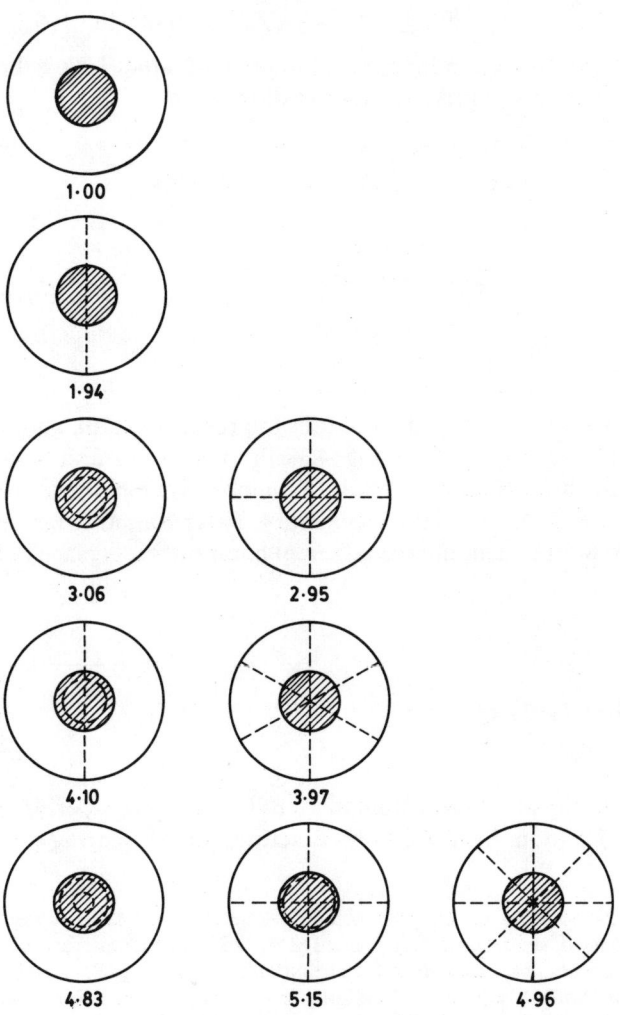

Figure 4.38 Nodal lines (shown dashed) corresponding to the first nine modes of vibration of the composite membrane. Also shown are the frequencies. The various modes are arranged in different rows according to the harmonic to which they closely correspond. (After ref. [4.34].)

$$\Psi_i(r,\theta) = \Psi_i(r_1\theta + 2\pi n) \quad i = 1, 2, \tag{4.74b}$$

$$\Psi_1(a,\theta) = \Psi_2(a,\theta), \tag{4.74c}$$

$$\delta\Psi_1(a,\theta)/\delta r = \delta\Psi_2(a,\theta)/\delta r, \tag{4.74d}$$

$$\Psi_2(b,\theta) = 0. \tag{4.74e}$$

Keeping (4.74a, b) in mind, the general solutions are written as

$$\Psi_1(r,\theta) = A_n J_n(\lambda_1 r)\sin(n\theta + \phi_n),$$

$$\Psi_2(r,\theta) = [B_n J_n(\lambda_2 r) + c_n Y_n(\lambda_2 r)]\sin(n\theta + \phi_n), \tag{4.75}$$

where Y_n denotes the Neumann function of the second kind. Using the boundary conditions (4.74c, d, e), Ramakrishna and Sondhi obtain

$$\sigma \frac{J_{n-1}(\sigma kx)}{J_n(\sigma kx)} = \frac{J_{n-1}(kx)Y_n(x) - J_n(x)Y_{n-1}(kx)}{J_n(kx)Y_n(x) - J_n(x)Y_n(kx)}, \tag{4.76}$$

where

$$x = \lambda_2 b = \omega b/c_2,$$

$$\sigma^2 = \rho_1/\rho_2 = \lambda_1^2/\lambda_2^2 = c_2^2/c_1^2,$$

$$k = a/b.$$

A graphical procedure is employed to obtain the roots of (4.76), and some of the results so deduced are displayed in Fig. 4.38. Ramakrishna and Sondhi further carried out an experimental study to obtain a quantitative measure of the degree of harmonicity and the closeness of the degeneracy. Later Ramakrishna also studied the left-hand *tabla* where the nonlinear adherent composition is placed eccentrically [4.35].

4.4 Whispering Gallery

In Chapter 3, we briefly mentioned Raman's studies on the whispering gallery in St Paul's Cathedral, London. Lord Rayleigh describes the whispering gallery phenomenon as follows:

> One of the most striking of the phenomena connected with the propagation of sound within closed buildings is that presented by 'whispering galleries', of which a good and easily accessible example is to be found in the circular gallery at the base of the dome of St Paul's Cathedral.... Judging from some observations that I have made in St Paul's whispering gallery, I am disposed to think that the principal phenomenon is to be explained somewhat differently [as compared to what others had done earlier]. The abnormal loudness with which a whisper is heard is not confined to the position diametrically opposite to that occupied by the whisperer, and therefore, it would appear, does not depend materially upon the symmetry of the dome. The whisper seems to creep round the gallery horizontally, not necessarily

along the shorter arc, but rather along that arc towards which the whisperer faces. This is a consequence of the very unequal audibility of a whisper in front of and behind the speaker, a phenomenon which may easily be observed in the open air. [4.36]

Lord Rayleigh also sketched a mathematical theory of the phenomenon, and one would have thought that the last word on the subject had been said. Not quite! In a footnote to his original paper as reprinted in his collected works [4.36], Lord Rayleigh commented that his theory should be equally applicable to electromagnetic waves. Raman and Bidhubhusan Ray put this idea to test, using a strip of mirror 100 cm long and 5 cm wide and on which a curvature of variable magnitude could be imposed. A razor edge was placed on the mirror at one end forming a very fine slit between it and the surface which was illuminated with light. The illumination at the other end was observed using an eyepiece. Describing their observations, Raman writes:

> When the surface was quite plane, there was only a very faint general illumination of the field, the edge of the mirror being, however, a perfectly black line. When a slight curvature is put on, there is a very rapid increase in the luminosity of the field, a bright band of light flashing out next to the surface of the mirror, which continues to be seen as a line of zero illumination. With further increase of curvature this band contracts in width, and is followed by a second bright band, separated from it by a dark band. Then a third bright band appears, preceded by another dark band, and so on, the number of bands and their sharpness increasing, and their width decreasing with the increase in the curvature of the mirror [4.37]

Rayleigh's theory predicted only one band, whereas Raman's experiment revealed the existence of many.

While in London, Raman recalled his Calcutta experiment and wondered whether multiple intensity bands occurred also in the whispering gallery in St Paul's. None of the previous observers had reported this and therefore a fresh experiment was called for. With the kind co-operation of the Cathedral authorities and with encouragement from Prof. Porter of the University College, London, Raman and Sutherland performed an extensive series of experiments with one of them producing the sound and the other tracking it [4.37]. It was found that as the observer moved along a radius, he heard a repeated waxing and waning of the sound intensity. In other words, there were multiple bands, exactly as found in the optical experiment in Calcutta.

Rayleigh's theory obviously needed an extension, but Raman and Sutherland did not attempt it. Raman later investigated several whispering galleries in India (like the *Gol Gumbaz* in Bijapur) [4.37a] and also used these ideas to explain iridescence in pearls.

4.5 Studies in Optics

If Raman appeared to excel in acoustics, he really sparkled where optics was concerned. As G. N. Ramachandran remarks:

The study of acoustics is intimately connected with the study of vibrations and waves, and it is not surprising that Raman's interests passed from his early love for acoustics on to a life-long devotion to optics, the other great domain of classical wave mechanics. In fact, if one may talk of a unifying trend in the scientific work of Raman, it may be said to reside in the study of wave phenomena. [4.38]

Some Preliminaries[14]

There can be little dispute in crediting Newton with being the father of modern optics. Newton relied on direct observation, describing his work as experimental philosophy. He strongly supported the corpuscular theory of light, but already there were suggestions that light had a wave-like character. However, Newton stoutly opposed any such notion, claiming it was incapable of explaining ray optics. Such was the sway of Newton's personality, that it stifled all advances in the wave theory for several decades. But the wave theory came back into its own in the nineteenth century, thanks in the first instance to Thomas Young and Jean Fresnel. In the

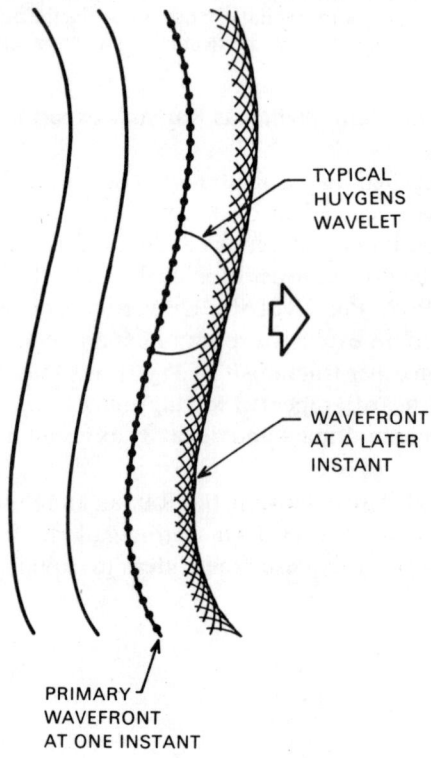

Figure 4.39 Propagation of a wavefront according to Huygens' principle.

meanwhile, the study of electricity and magnetism, particularly by Michael Faraday, was beginning to bear fruit. Later Maxwell added the crowning touch by proving that the electromagnetic field could propagate as a transverse wave, with a velocity equal to that measured in the laboratory for light. The conclusion was now inescapable that light was nothing but a propagating electromagnetic wave.

The question of how waves propagate had received attention as far back as 1690 when the Dutch physicist Christiaan Huygens enunciated a convenient working principle. To understand it, consider a point source emitting spherical waves. The spherical surface is an example of a *wavefront*. More generally, a wavefront is a surface over which the optical disturbance has a constant phase. Huygens hypothesized that the progress of the primary wavefront is due to the generation of secondary wavelets from every point of the primary wavefront. The primary wavefront at a later time is the envelope of these secondary wavelets (see Fig. 4.39).

The wave-like nature of light does not contradict our usual experience of

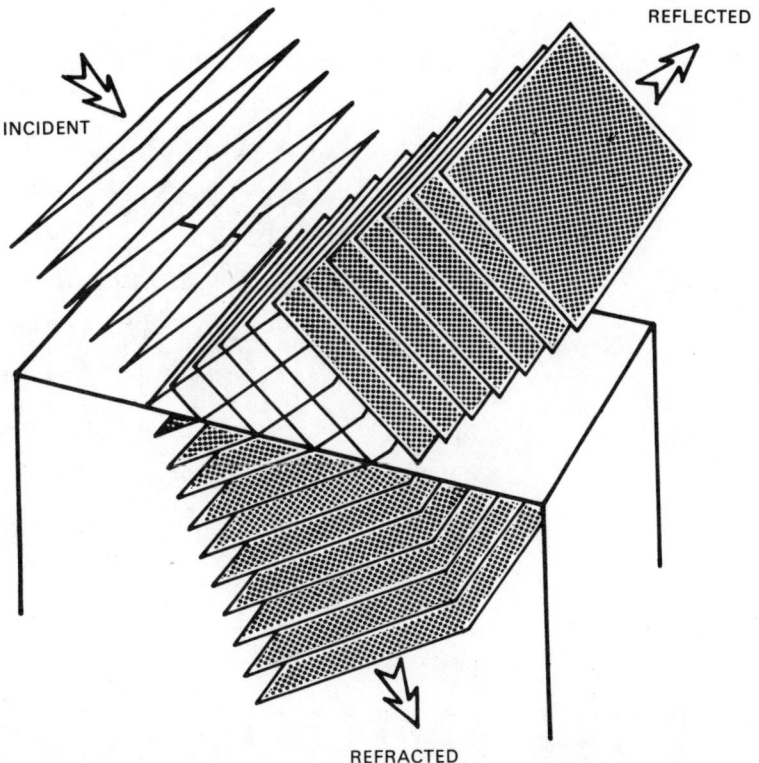

Figure 4.40 Reflection and transmission of plane waves. Shown here are plane wavefronts associated with the incident, reflected and refracted waves at various instants. The wave concept is no hindrance to understanding rectilinear propagation. The usual rays that one draws for the incident, reflected and refracted beams are just normals to the wavefronts.

reflection and refraction, which we are accustomed to explain in terms of rectilinear propagation of light (see Fig. 4.40). On the other hand, wave theory is particularly helpful in understanding certain phenomena like interference (see Fig. 4.41).

In the sixteenth century, Francesco Grimaldi observed that under certain conditions, light could be deviated from a straight path. Called by him *diffractio* and known to us as diffraction, this phenomenon also is a consequence of the wave nature of light. A convenient definition is provided by Sommerfeld who describes it as "any deviation of light rays from rectilinear paths which cannot be interpreted as reflection or refraction". There is really no significant difference between interference and diffraction. However, at the operational level one always talks of diffraction at an aperture or an edge.

There are two important limiting cases in diffraction, the difference between which

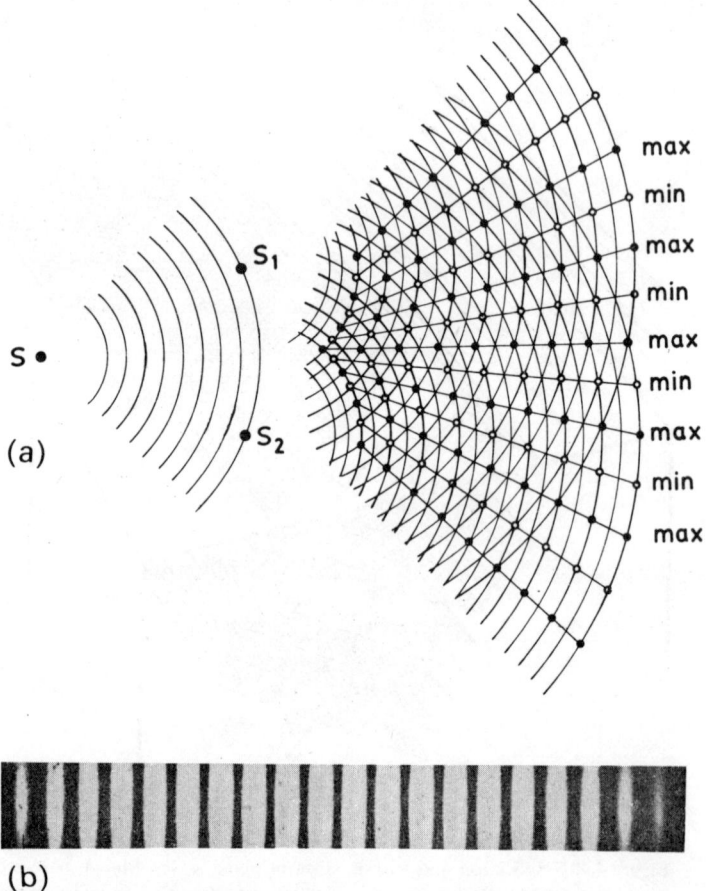

Figure 4.41 (a) Schematic of a standard interference experiment. Two slits S_1 and S_2 are illuminated by the same sources. The Huygens waves emerging from S_1 and S_2 then interfere to produce a pattern as in (b).

Glimpses from the Golden Era

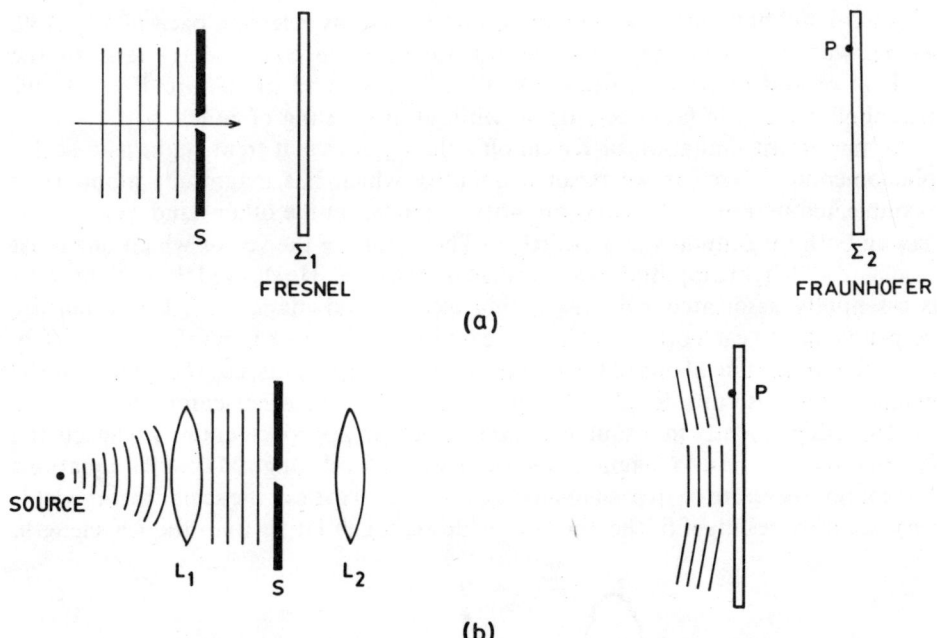

Figure 4.42 (a) Schematic of a diffraction experiment. The slit S is illuminated with plane waves and its image on a screen Σ is observed. Depending on whether Σ is close to or far away from S, one obtains the Fresnel or the Fraunhofer limits respectively. (b) Illustration of how Fraunhofer diffraction may be observed in practice. Lens L_1 renders light from the source into plane waves. The effect of putting Σ at infinity is simulated by a second lens L_2. Thus a typical point P on Σ receives plane waves.

may be understood by referring to Fig. 4.42. A slit S is illuminated with plane waves and its shadow is observed on a screen Σ. When Σ is sufficiently close to S, one would see a sharp shadow of the aperture. As Σ recedes away from S, the image of the aperture develops diffraction fringes, and the phenomenon observed is known as *Fresnel* or *near-field* diffraction. As Σ moves farther and farther away, the image on the screen spreads out, bearing little or no relationship to the actual aperture. At this stage, the movement of the screen does not change the shape but merely the size of the pattern appearing on the screen. This is *Fraunhofer* or *far-field* diffraction. The arrangement in Fig. 4.42b may be employed to realize Fraunhofer diffraction in practice. The source is placed at the focus of the lens L_1. Consequently, the slit S is illuminated with plane waves. The plane of observation is the focal plane of the second lens L_2.

The mathematical analysis of interference and diffraction phenomena, particularly the latter, is based on the wave nature of light. The foundations for such an analysis were laid by Kirchhoff, who, starting with the intuitive ideas of Huygens as modified by Fresnel, showed that the Huygens–Fresnel principle followed directly from a differential equation for wave propagation. The physical import of the

Fresnel–Kirchhoff modification may be appreciated by referring back to Fig. 4.39, where, wanting to portray a wavefront advance, we conveniently ignored the back wave and omitted to draw the other hemisphere! Thanks to Fresnel and Kirchhoff, we can in fact freely do so without any feeling of guilt.

An important limitation of Kirchhoff's theory is that it treats light as a scalar phenomenon. By scalar we mean a quantity which has magnitude alone. (For example, temperature is a scalar quantity; velocity, on the other hand, is a vector, having both magnitude and a *direction*.) The nature of the vector which one must associate with light amplitude was clarified by Maxwell. He showed that a light wave is essentially associated with oscillating electric and magnetic fields which are perpendicular to each other and in phase (see Fig. 4.43). In Kirchhoff's theory, only the scalar amplitude of one of the two transverse components, i.e., the electric or the magnetic field, is considered, it being assumed that the other component can be treated independently in a similar manner. Such an approach entirely neglects the fact that the electric and magnetic vectors are *coupled* through Maxwell's equations and cannot therefore be treated independently. Nevertheless, the scalar theory yields very accurate results if (i) the aperture width is much larger than the wavelength,

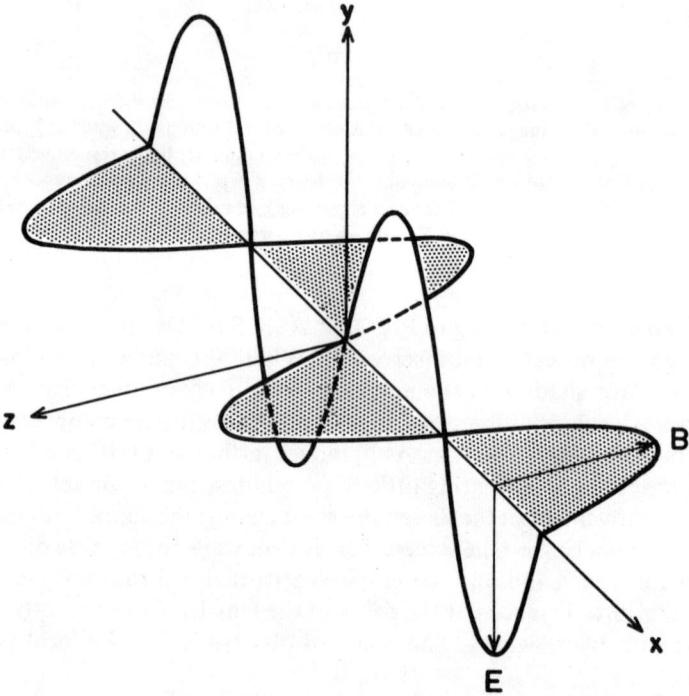

Figure 4.43 Shown here is an electromagnetic wave propagating along the x direction. The associated electric and magnetic fields (**E** and **B**) will then be perpendicular to x and, further, perpendicular to each other, e.g., as in the figure. At any instant, the fields will vary in amplitude along x as shown. If, on the other hand, the fields are observed at a fixed point x, they will vary sinusoidally in time.

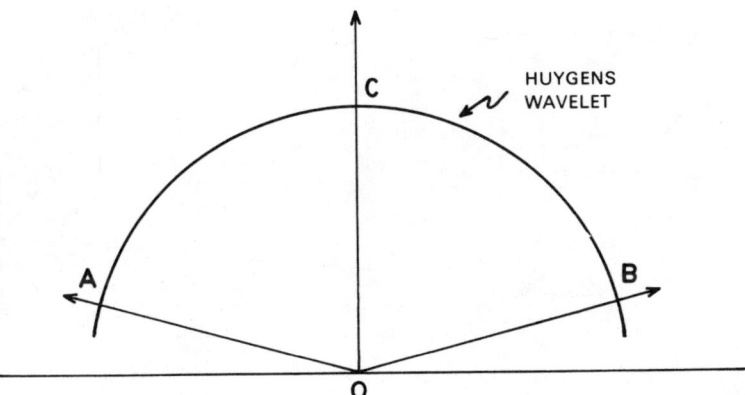

Figure 4.44 Illustration of the origin of the obliquity factor. Usually observations are made along OC but when they are made along OA or OB, the obliquity factor becomes important.

and (ii) the diffraction effects are not observed too close to the aperture. In practice both these requirements are often met, making Kirchhoff's theory quite useful. At the same time we must also record that the Kirchhoff treatment does not allow for the fact that the amplitude of the secondary wavelet may vary with angle. For instance, in Fig. 4.44 the rate of change of the amplitude near A and B is quite high in contrast to that near C where it is zero. Such amplitude variations lead to what is called an *obliquity factor* which becomes important in problems like the one that Raman studied in college. In fact Raman repeatedly studied various phenomena under oblique conditions to assess the effects produced by this factor.

Despite its widespread utility, Kirchhoff's theory is but an approximation. On the other hand, theorists are always attracted to problems capable of *exact* mathematical solution, even if the problem itself is idealized. The first truly rigorous solution to a diffraction problem was given in 1896 by Sommerfeld, who treated the diffraction of a plane electromagnetic wave by an infinitely thin and infinitely conducting screen with a perfectly sharp edge. Such exact solutions are useful in that one may compare them with those obtained by approximate methods, e.g., Kirchhoff's theory, in order to assess how good the latter solutions are (as has been done, for example, by Kottler). Raman and his colleagues drew much inspiration from this famous work of Sommerfeld.

In applying the Huygens principle to the problem of diffraction by apertures, we found it convenient to regard each point on the aperture as a new source of spherical waves (recall Fig. 2.1). However, such sources are merely mathematical conveniences and have no physical significance. A more physical viewpoint, first expressed qualitatively by Young, is to regard the observed field as a *superposition* of the *incident* wave transmitted unperturbed through the aperture, and *diffracted waves* originating at the *rim* of the aperture. It has been shown that the Kirchhoff diffraction formula can indeed be manipulated to this form. While explaining diffraction, Raman always preferred this approach to the Huygens principle.

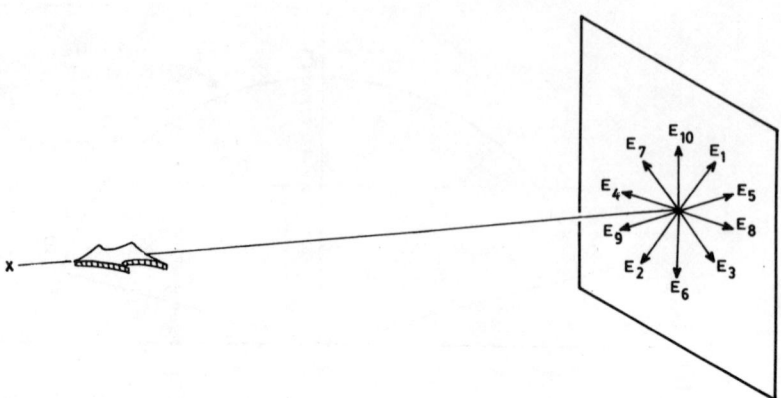

Figure 4.45 Concept of unpolarized light. Here light is propagating along the x-direction; therefore the electric vector **E** must be in the yz-plane. However, its direction need not remain fixed; rather it varies rapidly in a random fashion – from $\mathbf{E}_1$ to $\mathbf{E}_2$ to $\mathbf{E}_3$ etc.

We now turn to the phenomenon of *polarization*. The light we get from the usual sources of light is *unpolarized*. What it means is that the electric field vector varies its direction rapidly in a completely unpredictable fashion so that one cannot assign any one preferred direction to it (see Fig. 4.45). However, unpolarized light may be rendered polarized by a selection process. An optical device whose input is natural light and whose output is polarized light is called a *polarizer*. By suitably orienting the polarizer, one can make the electric vector E of the emerging light have any angle θ, as in Fig. 4.46a. How does one check that the emergent light is indeed polarized? For this purpose, a second optical element called the *analyser* is introduced, as in Fig. 4.46b. When the analyser is rotated keeping the polarizer orientation fixed (i.e., holding θ constant), the intensity varies as in Fig. 4.46c. When the polarizer and the analyser are *crossed*, light is extinguished. This can happen only if light emerging from the polarizer is polarized.

Many crystalline substances are optically anisotropic, i.e., their optical properties are not the same in all directions within any given sample. Such crystals exhibit a property called *double refraction* or *birefringence* meaning essentially that if an unpolarized ray of light is transmitted through them, it will in general be split into two rays which are polarized perpendicular to each other (Fig. 4.47). It is evident that birefringence can be exploited to make optical elements which function as polarizers and analysers. The so-called nicol prism invented by the Scottish physicist William Nicol was for many years the work-horse of all polarization analysis. In fact, Raman always carried a pair of nicol prisms. Thanks to progress in materials technology, we now have superior polarization devices.

With this background, we are now ready to survey the work done by Raman and his collaborators in the field of optics.

Optics saw Raman transform from a loner into a group leader. He became deeply

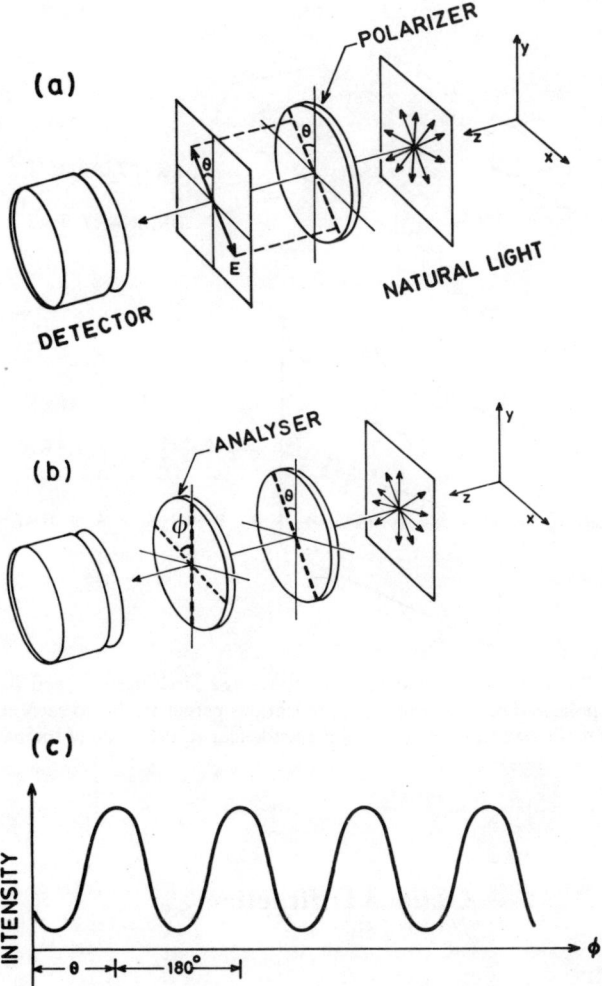

Figure 4.46 Illustration of (a) the action of a polarizer and (b) how the state of polarization may be analysed using a second element. If θ is kept fixed and ϕ, the orientation of the analyser, is varied, the intensity variation would be as in (c).

involved in optics only in the second decade of his stay in Calcutta by which time there were enough students and associates to pursue various leads for which he himself could not spare the time. Naturally, therefore, our survey must sample not only the work of Raman but of others as well, so as to provide a flavour of Raman's overall thrust. The experiments on the scattering of light were an offshoot of the investigations in optics but since they have a profound significance of their own, they are dealt with in a separate chapter.

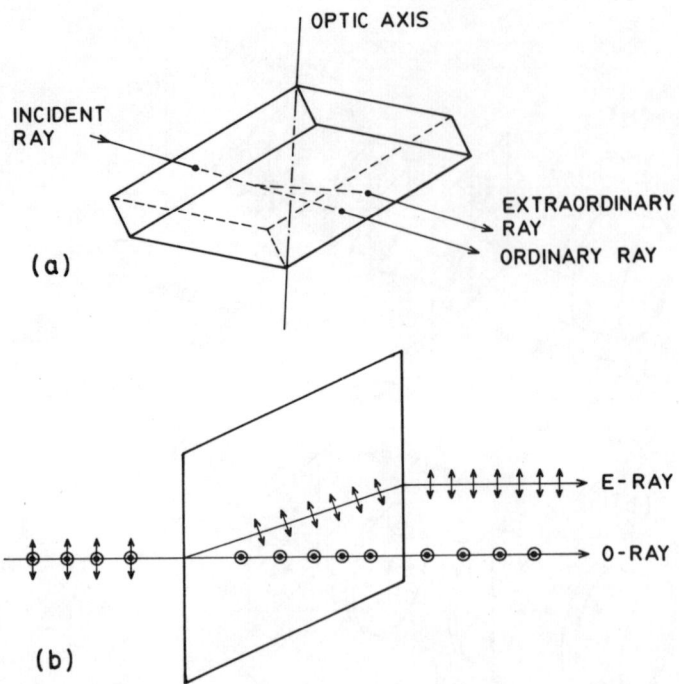

Figure 4.47 An unpolarized light beam traversing a birefringent crystal like calcite splits into two polarized beams with their polarizations perpendicular to each other. In (b), ⊙ means that the polarization vector is perpendicular to the plane of the paper.

Oblique Diffraction

Raman commenced his studies in optics in Calcutta practically at the same time as those on vibrations [4.39, 4.40], and it is not surprising that the first problem to receive his attention was that of oblique diffraction. In the paper with which he made his entry into the scientific world (see Chapter 2), he had already explained the basic asymmetry of the diffraction pattern, and the fact that the number of fringes on one side was limited. At the IACS he photographed the fringes and found that those on one side were fainter than those on the other. Analysing the problem mathematically, Raman came to the conclusion that the obliquity factor, justifiably ignored in the conventional treatment of diffraction, becomes significant in the present case. The next step was clearly an experimental verification of his theoretical prediction. Today there is no difficulty in measuring variations in light intensity, thanks to devices like photomultipliers. Raman had no such gadgets and resorted instead to a photometer (see Fig. 4.48). This consists essentially of two discs with various sectors

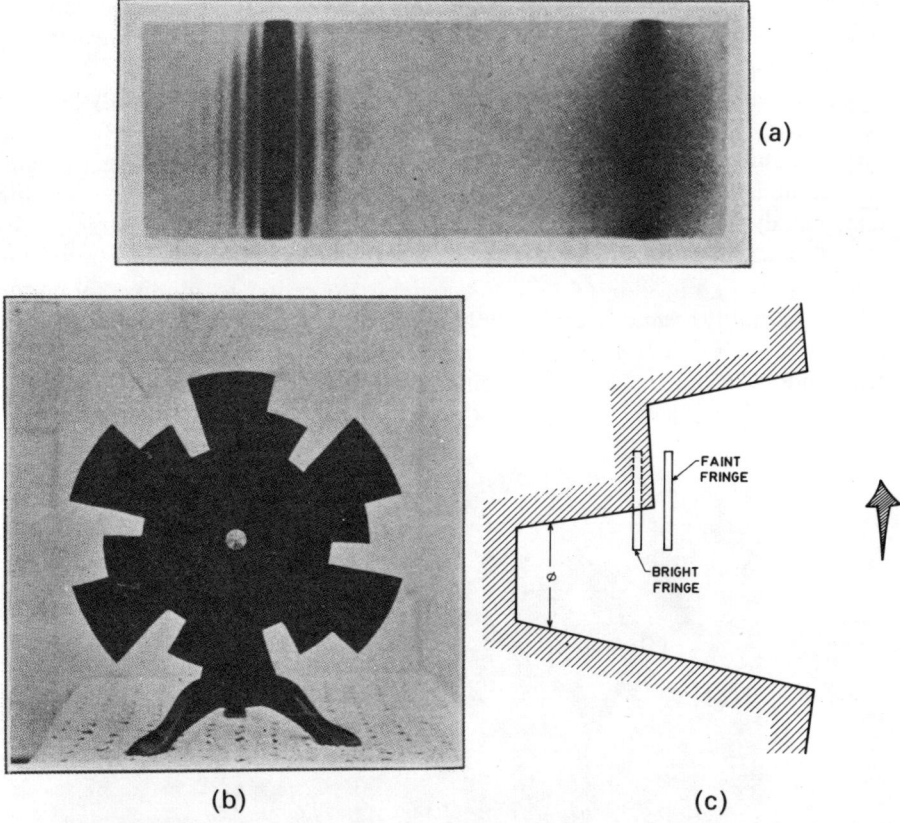

Figure 4.48 (a) Diffraction pattern observed by Raman. (b) Photometer disc used by Raman. (c) Scheme used by Raman to measure relative intensities. For explanations, see text. ((a) and (b) after ref. [4.40].)

cut out in them, as illustrated. The relative overlap of the discs can be varied as desired. To compare the intensities of the faint fringes with those of the bright ones, Raman first focused the diffraction pattern on the photometer disc so that the central bright band lay at the edge between the two annuli, and the two fringe systems were on either side. The disc was rapidly rotated with a motor and the appearance of the fringes on the two sides of the edge were compared. The overlap (i.e., the angle ϕ in Fig. 4.48) was then carefully adjusted so that the fringes on the two sides appeared *equally* bright. The ratio of the intensity of the dull fringe to that of the bright fringe is then equal to the ratio of the lengths of the chords of the small and large sectors. The experiment quantitatively confirmed Raman's analysis that the obliquity factor was indeed the one responsible for the observed anomalies in intensity.

While Raman made an experimental study of oblique diffraction by a rectangular aperture, his theoretical analysis also covered the case of oblique reflection from a rectangular slab. Mitra [4.41] pursued this question further, and examined both theoretically and experimentally the question of what happens when one has not one but two or more parallel reflecting surfaces.

Raman's analysis, while highlighting the importance of the obliquity factor, was nevertheless not rigorous enough. On the other hand, the systems studied experimentally by Raman and by Mitra were not easily amenable to rigorous mathematical analysis. Best suited for this purpose was of course the semi-infinite screen analysed by Sommerfeld; why not then investigate oblique diffraction by it? This is precisely what Mitra did, and some of the beautiful fringes he photographed can be seen in Fig. 4.49 [4.42]. Mitra then proceeded to modify Sommerfeld's analysis suitably to make it applicable to the case of oblique diffraction and in the process made an interesting discovery. Whereas Sommerfeld had found that the amplitude of the diffracted waves is different for light polarized in and at right angles

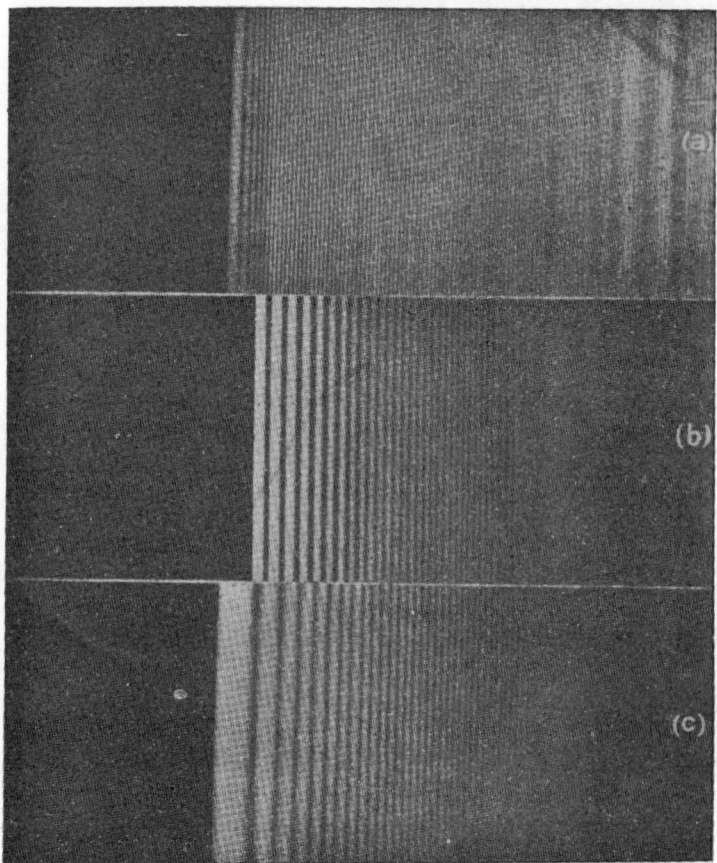

Figure 4.49 Diffraction fringes due to semi-infinite screen under conditions of oblique incidence. The different photographs correspond to different angles of incidence. (After ref. [4.42].)

to the plane of incidence, Mitra deduced theoretically that at oblique incidence this difference disappears, a finding which he confirmed by examining the fringes with a nicol. It was observed that the intensity of the fringes was independent of the plane of polarization of the incident light. Mitra also established photometrically that the ratio of the intensity of illumination at the various maxima and minima was in close agreement with his predictions.

Foucault's Test

Banerji also drew inspiration from Sommerfeld's memoir to extend the latter's analysis to apertures "of limited area and of specified form". Banerji's work is interesting for two reasons – firstly as it deals with diffraction at curvilinear boundaries, and secondly because it is a forerunner of modern Fourier optics. Of course Banerji himself was not aware of this, and he was merely following some earlier leads of Lord Rayleigh[15].

Long ago, Foucault recognized that good optical elements (e.g., lenses) must have a surface which is smooth to a fraction of the wavelength λ of the light employed. He also suggested a method to test for optical smoothness which Lord Rayleigh describes as follows:

> According to geometrical optics rays issuing from a point can be focused at another point, if the optical appliances are perfect. An eye situated just behind the focus observes an even field of illumination; but if a screen with a sharp edge is gradually advanced in the focal plane, all light is gradually cut off, and the entire field becomes dark simultaneously. At this moment any irregularity in the optical surfaces, by which rays are diverted from their proper course so as to escape the screening, becomes luminous; and Foucault explained how the appearances are to be interpreted and information gained as to the kind of correction necessary. [4.43]

Lord Rayleigh made a detailed mathematical analysis of the Foucault's test, considering for convenience a test object bounded by parallel straight edges. The object is illuminated with a fine slit parallel to the boundaries of the test object, and a knife-edge is placed at or near the focal plane and having its edge parallel to the boundaries of the object. It was known that when the illumination of the surface under observation is cut off by advancing the knife-edge in the focal plane, the edges remain bright. Repeating the experiment, Banerji found new phenomena not previously reported.

> I have found that as the knife-edge is gradually advanced in the focal plane, the surface lying between the boundaries does not continuously decrease in brightness but that the illumination of the entire surface undergoes large fluctuations, becoming alternately greater or less, finally tending, however, to zero when the knife-edge is advanced considerably. With white light some very remarkable colour-effects may be noticed. It is found that the boundaries of the surface appear luminous and white, but the region inside the boundaries shows colour, this being practically of the same tint throughout, but most marked midway between the

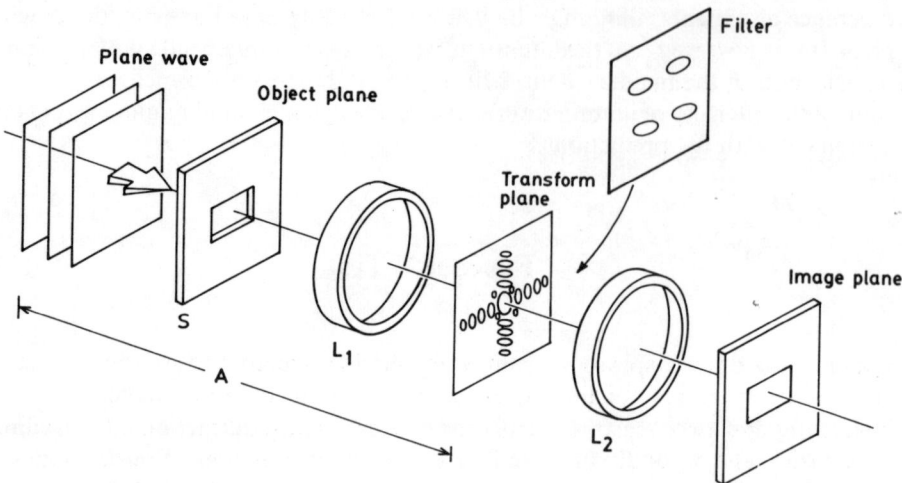

Figure 4.50 Geometry of Banerji's experiment. Without L_2, one would obtain a Fraunhofer pattern. Lens L_2 synthesizes an image. However, the image obtained could be doctored using the filter.

boundaries. The whole of the field between the boundaries passes through an interesting succession of colours as the knife-edge is slowly moved in the focal plane. The field outside the boundaries also shows a colour (though much less vividly) which is in general complementary to that observed between the boundaries. [4.44]

Banerji used Rayleigh's theory to explain the remarkable colour phenomena discovered by him[16]. Shortly thereafter, Banerji started his investigation of diffraction by "limited-area" apertures to which a reference has already been made [4.45]. The idea was to see if a Sommerfeld-type analysis could be extended to such cases.

The geometry of Banerji's experiment is illustrated in Fig. 4.50. The normal Fraunhofer experiment would involve only the components in part A in the figure. Banerji employed another lens L_2 to view the object. As he describes it, this was for viewing the aperture "solely by the diffracted light". One naturally expects to see an image of the aperture S on the screen. Besides a circular aperture, Banerji also experimented with apertures of polygonal shapes. The diffracted rays passed through a screen placed at or close to the focal plane where the diffraction pattern is formed. The screen had apertures in it and the image observed depended very much on the disposition of the openings in this screen. Banerji found that a remarkable transformation of the image occurred when the screen blocked out the region of maximum intensity in the Fraunhofer pattern. The image then formed was a delineation of the boundaries of the aperture, the area within the boundaries appearing more or less perfectly dark. However, the edges of the aperture did not

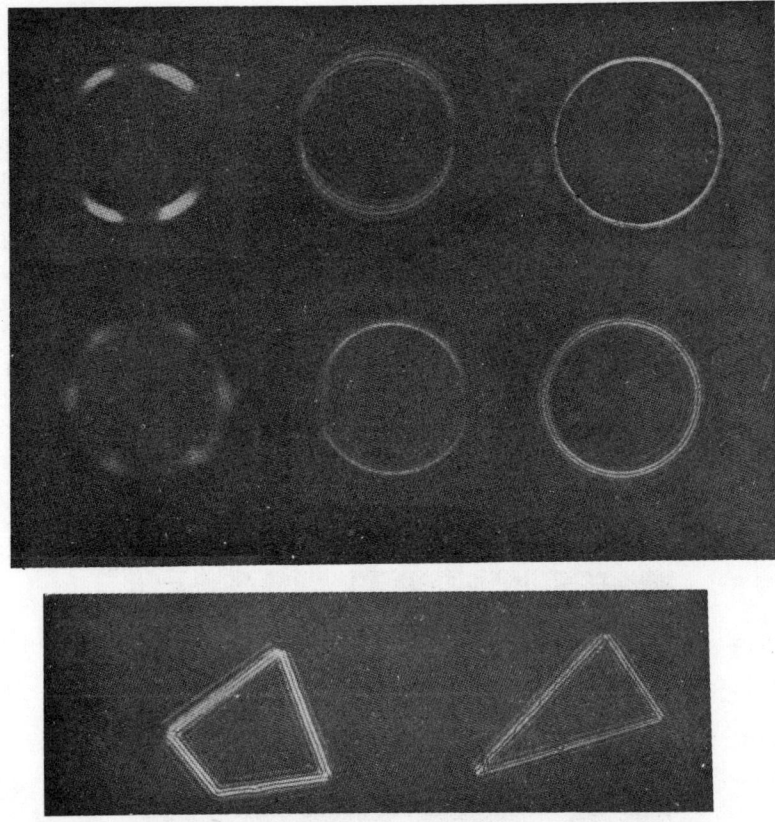

Figure 4.51 Some images obtained by Banerji. Observe the considerable manipulation achieved in the diffraction pattern of a circular aperture. (After ref. [4.45].)

appear as bright lines in the image so formed. On the contrary, they appeared as *perfectly dark lines* bordered by alternately bright and dark fringes on either side.

Strikingly different images were obtained by Banerji by manipulating the apertures in the screen (see Fig. 4.51), and these were painstakingly explained by him by drawing upon the works of Lord Rayleigh and Sommerfeld. Today of course one would crisply explain all his observations by saying that L_1 (see Fig. 4.50) produces a Fourier spectrum of the aperture S on the transform plane, which L_2 then synthesizes back into an image. The role of the screen is that of a filter. It then stands to reason that, depending upon what part of the Fourier spectrum is filtered out, variations would occur in the reconstructed image. Fourier optics is today a thriving field of activity, especially in image processing.

Curvilinear Apertures

An aperture of particular interest is one with an elliptic shape. It was known that the diffraction pattern of such a slit consists of a central spot of light surrounded by alternate dark and bright elliptic rings, the direction of these rings being rotated with respect to the aperture. What happens to the fringe system when one moves away from the ideal Fresnel geometry was, however, not known. Raman investigated the complete transition from the Fresnel to the Fraunhofer case (see Fig. 4.52), and made an interesting discovery, namely, the occurrence of a *diffraction caustic* [4.46].

Consider the curve S in Fig. 4.53a and let P be a point on it. Visualizing a small region near P to be a part of a circle, one may locate C the centre of the circular portion. The curve E is traced by C as the point P moves on S. Known as the *evolute*, E may also be regarded as the envelope of the normals to S drawn at various points along it, as illustrated in Fig. 4.53b. Raman found that if the eccentricity of the ellipse was large, there was a marked concentration of luminosity along four curves, which in fact constitute the evolute of the elliptic boundary (see Fig. 4.54). This line of luminosity is also referred to as a diffraction caustic.

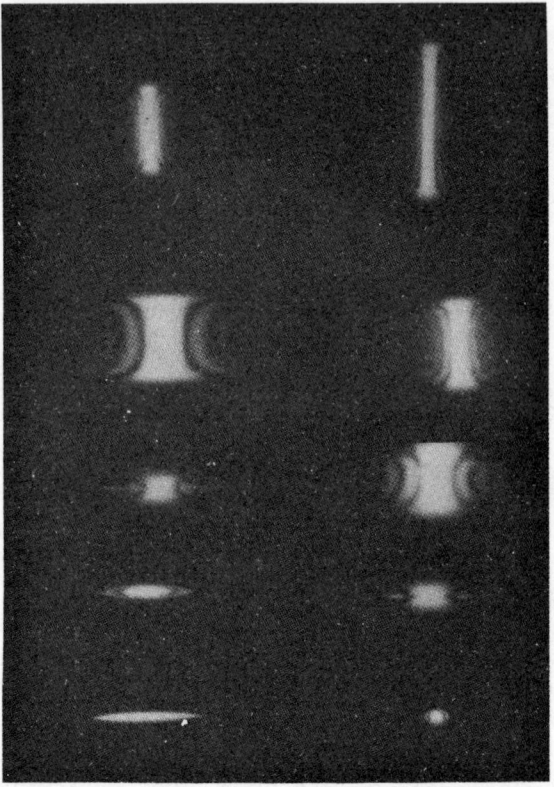

Figure 4.52 Diffraction fringes due to an elliptic aperture. (After ref. [4.46].)

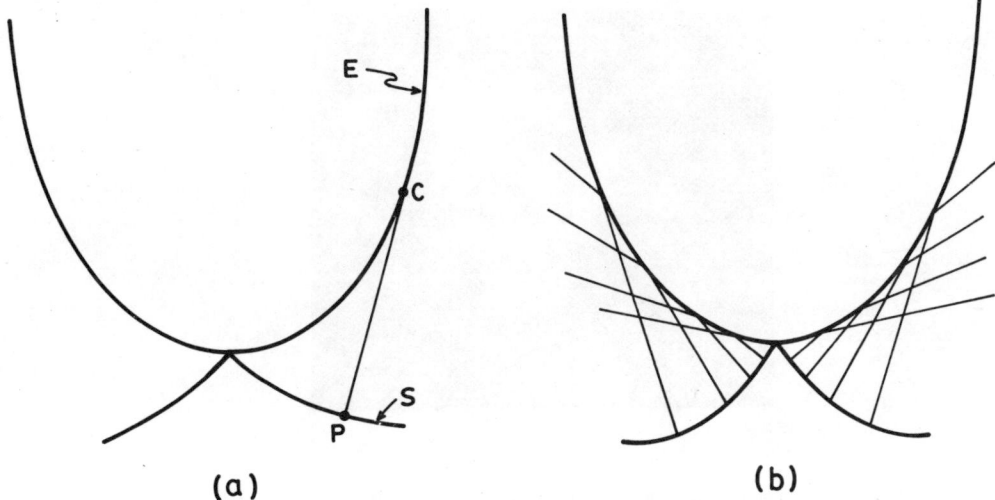

Figure 4.53 Concept of a caustic. For explanations see text.

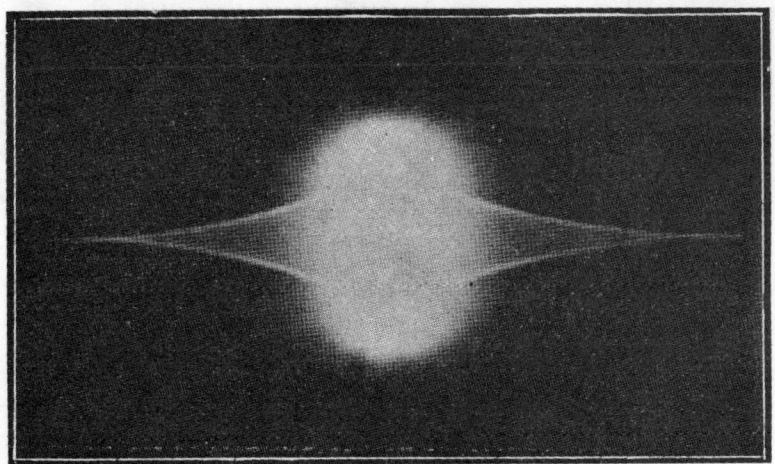

Figure 4.54 Diffraction caustic associated with an elliptic aperture. (After ref. [4.46].)

Mitra pursued the subject further [4.47], obtaining many beautiful patterns, as in Fig. 4.55. Diffraction caustics are strikingly exhibited in the shadow of a disc with an undulatory margin. Two photographs taken with a nickel one-anna coin are reproduced in Fig. 4.56 where the picture taken with a longer exposure clearly reveals the caustics. Their geometric form is that of the evolute of the undulating boundary.

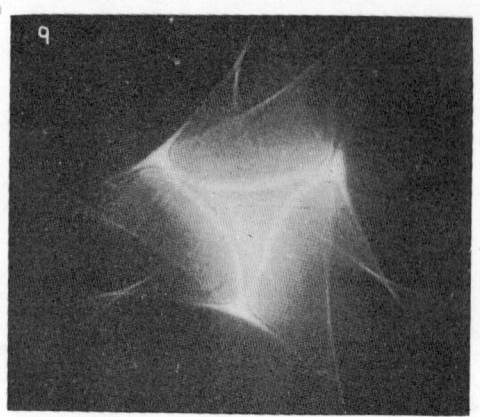

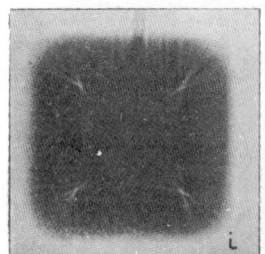
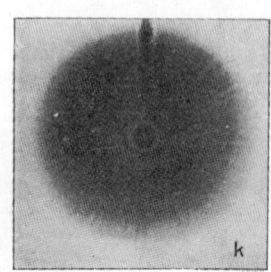
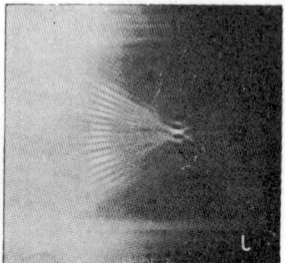

Figure 4.55 Some of the diffraction caustics observed by Mitra. (After ref. [4.47].)

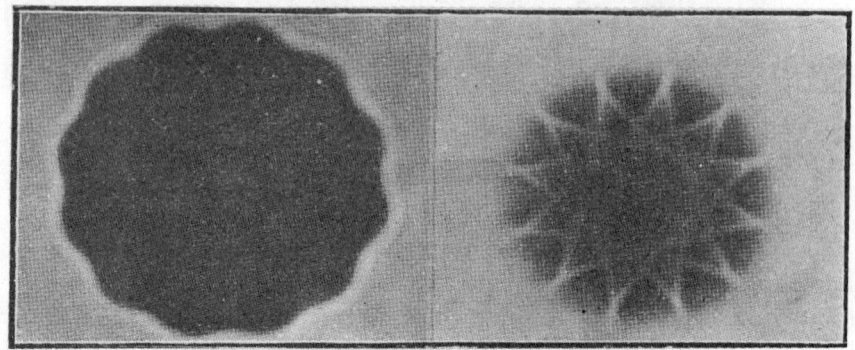

Figure 4.56 Diffraction caustics observed with a one-anna coin. (After ref. [4.66].)

Curved Surfaces

Diffraction from curved surface also received attention from various workers at the IACS. Earlier, Brush had examined the diffraction of light by the edge of a cylindrical obstacle. He observed the fringes formed within a few millimetres of the diffracting edge through a microscope, and found that they appeared brighter and sharper with every increase in the radius of the cylinder. To account for these phenomena, Brush suggested that the cylindrical diffracting surface may be regarded as consisting of a great many parallel elements, each acting as a diffracting edge and producing its own pattern, the observed fringe system being a superposition of such "elementary" patterns. However, Brush made no attempt to substantiate his conjecture by mathematical analysis.

Raman disagreed with Brush's speculations and asked Basu to explore further. For a start Basu [4.48] made observations of his own (see Fig. 4.57) and found that the phenomenon observed depended on the position of the focal plane with respect to the diffracting edge of the cylinder and that an interesting sequence of changes occurred as the focal plane was gradually moved towards the light up to and beyond

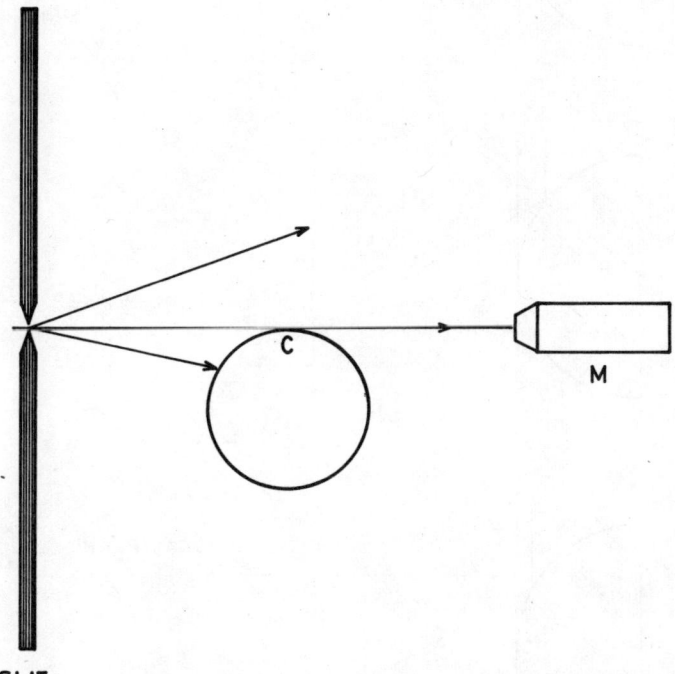

Figure 4.57 Light from a slit falls on a polished cylinder of glass and passes it tangentially. The axis of the cylinder is parallel to the slit. The fringes bordering the shadow of the edge are observed through a microscope objective M. Instead of glass, a polished metallic cylinder can also be used.

the edge (C in Fig. 4.57). This gave him a clue regarding how to construct an explanation.

Basu's viewpoint may be appreciated by considering Fig. 4.58. When the focal plane coincides with the edge of the cylinder as in Fig. 4.58a, the fringes may be regarded as formed by simple interference between the light that passes the cylinder unobstructed and the light that suffers reflection at the surface of the cylinder at various incidences. For instance, at the point P, there can be interference between the direct wavefront coming along RP and the reflected one coming along QP. If the fringes are observed in a plane which is farther from the source of light than the edge of the cylinder as in Fig. 4.58b, then, besides the mutual interference of the direct and the reflected rays, the effect of the diffraction at the edge must also be taken into

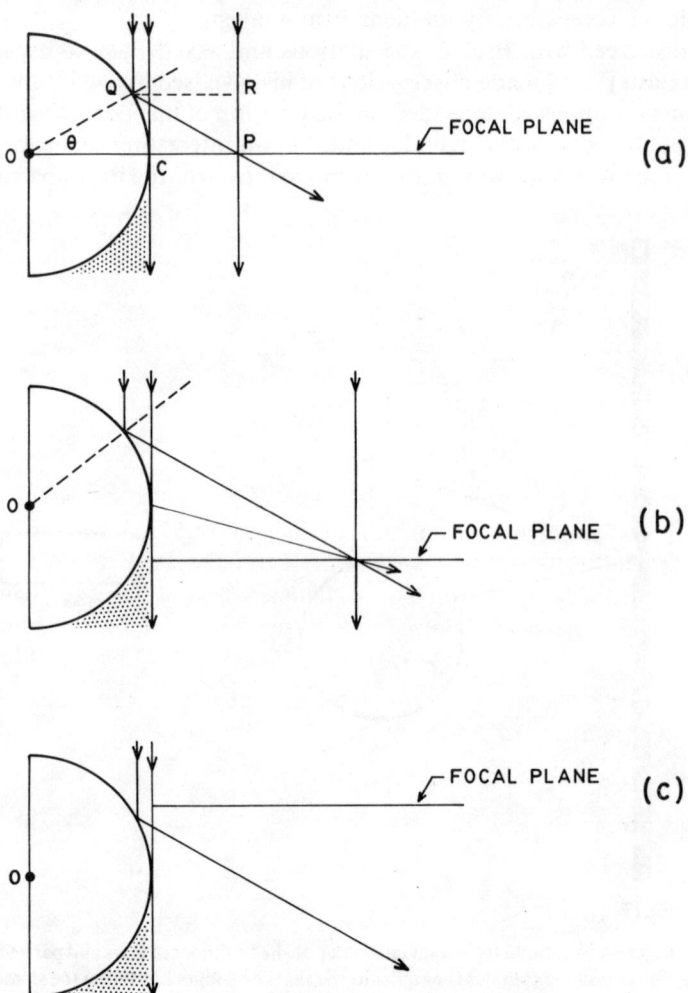

Figure 4.58 Illustration of Basu's analysis; explanations in the text.

account. On the other hand, if the focal plane is sufficiently towards the light as in Fig. 4.58c, the direct and reflected rays may not exactly cover the same part of the field and may even be entirely separated. Basu was able to quantitatively verify many aspects of his theoretical analysis.

Chinmayanandam [4.49] continued where Basu left off. Basu's analysis had, for example, indicated that the positions of the dark bands would vary with position, as in Fig. 4.58. But what about the *intensity* distribution in the fringe patterns? Chinmayanandam addressed himself to this question, and checked his theoretical estimates by direct photometric measurements. Interestingly, he used a polarization technique instead of the rotating disc method favoured by others. Chinmayanandam also made calculations about the actual flow of energy around the cylinder.

Sethi [4.50] too studied diffraction by a cylinder although he came to the problem by a different route, having been stimulated by some spectacular effects he had observed analogous to the Christiansen experiment (concerning the latter, more will be said in Chapter 12). Sethi immersed a glass cylinder in a parallel-sided cell containing a mixture of carbon disulphide and benzene whose refractive index could be varied. The cylinder and the cell were kept at a distance of about half a metre from a narrow slit which was illuminated with a powerful source of monochromatic light. The source was then viewed through the cylinder, whereupon a long band of light was seen extending to very large angles on either side and broken up into a number of fringes, the nature of which depended on whether the refractive index μ of glass was greater or less than μ' the index of the liquid mixture surrounding the cylinder. Sethi observed that when μ was greater than μ', there were interesting effects analogous to the rainbow.

One might wonder if diffraction from a sphere was studied; indeed it was. It was already quite well known that at the centre of the shadow of a spherical obstacle cast by a small source of light there is a bright spot similar to that found in the shadow of a circular disc. However, it was generally assumed by experimentalists that a sphere and a disc of equal radius would give practically identical results as far as the central spot was concerned. Raman and Krishnan [4.51] showed that contrary to such widely held notions, there were notable differences between the two cases.

They mounted side by side on a glass plate a spherical ball and a circular disc of equal radii, so that the bright spots at the centre of their shadows could be seen at the same time. The diffraction patterns within the shadows of the disc and the sphere were seen simultaneously through a lens of sufficiently wide field of view. It was noted that the general illumination within the geometrical shadow was much greater for the disc than for the sphere. A quantitative study of the relative intensities of the central white spots of the two diffraction patterns was made by the usual rotating photometer method, and the results obtained may be seen in Fig. 4.59.

Raman and Krishnan had a ready explanation for their observations.

> In the case of the disc the rays diffracted by the illuminated edge reach the point of observation directly. In the case of the sphere, however, the position is somewhat different. Drawing tangent cones enveloping the spherical obstacle, with the source and the point of observation as apexes [see Fig. 4.60], we see that they now touch the sphere at different circles of contact, X and Y respectively. Thus, the circle of

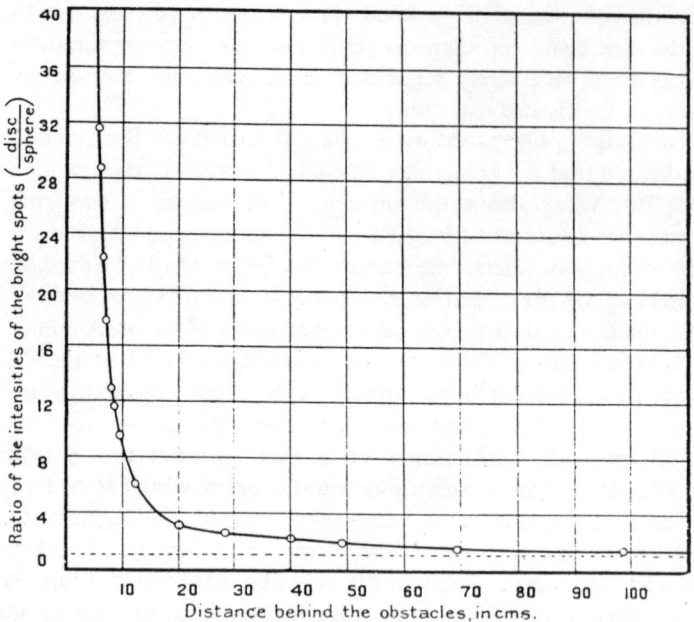

Figure 4.59 Comparison of the intensities of the central bright spots obtained with a disc and a sphere. (After ref. [4.51].)

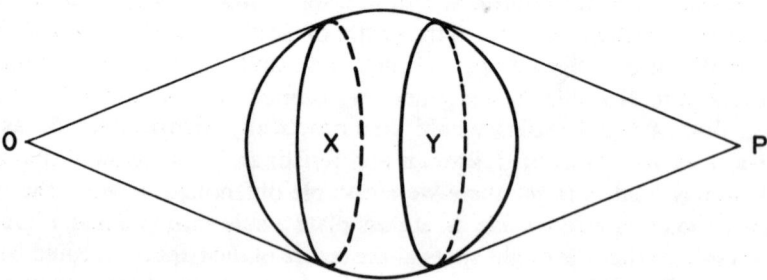

Figure 4.60 Diagram to amplify certain points related to diffraction by a sphere.

contact Y, from which diffracted rays originating at the surface can reach the point of observation, lies within the region of geometrical shadow, and not at its edge, as in the case of the disc. The disturbance incident on the surface of the sphere has to creep round it, as it were, over the arc XY before the rays diffracted out by the sphere can reach the point of observation, and must suffer a very considerable diminution in the process. Thus, we can see that the intensity of the central white spot in the diffraction pattern of a sphere will be less than in the case of the disc at the same distance behind by a quantity depending on the length of the arc XY between the circles of contact of the tangentially incident and diffracted rays. [4.51]

Glimpses from the Golden Era 149

Raman and Krishnan offer some more comments but recognize that their discussion is purely qualitative. Why did they not attempt a mathematical analysis? Even if Raman was impatient, was there not Krishnan who enjoyed such exercises? The year of the publication perhaps gives us a clue. Both Raman and Krishnan were very busy at that time, with light scattering occupying much of their attention. Besides, Raman also had the *Handbuch* article on his hands.

Although Raman and Krishnan do not offer a mathematical analysis of their own, they nevertheless make some pertinent observations:

> This problem [diffraction from a large sphere on the basis of electromagnetic theory] has been handled by Poincaré, Nicholson, Macdonald, Bromwich, G. N. Watson and others. The paper by Macdonald, on "The diffraction of electric waves round a perfectly reflecting obstacle", might in particular be referred to, as the analysis contained in it approaches most closely to the point of view from which we have explained our experimental results. The formulae given by Macdonald are, however, not in a form capable of immediate application to the problem without considerable labour.

As a hint of their preoccupations they also add,

> As the experimental work was completed last summer, and as we are at present engaged on other work, we have thought it best not to defer publication of the results any longer.

Sommerfeld Stretched

In 1886 Gouy discovered that when a metallic screen with a sharp and highly polished edge, e.g., a razor blade, is held in a beam of light, its boundary appears as a luminous line when viewed from within the region of shadow or from the region of light. The light diffracted by the edge is strongly polarized, but in perpendicular planes in the two regions. Gouy experimented with edges of various metals and discovered that both the colour of the light diffracted into the shadow and its state of polarization depended in a remarkable way on the material of the edge and on the extent to which it had been rounded off in the polishing process.

Gouy's results were discussed by Poincaré in two memoirs published in *Acta Mathematica*. Poincaré also dealt with the special case of an ideal screen, a problem subsequently solved exactly by Sommerfeld (reference has already been made to this famous work). However, all these discussions bypassed an essential element of Gouy's discovery, namely, the role played by the *material* of the edge. In 1927 Raman and Krishnan decided to remove this lacuna [4.52].

Consider a semi-infinite screen as in Fig. 4.61, and let light be incident on it at an angle ϕ as illustrated. The edge ray AO in the figure clearly defines three regions. Accordingly, three kinds of waves must be considered: (i) plane waves travelling towards the edge, (ii) waves reflected from the upper surface of the screen and receding from it, and (iii) cylindrical waves radiated from the edge of the screen. One

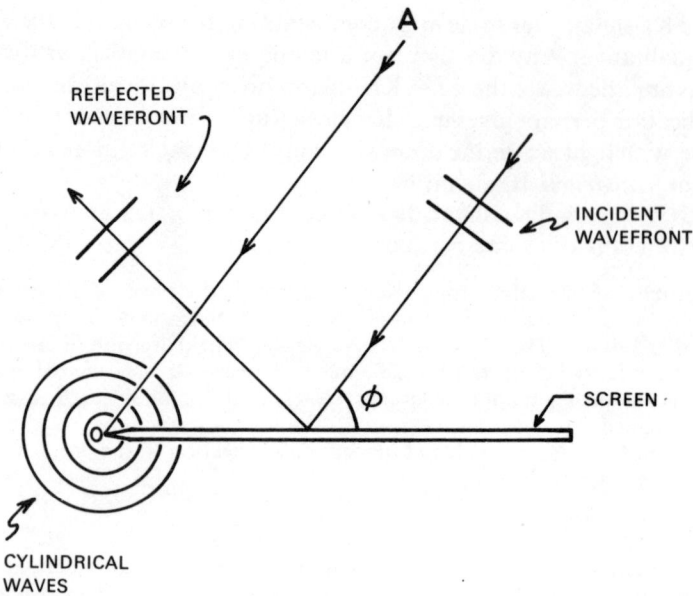

Figure 4.61 Features associated with diffraction by a semi-infinite screen.

must also take into account the fact that the electrical vector in the incident wave could either be parallel to the edge or be perpendicular to it. The effects observed by Gouy are the result of a complex interplay between all these different waves. In the treatment of Raman and Krishnan, the material properties determine the amplitude of the reflected waves; Sommerfeld assumed that the reflector was perfect, i.e., that the screen was a perfect conductor. The effect of imperfect conductivity is to make the intensity of the diffracted light less and less as the incidence becomes more and more oblique, i.e., as ϕ tends to zero. This diminution is least for wavelengths for which the reflection coefficient is largest. The colour effects noted by Gouy arise in this way, becoming more prominent at oblique incidences.

Raman and Krishnan studied in detail the diffraction from steel and gold edges under varying conditions of incidence, with due regard to the optical properties of these materials. Their analysis succeeded in explaining various observed phenomena then known, including polarization effects. The theory of Raman and Krishnan was put to a rigorous test by Savornin in France about a decade later, and was found to be entirely satisfactory.

Colours of Heated Metals

It is a matter of common experience that characteristic colours appear on a metallic surface when it is heated. The explanation usually put forward was that the

Glimpses from the Golden Era

colours were due to the interference of light reflected at the surface of a thin oxide film formed on the surface of the metal. Rejecting this view, Mallock argued instead that the colour was due to the material property of the oxide. Raman disagreed with Mallock, being of the opinion that the colours seen were basically due to an optical phenomenon, though not of the interference type as usually proposed. As he puts it, "The colours under discussion are in the nature of diffraction effects arising from a film which is *not continuous* but has a coarse-grained structure." Raman also described some observations of his own in support of this viewpoint [4.53].

A more detailed investigation was later undertaken by Chuckerbutti [4.54]. His first problem was that of obtaining highly polished plates. Chuckerbutti found that direct heating in a Bunsen flame, as one would normally tend to do, was not acceptable since it resulted in non-uniform films in addition to a change in coloration that was too fast for documentation. A scheme for slow and controlled heating had therefore to be devised; the phenomenon could then be observed conveniently, and he describes it as follows:

> In each case, colour starts at about a reddish or violettish tint which is rather difficult to distinguish on account of the surface colour of the metals. Next, the colour turns to violet in the case of copper, and indigo in the case of iron. Copper being further heated shows indigo, green and yellow in succession. On still further heating, rather at a high temperature the colour is almost white and then again it starts from red and ends in green and yellowish green. Further heating blackens the plate due to the formation of black oxide. The colours exhibited at this repetition are very rich and gorgeous. In this stage the formation of the surface-structure becomes visibly discontinuous and granular.... [4.54]

Chuckerbutti made quantitative studies, starting with a measurement of the

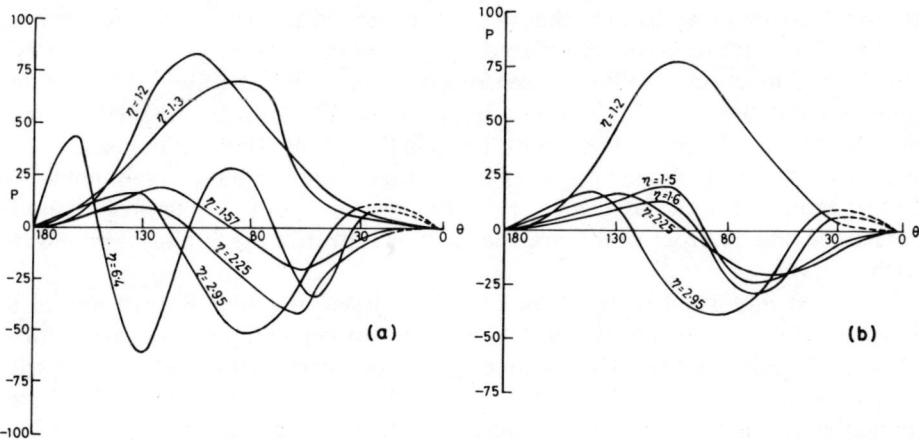

Figure 4.62 (a) Experimental data for heated copper plates. Light is incident obliquely, and what is plotted is the polarization as a function of the scattering angle θ. The latter is so defined that $\theta = 0$ corresponds to back scattering. η is a parameter characterizing the size of the particles formed on the surface by the heating process. (b) Theoretical curves. (After ref. [4.54].)

diameters of the particles forming the surface. Examining next the optical properties, he found that

> if a beam of white light be allowed to fall upon one of the [heated] metal plates... then the colour and polarization of the reflected light vary with the angle of incidence and the thickness of the film upon the surface in a very remarkable way.

Chuckerbutti also found that the metallic surfaces scattered light, and, what was more interesting, the colours of the scattered light were *complementary* to the colour of the reflected light. There were also strong polarization effects, and Fig. 4.62a shows some data for copper. In interpreting these results, Chuckerbutti assumed that all the observed effects can be understood in terms of the scattering by spherical particles (of the material concerned). Drawing upon earlier work by Lord Rayleigh, Mie and J. J. Thomson on scattering by spheres as well as published data on the optical properties of metals, Chuckerbutti succeeded in obtaining trends as in Fig. 4.62b which qualitatively compare quite favourably with those he had measured.

Coronae and Halos[17]

Diffraction by simple objects (especially of a geometrically well-defined shape) is of particular scientific interest since the problem is often amenable to theoretical analysis. On the other hand, many natural phenomena involve diffraction by a great many particles or obstacles simultaneously. The phenomena observed in all such cases may be included under the general description of "diffraction halos". At the practical level one distinguishes two subcategories, namely, halos and coronae.

Historically, the term corona referred to rainbow-coloured rings, usually of only a few degrees in angular width, concentrically surrounding the Sun, Moon or any other bright object when covered by a thin veil of cloud. Coronae owe their origin to minute spherical droplets of water contained in the clouds. Thin clouds consisting of small particles of crystalline ice also exhibit rings known as halos surrounding the Sun or Moon when the latter are seen through them. Coronae differ from halos in having the reverse order of colours; that is, blue is nearest the Sun say, and red the farthest away.

The most important work on coronae was done only after Raman moved to Bangalore but already in Calcutta the problem was beginning to receive attention. Mitra [4.55] obtained for the first time a series of corona photographs taken with droplets of uniform size. Observing the coronae with monochromatic light, Mitra found that the patterns he obtained were very different from the diffraction patterns due to opaque spheres. The effect of waves transmitted through the water droplets was clearly important but it was only much later that G. N. Ramachandran solved the problem completely (see Chapter 8).

Very beautiful phenomena are shown by heterogeneous films known as mixed plates. Spread over a period of seven years, five papers emerged on mixed plates in three of which Raman was an author. Raman describes how mixed plates may be prepared and what they exhibit. To obtain them, a few drops of egg albumen are spread between two plates of glass about ten centimetres square in size and a centimetre thick. The plates are then separated and put back together a few times and slid over each other with a circular movement. The material is thus worked up into a film of uniform thickness which, when seen under the microscope, appears as a thin layer of liquid enclosing a layer or air-bubbles. These vary in size and are irregularly arranged and sometimes depart from a circular shape, but, except in special circumstances, show no bias towards elongation in any particular direction.

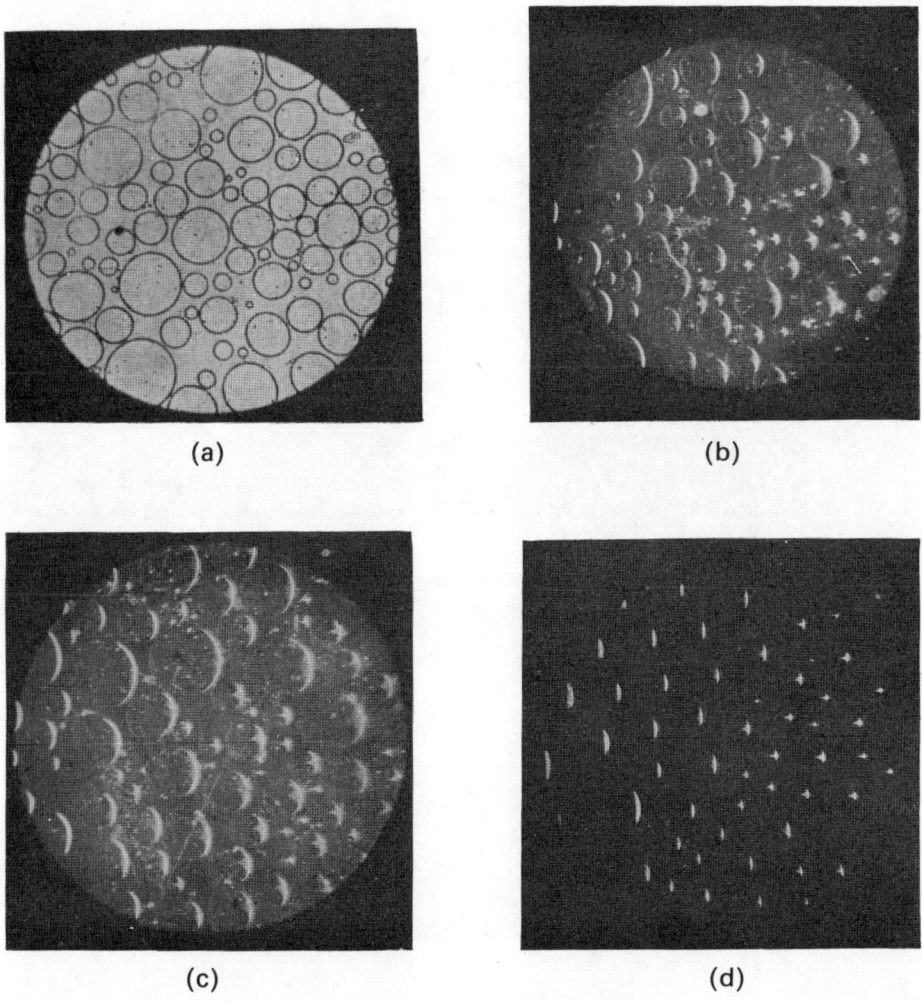

Figure 4.63 Structure of mixed plates observed (a) in normally incident light, and (b)–(d) at different obliquities. (After ref. [4.56].)

Gorgeous colours are shown by such films when they are freshly prepared and are not too thick. On being allowed to stand, the albumen in the film begins to dry up and forms hexagonal networks between the two plates. The character of the optical phenomena then completely alters.

Figure 4.63a shows the structure of mixed plates in direct light incident normally on the plates [4.56]. The edges of the air-bubbles appear dark while the rest of the field of view is illuminated. These edges reflect and diffract the light incident on them in directions different from that of the incident light, thus allowing no light to be directly transmitted through them. Hence they appear dark in direct light, while the remaining field of view appears illuminated. Figure 4.63b, c, d show the structures when plates are viewed at increasing angles from the direction of the incident light.

If the glass plates are pressed together and continuously moved over each other in any one direction, the bubbles in the film become distorted, assuming elliptical shapes as in Fig. 4.64a, the major axis being in a direction perpendicular to that of

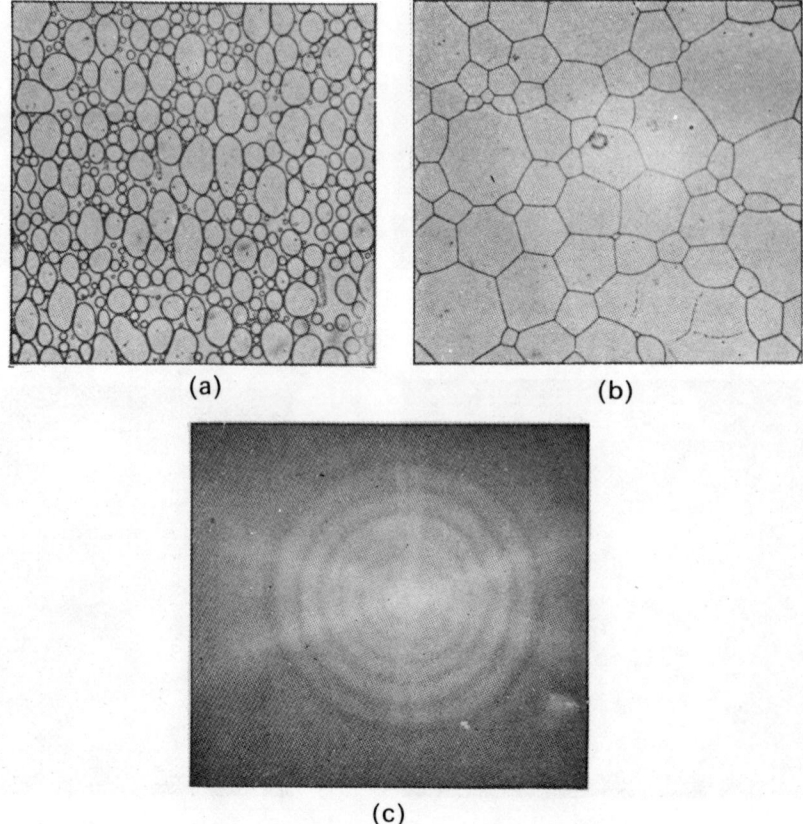

Figure 4.64 Mixed plates under different conditions of preparation. For explanations, see text. (After ref. [4.57].)

the movement of the plates (compare with Fig. 4.63a). When the mixed plate is allowed to stand for some time the bubbles start joining along the edges. After a day or two when the film has completely dried up, the albumen is confined to a number of very fine ridges, forming a random network as in Fig. 4.64b [4.57].

Optical effects in a mixed plate are studied by holding it *normally* in front of the eye, and viewing a distant point-source of light through it. Observations were made by Raman and others with both monochromatic and white light. The characteristic feature observed by them which demands explanation is the diffraction halo, one example of which is shown in Fig. 4.64c. When viewed in monochromatic light, the halo consists of a series of circular rings, alternately bright and dark, which concentrically surround the source. The rings are narrowest near the centre of the halo and widen as one proceeds towards its outer margin. If white light is used, the outermost ring is practically achromatic and is followed by coloured rings. A thick plate shows numerous close rings, while a thin plate shows fewer rings which are wide apart. The rings move inwards when the thickness of the film is reduced. Thus, the thinner the plate, the more striking are the colours shown by the rings nearest the centre of the halo.

All the effects mentioned above can be understood by considering in detail the propagation of light through the medium. Figure 4.65 represents a section of the mixed plate system, showing in particular the profile of the liquid meniscus. For convenience, the angle of contact of the liquid with the glass plate is taken to be zero. One can distinguish at least five types of rays: (1) rays which pass right through the air-bubble, (2) rays which are twice refracted at the liquid-air boundary, (3) rays which are totally reflected into the liquid, (4) rays which pass right through the

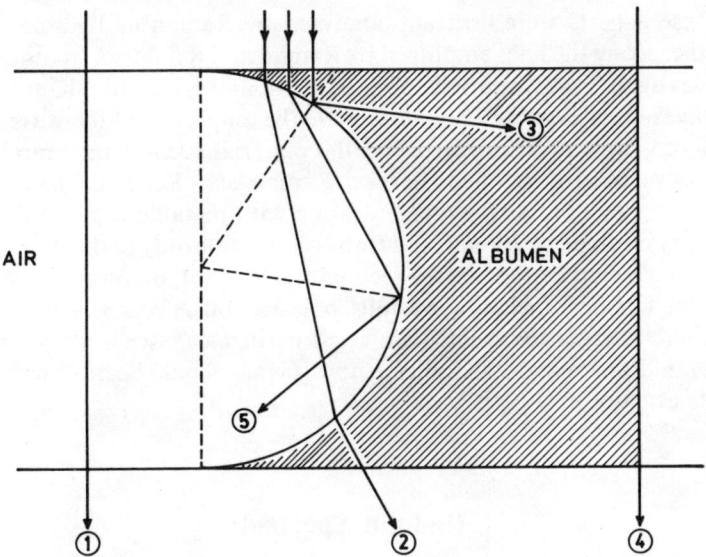

Figure 4.65 Section of a mixed plate, and the passage of various rays through it. For explanations, see text.

liquid, and (5) rays that are refracted once and thence reflected. In Fig. 4.65, the last-mentioned rays diverge to the left. If one now systematically explores various possibilities for interference amongst these five types of rays, one can not only account for all the observed phenomena qualitatively but also quantitatively. Naturally the peculiarities under conditions of oblique incidence were not overlooked!

Striae in Mica

As is well known, mica has a perfect cleavage and the sheets obtained exhibit a remarkable uniformity of thickness. However, on examining even the most regularly split and transparent pieces of mica by diffuse reflected light, a few fine, hair-like and rather irregular lines may generally be seen running along the surface. The sheet as a whole, being optically good, remains invisible, but the striae shine out as brilliant and vividly coloured lines of light. Raman and Ghosh who studied this phenomenon observe that

> a stria at normal incidence may appear crimson and, as the mica is rotated about an axis in its own plane, become successively purple, green, yellowish-green, yellow, orange, scarlet-red, green, yellow and red. [4.58]

Single striae are extremely sharp laminar edges and are therefore well suited for a critical study of the diffraction effects due to such edges. They diffract light through large angles, exhibiting colour and polarization effects which in some respects are analogous to and in other respects differ from those observed with sharp metallic edges. These aspects were critically analysed by Raman and Ramakrishna Rao [4.59] rather along the lines employed by Raman and Krishnan. In other words, the take-off was again from Sommerfeld's theory but in this case the shadow region can receive waves *transmitted through the screen*, the amplitude of the wave so received depending on the optical properties of the material even as the amplitude of the reflected wave is so controlled. Earlier, Raman and Krishnan had ignored the transmitted component as they were dealing with metallic edges.

The theory outlined by Raman and Rao is applicable only to the thinnest laminae, a restriction which they themselves recognize. However, it succeeded in explaining the essential features, including various polarization effects which, according to them, "would otherwise be unintelligible". Striae in mica often have multiple steps as in an echelon. Raman and Rao indicate how this case could be dealt with but do not themselves attempt it.

Radiant Spectrum

Closely related to halos and coronae is the phenomenon of "radiant spectrum". When a very small and intensely luminous source of light is viewed directly by the eye against a dark background, an enormous number of luminous streamers appear to emerge from the source and stretch out from it in radial directions. Illuminating a pin-hole with an electric arc and viewing it from about three to four metres, one would see the following:

> The luminous pin-hole appears surrounded in the first instance by a circular patch full of luminous streaks starting out more or less radially from it, and occasionally crossing each other. These streamers are generally white, but appear here and there tipped with streaks of colour. The circular patch is surrounded by a relatively dark ring, outside which again the streamers reappear passing radially through a luminous coloured halo surrounding the dark ring. The halo is, in fact, made up of short sections of the streamers which, here, are strongly coloured. Outside the halo the streamers emerge again, but are much fainter, and they form a broad and somewhat ill-defined ring of luminosity extending to a considerable angular width from the source. The inner margin of this luminous ring is greenish-blue, and the outermost visible periphery is of an orange-red colour, but some fainter fluctuations of luminosity and colour may be observed within it. [4.60]

With monochromatic light, radial streamers are *not* obtained.

> We see, instead, a circular area round the source filled with *granular* patches of light, and outside this, a relatively dark ring, followed by a well-defined circular halo, and some faint outer rings of luminosity.

Helmholtz had earlier studied this phenomenon and was of the opinion that the streamers were due to diffraction of light at the irregular margin of the pupil of the eye. Raman agreed it had something to do with the eye, but was not sure of Helmholtz's explanation. He therefore made some studies of his own. He records:

> The phenomenon being of a subjective character, the assistance of a number of independent observers with normal vision was obtained in order to confirm my personal observations. This appeared all the more necessary, as the conclusions arrived at as to the origin of the phenomenon differ from those of Helmholtz.

Raman found that the angular diameters of the circular patch containing the streamers and of the halos surrounding it were entirely independent of the aperture of the pupil of the eye. This was checked by effecting a considerable variation of the intensity of the light. Under these circumstances, the aperture of the pupil would have varied substantially, and if the aperture of the pupil was the controlling factor then the angular diameters of the halos would also have shown much variation, which, however, was not the case.

Raman made many tests and reported them in a paper published in *Philosophical Magazine* [4.60]. A short note on the subject was also sent to *Nature* [4.61] which drew an objection from one Hartridge but that was easily disposed of. Nevertheless the question remained as to what actually caused the streamers. A firm answer to this was furnished by Raman in a paper read by him before the Royal Society of Edinburgh in 1921 [4.62], in which he convincingly argued that the observed phenomena were due to diffraction by *randomly distributed* structures within the eye, possibly corneal corpuscles. In support of this hypothesis, Raman cited the type of halos one obtains when a glass plate is dusted with lycopodium powder. As we shall see in Chapter 8, this was the forerunner of the famous speckles phenomenon.

Haidinger's Rings

Haidinger's rings are interference patterns observed between plane parallel surfaces under diffuse monochromatic illumination. A scheme for observing them is shown in Fig. 4.66. The natural cleavages of crystals like mica enable one to obtain transparent plates with good optical surfaces well suited for the observation of these fringes. In fact Haidinger himself was the first to observe such rings in mica. Subsequently Lord Rayleigh made a brief study and pointed out the desirability of investigating the influence of the birefringence of mica on the ring system. This is precisely what Chinmayanandam did [4.63]. Using a good specimen from Burma obtained through the courtesy of Dr Hayden, FRS, of the Geological Survey of

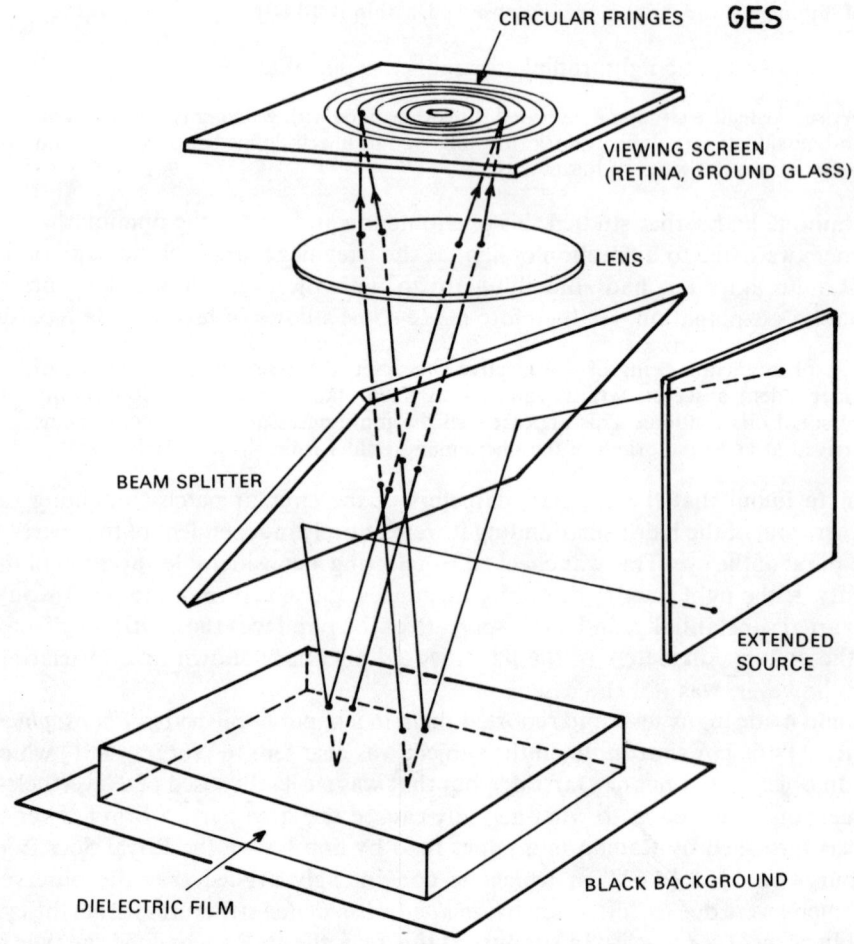

Figure 4.66 Arrangement for the observation of Haidinger's rings.

Glimpses from the Golden Era

Figure 4.67 Haidinger's rings in mixa. (After ref. [4.63].)

India, Chinmayanandam obtained excellent photographs of the ring system, an example of which is shown in Fig. 4.67. Some innovations were necessary to obtain pictures of good quality.

On an examination of Fig. 4.67 it will be found that the rings do not appear everywhere with the same brightness and that their visibility is a minimum along four arcs of roughly hyperbolic form. This was a new finding, and Chinmayanandam explained the occurrence of these arcs as follows. Any ray of light upon entering the crystal would be split into two rays (recall Fig. 4.47). Correspondingly, there would in reality be two ring systems. For a point to be one of minimum visibility, it is clear that it must *simultaneously* belong to the minima of *both* the ring systems. The problem of finding the loci of minimum intensity is now mathematically well defined. And when Chinmayanandam solved it, he got exactly the pattern he actually observed.

Quetelet's Rings

Today's students of physics would most probably have never heard of Quetelet's rings. They are somewhat similar to the more familiar Newton's rings, and Raman describes them for us:

> When a distant point-source of light is viewed by reflection from a plane mirror silvered on the back, the scattered light surrounding the reflected image of the source exhibits a system of coloured rings, the brilliancy of which is greatly enhanced by purposely dimming the front surface of the mirror, as, for instance, by breathing upon it. [4.64]

These rings are referred to as Quetelet's rings.

As always, Raman was interested in the case of oblique incidence. Sir George Stokes had earlier attempted to see the rings under this condition, but was not successful. Raman overcame the difficulty as follows. He first took a front-silvered plate of glass. Next he coated a second glass plate with a scattering film; a layer of

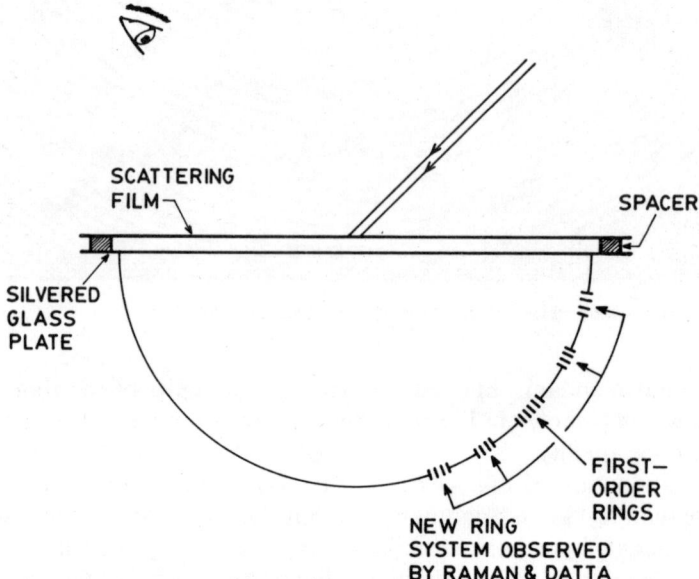

Figure 4.68 Observation of Quetelet's rings.

ammonium chloride deposited by volatilization produced the desired uniform surface. The silvered surface and the coated surface were then placed face to face, with thin strips of paper or mica at the corners to act as spacers. In this manner, very small separations between the mirror and the scattering film could be obtained. Gorgeously coloured rings were seen when the arrangement was held close to the eye and a small source of light was viewed by reflection.

Raman was particularly interested in the patterns obtained with white light at oblique incidence. He found that besides the usual system of rings, there were several other fringe systems.

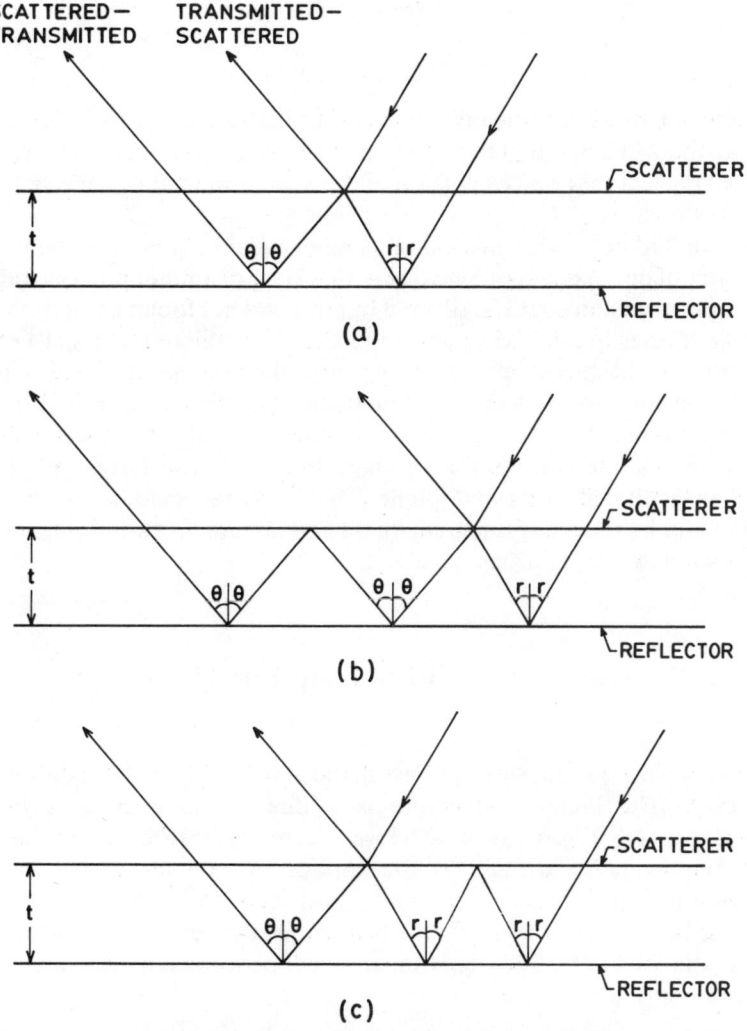

Figure 4.69 Origin of interference leading to Quetelet's rings. (**a**) The case considered by Stokes; (**b**) and (**c**) examples of additional cases considered by Raman.

There was already a theory for the rings put forward by Stokes, Lommel and Exner, but it could explain only what is referred to in Fig. 4.68 as first-order rings. The other ring systems are not addressed by it. The basic idea behind the theory of Stokes is illustrated in Fig. 4.69a. In essence, the rings appear owing to the interference of two sets of rays, one of which is scattered at entry and the other at exit, the interference condition being given by the equation

$$2t(\cos r - \cos \theta) = \pm n\lambda.$$

Raman argued that there are other possibilities, for example, as in Fig. 4.69b and c, governed by the equations

$$2t(\cos r - 2\cos \theta) = n\lambda$$

and

$$2t(2\cos r - \cos \theta) = n\lambda$$

respectively. By systematically following through all such possibilities, Raman was able to obtain a consistent explanation of all his observations. The theory of Stokes ignored the fact that a *given* portion of the wavefront could be affected *both* at entry and at emergence.

Raman had noted that instead of using two glass plates, one might as well use a thin sheet of mica coated on one side with a layer of ammonium chloride. The other side of the mica film could be silvered but this was not found necessary. The study of Quetelet's rings in mica was pursued further by Sethi and Sogani [4.65] who were interested in the special effects arising from the double-refracting nature of mica. Apart from the ring system, they found, as Chinmayanandam had in his study of Haidinger's rings, hyperbolic lines of minimum visibility. At oblique incidence, these loci of minimum intensity varied in shape and size, particularly when the mica sheet was slowly rotated in its own plane. Thanks to the spadework already done by Raman and by Chinmayanandam, Sethi and Sogani had no problem in explaining their own new observations.

Black Soap Films[18]

It was known from the work of Perrin and several others that stratified soap films can exist. The lamellar structure is a direct consequence of the molecular arrangement (see Fig. 4.70). In each layer, the molecules are packed side by side, with their lengths perpendicular to the surface. As a consequence of the regular orientation of the molecules, one may expect soap films to be doubly refracting, and it would be of interest to study the behaviour of soap films of various thicknesses under polarized light. Such experiments were performed by Raman, and here is his description:

> Some beautiful and interesting phenomena are noticed when a soap-bubble is placed between two crossed nicols or polaroids and viewed by transmission against

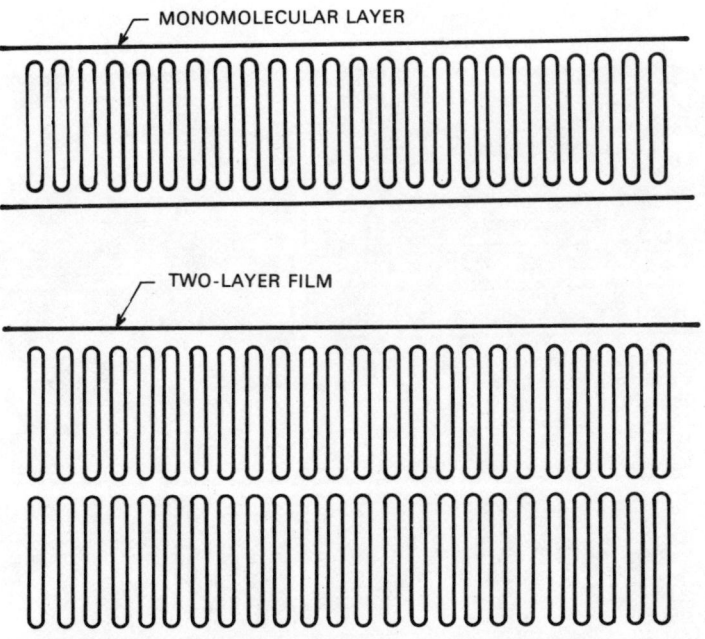

Figure 4.70 Molecular arrangement in a soap film. Observe the alignment of the molecules.

a bright source of light [see Fig. 4.71]. When only one of the nicols is present, the usual colours by transmitted light are noticed, except that they are slightly more vivid towards the end of one diameter of the sphere and slightly less vivid at the ends of the perpendicular diameter. With both nicols present, a black cross appears on the surface of the bubble with its arms parallel to the vibration planes of the two nicols. Elsewhere, the surface of the bubble exhibits striking colours which recall those seen by reflected light in their vividness and are, indeed, complementary to the usual transmission tints. With a monochromatic light source, the interferences are as striking as those ordinarily seen by reflected light. Near the margins of the bubble, the minima are sharp and much finer than the maxima, thus reversing the effects usually observed in transmitted light. As the bubble thins down, the interferences successively disappear so that in the penultimate stage its entire surface is bright except for the dark cross. When, finally, the soap-bubble goes "black", it retains a faint luminosity, while its spherical margin shines brightly as a crescent of light interrupted by the intersections with the black cross....

When the nicols are set with their vibration planes not exactly at right angles, the black cross breaks up into two curved arcs or isogyres. These shorten and approach the margins of the sphere rapidly as one of the nicols is further turned round. With a thick film, the isogyres are themselves the most vividly coloured parts of the bubble. With thinner films it is noticed that when the nicol is turned so that the isogyre moves across an area of the bubble, the colour of the same alters to the complementary tint. With monochromatic light, the isogyres show notable alternations of intensity and appear distorted where they cut the interference curves, while the latter exhibit dislocations at these points which may amount to as much as

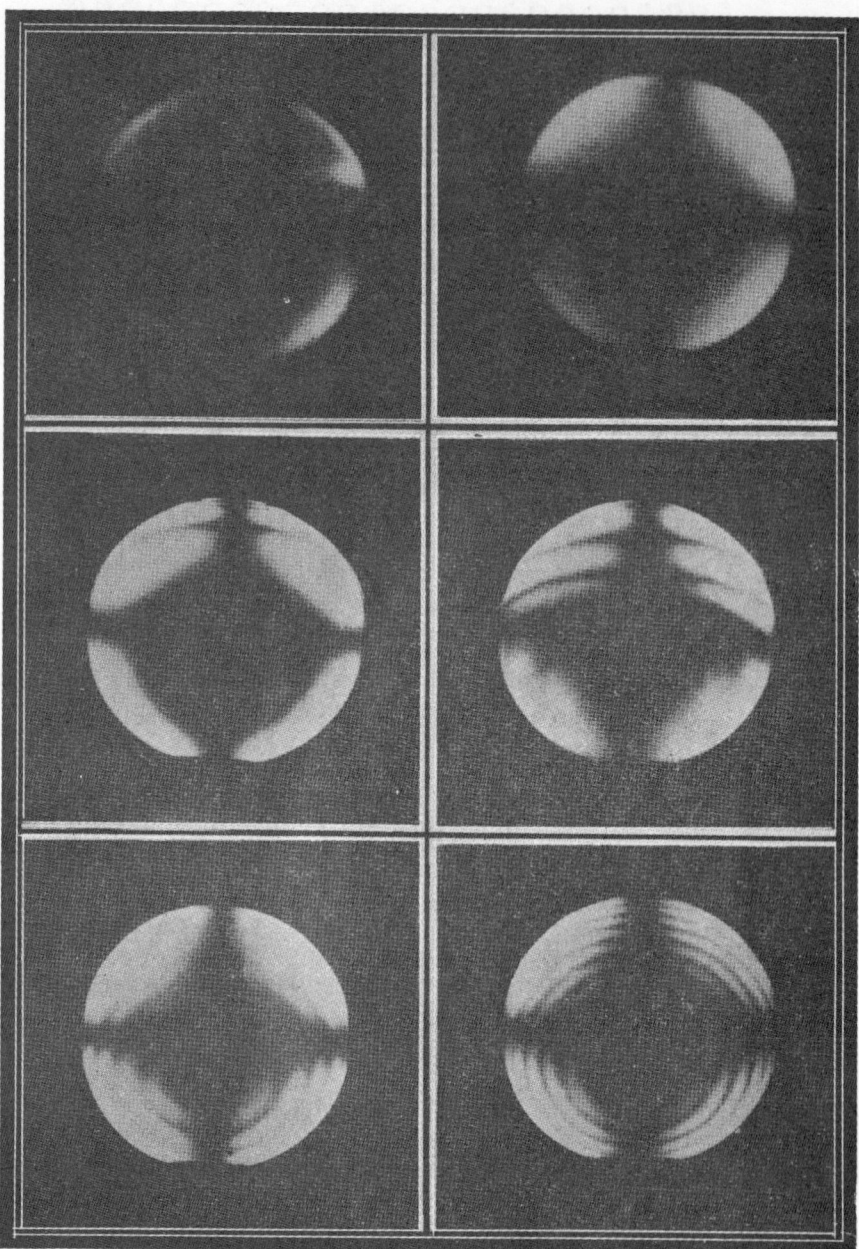

Figure 4.71 Soap-bubbles between crossed nicols. (After ref. [4.66].)

half a fringe. A "black spot" on the bubble usually appears as a dark area on a bright background. But when it passes over one of the isogyres, its optical character reverses and it is then seen as a *bright spot on a dark background*. [4.66]

Raman did not publish his observations; instead he left them for Krishnan to try and explain them if he could. Krishnan [4.67] started with the expectation that thick and thin bubbles showed different behaviours because the former were optically isotropic whereas the latter were anisotropic and therefore birefringent. However, a detailed mathematical analysis revealed that the observed effects could not be attributed to birefringence. Instead they were merely consequences of varying film thickness. Birefringence, even if present, was not the controlling factor.

Birefringence, Diamagnetism and Optical Anisotropy

As already seen, birefringence is basically a macroscopic optical phenomenon. However, its occurrence is related to the optical properties of the molecules in the substance concerned, in particular to the so-called optical anisotropy of the molecules. A molecule is said to be anisotropic when it has a non-spherical shape. Under these circumstances, it exhibits an anisotropy (i.e., angular variations) not only in the optical properties, but in the magnetic and electrical properties as well. As is only to be expected, these various effects are interrelated in some measure. For over a decade and a half, Raman and his students carried out extensive studies on optical anisotropy, electrical birefringence, magnetic birefringence and diamagnetism. The connecting thread running through these phenomena was brought out beautifully by Raman in a lecture delivered by him before the Physical Society in London in 1929 [4.68].

Consider a non-spherical molecule, and let us suppose it is placed in a *static* electric field. The molecule will then be (electrically) polarized, meaning that tiny electric dipoles will be induced along the three principal directions, as shown in Fig. 4.72. Let A', B', C' be the magnitudes of the three moments induced by unit fields acting respectively along these directions. A', B', C' will determine the static dielectric properties of the medium. Next we suppose that the molecule is subject to *oscillating* electric fields such as are present in a light wave. The moments A, B, C now induced (by unit oscillating fields along the three directions) will not necessarily equal A', B', C', since the high-frequency response could differ from the static response. The optical anisotropy alluded to earlier depends on the extent to which the quantities A, B and C differ from each other. In a light scattering experiment one often measures a quantity ρ referred to as the depolarization ratio (see Fig. 4.73). Lord Rayleigh has shown that for a gas or a vapour,

$$\rho = \frac{6[(A-B)^2 + (B-C)^2 + (C-A)^2]}{10(A+B+C) + 7[(A-B)^2 + (B-C)^2 + (C-A)^2]}.$$

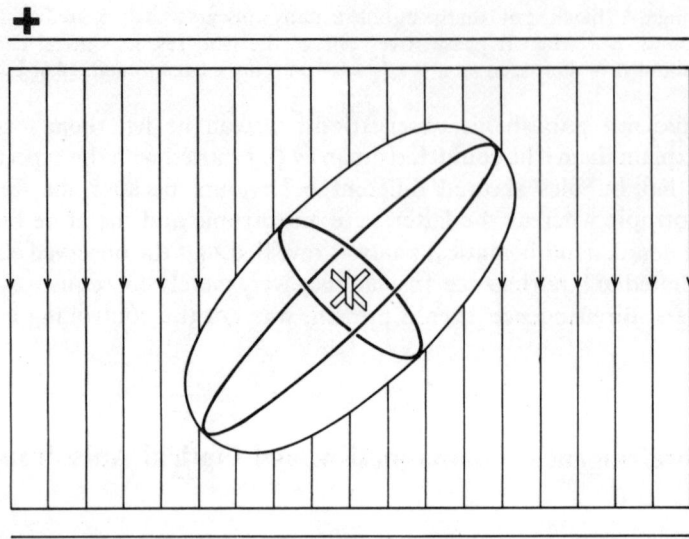

Figure 4.72 Illustration of the electric dipoles induced (along the three principal directions) when a molecule is placed in an electric field.

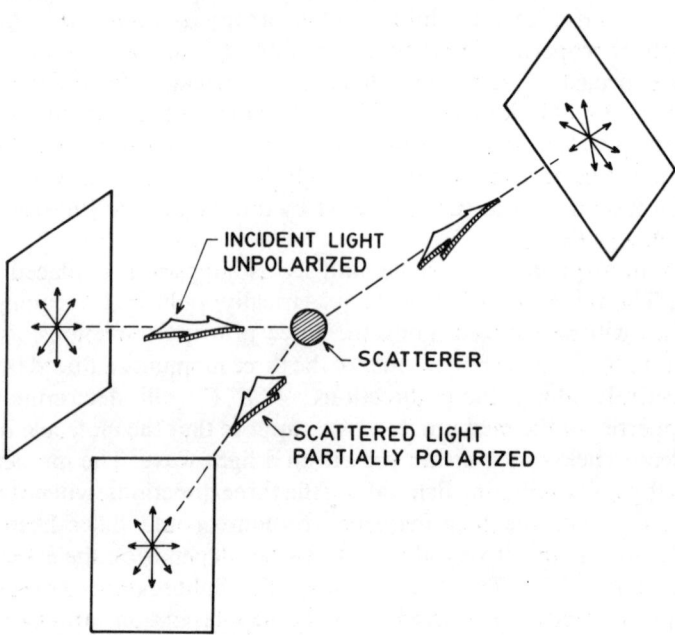

Figure 4.73 Illustration of the concept of depolarization. When unpolarized light is scattered by molecules of a gas and is observed at 90°, one expects the scattered light to be completely polarized. However, this is true only if the molecules are spherical. If the molecules are non-spherical, the transverse scattering is only partially polarized. In this case one measures a quantity called the depolarization ratio.

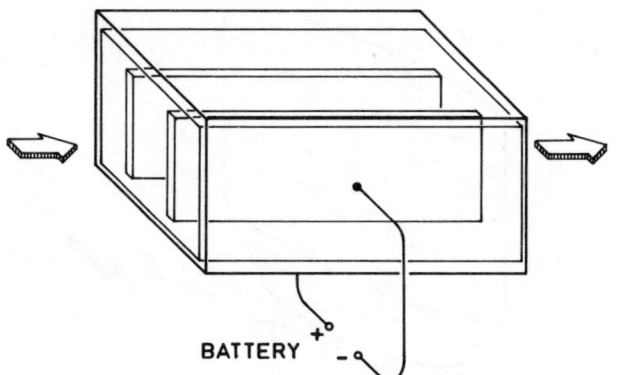

Figure 4.74 Diagram to illustrate the Kerr effect.

This expression vanishes when $A = B = C$, i.e., when the molecule is optically isotropic; thus ρ is a measure of optical anisotropy.

Sometime in the last century, Kerr showed that gases or liquids containing anisotropic molecules exhibit feeble birefringence when placed in an electric field (see Fig. 4.74). The electric field orients the individual molecules, causing the fluid as a whole to behave as if it were a doubly refracting crystal. A quantity K referred to as the Kerr constant measures the magnitude of the electrically induced birefringence, and is dependent on A, B, C as well as A', B', C'.

Magnetic polarization is analogous to electric polarization and occurs when the molecule is placed in a static magnetic field. The magnetic anisotropy is determined by the magnetic moments A'', B'', C'' induced along the three directions by unit magnetic fields acting respectively along the three directions.

The Cotton–Mouton effect is the magnetic analogue of the Kerr effect, and refers to the birefringence which occurs when a fluid containing anisotropic molecules is placed in a magnetic field, as sketched in Fig. 4.75. The quantity C_m which measures the magnitude of the effect is related to A, B, C and A'', B'', C''.

Let us now consider the work of Raman and his colleagues in this area. The broad objective was to see the interrelationships between experimental data relating to the various phenomena just referred to, and to learn, if possible, something about molecular arrangement in the condensed state. It was shown, for example, that optical anisotropy was stronger in the aromatic molecules than in the aliphatic ones, and that there was a corresponding enhancement of the Kerr effect and the Cotton–Mouton effect in the aromatics. Sircar studied the Kerr effect in radio-frequency fields (~ 300 MHz) and demonstrated that the Kerr effect gets quenched when the applied frequency matches one of the (microwave) absorption frequencies of the molecule.

The *tour-de-force* was the prediction of the orientation of the molecules in crystalline naphthalene and anthracene. In 1924 the Braggs (father and son) had proposed a structure for these crystals which is shown in Fig. 4.76. Raman called this to question, based on magnetic susceptibility measurements carried out by Bhagavantam. He remarked,

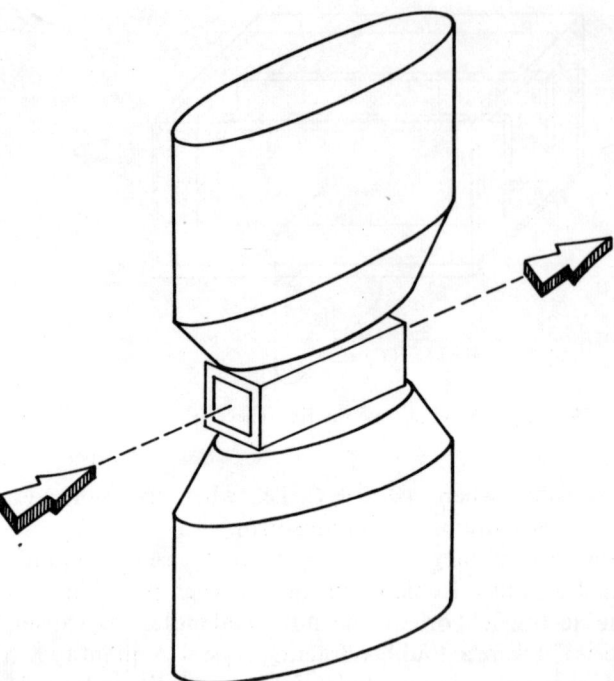

Figure 4.75 The Cotton–Mouton effect is the magnetic analogue of the Kerr effect, and occurs when a liquid containing anisotropic molecules is placed between the poles of a powerful electromagnet.

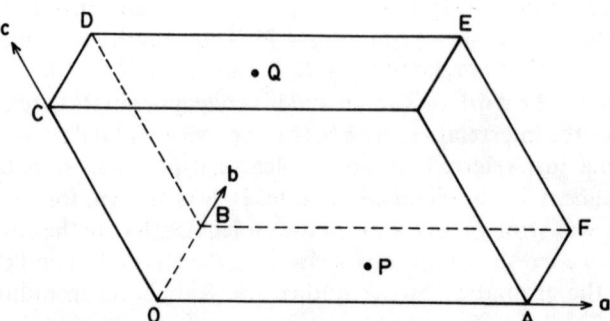

Figure 4.76 Unit cell of anthracene and naphthalene. In both cases, $BOC = AOB = 90°$. However, COA is obtuse. Bragg placed one molecule at each corner and one each at P and Q. Thus each cell contains two molecules, both of them similarly oriented. Further, Bragg proposed that the plane of the molecules was the ac plane, the length being along the c axis. Bhagavantam disagreed with Bragg, and was later proved correct.

I fear that in the case of naphthalene and benzene, the orientation which Sir William Bragg, in his book published some years ago, assigned as probable, cannot now be accepted, because it does not take into account the optical and magnetic aspects and there is evidence that more recent X-ray results will lead to a revision of their structures so as to fit in with the optical and magnetic characteristics [4.68].

Based on his magnetic measurements, Bhagavantam convincingly argued:

Two of the magnetic axes of the naphthalene crystal are found to coincide with two of the crystallographic axes, namely, b, and c, and the third is a line perpendicular to both of them.... Bragg supposes that the two molecules in the cell are similarly oriented with their planes parallel to the ac plane. Contrary to this, the foregoing results suggest that the molecules are so oriented as to have their planes parallel to the bc plane of the unit cell in the crystal. [4.69]

Preliminary X-ray measurements by Kedareshwar Banerjee provided support to the conjecture made by Bhagavantam. Full vindication came shortly thereafter when the structure of anthracene was completely solved.

In describing these various scientific findings, one should not overlook the fact that there was much innovation in the design of the experiments. Thus Ramanadham was able to measure magnetic birefringence "occurring in a great many common substances, including something like 35 compounds in which Cotton and Mouton failed to observe the effect, and in which it is exceedingly small". Vaidyanathan developed a technique for measuring the feeble diamagnetism of gases and vapours, while Bhagavantam designed another suited for crystals.

Our discussion of birefringence would not be complete without a brief reference to the elegant work of Raman and Krishnan on flow birefringence [4.70]. As early as the last century, Maxwell had observed that when a viscous liquid like Canada balsam is mechanically agitated, it exhibits optical anisotropy. The subject was later pursued by other investigators, and in 1925 Vorländer and Walter examined as many as 172 liquids, using an arrangement suggested by Maxwell himself. In their experiment, the liquid under investigation was contained in the gap between two coaxial cylinders, and the inner cylinder was rapidly rotated. A beam of polarized light was passed through the column of liquid parallel to the axis of the cylinders, and analysed for birefringence in the usual manner. Their work clearly brought out the fact that "the power to exhibit birefringence under mechanical flow is just as much a characteristic of pure liquids as, for instance, the power of exhibiting birefringence when placed in an electrostatic field".

Raman and Krishnan provided a comprehensive description of the experimental observation by regarding mechanical birefringence

as arising from the optical anisotropy of the molecules, taken together with a tendency for them to orientate under the mechanical stresses within the fluid. The effective cause of such orientation is taken to be the non-spherical shape of the molecules. [4.70]

Earlier, while discussing viscous flow, Sir George Stokes had shown that in the case of a simple sliding motion parallel to a plane, the tangential stresses acting along the

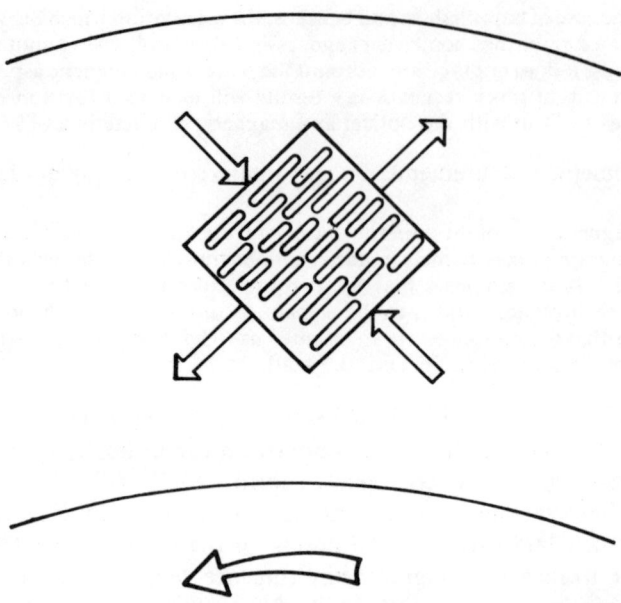

Figure 4.77 Schematic of a section of a fluid contained between two cylinders. Stokes showed that the tangential stresses acting may be replaced by two sets of stresses, one compressive and the other tensile, as shown. The outer cylinder is fixed and the inner one is rotated in the direction shown. Birefringence results because of molecular alignment.

plane may be replaced by two sets of stresses, one set consisting of tensions, and the other set of pressures, acting along two directions which are mutually perpendicular and inclined at 45° to the line of flow, as shown in Fig. 4.77. Raman and Krishnan observe:

> It is clear that if the molecules are highly asymmetrical in shape, the set of tensions and pressures pictured in Fig. 1 [our Fig. 4.77] would tend to cause them to orient in the fluid, in such manner that the longest dimension of a molecule lies along the axis of tensions and the shortest one along the axis of pressures.... This orientative tendency of the molecules is, however, opposed by their thermal agitation, which tends to throw them into disarray. The resulting state of statistical equilibrium can be found by an application of the Boltzmann principle.

This is precisely what Raman and Krishnan did, following which they computed the Maxwell constant ∇ (the analogue of K and C_m introduced earlier) which is a measure of the flow-induced birefringence. They find it significant

> that all the low values of the Maxwell constant belong to the aliphatic or hydroaromatic compounds, and all the high values to the aromatic compounds, and that the average values for the two sets of compounds differ just in the way we should expect in view of the greater refractive index and optical anisotropy of benzene and its derivatives.

Einstein questioned

Early in this century, the nature of light became a controversial issue. Raman pondered over this question in the context of light scattering experiments, as we shall see in the next chapter. We bring it up here in a different connection, namely, with reference to an experiment that Einstein proposed in order to resolve the controversy. Raman pointed out that the proposed experiment would be inconclusive as there was a flaw in Einstein's argument.

As early as 1905, Einstein had argued that light consisted not of waves but of packets of energy (later named photons). Despite Einstein's outstanding success in explaining the photoelectric effect, there was much reluctance in accepting his hypothesis concerning the nature of light. It almost seemed disloyal to the wave theory which for nearly a century had enjoyed a progression of triumphs. Strong in his conviction, Einstein felt quite perturbed by this rejection and in 1921 proposed in the Berlin Academy an experiment that could provide a definitive means of deciding between the widely-held wave theory, and the corpuscular theory advocated by himself. Raman, in two papers published in quick succession of each other [4.71, 4.72], showed that Einstein's reasoning was at fault and that a null effect would result from *both* theories; as such, a resolution of the controversy was not possible as Einstein had proposed.

Einstein's proposed experiment is based upon J. Stark's observation that the light emitted by moving ions (as in *Kanalstrahlen*) exhibits the Doppler effect, i.e., the wavelength of the emitted light depends on the angle between the line of movement of the atoms and the direction of observation. With this background, let us see what Raman himself has to say about Einstein's experiment.

> Fig. 1 [our Fig. 4.78a] illustrates the proposed experiment. K is a stream of canal rays, L_1 is a focusing lens, S is a screen containing a slit which serves to isolate a definite pencil of light, and the lens L_2 renders the emergent beam parallel. The emergent pencil is observed through a telescope focused for infinity, so that the image of the slit in the screen S would be seen sharply focused in the field of view. Since the atoms in the canal rays emitting light are in motion, the Doppler effect comes into evidence, and the rays proceeding at any instant from individual luminous atoms in different directions should, according to the wave theory of light, be of different frequencies. Einstein suggests that the rays passing through the slit S and incident on the upper and lower parts of the lens L_2 should consequently be of different frequencies. If, therefore, a layer of a dispersing medium such as carbon disulphide be placed between the lens L_2 and the observing telescope, the different rays would travel through it with different velocities. Hence the wavefront should suffer an aberration and the image of the slit seen in the focal plane should shift through an extent proportionate to the thickness of the dispersing layer introduced. Einstein conceives that according to the quantum theory of light, on the other hand, such displacement should not occur, and he believes that the proposed arrangement furnishes an *experimentum crucis* to decide between the rival theories. [4.71]

Paraphrasing, Einstein argued that if his quantum hypothesis concerning light were true then the image of the slit would remain unaffected by the introduction of

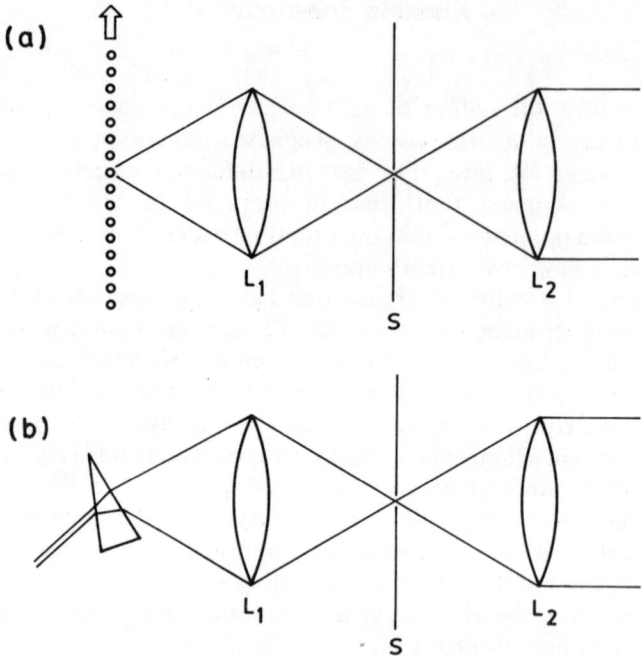

Figure 4.78 (a) Schematic of the experiment proposed by Einstein. (b) Equivalent arrangement where a prism simulates the input conditions.

the carbon disulphide layer whereas if the wave theory were true, the image would shift to an extent proportional to the length of the dispersion layer. According to Einstein such a shift was predicted by the wave theory because the wavefront emerging after L_2 would have different wavelengths at the top and the bottom.

Raman disputed Einstein's reasoning first in a letter communicated to *Nature* on 16 March 1922, and then again in a slightly longer paper sent to the *Astrophysical Journal* barely four days later. Several arguments were advanced by him, all to show that no shift of the slit image was expected in the wave theory also! He explains, "The problem is exactly similar to that which arises in connection with Michelson's determination of the velocity of light in dispersive media by the revolving mirror method, and was very clearly dealt with by Willard Gibbs." Following this line of reasoning, Raman established that no shift of the slit image would occur. Perhaps the most appealing argument employed by him is indicated in Fig. 4.78b. Here the stream of canal rays is replaced by a stationary source of light in the form of a prism from which a spectrum diverges, illuminating the top of L_1 with one colour (i.e., one particular wavelength) and the bottom with another colour, exactly as Einstein proposes. It is then straightforward to see (and this is easily confirmed by direct experimentation) that no shift may be expected in the wave theory when the region behind L_2 is filled with carbon disulphide. Let us suppose the slit is narrow. In accordance with the Huygens principle, cylindrical secondary waves would diverge

Glimpses from the Golden Era

from it. Since the slit is being illuminated with *all* colours, secondary waves corresponding to all wavelengths in the incident spectrum would be produced and *both* the top and the bottom of L_2 would be illuminated by *all* colours. Under these circumstances, a shift of the image is not possible.

Raman remarks that

> the image of the slit would be formed in the focal plane of the observing telescope according to the ordinary laws of geometrical optics. No difference would therefore be made by the interposition of a plane-parallel layer of dispersing medium. [4.72]

Raman concludes by saying,

> The error in Einstein's reasoning lies in his having ignored the vitally important part which diffraction plays, according to the wave theory of light, in the theory of the formation of images of illuminated apertures by optical instruments. [4.7]

4.6 Studies on Impact

Raman's studies on impact illustrate that he was no narrow specialist, and that he allowed his attention to be captured not only by beautiful phenomena but by interesting questions of physics as well.

The first worthwhile statement as to what happens when one body impacts against another seems to have been made by Newton, who observed that impact can be divided into two categories, namely, perfectly elastic and imperfectly elastic, the latter being characterized by energy dissipation.

Quantitative studies on impact started with Hertz, who confined himself to the first of the two categories mentioned above, which clearly rules out cases like impact by bullets, for example. Hertz assumed that the stresses set up are the same as those

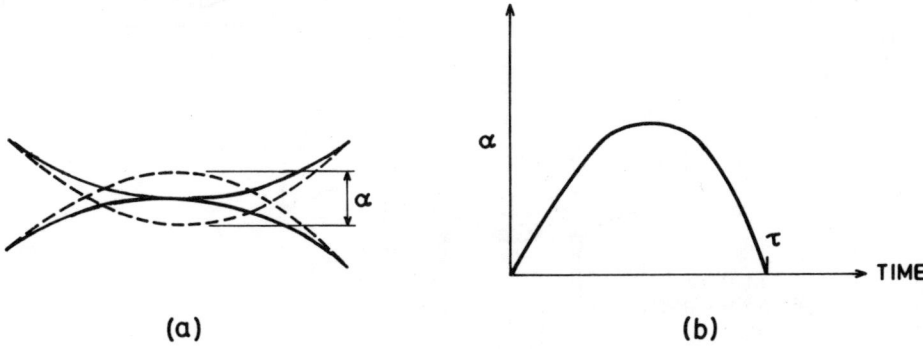

Figure 4.79 (a) Contact parameter α for two colliding spheres. The solid lines indicate the distorted profiles, while the dashed lines indicate the profiles in the absence of elastic deformation. The latter imply interpenetration, which is obviously not possible. However, they permit the definition of α. At the instants when the contact is first made and when it is broken, the solid and dashed lines coincide, i.e., $\alpha = 0$. (b) The variation of α with time τ is the contact time.

that would be produced if the load was turned on slowly. Given that the material suffers compression on impact, the collision is not instantaneous but of finite duration. To estimate this duration, we need to define a contact parameter α; this is done in Fig. 4.79a for two colliding spheres. The quantity α is non-zero during contact, and has a time variation as in Fig. 4.79b from which the duration time τ is easily defined. To the experimentalist, the quantity of interest is *the coefficient of restitution* $e = -(v(\tau)/v(0))$, where $v(0)$ is the impact velocity (for the purpose of definition we suppose the sphere is impacting on an infinite slab) and $v(\tau)$ the velocity on rebound. It should be intuitively obvious that for small impact velocities, the sphere would rebound with the same velocity as it originally had, meaning that e would equal unity. As the impact occurs with greater and greater velocities, one would expect the value of e to decrease since on the one hand there would be energy dissipation, and on the other hand elastic deformation. While there were general theoretical expectations that e would decrease with increasing $v(0)$, detailed experimental data were not available. Raman therefore decided to follow the velocity variation of e, particularly for small impact velocities [4.73].

The experimental arrangement used by him is shown in Fig. 4.80. Two spheres hung by bifilar suspensions were held apart and dropped electromagnetically by the aid of small iron washers firmly fixed near the point of suspension. The motion of the balls before and after impact was recorded photographically by what may be

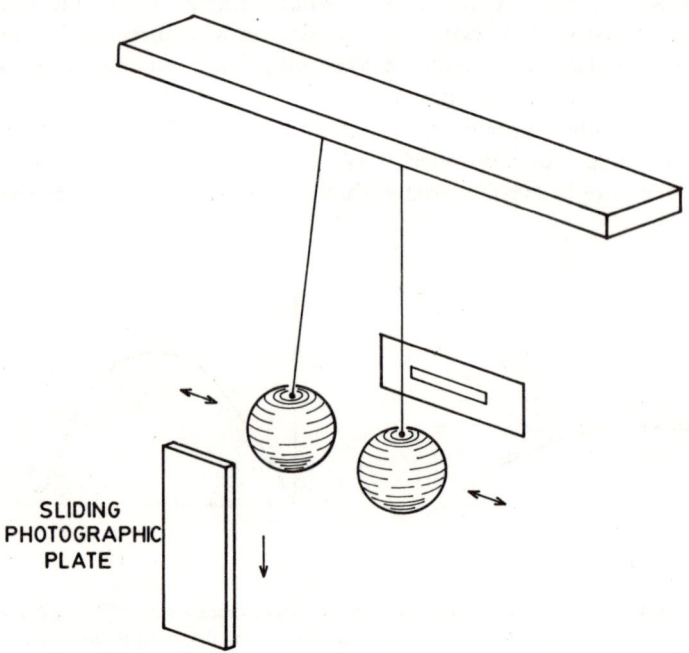

Figure 4.80 Experimental arrangement used by Raman to study colliding spheres. For explanations, see text.

The ancestral house of Raman.

The house at Tiruvanaikkaval where Raman was born.

Raman's father, Chandrasekara Iyer.

Raman's mother, Parvati Ammal.

Raman's elder brother, C. S. Iyer.

A group photograph taken in 1905 in the Mrs A.V.N. College, Visakhapatnam. Raman's father is seated in the third row from the top, second from the right.

A view of the physics laboratory of the Presidency College, Madras.

Group photograph taken in the Presidency College, Madras, of the BA Class of 1903. Raman is in the second row from the top, fourth from the right.

Raman and Lokasundari in the early Calcutta days.

Mahendra Lal Sircar, founder of the Indian Association for the Cultivation of Science.

Asutosh Dey, popularly known as Ashu Babu, and a pillar of strength to the researchers at the Association.

A view of the Vizianagaram Laboratory.

Raman as Assistant Accountant-General in Calcutta.

Group photograph taken during Raman's stay in Rangoon. Raman is seated in the second row from the bottom, third from the left.

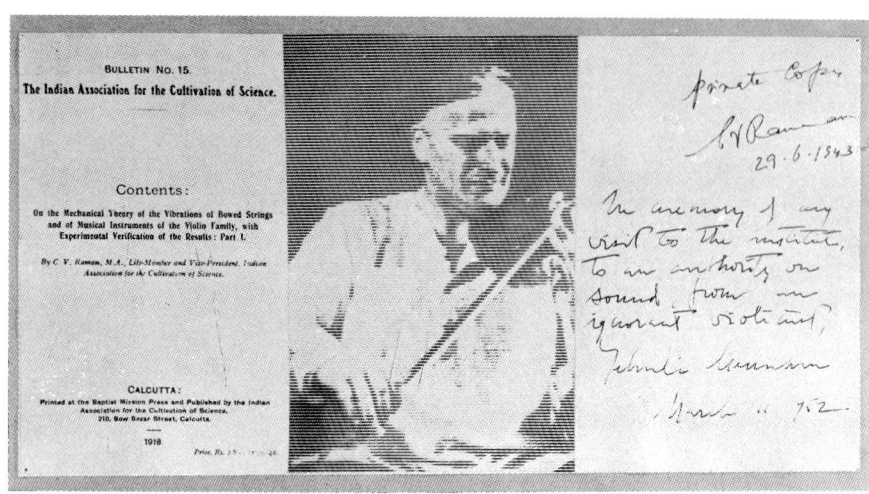

Raman's monograph on the violin autographed by Yehudi Menuhin who declares himself as an "ignorant violinist"!

Letter from the Registrar of Calcutta University, discreetly waiving the requirement of foreign training.

During a seminar at the Association.

Some of the research scholars of the Association. Sir K. S. Krishnan is seated in the middle.

Raman with his baby quartz spectrograph in Calcutta.

Some of the first spectra taken by Raman. These were shown by him during his lecture on March 16, 1928, in Bangalore.

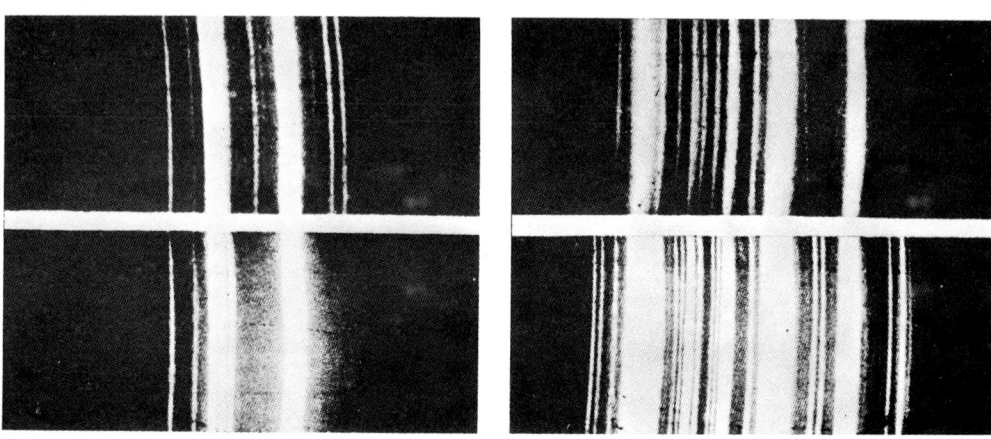

With fellow Nobel Laureates. That year (1930), H. Fischer of Germany received the prize for Chemistry, K. Landsteiner of Austria for Medicine, and Sinclair Lewis of the US for Literature.

Lady Raman with Princess Ingrid at the Nobel ceremony.

Lady Raman (*left*) with friends in Calcutta.

A souvenir from the American tour of 1924.

Glimpses from the Golden Era

called the eclipse technique. A fine horizontal slit was illuminated by an electric arc and an image of it was focused on the lowest point of the circle on which the centres of the balls moved when dropped. As long as the balls are in contact, the image of the slit when viewed from behind is completely obscured. The image appears only when the balls draw apart. Raman arranged a sliding photographic plate behind the balls

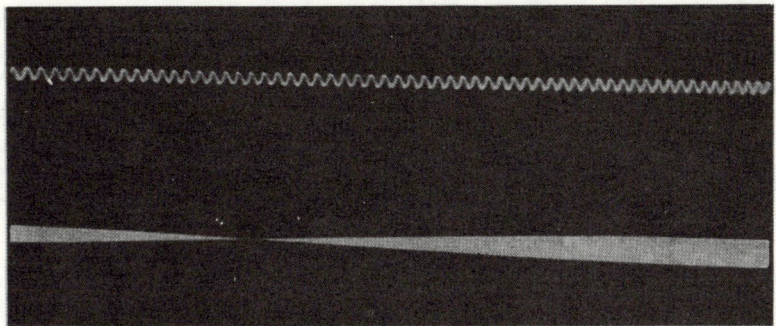

Figure 4.81 Specimen photograph showing the eclipse observed with the apparatus of the previous figure. (After ref. [4.73].)

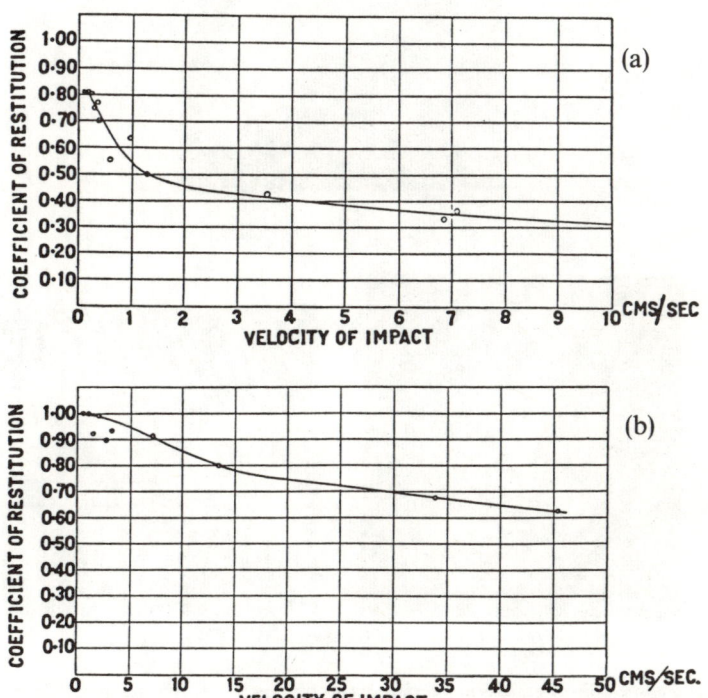

Figure 4.82 Variation of coefficient of restitution e with velocity for (a) lead and (b) aluminium spheres. (After ref. [4.73].)

and recorded the view of the slit as a function of time. For calibration, he also recorded the vibrations of a tuning fork. A typical photograph is shown in Fig. 4.81.

The variation of the coefficient of restitution with impact velocity was studied for various materials, and a selection of these results is offered in Fig. 4.82. Observe how rapidly e departs from unity in the case of lead spheres.

An important assumption underlying Hertz's theory is that the energy of the colliding bodies remains as translational energy even after impact. If such were indeed the case, then e should not decrease, whereas experiment indicates just the opposite, clearly implying that Hertz's assumption was invalid. In other words, one must allow for energy dissipation occurring either via the excitation of elastic waves in the impacting bodies, and/or the introduction of plastic deformation. Raman extended Hertz's theory by allowing for the former possibility, and his predictions were confirmed by experiments performed by Venkatasubbaraman [4.74] and by Seshagiri Rao [4.75].

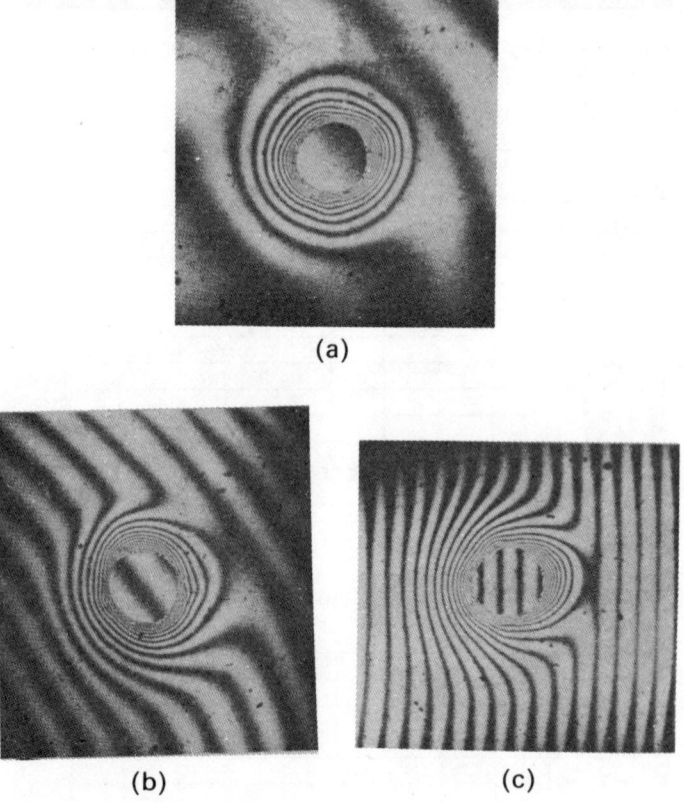

Figure 4.83 Percussion interferograms obtained by laying an optical flat on the deformed surface and forming interference fringes with the wedge. In (a) the test plate is nearly parallel to the surface of the glass block; (b) and (c) are obtained with some tilt. (After ref. [4.76].)

Glimpses from the Golden Era

Raman also found that when steel balls were dropped on glass plates, they produced beautiful percussion figures. A circular crack starts from the surface of the plate and spreads obliquely inwards in the form of a surface of revolution, revealing itself by the light which it reflects. The deformation of the external surface of the glass plate was studied by laying another flat glass plate on it and observing the interference fringes formed by the wedge made up by the two surfaces. Figure 4.83 shows such percussion interferograms. Raman notes that there are

> three distinct regions in the percussion figures [Fig. 4.83]. Firstly, there is a central area, nearly circular, which, apparently, is unaffected by the impact, as is shown by the fringes passing through it [in Fig. 4.83c] being straight and parallel. Secondly, there is a narrow annular region of fracture full of a network of irregular fringes, showing severe injury to the surface. Thirdly, and just beyond this, there is a sudden elevation of the surface which slopes down, first quickly and then more slowly, to the original level of the surface at the edge of an area which sets the limit to the percussion figure. Closer examination reveals another remarkable feature, namely, that the central area of the percussion figure, though it remains plane and apparently undisturbed, has, in reality, been *depressed below* the original level of the surface by an appreciable fraction of a wavelength, as shown by the fact that the courses of the fringes outside the percussion area and within the central circle are distinctly out of register. [4.76]

Later, Smith [4.77] studied percussion figures in crystalline rock-salt using the same technique.

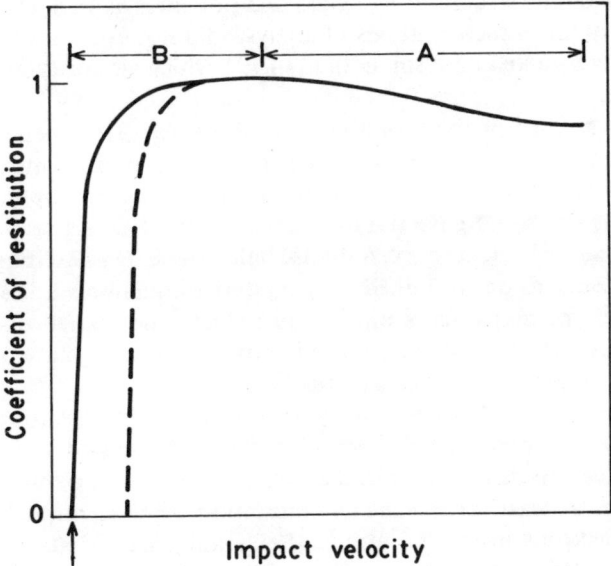

Figure 4.84 Schematic plot of *e* versus impact velocity. Region A is that explored by Raman while B is that proposed for space experiments. The arrow shows the threshold velocity by measuring which the surface energy is obtained. The dashed curve would result if gravitational energy is comparable to surface energy. In this case the threshold velocity will not be determined by surface forces alone.

It is also worth making a brief reference to some recent developments which, in a sense, represent an extension of Raman's work [4.78]. All the older theories including that of Raman predict that e approaches unity as $v(0)$ tends to zero. However, this neglects surface and gravitational forces. It should be clear from a little reflection that inside a spacecraft where gravity effects are small, a small ball impacting a flat surface with low velocity may be prevented from rebounding by forces of adhesion. Such forces could modify e as shown in Fig. 4.84. An experiment was in fact proposed to measure adhesive forces by observing rebound in a spacecraft environment.

4.7 X-rays

In 1911 von Laue found that X-rays were diffracted by crystals and produced a well-defined interference pattern. This great discovery (duly rewarded with the Nobel Prize) marked the beginning of the use of X-rays for probing atomic arrangements in condensed matter. Now the use of X-rays for unravelling the architecture of crystals is readily understable, given their highly geometric structures. But such is the capacity of human ingenuity that today X-ray diffraction (together with electron and neutron diffraction) is routinely employed to explore a variety of non-crystalline structures like glass and polymers as well. However, way back in the early twenties such methods of analysis did not exist and therefore the pioneering work by Raman and Ramanathan [4.79], which we are about to discuss, acquires significance.

The first X-ray studies of the liquid state go back to 1916 when Debye and Scherrer discovered that when a narrow pencil of monochromatic X-rays was made to pass through a thin layer of liquid and received on a photographic plate, the X-rays scattered by the liquid gave rise to a fuzzy circular halo separated from the central spot by a relatively clear space. A diffuse halo not being as exciting as a set of well-defined spots such as one obtained with crystals, diffraction studies on liquids remained neglected for many years until in 1921 Debierne carried out extensive investigations of several liquids as well as liquid mixtures. In every case one or more diffuse halos were observed, and the question was why.

Being electromagnetic in character, X-rays interact basically with the electrons in the medium. Thus the observed diffraction pattern of a liquid is in essence a signature of the electron distribution in the system. In discussing this distribution one must recognize at least three levels of description, relating respectively to the arrangement of electrons inside the atoms, the arrangement of atoms inside the molecules (if any), and finally the arrangement of the molecules/atoms themselves in the liquid. While today we have a comprehensive theory which takes adequate account of these features, in the early twenties there was none such.

In 1923 Raman and Ramanathan proposed a theory of X-ray diffraction in liquids. Preoccupied as they were at that time with light scattering in liquids, they naturally approached the subject from the standpoint of optics. If light could be scattered by fluctuations (see Chapter 5) then it stood to reason that so must X-rays. But there is a complication in borrowing the Einstein–Smoluchowski theory (see Chapter 5) developed earlier for the scattering of light. Unlike in the case of light the wavelength of X-rays (~ 1 Å) is comparable to interatomic spacing. However, Raman and Ramanathan noted: "In one case, the Einstein–Smoluchowski theory may be applied as it stands to the problem of X-ray scattering. This is when the angle of scattering is small." Their reasoning may be understood as follows. A good thumb rule in all scattering experiments is that the radiation (of wavelength λ) used probes the structure of the scatterer on a length scale L of the order of $\lambda/2\sin(\theta/2)$ where θ is the angle of scattering. Taking $\lambda = 0.71$ Å as typical of X-rays, one finds $L \sim 1$ Å for $\theta \sim 90°$, meaning that X-rays probe on a length scale of the order of interatomic spacing. However, if θ is reduced to 10 minutes of arc, then L increases to about 240 Å, which is sufficiently large compared to interatomic spacing as to make the Einstein–Smoluchowski formula applicable. Indeed, in modern theories of X-ray scattering by liquids, it is a necessary requirement that the scattering at small angles must go to the Einstein–Smoluchowski limit, or the compressibility limit as it is better known. This important result was first deduced by Raman and Ramanathan but unfortunately that fact is hardly known today.

Turning next to their description of the diffraction halo (which occurs at slightly larger angles of scattering), this too was based on the concept of fluctuations, but, as we shall soon see, did not go far enough. Raman and Ramanathan start with the notion that there is in a liquid a characteristic distance λ_0 analogous to the lattice spacing d in a crystal. As a consequence, there is a strong diffraction of X-rays at an angle θ_0 determined by the Bragg relation

$$\lambda = 2\lambda_0 \sin(\theta_0/2). \tag{4.77}$$

However, one knows that liquids give halo-type diffraction maxima rather than sharp rings as polycrystals do. Raman and Ramanathan explain this blur by invoking what they call a "structural spectrum". In essence what it means is that owing to fluctuations, diffraction occurs not only at a particular scattering angle θ_0 but over a range of angles centred around it.

☆Raman and Ramanathan tackle the problem of diffraction by the structural spectrum as follows.

> Imagine a cube in the fluid, which is normally of edge-length λ_0, distended or compressed into a cube of edge-length λ_1. The work done in the process is given by the expression
>
> $$\frac{1}{2}\frac{\lambda_0^3}{\beta}\left(1 - \frac{\lambda_1^3}{\lambda_0^3}\right)^2,$$
>
> where β is the isothermal compressibility of the fluid. Actually as a result of thermal agitation, the cube might change shape as well as volume. If we take the cube to remain always a rectangular parallelopiped, the three edge-lengths may each be either greater or less than λ_0. It is only one chance in eight that all the edge-lengths would be *greater* (or less as the case may be) than λ_0. The average work

corresponding to a change of one of the edge-lengths from λ_0 to λ_1 may thus be taken to be

$$\frac{1}{16}\frac{\lambda_0^3}{\beta}\left(1-\frac{\lambda_1^3}{\lambda_0^3}\right)^2$$

and its thermodynamic probability may in accordance with Boltzmann's principle be written as

$$A\exp\left[-\left\{\frac{1}{16}\frac{N}{RT\beta}\lambda_1^3\left(1-\frac{\lambda_1^3}{\lambda_0^3}\right)^2\right\}\right]d\lambda_1,$$

where A is a constant. [4.79]

Using the above result, Raman and Ramanathan obtained the diffracted intensity $I(\theta)$ as

$$I(\theta) \sim \text{const.} \times \exp\left[-\frac{1}{16}\frac{N}{RT\beta}\lambda_0^3\left(1-\frac{(\lambda/2\sin(\theta/2))^3}{\lambda_0^3}\right)^2\right] \times \cos(\theta/2)d\theta.$$

As is readily seen, when θ becomes equal to θ_0 defined by (4.77) the intensity attains a maximum.

The problem now remains of specifying λ_0. For this purpose Raman and Ramanathan argue that no appreciable error arises "if we take θ_0 to be identical with the mean distance between neighbouring molecules in the fluid". After some analysis they conclude that "in liquids under ordinary conditions, λ_0 is of the same order of quantities as σ [mean linear dimension or diameter of the molecule] but may be some 10% to 20% greater".　　　　　　　　　　　　　　　　　　　　　　　　　　　☆

The performance of the Raman–Ramanathan theory may be judged from Fig. 4.85 taken from their paper. Here (a) is the diffraction pattern calculated using a formula derived by Ehrenfest, assuming that the dominant factor is the "interference of the effects of two neighbouring molecules". Brillouin (see also Chapter 9), who at that time was engrossed in the interaction of light with sound waves, extended his analysis to the case of X-rays, with consequences as sketched in (b). Figure 4.85c shows the data for liquid benzene as obtained by Hewlett, while (d) gives the curve computed by Raman and Ramanathan. As they observe,

> When account is taken of the imperfect homogeneity of the X-rays used by Hewlett, it will be seen that his experimental curve reproduces with remarkable fidelity the indications of theory.

Some years later, Kedareshwar Banerjee extended the theory of Raman and Ramanathan by regarding the liquid as

> a degeneration of the crystal structure brought about by the thermal agitations and hence some of the intense crystal diffraction lines give rise to liquid diffraction maxima while others are quenched out.

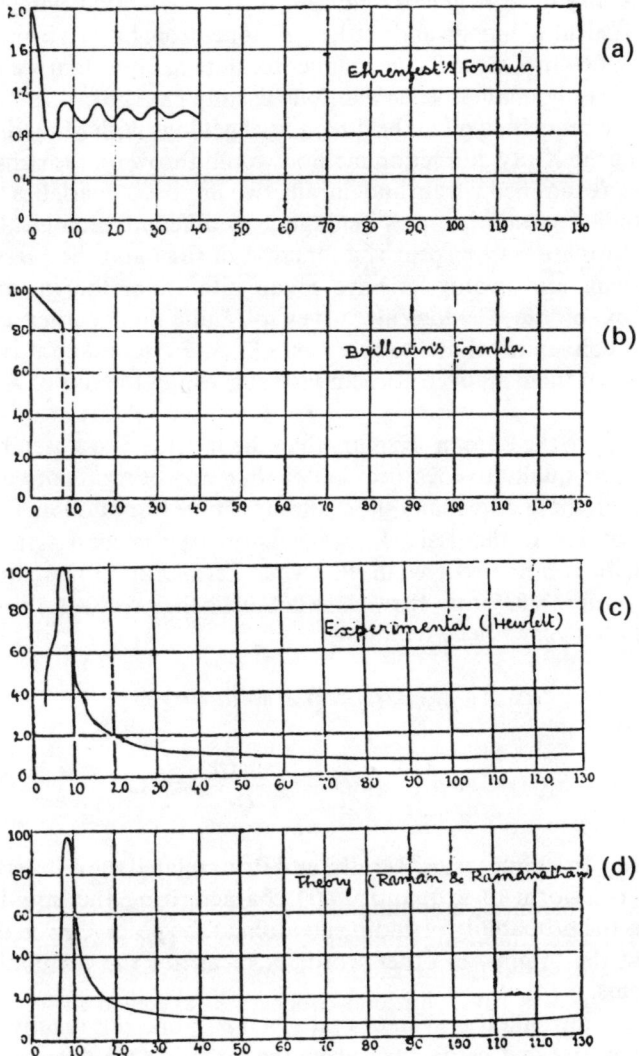

Figure 4.85 X-ray diffraction in liquids. For explanations, see text. (After ref. [4.79].)

In effect, Banerjee was viewing a liquid as a highly disordered crystal with certain lattice spacings alone surviving, each associated with a particular λ_0.

It was not merely theorising; a fair amount of experimental work on X-ray diffraction was also carried out at the IACS. Kedareshwar Banerjee himself studied several liquid alloys of sodium and potassium while Sogani examined liquid hydrocarbons of varying complexity. Noteworthy is the fact that molecules of biological interest like dextrin, glucose, etc. also received attention, with one student studying "vesical calculus (human and cattle), gallstones from human beings, uric acid and cholestrol". Of course, since the techniques for data analysis that we have today did not exist then, only qualitative conclusions usually emerged.

There was also a branching off in the direction of geology with Mahadevan trying "to see what light the X-ray diffraction method would throw on the constitution of coal". Mahadevan examined vitrain and durain, two important varieties of coal, and in particular studied vitrain samples belonging to different geological ages. The process of coal formation from peat is a function of time and the pressure of the overlying sediment, and geologists have much interest in the metamorphosis. Fermor and Fox of the Geological Survey of India had earlier carried out many studies by conventional methods. Raman knew Fermor, and it is likely that discussions between them spurred Raman into suggesting the use of X-rays as an analytical tool.

Returning briefly to the Raman–Ramanathan theory: it served a useful purpose in that it predicted the qualitative features better than any other theory available at that time. However, there were many shortcomings in the formulation of the theory, and it could not stand the test of high-quality experimental data. The real breakthrough in the subject occurred in 1927 when Zernike and Prins [4.80] applied the method of Fourier transforms. In modern language their result (for a monatomic liquid) can be stated very briefly as follows.

$$I(Q) = \text{const. } S(Q), \quad Q = 4\pi \sin(\theta/2)/\lambda$$

where

$$S(Q) = 1 + \rho \int_0^\infty [g(r) - 1] \frac{\sin Qr}{Qr} 4\pi r^2 \, dr,$$

ρ being the density. In simple terms, Zernike and Prins related the diffraction pattern to the Fourier transform of a quantity $g(r)$ characterizing the liquid structure. Physically $g(r)$ is the probability of finding an atom at the point r, given that there is another atom at the origin – in other words, it measures the correlations in the positions of atoms.

From the above formula it is evident that given $g(r)$, one can readily predict the diffraction pattern, and that precisely is where the catch is! One does not yet have a simple prescription for calculating $g(r)$ for a liquid, and indeed that is one of the important problems of liquid state physics. Experimentalists, on the other hand, Fourier-invert the measured diffraction pattern to obtain $g(r)$, rather in the manner crystallographers use X-ray data to solve crystal structures. Such information is of much value to theorists dealing with the liquid state.

Commenting on Raman's work Ramaseshan [4.81] records that once Raman wistfully said,

> We were so preoccupied with light scattering that we did not apply the idea of Fourier transforms to X-ray scattering in liquids although we were so close to it.

An unfortunate miss indeed for Raman.

4.8 An Appraisal

Calcutta saw Raman graduate from an amateur to a professional. Starting as a mere hobbyist, he soon became a full-time physicist involved in all the aspects of science – research, teaching, building up a school and establishing contact with peers.

That aesthetics of natural phenomena was the principal drawing magnet becomes abundantly clear from the problems he chose – the colours of mixed plates, the radiant spectrum, and so on. It is remarkable that at a time when one would have thought that interference and diffraction were thoroughly studied subjects, Raman and his coworkers found enough to keep themselves busy for over a decade and a half. The notion that such work was trivial is dispelled not only by the reputation of the journals where the results were published but also by the recognition conferred on Raman by the Royal Society when it elected him a Fellow. In fact, Sommerfeld knew about Raman mainly from his work on optics, and well before the discovery of the Raman effect.

It should also not be hastily concluded that Raman experimented with various phenomena merely because they appeared nice. His perception of beauty and aesthetics was far more subtle, being not confined just to the visual aspect. It lay in a comprehension of the way Nature operates to produce something that is, among other things, pleasing to the eye. Thus it was a total experience, of which seeing directly was just the first part.

Raman's style was largely that of a bygone era. This is particularly reflected in his various notes to *Nature*, some of which were transmitted by cable! While to the present generation they might sometimes read rather strange, one would find nothing strange about them if one takes the trouble to browse through publications of the late nineteenth and early twentieth centuries.

Two aspects of Raman's writings are, in particular, worth commenting upon. Firstly, they often give the impression that the author is thinking aloud. Secondly, there is a distinct preference for words over *formal* mathematics[19]. It is not as if Raman did not *know* the mathematics involved. From the various subtleties that he explains, it is abundantly clear that he understood the delicate mathematical nuances which can flow only from full analytical comprehension. But at the same time, he gives a distinct feeling of impatience at being tied down by a systematic development of the algebra; his mind was always racing ahead, or, as Born was to put it later, his quick mind leapt over mathematics.

Raman's basic strength lay in classical physics – in particular, in vibrations, waves and optics. But he was also conscious of new and emerging trends in physics as his studies on magnetism and X-rays illustrate. It was alleged then (and indeed sometimes even now) that Raman had no appreciation of the revolutionary changes that swept physics in the twenties. This is a crude distortion. If it was so, how then was he aware of the very important paper by Kramers and Heisenberg (see Chapters 5 and 6)? (Born is reported to have told Nagendra Nath that he scarcely thought Raman would have understood the Kramers–Heisenberg process but upon meeting

Raman found that he had understood the process much better than many theoreticians had done at the time the paper was published.) Did he not worry about the quantum nature of light and the spin of the photon (see Chapters 5 and 6)? What about the lectures on wave mechanics by Sommerfeld which he arranged? One should also not forget the efforts Raman made to get distinguished theoretical physicists like Max Born, Schrödinger, Peierls and others to Bangalore (see Chapter 8). Why would he want to do that if he did not appreciate the new and emerging trends in physics?

No, Raman was not out of touch (at least at that time) as his detractors made him out to be. What could be said with fairness is that he did not absorb all the subtle nuances of quantum mechanics. But then so was the case with most physicists of his generation, particularly experimentalists (as we shall see to a certain extent in Chapter 6). There is no special need to single out Raman on this score.

Apart from his brilliant research, the Calcutta period of Raman is also notable for the inspiring leadership he provided and the vigorous school of physics that he built up. The IACS became a truly exciting place to be in, praised by all who visited it, both Indian and foreign. For the aspiring young scientist in India, there was at last a place of excellence whose doors were open, unlike the Government-run establishments which still tended to look upon Indians largely as a service pool. Bangalore too grew under Raman, but one wonders if it ever reached the dizzy heights that the IACS did.

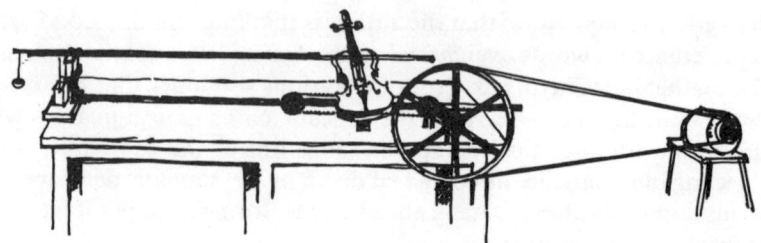

5 *Elementary, My Dear Watson!*

Watson, you can see everything. You fail, however, to reason from what you see. You are too timid in drawing your inferences.
                                    The Adventure of the Blue Carbuncle

Conceptually the Raman effect is quite simple. Suppose a beam of light of frequency v is incident on a liquid (or, for that matter, on a solid or even a gas). The incident light will either be absorbed by the molecules in the system or it will be scattered. We ignore absorption and consider scattering for which there are two possibilities. Either the scattered light has the same frequency v as the incident light, or it has a frequency v' which is different. In the former process, which is referred to as Rayleigh scattering, there is no change of colour. The term Raman scattering is applied to the latter process, and in it the scattered light appears with a different colour[1]. Simple though it all sounds, unravelling these facts took a long time involving many people in many places. As it turned out, there was an exciting race as well. But that comes later.

5.1 The Beginning

For Raman, the story begins in Calcutta in the year 1919. He is starting to develop a new interest, molecular diffraction of light as he calls it, and signals this curiosity with a small note in *Nature* [5.2] discussing the Doppler effect in scattering[2]. What he is really seeking is the molecular basis of the macroscopic phenomenon of refraction. In other words, he is beginning to look at optics from a *microscopic* point of view.

It is two more years before he expresses himself again on this subject, this time in the *Proceedings of the Royal Society* along with Bidhubhusan Ray [5.3]. He does not examine molecular scattering *per se*, but a scattering problem nevertheless. Keen and Porter had earlier done an interesting experiment. A beam of light was passed through a solution in which particles formed. As was to be expected, the intensity of the transmitted light decreased as the solution became more and more turbid consequent to the generation of suspension particles. But there was a surprise.

> Keen and Porter observed, however, that after further lapse of time, light begins again to be transmitted by the suspension, the colour of the light which passes through being at first indigo, then blue, blue-green, greenish-yellow, and finally again white [5.3].

Lord Rayleigh had addressed himself to this remarkable phenomenon and

> attempted to investigate the effects on the basis of the mathematical theory of the scattering of light by small transparent spheres. The explanation of the phenomenon observed in the earlier stages of the experiment presented no difficulty.... Lord Rayleigh did not, however, find it possible to explain the reappearance of the transmitted light in the later stages studied by Keen and Porter, and he went so far as to suggest that there might be some doubt whether the effect was really due to transmitted light in the technical sense of the term.

In Raman's view, the explanation is unsettled and needs sorting out. His approach is simple and clear. Consider the passage of a plane wavefront through the medium. Now consider a thin layer of the medium. In accordance with the Fresnel–Huygens principle, secondary wavelets would be produced by the particles in the slice under consideration. As far as the primary wave itself is concerned, there is a reduction in amplitude due to the obstruction produced by the suspended particles. In addition, interference of the light scattered in the forward direction with the primary wave can alter the intensity of the transmitted beam. Thus two factors control the transmitted intensity: (i) reduction in amplitude due to particle obstruction and (ii) interference in the forward direction. Initially effect (i) dominates, causing a continuous decrease in the transparency of the medium as the size of the suspension particles grows. But growth in size enhances forward scattering, and a stage is reached when effect (ii) starts becoming important, whereupon the transparency starts increasing again.

These ideas are neatly translated into a simple theory, and the expectations of the theory are quantitatively verified with an experiment on sulphur suspensions.

5.2 On Board the SS *Narkunda*

It is now September 1921, and Raman is returning from his first trip abroad. It is a long sea voyage (those were the days before the jets) taking about fifteen days from

Southampton to Bombay, and he cannot remain idle during that period. Like a doctor, he has his little kit consisting of pocket nicol prisms, a small telescope to which polarizers and analysers could be attached, and even a diffraction grating! He is fascinated by the deep blue waters of the Mediterranean and spends hours studying it. From Aden he mails, on September 18, a small note to *Nature* [5.4] describing how the visibility of indistinct objects (in particular the horizon at sea during a haze) could be improved by viewing through a nicol.

The last lap of the journey is used to generate another letter to *Nature* [5.5], this time on the colour of the sea, a theme to which he is to return several times. Mailed on September 26, the byline reads SS *Narkunda*, Bombay Harbour! A quaint, old-world touch. Incidentally, Raman wrote one more paper [5.6] on board a ship. That was in 1924, on board the SS *Kaisar-i-Hind* near Marseilles.

5.3 Why is the Sea Blue?

Raman begins his Bombay Harbour paper by remarking,

> The view has been expressed that "the much-admired dark blue of the deep sea has nothing to do with the colour of water, but is simply the blue of the sky seen by reflection" (Rayleigh's Scientific Papers, vol. 5, p. 540, and *Nature*, vol. 83, p. 48, 1910). [5.5]

He then questions this view describing his own experiments on board the ship, and adds:

> Observations made in this way in the deeper waters of the Mediterranean and Red seas showed that the colour, so far from being improverished by suppression of sky-reflection, was wonderfully improved thereby.... It was abundantly clear from the observations that the blue colour of the deep sea is a distinct phenomenon in itself, and not merely an effect due to reflected sky-light. When the surface-reflections are suppressed the hue of the water is of such fullness and saturation that the bluest sky in comparison with it seems a dull grey.... The question is: What is it that diffracts the light and makes its passage visible? An interesting possibility that should be considered is that the diffracting particles may, at least in part, be the *molecules* of the water themselves.

For the next few years, the scattering of light by molecules was to be an obsession.

Soon after reaching Calcutta, Raman dashes off another letter to *Nature* [5.7] on October 15, this time drawing attention to the connection between the colour of the deep waters and the Einstein–Smoluchowski formula. A month later his thoughts are properly organized, and he communicates them in a long paper to the Royal Society [5.8].

Naturally he must start with Lord Rayleigh to whom we owe the explanation of the colour of the sky. Raman observes that, as noted by Rayleigh, the sky is blue because of the scattering of light by the molecules in the upper atmosphere, a fact brilliantly confirmed by experiments. The colour of the sea is a different matter. Rayleigh of course believed it was all due to reflection, but Raman remarks:

On the other hand, observers familiar with the sea, such as J. Y. Buchanan, of the "Challenger" expedition, who have had very wide opportunities for study, have published detailed descriptions which support an entirely contrary view. An admirable *précis* of the literature on the whole subject has been recently published by Prof. W. D. Bancroft. From a perusal of this very convenient summary, and from the account given in Kayser and Runge's "Handbuch", it would appear that the general trend of opinion is that, so far as there is any real effect to be explained at all (that is, apart from reflected sky-light) the colour of water is due to absorption, the return of the light from the depths of the liquid being due to suspended matter in it. [5.8]

Raman wishes to

urge an entirely different view, that in this phenomenon, as in the parallel case of the colour of the sky, *molecular diffraction* determines the observed luminosity and in great measure also its colour.

Assuming that the colour of the sea is indeed due to scattering by water molecules, one needs a quantitative theory. Rayleigh's theory of scattering is applicable only to gases. In a gas the relative positions of adjacent atoms and molecules would not be correlated, and the scattered waves originating from the different molecules would have no definite phase relationships to each other – random-phase approximation in modern parlance. The total scattered intensity can therefore be obtained as the sum of the intensities contributed by the individual molecules.

On the other hand, in the case of liquids, the spacing of the molecules is far closer and their freedom of movement much less, and we should no longer be justified in making the same assumptions. A gramme-molecule of steam occupies at 100°C over 1600 times the volume which an equal mass of water would occupy, and it is clear, *prima facie*, that volume for volume, water would not scatter light 1600 times as strongly as pure steam, but only in a lesser degree. The question is, by how much less?

Raman does not answer this question in terms of the scattering by individual molecules and their relative positions, but turns instead to a different approach due to Einstein and Smoluchowski (E–S).

5.4 The Einstein–Smoluchowski Formula

It has been known since the last century that fluids scatter light strongly when their temperature is close to the critical temperature. The first clue as to why this happens came from Smoluchowski and Einstein.

In the early years of this century both Einstein and Smoluchowski had independently been concerned with fluctuations. Smoluchowski in particular studied particle number fluctuations in an ideal gas, and deduced the following result for $\overline{\Delta\sigma^2}$ the mean square fluctuation in density:

$$\frac{\overline{\Delta\sigma^2}}{\sigma^2} = \frac{RT\beta}{N_A V}.$$

Here R is the gas constant, T the temperature, β the isothermal compressibility, N_A the Avogadro number and V the volume. What the above formula essentially says is that if we consider a small volume of the gas, small compared to V but large enough to contain many molecules, then the number of molecules in the volume considered would fluctuate. Usually, $\overline{\Delta\sigma^2}$ is rather small but near the critical temperature it increases anomalously since β tends to infinity. Smoluchowski argued correctly that critical opalescence is due to the abnormal increase of $\overline{\Delta\sigma^2}$ near the critical temperature.

Einstein completed the picture. Particle number fluctuations means density fluctuations. A medium in which density is varying spatially is optically inhomogeneous and must therefore scatter light. The scattering coefficient[3] $\mathscr{R}$ was obtained by him (for the case of transverse scattering) as

$$\mathscr{R} = \frac{\pi^2}{18} \frac{RT\beta}{N_A \lambda^4} (n^2 - 1)^2 (n^2 + 2)^2,$$

where λ is the wavelength of light and n the refractive index of the medium.

For an ideal gas, $n \approx 1$ and $\beta = 1/p$, where p is the pressure, whence one obtains

$$\mathscr{R} \text{ (ideal gas)} = \frac{\pi^2}{2\lambda^4} \frac{RT}{N_A p} (n^2 - 1)^2.$$

This is the same as the formula obtained earlier by Lord Rayleigh for scattering by gas molecules, although by a somewhat different route involving the scattering by individual molecules.

Observe the $1/\lambda^4$ factor in both the expressions for $\mathscr{R}$. It is because of this that blue light is scattered much more than any other colour in the visible spectrum.

5.5 Back to the Blue Seas

Let us return to Raman and the colour of the sea. Using the E–S formula Raman estimates that water at 30°C "should scatter light 159 times as strongly as dust-free air under standard conditions". To test this estimate, he sets up an experiment to measure the intensity and the polarization of light scattered by water in a direction transverse to the incident beam. He begins with "ordinary town-supply water" which shows strong scattering when a beam of light is sent through it.

> The track [of the beam] was practically white and showed innumerable motes floating about in the water. Repeated filtration through several thicknesses of Swedish filter paper made an improvement.... A somewhat casual attempt was then made to clear the water by adding alkali and alum and thus throwing out a gelatinous precipitate of hydroxide. This made a further improvement.... The next attempt was made with ordinary distilled water.... This gave immediately a much smaller intensity of light scattering than the tap-water had done after several attempts at filtration. [5.8]

In this way, step by step the quality of the water sample was progressively improved until "the track of light was hardly conspicuous unless a dark background was provided for it to be viewed against". Raman is now ready for quantitative measurements which are performed using saturated ether vapour as a standard, as had been done earlier by Lord Rayleigh in his experiments on gases. Finally

> the observations showed that the scattering power of the sample of water used was 175 times that of dust-free air under standard conditions. This, though not agreeing exactly with the theoretical value, is only slightly in excess, and the difference is not more than can be reasonably explained as due to residual suspended matter present in the sample of water used.

He would like more experiments with still purer samples of water. Meanwhile the experimental results give credibility to his estimate based on the E–S formula. He is now able to assert that a "layer of water 50 metres deep would scatter approximately as much light as 8 kilometres of homogeneous atmosphere; in other words, it should appear nearly as bright as the zenith sky". Of course this rough calculation requires to be touched up with suitable corrections. These are duly made, and the scattering by water at different wavelengths is computed in units of kilometres of dust-free air. Once again, thanks to the $1/\lambda^4$ factor, the colour of water is also blue. Thus, "the blue colour of the scattered light is really due to diffraction, the selective absorption of the water only helping to make it a fuller hue".

All this is fine, but how may one assert that ocean water is clear and transparent so as to permit light scattering at great depths? This is no problem, for:

> There is ample evidence regarding the transparency of oceanic waters.... Writing of the mid-Pacific, J. Y. Buchanan states that a metal plate only 4 inches [~ 10 cm] by 4 inches, painted white and suspended at a depth of 25 fathoms (45 metres), was distinctly seen with sharply defined edges. The plate became indistinct at greater depths, but this was only on account of its smallness and want of steadiness owing to the movement of the boat from which the observations were made. The plate was seen through the glass bottom of a floating tub, in order to eliminate the effect of ripples on the surface of the water. Buchanan adds that the colour of the column of the liquid, 25 fathoms in length, resting on the plate was a pale but pure ultramarine.... Buchanan's observations clearly indicate a high degree of transparency, and show that the colour of the sea really arises from a scattering of light upwards from within the water.

The story is not over yet, for there are the objections of Lord Rayleigh which still need to be taken care of. According to Rayleigh:

> 'When the heavens are overcast the water looks grey and leaden, and even when the clouding is partial, the sea appears grey under the clouds.... One circumstance that may raise doubts is that the blue of the deep sea often looks purer and fuller than that of the sky. I think the explanation is that we are apt to make comparison with that part of the sky which lies near the horizon, whereas the best blue comes from near the zenith. In fact, when the water is smooth and the angle of observation such as to reflect the low sky, the apparent blue of the water is much deteriorated. Under these circumstances a rippling due to wind greatly enhances the colour by reflecting light from higher up. Seen from the deck of a steamer, those parts of the waves which slope towards the observer show the best colour for a like reason.'

Elementary, My Dear Watson!

Raman comments, "The *facts* indicated in this quotation are, of course, quite correct; but the *explanations* given may well be questioned, and a closer examination shows that they require to be modified considerably." What follows is a masterly disposal of the various objections raised by Lord Rayleigh, in a manner that would do credit to a famous trial lawyer!

Raman must have worked really hard on board the ship when all his copassengers were probably amusing themselves with parties and deck games. And no rest afterwards either, for this long paper is completed within a month of return. Incredible speed indeed!

5.6 A Dash up the Mountains

Even as the manuscript is on its way, Raman is already thinking about how solids would scatter light [5.9]. According to Debye's theory of specific heats, there must be elastic waves in a solid due to the thermal agitation of the atoms. In turn these would lead to density fluctuations, whence, by the E–S argument, one can expect light scattering. However, the E–S formula would not predict the intensity correctly since it uses the equipartition principle which is not valid at ordinary temperatures, especially for solids with high "characteristic temperature" (or Debye temperature as we would refer to it today). Hence the scattering predicted by the E–S formula "must be diminished in the ratio which the actual heat content at the temperature of observation bears to the heat content indicated by the equipartition principle". Thus, for example, diamond and quartz would scatter less than would be predicted by the E–S formula.

The reference to quartz is interesting. The Soviet physicists Landsberg and Mandel'shtam were studying light scattering by quartz for many years. Eventually, they discovered the Raman effect independently, and almost at the same time as Raman did. But Raman had the priority of publication and all the honours went to him, causing anguish and some bitterness in the Soviet camp. That story we shall pick up later.

Many experiments on light scattering had shown that the molecules of gases (like nitrogen and oxygen) are non-spherical. Photographic recording was used in all these experiments. Raman wonders whether a direct *visual* observation of the effects of anisotropy is possible. It is noteworthy that Raman attached much importance to visual observation in his experiments on optics. Indeed, as we shall discuss later, even the Raman effect was first detected through visual observations!

Getting back to our story: Raman decides to establish the anisotropy of gas molecules by making visual observations on the polarization of sky-light. There are masking influences like dust and low-lying mists, but to beat them all he goes to Dodabetta, the highest point (8750 feet or about 3000 m above mean sea level) on the Nilgiri mountains in the Western Ghats. On December 4 he makes the observations, and on December 19 he mails a paper to *Nature* [5.10] from Calcutta. Presumably he made a quick trip to South India in the early part of December. One wonders why he did not go to Darjeeling on the Himalayan slopes, especially considering its closeness to Calcutta. Did he regard the altitude as insufficient or was it that he had some other personal business in the South? We do not know. But we do know that the Dodabetta results are "in agreement with the laboratory determinations of Lord Rayleigh".

5.7   Some Reflections

The year had been a hectic one, and a pause was required to gather one's thoughts. Raman does this through an essay entitled *Molecular Diffraction of Light* [5.12], published by the Calcutta University (in 1922). Not surprisingly, he dedicates it to Sir Asutosh Mookerjee with his "warmest admiration and esteem". As Ramaseshan [5.11] comments, "The monograph bristles with ideas, but it is obviously hastily written...."

Raman goes through all the known ideas concerning scattering, his own experiments, and diverse natural phenomena like the polarization of sky-light, twilight, and afterglow, all related to light scattering. But he also looks beyond. Experiments had already revealed that when light is scattered by molecules, there is a certain amount of depolarization. Born and Gerlach had calculated these effects using the Bohr–Sommerfeld model, but their results did not agree with experiments. One more piece of evidence concerning the unsatisfactory nature of the then-existing quantum theory. Raman goes further.

Consider a cavity with perfectly reflecting inner walls. The radiation energy within the cavity would be distributed amongst the different wavelengths according to Planck's law of radiation.

> We may assume further that the enclosed space contains a few molecules of a gas at the same temperature, and ... that the molecules ... merely scatter the radiations incident on them in accordance with the Rayleigh law of scattering.... If we assume that the molecules scatter the waves incident on them continuously, the mechanism provided for the interchange of energy would operate according to the classical laws of electrodynamics, and the final distribution of energy in the enclosure would not be that given by Planck's law but would necessarily be that consistent with the principle of the equipartition of energy, viz., $f(\lambda)\,d\lambda = 8\pi RT\lambda^{-4}d\lambda$. In other words, the distribution of energy in the enclosure which was postulated in the first instance would be altered, and the thermodynamic equilibrium of the system would be upset. As the system was assumed to be initially at the same temperature throughout, such a conclusion is *prima facie* unacceptable, and we must therefore draw the inference either that the Rayleigh law of scattering is not valid or that the molecules do not scatter the radiations incident on them continuously. Since the Rayleigh law of scattering is supported by experiment, at least over a considerable range of wavelengths, it seems more reasonable to accept the latter conclusion, and to infer that molecular *scattering of light* cannot take place in a continuous manner as contemplated by classical electrodynamics. It seems to be difficult, however, to

reconcile this with the hypothesis that light is propagated through space in the form of continuous waves, and we are apparently forced to consider the idea that light itself may consist of highly concentrated bundles or quanta of energy travelling through space. [5.12]

From today's perspective one might wonder what is novel about this last speculation, considering that as early as 1905 Einstein had invoked light quanta while explaining the photoelectric effect. As already mentioned briefly in Chapter 4 and as Pais [5.13] records in detail, the concept of light quanta did not find universal acceptance during those days. When Planck ushered in the (old) quantum theory in 1900 with his theory of black-body radiation, he did so by applying quantization to matter, i.e., his material oscillators. He was unaware of the fact that his proposal implied any need for revising the classical theory of radiation. By contrast, Einstein argued that radiation energy itself was quantized, and that it consisted of independently moving point-like particles of energy $h\nu$. This was a challenge to the wave theory of light enshrined in Maxwell's celebrated equations. Many leading physicists of that era were naturally concerned, since Maxwell's theory was considered irrefutable, especially after Hertz's experiments on radio waves. Thus, seeking reassurance, Planck wrote to Einstein in 1909, "I assume that what happens in the vacuum is rigorously described by Maxwell's equations."

Planck was not the only one to be disturbed. There were several others, like Bohr, for example. Pais remarks, "... even after Einstein's photoelectric law was accepted, almost no one but Einstein himself would have anything to do with light quanta." In fact, as late as 1922 (the year in which Raman wrote his essay), when Einstein was awarded the Nobel Prize, the citation said, "For his services to theoretical physics and especially for the discovery of the photoelectric effect". Of the quantum of the electromagnetic field – not a word!

Raman does not belong to the crowd of doubters. Instead he observes:

> The tendency has therefore been to regard the propagation of light in space as determined by Maxwell's equations, but that these equations for some reason or other fail when we have to deal with the emission or absorption of energy from atoms or molecules.... the idea that emission and absorption are discontinuous while the propagation of light itself is continuous belongs to the class which Poincaré has described as "hybrid hypotheses"... Such hybrid hypotheses... must ultimately make way for a more consistent system of thought.... Newtonian dynamics as applied to the ultimate particles of matter has received a rude shock from quantum theory... and there seems no particular reason why we should necessarily cling to Newtonian dynamics in constructing the mathematical framework of field equations which form the kernel of Maxwell's theory. Rather, to be consistent, it is necessary that the field equations should be modified so as to introduce the concept of the quantum of action.

That step had to wait for six more years till Dirac showed the way.

Raman's own interest is in finding "some experimental support for Einstein's conception that light itself consists of quantum units". He would like an experiment that does not involve "catastrophic changes in atoms and molecules", like the expulsion of an electron, for example. A scattering experiment appeared ideal.

5.8 On with the Charge

The goal had been sharpened. For the next two years, 210 Bow Bazar Street is bustling with activity, with people studying light scattering in all sorts of systems – water, ether, alcohol, benzene, chloroform, other liquids, various liquid mixtures, and also some solids. A comprehensive picture emerges from all this.

1) The scattering at 90° (the geometry invariably employed) has a mixture of a polarized and an unpolarized components, the former arising from density fluctuations and the latter from molecular anisotropy[4].
2) In liquid mixtures, fluctuations of concentration make an additional contribution to the scattering, outweighing, in fact, that due to density fluctuations.
3) Density fluctuations produce scattering not only near the critical temperature but even at temperatures far removed from it.

Of course, not all these are original findings of Raman and his group. But the extensive work done in Calcutta, in conjunction with results of other workers elsewhere, did clarify matters.

There are some useful fall-outs too, like a value for the Avogadro number[5] [5.14]. There is also an explanation for the colour of ice based on results for scattering obtained in the laboratory.

> In his lecture on ice and glaciers, Helmholtz describes very vividly the experience of the Alpine traveller who, traversing the broken surface of the glacier along a narrow ridge, looks down into the crevasses on either side and views with mixed feelings of pleasure and awe their dark blue walls going down to the depths. It is obvious that in such a case as this, the light filtering *down* into the solid mass of transparent ice forming the glacier through the superficial layers or otherwise, has no possibility of returning to the observer above except as the result of internal scattering. [5.15]

We are now in the year 1924, and one would have thought that all that was to be learnt about light scattering had in fact been learnt. Not quite. Many students continue to be put on scattering studies but Raman himself takes time off to glance sideways. Along with Krishnan he examines how the optical anisotropy of molecules as deduced from light scattering experiments could be utilized to interpret the optical and dielectric behaviour of fluids and also the electric, magnetic and mechanical birefringence exhibited by them. In fact he is even able to establish correlations between the optical, electric and magnetic properties of solids in the crystalline state (recall Sec. 4.5).

5.9 Feeble Fluorescence

We have seen that as early as 1922 Raman had envisaged "the possibility that the corpuscular nature of light might come into evidence in scattering". Yet, starting with the blue of the sea all the experimental results were being interpreted in terms of the E–S formula (or modifications of it), and this formula had been derived using classical electromagnetic theory. Did it mean that light quanta were forgotten? Not quite.

In fact, with the discovery of the Compton effect in 1923, Raman begins to think of an optical analogue of the Compton effect, i.e., scattering of light with a *change* of frequency – in other words, a *new* type of light scattering process. Already there were hints that something like this might be happening. Apart from scattering due to fluctuations (which was polarized) and an extra contribution due to molecular anisotropy (which was depolarized), there appeared to be *yet another* (i.e., a third), though exceedingly feeble, component in the scattering.

> At a very early stage in our investigations, we came across a new and entirely unexpected phenomenon. As early as 1923, it was noticed that when sunlight filtered through a violet glass passes through certain liquids and solids, e.g., water or ice, the scattered rays emerging from the track of the incident beam through the substance contained certain rays not present in the incident beam. The observations were made with colour filters [see Fig. 5.1]. A green glass filter was used which cut off all light if placed between the violet filter and the substance. On transferring the glass to a place between the substance and the observer's eye, the track continued to be visible though feebly. This is a clear proof of a real transformation of light from a violet into a green ray. The most careful chemical purification of the substance failed to eliminate the phenomenon. [5.16]

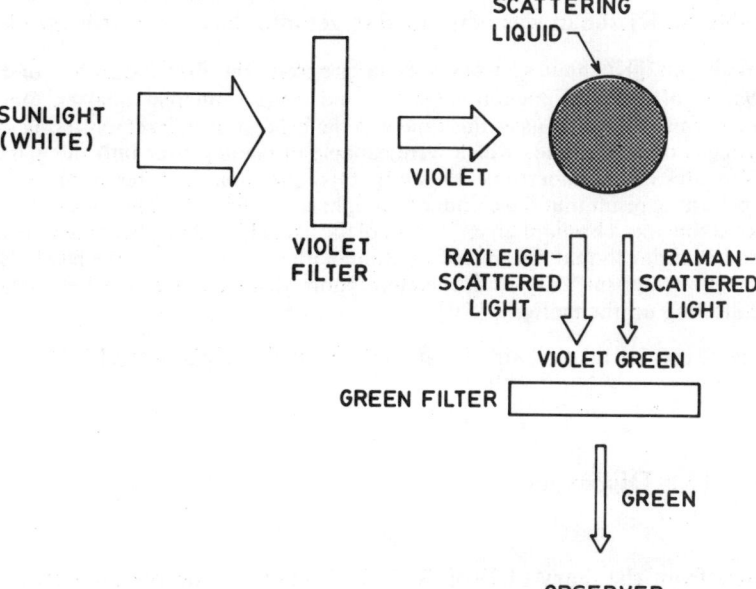

Figure 5.1 System of complementary filters devised by Raman to detect light scattering with a change of frequency. The observation is made visually, the advantage being that a number of scattering substances can be studied rapidly. For quantitative studies Raman used a mercury arc as the source of light and a quartz spectrograph for recording the spectra.

Ramanathan [5.17] noticed this phenomenon as far back as 1923 but he dismissed it as a "trace of fluorescence". In 1925 Krishnan [5.18] noticed it again during an extensive series of experiments. From time to time various efforts were made to study this "feeble fluorescence" and on one occasion even a spectroscopic study was attempted but they all failed owing to the lack of a sufficiently powerful light source. As Raman recounted later:

> Its real significance as a twin brother to the Compton effect first became clear to me at the end of 1927 when I was preoccupied with the theory of the subject[16]. I regarded the ejection of the electron in the Compton effect essentially as a fluctuation of the atom of the same kind as would be induced by heating the atom to a sufficiently high temperature, and the so-called directed Compton effect as merely an unsymmetrical emission of radiation from the atom which occurs at the same time as the fluctuation in its electrical state. The conception of fluctuation is a very familiar one in optical and kinetic theory, and in fact all our experimental results in the field of light scattering had been interpreted with its aid. There was, therefore, every reason to expect that radiations of altered wavelength corresponding to fluctuations in the state of scattering molecules should be observed also in the case of ordinary light. [5.16]

A follow-up was needed, and the person selected to do it was Krishnan. According to Ramaseshan [5.11], Krishnan was at that time preoccupied with various theoretical investigations. It was more than a year since he had been near experiments, and Raman felt that prolonged isolation from experiments was not desirable. So Krishnan was persuaded to get into the act towards the close of 1927.

> While his [Krishnan's] work was in progress, the first indication of the true nature of the phenomenon came to hand from a different quarter. One of the problems interesting us at this time was the behaviour in light scattering of highly viscous organic liquids which were capable of passing over into the glassy state. Venkateswaran undertook to study this question, and reported the highly-interesting result that the colour of sunlight scattered in a highly purified sample of glycerine was a brilliant green instead of the usual blue. The phenomenon appeared to be similar to that discovered by Ramanathan in water and the alcohols, but of much greater intensity, and, therefore, more easily studied. No time was lost in following up the matter. [5.20]

At this point, let us pick up the story from Krishnan's diary [5.21].

5.10 The Discovery

Extracts from the diary of Prof. K. S. Krishnan for the period February 5 to 28, 1928:

> February 5, 1928
>
> For the last three or four days, I have been devoting all my time to fluorescence. The subject promises to open out a wide field for research, since at present there is no theory of fluorescence which could explain even the outstanding facts.

Studied anthracene vapour. It exhibits strong fluorescence which does not show any polarization when viewed through a double image prism. Prof. [Raman] has been working with me all the time.

Recently. Prof. has also been working with Mr Venkateswaran on the fluorescence exhibited by many aromatic liquids in the near-ultraviolet region present in sunlight and the fluorescence of some of the liquids is found to be strongly *polarized*. However, in view of the fact that the fluorescence of anthracene vapour does not show any polarization, Prof. has asked me to verify again his observations on the polarization of liquids.

February 7, Tuesday

Tried to verify the polarization of the fluorescence exhibited by some of the aromatic liquids in the near-ultraviolet region. Incidentally, discovered that all pure liquids show a fairly intense fluorescence also in the visible region, and what is much more interesting, all of them are strongly polarized, the polarization being greater for the aliphatics than for the aromatics. In fact, the polarization of the fluorescent light seems in general to run parallel with the polarization of the scattered light, i.e., the polarization of the fluorescent light is greater the smaller the optical anisotropy of the molecule.

When I told Prof. about the results, he would not believe that *all liquids* can show *polarized* fluorescence and that too *in the visible region*. When he came into the room, I had a bulb of pentane in the tank, a blue-violet filter in the path of incident light, and when he observed the track with a combination of green and yellow filters, he remarked, "You do not mean to suggest, Krishnan, that *all that* is fluorescence." However, when he transferred the green-yellow combination also to the path of the incident light, he could not detect a trace of the track. He was very much excited and repeated several times that *it was an amazing result*. One after another, the whole series of liquids was examined and every one of them showed the phenomenon without exception. He wondered how we missed discovering *all that* five years ago.

In the afternoon, took some measurements on the polarization of fluorescence.

After meals at night, Venkateswaran and myself were chatting together in our room when Prof. suddenly came to the house (about 9 p.m) and called for me. When we went down, we found he was much excited and had come to tell me that what we had observed that morning must be the Kramers–Heisenberg effect we had been looking for all these days. We therefore agreed to call the effect *modified scattering*. We were talking in front of our house for more than a quarter of an hour when he repeatedly emphasized the exciting nature of the discovery[7].

February 8, Wednesday

Took some preliminary measurements of the polarization of the modified scattering by some typical liquids.

February 9, Thursday

Set up this morning the long telescope and made preliminary arrangements for observing the effect with vapours. Before the arrangements were completed, Prof. left for the college for his lecture.

In the afternoon, tried ether vapour and it was surprising that the modified radiation was very conspicuous. Tried a number of others in quick succession without, however, the same success.

When Prof. came from the college at about three, I announced to him the result, and there was still enough sunlight for him to see for himself. He ran about the place, shouting all the time that it was a first-rate discovery, that he was feeling miserable during the lecture because he had to leave the experiment, and that, however, he was fully confident that I would not let the grass grow under my feet till I discovered the phenomenon in gases. He asked me to call in everybody in the place to see the effect and immediately arranged in a most dramatic manner with the mechanics to make arrangements for examining the vapours at high temperatures.

Evening was busy and when Prof. returned after his walk he told me that I ought to tackle big problems like that and asked me to take up the problem of the experimental evidence for the spinning electron after this work was over.

February 10 to 15

Studied a number of vapours, [and] though a number of them showed the effect, nothing definite could be said regarding the polarization of the modified scattering.

February 16, Thursday

Studied today pentane vapour at high temperature and it showed a conspicuous polarization in the modified scattering. We sent a note today to *Nature* on the subject under the title "A new type of secondary radiation".

February 17, Friday

Prof. confirmed the polarization of fluorescence in pentane vapour. I am having some trouble with my left eye. Prof. has promised to make all observations himself for some time to come.

February 19 to 26

Studied a number of other vapours.

February 27, Monday

Religious ceremony in the house. Did not go to the Association.

February 28, Tuesday

Went to the Association only in the afternoon. Prof. was there and we proceeded to examine the influence of the wavelength of the incident light on the phenomenon. Used the usual blue-violet filter coupled with a uranium glass, the range of wavelengths transmitted by the combination being much narrower than that transmitted by the blue-violet filter alone. On examining the track with a direct vision spectroscope[8], we found to our great surprise [that] the modified scattering was separated from the scattering corresponding to the incident light by a dark region.

The so-called feeble fluorescence was not fluorescence at all but demonstrated to be something totally different: the Raman effect[9] had been discovered. A newspaper announcement was made the following day, i.e., February 29 (it was a leap year!). In addition to a brief account of the discovery, the news item stated that Prof. Raman would deliver a lecture demonstrating the phenomenon on March 16 in Bangalore. We shall discuss that lecture shortly.

Why was this discovery not made earlier? As Raman himself has pointed out, it was for lack of a sufficiently strong source of illumination. In all the early studies the excitation source was sunlight, of which, as Raman once remarked, there was a plentiful supply in Calcutta. But the study of the "feeble fluorescence" required an even more powerful source, which, however, was then not available. In early 1927 a seven-inch (~ 18 cm) refracting telescope was acquired by the IACS. Raman promptly coupled this to a short-focus lens to condense sunlight, thereby obtaining a more powerful source of illumination suitable for visual observations. Once the phenomenon was clearly seen through the use of complementary filters, use of the spectroscope was the next obvious step, which, as we have seen, was taken on February 28.

5.11 The Mechanism

This is an appropriate juncture to say a few words about the mechanism of the Raman effect. Further details likely to be of interest to physicists may be found in Chapter 6.

The Raman effect is a quantum-mechanical phenomenon, and to understand it we must consider the energy levels of the scattering system. Let us, as Smekal [5.22] first did in 1923, consider the energy levels of an atom. The usual optical processes one is interested in are emission and absorption, both of which are easily understood in terms of transitions between various atomic levels. Thus in Fig. 5.2 a transition from state l to state k, wherein the atom goes from a higher to a lower energy level, is accompanied by the emission of radiation. Likewise, an absorption of radiation energy is necessary to produce the transition $k \to l$.

Smekal considered scattering and showed that an atom of mass M in an energy state E_l moving with a velocity v can, upon collision with a light quantum of energy hv, pass on to another energy state E_k and experience as well a change of velocity to v'. From the principle of conservation of energy we have

$$\tfrac{1}{2}Mv^2 + E_l + hv = \tfrac{1}{2}Mv'^2 + E_k + hv',$$

where h is Planck's constant and v' the frequency of the scattered light. Smekal showed that the change in velocity was negligible so that

$$E_l + hv = E_k + hv' \quad \text{or} \quad v' = v - (E_k - E_l)/h,$$

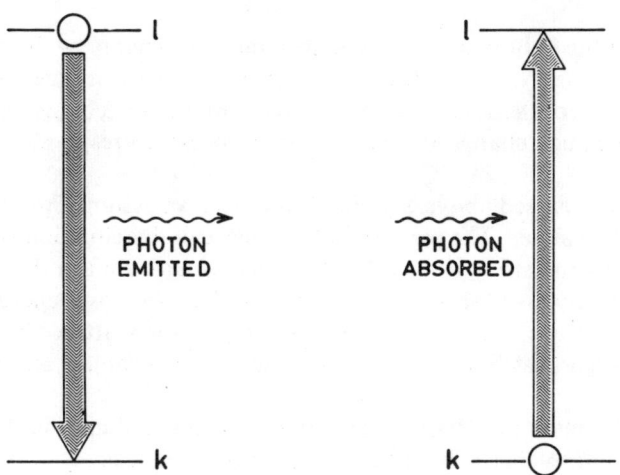

Figure 5.2 Emission and absorption of light by an atom can be understood in terms of transitions between the various energy levels of the atom. When the atom makes a transition from a higher energy state l to a lower one k, a light quantum, i.e., a photon, is emitted. Conversely, absorption causes the atom to go to a state of higher energy.

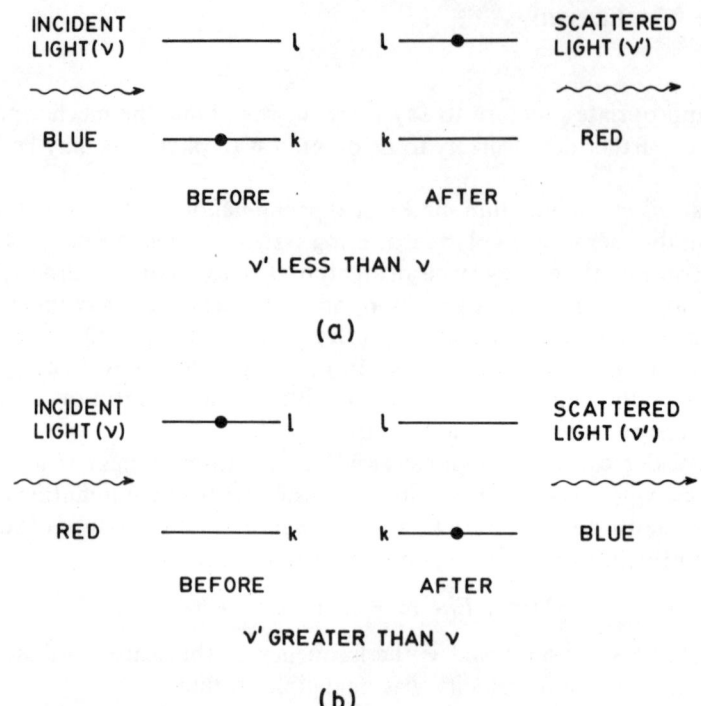

Figure 5.3 Schematic illustration of the changes in the atomic states following Raman scattering. There are two possibilities, (**a**) and (**b**). For further explanations, see text.

which means that the light quantum gives up some of its energy to the atom and is scattered with less energy. This is Raman scattering. If Δv the frequency difference between v' and v is zero then the light is scattered without any change of frequency, and the atom does not change its state l. This process corresponds to Rayleigh scattering.

In the example discussed above (see Fig. 5.3a), we have assumed the state k to be lower in energy than state l. The practical consequence of this situation is that if blue light is scattered by the atom, it would suffer a wavelength shift in the direction of red. The converse process is also possible in which the atom loses energy, as in the scattering process shown in Fig. 5.3b. Correspondingly, the scattered light is shifted towards blue. Such a possibility was first pointed out by Kramers and Heisenberg [5.23] in 1925.

The paper of Kramers and Heisenberg is an important landmark in the history of quantum mechanics. As far as light scattering with a frequency shift is concerned, this paper considerably amplified Smekal's rudimentary ideas. In particular, Kramers and Heisenberg made the following point:

> Under the influence of irradiation with monochromatic light, an atom not only emits coherent spherical light; it also emits systems of incoherent spherical waves, whose frequencies can be represented as combinations of the incident frequency

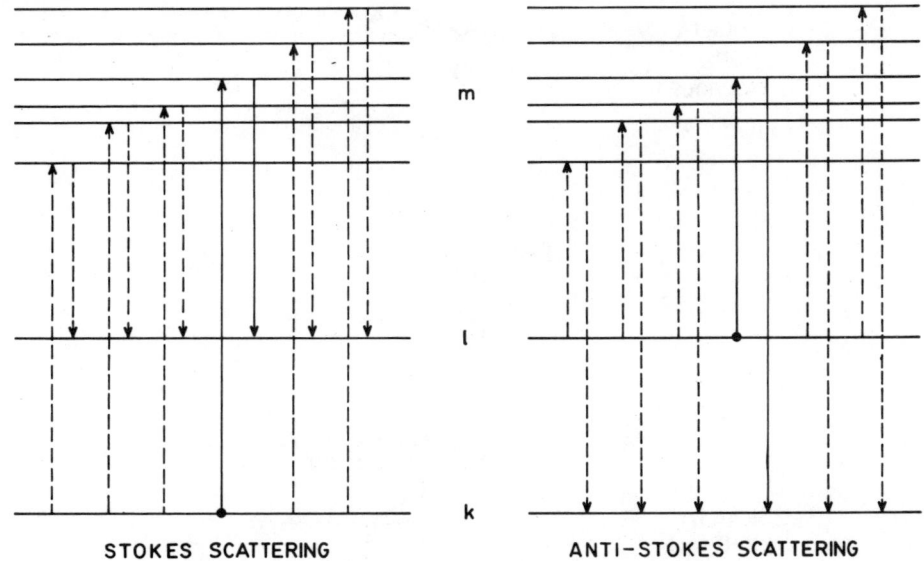

Figure 5.4 In Raman scattering, the atom changes its state, the transition occurring via an intermediate state m. The bold lines indicate the two-step process. But there are many choices for m and therefore one must consider also all the additional possibilities shown with dotted lines. Stokes and anti-Stokes scattering are explained in the text.

with other frequencies that correspond to possible transitions to other stationary states [5.23].

Apart from introducing the concept that light scattering with frequency change is incoherent, the Kramers–Heisenberg paper is important also for the reason that it introduced the idea of the virtual intermediate state. We have already seen that in Raman scattering, a change of state $l \to k$ or $k \to l$ is involved. Kramers and Heisenberg pointed out that such changes do not occur directly. Rather the atom takes a roundabout route via an intermediate state m. But there are so many states that are candidates for the intermediate state m. Which of these is relevant? All possibilities must be considered! See Fig. 5.4.

Where does fluorescence fit into all this? Certain substances, when illuminated, emit radiation having wavelengths different from that of the light falling on them. Observation of this phenomenon dates back to the sixteenth century but the first real study of the subject was made by Herschel and Sir David Brewster in the middle of the last century. Sir George Stokes was of the opinion that the wavelength of the emitted radiation was always greater than that of the exciting light. This is known as Stokes law[10].

The phenomenon of fluorescence can be qualitatively understood in terms of the schematic diagram of Fig. 5.5. The incident light is absorbed and excites the atom from level k to some level l, say. The atom then loses energy by some mechanism not involving the emission of light, and arrives at the level m. From m the atom makes a

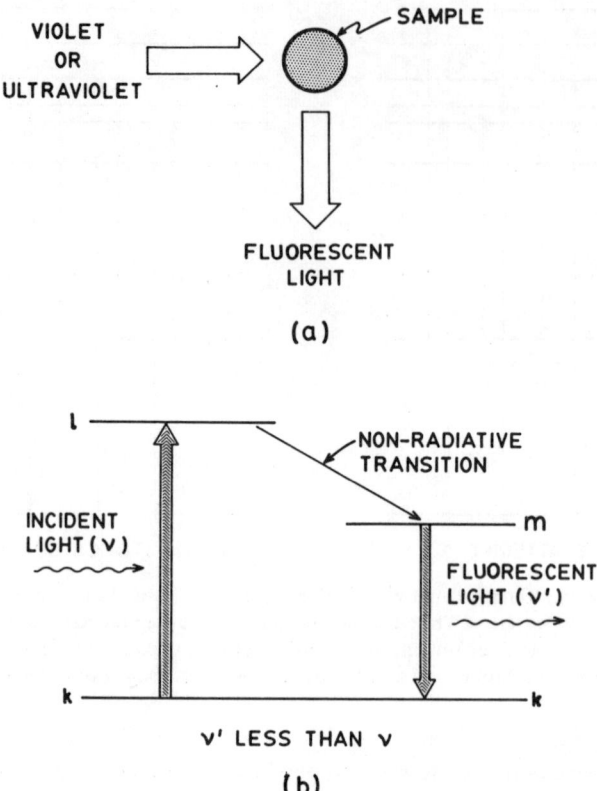

Figure 5.5 (a) Schematic of a fluorescence experiment. (b) The changes in the energy level. First the absorption of the incident light causes the atom to go from the state k to a higher energy state l. The atom then makes a transition $l \to m$ *without* emitting any radiation. Finally the atom makes the transition $m \to k$, this time emitting light which appears as the fluorescent radiation.

jump back to the original level k, emitting light in the process. Since the energy difference $(E_m - E_k)$ is less than the difference $(E_l - E_k)$, the emitted light or the *fluorescent radiation* as it is called, has a lower frequency than the incident light. By analogy, if the spectral line in the Raman spectrum has a frequency less than that of the incident line, one refers to the spectral line concerned as a Stokes line. Conversely, Raman lines with frequencies greater than that of the incident light are referred to as anti-Stokes lines.

A few words now about polarization effects in Raman scattering. A characteristic feature of the scattered radiation in the Raman effect is that it is strongly polarized. In this respect, therefore, Raman scattering is somewhat like Rayleigh scattering. However, it differs from the latter in that the scattered radiation is frequency-shifted. As far as frequency changes are concerned, the Raman effect therefore has something in common with fluorescence. However, fluorescent radiation is *not*

Elementary, My Dear Watson! 203

Table 5.1 Comparison of Rayleigh scattering with fluorescence and the so-called feeble fluorescence (later identified as the Raman effect).

Phenomenon	Polarization of emerging radiation	Frequency change
Rayleigh scattering	Yes	No
Fluorescence	No	Yes
Feeble fluorescence	Yes	Yes

polarized. Thus Raman scattering is like Rayleigh scattering with respect to polarization effects and like fluorescence as regards frequency shifts (see Fig. 5.6 and Table 5.1). As we shall see in Chapter 6, the physical processes leading to the Raman effect are quite different from those responsible for either Rayleigh scattering or fluorescence.

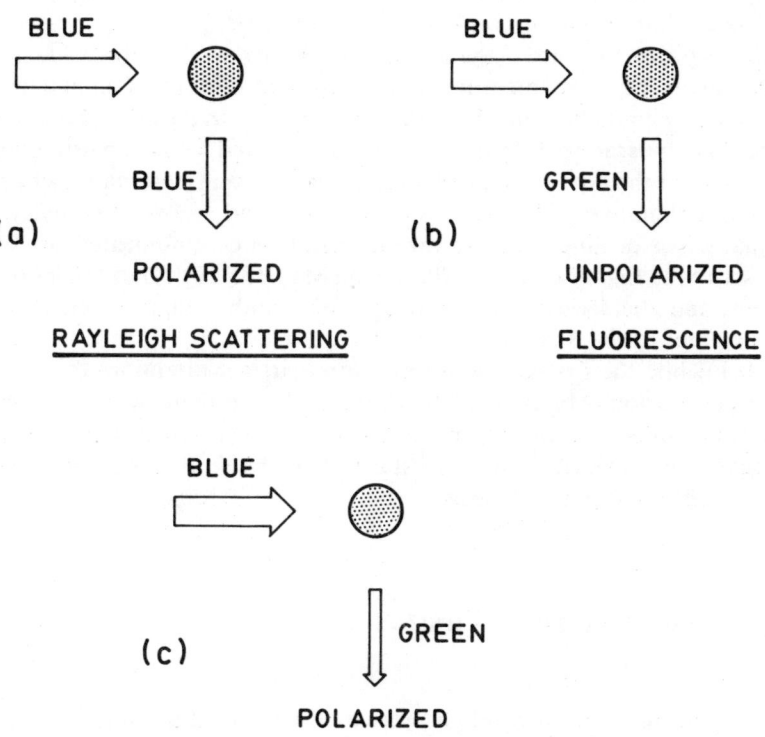

Figure 5.6 Schematic illustration of the differences between (a) Rayleigh scattering, (b) the usual fluorescence and (c) the mysterious feeble fluorescence. In 1928 Raman and Krishnan recognized that (c) was a new phenomenon quite distinct from (a) and (b). Subsequent spectroscopic examination revealed that the frequency shifts in (c) are related to the vibrational frequencies of the molecules.

The discovery was thus something completely new. In late 1928 Raman and Krishnan investigated in detail the polarization characteristics of Raman scattering.

5.12 The Follow-up

Events now move rapidly. On March 8, 1928, Raman communicates a letter to *Nature* [5.25] entitled "A change of wavelength in light scattering", being a sequel to the letter [5.26] mailed earlier on February 16 by Krishnan and himself (see Krishnan's diary entry for that date). A week later, he inaugurates the South India Science Association at Bangalore with an address titled "A new radiation". This lecture was printed and distributed to many, including some abroad – a forerunner of today's preprint[11]. The text of the lecture was also formally published in the *Indian Journal of Physics* on March 31, 1928 [5.27].

In this paper, the third in the series, Raman not only describes in detail the various experiments but also presents photographs of spectra of monochromatic light scattered by liquid benzene. Raman also points out that the frequency change which occurs in the scattered light is related to infra-red vibrational frequencies of the molecules of the liquid. He further reports that visual observations had been made on a large number of liquids, some gases, ice and amorphous materials. A more formal paper dealing with the various results is communicated by Krishnan and himself to the *Indian Journal of Physics* on May 7 [5.28]. A few other notes to *Nature* follow, and the series is wrapped up with another long paper, this time in the *Proceedings of the Royal Society*[12] [5.29].

Meanwhile the first two notes in *Nature* attract attention almost immediately after publication (March 31). The French, who, in their own way, were also after frequency shifts in scattering, move in swiftly with two papers in April and as many as seven in June! It is at this stage that the first paper of Landsberg and Mandel'shtam appears in print.

5.13 Who Got There First?

Studies on the scattering of light were avidly pursued not only in Calcutta, but in several other places including the Soviet Union where Landsberg and Mandel'shtam were active in this area. Mandel'shtam was the senior partner of the Soviet pair, and as early as 1913 he had, following Einstein and Smoluchowski, made a theoretical study of the scattering of light by the interface between two media, the interface being disturbed by thermal fluctuations[13]. Mandel'shtam expanded the interface roughness as a two-dimensional Fourier series and regarded each term as if it were a diffraction grating. However, the gratings were treated as being static.

Around the same time, Debye published his famous theory of specific heats[14] in which the heat motion of the atoms was visualized in terms of elastic waves propagating in the medium. However, Mandel'shtam failed to see the connection between his Fourier components and Debye's elastic waves. That connection became clear only five years later, whereupon he prepared a note entitled "Light scattering in an inhomogeneous medium". But for some reason the publication was delayed till 1926 by which time Brillouin, working independently, had published his now famous paper on the subject.

The highlight of the Brillouin theory is that when light is scattered by a crystal, it would not only be diffracted by the Debye elastic waves acting as gratings but in addition there would be frequency shifts in the scattered light consequent to the elastic waves propagating with the velocity of sound. Thus the scattered light would contain frequency components

$$v' = v \pm \Delta v,$$

where

$$\Delta v = 2nv(v/c)\sin(\theta/2),$$

where n is the refractive index, v the velocity of elastic waves, θ the scattering angle and c the velocity of light.

The elastic waves postulated by Debye do not exhaust all the possible wave motions of atoms in a solid consequent to heat motion. Other types of waves are possible (in some solids), associated with the so-called optic branches[15]. That optic branches can exist had been known from the pioneering work of Max Born and Th. von Karman in 1912. Nevertheless, neither Mandel'shtam nor any one else at that time applied his mind to the scattering of light by atomic vibrational waves corresponding to the optic branches. As it turns out, it is such scattering that corresponds to the Raman effect in crystals[16].

Light scattering studies in the Soviet Union commenced in 1925 at Mandel'shtam's initiative. Solids were chosen as the scattering medium and the idea was to look for Brillouin scattering. Quartz was selected for a start since large single crystals could be had. But it turned out that crystals of good quality were not all that easy to obtain. In a historical review of the Soviet work, Fabelinskii [5.31] describes how Landsberg combed shops trading in antiques so that he could pick up a good quartz piece from seals of Russian nobility (by then extinct!). However, such a quest was not without mishaps since several dishonest dealers palmed off clever imitation seals made in cut glass. In the laboratory one can easily distinguish quartz from glass but in a shop this was not easy, and Landsberg was cheated several times. Fabelinskii remarks, "If these shop dealers could have known what injury they would be inflicting on science in general and on the personal finances of one scientist in particular, they might have acted differently."

One problem with quartz is that it tends to have inclusions and various other defects all of which produce background fog (i.e., Rayleigh scattering) masking the scattering due to thermal fluctuations. Nevertheless, by 1928 Landsberg and Mahdel'shtam had obtained a specimen suitable for experimentation. Their approach to experiments was conventional. No sunlight, no complementary filters and no visual observations. A mercury lamp was used as the source, and the scattered light was analysed with a quartz spectrograph. Long exposures were necessary but they were rewarding for they revealed spectral lines distinct from that corresponding to the exciting frequency. However, Δv was much larger than expected from the Brillouin theory. This was puzzling, and several tedious tests were made to confirm the finding. Fabelinskii observes, "Landsberg and Mandel'shtam in their delicate and laborious experiments definitely proved the existence of a new optical effect and recorded combinational lines in the spectrum on February 21, 1928." By way of substantiating this, Fabelinskii quotes from a letter written by Mandel'shtam to another Soviet physicist in which Mandel'shtam says, "We first noted the appearance of the new lines on February 21, 1928. On a negative from an experiment of February 23–24 (exposure time of 15 hours), the new lines were clearly visible." Febelinskii reproduces this spectrogram which also carries the dates.

Landsberg and Mandel'shtam did not publish their results but in April of that year, Landsberg presented the work at a Moscow seminar. "Landsberg's report... made a very strong impression on the audience. In fact, one of the physicists present said, 'This cannot be correct, because it would mean that we were hearing a molecule talk'. [5.31]" It was only on May 6, 1928, that Landsberg and Mandel'shtam ventured to communicate their findings in a small note to *Naturwissenschaften*. But by then they were late and their claim to priority was lost. Naturally there was much disappointment and even some bitterness in Soviet circles. Landau and Lifshitz, for example, refer to Raman scattering as Smekal–Mandel'shtam–Raman scattering [5.32] while several others in the Soviet Union ignored Raman's contribution altogether, calling it combinational scattering. In fact, in the historical review already alluded to, Fabelinskii consistently avoids using the words Raman effect or Raman scattering, although *fifty* years had passed since the discovery of the effect! The abstract of his article begins with the words, "Soon after research on molecular scattering of light began in the Soviet Union, Landsberg and Mandel'shtam made one of the most important discoveries in physics in this century: combinational scattering of light." At this point, Dave Parsons who has translated this article into English (for the translation journal *Soviet Physics Uspekhi*, published by the American Institute of Physics) is forced to comment in parentheses, "or the Raman effect as it is known outside the Soviet Union".

Fabelinskii is a student of Landsberg and his chagrin in understandable but not so are some of the comments he makes which appear to be a product of both prejudice and inadequate researching. To set the matter in perspective, we first reproduce, as Fabelinskii does, some of the early short notes published by Raman as well as by the Soviets.

A New Type of Secondary Radiation[17]

If we assume that the X-ray scattering of the 'unmodified' type observed by Prof. Compton corresponds to the normal or average state of the atoms and molecules, while the 'modified' scattering of altered wavelength corresponds to their fluctuations from that state, it would follow that we should expect also in the case of ordinary light two types of scattering, one determined by the normal optical properties of the atoms or molecules, and another representing the effect of their fluctuations from their normal state. It accordingly becomes necessary to test whether this is actually the case. The experiments we have made have confirmed this anticipation, and shown that in every case in which light is scattered by the molecules in dust-free liquids or gases, the diffuse radiation of the ordinary kind, having the same wavelength as the incident beam, is accompanied by a modified scattered radiation of degraded frequency.

The new type of light scattering discovered by us naturally requires very powerful illumination for its observation. In our experiments, a beam of sunlight was converged successively by a telescope objective of 18 cm aperture and 230 cm focal length, and by a second lens of 5 cm focal length. At the focus of the second lens was placed the scattering material, which is either a liquid (carefully purified by repeated distillation *in vacuo*) or its dust-free vapour. To detect the presence of a modified scattered radiation, the method of complementary light filters was used. A blue-violet filter, when coupled with a yellow-green filter and placed in the incident light, completely extinguished the track of the light through the liquid or vapour. The reappearance of the track when the yellow filter is transferred to a place between it and the observer's eye is proof of the existence of a modified scattered radiation. Spectroscopic confirmation is also available.

Some sixty different common liquids have been examined in this way, and every one of them showed the effect in greater or less degree. That the effect is a true scattering and not a fluorescence is indicated in the first place by its feebleness in comparison with the ordinary scattering, and secondly by its polarization, which is in many cases quite strong and comparable with the polarization of the ordinary scattering. The investigation is naturally much more difficult in the case of gases and vapours, owing to the excessive feebleness of the effect. Nevertheless, when the vapour is of sufficient density, for example with ether or amylene, the modified scattering is readily demonstrable.

C. V. RAMAN
K. S. KRISHNAN

210 Bowbazar Street
Calcutta, India
Feb. 16

A Change of Wavelength in Light Scattering[18]

Further observations by Mr Krishnan and myself on the new kind of light scattering discovered by us have been made and have led to some very surprising and interesting results.

In order to convince ourselves that the secondary radiation observed by us was a true scattering and not a fluorescence, we proceeded to examine the effect in greater detail. The principal difficulty in observing the effect with gases and vapours was its excessive feebleness. In the case of substances of sufficient light-scattering power, this difficulty was overcome by using an enclosed bulb and heating it up so as to secure an adequate density of vapour. Using a blue-violet filter in the track of the incident light, and a complementary green-yellow filter in front of the observer's eye, the modified scattered radiation was observed with a number of organic vapours, and it was even possible to determine its state of polarization. It was found that in certain cases, for example, pentane, it was strongly polarized, while in others, as for example naphthalene, it was only feebly so, the behaviour being parallel to that observed in the liquid state. Liquid carbon dioxide in a steel observation vessel was studied, and exhibited the modified scattering to a notable extent. When a cloud was formed within the vessel by expansion, the modified scattering brightened up at the same time as the ordinary or classical scattering. The conclusion is thus reached that the radiations of altered wavelength from neighbouring molecules are coherent with each other.

A greater surprise was provided by the spectroscopic observations. Using sunlight with a blue filter as the illuminant, the modified scattered radiation was readily detected by the appearance in the spectrum of the scattered light of radiations absent from the incident light. With a suitably chosen filter in the incident light, the classical and modified scatterings appeared as separate regions in the spectrum separated by a dark region. This encouraged us to use a mercury arc as the source of light, all radiations of longer wavelength than 4358 Å being cut out by a filter. The scattered radiations when examined with a spectroscope showed some sharp bright lines additional to those present in the incident light, their wavelength being longer than 4358 Å; at least two such lines were prominent and appeared to be accompanied by some fainter lines, and in addition a continuous spectrum. The relation of frequencies between the new lines and those present in the incident light is being investigated by photographing and measuring the spectra. The preliminary visual observations appear to indicate that the position of the principal modified lines is the same for all substances, though their intensity and that of the continuous spectrum do vary with their chemical nature.

C. V. RAMAN

210 Bowbazar Street
Calcutta, Mar. 8

Concerning this paper, Ramaseshan [5.11] has this to say:

> Story goes that the note sent to *Nature*... on March 8th was rejected by a referee but published anyway by the editor.
>
> The following is an extract of the cable to *Nature* from Prof. R. W. Wood, the distinguished optical physicist of Johns Hopkins University, USA:
>
> "Prof. Raman's brilliant and surprising discovery; I have verified his discovery in every particular. Raman's discovery thus makes it possible to investigate remote infra-red regions hitherto unexplored. It appears to me that this very beautiful discovery which resulted from Raman's long and patient study of the phenomenon of light scattering is one of the most convincing proofs of the quantum theory."

I wrote to John Maddox, the present editor of *Nature*, to check on the story of the rejection. His reply said:

> I am very sorry we cannot help with your request. The difficulty is that our records from this period were destroyed mostly during the Second World War! But I would not at all be surprised if what you say about Raman's paper is in fact correct. This was at the time when we rejected Krebs' paper on the Krebs cycle and (less well-known) Fermi's paper on beta decay.

There are of course many other examples of epoch-making papers which were first rejected. Anyway, Raman's paper did make it into print, even if it faced problems initially.

Parenthetically it may be added that after seeing the spectral shifts with a pocket spectroscope, Raman photographed the spectrum with a baby Hilger quartz spectrograph. In 1929 a slightly bigger spectrograph was purchased "through the kindness of Mr G. D. Birla and Mr Sajan Kumar Chowdhuri who each gave a donation of Rs. 1,500". Much of the subsequent work at the IACS was done using this spectrograph.

Let us now turn to the first paper of Landsberg and Mandel'shtam.

A New Effect in the Scattering of Light in Crystals[19]

In a study of the molecular scattering of light in solids which we undertook in order to determine whether there is a change in wavelength, as could be expected on the basis of the Debye theory of specific heats, we found a new effect, which we believe will be of definite interest.

This effect consists of a change in wavelength, but the change is of an order of magnitude different from what we expected, and it is of an entirely different origin.

In the experiments, an intense light beam from a quartz mercury lamp was passed through a quartz crystal. The light reflected at an angle of 90° with respect to the incident beam was studied with a quartz spectrograph. The standard measures were taken to combat spurious light signals[1]. A reference spectrum was obtained by reflecting light from black velvet. The exposure time was 2–14 h.

Experiments were carried out with two different quartz samples. On all the spectrograms, all the mercury lines were accompanied by clearly defined satellites at a slightly higher wavelength. In addition, near each line we found a hint of two or three less well defined lines.

No traces of these satellites were found in the reference spectrum.

One of the spectrograms is shown in the figure [not reproduced here]. Approximate measurements from the spectrograms show that the wavelength for the brightest satellite behaves as indicated in the accompanying table. We carried out a variety of control experiments to establish firmly that the observed lines were not due to spurious light signals.

We believe that the following experiment was decisive. Between the scattering quartz crystal and the spectrograph slit we placed a quartz vessel filled with mercury vapour, which absorbed all the light at 2536 Å. On the spectrogram we did not find this line – only the satellites.

This experiment was solid proof that the satellites were in fact at a wavelength different from that of the fundamental line.

At this point we think it is too early to give a definite explanation of the observed effect.

One of the possible theoretical explanations runs as follows: Certain natural infra-red frequencies of quartz may be excited, at the expense of the energy of the scattered light. The energy of the scattered quantum and thus its frequency will decrease by amounts corresponding to an infra-red quantum.

When we start from the frequency corresponding to the wavelength[2] $\lambda = 20.7 \mu$ we find good agreement between the calculated and measured values (see the accompanying table).

At present we cannot say how closely this effect is related to the effect described by Raman[3], because Raman's description was so brief.

G. Landsberg and L. Mandel'shtam, Institute of Theoretical Physics, First University, Moscow, May 6, 1928.

Note added in proof. More recently, we have studied light scattering in Iceland spar and have observed the same effect. The change in the wavelength is greater in

λ(Å)	$\Delta\lambda$(Å)	
	Observation	Calculation
2536	About 30	30.8
3126	About 47	47
3650	About 63	64

this material than in quartz. The change corresponds to an infra-red frequency, for which the corresponding wavelength is $\lambda = 9.1\,\mu$. [See Note 20.]

1 Landsberg, G. Zs. Phys., 1927, Bd. 43, S. 773; Bd. 45, S. 442
2 Rubens, Nichols. Ann. d. Phys., 1927, Bd. 60, S. 418. The value 20.7 is used uncorrected, since this refers to our preliminary measurements.
3 Raman, C. V., Krishnan, K. S. Nature, 31 March 1928, v. 121, p. 501; 21 April 1928, ibid., p. 619.

Having reproduced the papers, Fabelinskii makes several observations and comments critical of Raman. In a nutshell, he alleges inadequate experimentation and incomplete understanding of the effect on the part of Raman. Criticism concerning the experiment centres around several statements made in the two letters to *Nature*, all of which stem in a sense from the use of the visual observation technique. Visual observation was Raman's style, and it has some merits to commend it. Provided it is defect-free, the human eye is a remarkably sensitive photon detector. Let us not forget that the Rutherford school made skilful use of visual observations even in nuclear physics experiments.

Raman's reliance on visual observations is closely linked to his keenness to rapidly try out various ideas, besides experimenting with various systems. Of course Raman had the advantage of working with liquids, which generally have greater scattering power than solids. Recall that Raman and Krishnan were able to *see* clearly even the spectral shift!

One disadvantage of the visual method is that it does not afford a quantitative measurement of the spectral shift. Raman was well aware of that. While remarking [5.25] that "the position of the principal modified lines is the same for all substances", he qualifies the statement by pointing out that (i) the observations are *preliminary*, and (ii) the frequency shifts *are* being studied by photographing the spectra. This in fact was soon done, and Raman *did* present such spectra in his public lecture in Bangalore on March 16, 1928. One feels that the preliminary comments of Raman (including the one on whether the scattered radiation is coherent or incoherent) have been overplayed by Fabelinskii. Certainly they were not as much off the mark as Fermi's speculation [5.34] that the radioactivity he observed on bombarding uranium with slow neutrons was due to the formation of transuranic elements[21]. In any case, Raman suitably revised his preliminary observations barely two weeks later, after the spectra had been photographed.

Let us now turn to Fabelinskii's allegation of incomplete understanding on the part of Raman. Essentially it is based on (i) Raman's references to the analogue of the Compton effect, and (ii) Raman not identifying the frequency shift with the vibrational frequencies of the molecule. Fabelinskii remarks:

> The first prediction of this new effect – combinational scattering of light – was offered by Smekal (1923) on the basis of quantum-mechanical considerations. Credit for the complete quantum-mechanical theory goes to Kramers and Heisenberg whose paper appeared two years later. It turns out that none of the people doing the experiments paid any attention to these theoretical papers. [5.31]

He also adds,

> It is difficult for us to determine just when Raman stopped asserting that he had found an optical analogue of the Compton effect, and it is difficult to say whether he reached this conclusion independently or was influenced by the papers of the French authors, by other papers, or by some other factors, but already in his paper in *Nature* on July 7, 1928, and later, he gave the observed effect the same explanation as given by Landsberg and Mandel'shtam and by the French authors in their papers in *Comptes Rendus*....

In making his various observations, Fabelinskii consistently ignores the views expressed by Raman in his address to the South India Science Association, reprinted in the *Indian Journal of Physics* in its issue of March 31, 1928 [5.27]. One fails to understand why.

Let us first consider Raman's use of the words "the analogue of the Compton effect". In a footnote Fabelinskii points out that a frequency shift due to the same mechanism as the Compton effect would be too small to be observed in a light scattering experiment (at least in those days). It is difficult to believe that Raman was not aware of this elementary fact, especially since he had himself done both theoretical and experimental studies on the Compton effect[22]. According to the Chambers Dictionary, analogy means "correspondence in certain respects between things otherwise different". Raman evidently used the word analogue in this sense and none other. The correspondence was with respect to the occurrence of a frequency shift and not with respect to the specific microscopic mechanism[23]. As far as the latter is concerned, it is clear from his remarks to Krishnan (see Krishnan's diary entry dated February 7) prior to the observation of the line spectrum that he envisaged a process implied by the Kramers–Heisenberg theory. This conclusion is further reinforced by the following remarks made during his Bangalore address on March 16:

> ...such a possibility [scattering with a frequency shift] is already contemplated in the Kramers–Heisenberg theory of dispersion. If we accept the idea indicated above, then the difference between the incident and scattered quanta would correspond to a quantum of absorption by the molecule. The measurement of the frequencies of the new spectral lines thus opens a new pathway of research into molecular spectra, particularly those in the infra-red region.

As a final comment on the question of analogy to the Compton effect, attention is also drawn to Raman's own succinct account of the sense in which he drew parallels, quoted earlier.

Fabelinskii stresses the deductions made by Landsberg and Mandel'shtam concerning the observed frequency shifts. But besides frequency shifts there is also the nature of the polarization of the scattered radiation. Somehow Fabelinskii does not refer to the fact that Raman and Krishnan had also studied this aspect whereas the Soviet physicists had not.

Table 5.2 Early papers on the Raman effect (based on reference [5.35]).

No.	Authors	Reference	Date of publication
1.	C. V. Raman and K. S. Krishnan	*Nature* **121**, 501 (1928)	31 March 1928 (communicated on 16 February)
2.	C. V. Raman	*Indian J. Phys.* **2**, 387 (1928)	31 March 1928
3.	C. V. Raman	*Nature* **121**, 619 (1928)	21 April 1928 (communicated on 8 march)
4.	Y. Rocard	*Comptes Rendus* **186**, 1107 (1928)	23 April 1928
5.	J. Cabannes	*Comptes Rendus* **186**, 1201 (1928)	30 April 1928
6.	C. V. Raman and K. S. Krishnan	*Nature* **121**, 711 (1928)*	5 May 1928 (communicated on 22 March)
7.	A. Cotton	*Comptes Rendus* **186**, 1475 (1928)	4 June 1928
8.	J. Cabannes and P. Daure	*Comptes Rendus* **186**, 1533 (1928)	4 June 1928
9.	C. V. Raman and K. S. Krishnan	*Indian J. Phys.* **2**, 399 (1928)	5 June 1928 (communicated on 5 May)

*This paper discusses analogy with the Compton effect and is criticized by Fabelinskii.

A few dates are worth noting, and for convenience they are gathered in Table 5.2 [5.35]. We see that the first papers published by a group outside India appeared on April 23. By that time, Raman's Bangalore lecture had already been published in the *Indian Journal of Physics* (on March 31). We have already seen that in this paper Raman had clearly identified the frequency shift with infra-red frequencies of molecules. Incidentally, the preprint of this paper was available to the French authors Rocard, Cabannes and others, who quote it in their papers. Even if credit is not given on the ground that actual numbers for the frequency shifts had not been quoted by Raman in his March 16 address, one has to concede that this as well as explicit identification with previously measured vibrational frequencies were both done in the paper communicated on May 5 (item 9 in table above). There is a time span of 11 days between the paper of Rocard [5.36; item 4 in table] and the receipt by the *Indian Journal of Physics* of the just mentioned paper of Raman and Krishnan. In the case of the paper of Cabannes [5.37], the gap is even smaller since his paper appeared on April 30. Given the poor communication facilities of those days, it is impossible that Raman would have seen the papers of the French authors *before* he communicated his own in early May. Indeed even today, journals published in the West take several weeks and at times months to reach India by surface mail. One must regretfully conclude that Fabelinskii has been swept by emotion.

Looking back, it is clear that several groups in several countries were roughly on the same track, namely, detection of the scattering of light with a frequency shift. But their motivations were somewhat different. In fact as Rocard, Cabannes and others have recorded [5.36, 5.37], what Raman discovered was what they themselves were looking for, i.e., scattering with a frequency change corresponding to molecular vibrations. They failed in their search because they experimented with gases where the scattering is quite weak compared to that in liquids. Fabelinskii is certainly right to the extent that none of these groups understood *in detail* the full mechanism of scattering. But that is not surprising since a proper theory of the Raman effect did not exist. (Such a theory came later, interestingly from the Soviet Union [5.38; see also 5.39].) However, each group *did* have *some* perception of the theory even at that time, and there is documentary evidence to support the belief that Raman independently understood, possibly even ahead of the others, the essential significance in terms of the Kramers–Heisenberg theory of dispersion. One must also accept that Landsberg and Mandel'shtam had in fact made an independent discovery, although their publication was somewhat delayed.

Having said that, one also wishes that, at least for the sake of posterity, Raman had composed his letters somewhat differently, in a more pedantic style, and citing due references, particularly the Kramers–Heisenberg paper, of which he was well aware. But then Raman's papers reflect his personality. When he beholds something interesting, his heart leaps like that of Wordsworth. He becomes excited and must immediately call the attention of others to what he sees. As remarked in an earlier chapter, his papers (especially letters), often tend to read like the report of a commentator, a style not uncommon in an earlier era.

Landsberg and Mandel'shtam are not any less distinguished because they did not share the Nobel Prize with Raman. Both made many significant contributions during their careers. Conversely, Raman would not have been any less distinguished himself had he lost the race. For science, it was good that the discovery had been made, for it was to prove so useful[24].

5.14 A Mild Confrontation

The story is almost over, barring a few historical footnotes. In August 1928 the Association of Russian Physicists held in Sixth Congress, which was attended among others by Born, Brillouin, Debye and Dirac. Naturally the work of Landsberg and Mandel'shtam was the centre-piece, and on return to Western Europe, Born and Darwin wrote accounts of the highlights of the meeting in the Soviet Union. Darwin's note provoked a characteristic response from Raman [5.40].

Investigations of the Scattering of Light

Prof. C. G. Darwin, in his interesting account in *Nature* of Oct. 20, 1928 (p. 630), makes a reference to recent work on the scattering of light. It appears desirable in this connection to point out that the existence in the light scattered by liquids and solids of radiations of modified wavelength was established so early as 1923 by investigations made at Calcutta. Dr K. R. Ramanathan showed (*Proc. Indian Assoc. Cultiv. Sci.*, vol. 8, p. 190; 1923) that when violet rays pass through carefully purified water or alcohol there is an appreciable quantity of radiations in the green region of the spectrum present in the scattered light. Further studies of the effect in other substances are described by Mr K. S. Krishnan in the *Philos. Mag.* for October 1925 and by me in *J. Opt. Soc. Am.* for October 1927. These investigations were of course well known to workers in this field.

In a lecture delivered at Bangalore on Mar. 16, 1928, and published and distributed on Mar. 31, investigations were described showing *first*, the universality of the effect, namely, that it is observed in the widest variety of physical conditions (gas, vapour, liquid, crystal, or amorphous solid) and in the largest possible variety of chemical individuals (more than eighty different substances); *secondly*, that the modified radiation is strongly polarized and is thus a true scattering effect; *thirdly*, that each incident radiation produces a different set of modified scattered radiations; *fourthly*, that the scattered radiations consist in many cases of fairly sharp lines in displaced positions; and *fifthly*, that the frequency differences between the incident and scattered radiations represent the absorption frequencies of the medium. These observations established and emphasized the fundamental character of the phenomenon in a manner which any isolated observation with a single substance would have quite failed to achieve.

The Russian physicists, to whose observation on the effect in quartz Prof. Darwin refers, made their first communication on the subject after the publication of the notes in *Nature* of Mar. 31 and Apr. 21. Their paper appeared in print after *sixteen* other printed papers on the effect, by various authors, had appeared in recognized scientific periodicals.

C. V. RAMAN

210 Bowbazar Street
Calcutta, Nov. 13

Around the same time, Raman also wrote to Mandel'shtam. The letter was dated August 7, 1928, and after the standard greeting, Raman wrote as follows:

My attention has been brought to two recent papers on light scattering in crystals which you have published together with Dr Landsberg in *Naturwissenschaften* on July 13, 1928, and in *Comptes Rendus* on July 8, 1928, which were received in Calcutta in the last mail delivery. In *Naturwissenschaften* I found a reference to two letters to the editor of *Nature* in which the discovery of a new type of secondary radiation and its spectral composition were reported from Calcutta. But in none of your communications have I found any mention of my speech of March 15, 1928, entitled "A new radiation" and published in the *Indian Journal of Physics*. In this speech, which was published on March 31, 1928, the phenomenon of a change of wavelength upon scattering in crystals was clearly described and explained. Your university library regularly receives the *Indian Journal of Physics* by exchange, and over the past three months this paper has been cited several times by Cabannes and Daure and other authors who had published in *Comptes Rendus* and with whose articles you are evidently familiar. Therefore the absence of any reference to prior publication of the discovery of this effect in crystals, made in Calcutta, is somewhat unexpected. I trust that this omission will be corrected in your future publications on the subject.

Thanking you in advance,

Respectfully yours,
C. V. Raman

Fabelinskii reproduces this letter and adds, "... this letter is unique and constitutes the entire correspondence between the two groups...."

5.15 The Honours

It was clear right from the beginning that Raman's discovery was a major one, both in terms of its implications for quantum physics as well as in terms of its applications. While the possibility of widespread application had been foreseen by Raman himself, Rutherford had this to say in his presidential address to the Royal Society in November 1929:

> It is clear that this new effect may be of great importance in determining the slow characteristic frequencies of molecules in the infra-red, which may be difficult to measure by other methods. This new discovery, of great interest in itself, thus promises to open up a new field of experimental enquiry and throw valuable light on the modes of vibration and constitution of the chemical molecule. This discovery has attracted much attention, and a number of papers dealing with it have been published in all parts of the scientific world. [5.41]

Indeed it is true. By August 1929, Ganesan [5.35] was able to compile a bibliography of 150 papers!

It is not surprising that various honours followed in rapid succession. The Italian Society of Sciences, Rome, awarded him the Matteucci Gold Medal for "the most important physical discovery of the year". The British Government conferred a knighthood. The Faraday Society of London invited Raman to deliver the opening address at a meeting on "Molecular spectra and molecular structure", and to lead the discussion. The University of Freiburg awarded a doctoral degree *honoris causa*, while the Physical Society of Switzerland elected him an Honorary Member. Invitations to lecture were received from numerous quarters both in England and on the Continent.

The plum came in 1930[25]; Ramaseshan describes it as follows:

> Nobel Prizes are announced in the second or third week of November[26]. The meetings of the Nobel Committee are held in the highest secrecy and the awards are announced in November about a month before the prize-giving ceremony in mid-December. It would have been surprising enough that Raman could leave by steamship after receiving the news by telegram to reach Stockholm in time for the ceremony. It is now a historical fact however that Raman had booked two tickets for himself and his wife in July that year to enable them to reach Stockholm in early December! [5.11]

Bhagavantam too describes the incident and his account is not only first-hand but a bit more colourful!

> I had the privilege of being one of his active collaborators at the time when he was awarded the Nobel Prize and I vividly recall his reactions when I communicated to him the first news of the award after knowing it on telephone from one of the Indian

news agencies in Calcutta. He asked if he was the sole awardee or was he to share the bed with other strangers!... Two months before he knew he was awarded the Nobel Prize, he had the supreme audacity of booking his steamer passage to be in time for the ceremony at Stockholm. That not only did he take such a step but went further and declared publicly that he did so are both interesting facts of his life. [5.42]

The award was made on December 10, 1930. Lady Raman has given a vivid account of that exciting week in Stockholm in the Calcutta Municipal Gazette:

> We arrived by train at Stockholm at 8 o'clock in the morning of 9th December. The platform was crowded with people waiting to receive the guests.... Not being white-skinned, we could not mix with the natives of the country without being recognized.... It was very difficult to get rid of the newspapermen. They must hear about Indian politics.... The award of the Nobel Prizes took place on the 10th between 4 and 7 p.m. If I shut my eyes, I can still see as in a dream the great concert hall of Stockholm, decorated with flowers and flags, filled with more than 4000 people, the King and the Queen of Sweden and the royal family occupying the first seats. The Nobel Laureates then entered the hall.... The secretary of the Academy of Sciences read a report and gave a brief account of Nobel's life. Dr Pleijel, Professor of Electro-Technics in the University of Stockholm, then rose and spoke for twenty minutes on my husband's investigations on the scattering of light and the new effect that had been discovered by him, and addressing him said: Sir Venkata Raman, the Royal Academy of Sciences has awarded you the Nobel Prize in Physics for your eminent researches on the diffusion of light and for your discovery of the effect that bears your name. The Raman effect has opened new routes and has given most important results. I now ask you to receive the prize from the hands of His Majesty.
>
> On Sir Chandrasekhara rising to receive the Prize, the whole audience including the King stood up and the British flag was held aloft. The Laureate then approached the royal seat and bowed before the King who took him by the hand and presented him the Nobel Medal, Prize and Diploma. This was attended with loud cheering and was followed by orchestral music[27]....
>
> The ceremony... was followed by the Nobel banquet.... The Nobel Laureates sat at the royal table. The dinner was of the most lavish scale. Wine flowed freely[28]. A few vegetarian dishes had been considerately provided for us. When it came to the drinking to health, we had our cups filled with water. In replying to the toast, Sir Raman[29] spoke of the glories of ancient India. He spoke of the great renunciation of Buddha, the royal ascetic and world teacher, and of his message of non-violence and love which embraced all living creation. [5.43]

After participating in the Nobel ceremony and the festivities which followed, Raman toured Europe lecturing at Uppsala, Goteborg, Oslo, Copenhagen, Munich, Strasbourg and London. He attended the special convocation of the Glasgow University where an honorary LL D degree was conferred upon him. The Hughes Medal for 1930 was also awarded to him by the Royal Society. On that occasion Lord Rutherford observed:

> The Raman effect must rank among the best three or four discoveries in experimental physics in the last decade; it has proved and will prove [to be] an instrument of great power in the study of the theory of solids. In addition to important contributions in many fields of knowledge, he [i.e., Raman] has developed an active school of research in physical sciences in the University of Calcutta.

It was the hour of glory. One of the problems of reaching the top is to stay there thereafter. Almost all winners of the Nobel Prize have faced this difficulty to some extent or other, especially when the honours came to them early in life. Raman was 41 when he won the Prize – not too young nor too old either. What he faced subsequently was not merely a professional gradient but a traumatic experience – in fact twice in quick succession, first in Calcutta and then in Bangalore. The Calcutta story has already been told. We now shortly move with Raman to Bangalore to see what happened there.

> The king is set from London: and the scene
> Is now transported, gentles, to Southampton:
> There is the playhouse now, there you must sit....

(Henry V, Act II)

☆6 *I say, What is this Raman Effect?*

> We see a number of sophisticated, yet uneducated, theoreticians who are conversant in the LSZ formalism of Heisenberg field operators, but do not know why an excited atom radiates, or are ignorant of the quantum-theoretic derivation of Rayleigh's law that accounts for the blueness of the sky.
>
> J. J. SAKURAI

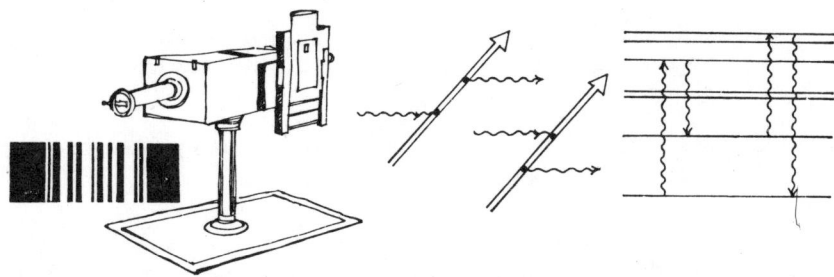

A bright young theorist (needless to say a particle physicist!) once asked a colleague of his: "I say, what is this Raman effect?" Upon being given a brief qualitative answer, his next question was: "Is there an explanation for it in quantum electrodynamics?" When assured there was, he felt relieved and lost further interest!

The purpose of this chapter is not to instruct those with gaps in their education, but rather to highlight the basic physics underlying the Raman effect, and also to provide an anchor to some of the descriptive material of the preceding chapter.

6.1 Optical Inhomogeneity and Light Scattering

A perfectly homogeneous medium does not scatter light. To understand this, consider a block ABCD of perfectly homogeneous matter, illuminated as sketched in Fig. 6.1. Consider now an elementary cube P whose linear dimensions are much smaller than the wavelength of light. The different portions of P will then scatter light in phase with each other, and, at a (faraway) point at distance x from P, we may write the scattered wave amplitude as

$$A \cos \{\omega(t - x/c) + \theta_x\},$$

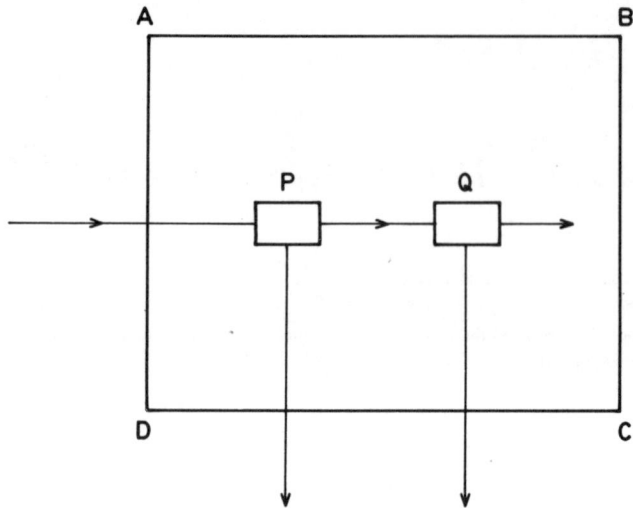

Figure 6.1 Illustration of the argument that a homogeneous block of matter will not scatter light except in the forward direction. For explanation see text.

where the various symbols have their usual and obvious meanings. Another block Q will similarly produce at x an amplitude

$$A \cos \{\omega(t - x/c) + \theta_x - \phi\},$$

where ϕ is the phase difference due to the passage of the incident light from P and Q. If the block ABCD is big enough, we can always choose Q such that $\phi = \pi/2$ in which case the waves scattered by Q and P would cancel each other at x. In this way, all the secondary waves coming from the various elementary cubes may be paired off, leaving no aggregate scattering in the x-direction (boundary scattering neglected). A similar argument may be extended to scattering in other directions. The only exception is the forward direction, where the various secondary wavelets reinforce each other; in fact, it is this which emerges as the refracted beam.

It should be evident from the foregoing discussion that both gases and liquids would scatter light to a much greater extent than crystals. According to our argument, a *perfect* single crystal should not scatter light but then no crystal is truly homogeneous, thanks to the presence of inclusions and various other structural defects. That is why Landsberg and Mandel'shtam had to reject many specimens of quartz since, like Raman, they too were looking for scattering produced by thermally-induced inhomogeneity.

It was known well before Raman's time that optical inhomogeneity was responsible for (Rayleigh) scattering from gases. Einstein and Smoluchowski used the same idea to explain critical opalescence in fluids. Raman recognized that *normal thermal motion of atoms in solids and liquids also produces optical inhomogeneities* and should therefore lead to the scattering of light. For almost a decade from 1920 onwards, this was a constant refrain in his papers on scattering.

6.2 Light Scattering and the Dipole Approximation

The dipole approximation plays an important role in the theoretical treatment of Rayleigh and Raman scattering – not surprising since the dimensions of the scattering agency, i.e., the molecule, are much less than the wavelength of light.

In elementary treatments, light scattering is visualized as a two-step process. The incident light first induces an oscillating dipole within the molecule. Next the induced dipole radiates (as it is required to do by Maxwell's equations), and it is this radiation which is interpreted as the scattered light.

Given the above picture, one can readily understand the nature of the angular distribution of the scattered radiation. Suppose unpolarized radiation is incident on the molecule along the z-direction, as illustrated in Fig. 6.2. Dipoles would then be induced with oscillations along the x and y directions. In the scattering plane yz, the dipole A contributes equal intensity at all scattering angles θ while the intensity associated with dipole B varies as $\cos^2 \theta$. Thus the net effect is $\sim (1 + \cos^2 \theta)$. If observations are made along the y-direction, the $\cos^2 \theta$ contribution drops out. This is the well-known transverse scattering geometry highly popular not only with the Calcutta investigators, but also with others elsewhere.

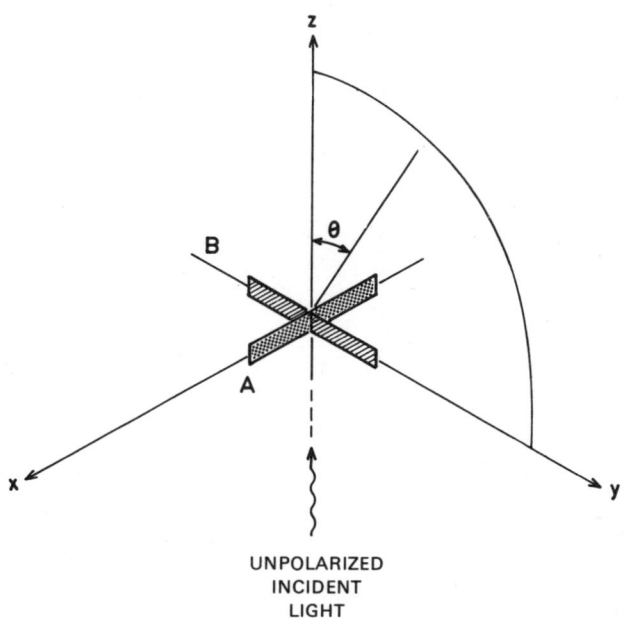

Figure 6.2 Unpolarized light incident along the z-direction will induce electrical dipoles having their moments in the xy plane. By resolving, the cumulative effect can be discussed in terms of the dipoles A and B.

6.3 Rayleigh Scattering

There are various ways of deriving the formula for Rayleigh scattering by gases. We outline one which emphasizes the random-phase approximation.

Consider an oscillating dipole $P = P_0 \cos \omega t$ oriented as in Fig. 6.3. By virtue of Maxwell's equations this dipole would radiate, the magnitude of the average value of the Poynting vector being given by

$$\langle |S| \rangle = \frac{\omega^4 P_0^2 \sin^2 \phi}{8\pi c^3 r^2}, \tag{6.1}$$

where r and ϕ are defined in Fig. 6.3. We now suppose that the moment P is induced by an electric field $E_0 \cos \omega t$ incident along the x-direction on the molecule (here assumed to be spherically symmetric). If α is the polarizability, then $P = \alpha E_0 \cos \omega t$. Taking $\phi = 90°$ for convenience, we then obtain

$$\langle |S| \rangle = I_0 \frac{16\pi^4 \alpha^2}{\lambda^4 r^2}, \tag{6.2}$$

where $I_0 = (cE_0^2/8\pi)$ represents the incident intensity. We now note that the polarizability of the molecule is given by

$$\alpha = \frac{(\varepsilon - 1)}{4\pi N}, \tag{6.3}$$

where ε is the dielectric constant of the gas and N is the number of molecules per unit

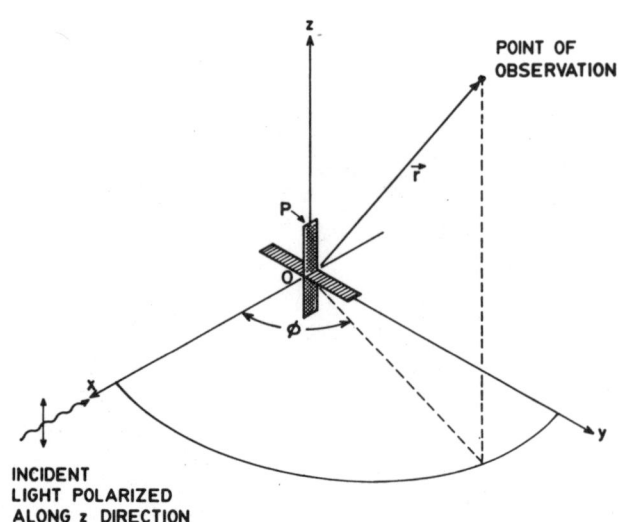

Figure 6.3 Illustration of the various quantities involved in calculating the power radiated by a dipole.

volume. By substituting (6.3) into (6.2) we obtain the intensity scattered by one molecule. In a volume V there would be NV molecules and since their positions are not correlated we may assume that the intensities contributed by them add, obtaining the net scattered intensity as

$$I = I_0 \frac{\pi^2 V (\varepsilon - 1)^2}{\lambda^4 r^2 \; N}. \tag{6.4a}$$

If the incident polarization was along the y-axis, we would similarly obtain

$$I = I_0 \frac{\pi^2 V}{\lambda^4 r^2} \frac{(\varepsilon - 1)^2}{N} \cos^2 \theta, \tag{6.4b}$$

where θ is the scattering angle. Thus, for unpolarized light incident along the x-axis, the total scattered intensity would be

$$I = I_0 \frac{\pi^2 V}{\lambda^4 r^2} \frac{(\varepsilon - 1)^2}{N} (1 + \cos^2 \theta). \tag{6.5}$$

The scattering coefficient $\mathscr{R}$ introduced in Chapter 5 is defined by

$$\mathscr{R} = (Ir^2 / 2I_0 V), \tag{6.6}$$

where V now denotes the scattering volume. Remembering

$$N = N_A p / RT, \tag{6.7}$$

we then obtain (for transverse geometry)

$$\mathscr{R} = \frac{\pi^2}{2\lambda^4} \frac{RT}{N_A p} (\varepsilon - 1)^2, \tag{6.8}$$

which reduces to the formula quoted in the earlier chapter when we remember $\varepsilon = n^2$, where n is the refractive index.

6.4 Polarization Effects in Rayleigh Scattering

A few remarks next about the polarization of the scattered light. Reference has already been made to the fact that even if the incident light is unpolarized the light transversely scattered by a gas is completely polarized, provided the scattering molecules are spherical. However, some depolarization may be expected if the molecules are non-spherical (i.e., optically anisotropic). Taking this effect into account, Lord Rayleigh showed that (6.8) becomes modified into

$$\mathscr{R} = \frac{\pi^2}{2\lambda^4} \frac{RT}{N_A p} (\varepsilon - 1)^2 \frac{6 + 6\rho}{6 - 7\rho}, \tag{6.9}$$

where ρ is the depolarization ratio introduced earlier (see Sec. 4.5).

Experiments by Rayleigh, Cabannes and others on dust-free air in the laboratory not only provided a basis for the explanation of the blue of the sky, but also showed through the existence of a depolarization effect that the molecules of air were non-spherical. All these workers had employed the photographic technique for making observations. Raman wanted to establish that the molecules of air are anisotropic by direct observations on depolarization effects in sky-light. To obtain reliable results he climbed the Dodabetta, converting the sky itself into a laboratory!

6.5 Scattering of Light by Liquids

Compared to a gas, a liquid is a very much denser medium. The positions of adjacent atoms would therefore be correlated to some extent, and one cannot glibly make the random-phase approximation.

The problem of calculating the scattering of light by liquids with due regard to the intermolecular correlations is a non-trivial one. However, the problem may be circumvented to a certain extent by following what Einstein and Smoluchowski did in the case of critical opalescence. In this approach, one disregards the details of the atomic structure and treats the medium as a continuum. However, there could be local density fluctuations leading to optical inhomogeneity and hence to light scattering. To compute the scattering, let us suppose that a given volume element V has a mean dielectric constant ε which attains an instantaneous value $(\varepsilon + \Delta\varepsilon)$ consequent to a fluctuation. An electric field of amplitude E_0 will then induce in the volume V a dipole moment of magnitude $(\Delta\varepsilon E_0 V/4\pi)$, whence the transverse scattering is given by

$$\frac{\omega^4}{c^4}\left(\frac{\Delta\varepsilon V}{4\pi}\right)^2. \tag{6.10}$$

The process of obtaining the average effect from all the volume elements at a given instant is equivalent to obtaining the time-average effect from a given element which means that $(\Delta\varepsilon)^2$ in (6.10) must be replaced by $\overline{\Delta\varepsilon^2}$, where the bar denotes an average.

Now according to the Clausius–Mossotti formula,

$$\frac{\varepsilon - 1}{\varepsilon + 2} = K\sigma, \tag{6.11}$$

where σ is the density and K an appropriate constant. From this it is easily seen that

$$\overline{\Delta\varepsilon^2} = \frac{(\varepsilon - 1)^2(\varepsilon + 2)^2}{9} \frac{\overline{\Delta\sigma^2}}{\sigma^2}. \tag{6.12}$$

Using now the Smoluchowski result

$$\frac{\overline{\Delta\sigma^2}}{\sigma^2} = \frac{RT\beta}{N_A V} \tag{6.13}$$

I say, What is this Raman Effect?

in (6.12), one obtains after some elementary algebra the Einstein–Smoluchowski (E–S) result (for transverse scattering)

$$\mathcal{R} = \frac{\pi^2 RT\beta}{18 N_A \lambda^4} (\varepsilon - 1)^2 (\varepsilon + 2)^2. \tag{6.14}$$

The corresponding formula in Chapter 5 is recovered when one remembers $n^2 = \varepsilon$.

Two comments are worth making at this stage. Firstly, intermolecular correlations are *implicitly* taken into account while writing the Clausius–Mossotti formula (6.11). For a gas, on the other hand, it would be more appropriate to use $\varepsilon - 1 = K'\sigma$ where $K' = K/3$, a result which neglects the effects of correlations. Following through, one would then obtain the Rayleigh formula (6.8) instead of the E–S one. Our second comment is that a better estimate of $\overline{\Delta\sigma^2}$ than given in (6.13) is possible by using thermodynamics.

In deducing (6.14), no account has been taken of the possibility that molecules in the system might be non-spherical and that there could be consequent depolarization. When this is done, one arrives at the following result:

$$\mathcal{R} = \frac{\pi^2 RT\beta}{18 N_A \lambda^4} (\varepsilon - 1)^2 (\varepsilon + 2)^2 \frac{6 + 6\rho}{6 - 7\rho}. \tag{6.15}$$

Now depolarization also is affected by intermolecular correlations; thus $\rho_{gas} \neq \rho_{liquid}$. In fact

$$\frac{RT\beta N}{N_A} \frac{\rho_{liquid}}{6 - 7\rho_{liquid}} = \frac{\rho_{gas}}{6 - 7\rho_{gas}}, \tag{6.16}$$

where, as before, N is the number of molecules per unit volume. However, Raman and Krishnan [6.1] found that ρ_{liquid} as computed from (6.16) using measured values for ρ_{gas} was very much at variance with experiment (see Table 6.1). To explain this discrepancy they then argued that when a liquid is subject to an external electric

Table 6.1 Comparison of depolarization ρ_{liquid} observed in various liquids (at 30°C) with computed values. The entries in the third column are based on (6.16), while those in the fourth are due to Raman and Krishnan using an anisotropic polarization field.

Substance	ρ_{gas}	ρ_{liquid} (Isotropic polarization field)	ρ_{liquid} (Anisotropic field, calculation by Raman and Krishnan)	ρ_{liquid} (Observed)
Pentane	0.0136	0.21	0.074	0.075
Hexane	0.015	0.31	0.087	0.100
Heptane	0.0158	0.38	0.083	{ 0.127 / 0.100
Octane	0.0186	0.46	0.105	0.129

field E, the molecules in the liquid do not see an isotropic effective field

$$E' = E + (4\pi P/3), \quad P = \text{polarization/unit volume},$$

as is customarily assumed in the derivation of (6.14) and (6.15), but rather an anisotropic field. Their modified analysis results in better agreement with experiment, as Table 6.1 shows.

6.6 Raman Effect in the Polarizability Picture

The Raman effect is completely quantum-mechanical in origin, which is why it is remarkable that Raman, a thoroughbred classical physicist, was able to track it and finally discover it. However, as we shall presently see, one can in fact qualitatively understand Raman scattering by close analogy with Rayleigh scattering, provided one makes some minor conceptual concessions.

It will be recalled that Rayleigh scattering is explained by assuming that the electric field E of the incident radiation induces a dipole $P = \alpha E$, and that the scattered light is essentially the radiation emitted by P. Likewise, Raman scattering may be visualized as due to an induced *transition* dipole

$$\mathbf{P}^{fi} = \langle f|\hat{\alpha}|i\rangle \mathbf{E} \equiv \alpha^{fi}\mathbf{E}, \tag{6.17}$$

where $\hat{\alpha}$ is a suitably-defined polarizability operator. In this language, Rayleigh scattering is determined by dipoles $\mathbf{P}^{ii}$.

What about the change in frequency on scattering? This too can be qualitatively understood as follows: Let us first write

$$E = E_0 \cos \omega_0 t \tag{6.18a}$$

$$\alpha = \alpha_0 + \alpha_k \cos(\omega_k t + \delta_k). \tag{6.18b}$$

Equation (6.18b) implies that the polarizability α of the molecule is modulated in time with a frequency ω_k, associated, say, with a molecular vibration. Using the above,

$$P = \alpha E$$

where
$$= P_0 + P_1, \quad \text{say}, \tag{6.19a}$$

$$P_0 = \alpha_0 E_0 \cos \omega_0 t \tag{6.19b}$$

$$P_1 = \frac{1}{2}\{\alpha_k E_0 \cos[(\omega_0 - \omega_k)t - \delta_k]$$

$$+ \alpha_k E_0 \cos[(\omega_0 + \omega_k)t + \delta_k]\}. \tag{6.19c}$$

Clearly P_0 contributes to Rayleigh scattering while the two terms on the r.h.s. of (6.19c) produce Stokes and anti-Stokes Raman scattering.

6.7 Time-dependent Perturbation Theory and Raman Scattering

As noted in the previous chapter, Raman scattering was theoretically anticipated first by Smekal and then by Kramers and Heisenberg whose paper [6.2] is a landmark not only in the history of the Raman effect but indeed in the history of quantum mechanics itself.

Kramers and Heisenberg employ time-dependent perturbation theory. The scattering system is treated quantum-mechanically but the radiation itself is treated classically so that in today's parlance, one would describe the overall treatment as semi-classical.

Let us write the perturbed wavefunction Ψ_i as

$$\Psi_i \approx \Psi_i^{(0)} + \Psi_i^{(1)}, \tag{6.20}$$

where the unperturbed functions are given by

$$\Psi_i^{(0)} = \psi_i \exp(-i\omega_i t). \tag{6.21}$$

The time-independent wave function ψ_i includes an arbitrary phase factor $\exp(i\delta_i)$.

Introduce now the dipole moment operator **M** defined by

$$\mathbf{M} = \sum_j e_j \mathbf{r}_j, \tag{6.22}$$

where e_j is the charge of the jth particle and $\mathbf{r}_j$ its position vector. The transition dipole moment for the transition $\Psi_i \rightarrow \Psi_f$ may then be written [6.3]

$$\int \{\Psi_i^* \mathbf{M} \Psi_f + \Psi_f^* \mathbf{M} \Psi_i\} d\tau. \tag{6.23}$$

One now uses standard perturbation theory, taking the perturbation Hamiltonian to be

$$\mathcal{H}' = -\mathbf{M} \cdot \mathbf{E}, \tag{6.24}$$

where

$$\mathbf{E} = \mathbf{E}_0^- \exp(-i\omega_0 t) + \mathbf{E}_0^+ \exp(i\omega_0 t), \quad \mathbf{E}_0^- = (\mathbf{E}_0^+)^*. \tag{6.25}$$

The following result is then obtained [6.3] for the matrix elements of the polarizability tensor:

$$\alpha_{\mu\nu}^{fi} = \frac{1}{\hbar} \sum_r \left\{ \frac{\langle i|M_\mu|r\rangle \langle r|M_\nu|f\rangle}{\omega_{rf} + \omega_0} + \frac{\langle i|M_\nu|r\rangle \langle r|M_\mu|f\rangle}{\omega_{ri} - \omega_0} \right\} \tag{6.26}$$

where

$$\omega_{ri} = \omega_r - \omega_i \quad \text{etc.} \tag{6.27}$$

represent the frequencies associated with the transitions $|i\rangle \rightarrow |r\rangle$ etc.

Two points are worthy of note here, both first emphasized by Kramers and Heisenberg themselves. Firstly, the transition $|i\rangle \to |r\rangle$ occurs via intermediate states $|r\rangle$ (as already mentioned in Sec. 5.11). In fact, transitions via virtual intermediate states have now acquired the generic name of Raman processes (at least in certain branches of condensed matter physics). Our second point relates to the incoherence of the scattered radiation. In Rayleigh scattering, the scattered wavelet emerging from every molecule in the system is in phase with the incident radiation. Thus interference of waves scattered by *different* molecules is possible, and the angular distribution of the scattered radiation depends entirely on the spatial arrangement of the atoms. In Raman scattering this is not so. The scattered wave has an arbitrary phase relationship with the incident wave, and consequently the waves scattered by *different* molecules *cannot* ever interfere with each other. In other words, the scattering is incoherent. One can see this by considering the phase $\exp\{-i(\delta_f - \delta_i)\}$ associated with the transition moment $\mathbf{P}^{fi}$. Since the phases are arbitrary, the phase factor $\exp\{-i(\delta_f - \delta_i)\}$ will fluctuate randomly from one molecule to another, leading in effect to the incoherence of the Raman scattering. Kramers and Heisenberg make this quite clear. One is therefore a bit puzzled by some slightly confusing statements concerning the coherence aspects made by Raman in his early papers, especially in view of his familiarity with the paper of Kramers and Heisenberg. But his remarks are tentative, and it would seem that he is intrigued by some experimental observations made with carbon dioxide vapour. Certainly he offers room for Fabelinskii to score a debating point. (See, however, the remarks at the end of Sec. 6.10.)

6.8 Raman Scattering by Molecules

Kramers and Heisenberg considered the inelastic scattering of light by atoms, and it is not immediately obvious that their arguments could be applied to molecules as well. Indeed, the extension of their treatment to the molecular case is non-trivial, and we owe our present understanding of this aspect to a famous paper by Placzek [6.4].

Unlike an atom, a molecule can have rotational and vibrational states in addition to the usual electronic states. To deal with this complication, one first makes the so-called Born–Oppenheimer or adiabatic approximation

$$\psi_i \approx \chi_i^{el} \Phi_i^{VR}, \tag{6.28}$$

where χ is the electronic wave function and Φ^{VR} the wave function for vibrations and rotations combined. Next one supposes that the vibrations and rotations are decoupled so that

$$\Phi^{VR} = |v\rangle |\Theta\rangle, \tag{6.29}$$

where the two bras on the r.h.s. denote respectively the vibrational and rotational states.

I say, What is this Raman Effect?

One more detail has to be attended to before considering molecular scattering. Let x, y, z and x', y', z' denote space-fixed and body-fixed frames of reference respectively. Two frames are necessary since molecules can rotate. Tensor operators in the two frames are related by

$$\hat{O}_{\mu\nu} = \sum_{\mu'\nu'} \hat{O}_{\mu'\nu'} \cos(\mu\mu') \cos(\nu\nu'), \tag{6.30}$$

where $(\mu\mu')$ and $(\nu\nu')$ are the angles between the concerned pairs of directions.

Placzek showed that

$$\alpha^{fi}_{\mu\nu} = \sum_{\mu'\nu'} \langle v_f|\hat{\alpha}^{el}_{\mu\nu}|v_i\rangle \langle \Theta_f|\cos(\mu\mu')\cos(\nu\nu')|\Theta_i\rangle, \tag{6.31}$$

where $\hat{\alpha}^{el}$ is the electronic polarizability in the state χ^{el}_i. From this we see how the transitional polarizability is related to vibrational and rotational transitions.

Obviously Raman did not go through such detailed mathematical reasoning before concluding that what he saw was the molecular case of the Kramers and Heisenberg dispersion. As Born once remarked, Raman's quick, intuitive mind was always leaping over mathematics!

6.9 Intensity Formulae

Our discussion thus far has been a bit sketchy, stressing mainly the concept of the transition dipole in Raman scattering. We now consider the formula for the intensity of Raman scattering as this is necessary for discussing certain aspects.

We begin by assuming that the incident electric field is of the form

$$\mathbf{E}(t) = \mathbf{E}^- \exp(-i\omega_i t) + \mathbf{E}^+ \exp(i\omega_i t). \tag{6.32}$$

The induced moment pertinent to Rayleigh scattering is then given by

$$\mathbf{m}(t) = [\alpha^{ll}(\omega_i)]^* \mathbf{E}^- \exp(-i\omega_i t)$$
$$+ [\alpha^{ll}(\omega_i)] \mathbf{E}^+ \exp(i\omega_i t), \tag{6.33}$$

where it is assumed that the molecule is in the state $|l\rangle$. Likewise, the induced moment for the Raman transition $|l\rangle \to |m\rangle$ is given by

$$\mathbf{m}(t) = [\alpha^{ml}(\omega_i)]^* \mathbf{E}^- \exp[-i(\omega_i + \omega_{lm})t]$$
$$+ [\alpha^{ml}(\omega_i)] \mathbf{E}^+ \exp[i(\omega_i + \omega_{lm})t]. \tag{6.34}$$

Given $\mathbf{m}(t)$, one may calculate the Poynting vector, its scalar magnitude, and then its time average. The result is:

$$\langle |\mathbf{S}|\rangle = \frac{\omega_s^4}{2\pi c^3 R^2}(\boldsymbol{\eta}^s \cdot \alpha^{ml}(\omega_i)^* \cdot \mathbf{E}^-)$$
$$\times (\boldsymbol{\eta}^s \cdot \alpha^{ml}(\omega_i) \cdot \mathbf{E}^+), \tag{6.35}$$

where $\omega_s = (\omega_i + \omega_{lm})$, **R** is a vector joining the dipole to the point of observation, and η^s is a unit vector along the direction of the scattered electric field. The differential scattering cross-section for the transition $|l\rangle \to |m\rangle$ is

$$\frac{\mathbf{R}^2 \langle |S| \rangle}{I_0} = \frac{\omega_s^4}{c^4} |\langle m|\eta^s \cdot \hat{\alpha} \cdot \eta^i|l\rangle|^2, \tag{6.36}$$

where η^i is a unit vector along the direction of the incident field. Our interest is in the cross-section $(d^2\sigma/d\Omega d\omega_s)$ obtained by summing (6.36) over all final states $|m\rangle$ and averaging over all initial states $|l\rangle$, subject to energy conservation, i.e.,

$$\omega_s - \omega_i = \omega_l - \omega_m.$$

Performing the various operations indicated above,

$$\frac{d^2\sigma}{d\Omega d\omega_s} = \frac{\omega_s^4}{c^4} \sum_m \rho_m \sum_l |\langle m|\eta^s \cdot \hat{\alpha} \cdot \eta^i|l\rangle|^2$$
$$\times \delta(\omega_s - \omega_i + \omega_m - \omega_l), \tag{6.37}$$

where ρ_m is the density of final states.

The structure of the above formula is a familiar one. Using now a trick due to van Hove [6.5], Gordon [6.6] recast (6.37) into the following form, specializing to the case of rotational-vibrational scattering:

$$\frac{d^2\sigma}{d\Omega d\omega_s} = \frac{\omega^4}{c^4} \frac{1}{2\pi} \int_{-\infty}^{+\infty} \exp(-i\omega t) \langle [\eta^s \cdot \alpha^v(0) \cdot \eta^i][\eta^s \cdot \hat{\alpha}^v(t) \cdot \eta^i] \rangle dt. \tag{6.38}$$

In the above, $\hat{\alpha}^v$ is the polarizability operator associated with a particular vibrational transition, and the average $\langle \rangle$ is performed over all rotational states. The physical content of the above formula is that the Raman vibrational line is influenced by rotational correlations. More explicitly, the line-shape is a Fourier transform of the orientational correlations of the molecule. Gordon has successfully used such an approach to analyse the rotational relaxation of various molecules in the liquid state.

It turns out that Rayleigh scattering also can be expressed as the Fourier transform of a temporal correlation, i.e., as

$$\sim \int_{-\infty}^{+\infty} \exp(-i\omega t) \langle \Delta\varepsilon(\mathbf{0}, 0) \Delta\varepsilon(\mathbf{r}, t) \rangle dt, \tag{6.39}$$

where $\Delta\varepsilon(\mathbf{r}, t)$ is the fluctuation in the dielectric constant at the point **r** at time t. The simple, schematic treatments of Sec. 6.3 and Sec. 6.5 ignored the intricacies of temporal fluctuations, whence, according to them, the scattered spectrum would be $\sim \delta(\omega)$ (see Fig. 6.4). Inclusion of temporal fluctuations leads to line-broadening in Rayleigh scattering.

In 1934 Landau and Placzek [6.7] used linear hydrodynamics in conjuction with irreversible thermodynamics to evaluate the power spectrum (6.39), and their results are sketched in Fig. 6.4c. Here the central component is the Rayleigh peak while the frequency-shifted lines are the Brillouin peaks arising from the scattering

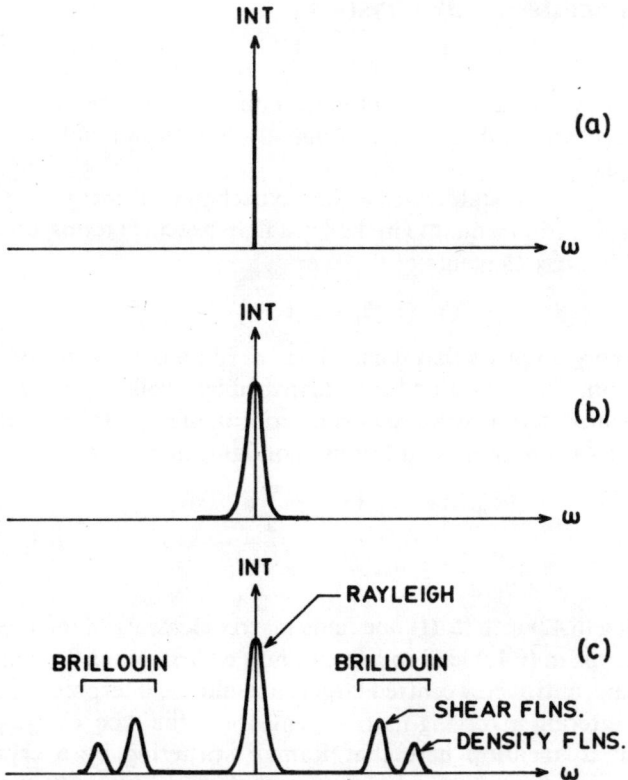

Figure 6.4 Spectrum of light scattered by a fluid according to various models. If the inhomogeneities are *static*, the spectrum would be as in (a). Naive allowance for the temporal fluctuation of the inhomogeneities leads to the broadening of the Rayleigh line, as in (b). The Landau–Placzek theory not only predicts a broadening of the Rayleigh line but also a Brillouin peak due to longitudinal sound waves (manifestation of density fluctuations) as in (c). In highly viscous liquids, a separate peak due to shear modes is possible. Such a peak was first discovered in Bangalore (see Chapter 8).

due to propagating sound waves. Landau and Placzek showed that the ratio of the intensity of the central component to that of the two (longitudinal) Brillouin lines combined should be

$$\frac{I_C}{2I_B} = \frac{C_p - C_v}{C_v}, \qquad (6.40)$$

where C_p and C_v are the specific heats at constant pressure and constant volume respectively. This is the famous Landau–Placzek ratio. In Bangalore Raman's student C. S. Venkateswaran made many experimental studies relating to the experimental verification of the Landau–Placzek theory [6.8] (see also Chapter 8).

6.10 Raman Scattering by Crystals

Raman scattering by crystals is relevant in the context of the famous Born–Raman controversy, which will be discussed in Chapter 10. Here we will introduce a few pertinent concepts.

Raman scattering by crystals arises owing to exchange of energy by the photons with phonons or vibrational quanta in the crystal. In practical terms, this means one has to deal with matrix elements of the type

$$\langle v'|\hat{\alpha}^{el}_{\mu\nu}(X)|v\rangle \tag{6.41}$$

(see (6.31)). One now supposes that the nuclear coordinates collectively represented by X can vary in time, the vibration being described by a collection or normal mode vibrations, associated with which are normal coordinates q. One may thus expand the operator in (6.41) about an equilibrium configuration X_0 as

$$\hat{\alpha}^{el}_{\mu\nu}(X) = \hat{\alpha}^{el}_{\mu\nu}(X_0) + \sum_j a_{\mu\nu}(j)q_j$$
$$+ \sum_{jj'} a_{\mu\nu}(jj')q_j q_{j'} + \cdots. \tag{6.42}$$

Upon substituting (6.42) into (6.41) one finds matrix elements of the type $\langle v'|q_j|v\rangle$ etc. The linear terms in (6.42) lead to one-phonon or first-order Raman scattering. The Born–Raman controversy centred largely around the interpretation of second-order Raman scattering involving matrix elements of the type $\langle v'|q_j q_{j'}|v\rangle$.

It is pertinent to mention here that Raman scattering by a crystal is also incoherent in the sense that scattering from two *independent* systems cannot lead to interference effects. However, nothing prevents one from regarding the crystal itself as a "giant molecule", and portions of the scattered wavefronts emerging from the different parts of this "giant molecule" *can* interfere and exhibit directional effects. Indeed this happens in Brillouin scattering (which is another name for Raman scattering by acoustic phonons – see Chapter 10).

6.11 Quantum Theory of Light Scattering

A few words next about a full-fledged quantum-mechanical treatment. The Hamiltonian for a charged particle moving with a velocity **v** and interacting with a vector potential $\mathbf{A}(\mathbf{r},t)$ is given by

$$\mathcal{H}_A + \mathcal{H}_{AA} \equiv \frac{e}{c}\mathbf{v}\cdot\mathbf{A} + \frac{e^2}{2mc^2}\mathbf{A}^2. \tag{6.43}$$

When dealing with a molecule, we should write

$$\mathcal{H}_A = \sum_j \frac{e_j}{c}\mathbf{A}(\mathbf{R}_j)\cdot 0_j, \tag{6.44}$$

I say, What is this Raman Effect?

where e_j is the charge of the jth ion of the molecule, $\mathbf{R}_j$ being the position of the ion. A similar modification is required for $\mathcal{H}_{AA}$.

In the second-quantized notation $\mathbf{A}(\mathbf{r}, t)$ is expressed as

$$\mathbf{A}(\mathbf{r}, t) = \sum_{qj} (\text{const}) \{ a_{qj} \exp[i(\mathbf{q} \cdot \mathbf{r} - \omega t)] + h \cdot c \}. \tag{6.45}$$

where a_{qj} and a_{qj}^+ are the annihilation and creation operators for photons.

In a scattering event the scattering system goes from a state $|l\rangle$ to a state $|m\rangle$ while the radiation field goes from the state $|n_i, n_s\rangle$ to the state $|n_i - 1, n_s + 1\rangle$. Here n_i is the number of photons in the incident radiation field having a frequency ω_i, wavevector $\mathbf{q}_i$ and polarization $\mathbf{\eta}^i$. The quantity n_s refers to the field of the scattered photon, and is defined similarly.

The transition probability per unit time is given by the Golden Rule:

$$W_{ml} = \frac{2\pi}{\hbar} \delta(\omega_s - \omega_i + \omega_m - \omega_l) \times$$

$$\times \Bigg| \langle m, n_i - 1, n_s + 1 | \mathcal{H}_{AA}(\omega_i - \omega_s) + \mathcal{H}_{AA}(\omega_s - \omega_i) | l, n_i, n_s \rangle$$

$$+ \sum_r \frac{\langle m, n_i - 1, n_s + 1 | \mathcal{H}_A(-\omega_s) | r, n_i - 1, n_s \rangle \langle r, n_i - 1, n_s | \mathcal{H}_A(\omega_i) | l, n_i, n_s \rangle}{(\omega_{lr} + \omega_i)}$$

$$+ \sum_r \frac{\langle m, n_i - 1, n_s + 1 | \mathcal{H}_A(\omega_i) | r, n_i, n_s + 1 \rangle \langle r, n_i, n_s + 1 | \mathcal{H}_A(-\omega_s) | l, n_i, n_s \rangle}{(\omega_{lr} - \omega_s)} \Bigg|^2. \tag{6.46}$$

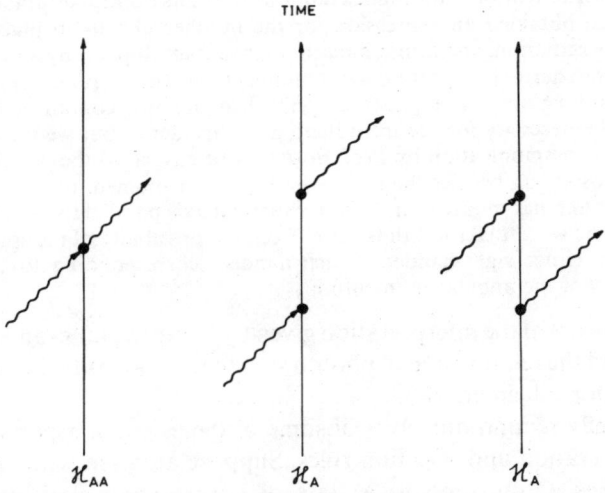

Figure 6.5 Feynman diagrams associated with the various terms in (6.46).

The Feynman diagrams associated with the various terms in (6.46) are shown in Fig. 6.5. Thus, as indicated in the beginning, Raman scattering has a firm basis in quantum electrodynamics also! In passing we observe that the Feynman diagrams for Compton scattering are very similar, which justifies Raman's use of the analogy with the Compton effect.

6.12 Spin of the Photon

It is a little-known fact that Raman and Bhagavantam [6.9] tried to use Raman scattering to establish that the photon had an intrinsic spin. The association of an angular momentum with light was known even in classical physics where it was discussed in terms of the torque generated following the absorption of radiation. After Dirac's monumental work, the idea that the photon had an *intrinsic* spin began to gain currency. According to this view, the intrinsic spin angular momentum of a photon is either $+\hbar$ or $-\hbar$, and a beam consisting of photons of one or the other type corresponds to right- or left-circularly polarized light respectively. Commenting on these developments Raman and Bhagavantam observe:

> The hypothesis that the light-quantum possesses an intrinsic spin in addition to energy and momentum has the merit of enabling the corpuscular concept of radiation to become a complete and intelligible picture.... [But] the spin of the photon...remains at the present time a somewhat nebulous mathematical abstraction without any convincing experimental support. [6.9]

This they now wished to provide, using Raman scattering as a tool. That Raman was well informed about the concept of photon spin is evident from the following:

> In his well-known derivation of the Planck radiation formula from quantum statistics, Prof. S. N. Bose obtained an expression for the number of cells in phase-space occupied by the radiation, and found himself obliged to multiply it by a numerical factor 2 in order to derive from it the correct number of possible arrangements of the quantum in unit volume. The paper as published did not contain a detailed discussion of the necessity for the introduction of this factor, but we understand from a personal communication by Prof. Bose that he envisaged the possibility of the quantum possessing besides the energy $h\nu$ and linear momentum $h\nu/c$ also an intrinsic spin or angular momentum $\pm h/2\pi$ round an axis parallel to the direction of its motion. The weight factor 2 thus arises from the possibility of the spin of the quantum being either right-handed or left-handed, corresponding to the two alternative signs of the angular momentum.

Raman was also aware of the interpretation given by Dirac to plane- and elliptically-polarized light, and the relationship of photon spin to selection rules in spectroscopy as discussed by Oppenheimer.

Let us now briefly remind ourselves of some of the essential aspects of angular momentum conservation and selection rules. Suppose an atom with spin 1 in the excited state makes a transition to a state of lower energy with zero angular momentum. Depending on the m-value of the initial state, the emitted photon will be

I say, What is this Raman Effect?

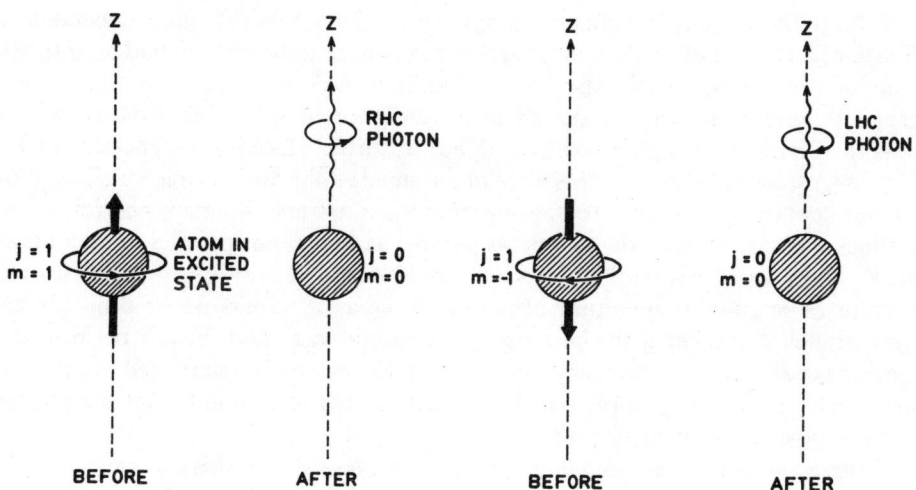

Figure 6.6 Illustration of the conservation of angular momentum in photon emission.

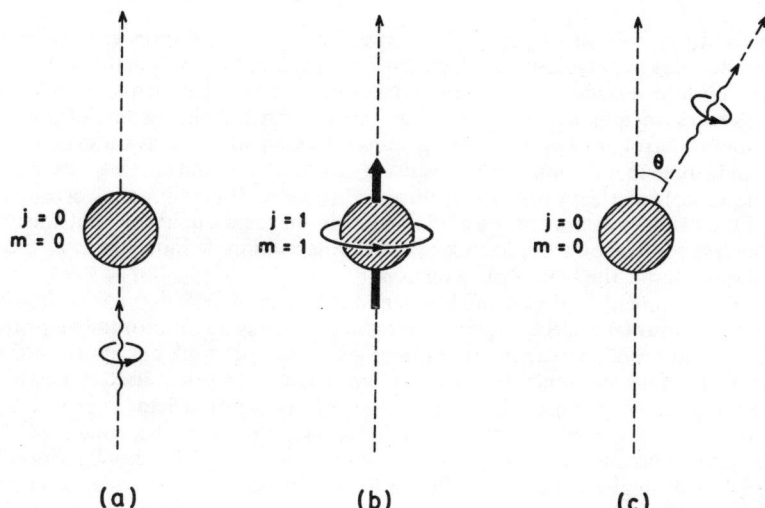

Figure 6.7 Angular momentum conservation in Rayleigh scattering.

either right- or left-circularly polarized, as illustrated in Fig. 6.6. The point to note is the conservation of angular momentum.

Slightly more complicated is the case of Rayleigh scattering illustrated in Fig. 6.7 where a right-circularly polarized photon is scattered by an atom in its ground state. The process may be regarded as involving two steps as shown in Fig. 6.7b, c, i.e., first absorption and then emission. At each step angular momentum is conserved, and, as a result, a right-circular photon is emitted.

Prior to the work of Raman and Bhagavantam, Frisch [6.10] and, independently, Kastler [6.11] had attempted an experiment to check if the photon had an intrinsic spin or not. In Kastler's experiment a sodium vapour source was placed in a magnetic field H_1 so that the spectral lines were Zeeman-split. Light from the source was then passed through a sodium vapour absorber placed in magnetic field H_2. The magnitudes of the two fields were maintained equal, and resonance absorption of the circularly-polarized (σ) components of the transverse Zeeman spectrum of the D-lines was studied with the two fields parallel as well as antiparallel to each other. In Kastler's view, resonant absorption would occur in both cases if the classical picture of angular momentum obtained, whereas it would occur only for the antiparallel alignment if the photon spin concept was valid. Frisch performed a similar experiment with mercury vapour. Both Frisch and Kastler failed to detect an asymmetry in the absorption, which then led Kastler to conclude that the photon did not possess an intrinsic spin.

Raman and Bhagavantam [6.12] were not impressed with this evidence. Instead, they were of the opinion that the experiments of Bär [6.13] on forward Raman scattering of circularly-polarized light by several liquids strongly suggested the existence of photon spin. Commenting on these developments, Saha and Bhargava wrote:

> Recently a number of papers have appeared on the question as to whether the phenomena of polarization of light can be explained by the assumption of a 'spin' of the photon. Kastler and Frisch deduce from their experiments that the photon possesses no spin, and Kastler argues further that the phenomena of polarization should be explained on statistical grounds. Raman and Bhagavantam, on the other hand, argue that the interesting results obtained by Bär and Hanle on the reversal of the state of polarization of Raman lines when observed in the direction of propagation of the primary beam can be explained only on the assumption that the photons possess spin. They seem to link circular polarization definitely with a spin of the photon about the line of propagation.
>
> The arguments of Frisch and Kastler are based upon the Sommerfeld–Rubinowicz explanation of the selection principle for the azimuthal quantum number (principle of conservation of angular momentum of atom plus photon), but applying the same principle, and the principle that the atom-magnet can orient itself in any direction making certain definite quantized angles with the external field (as proved by Stern and Gerlach's experiment), it can be shown that the absorption of Zeeman components can never disappear with reversal of the field, but it will be modified on passing through two fields, whether parallel or antiparallel. Hence the experiments of Frisch or Kastler cannot be interpreted in the way supposed by them and show no light on the question of the spin. Secondly, and this is more important, a discussion of the Zeeman effect of the π-components of the D_1 line, assuming that the principle of conservation of angular momentum holds during radiation, shows that there may be photons without any 'spin' whatsoever, although they may show polarization. It therefore seems unjustifiable to describe polarization with the aid of a 'spin'. It appears that Bär and Hanle's results should be explained in some other way than that proposed by Raman and Bhagavantam. [6.14]

Obviously Saha was dissatisfied with the interpretations of Kastler and of Raman as well.

We now turn to the work of Raman and Bhagavantam who, unmindful of all

I say, What is this Raman Effect? 237

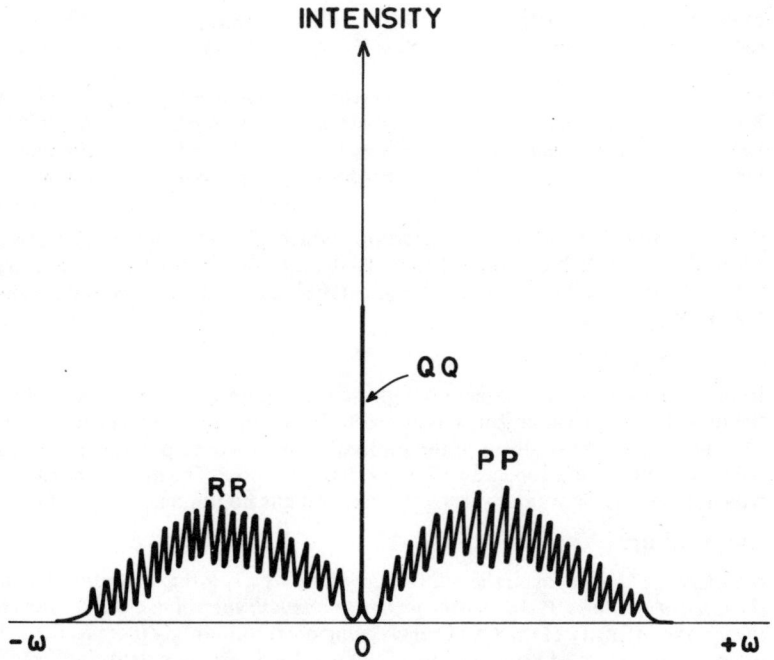

Figure 6.8 Schematic of the spectrum of scattered light, showing the Rayleigh line (QQ branch) and the rotational wings on either side (PP and RR branches).

previous experiments as well as comments, were determined to look for the spin of the photon. In considering their investigation we note first that the spectrum of scattering by molecules of a gas has three components in the vicinity of the incident line. Besides the unshifted Rayleigh line, there are the rotational bands on either side (see Fig. 6.8). While in hydrogen the rotational lines could be easily resolved, in the case of the other gases one usually saw only smeared wings (with high resolution, however, the line structure could be resolved). Concerning the selection rules for rotational scattering, Raman and Bhagavantam comment as follows:

> If we regard radiation as a particle possessing an intrinsic spin and impinging on a rotating molecule, the *sense* of the spin which the molecule possesses in relation to that of the photon should obviously enter into the problem in a fundamental way. If the molecule is spinning in a right-handed sense, it is obvious that if it gains energy of spin from the encounter, it must also gain angular momentum from it in a right-handed sense; this obviously it cannot if the photon before the encounter has a left-handed spin, since the latter can take up but not give up right-handed spin. Similarly, a molecule with left-handed spin cannot gain both energy and spin from a right-handed photon. We may in fact make out the following scheme of forbidden and allowed transitions:

Molecule spin	Photon spin	PP K→K−2	QQ K→K	RR K→K+2
Right	Left	Allowed	Allowed	Forbidden
Right	Right	Forbidden	Allowed	Allowed
Left	Right	Allowed	Allowed	Forbidden
Left	Left	Forbidden	Allowed	Allowed

It will be seen that while the transitions which give the QQ branch are always allowed, those which could contribute to the intensity of the PP and RR branches are forbidden in half the total number of possible cases and allowed only in the other half. [6.9]

What happens in the classical case?

In the treatment of the problem of light scattering on the Maxwellian field theory, the function of the radiation is assumed to be merely that of furnishing a periodic electric force at the position of the molecule which would polarize it and cause to emit secondary radiation as a rotating Hertzian dipole. The question of the sense in which the molecule rotates does not arise, and has no influence on the final results.

To sum it all up,

the effect of photon spin on the scattering of light by a rotating molecule is to diminish the intensity of the PP and RR branches to one half the value given by the classical theory, the intensity of the QQ branch being correspondingly increased, so that the intensity and state of polarization of the total scattering remain unaffected.

Raman and Bhagavantam also state,

The PP and RR branches of the scattering in the forward direction would in this view [conservation of angular momentum] consist of photons which have experienced a reversal of spin. In other words, if the incident light be circularly polarized, Rayleigh or undisplaced scattering would be circularly polarized in the same sense, and the rotational [Raman] scattering would be *reversely* circularly polarized.

In their opinion, the experiments of Bär essentially support this point of view.

Raman and Bhagavantam rely on depolarization studies for obtaining evidence concerning photon spin. Their arguments are not entirely clear but in essence they say that since the QQ branch consists mainly of polarized light while the PP and RR branches consist of unpolarized light,

the defect in polarization [ρ_0] of the QQ branch when spectroscopically separated from the PP and RR branches would be *smaller* than that of the total unresolved scattering consisting of all the three branches [i.e., ρ_∞].

According to them, $(\rho_\infty - \rho_0)_{\text{classical}} > (\rho_\infty - \rho_0)_{\text{spin theory}}$, and "this fact furnishes us with a convenient *experimentum crucis* for the existence of photon spin".

The actual experimental is performed by them as follows. Light from a mercury lamp is transversely scattered by the gas, allowed to pass through a nicol, and then focused on the slit of a spectrograph. Two photographs are taken of the spectrum of the scattered light with the nicol oriented in two perpendicular positions, and the state of polarization is determined by varying the relative exposures until the horizontal and vertical components are obtained with equal intensity. The slit is kept sufficiently narrow so that only the undisplaced Rayleigh line is recorded. In this way, the depolarization ρ_0 of the QQ branch is determined. A second set of measurements made with the slit open as wide as possible gives the depolarization ρ_∞ of the unresolved scattering, the comparison of which with the accepted value serves as a check of the perfection of the experimental arrangements.

The experiment itself is quite delicate, and, deeply conscious of the various possible sources of error, Raman and Bhagavantam took all the feasible precautions. In practically every case, the depolarization observed for the QQ branch itself was significantly less than that for all the branches combined. The observed values were found to be in accord with theoretical estimates (which I am afraid I do not quite understand), and it was finally concluded that the photon does have spin, as in fact it should.

While the theoretical reasoning of Raman and Bhagavantam might appear quite persuasive at first sight, a careful reading raises doubts. Can one argue as they do that "the results of an encounter between a spinning photon and a rotating anisotropic molecule must depend upon the relative sense of their spins before impact"? In other words, can one talk of molecules spinning to the right and molecules spinning to the left? It is obvious that Raman and Bhagavantam were misled into believing that they could do so from the example of the photon where the spin states and states of circular polarization are related.

Looking back, there seems to have been much confusion all round. Kastler[1] was wrong in his analysis, and Saha seems no better if one goes by his statement that "it therefore seems unjustifiable to describe polarization with the aid of a 'spin'". As for Raman, he obviously could not completely shake off his classical training before dealing with this problem. Altogether, people had not yet adjusted to the concept of quantum-mechanical spin, and the prevailing widespread misconceptions and misinterpretations are not surprising.

6.13 Anti-Stokes Scattering

The concept of anti-Stokes scattering has already been explained in Chapter 5. Here we make a brief reference to that topic for historical reasons.

In May 1928 Raman and Krishnan published a short note in *Nature* [6.15] reporting anti-Stokes lines in Raman scattering. However, leaning on the paper of Kramers and Heisenberg they identified these lines as due to "negative absorption of radiation". Translated into modern language, this would read stimulated Raman scattering (see Chapter 7). From our present experience we know that to observe stimulated Raman scattering one needs a powerful excitation source like a laser, which means that Raman and Krishnan were in error with their identification. Shortly thereafter, Saha, Kothari and Toshniwal [6.16] pointed out correctly that the new lines reported by Raman and Krishnan were just anti-Stokes lines. It is curious that Saha *et al.* make no reference to the work of Kramers and Heisenberg. Indeed one even wonders whether they were aware of this paper at that time, as the following observation of theirs suggests:

> We wish further to point out that though the phenomenon [of Raman scattering] has been described as one of 'scattering', it seems to be intermediate between pure scattering (e.g., by fog particles in which the agent responsible for scattering does not suffer physical change) and pure absorption (e.g., the absorption of the sodium line by the sodium atom, resulting in the utilization of the total energy of the energy-particle in lifting the electron to the higher orbit and production of a new system). This phenomenon is just intermediate between the two.... [6.16]

No reference whatsoever to the virtual intermediate states.

While Saha might not have been familiar with the work of Kramers and Heisenberg, he certainly saw interesting extensions of the concept of anti-Stokes scattering to free electrons which would "probably afford an easy explanation of the origin of bright and broad bands in the spectra of novae, and of winged lines in the solar spectrum".

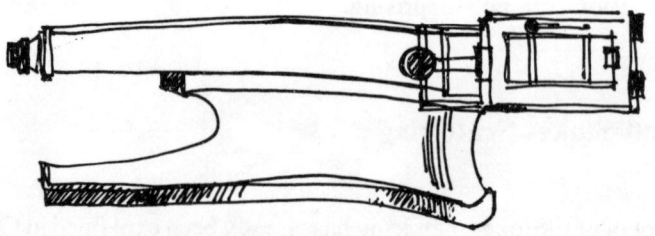

7 Brighter than a Thousand Suns

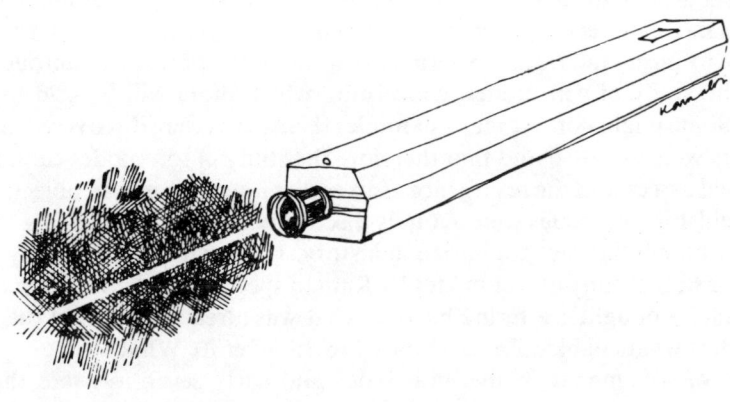

There is a popular saying in high energy physics that yesterday's discovery is today's calibration and tomorrow's background. The Raman effect did not quite pass into oblivion in this fashion although, as we shall see in Chapter 8, it was quickly transferred by physicists to the care of chemists who exploited the effect largely as an analytical tool. While a slow fade-out of discoveries is the norm, it occasionally happens in physics that there is a breakthrough (often of a technical nature) thanks to which a subject which had earlier reached a plateau experiences a dramatic revival. This is precisely what happened in the case of Raman scattering when the laser was invented.

The impact of the laser on light scattering is best appreciated by considering some numbers. The intensity I of light is usually expressed in watts per square metre (W/m^2). Since light is electromagnetic in nature, there is an associated electric field E which, when expressed in units of volts per metre (V/m), leads to the following relationship between I and E:

$$I(W/m^2) = 1.327 \times 10^{-3} E(V/m).$$

The irradiance produced by solar radiation at the earth's surface is about 1.4×10^3 W/m^2, corresponding to an electric field intensity of around 10^6 V/m.

The laser produces a highly directed (and near-monochromatic) beam of cross-section typically around 10^{-5} m^2 which can be further reduced by focusing. Thus, with a continuous wave laser delivering an average power of just a few watts, irradiances in the range 10^4 to 10^6 W/m^2 can be readily obtained with suitable optics. With giant pulse lasers, irradiances of the order of 10^7 W/m^2 are quite easy to achieve, the corresponding electric fields being in the neighbourhood of 10^{10} V/m. Truly mind-boggling irradiances and electric fields are produced in laser fusion experiments. It is not surprising therefore that the advent of the laser led not only to the discovery of processes far beyond what Raman did, but also to the emergence of several new techniques and devices as well. We offer below a flavour of this new development.

The laser is basically a highly monochromatic (coherent[1]) light source capable, in some instances, of very high intensity. Given such a source, one's first inclination would be to push traditional experiments to domains previously untouched. One such is the study of soft modes, concerning which more will be said in the next chapter. Some might contest that soft modes themselves were discovered around the time lasers were invented and that therefore the study of soft modes cannot strictly be regarded as a case of the resurgence of an earlier interest. Such an objection would not be valid, for soft modes were actually discovered much earlier, in fact by Raman himself, although that fact got buried in history. That apart, from a purely technical point of view, the study of soft modes by Raman spectroscopy does not require any new principles of light scattering beyond what was already known in the pre-laser era (which is what enabled Raman himself to discover it). What strongly promoted the study of soft modes in the mid-sixties and early seventies were the intense scientific interest on the one hand (both on account of their relationship to the phenomenon of ferroelectricity and because of the larger implications for the study of phase transitions), and the rapid advances in instrumentation on the other. The quality of data obtainable at present may be judged from Fig. 7.1 which is to be compared with Fig. 8.34.

While many examples of such "extensions" exist, far more interesting are the study of new phenomena and new types of experiments previously inconceivable. Our first example in the latter category will be resonance Raman spectroscopy. In the pre-laser era, there was not much choice with respect to ω_0 the frequency of the excitation source. Raman himself used the 4358 Å radiation from a quartz mercury lamp in his very first experiment; other workers used similar sources but invariably they were spectral lines emitted by suitably constructed lamps. Not only was the total set of excitation frequencies so available rather limited, but an even more severe restriction was that the frequency of the source could not be *continuously varied*. One might wonder why one would want to continuously vary ω_0 since, *per se*, the Raman effect does not call for any such tuning. The answer lies in the celebrated paper of Kramers and Heisenberg who had pointed out that the intensity of (Raman) scattering would be *considerably* enhanced if the frequency of the incident radiation were to coincide with one of the electronic transitions of the scattering medium (see

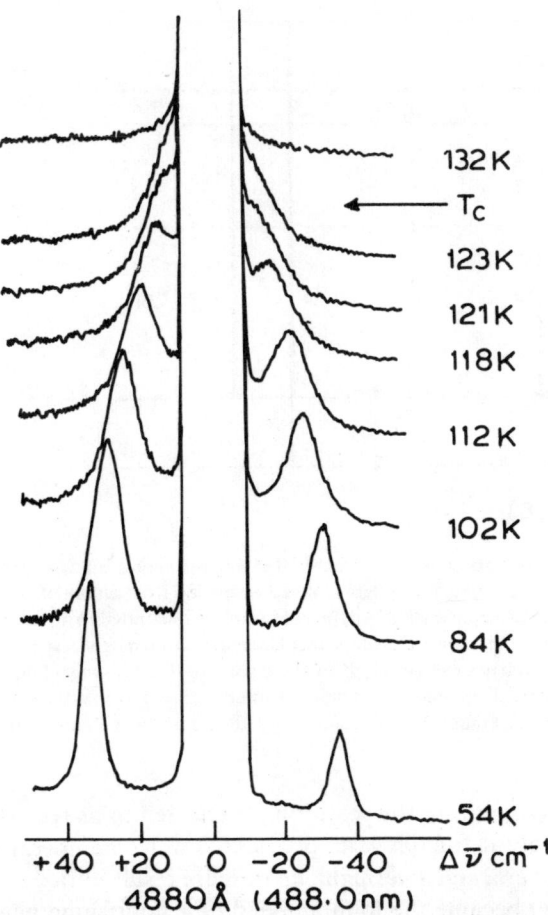

Figure 7.1 Raman spectra illustrating the soft mode in a single crystal of potassium selenate. The strong peak with zero frequency shift arises owing to Rayleigh scattering. On either side of it may be seen Raman lines whose frequency is strongly temperature-dependent, in marked contrast to the usual behaviour. In fact, the frequency of this mode moves from $35\,\text{cm}^{-1}$ at low temperature towards zero as the temperature approaches 130 K. For this reason the vibration is referred to as a *soft mode*. The study of soft modes is most conveniently carried out using Raman spectroscopy. Further remarks on the subject are made in Chapter 8. (After ref. [7.1].)

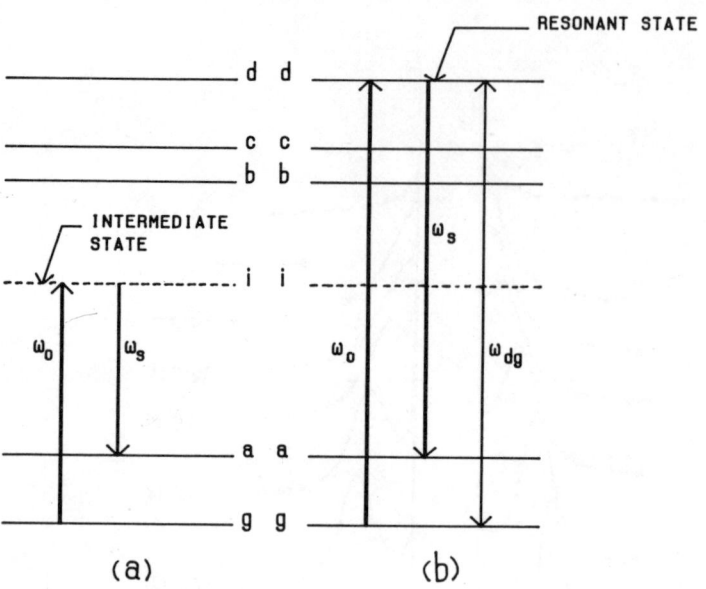

Figure 7.2 Illustration of the concept underlying resonance Raman spectroscopy. In the usual Raman scattering, the system makes a transition from an initial state to a final state via a virtual intermediate state. A typical example is illustrated in (**a**) for a Stokes process, ω_0 and ω_s being the frequencies of the incident and scattered light respectively. However, if the incident frequency can be varied and be made equal to a *resonant* frequency ω_{dg}, say, as in (**b**), then state d becomes the dominant intermediate state and is said to be resonant instead of virtual. Under these circumstances, the intensity of Raman scattering is greatly enhanced.

Fig. 7.2). Under these conditions, the scattering is referred to as resonance Raman scattering. The arrival of the tunable laser provided a welcome liberation from the constraints of an earlier era, and, overnight, an obscure result buried in the pages of an old German journal became the fountainhead of a flourishing new technique.

A good example of the power of resonance Raman scattering is provided by Fig. 7.3. Figure 7.3a shows a conventional Raman spectrum of K_2CrO_4 (potassium chromate) obtained with 6328 Å radiation (from a helium–neon laser). The absorption spectrum of the scattering material is shown in Fig. 7.3b from which it is clear that the excitation frequency of the helium–neon laser source lies outside the absorption band. Suppose now we have a laser whose frequency can be varied in the direction of the fat arrow. If, by so doing, the excitation frequency is brought into the absorption band, resonance Raman scattering would occur. A spectrum thus measured is shown in Fig. 7.3c, this particular one having been obtained with 3638 Å excitation radiation. Instead of a sparse spectrum as in Fig. 7.3a, there is a rich one with several overtones.

Overtones are no doubt interesting; nevertheless resonance Raman spectroscopy is usually employed not for the study of overtones but for trace detection, exploiting the fact that at resonance the intensity of the fundamental is enhanced by several

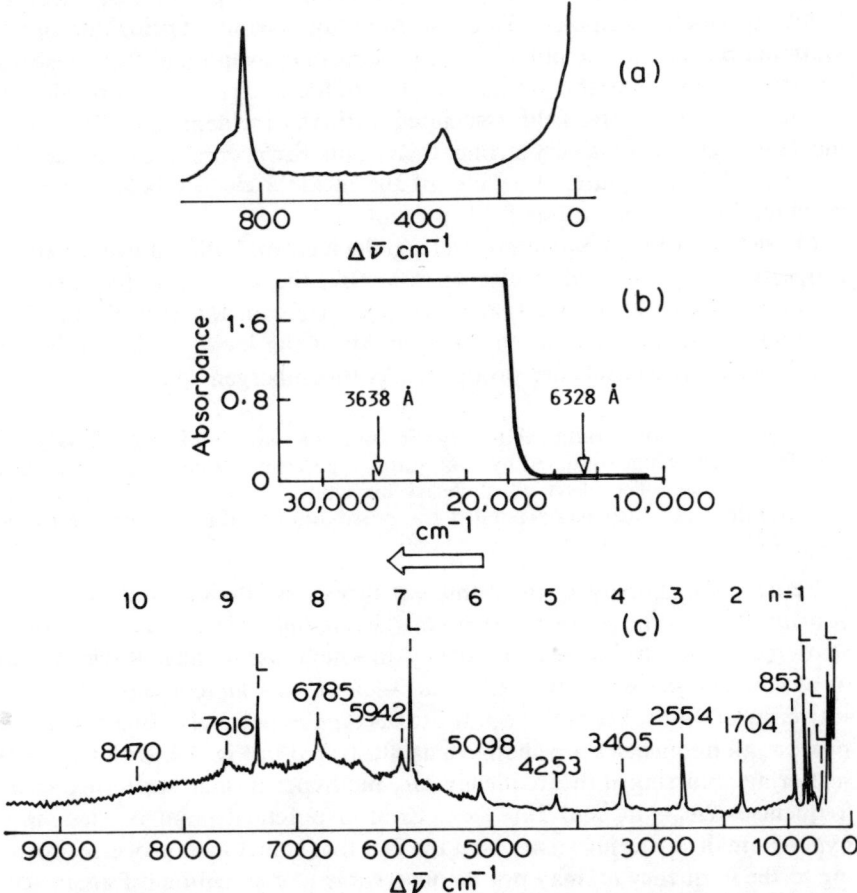

Figure 7.3 Resonant Raman scattering (RRS) in potassium chromate. In (a) is shown a conventional Raman spectrum obtained using a helium-neon laser. The absorption spectrum of the material (in aqueous form) is shown in (b). If the laser light can be tuned so that its frequency ω_0 can be brought within the absorption band, then RRS will occur. Tuning is not possible with the helium-neon laser, but there exist lasers with this capability. A resonant spectrum so obtained is shown in (c) where the lines márked L are laser lines. Observe the rich detail; as many as ten overtones of the fundamental at $853\,\text{cm}^{-1}$ can be seen. Note, however, that the overtone frequencies are not exact integer multiples of 853. This is due to anharmonicity. (After ref. [7.1].)

orders of magnitude. Thus chemists and biologists use the technique to detect molecules present in very low abundance.

The phenomena we consider next belong to the general realm of nonlinear optics, and arise specifically on account of the very high electric fields present in the incident radiation. It will be recalled that the scattering process which Raman discovered can be visualized as a two-step affair in which the electric field of the incident light first polarizes the molecule of the scattering medium to produce an electric dipole following which the dipole radiates, the radiation so emitted being interpreted as the scattered radiation. The polarization process in conventional Raman scattering is linear, meaning that the strength of the induced dipole is proportional to the strength of the electric field associated with the incident light. If, however, the incident electric field is very strong, then nonlinear processes can occur, involving, for example, the square, cube, etc. of the incident electric field, not to mention combinations of the various fields present.

In brief, in ordinary Raman scattering, the incident light is a probe exploring the properties of the scattering molecule, rather like the press faithfully reflecting public opinion. But, even as the press can under certain circumstances itself influence public opinion, it happens that when the strength of the incident electric field is high, dramatically new effects are produced. As Bloembergen[2] puts it:

> There is a close relationship between some of these nonlinear effects and the frequency change observed by Prof. Raman in the scattering of light by molecules and crystals. The effect which bears his name has a nonlinear or stimulated counterpart which has expanded the possibilities of Raman-type spectroscopy. [7.2]

Perhaps the simplest of the nonlinear processes alluded to above is the *hyper-Raman effect*. In the process earlier considered, light of frequency ω_0 can either be scattered without a change of frequency (in which case we have Rayleigh scattering) or be modified in frequency to either $\omega_0 - \omega_v$ or $\omega_0 + \omega_v$ (corresponding to Raman scattering of the Stokes and the anti-Stokes type respectively). In a hyper-scattering process, all frequencies are changed, as illustrated in Fig. 7.4, with hyper-Rayleigh scattering occurring at the frequency $2\omega_0$ and hyper-Raman scattering occurring at frequencies $2\omega_0 - \omega_v$ and $2\omega_0 + \omega_v$. From a practical point of view, interest in hyper-Raman scattering stems from the fact that sometimes, processes corresponding to the frequency ω_v may not be observable in conventional Raman scattering. More technically, they could be *forbidden*. However, forbidden processes could reveal themselves in hyper-Raman spectroscopy, as in the example of Fig. 7.4c.

Undoubtedly the most spectacular nonlinear effect of the type we are considering relates to stimulated Raman scattering (SRS). Key to an understanding of this process is *stimulated emission*, a concept going back to Einstein. Consider a system like, say, a collection of molecules with energy levels as in Fig. 7.5. While the natural tendency would be for all molecules to be in the state of lowest energy (ground state, as it is often called), higher levels (or excited states) like a, b, c, etc. can also be occupied, especially at high temperatures. According to Bohr, a molecule or an atom in an excited state such as a would make a downward transition emitting electromagnetic radiation, the process itself being called *spontaneous* emission. In the twenties Einstein made the interesting discovery that if a system is in an excited state such as a and if radiation of frequency $\omega = 2\pi(E_a - E_g)/h$ (where h is a quantity called the Planck's constant) is incident upon the system, then the latter can be *induced* or *stimulated* to make the jump from level a to level g. Such a process is

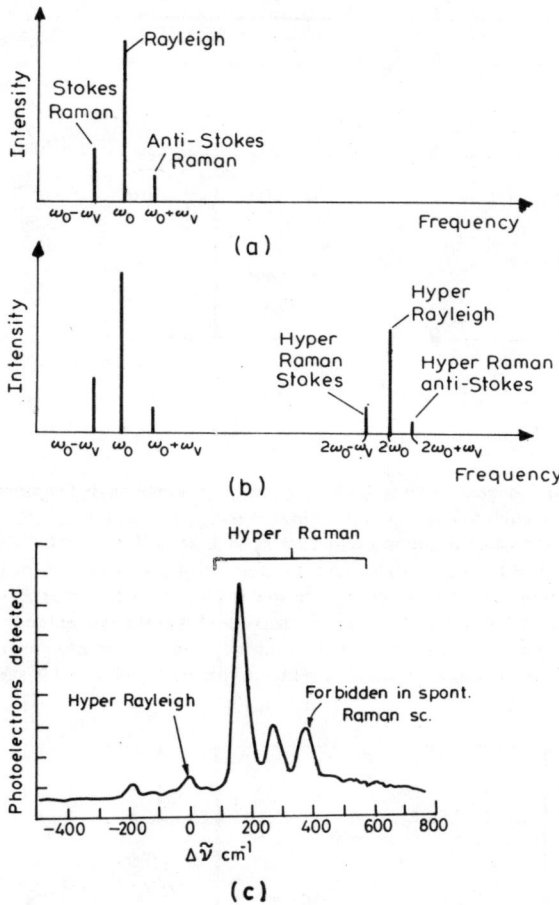

Figure 7.4 (a) Schematic of the spectrum of normal Raman scattering. If nonlinear processes are possible then hyper-scattering can also occur, as in (b). A typical hyper-Raman spectrum is shown in (c), the material in this case being ammonium chloride. The band at 360 cm^{-1} cannot be seen either in the infra-red spectrum or in the conventional Raman spectrum. This particular mode was once of much interest in connection with a structural phase transition in ammonium chloride. ((c) After ref. [7.1].)

referred to as stimulated emission and occurs at a rate greater than that for spontaneous emission.

Although the concept of stimulated emission had been known for a long time, it remained largely unexploited until, of course, the laser came along. And once the laser became available, stimulated emission was exploited in diverse ways leading, in fact, to a new device, namely, the Raman laser.

At the heart of the Raman laser is the idea of stimulated Raman scattering to understand which a reference to Fig. 7.6 is necessary. In Fig. 7.6a is illustrated the process involved in normal Raman scattering, or spontaneous Raman scattering as

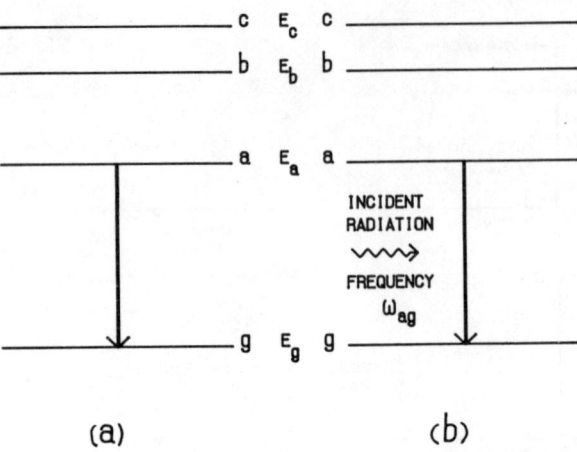

Figure 7.5 (a) Schematic energy level diagram of an atom (say). The normal tendency would be for the atom to be in the state of lowest energy, i.e., the ground state g, but means exist to make the atom reside in excited states such as a. Bohr pointed out that atoms cannot live forever in excited states and that they must eventually return to the state g, emitting radiation whose frequency ω_{ag} is related to the energy difference $(E_a - E_g)$ between the excited and the ground states. Einstein discovered that the rate of the transition $a \to g$ can be greatly enhanced if radiation of frequency ω_{ag} is simultaneously present, as in (b). In such a case, the emission associated with the atomic transition $a \to g$ is stimulated.

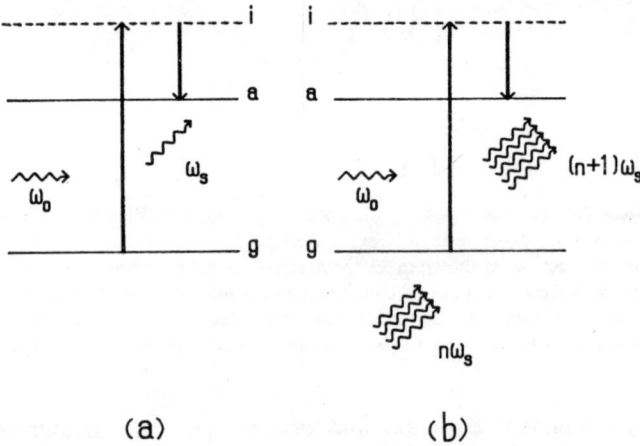

Figure 7.6 Schematic illustration of the principle underlying stimulated Raman scattering (SRS). In (a) are shown the transitions associated with the familiar spontaneous Raman scattering. In (b), the atomic transitions are similar, except that they occur in the presence of radiation of frequency ω_s (Stokes line). The transition $i \to a$ is therefore stimulated. In the practical realization of SRS, one cleverly arranges the situation so that starting from (a) one goes over to the case (b), i.e., the solitary photon that is first emitted in spontaneous Raman scattering is recycled to promote SRS, emitting more Stokes photons. In turn, these are recycled, leading to a net *coherent* amplification of the Stokes line.

it is nowadays called. Before the scattering event there is no photon of frequency ω_s while after the event there is a single photon of that frequency. In Fig. 7.6b is depicted the scattering process when radiation of frequency ω_s is already present; naturally, this radiation promotes stimulated emission of photons of frequency ω_s. At first sight one might not notice any material difference between cases (a) and (b) since in both instances the number of photons of frequency ω_s changes by just one. There is, however, a subtle difference. If conditions can be so arranged that the scattered photons are trapped and made available to contribute to stimulated emission, one can envisage a build-up of the scattered intensity. It is natural then to ask why such a recycling-cum-multiplication at frequency ω_s was not accomplished by Raman himself or anyone of his generation. The answer is that such a multiplication is essentially a nonlinear process and was simply not possible in the pre-laser era; in fact, even the conceptual visualization of such processes did not exist since there did not seem to be any need for such flights of fancy. Actually, stimulated Raman scattering was discovered accidentally [7.3], the theoretical explanation itself coming later.

A popular scheme to demonstrate SRS is shown in Fig. 7.7. Excitation radiation from a ruby laser is allowed to be incident on a cell containing a good scattering liquid like benzene. Raman scattering will then occur in all directions but the

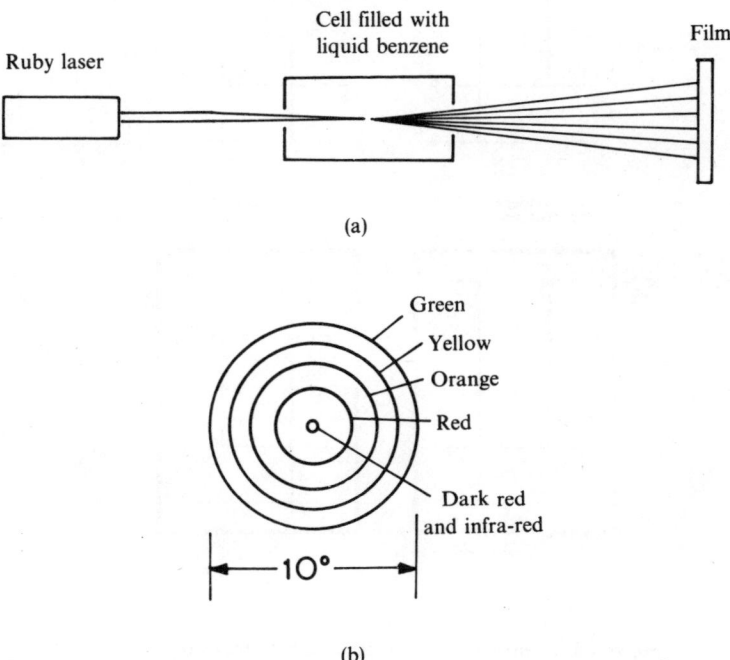

Figure 7.7 (a) Experimental arrangement to demonstrate SRS. The ring pattern is shown separately in (b). The angular width of the pattern is about 10°.

forward-scattered radiation can bounce back and forth in the cavity promoting stimulated emission at ω_s. The occurrence of SRS is established by placing a colour-sensitive film some distance away, whereupon a striking pattern of coloured rings is observed. A notable feature is that a substantial fraction of the incident radiation appears as SRS radiation. Interestingly this radiation is coherent, in marked contrast to what happens in spontaneous Raman scattering. Just to give a quantitative feel, in benzene, about 50% of the radiation incident at ω_0 is converted into Stokes radiation at frequency $\omega_s = (\omega_0 - \omega_v)$. In other words, one has a new light source at the Stokes frequency ω_s; not surprisingly, such a source is referred to as a Raman laser.

Reference was just made to the coherence aspect of nonlinear scattering. Indeed, there is a strong undercurrent of coherence in almost all nonlinear Raman processes, as an illustration of which we will briefly consider coherent anti-Stokes Raman spectroscopy (CARS). It will be recalled that in spontaneous Raman scattering, although the Raman lines occur symmetrically on either side of the unshifted frequency, the anti-Stokes line at $(\omega_0 + \omega_v)$ is very much weaker than the Stokes line at $(\omega_0 - \omega_v)$, except of course at high temperatures. In nonlinear optics, on the other

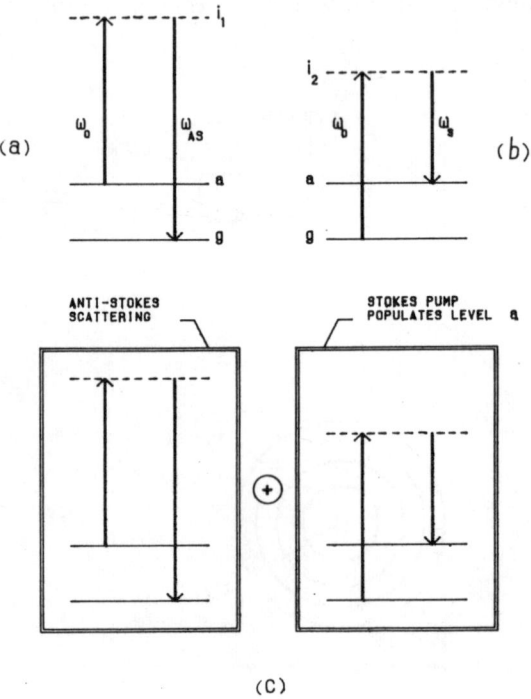

Figure 7.8 Transitions in the usual anti-Stokes (**a**) and Stokes (**b**) processes. When nonlinearities are present, the two can couple (**c**) and reinforce coherently to produce a strong anti-Stokes line of frequency $\omega_{AS} = (2\omega_0 + \omega_s)$. As is to be expected, resonances are possible when ω_0 or ω_{AS} matches a resonant frequency.

hand, both the Stokes and the anti-Stokes processes can be of comparable intensity. Bearing this in mind, we turn to the principle underlying CARS which may be understood by referring to Fig. 7.8. In Fig. 7.8a, b are shown the transitions associated with anti-Stokes and Stokes processes in spontaneous Raman scattering. Let us now consider the anti-Stokes process in the presence of a strong incident electric field. Even as the ground state g is being populated by the anti-Stokes process, the laser light can pump the system upwards from where it can descend to the state a by a Stokes process. In other words, there are two things happening simultaneously – an anti-Stokes process, and a Stokes pumping which feeds the former, coherently enhancing its intensity. However, such a coupling is possible only in a nonlinear regime. By tuning the incident frequency, various kinds of resonances can also be achieved in CARS as in resonance Raman spectroscopy. These ramifications have been skilfully exploited to generate a variety of novel applications.

Enough has been said (despite the brevity of our description) to convey the fact that the world of nonlinear processes is both vast and fascinating. Not only has Raman spectroscopy acquired a new dimension with the arrival of the laser, but new words like SRS, CARS, Raman laser, etc. have also been added to the scientific literature. Bloembergen comments,

> As spontaneous Raman spectroscopy has blossomed and grown during one half-century, it may be predicted with some confidence that coherent nonlinear Raman spectroscopy will yield many new results in the next half-century. [7.2]

Indeed nonlinear optical processes play an important role in fields as diverse as laser fusion and Star Wars! Raman's original discovery exposed but the tip of the iceberg – and what an iceberg it was!!

☆A brief interlude for the physicist to indicate the relationship of nonlinearities to some of the processes outlined above. In the interests of simplicity we will restrict ourselves to a classical discussion.

Earlier we have seen that spontaneous Raman scattering may be understood in terms of an induced dipole. Restricting ourselves to linear processes, an induced dipole moment **P** is related to the applied field **E** by

$$\mathbf{P} = \chi \cdot \mathbf{E}, \tag{7.1}$$

where χ is the susceptibility tensor. If nonlinearities are present, one must generalize (7.1) as:

$$\mathbf{P} = \mathbf{P}^{(1)} + \mathbf{P}^{(2)} + \mathbf{P}^{(3)} + \ldots, \tag{7.2}$$

where

$$\mathbf{P}^{(1)} = \chi^{(1)} \cdot \mathbf{E}, \tag{7.3a}$$

$$\mathbf{P}^{(2)} = \chi^{(2)} \cdot \mathbf{EE}, \tag{7.3b}$$

$$\mathbf{P}^{(3)} = \chi^{(3)} ::\mathbf{EEE}, \quad \text{etc.}, \tag{7.3c}$$

with $\chi^{(2)}$, $\chi^{(3)}$, etc. denoting the various nonlinear susceptibilities. While (7.3) gives the structural relationship between the induced dipole and the applied electric field, one must be careful enough to indicate the various wavevector and frequency dependences. Thus (7.3b) should be written

$$\mathbf{P}^{(2)}(\mathbf{k}, \omega) = \chi^{(2)}(\mathbf{k} = \mathbf{k}_i + \mathbf{k}_j; \omega = \omega_i + \omega_j) : \mathbf{E}(\mathbf{k}_i, \omega_i)\mathbf{E}(\mathbf{k}_j, \omega_j), \tag{7.4a}$$

and (7.3c) as

$$\mathbf{P}^{(3)}(\mathbf{k}, \omega) = \chi^{(3)}(\mathbf{k} = \mathbf{k}_i + \mathbf{k}_j + \mathbf{k}_l; \omega = \omega_i + \omega_j + \omega_l) : :$$
$$\mathbf{E}(\mathbf{k}_i, \omega_i)\mathbf{E}(\mathbf{k}_j, \omega_j)\mathbf{E}(\mathbf{k}_l, \omega_l), \tag{7.4b}$$

and so on.

The above formulae provide a wide enough umbrella to encompass a host of nonlinear phenomena. For instance, using the subscript 0 to denote the incident radiation, the term

$$\mathbf{P}^{(2)} = \chi^{(2)}(\mathbf{k} = \mathbf{k}_0 + \mathbf{k}_0; \omega = 2\omega_0) : \mathbf{E}_0^2$$

describes hyper-scattering. Similarly,

$$\mathbf{P}^{(3)} = \chi^{(3)}(\mathbf{k} = \mathbf{k}_s; \omega = \omega_s) : : \mathbf{E}(\omega_s)|\mathbf{E}(\omega_l)|^2$$

describes SRS, and

$$\mathbf{P}^{(3)} = \chi^{(3)}(\mathbf{k} = 2\mathbf{k}_0 + \mathbf{k}_s; \omega = 2\omega_0 - \omega_s) : : \mathbf{E}(\omega_s)\mathbf{E}^2(\omega_0)$$

describes CARS.

As is only to be expected, symmetry provides some simplification of the structure of the various susceptibility tensors. Nevertheless, these are complicated objects. $\chi^{(3)}$, for example, has 81 elements, in general complex, and each element consisting in general of 48 terms! An explicit study of $\chi^{(3)}$ has been carried out by Bloembergen in an article contributed on the occasion of the golden jubilee of the Raman effect [7.4].

Relations (7.3, 7.4) are in the nature of constitutive equations. For a full description of the complex wave phenomena associated with various processes, one must solve the coupled wave equations with proper boundary conditions. Quantum extensions to these treatments are also available. ☆

Many are the novel applications of Raman spectroscopy that have emerged after the advent of lasers, but of these, it suffices to describe just one to give an idea of the new things possible. The system we select for description is the Raman microscope, or Raman microprobe as it is often called [7.5]. It is somewhat similar to the electron microprobe, but there are also important differences. The basic principles of the two systems are compared in Fig. 7.9a. In the electron microprobe, an electron beam is focused (in vacuum) onto a spot on the surface of the specimen under examination, causing characteristic X-rays or electrons to be emitted. The beam is rastered across the sample, producing a profile of the atomic abundances. In the Raman microprobe, photons are used instead of electrons, and instead of characteristic X-rays, one has scattered Raman radiation, bearing a signature of the *molecular* species on the surface. Once again by a rastering technique, one can obtain a profile of the molecular abundance on the surface. A schematic of such a profile is

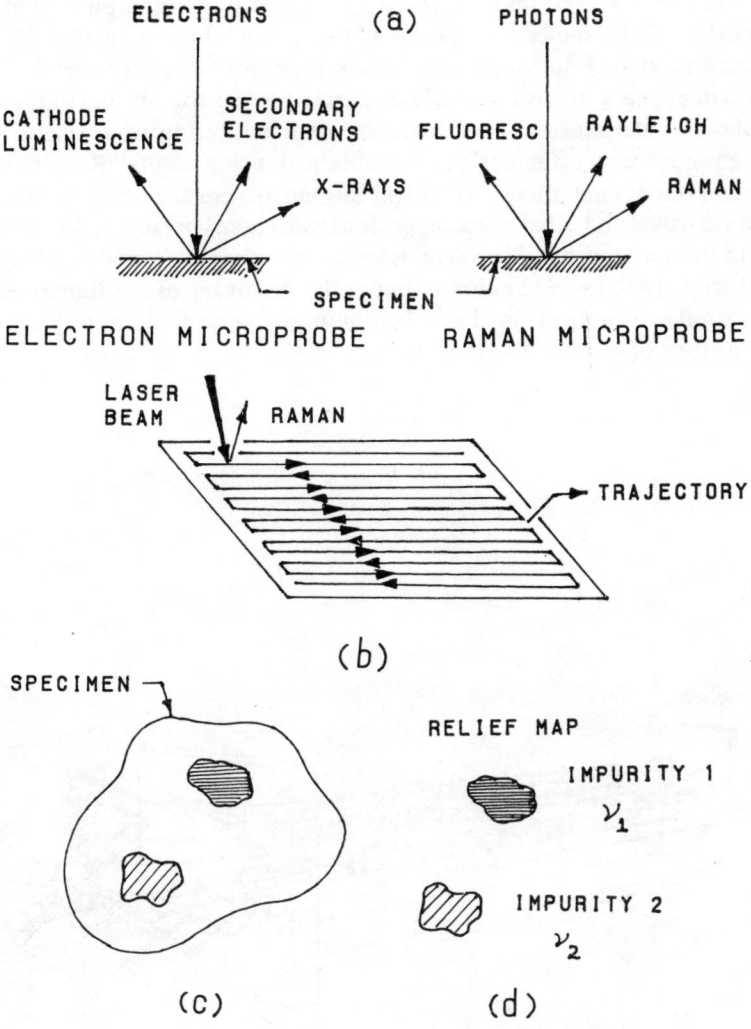

Figure 7.9 (a) Comparison of the basic principles of the electron microprobe and the Raman microprobe. In both cases, there is an incident radiation which, when it falls on the sample, produces secondary radiations. Those associated with the electron beam yield atomic signatures while the Raman scattered radiation associated with the photon beam yields molecular signatures. To detect molecules of a particular species, the incident laser beam is swept in a raster across the specimen surface as in (b). At each spot, the intensity of the Raman line corresponding to a favourable molecular vibration of the molecular species of interest is measured. Schematics of intensity maps so obtainable are shown in (d). They reveal the presence of impurities in the specimen, as shown in (c). The embedded impurities may not often reveal themselves to a conventional optical examination.

shown in Fig. 7.9c. For many applications (e.g., a study of impurities in crystals), a knowledge of the molecular species is more important than that of the atomic species, and in such instances the Raman microprobe is very useful.

Raman appears to have seen a laser during his last trip abroad in the mid-sixties. In 1968 an international conference on light scattering was held in New York highlighting the spectacular results obtained using lasers. Raman received an invitation but did not attend. As Joseph Birman, the conference secretary, wrote, "the rigours of travel and other pressing professional claims prevented Professor Sir C. V. Raman from attending". However, Raman sent a message, essentially recalling the events from 1922 to 1928 culminating in the discovery of the Raman effect. There was a similar message from Leon Brillouin giving an illuminating account of the early history of Brillouin scattering.

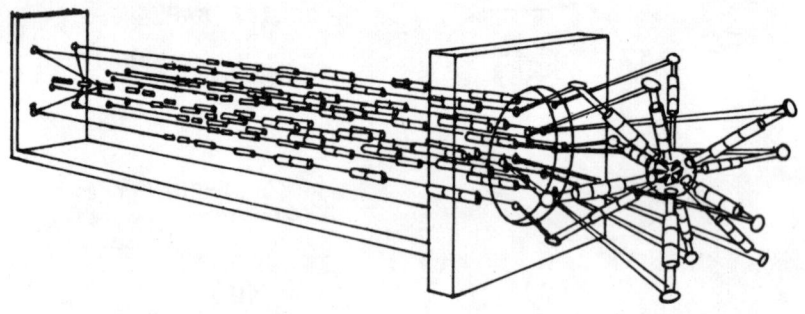

8 On to Bangalore

What is highest is Envy's mark

OVID

The transition to Bangalore did not end Raman's difficulties. Outwardly the problems he faced in Bangalore were different in character from those he had experienced in Calcutta, but in a generic sense they too were related to the question of developing and maintaining scientific excellence in a backward country. Once again the forces rallied against him were able to dislodge him from a position of authority but this time he did not completely leave the organization. Instead, he reluctantly reconciled himself to a lower post and continued till superannuation. But the replay of Calcutta was to leave deep scars.

8.1 The IISc – Its Origin

The origin of the Indian Institute of Science[1], while interesting in its own right, also offers many contrasts to that of the IACS. Founded in 1911, the Institute owes its existence to the foresight, patriotism and munificence of Jamshedji Nusserwanji Tata. Even today, it is better known in Bangalore as the Tata Institute than by its official name.

J. N. Tata was a distinguished member of the Parsi community, to which a reference has already been made. Born in 1839, he rose to a position of great eminence not only in his own community but in the entire nation. The city of Jamshedpur in Bihar named after him is a testimony to his vision in introducing modern steel technology into the country But Tata will be remembered for something more than merely the founding of the Tata Iron and Steel Co., or the Tata Power Stations or even the legendary House of Tatas. Tata was convinced, especially after a careful study of the European and American scenes, that India's problems could be solved in the long run only by the large-scale introduction of science and technology. But factories alone do not produce science and technology; more important was the development of human resources. No doubt the British had set the ball rolling by introducing Western education, but the prevailing educational system was quite disappointing, functioning as it did largely as a degree mill. The existing universities provided affiliation to colleges scattered all over, and acted mainly as examining bodies rather than as residential schools. While this suited the Indian parent who preferred his son staying at home, it certainly did not permit the universities to emerge as the centres of learning they were supposed to be. Thus the noted chemist Sir William Ramsay wrote: "The colleges are wretched places as a rule ... and the whole system is rigidly examinational like the London University. They are reaping the fruits of it in a number of cramming shops, miscalled colleges." One wonders what Ramsay would have to say about today's colleges!

In 1889 Lord Reay, the Governor of Bombay Presidency, speaking as the Chancellor of Bombay University, pleaded that "we establish in this old home of learning, real universities which will give a fresh impulse to learning, to research, to criticism, which will inspire reverence and impart strength and self-reliance to future generations of our and your countrymen".

Jamshedji Tata was deeply impressed by Lord Reay's speech, and vague plans which were then churning in his mind now began to take concrete shape. For a start, Tata decided to send, every year, a few chosen students to England for advanced training as well as exposure to a better educational system. The expenses involved were considerable but Tata was shrewd and offered the money as a loan on a very low rate of interest rather than as charity. Since the persons selected were brilliant they invariably managed to secure good jobs on return, and loan repayment was never a problem. By this means the fund was conserved and made available to several generations of students. Tata also regarded his scheme as a good investment for his country and said, "Every Indian that gets into the Civil Service ... effects a saving of two lakhs of rupees: that is what a civilian's pay, allowance and pension come to, most of which usually goes to Britain."

Tata was justly proud of his scholars and pleased with their success. But he would not rest content with merely sending scholars abroad for study. Why not go further and bring to India that kind of education, especially in science and technology? The steady increase of his fortune certainly permitted Tata to think in such grandiose terms. But to give these plans a practical shape much work was required for which Tata had no time. So he sought the help of one Mr Burjorji Padshah and sent him on a tour of Europe. Mr Padshah travelled across the Continent for eighteen months meeting many educationists, each of whom had his own recipe – one favoured a school of medical research, another a research institute for the chemistry of tropical plants, and so on. Meanwhile, there were two opinions about how the proposed centre of higher education should be organized, whatever its specialization. Some wanted it to be affiliated as a post-graduate centre to one of the existing universities, preferably the Bombay University since Tata himself had studied there. But there was also another view favouring a separate university or institute of research open to the graduates of all the existing universities. Even as this debate was going on, the Tata plan was made public in 1898 and immediately brought forth a chorus of praise. Thus *The Hindu* wrote, "Mother India has long been crying for a man among her children, and in Mr Tata she has found the son of her heart." Nevertheless, within the Parsi community itself there was some heart-burn about the diversion of a great fortune for the support of non-Parsi causes. But Tata was not much impressed with the clamour for further contributions to Parsi charities. He said:

> What advances a nation or community is not so much to prop up its weakest and most helpless members, as to lift up the best and most gifted so as to make them of the greatest service to the country. I prefer this constructive philanthropy which seeks to educate and develop the facilities of the best of our young men. And if this is to be done, what I ask my fellow Parsis is, "What difference is it to them whether it is exclusively to their benefit or open to all?" If able professors and specialists are to be obtained, the cost will be the same, whether it is only a few Parsis alone that attend their lectures or young men of all communities. The Parsis cannot supply more than a few students for each post-graduate class, and it would be foolish to have costly professors to lecture to only two or three Parsis to the exclusion of hundreds who are anxious to benefit likewise.

A committee was now formed to bring the proposed institute into existence. It would be all-India in character, called the Imperial University of India, and modelled after the Johns Hopkins University in Baltimore, USA. The location of the university also began to receive some attention. Tata favoured a place with a salubrious climate, being of the view that hot weather, so common throughout India, was enervating and not conducive to intense creativity. He then went on a site-selection trip through the country, and his tour brought him to Bangalore where his friend Sir Seshadri Iyer pressed the claims of that city. There was also a strong Bombay lobby, but it failed to find the required large plot of land within the city. Trombay, a suburb of Bombay where Tata owned some property, was briefly considered but ruled out owing to lack of proper communication[2]. Meanwhile the Maharajah of Mysore furthered Bangalore's claim by offering a gift of 300 acres of land, Rs. 5 lakhs towards the building and an annual subsidy of Rs. 1 lakh, a princely offer which could hardly be ignored! Tata now began to swing in favour of Bangalore, especially since a marriage alliance had been forged between the Tata family and the Bhabha family[3] of Mysore.

One could hardly start a venture of such a magnitude without the blessings of the Government. Tata was able to prevail upon Lord Curzon, the Viceroy, and soon, conceptual approval of the Government was forthcoming after some administrative details were sorted out. At this stage it was suggested that the new university should be called the Tata University but Tata rejected the idea. He now invited Sir William Ramsay to visit India and work out the details. Ramsay approved the choice of Bangalore on various practical considerations. Besides the pleasant weather and the mineral and hydroelectric power potential of the state of Mysore, there was the all-important factor of free land and the annual subsidy offered by the Maharajah. Ramsay also found a "certain nucleus of scientific society" congenial to the staff and the students. As for the name, Ramsay was not in favour of calling the proposed centre a university – not surprising in view of the low opinion he held of Indian universities! He was also practical. If the new centre of learning were to be called a university, it would promptly arouse a certain amount of jealousy in the already existing ones. But there was an even more important objection. "Such work as is here proposed" he wrote, "is not, as far as I am aware, carried out at any university.... It is not 'universal' ..., i.e., it does not cover research in all branches of knowledge, at least at present." It was finally decided that the new centre would be called the Indian Institute of Science (the name by which it continues to be known).

The publication of the Ramsay report generated some protests, especially from Calcutta which felt that its own claims were superior to those of Bangalore. As far as the Government of India was concerned, Ramsay's words were not enough, and it appointed a committee of its own consisting of Col. Clibborn of the Roorkee Engineering College and Prof. David Manson of Melbourne to "examine" the Ramsay report. Not unexpectedly, this committee felt that Roorkee was a better choice than Bangalore (!), but who in Roorkee could match the donations of the Maharajah of Mysore? Meanwhile, the Government of India began to drag its feet by not firmly committing its share of funds, and Tata became desperate as he was getting old. He was most anxious to see his dream come true but, as it turned out, he was denied that pleasure since death snatched him away in 1904. Finally, in 1905, the Government of India accepted the choice of Bangalore and the broad framework of Ramsay's recommendations with, however, the inevitable financial pruning. A Committee of the Royal Society was then asked to select a nominee for the post of Director and its choice fell on Dr Morris Travers, FRS. Travers sailed for India in 1906. But things still moved quite slowly and all that happened during the next three years was that the constitution of the Institute was approved by the Viceroy (now Lord Minto), and the Council of the Institute came into existence. It was only in 1911 that the Institute finally started functioning with three departments – General and Applied Chemistry, Organic Chemistry and Electro-Technology.

8.2 The Institute Before Raman

Prior to Raman's entry, the Institute consisted of the departments just mentioned and one more, Biochemistry, added subsequent to the founding. All of these engaged in teaching but in addition the various chemistry departments also did some research, mostly applied in character. Thus there were studies in ore beneficiation, manufacturing routes for organic chemicals, and sewage disposal, to name just a few topics.

The management of the Institute was vested in the Visitor (then the Viceroy) and in the following bodies – the Court, the Standing Committee of the Court, the Council and the Senate. The Court had eighty odd (!) members, its sole function being to elect two members to the Council. About the Court one Review Committee wrote, "Many of these gentlemen [members of the Court] on being addressed replied that they had no knowledge of the Institute." The job of the Standing Committee was to review the working of the Institute and advise the Council regarding the expenditure, while the latter body was charged with overseeing the Institute management. The Senate was essentially a Faculty body which made recommendations to the Council.

The working of the Institute was reviewed from time to time by committees appointed by the Visitor. Thus there was a Review Committee in 1921 headed by Sir William Pope, FRS, and another in 1930 under Lt. Col. Sewell. The Pope Committee was apprehensive that the Institute was veering far too much towards industrial application, and made a plea that basic research be made a necessary part of the Institute's programme. It said:

> We would point out that great numbers of scientific industries have originated in purely academic scientific work; the same process of evolution will certainly lead to the establishment of important branches of technology which are not yet foreseen. Apart, however, from the fact that applied science is largely founded on discoveries in pure science, it may be remarked that the exclusion of pure scientific research will inevitably narrow the interests of the workers and check the development of that scientific atmosphere which is so essential in such an organization as the Institute. It is desirable that workers in pure science should be encouraged in the Institute....

And when the opportunity arose for doing precisely this, the Institute management did exactly the opposite!

Lt. Col. Sewell who headed the next Review Committee in 1930 was the Director of the Zoological Survey of India. He was also the President of the Anthropology Section of the Indian Science Congress session of 1929 (which met in Madras and over which Raman had presided). The other members of the Committee were Meghnad Saha and a Major Howard, an electrical engineer from Madras. If the Pope Committee saw a danger of the Institute becoming industrialized, the Sewell Committee leaned the other way worrying whether applied research would be overshadowed by purely scholastic enquiries! With Raman's entry, this debate of pure versus applied research snowballed into a major controversy.

The Sewell Committee made a number of observations, many of which are relevant to subsequent developments involving Raman. Concerning finances it noted that whereas in the early twenties the expenditure matched receipts, things had subsequently changed with expenditure for 1929–30 being (estimated as) Rs. 5.94 lakhs, as against an opening balance (including funds held in suspense for retiring allowances and depreciation) of Rs. 9.08 lakhs, and an estimated total receipt of Rs. 6.68 lakhs. The Committee observed,

> Although in some undertakings such figures would indicate a most satisfactory state of affairs, we view with some apprehension the accumulation of funds to this extent in an institution which is purely educational.

In fact the Committee recommended an expenditure up to the possible limit. When Raman became the Director, he did not allow the pointless accumulation of funds. Instead he tried to develop the Institute and in so doing he overspent, promptly landing into trouble!

The Sewell Committee was quite clear about what in its view the character of the Institute should be. It said:

> The Institute should do what no other institution can do. It should maintain a position of pre-eminence; it should acquire a national and even a world reputation; it should become a place of reference.... It is a well-known fact that, in more cases than one, the reputation of a university has been built up round the work of some pre-eminent man. Students are attracted by the reputation of the man under whom they hope to work.... A Nernst or a Ramsay would draw men to any institution to which he happened to be attached. We are of the opinion that the chairs in the Institute should be filled by men of the highest eminence, irrespective of nationality....

How soon was all this to be forgotten!

The Committee had some interesting things to say about the post of Director. Sir Dorab Tata (a member of the Council) had suggested that the directorship should be abolished as a full-time post, and that the administrative duties of that office be discharged by heads of departments in rotation. To deal with routine work, Sir Dorab recommended that a Registrar be appointed (at that time that post did not exist). The Sewell Committee while agreeing that a registrar was necessary felt nevertheless "that the post of the Director, as a whole-time appointment, should be retained". As it turned out, a Registrar was not appointed when Raman became the Director. However, an appointment to that post was made after his resignation.

The Sewell Committee also examined the salary structure. The majority opinion was that the (rather high) salary of Rs. 3,000 per mensem paid to the Director was justified as there was a need to attract men of calibre. But, in a minute of dissent, Saha argued that "the idea that the provision of a very high salary will enable the Institute to secure a proportionally superior type of man is entirely misleading"[4]. There was a similar disagreement concerning the salaries of professors. The report says:

> The present scale of remuneration of professors is Rs. 1,250–50–1,500 per mensem, plus an overseas allowance of Rs. 500 per mensem for European professors. Professor Saha is of the opinion that this scale of remuneration is too high.... He refers in this connection to the opinion of Dr Bhabha, one of the representatives of the Tata family on the Council of the Institute, that Indian professors should be patriotic enough to serve at a lower scale of pay than their European colleagues[5].

Saha's basic complaint about high salaries for the top posts is understandable, for in those days these posts invariably went to Europeans who also received additional overseas allowance. Unbelievable though it might sound, the rupee was a hard currency and Englishmen in India could freely repatriate their earnings back to England. Besides, the professor's salary in the Institute was higher than what could be commanded back home. Given this background, one can readily understand the majority view of the Sewell Committee!

The recruitment practices of those days are also interesting to read about. Whenever the directorship or a professorship fell vacant, the Council appointed *two* Selection Committees, one in Britain and the other in India. Ultimately, the recommendations of the two Committees came before the Council but since the two sets of recommendations seldom coincided, protracted correspondence with the two Committees was frequently needed before a final decision could be reached.

The observations of the Sewell Committee on the Department of Electrical Technology are pertinent in the context of subsequent developments. It said, "No important research work had been undertaken by the department, and the facilities for this class of work are greater than the demand." The Committee noted that physics was yet to find a place in the Institute and, stressing the importance of physics to technology, observed:

> The Pope Committee Report envisaged the establishment of the department [of physics] ten years ago.... We wish further to point out that physics forms the basis of such important industries as wireless, refrigeration, the glow lamp industry and metallurgy, and of the theory and design of prime movers, pumps, windmills, etc.

The Institute had enjoyed twenty-two years of quiet and uneventful existence before Raman was appointed its Director, and there was nothing to suggest that it was making a dramatic impact on the industrial development of India. A popular saying then was that the Institute was a sanatorium to which a few laboratories were attached! It is not surprising then that Raman tried to change this stagnant state of affairs. Earlier he had worked that miracle in Calcutta but here in Bangalore he failed. In Calcutta he had the advantage of starting with a clean slate and therefore he encountered no opposition for quite some time. In Bangalore, on the other hand, there were already many well-entrenched persons with vested interests to whom Raman was an unwelcome transplant. And it did not take very long for the transplant to be rejected.

8.3 Enter Raman

The Institute that Raman inherited was rich in prestige but short on academic distinction, a situation of course not to his liking. He entered the Institute wearing two hats (or two turbans as someone put it!), one as the Director and the other as the Professor of Physics, but pretty soon he lost the former. One has heard of new brooms sweeping clean but Raman, it would seem, was more like a tornado. Little did he understand that "hell hath no fury like mediocrity scorned", and the rebellion he faced left him shattered to a far greater extent than was realized at that time. As his resignation from the Institute directorship marks a turning point in his life, we reserve for later discussion the scientific work he did at the Institute, concentrating for the present on the circumstances which led to this unfortunate event.

Raman's association with the Institute commenced even in the late twenties when he was nominated to the Council as a representative of the Eastern Group of Universities. Twice during this period he was warmly felicitated by the Council, once when the knighthood was conferred upon him, and later when he won the Nobel Prize. Dr Martin Forster was the Director at that time, and his term of office (after two short extensions) was coming to a close on April 1, 1933. In anticipation of Dr Forster's retirement, the Council appointed in July 1931 two Selection Committees which would submit names of suitable candidates for a successor. The Committee in England was to be chaired and convened by Sir William Bragg, and have Sir William Pope and Sir Robert Robertson as the other two members. The Indian Committee was to consist of Sir Samuel Christopher, Sir T. Vijayaraghavachariar and Sir M. Visveswarayya, the last mentioned being the Chairman and Convener. Both Committees unanimously favoured Raman for the post, and in July 1932 the Council recommended to the Viceroy that Raman be appointed Director[6]. It was also agreed that, as recommended by the Sewell Committee, a department of physics would be created to enable Raman to continue with his research.

There was trouble right from the beginning. In December 1932, i.e., a few months prior to taking over, Raman requested the Council to add to the staff one personal assistant and one scientific assistant to help him in administrative and scientific matters respectively. It will be recalled that the Sewell Committee had favoured the appointment of a Registrar to assist the Director. However, the Viceroy advised the Council that "the question of creating a post of Registrar for carrying on the ordinary routine work of the Director should be held over pending the appointment of the next Director so as to enable him to express an opinion in the matter". The Council accepted Raman's preference for a personal assistant, and approved the appointment of Dr P. Krishnamurthy who had earlier provided assistance to Raman in Calcutta. Raman also wanted Bhagavantam as his scientific assistant (actually he preferred the term research physicist) but a minority group in the Council (which included Sir P. C. Ray and Saha) opposed the appointment on the ground that the by-laws did not give the Council the power to make such an appointment. Saha further wanted a reduction in pay by half which was not to Raman's liking.

Upon assuming office in April 1933, Raman did three things, namely, bring a new physics department into existence, restructure some of the existing departments, and reorganize the workshop (the changes being, in his opinion, for the betterment of the Institute). But every one of these actions boomeranged!

To organize a new department one needs money, staff and students. Raman had no problem in attracting students but he did have difficulties in finding money and in making staff appointments. As far as money was concerned, he had been given a capital grant of Rs. 1 lakh and a recurring sum of Rs. 25,000 which he found woefully inadequate since salaries, studentships, journals, books and equipment had all to be accommodated within this sum. He therefore re-apportioned the budget to aid the fledgling Physics Department, an act which invited charges of embezzlement!

Raman was firmly wedded to the view that excellent work comes from excellent people. He was quite dissatisfied with the performance of the existing departments and strongly felt that the Institute needed fresh blood. Luckily, an opportunity for inducting new talent was presenting itself since many eminent scientists were fleeing Hitler's Germany just then. Why not bring some of them over to the Institute? As we shall soon see, one particular appointment which he pursued with enthusiasm created a huge problem!

The reorganization previously alluded to antagonized both the Professor of Chemistry and the Professor of Electrical Engineering. Raman found that the Physical Chemistry Section (attached to the Inorganic Chemistry Department) was engaged mainly in studies relating to magnetism. At the IACS, magnetism had been one of the strong points of his research group (see Chapter 4). Considering the isolation of the Physical Chemistry Section in relation to the other activities of the Chemistry Department, it seemed appropriate to Raman to make the section a part of the new Physics Department, especially since the merger would provide the chemists concerned with opportunities for constant and profitable interaction with other colleagues having allied interests. Professor Watson, under whose overall care the Physical Chemistry Section had earlier functioned, was deeply offended and he resigned. Likewise, Professor Mowdawalla of the Electrical Technology Department opposed Raman's idea that the Institute workshop, instead of merely training students, also assist the research workers not only by carrying out repairs but further by fabricating new equipment as desired etc. Mowdawalla also became resentful and chose to leave! The sleepy campus was coming alive with controversy, and pretty soon it would be time for the Council to sit up and take notice, which is exactly what the opposition wanted.

The Born episode brought things to a boil. Like many others, Max Born left Germany in the early thirties and found for himself a temporary berth in Cambridge. At that time he received a letter from Raman asking for the names of bright theoretical physicists wanting to leave Germany who could be considered for appointment at the Institute. Born regretted his inability to recommend names without knowing about the conditions in India. Raman understood Born's position; so could Born come to Bangalore for a while and see for himself how things were? The Institute Council approved a temporary readership for six months (as it did also for Prof. George Hevesy), and Born accepted the offer especially as his Cambridge appointment was drawing to a close. Further, Rutherford advised him to try out Bangalore as the salary was better!

In the autumn of 1935 Born and his wife Hedi sailed for India and landed in Cochin, where they got their first taste of the Sahibs. As Born describes:

> We paid a visit to the British Club where we had our first impression of the separation between the ruling British and the Indians. It was a perfectly English atmosphere; everyone knew everyone else even if he had just arrived from a distant corner of the world, and they talked about schoolmates and what had become of them. In spite of the heat, a considerable amount of alcohol was consumed. We rather felt ourselves foreigners and outsiders, though they were all very friendly to us. [8.1]

From Cochin the Borns travelled to Bangalore by train. They were received by Mrs Metcalf, the wife of the Vice-Chancellor of Mysore University, and Lady

Raman. The latter took them to their bungalow in the Institute campus, describing which Born writes:

> [It] was actually a big two-storey house with numerous rooms, electric light, and two WCs, at that time a rare convenience in India. A few rooms had been furnished for us by Mrs Metcalf, quite cosy and comfortable.
>
> We had a large garden with beautiful trees and flowers and two tennis courts which were screened off by marvellous bougainvillaea shrubs. The Raman family lived in a similar house just across the road.
>
> We liked Lady Raman right from the beginning. Her husband was absent and appeared a few days later. We were fascinated by his appearance and talk. Hedi said he looked in his Indian dress and turban like a prince from the *Arabian Nights*.

Born lectured extensively and spent his time pondering about odd problems he had brought with him from Europe. He recalls that he used to work in the Institute verandah and that monkeys used to come and watch. Once they stole his glue and ate it!

Soon after Born arrived, a Professor of Electrical Engineering named Aston came from England. He was Mowdawalla's replacement, duly selected by two Committees. The Astons stayed with the Borns till their own bungalow was ready. Later Aston actively worked against Raman and also attacked Born, opposing his permanent appointment at the Institute.

Raman developed a great liking for Born despite the difference of opinion he had concerning theories of lattice dynamics. (As we recount in Chapter 10, these differences later snowballed into a controversy.) He was very keen that Born should continue in the Institute as a permanent member. But first he had to persuade the Faculty to accept the idea. He proposed to the Senate of the Institute that it recommend the appointment of Max Born as Professor of Mathematical Physics. Speaking in support of his motion, Raman strongly eulogized Born but all the same, Raman's motion was not well received by many. As Born (who was present) describes:

> Aston went up and spoke in a most unpleasant way against Raman's motion, declaring that a second-rank foreigner driven out from his country was not good enough for them. This was particularly disappointing since we had been kind to the Astons.... I was so shaken that when I returned to Hedi I simply cried.

Meanwhile the turmoil in the campus continued to grow, with Raman finding himself increasingly in isolation. A review of the Institute affairs seemed appropriate, and in July 1935 the Council recommended to the Viceroy that a Review Committee be appointed. It also submitted a panel of names which included Sir William Bragg, Sir L. L. Fermor, Sir M. N. Mukerji, Sir Courtney Terrell, Dr E. F. Armstrong and Prof. J. C. Ghosh.

More important, the Council, unlike in the past, spelt out in detail the various items the proposed Committee should address itself to. The list was heavily loaded against Raman but his protests were overruled.

Despite these reverses, Raman continued to steadfastly campaign for Born's appointment, and in November 1935 the Council accepted his suggestion that a professorship of mathematical physics be created. It also appointed as usual two

Committees. The Committee in India had Raman as the Convener and Prof. G. S. Mahajani and Dr S. K. Banerji as the two other members while the Committee in England consisted of Prof. T. M. Lowry, Prof. Dirac and Lord Rutherford, the last mentioned being the Convener. As it turned out, all this was an infructuous exercise, for the Review Committee completely upset Raman's calculations. Most tragically, the Institute lost Born.

It was not merely Born but many others whom Raman wanted to bring to Bangalore. Concerning these attempts Ramaseshan observes:

> There is a letter from Schrödinger, the originator of wave mechanics, saying that Raman's offer had arrived a bit too late as he had just accepted an offer from Dublin [and] regretting [the fact] that he could not settle in the land of the *Upanishads*. Ewald, Peierls, Kuhn and many others were also in Raman's list. Years later I once asked him about his attempts to bring these scientists to India. He replied that he had always been against young Indians going abroad to be initiated in scientific research because it would have been done in an environment so completely different from what exists in India. This type of training could have made them useless in our country. If great minds like Born and Schrödinger who were seeking a country to adopt had been provided with a home here, a real scientific movement could perhaps have been started in the country. [8.2]

What a great setback it was for our science, especially when one recalls the tremendous advantage gained by America from immigrant scientists!

8.4 The Irvine Report

The Review Committee (or rather the Second Quinquennial Review Committee as it was officially called) appointed by the Viceroy did not contain any of the persons proposed by the Council. Instead, the Viceroy nominated Sir James Irvine, FRS, Principal and Vice-Chancellor of St Andrews University, to head the Committee. Besides Sir James there were Dr A. H. Mackenzie, Pro-Vice Chancellor, Osmania University, and Dr S. S. Bhatnagar.

Irvine was born in Glasgow in 1877 and practically grew up with St Andrews, being associated with it successively as a student, lecturer, professor, Dean of the Faculty of Science, principal, and finally vice-chancellor. His biographer describes his outlook as that of a lowland Scot, strongly Calvinistic. Professionally, Irvine was an organic chemist with some work on carbohydrates to his credit. It is noteworthy that his research was highly slanted towards industrial applications, and at one stage he ran practically a research factory with as many as sixty investigators organized in nine departments.

Bhatnagar was at that time Professor of Chemistry in Punjab University. Later of course he rose to fame by founding a chain of National Laboratories, with blessings and support from Nehru. As for Dr Mackenzie, he appears to have been included more on the strength of his administrative background than on the basis of his scientific credentials—earlier he had served as the Director of Public Instruction in the United Provinces.

The appointment of the Committee was formally announced on January 20, 1936, but Sir James Irvine must have had some advance intimation for otherwise it is difficult to see how he was able to immediately reschedule all his activities, sail for India and hold the first meeting of the Committee in Bangalore on February 24. The Committee submitted its report a few months later. Running to 42 pages and priced at six annas or eight pence, it is a masterpiece in sugar-coating. Anyone reading it without background knowledge would regard it as fair and justly critical (in mild language) of Raman's actions. However, as Jeeves would put it, there are wheels within wheels, and the reader has to be a little patient to obtain an insider's view of the whole affair. But first let us interrupt our narration and examine the report at its face value.

The Irvine Committee report begins with an analysis of the aims and objectives of the Institute. As there had been

> a period of unusually rapid changes, it is more than ever necessary to secure that the policy pursued is consistent with the wishes of the Founder and of the contributing bodies.

The Committee feels that there have been too many "fluctuating and individual interpretations" of what the Institute is supposed to do, and claims that "the purposes for which the Institute was created have never been clearly defined". On the other hand, the heirs and the executors of Tata had laid down the following:

> The object of the Institute shall be to establish chairs and lectureships in science and arts especially with a view to the promotion of original investigations in all branches of knowledge and their utilization for the benefit of India, and to provide and to assist in the provision of suitable libraries, laboratories, and all necessary appliances.

Commenting on this, the Committee observes:

> In our opinion, the phrase "the benefit of India" strikes the keynote of the policy to be followed by the Institute.... It may be argued with good reason that the prosecution in India of research of high merit is a benefit to the country through the prestige it confers. Similarly, the provision of a steady succession of young men whose minds have been developed through the discipline of research in pure science is another result which comes within the compass of the phrase "the benefit of India". But, whilst in full agreement with such arguments, we believe that the initial conception of the Founder was that the activities of the Institute should be devoted primarily to securing for India the material benefits expected to follow from the close association of scientific research with the industries of the country.

Having read the mind of J. N. Tata, now long gone, and decided what is good for India, the Committee's first recommendation is that the objectives of the Institute be modified from what had been laid down by the heirs of Tata to the following:

> The object of the Institute shall be to establish chairs and lectureships in science and arts for the purpose of providing advanced instruction and conducting original investigations in all *branches of knowledge and particularly in such branches of*

> knowledge as are likely to promote the material and industrial welfare of India [italics mine]; to provide suitable libraries, laboratories....

According to the terms of reference, the Committee was only supposed to review the working of the Institute "with special reference to the purposes for which it was founded", and not to modify the objectives. But the Committee does that anyway, and, as a natural corollary, declares that

> the major part of the resources of the Institute should be applied to those investigations which are likely to be of direct benefit to industry in India.

The Committee next examines the research work done at the Institute and observes,

> In the course of our inspection, we have elicited evidence of the existence of a genuine desire on the part of several responsible members of the staff to co-operate energetically in applied research if given the opportunity.

Then, without giving a shred of reasonable evidence, the Committee simply says,

> It appears to be well established that applied research does not receive the sympathetic support of the Director

and goes on to add,

> It is not an exaggeration to say that the resources of the Institute are now being concentrated in the direction of research in pure physics, and that an atmosphere has been created within the Institute and, more publicly, through the agency of the press, which is designed to place such research at a premium.

The reference to the press is interesting, for we shall soon see how the opposition used the press, not to mention other tactics, to malign Raman.

The Committee disapproves of the way Raman ran the Institute, but first the sugar-coating:

> Our first impression, gained from a study of annual reports and published papers was that the Institute had made remarkable progress during the past five years, and particularly since the appointment of the present Director. A notable gap in the facilities provided by the Institute had been filled by the creation of a Department of Physics...and from it...had flowed a steady succession of original papers.

And now the pill:

> The Committee appreciated highly the progress made in physics, but evidence [nothing is cited] quickly revealed the fact that physics was in the process of becoming the dominant feature of the Institute. ...

One wonders whether the other affected departments were doing wonderful work for the "benefit of India" before Raman's entry stopped them dead in their tracks. Certainly the Committee does not cite any evidence to this effect. How can it?

While the Committee is not enchanted by Raman's researches, it is even less enamoured by Born's work and decides that "modern mathematical physics, with its attractive fields of speculation and experiment, has little direct contact with industry", and therefore not "likely to be of service to India". Reading this today certainly makes one's mind boggle.

The Committee is not finished with Raman yet. It says:

> As we read the situation, and there is a remarkable uniformity in the evidence [once again nothing is cited], the Director's policy is to make the Institute a centre for physical and mathematical studies. True, there has been no suggestion that other departments should be suppressed, but as time goes on they will inevitably become more and more subordinated to physics until their individuality is to all intents and purposes extinguished.

Next come some sketchy references to alleged irregularities on the part of Raman, and the remarks:

> Making full allowance for possible exaggeration, we greatly fear that there is much truth in these allegations. The resignations of Professor Watson and Professor Mowdawalla can now be readily understood.

An incredible distortion of facts, as we shall soon see. Watson, in particular, had on two earlier occasions temporarily served as the Institute Director during periods when Martin Forster was absent. Clearly, he had entertained hopes that he would be the successor. Not only was he denied that pleasure but Raman made things worse by taking away the Physical Chemistry Section. The fact is that both Watson and Mowdawalla were complete nonentities in the academic world, their only claim to fame being their resignations!

According to the Irvine Committee, Raman was guilty not only of administrative irregularities, but of financial lapses as well. The Committee cites figures to show that lately there had been a systematic deficit in the Institute budget, around Rs. 50,000 in 1932–35, touching a peak of Rs. 1.1 lakhs in 1934–35, and declining somewhat thereafter. There can be no dispute about these figures, and indeed there was a deficit from around the time Raman entered the Institute. But then how could it be otherwise, especially since the income remained steady between Rs. 5.5 and Rs. 6 lakhs? There was no physics department before Raman came, and it had been agreed that one would be created for him. How was he supposed to build a totally new department without dipping a little into the reserves, especially when the income of the Institute remained steady? And did he not make up for it by showing performance?

Nobody bothered to examine the root cause of the problem, part of which was the paucity of funds. True, Bangalore was no Cambridge, but one cannot help comparing the apathy Raman received with the patronage Rutherford enjoyed. The latter got all the money he needed whereas here in India, with all his credentials, Raman got precious little.

To combat the deficit in the Institute budget, the Economy Committee appointed by the Council had suggested several measures which, all taken together, were to lead to a saving of Rs. 52,000. The Irvine Committee is critical of Raman's efforts in implementing the suggested economy measures.

> The unsatisfactory manner in which these recommendations have been implemented may be illustrated by reference to the proposal to effect economy in the

expenditure on water supply by the installation of tube wells. The [Economy] Committee estimated that this measure would result in a net saving of Rs. 9,000 per annum and this was accordingly included by them in the savings of Rs. 52,000 mentioned above. But the Council, on the recommendation of the Director, decided to utilize the saving on the water scheme towards the cost of creating a new chair of mathematical physics.

Reading these words one cannot believe one's eyes. Raman was trying to get *Max Born* appointed to the chair mentioned. The Sewell Committee had explicitly recommended the hiring of men of the highest eminence. And yet, according to the Irvine Committee, the saving of Rs. 9,000 was more important. One can understand an administrator nit-picking in this fashion, but a scientist, especially one coming from Europe? Could Sir James have really been unaware of the great revolution ushered in by quantum mechanics, in the founding of which Max Born had played such an important role? Did not Sir James know that quantum mechanics had even penetrated chemistry (the famous book by Pauling and Wilson, on which generations of chemists have been brought up, had already appeared by then)? We do not know the answers to these questions. All we know is that the Irvine Committee wanted Born's appointment to the post of Professor of Mathematical Physics to be cancelled, and that the Council (which earlier had accepted Raman's plea for the creation of the post) agreed to the cancellation.

The Irvine Committee was apparently genuinely concerned about economy and suggested many measures, one of which is quite astonishing by today's standards–the *reduction* in pay of clerks, servants and menials! The amount saved was Rs. 8,436. Dare anyone suggest a salary cut today? But the Irvine Committee got away with it! In passing, it is curious that the Committee did not apply its mind to the question of how the resources of the Institute could be enhanced. It seemed to accept shortage of funds as inevitable.

In the final part of its report, the Irvine Committee comments on the various departments and offers its views on how the Institute should be run. To start with it reiterates its desire to axe the proposal for a professorship in mathematical physics, observing:

> There can be no doubt that a department of mathematical physics is a necessary adjunct to a successful school of experimental physics, physical chemistry, or engineering. The creation of such a chair in the Institute is therefore to be commended, and we are certain that the presence of an eminent mathematician such as Dr Born would have a stimulating effect on the activities of the Department of Physics. If we have not seen our way to include in our financial proposals the stipend of a professor of mathematical physics, this does not mean that we fail to appreciate the desirability of adding to the strength of the Institute men of the reputation and calibre of Dr Born. But under the present financial conditions it appears to us imprudent to add this new charge....

The Committee then castigates Raman for various alleged sins against chemistry, dismissing Raman's argument that lately "the universities of India have been thoroughly aroused to a sense of their responsibility to provide instruction and research facilities in all branches of chemistry, including applied chemistry". Nor does Irvine accept the findings of the special Committee appointed by the Council regarding the reorganization of the Chemistry Department. And to rub it all in it says,

> In our opinion a strong representative school of chemistry, capable of playing a significant part in pure and applied research, requires chairs in the following divisions of the subject: (a) organic chemistry, (b) physical chemistry, (c) inorganic and mineral chemistry, (d) technical chemistry, (e) pharmacological and medicinal chemistry.

All well-meaning of course, but *five* professors in chemistry when one in physics was too much?! At this stage the Irvine Committee realizes it is running rather wild and curbs itself by settling for one professor and four assistant professors.

The Committee also has things to say about the Central Workshop. Its analysis of the departmental workshops versus a centralized workshop reads familiar since the question is a topic of perpetual debate in our scientific establishments. The Committee is of the view that the prime purpose of the workshop should be to provide routine training in workshop practice to the engineering students.

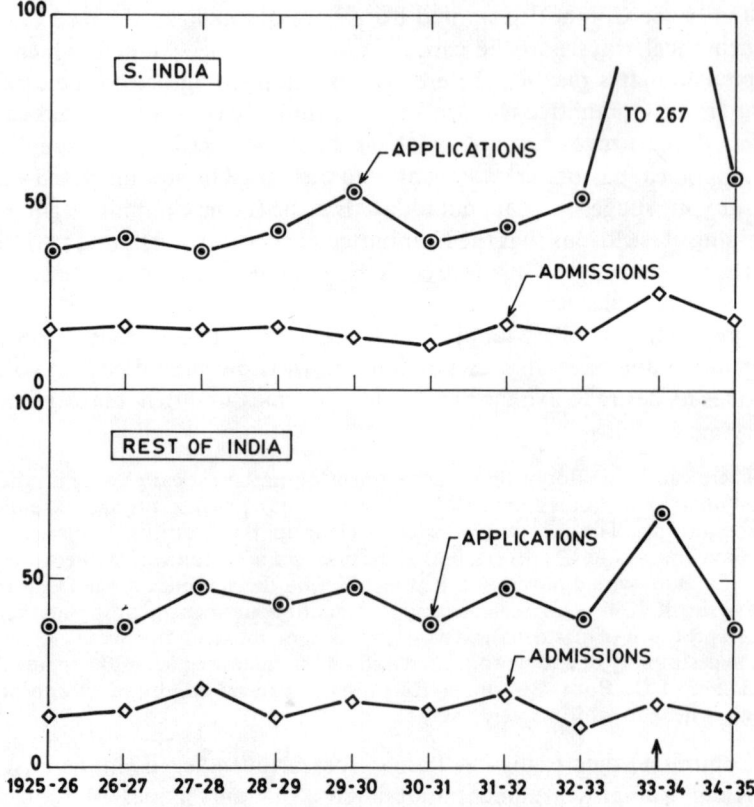

Figure 8.1 Time variation of student population in the Institute. The şolid line is just a guide to the eye. The arrow indicates the year of Raman's entry. Observe the sharp increase in the applications received. (Data derived from the Sewell and Irvine Committee reports.)

Apparatus fabrication, which Raman wanted to promote, should take a lower priority. How unimaginative!

Finally the student population. In Calcutta Raman was accused of collecting a South Indian coterie. He was now in South India, but once again the same charge! "Yes," one might argue, "was not the Institute supposed to be an all-India one?" Yes indeed, but then did not Mahendra Lal Sircar also dream that his Association (the IACS) become a national centre? Actually, this debate is irrelevant. If one examines the records (see Fig. 8.1), one does not find any statistically significant change in the demographic composition of the student population after Raman entered the Institute. Even if the charge that "Physics Department has drawn its students entirely from South India" was true, what does it prove? Just that these students were rash enough to follow a great master, though the job opportunities were far less than those offered by a pursuit of either chemistry or electrical technology[7]!

There can be little question that the Irvine Committee did what it was supposed to do – slay Raman. A gross accusation, one might say. Well, the reader is urged to continue and find out.

8.5 The Inside Story

As hinted previously, there was more in the entire episode than met the eye. Fortunately, the historian's job of digging that out is made easier by six letters exchanged by Born and Rutherford during October 1936. These letters were written after Born, realizing he had no place in Bangalore, returned to the United Kingdom and took up a position in Edinburgh. On his way, Born stopped briefly at Cambridge but could not meet Rutherford as the latter was away on a holiday. From Scotland Born wrote to Rutherford expressing his disappointment, and then went on to describe the conditions in Hitler's Germany which he had visited *en route*. No reference either to Bangalore or to the circumstances which compelled him to leave. Rutherford wrote a prompt reply, beginning,

> I have received your note this morning and make haste to reply as I have been intending to write to you about the matter of Raman, but have been pressed for time.

After referring to various other matters he says:

> Now about the Raman matter. When I was on holiday in the country, I received an SOS letter from Raman asking me to write to either the Secretary of State for India or to the Viceroy, pointing out his eminence as a physicist. This put me in a very awkward position, for it is difficult for me to interfere in the question of the management of the Bangalore Institute. My information on this subject is vague and uncertain and I am not in a position to make any pronouncement on the rights or wrongs of the present situation. Someone has since sent me several statements published in the Indian press on the situation at Bangalore and they are certainly rather unpleasant reading. Obviously if I barge in on the matter it looks as if I were expressing an opinion on a question still *sub judice* before the report of the

Commission [the reference here is obviously to the Irvine Committee] has been made public. I am very sorry for Raman, and I have no doubt about his great ability and enthusiasm for research. I gather, however, that everybody recognizes that, but that complaints arise from his failure to co-operate with the staff and troubles of administration. If you had been here in Cambridge I would have been very pleased to have a long talk with you on the subject. I do not myself know what were the essential views and plans of Raman that have precipitated this crisis. I do not know what you think about the matter except that you feel that Raman has many points in his favour. I feel great difficulty in moving in the matter as it is an implied criticism of a report I have not seen. If you know of any special points about which you would like to draw my attention I should be glad to hear them. I am myself afraid that Raman has got into such a devil of a mess that he may be driven to resign to save his face.

Born, who until then was quite reticent, now decided to speak out. In a lengthy reply written within days of receiving Rutherford's letter, he says:

If you think that my impressions of Raman's situation are of any value to you I shall be glad to tell you what my wife and I have seen and experienced at Bangalore.... Raman is a very able physicist, full of enthusiasm, as you say yourself. There is really no other Indian physicist who is of his rank.... [His] European intensity alone would be enough to make Raman suspicious to the average Indian professor. Now Raman, far too much aware of his own superiority, likes to make other people seem small in his presence....

Raman came to the Institute with the idea of making it a centre of science of international standard. What he found was a quiet sleepy place where little work was done by a number of well-paid people. My wife and I met an English couple – the man was a retired official – in Ootacamund. When I said that I was at the Indian Institute at Bangalore, this man said laughing, "That's a nice sinecure where people draw high salaries...." Similar expressions we have heard on other occasions. Raman's mere speeding up of the entire pace at the Institute was bound to look like criticism on the former work. Add to this that he made a heavy mistake in not waiting a year or two before starting actual reforms. Naturally he got into troubles with the professors who were at the Institute before him. Two of them left the Institute during the first year: an Indian, Mowdawalla, Professor of Electrical Engineering, and Watson, Professor of Physical Chemistry. The latter case seems to me one of the main sources of difficulties Raman was to encounter later. Watson's friends and he himself may have expected that he was to be the new Director after Sir Martin retired. Certainly Watson did not like to continue as a professor under an Indian Director. I was told this by some of his English friends. It is easy now to make the loss of Prof. Watson a point against Raman, but it is certainly not just. Openly the real reason for Watson's leaving the Institute were not known; only the given reason was known, namely that Raman's manners had driven him away. I know that Raman's manners *can* cause serious grievances but in Watson's case they were but a pretext....

Shedding more light on the Indo-European angle, Born observes:

To me the deepest sources of the trouble are: The English group resented an Indian Director, who, as a political principle, was wanted and pushed through by the powerful Tata group. The Tatas knew quite well that they offended Watson who could have been the successor of Sir Martin Forster and who seems to be a charming personality. Watson's leaving was in fact a protest to the address of the Tatas, and these – who as big merchants cannot afford to be on bad terms with the English –

may have found it the easiest way out of their difficult situation, to drop their own former favourite (I assume that they had expected that a Nobel Prize winner like Raman was the best chance to apply the principle of Indian directorship without difficulties arising from it). So when this case turned against the *person* of Raman, they took it up in the same sense. As I have said before Raman gives plenty of room for criticism. But the complicated situation has certainly not risen from his personal faults but from much deeper feelings and principles. Otherwise it would *never* have grown to this tenacity and heat. They will probably fight till Raman is 'slain' (if your word will not help him) and will make an outward peace by electing a new Director. But the Tatas *will* preserve a feeling of defeat which one day will burst out again.

I want to show you by a few examples that all this is not a matter of mere assumption. Three weeks after us arrived the new Professor of Electrical Engineering, Aston, at the Institute. Immediately after his arrival the open revolt amongst staff and students began and he became a centre for collecting ever so silly complaints against Raman. We wondered very much till one day Mrs Aston said to my wife that her husband had been made to accept the post by English colleagues (Aston had not got a professorship at London for which he had hoped, so Mrs Aston told my wife) in charging him with the definite "mission to clear up the Institute". Aston had been received in Bombay by the Tatas, had been their guest and had got instruction. Nothing can be easier in India than to rouse discord and to stir it.

Born then comments on the Saha–Raman conflict and its extension to Bangalore via the Council. Guha, an assistant professor of chemistry, provided Shyama Prasad Mookerjee, P. C. Ray and others with all the local information they needed. Commenting on Guha, Born writes:

Raman considers him an inferior scientist and wants to get rid of him[8]. And in this particular matter Raman has acted in a way which cannot be excused. He certainly does not regard the means by which he reaches his goal. He does not know and therefore does not respect other people's feelings. He is not a bad character, he is rather an eternal and badly educated child. When he became Director he had nothing in view but to raise the general standard by all means regardless of what he found. In his proper goal he enjoys the full assistance of the Mysore Government which is a far-sighted body.

Turning then to the Irvine Committee, Born says:

I have no right to criticize the attitude and the proceedings of the Quinqn. Comm. but I must say that it seemed to me rather surprising. Instead of visiting the Institute and carefully studying the work done in the laboratories, they settled in a Government building [this was the Residency] some four miles away, where they behaved like a law court. It was evident to me from the beginning that they had received instructions beforehand. They examined chiefly Raman's opponents, even students. Men like Mr Jatkar [a lecturer in physical chemistry] of whose untrustworthiness I got personal proofs were allowed to "give evidence" and be *persona grata*. All the dirty affairs were treated in detail but no voice raised to take into account the good intentions of Raman or his achievements at the Institute. His enemies the Tatas and the Bengali members of the Council had made up their minds to get rid of Raman, and the Irving [Born repeatedly gets the name wrong!] Comm. listened to them. The record of evidence of the physicist Prof. Venkatesachar given to the Quinqn. Comm. was so compressed that it got an altogether different meaning, namely an unfavourable one. Whereas Prof. Venkatesachar had admitted Raman's lack of tact and of administrative gifts, but had said many important good things in his favour. It was not seen with pleasure that Prof. Venkatesachar insisted on having the record corrected.

To Born, the situation was clear. Raman's intentions were completely honourable although many of his tactics were ill-advised. On the whole, Raman deserved sympathy, and so he says,

> There is always a *real* motive and a *pretended* one, and why in this case I wish to intercede for Raman is that – whatever his weak points may be – in this fight he is the *pretended* one, and he has become a sort of victim.

Born is particularly worried about the students.

> What I am sincerely sorry for is the fate of all the young students of the Institute. The *clever* boys are very devoted to Raman, for he is most interested in their progress and asks very much of them.

And so Born makes a direct appeal:

> If *you* could help Raman not being 'slain', and to find a way out so that he can further work and live in peace wherever it be, without being dishonoured and discriminated as a man and a citizen, you would act in the interest of justice.

A couple of days after dispatching this long letter, Born received from India a scurrilous anti-Raman pamphlet. This made him dash a hastily written note to Rutherford in which he says:

> I do not like to bother you any more with the Indian affair. But I think I should send you a pamphlet which I received today. (Perhaps you have got it too.) It is anonymous and full of misprints; it shows you, better than my words, the ugly methods used by Raman's opponents. I have marked sentences with red pencil, showing examples of the political background of the fight, or sustaining particularly fine lies. I wish to comment only upon two points concerning myself. The words quoted on p. 36 "Sir Raman's methods are hopeless etc...." are of course invented; my opinion is just the opposite: *only* the eminent men can get on with him. Further it is *true* that Raman tried to get Goldschmidt. I know this as myself have told Raman first that Goldschmidt had left Göttingen and accepted his old position in Oslo. Raman was very eager to have him in Bangalore, and Goldschmidt was inclined to come for a year or two. I thought that a man of such outstanding knowledge of materials would be very useful for India, and the Mysore Government understood this argument. But the Council turned it down....

And Raman was supposed not to have acted for the "benefit of India". Promptly acknowledging Born's letter, Rutherford wrote back that he too had received the pamphlet "The Raman Effect" (!). Rutherford adds:

> I trust you will not be worried by these gross perversions of the truth.... The amount of intrigue and jealously displayed in these pamphlets is almost unbelievable....

Viewing these sordid events half a century later, one perceives something more than mere intrigue, rivalry and petty jealousies. One sees a struggle between forces impatient for progress on the one hand, and vested interests threatened by that progress on the other. How often has this drama been re-enacted, though in slightly different terms! One sees also an early example of the tyranny of mediocrity which had successfully penetrated into a domain where it had no right to be. Raman lost the battle not only for himself but for excellence as well. Excellence, always difficult to sustain in a backward environment, has today become such a scarce commodity in our academic world that we are compelled to hold national seminars to hunt for it[9].

As for Sir James, he was an early proponent of what is now known as appropriate technology. His view was that India, being an agricultural country, did not require anything other than research relating to biochemistry, chemistry of natural products, and the like. Superficially, this advice appears sensible as do indeed all such arguments made subsequently in favour of appropriate technology. But then, does a vast and ancient country like India with such a glorious past live on bread alone? What about the dreams of her own sons, like Mahendra Lal Sircar, for example? Did he not understand India's problems and was he not concerned about "benefit to India"? A good fifty years earlier he had the vision to look beyond the immediate horizon. And then, what about J. N. Tata himself and the understanding reached in his lifetime that the Institute would be modelled after Johns Hopkins? For that matter, would Tata have complained if IISc had emulated Caltech or MIT instead? Caltech had built all the equipment for the famous Mt Wilson Observatory, and Raman, more modestly, wanted the Institute workshop to build equipment needed for basic research. But Irvine ruled that the workshop's prime duty was to train students. All one can say is: Thank God Sir James was the Vice-Chancellor of St Andrews and not of Calcutta University!

It is quite likely that Irvine sincerely believed in all that he said, in which case he is more to be pitied than censured. In his view the Institute must rest content with limited ambitions. Just around that period, Stalin, in an effort to retain Kapitza in the USSR, paid as much as £30,000 to buy all the equipment Kapitza had built at Cavendish[10]. But Raman had no godfather (it was more than a decade since Sir Asutosh had passed away), although he did not lack critics. One is almost certain that the report would not have been so damaging had a physicist, say Rutherford himself, headed the Review Committee. One would have at least felt satisfied if, as a result of the report, the Institute had in subsequent years blazed a new trail in applied research bringing great benefits to the country. It simply did not. In fact the flow of technology from the laboratory to industry has always been a problem in India and has nothing to do with Raman and his style. Years later Nehru complained about this lack of impact. Of course, Irvine could not have known all this. But if only he had been a little more sympathetic and not played into the hands of reactionary forces...

8.6 The Resignation

Resuming the narration, we now follow the events leading to the resignation. The Irvine Committee submitted its report to the Viceroy around May 1936, and two months later the Council, under the chairmanship of Lt. Col. Plowden, Resident in Mysore, held an extraordinary meeting to discuss the report. Besides Raman, twelve others participated in the meeting. The Council recognized that the business before it was of a contentious nature. Barring Raman and three others, everyone present supported the report with Shyama Prasad Mookerjee remarking:

> With regard to the personal indictment of the Director, I do not think it is necessary for us to go into the evidence. So far as the conclusions are concerned, we accept them.

The acting Dewan of Mysore, Rao Bahadur S. P. Rajagopalachari, however, had some reservations. Having read both the report and the Director's reply, he felt that "the Committee have drawn a darker picture than the circumstances warranted". The Irvine Committee had recommended that a Registrar be appointed who would be a full-time official of the Council with independent status, who would owe allegiance to the Council and not to any department. Commenting on this Mr Rajagopalachari observed:

> The object seems to be to dissociate the Director entirely from the work of the Institute excepting of course his professorship of physics and the chairmanship of the Senate. We consider this scheme will not work satisfactorily. There should be a Registrar with defined and specific powers but it will not do to set him up as an authority independent of the Director. It is sure to cause friction and further trouble. So long as there is a Director of the Institute, the Registrar must work under his general control except in regard to certain specified routine matters on which he can take action *suo moto*.

Professor B. Venkatesachar read out a memorandum in which he said:

> On a careful reading of the report and the memorandum thereon circulated by the Director, we are of the opinion that the Committee have failed to discover the true cause of what they call the 'unhappy situation' at the Institute. Anyone who tries to effect reforms and introduce changes in an institution with a view to increase its efficiency is sure to meet with opposition.

Dr Bawa Kartar Singh associated himself entirely with the memorandum read out by Venkatesachar.

Raman made a lengthy statement of his own in which he was, naturally, quite critical of the report. Interestingly, he *supported* the appointment of professors in the chemistry departments recommended by the Irvine Committee. He said:

> An Institute devoted mainly to scientific research must be judged by the quality, volume and usefulness of the results achieved in its laboratories. In spite of the large sums of money being spent annually on the three departments of chemistry at this Institute, their output has been unsatisfactory and disappointing. The reason for this is known to every man of science in India, viz., that the higher staff of these departments do not possess the requisite high qualifications. No improvements can be expected until the three professorships of chemistry at the Institute are filled by eminent men of science of established reputation.... It is the insistence of the Director on the necessity for a change in the personnel of the higher teaching staff in chemistry that has been largely responsible for the resentment borne towards him by a considerable section of those connected with the Institute, and which is reflected in the report of the Committee.

Raman was also critical of the Irvine Committee's suggestion that the budget be processed "through no fewer than four bodies in succession before it is finally sanctioned". He warned that

> it would place insuperable obstacles in the path of progressive administration such as is essential for the success of a scientific institution.

Prophetic words indeed, considering the bureaucratic ladders which have become the norm today!

There was no suggestion as yet that Raman should leave. It was generally believed that Raman would continue as Director but with his powers severely curbed. By way of softening the blow, Dr Gilbert Fowler, a member of the Council, declared:

> It is not always possible to find a galaxy of talent to fill all the appointments at the Institute. True wisdom will endeavour to obtain the best results from the material available. The Director under the new arrangement proposed will still be free to lead and encourage by the example of his energy and achievement rather than to dominate by the status of his position.

Meanwhile the pressure against Raman was stepped up in various ways. On August 24, there was a regular meeting of the Council at which a memorandum submitted by Aston, Guha and Subramanyam (of the Biochemistry Department) came up for discussion. The press gave full publicity to the charges in the memorandum. At the Council meeting itself, Shyama Prasad Mookerjee attacked Raman on various counts and moved a resolution that the Council declare its want of confidence in the Director in the matter of preparing the Council minutes. The Council did not go that far but did record its "sense of strong disapproval". It also appointed a Special Sub-Committee under Lt. Col. Plowden to enquire into the charges contained in the Aston memorandum.

The split now spread to the student community, and on September 16, the *Times of India* correspondent wrote:

> There is an atmosphere of unrest both among the students and the members of the staff. There are two factions among the staff of the Institute, those supporting Sir C. V. Raman and those against him. The students have also taken sides with these two groups.... there are endless wrangles... and the net result of all this is that actual research work has suffered a great deal. Unless drastic steps are taken to set matters right, a time may soon come when it will become impossible to pull the Institute out of its unsatisfactory condition.

Even sports activities were not spared. During the Annual Day celebrations of the Institute Gymkhana, Aston and an assistant of his staged a walk-out on a silly pretext, and this became a big incident.

There was also friction between Raman and the Special Sub-Committee. For instance, Raman objected to the Sub-Committee considering the question of the future of the Institute journal. Writing to Gilbert Fowler, the Secretary of the Sub-Committee, he observed:

> I have to point out that the journal is conducted by the Senate of this Institute, and as it is a research publication, its future is a matter to be decided by the Senate under Regulation 37 and not by the Council. If the Council wishes to review the Senate's policy with regard to the journal, the correct procedure would appear to be for the Council to invite the Senate as a body to submit its considered views. To address the

individual members of the Senate and then for the Council itself to deal with the matter is not in accordance with the Regulations.

The Council was naturally not pleased, and once again in November 1936 it expressed its strong disapproval etc.

With things going from bad to worse, and with the Special Sub-Committee's report being adverse (as expected!), Raman finally decided to step down from the directorship. On June 1, 1937, he wrote to the Chairman of the Council,

> Having considered all the circumstances I feel it would be best that I offer to terminate my contract of service with the Institute as its Director.

It seemed he wanted to leave the Institute completely for he asked for a special retiring allowance of Rs. 1 lakh. There were reports in the press that he planned to settle abroad. Along with his resignation letter, he submitted a memorandum regarding his work at the Institute. In it he said:

> Under the terms of my contract, I was required to undertake the creation of a new Department of Physics and the supervision of the teaching and research therein. Except during a period of six months when Dr Max Born was in residence at the Institute, there has been no physicist on the teaching staff of the department other than myself, and in consequence the whole of the work involved in the creation of the new department has fallen upon me personally. The aims of the department have been set very high. No work of the MSc type is undertaken, and in fact all the regular students of the department have worked or are working for the doctor's degree in science.... A special feature of this department is that all its students, including even beginners in research, are required to work independently and to publish their results in their own names. Work under this system calls out the highest qualities inherent in the students, and it is not surprising that in spite of its comparative infancy, the department has already produced at least one man who has achieved a first-class reputation, namely, Mr N. S. Nagendra Nath.... On the experimental side, the names of the more senior students, namely those of Mr R. Ananthakrishnan, Mr R. S. Krishnan and Dr S. Parthasarathy, are already familiar to workers all over the world in their respective fields of research.... Numerous textbooks and monographs on physics published abroad during the last two years contain detailed and highly appreciative references to the work of the new Bangalore school. The results already obtained leave no doubt that if the head of the department had more leisure to devote to his scientific work and is given adequate staff and money to assist him in his work, the Institute would rapidly develop into a world-factor for the progress of science.

About the Central Workshop he said:

> It was intended to avoid the multiplication of workshops in the Institute, and in fact, to substitute one well-equipped and efficiently conducted organization for a number of ill-equipped ones.... Valuable scientific apparatus has been manufactured at a fraction of the cost at which it would in the past have had to be imported from abroad. The experience gained in the fabrication of such apparatus has afforded lessons of immense value to the staff and students of the Institute, and the pursuit of applied science at Bangalore has thus gained contact with reality on a new plane.... The centralization under a competent staff has enabled valuable equipment previously thought fit only to be scrapped, namely, a liquid-air machine, a refrigerating plant for cold-storage, and a tilting furnace, to be put into working order and given a new lease of life and efficient operation.... During the past year,

the workshop succeeded in manufacturing electrical motors and other commercial products thus showing the way to the future industrial development of India. The creation of this new organization was not achieved without meeting a great deal of ill-advised opposition and even obstruction, but time and experience will undoubtedly prove that its initiation should be considered a landmark in the history of the Institute.

Raman gave us our first lesson in self-reliance. If only his lead had been followed we would perhaps not be in the sorry state we are in today with regard to scientific instrumentation. Roughly a decade later Bhabha emulated Raman on a much larger scale, and the results were there for all to see.

Raman also pointed out that the Physics Department acted as a pace-setter, generally stimulating an all-round elevation in the quality of student output, which in turn enabled the students to get jobs somewhat more easily. He said:

> As will be seen by a study of the data given in the annual reports of the Institute, there has been a considerable increase in the number of appointments received in my term of office. The published annual report for 1935–36 indicates that during that year as many as 42 students from the Institute obtained employment as compared with 25 students in the year 1932–33 just preceding that in which I took charge as Director. Considering the unemployment situation in the country, and the fact that the total number of students who, during the same year, worked at the Institute was 180, the figure of 42 appointments must be considered highly satisfactory.

The Irvine Committee had branded Raman as inimical to applied science. Raman defended himself against this charge by citing several instances of his efforts "to promote the applications of science for the benefit of India". Describing these he said:

> Amongst my efforts in this direction may be mentioned the following: (a) my work as Chairman of the Railway Board Committee on the Chemical Preservation of Timber; the report of this Committee has, to my knowledge, resulted in a considerable extension of the use of treated timber for various useful purposes throughout India; (b) the establishment of a flourishing new chemical industry at Bangalore devoted to the manufacture of rare earth chemicals and gas mantles of high quality; (c) my work as Adviser to the Government of Hyderabad on the work of the Hyderabad Industrial Research Laboratory, and as Adviser to the Government of Baroda on the organization of the Sayajee Jubilee Technical Institute; my reports have, I learn, been adopted as the basis for the further development of these institutes; (d) my work as Member of the Board of Industries of Mysore Government and Chairman or Member of several of its Committees; it is not irrelevant in this connection to refer to the establishment recently in Mysore of several new industries; (e) my work in fostering the development of the new school of scientific and technological study and research at the Andhra University in Waltair; (f) the course of study in chemical engineering and chemical technology at the **Institute initiated by me in 1933.**

It is easy to pick holes in Raman's defence by arguing that barring the institution of a course in chemical technology, all the contributions catalogued by Raman refer to his personal efforts in relation to organizations *other* than the Institute, rather than to the Institute itself. True, but such would be a narrow interpretation of intentions. What is noteworthy is that Raman, despite his uncompromising attachment to basic science where his own research was concerned, was able to bring a wider perspective to applied research in forums where such applied work was relevant and meaningful. Years later Bhabha adopted a similar strategy by creating an institute for *fundamental* research which served as a fountainhead for an applied programme in atomic energy.

Raman obviously believed in the tenet that superior skills are developed by facing basic challenges and that such skills are always useful when applications are demanded. Clearly this was a radical philosophy, perhaps ahead of its time as far as this country was concerned, but its merit may be judged in an overall context by considering examples like Caltech (where, remember, Raman spent four months).

After studying Raman's explanations, the Council resolved that his resignation be accepted and his request for a special retiring allowance granted. It also recorded that it was mutually agreed that the settlement should be regarded as final and amicable. As events transpired, neither was true!

Raman as Director forwarded the Council resolution to the Viceroy and along with it sent a letter of his own. The Viceroy was not quite pleased with the idea of Raman resigning from the directorship nor with the Council's recommendation that the resignation be accepted, and he wanted to know what specific charges the Council had against Raman. Another extraordinary meeting of the Council was summoned on July 19, 1937, to discuss the Viceroy's letter, and the Viceroy sent the Education Commissioner, Mr J. E. Parkinson, to attend the meeting. Meanwhile, the members of the Council were incensed that by writing a letter to the Viceroy, the "Director had reopened without the knowledge of the Council a settlement which was mutually agreed upon at the meeting as final and amicable". An informal meeting of the Council members took place on July 17 at the Residency under the chairmanship of Lt. Col. Plowden. The original offer and settlement were withdrawn, and Raman was summoned and told that he was "unfit to continue any longer as Director". He was offered two choices – either to continue as Professor of Physics or to resign with effect from April 1, 1938, on such retiring allowances as he may be entitled to according to the rules. Raman was also warned that if he declined both options, he would be suspended!

There was practically no support from any quarter. Raman saw the writing on the wall, and at the formal extraordinary meeting of the Council two days later, he agreed to continue as the Professor of Physics. There was a consoling letter from Rutherford who said:

> I am pleased to hear that you will be able to continue your work in physics in Bangalore without all the worries and distractions involved in acting as Director of the Institute. Now that the matter is settled, I trust that you will be able to carry on with your personal work and let bygones be bygones. It seems to me highly important that the staff at Bangalore should all pull together for the good of the Institute.
>
> I note that the tour arranged in India includes Madras and Bangalore on our return journey. I am not yet quite sure of my plans, but it may be that I shall have the opportunity of meeting you on my travels[111].

That was about it.

Where science was concerned, Raman still appeared to retain his old enthusiasm and spirit. It is said that on the day following his laying down of the directorship, he was in the lab as usual at 7 a.m. asking his students in a booming voice about the progress of their work. Was it mere outward bravado? We do not know. There can be no question that Raman was lion-hearted but it is also more than possible that deep inside he was very much shaken. He was never again the same man, increasingly prone to cloudy judgement where both persons and scientific matters were concerned.

8.7 Some Reflections

In its day, Raman's resignation was widely seen simply as the consequence of "maladministration" by an eminent and a well-meaning scientist lacking, however, in administrative ability. Thanks to Einstein, the image that the public had (and perhaps continues to have!) of a great scientist was that of an absent-minded, dreamy and somewhat disorganized person lost in his thoughts and deeply immersed in his own abstract world. Given this picture, it was but inevitable for everyone to assume that Raman was a misfit as Director, his natural calling being a professorship. For Raman this was like adding salt to injury, especially considering the encomiums paid to his administrative skills while he was in Government service. But perhaps one can excuse the public for its ignorance. The historian, on the other hand, cannot seek refuge under such facile explanations and has the responsibility of reflecting on the significance of the events just described. As Born correctly gauged, there was an apparent reason for Raman's problems (like his haste, manners, ego, etc.), and a real reason as well. However, one must go a little deeper than Born, whereupon one finds that the superficial incidents and even the personalities themselves fade away, leaving behind some basic issues.

All the early advances in science were made by isolated individuals but by the nineteenth century one can see several strong focal points in Europe – Cambridge, Göttingen, the French Academy, and so forth. The growth and development of science had not only become polycentric but the evolution itself was according to a dynamics of its own, according to unwritten but well-understood laws. A centre of learning had only one commitment, namely, to excellence, and there was no place in it for any but the best. Those who did not belong simply did not go there and if by chance they strayed in, peer pressure compelled them to withdraw. The aristocracy of excellence reigned supreme and there was no interference from outside. Society in Western countries tacitly understood the value system and did not regard the rejection of the mediocre as either undemocratic or unjust (a situation that obtains even today, perhaps more so).

Contrast this with the situation at the Institute. Here was a centre of learning founded with the best of intentions to foster excellence in the tradition of the great institutions of the West. But thanks to the general backwardness of the country and its geographical location, it could be populated only with average talent. In walks a spirited giant who finds this set-up all wrong. His mind is full of visions of Cambridge and Caltech, and he wants to recreate their atmosphere in his backyard. Noble objectives no doubt but are they easily accomplished? Cambridge and Caltech had men of eminence from day one to steer their destinies, and mediocrity was always severely excluded. The Institute, on the other hand, had accumulated much dead wood.

How does one create a beautiful garden without first clearing the weeds? It was Raman's unfortunate lot that not only was he faced with the job of cleaning up but also that he had to do it all alone. To make matters worse, there was no strong community of peers in the country to support him. The handful of scientists around were either too meek to speak out or were already lined up in opposition for personal reasons. When a staff member is refused tenure in one of the Big Ten universities in the US, people just shake their heads and mutter apologetically, quietly accepting the judgement as being in the best interests of the University; no cries of injustice or whatever. It is no accident that societies in which such enlightened views prevail are also those where some measure of full employment obtains – at least a person who cannot make it in the best of places can find alternative employment.

In a backward country the situation is altogether different. Jobs, being scarce, assume special significance in the struggle for survival. Thanks to wide disparities, there is also a considerable sensitization to social wrongs, and threats to jobs, job opportunities and promotions are not taken kindly. Further, legalities, even petty legalities, assume importance, for they create the illusion of an invincible sword against injustice. The legal form becomes supreme, and things must always *appear* to be done correctly, even if at times they are totally wrong!

Where weeds are concerned there are no soft options but society was blind and Raman had to pay a price. All the minor infringements of procedure he was guilty of were magnified beyond proportion, becoming points for heated debate in the Council. The legal types, who were in the majority, naturally saw Raman as a dictator rather than as the crusader he really was. Such mishaps are inevitable if academic destinies and management, indeed even micromanagement, are controlled by people lacking in academic perspective, values and vision, however distinguished they might otherwise be. (Most regrettably, the situation has not changed significantly since then.)

Superficially, Calcutta was a different story, but in a sense the Calcutta period also symbolized the hazards of growing science in a backward environment, especially from the resource angle.

What then is a good model for growing excellence in a backward society, given the need for a ruthless exclusion of mediocrity on the one hand and the demand for parity on the other (irrespective of quality), with the eternal problem of funds thrown in? The question does not appear to have been debated in depth; in fact it is avoided, being considered an explosive issue. Be that as it may, there are the odd success stories. Typically they consist of small units, well insulated by fortuitous circumstances from external interference, with a low national profile perhaps but enjoying high reputation in international scientific circles. Can this insulation last indefinitely against the continuing onslaught of outside social pressures? Is this true success, ask others. Should there not be a national impact of some kind – on teaching, training, education, and so on? How does one describe success? Is it international reputation at the expense of national involvement or is it acquiring an (apparent) image of national usefulness at the expense of quality? Are the demands for social justice and the quest for excellence perforce mutually exclusive in a developing country and is it only in advanced societies that they are reconcilable? If so, which of the two options does a developing country follow? How does one fix priorities? What kinds of problems does one study–does one follow fashions or does one explore one's own ideas at the risk of being ignored by peers outside? Can one compete with the advanced countries in experimental science if the large funding, which seems mandatory, is not available? On the other hand, can one ignore experiments and concentrate instead on theoretical science alone? How does one attract and not lose talented scientists always on the look-out for greener pastures abroad?

Endless questions. They were asked in Mahendra Lal Sircar's time, they were asked in Raman's time, and they are still with us – important questions which cannot be dodged if one is trying to foster excellence, which Raman was doing in his own way. Facing them squarely and surviving can be quite tricky, for the countercurrents are strong. It was unfortunate for Raman that while groping for answers he created a whirlpool, only to become a victim of it later. Perhaps it was inevitable.

8.8 Science in Bangalore

After a lengthy discussion of events painful, it is a relief now to turn to Raman's scientific contributions. Raman might have entered the Institute as the Director, but that did not mean physics research took a back seat. While the vacuum in physics which existed at the time of his entry was no doubt a blessing it also meant, on the other hand, an extra load compounded by difficulties due to shortages of staff, equipment and funds. However, these were not insurmountable hurdles, and soon there was a crowd of enthusiastic students. Physics research was thus in full swing practically from the time of Raman's arrival, although it took some time for things to stabilize. The lessons learnt in Calcutta on dividing time between research and other activities (be it teaching or office work) came in handy but, as earlier, they implied a heavy schedule. Pisharoty describes the latter:

> Professor Raman was at his desk at 6 a.m. every day. Till about 9 a.m. he would be there scrutinizing the work done by his students, or going round the laboratory. During these visits he would offer his suggestions for improvement in the experimental set-up, in the analysis of data or in the mathematical solution of problems, as the situation demanded. He had a prodigious memory regarding all the scientific material which had appeared in print in the English and German periodicals or books. He would go home for a quick breakfast at about 9 a.m., and would be back in his Director's Office by about 10 a.m. Twice or thrice a week there

would be seminars, where one of the students would report the work that he had been doing, experimental or theoretical, the difficulties encountered by him, the results obtained, etc. During these seminars, Professor Raman would think aloud.... In the evening, after discharging the onerous responsibilities of the Director, he would be back at his research desk at the service of his students, or digging away at his own research work, almost till 8.30 p.m. everyday. Occasionally, he would go out for a drive in the evening and would usually take along with him one of his students whom he could locate in the laboratory as he was leaving.... If any one [student] came to the laboratory after 7 a.m. he would miss the opportunity of having any long discussion with Professor Raman, for some other student would have taken up that opportunity. [8.3]

Ananthakrishnan adds:

He would give free expression to his joy when a new result was brought to his notice. In his public lectures, he would refer to his students by name and talk about their work. All this was a thrilling experience to young students and a powerful incentive for sustained hard work and endeavour. [8.4]

Recounting his experience, Thosar, another veteran of those days, offered to the author further glimpses into Raman's personality. Having obtained a scholarship, Thosar arrived in Bangalore. After establishing himself in the hostel, he called on Raman. The Director's office was on the ground floor of the main building (it is even now) and at that time the Physics Department also was located in the same building (the department now has a building of its own). Apparently Raman believed in leaving the doors of his office wide open so that people could walk in freely. As the young student entered timidly, he saw the Director busy discussing various files with his assistant Dr Krishnamurthy. When Raman discovered that his visitor was his latest student, he dropped his administrative work, and after enquiring whether he (i.e., Thosar) was properly settled and all that, took him on a tour of the laboratory. This done, the next job was to find a desk and a place to sit. Thosar told me that he was astounded when Raman personally helped to move the furniture. Indian society then was (and one regrets to say still is!) highly feudal, and even getting an audience with a minor dignitary protected by layers of pettier officials was usually quite an achievement. But here was the Director of a prestigious institute, and an eminent scientist to boot, not only freely accessible but also nonchalantly helping a raw student shift furniture. Westerners may see nothing great in all this but only a person who has had a personal taste of a feudal atmosphere can really appreciate how refreshingly different (and therefore a misfit!) Raman was.

After some discussion, the study of fluorescence in ruby was suggested as a problem to Thosar and he was handed a sample in the form of a ruby necklace! Apparently it had been purchased as a gift for Lady Raman, but rejected by her as it was not up to the mark!! In the course of his investigations, the young scholar damaged a mercury lamp owing to lack of experience. Everyone around was aghast; Professor would be furious, for money was scarce; he must not know. The situation was saved by raising a collection for a replacement.

We are talking here of the mid-thirties, and nuclear physics was then the new frontier in physics. Cambridge was throbbing with excitement. Raman wanted very much to enter this area but he could not; there was neither the money nor the support. And so it was back to good old optics, light scattering, fluorescence and, of course, a bit of Raman effect. Not mainstream physics obviously, but some good work came out nevertheless, including a few important contributions like the Raman–Nath theory and the soft mode concept. The accent was on experiments, inevitable after the squashing of efforts to build up a viable theory group. An unfortunate consequence was that the thinking of the Bangalore group remained largely classical. Quite possibly, if nuclear physics had been pursued, it might have forced a reorientation in outlook, Born or no Born. But events did not transpire that way and the urgent compulsions to absorb new concepts were largely lacking.

In what follows, we shall, as in Chapter 4, sample some of the scientific highlights of the Bangalore period, extending our coverage to the work of Raman's students as well. The study is interesting, for it shows how curiosity, even when it is not about the hottest problem on earth, can, if intense enough, lead to intellectual satisfaction besides occasional rewards. Thus many good contributions emerged, a few, in fact, with long-range impact.

Optics of Stratified Media

Optics continued to remain Raman's first love, and as in Calcutta he excelled in it. But increasingly, it was optics associated with natural phenomena, and so involved did he become with aesthetics that a casual reader of his papers is likely to miss altogether their physics content.

His very first publication from Bangalore (which, incidentally, was also the opening paper in the journal *Proceedings of the Indian Academy of Sciences* founded by him – see Chapter 11) signals his new focus. Entitled "The origin of the colours of the plumage of birds", it begins thus:

> Great interest naturally attaches to the investigation of the colours that form a striking feature of the plumage of numerous species of birds. Even a cursory examination, as for instance the observation of the feathers under a microscope, shows that the distribution of colour in the material and its optical characters are very different in different cases, indicating that no single explanation will suffice to cover the variety of phenomena met with in practice. It is usual to distinguish between those cases in which the colour is chemical in origin and those in which it is physical or structural. It must not be overlooked, however, that in any particular instance both physical and chemical colouration may be jointly operative. Then again, it is usual to divide the structural colours of feathers into two classes: those of the *iridescent* type in which the colour changes very obviously with the angles of incidence and observation, and the *non-iridescent* class in which such change, if any, is not very patent. [8.5]

The feathers of the peacock and the plumage of the Himalayan pheasant[12] provide striking examples of such iridescent colours but Raman's interest in this paper is directed to *Coracias indica*.

> This is a species of jay, very common in Southern India, which furnishes readily accessible material for the investigation of this type of colouration of birds.
>
> Seen sitting with its wing folded up, *Coracias indica* is not a particularly striking bird, though even in this posture its head, sides and tail show vivid colouration. It is when in flight that the gorgeous plumage of this bird is most strikingly seen, and museum specimens of the bird are therefore best mounted with the wings outstretched. The wings then exhibit a succession of bands of colours alternately a deep indigo-blue and a light greenish-blue; the tips of the wings show a delicate mixture of both colours.

This is not just a rambling description. There is a reason behind it which Raman reveals only rather slowly.

> A remarkable feature is the striking variation in the appearance of the wings with their position relatively to the source of light and the observer. Held between the observer and the source of light, such as an open door or window, the wings appear dark and dull, while when observed with the light behind the observer, they have a brilliant sheen, and at some angles an enamel-like lustre. There is also a distinct difference in the colours exhibited in the two positions. With the light behind the observer, the predominant colours are deep blue and a light greenish-blue, while with the light facing the observer, the same regions exhibit respectively a dark indigo colour and a light blue tint.
>
> Very noticeable changes in colour take place when the feathers are immersed in water and after some soaking the superfluous water is shaken off. The indigo-blue portions of the feather then appear green in colour, while the greenish-blue portions appear pinkish red.

More description follows, and finally (!) Raman makes the point that a down-to-earth physicist would be interested in:

> The foregoing observations indicate very clearly that the cause of the colouration is mainly physical.

Next follows a detailed record of observations made with an "Ultra-opak" microscope of Leitz make. In this instrument, the illuminating beam has the form of a hollow cone which is reflected downwards and converges on the surface of the object. If the axis of the cone is normal to the surface, one has "dark-field" illumination, the surface becoming visible by reason of the light scattered or diffracted by it. On the other hand, by suitably tilting the specimen, it is possible to arrange that the light reflected by the sample enters the microscope, in which case one has the ordinary or "bright-field" observing conditions. Raman finds that the characteristic colouration is confined entirely to the barbs of the feather. But the colour of the barbs is by no means uniform; rather, there appeared to be a series of cells, polygonal in shape, with the colour varying from cell to cell. Further investigations follow on wetted barbs as well as on the polarization characteristics.

The whole appearance is quite picturesque no doubt, but what is it all due to? Raman comments:

> Bancroft and others, including especially C. W. Mason, have put forward the theory that the non-iridescent blue colour exhibited by the feathers of numerous birds is a Tyndall effect due to the scattering of light by very fine air-bubbles or cavities

contained in the substance of the barbs. Mason includes *Coracias indica* in the list of birds examined by him in the investigations which are claimed to support the theory.

Raman finds that his own observations are not in accord with Mason's theory. He agrees that the cavities referred to by Mason have something to do with the colouration, but not via the Tyndall effect. Recalling various related experiments performed earlier in Calcutta, he narrows down the choice to one between diffraction by the cavities and interference from the surfaces of the cavities. He is not, however, able to decide between these two. He acknowledges that further experiments were necessary, but apparently never got round to doing them.

If the reader finds both the style and the tone of Raman's paper quite strange, then, just to establish a comparison, his attention must be directed to a series of papers by the younger Lord Rayleigh which had appeared just a few years earlier. The latter (following up some of his father's work) also was interested in iridescent colour and optical structures producing it, and one paper of his [8.6] is devoted to iridescent beetles!

Like the two Rayleighs, Raman became much interested in iridescence. Earlier in Calcutta, Ramdas [8.7] had investigated feeble iridescence from potassium chlorate crystals. Potassium chlorate belongs to the monoclinic class of crystals, and in its natural occurrence takes the form of flat plates containing many twins. When a crystal plate is held so as to reflect light obliquely and is turned around in its own plane, colours alternately appear and disappear twice in each complete revolution. The spectral character of the reflected light varies also with the angle of incidence.

Figure 8.2 shows some of the results obtained by Ramdas. Not only does one see that different wavelengths are selectively reflected at various angles of incidence, but also that corresponding to the selective reflection, there are extinctions in the transmitted spectra indicated by the presence of dark lines or bands. The spectral effects observed when the angle of incidence is fixed and the crystal is rotated in its own plane are shown in Fig. 8.3. All these effects are associated with stratified media.

In Bangalore Raman picked up this theme, using various specimens of natural origin to provide examples of stratified media. His study of the plumage of birds is the first of a series of publications, soon followed by two papers on iridescent shells [8.8, 8.9]. The first paper on shells is largely descriptive, being in the style of a naturalist's. Several specimens had been kindly loaned by Dr Baini Prashad, Director, Zoological Survey of India, himself an expert on shells, and there are delightful descriptions of these ("the Indian and Ceylon pearl-oyster is amongst the smallest of its kind, and compares unfavourably in size with the huge gold-lip pearl-oyster from Mergui and the South Seas, a species with shell large enough for a dinner plate"). The paper also abounds in Latin names seldom seen in a physics journal! The text is profusely illustrated with photographs (as indeed all the papers on iridescence were), two of which are reproduced in Fig. 8.4.

After setting the stage Raman quotes extensively from various zoologists and conchologists, leading up finally to the fact that iridescent shells are essentially laminates of calcium carbonate in the aragonite form cemented together by an organic substance, the so-called conchin.

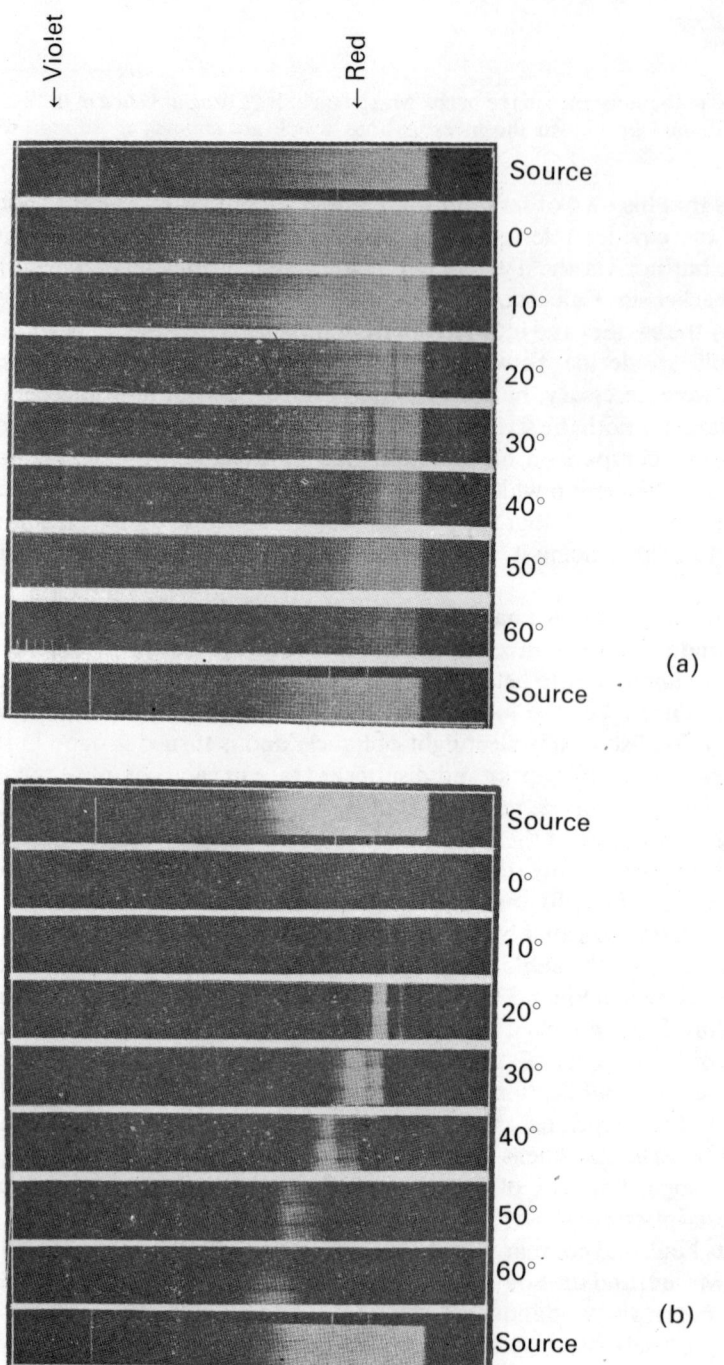

Figure 8.2 (a) Reflection spectra of potassium chlorate crystals for a sequence of incidence angles; (b) the corresponding transmission spectra. The two are complementary, i.e., a sharp band in reflection corresponds to an equally sharp dark band in transmission. (After ref. [8.7].)

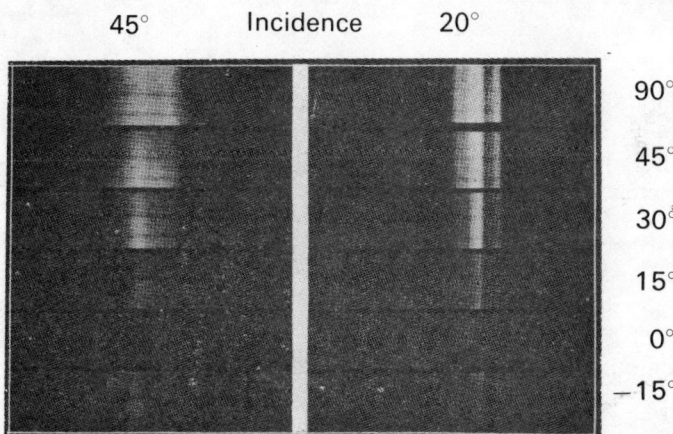

Figure 8.3 Reflection spectra of potassium chlorate crystal obtained by holding the angle of incidence fixed at 20° and 45°, and rotating the crystal in its plane. (After ref. [8.7].)

Raman's second paper on iridescent shells [8.9], which immediately follows the first, concentrates on the physics aspects. He begins by recalling some known examples of iridescence such as that of potassium chlorate crystals, wings of beetles, etc. But shells have some distinctive features of their own. Firstly, the internal laminations are nearly always inclined to the external surface instead of being parallel to it. Secondly, the laminae are granular and, further, the material is thick and not a thin film (as in the case of the wing of the beetle, for example). Concerning earlier studies on the general aspects, Raman observes:

> The theory of the reflection of light by a stratified film has been discussed by Rayleigh. His investigation indicates that the permissible cases fall into one or another of two classes, the first in which the reflection tends to become complete as the number of laminae becomes large, and the second in which the reflection and transmission remain fluctuating however great this number may be. Whether the one or the other state obtains is determined by the relation between the reflecting power of the individual laminae and the phase relation between the reflections occurring at the successive laminae.

Raman devotes much attention to the iridescence of mother-of-pearl. Earlier Sir David Brewster had investigated the colours and classified them as what he called transferable and non-transferable colours. The former could be transferred to gelatin or other soft substances employed to take an impression of the surface of the material, and hence it was inferred that they were due to diffraction by grooves on the exterior of the shell. The non-transferable colours on the other hand were regarded as interference or thin film colours due to the internal reflecting layers. Raman disagrees with Brewster, being of the opinion that diffraction effects would occur at the surface not only when light is incident externally on it, but also when it emerges at the surface after suffering reflection at the internal laminations. An integrated approach to the phenomenon therefore seemed preferable to a bifurcated

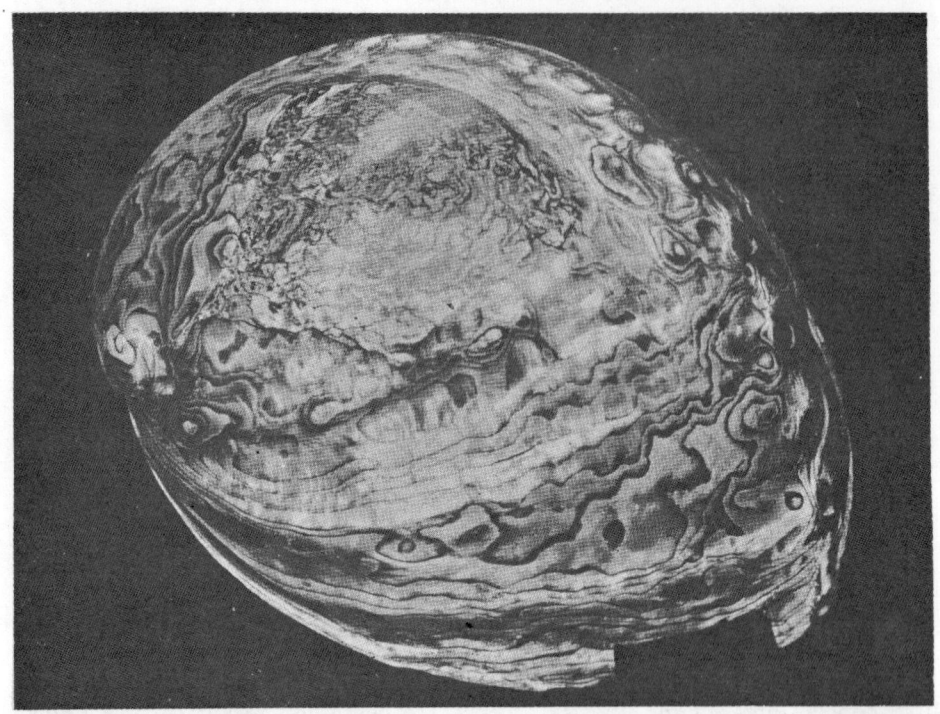

Figure 8.4 Photographs of abalone shell. (After ref. [8.8].)

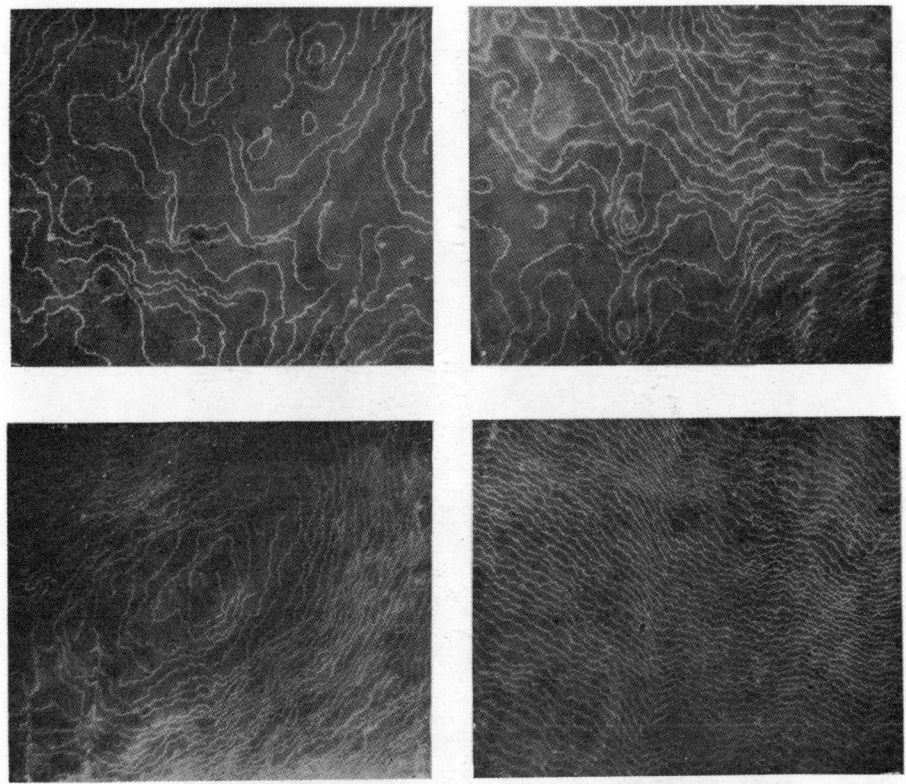

Figure 8.5 Abalone shell under hollow cone illumination. (After ref. [8.9].)

one. Clearly, fresh studies were called for. Figure 8.5 illustrates a typical outcome of these investigations. One sees numerous sharp, bright lines against a dark background. Raman finds that the lines represent the intersections of the internal conchin layers with the surface of the shell. These layers are quite thin compared to the aragonite layers which they separate.

The diffraction spectra were also studied. A pencil of light was allowed to fall on the surface and be reflected from it. It was then found that the reflected pencil was accompanied by diffraction spectra. There were three conspicuous spectra on one side of the reflected light, and only one on the other (see Fig. 8.6). The first-order spectra on both sides showed the complete set of colours present in white light. But the second-order and third-order spectra, which were present only on one side, did not show all the colours. The third-order spectrum in particular, though very intense, showed prominently only the characteristic iridescence colour for the particular angle of incidence.

An interesting variation of the experiment was made by cementing a thin microscope cover-slip of glass with a little Canada balsam on the surface of the shell.

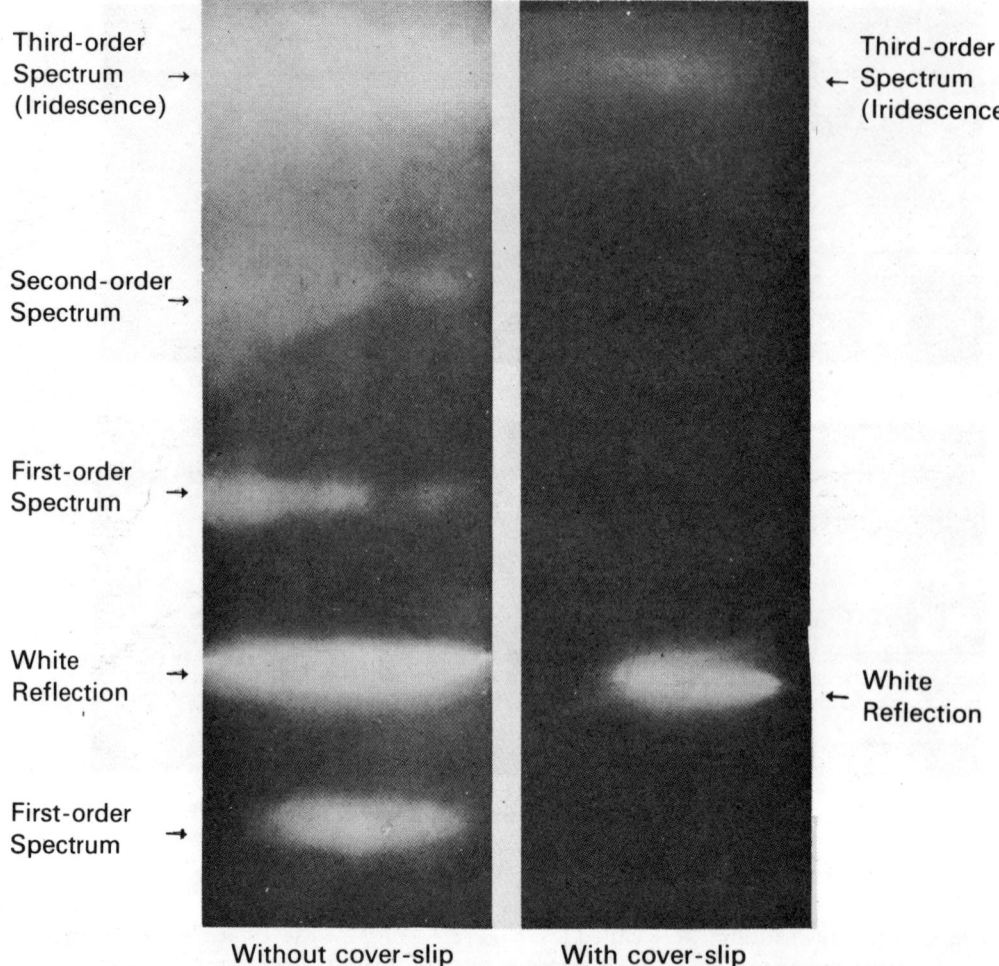

Figure 8.6 Diffraction spectra obtained with iridescent *Turbo* shell. (After ref. [8.8].)

The effect of the Canada balsam is to suppress the grooves present on the surface of the shell. Raman found that all the diffraction lines disappeared, except the one in the third-order spectrum. He thus concludes, quite reasonably, that the characteristic iridescence appears as the third-order diffraction spectrum and is primarily due to the *internal periodic structure*.

Further studies were carried out by polishing the surface to various degrees. From these and various other related observations Raman concludes that the edges of the terraces formed by abrasion on the external surface of the shell coincide with the curves along which the internal laminations meet the surface, that the interference arising from the internal laminations and the diffraction arising from the surface are linked, that iridescence is a diffraction effect produced by the internal periodic structure (as distinct from surface corrugations – see Fig. 8.7), and finally that the

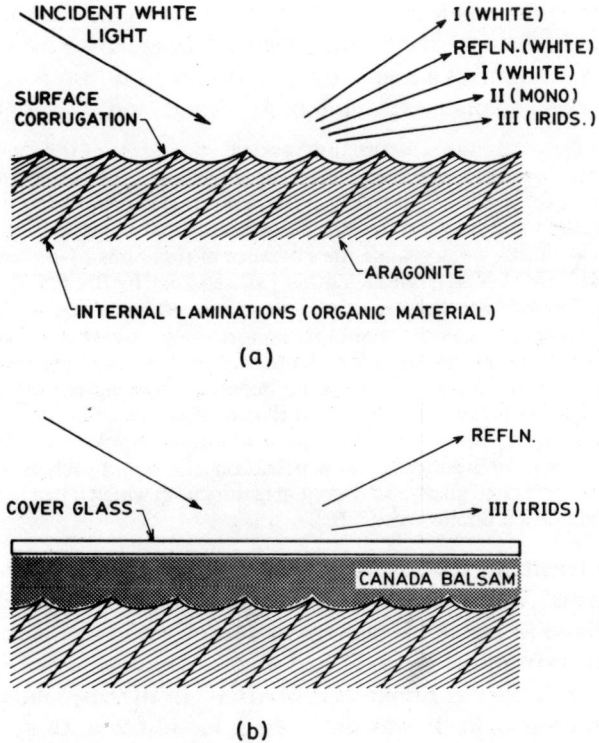

Figure 8.7 (a) Schematic of a section of an iridescent shell. The optical effects are due to both the surface corrugations and the internal laminations. When a glass cover-slip is used as in (b), the surface effects are suppressed but iridescence still survives (see the previous figure).

optical observations support the view of the noted zoologist Schmidt that mother-of-pearl (the optically interesting innermost layer of iridescent shells) consists of layers of aragonite separated by immeasurably thin strata of organic matter.

Iridescence in glass was studied next. Now glass by itself is not an iridescent material. However, ancient glassware excavated from archaeological sites often exhibit beautiful iridescence, and it is not surprising that Raman became interested in the phenomenon. Sir David Brewster who was the first to study it concluded that the iridescence was due to a lamellar structure resulting from decomposition.

How exactly does such a structure arise? The details were not known but it was clear that wind, water and chemical reagents played a role. Marcel Guillot, for example, had demonstrated that when a saturated solution of sodium bicarbonate was kept in contact with alkaline glass, marked iridescence resulted after some weeks or months. He in fact went so far as to suggest that the iridescence was due to the formation of stratified films by a chemical process analogous to the Liesegang phenomenon. (A description of this phenomenon and Raman's studies on it are described in a later subsection.) Raman was not interested in the chemical origin of the stratification but in the optical effects it produced. Brewster's view was that the medium consisted of thin films of air separating the decomposed layers of glass. This, Raman felt, was not an adequate explanation. As Raman and Rajagopalan put it:

> Brewster's views regarding the structure and optical characters of iridescent glass are not altogether free from obscurity. The refractive index of the decomposed material presumably differs little from that of the original glass. Assuming this to be the case, it is not easy to understand how the coloured reflections arise. If in order to explain the optical effects, we postulate the existence of thin films of air separating the decomposed layers of glass from each other [as was done by Brewster], it is not understood why the films are actually adherent to each other and to the glass, and why they exhibit in many cases a remarkable uniformity of colour and brightness over considerable areas. If, on the other hand, we assume that the cementing material is some solid substance, it would be necessary that its refractive index should be considerably higher or lower than that of glass, in order that sensible reflections should result; moreover, such an assumption would be difficult to reconcile with the astonishing rapidity of penetration of a liquid such as water or alcohol into the decomposed glass, and the equal facility with which it can evaporate and leave the iridescence unaffected. [8.10].

A re-examination seemed called for, which Raman and Rajagopalan carried out. But what about specimens? The museums in India could not provide much help but, undaunted, Raman and Rajagopalan manage to start off with "specimens of glass of no great age picked up from the ground", and plates "which were picked up at the site of an extinct glass factory at Ennur near Madras". Even these, under thorough probing, offered much insight. It was discovered, for instance, that the internal structures leading to the observed optical effects could be quite complex (as opposed to what Brewster had envisaged), consisting of cavities, films without boundaries, films with sharply defined boundaries, and so on. One example of such a structure may be found in Fig. 8.8 which shows cavities as they appear under microscopic examination. The Leitz Ultra-opak microscope described earlier was used for this purpose. The circular rings in the figure are the step-like walls of the cavity which correspond to the decreasing diameter of the successive laminae in the decomposed material, the number of such laminae being greatest at the centre and least at the margin of the cavity. Another example of internal structure can be seen in Fig. 8.9 where one notices "networks of the most varied forms". Raman and Rajagopalan present many photographs; they regret their inability to offer colour reproductions but make up for it with vivid descriptions. Some polarization studies were also carried out.

There are some interesting questions concerning what happens when the observations are made with the specimen immersed in a liquid. As Raman and Rajagopalan observe:

> If a piece of iridescent glass is breathed upon when cold, moisture condenses on it and the colour of the film apparently vanishes immediately; it however reappears when the plate is warmed up. Brewster noticed the apparent suppression of colour produced by placing a drop of water or alcohol on the film, as also its restoration when the fluid has evaporated. He noticed further that oil or balsam penetrates the film slowly and unequally, producing a succession of tints on the plate during its

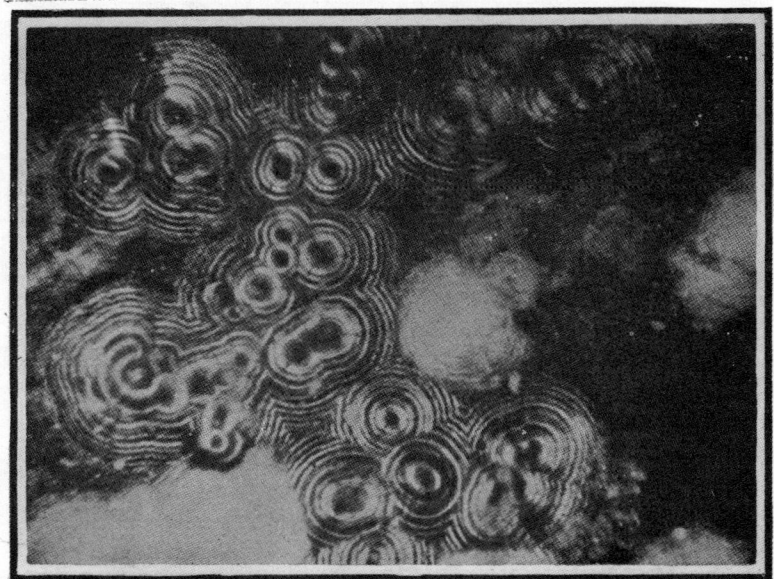

Figure 8.8 Cavities in decomposed glass. The white patches with indistinct rings indicate material produced by the decomposition, while the concentric rings with dark centres are due to the cavities formed by the removal of the decomposed material. (After ref. [8.10].)

advance. A careful study of these and other effects produced by contact with liquids is evidently of importance in order to elucidate the problem of the structure of these films and their optical characteristics.

It should be remarked in the first place that the observations made in the rather crude way mentioned above are deceptive. *In reality, the iridescence of decomposed glass does not disappear on immersion in a liquid; on the other hand, the colours actually become more vivid, though their intensity is considerably diminished.*

Raman and Rajagopalan demonstrated that the many new effects observed are due to the penetration of the liquid into the various cavities present. One example of the result of such penetration may be seen in Fig. 8.10. A glass film exhibiting uniform iridescence was taken (as the authors describe, "a reagent bottle which had developed a strikingly uniform internal iridescence was broken up and furnished suitable material for these investigations"), and a drop of monobrom-naphthalene was placed on it. The sharply bounded black central area (with a bright overlying ring due to reflected light) is the portion of the film actually covered by the liquid. The moderately dark circle surrounding the drop is the area of the film internally saturated with liquid, while beyond the same is seen a succession of dark and bright rings which indicate a variation in the quantity and distribution of liquid absorbed by the film.

Subsequent to this study, samples of ancient glass were obtained which permitted further investigations. Raman and Rajagopalan describe the circumstances leading

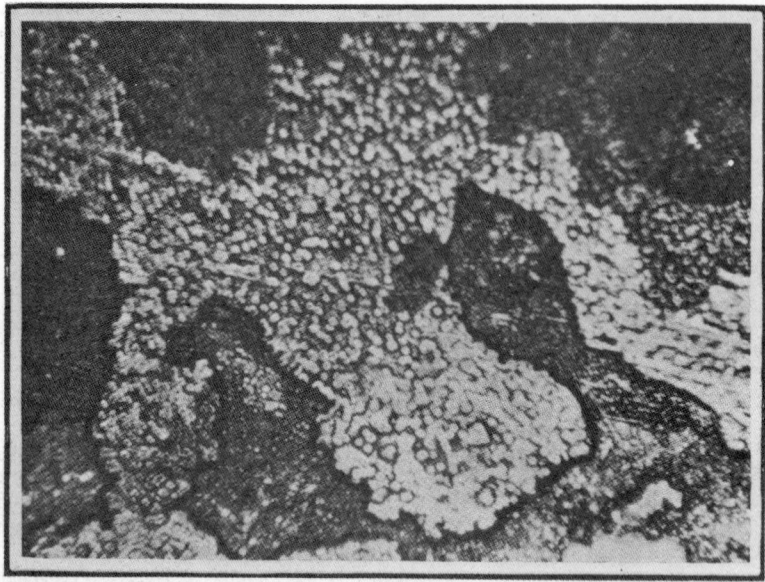

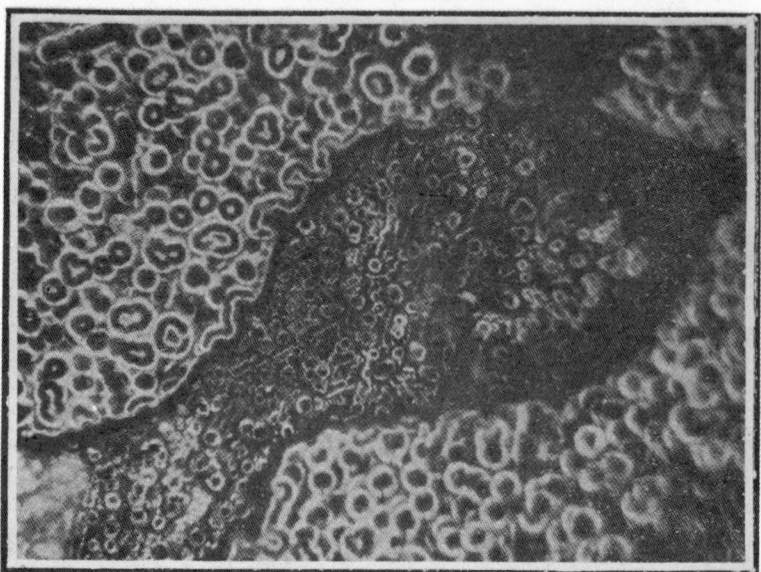

Figure 8.9 Dark-field pictures of decomposed glass. The luminosity is largely due to hollow cavities of varying sizes, some discrete and others running into each other to form patterns. (After ref. [8.10].)

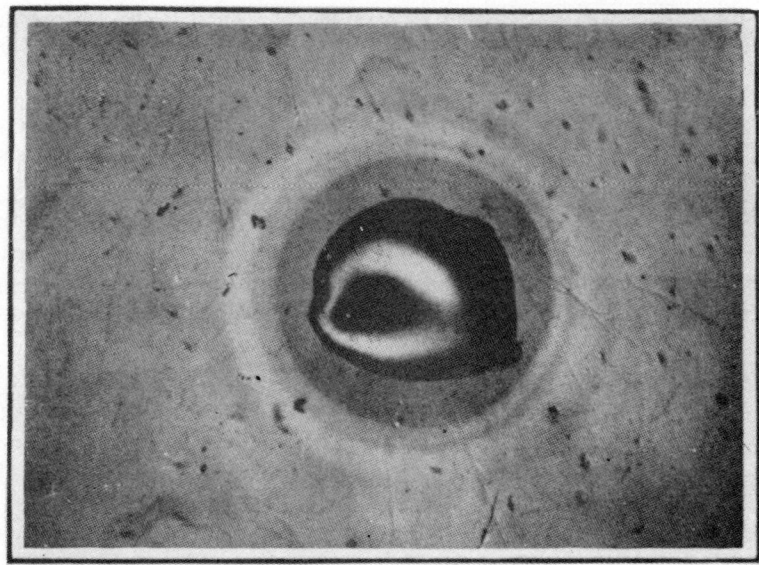

Figure 8.10 Photomicrograph of a plate of decomposed glass with a drop of liquid placed on it. (After ref. [8.10].)

to this:

> At the Palais de La Découverte in the Paris Exposition of 1937, one of us saw in the section of Optics a very striking exhibit of an ancient vase of glass which had been excavated in Syria by the French archaeologist M. Pupil. The remains of the vase together with numerous iridescent flakes resulting from its disintegration were placed in a plate-glass cabinet which was provided with viewing mirrors inclined at 45° to the vertical, and illuminated both from above and below, so that they could be seen simultaneously by transmitted and reflected light. The brilliance of the colours and the complementarity of the same as seen in transmission and by reflection were thus beautifully made evident. Through the kind offices of Prof. A. Cotton, a few flakes of glass from this exhibit were presented to us by M. Pupil. This gift enabled us to undertake the present investigation, which indeed we were desirous of doing, to supplement and complete our earlier work on the optical behaviour of decomposed glass of modern origin. [8.11]

Even a superficial examination showed a considerable difference in the character of the optical effects observed with ancient and modern decomposed glass. Raman and Rajagopalan give a graphic description of this difference:

> The flakes derived from the Syrian vase reflect light strongly, exhibiting an almost metallic lustre, the colour of which varies greatly. The thicker flakes amongst those given to us exhibit a bluish-white silvery lustre, while the thinner ones exhibit other tints in which greens and oranges are the most striking colours observed at normal incidence. The flakes also exhibit vivid colours by transmitted light, being in this respect much superior to the specimens of decomposed glass of modern origin considered in our [earlier] paper. The latter show scarcely any perceptible tints

when observed in transmission, while on the other hand, the flakes from the Syrian vase show colours in transmitted light which in many cases are more striking than those seen by reflected light. The transmission colours for the thicker specimens tend towards a rich red, while the thinner flakes showed colours ranging over the whole spectrum from violet to red. It is thus evident that the development of colour occurs in antique glasses to a greater depth and in a more uniform manner than in the modern specimens.

Clearly there was much to be studied. To start with a microscopic examination was carried out as before, and Fig. 8.11 shows a commonly observed feature. It was found that the laminae instead of being perfectly plane, consisted of shallow cups shaped like watch-glasses fitting together and dividing the surface of the flake into a large number of irregular polygons bounded by straight lines. The rings seen in the figure arise from an interference effect associated with the placement of a plane sheet of mica over the specimen. The uniformity of curvature of the surface of the cups is brought out by the perfectly concentric arrangement of the interference rings.

The hollow cup-like forms are naturally convex on one side of the lamina and concave on the other. To demonstrate this, the lamina was turned over on the stage of the Ultra-opak and observed by reflected light. Each cup-shaped area then showed an image of the illuminating annulus of the microscope, the surfaces themselves appearing dark (see Fig. 8.12). Some of the specimens were also examined under a polarizing microscope between crossed nicols, and Fig. 8.13 is a beautiful example where the cavities are ellipsoidal and do not meet to form polygonal figures. (Observe the crosses, and recall also Fig. 4.71.)

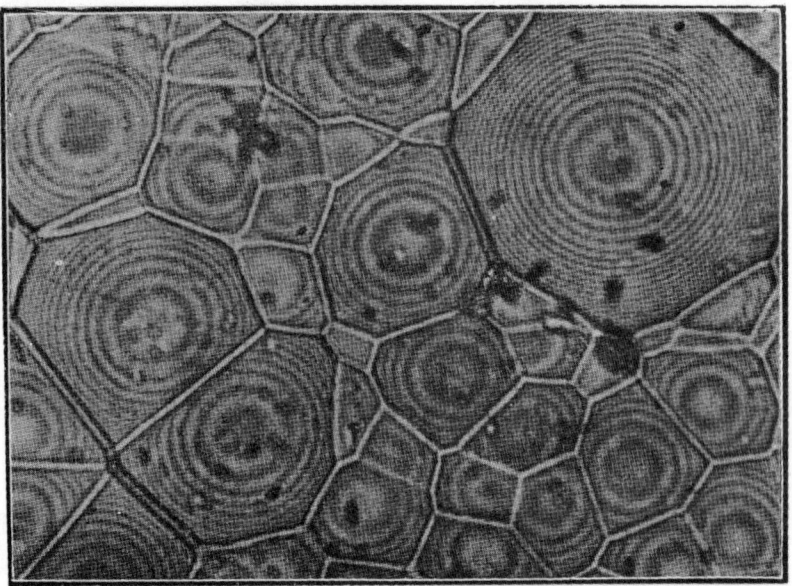

Figure 8.11 Photomicrograph of decomposed glass in monochromatic light, showing network of laminae. (After ref. [8.11].)

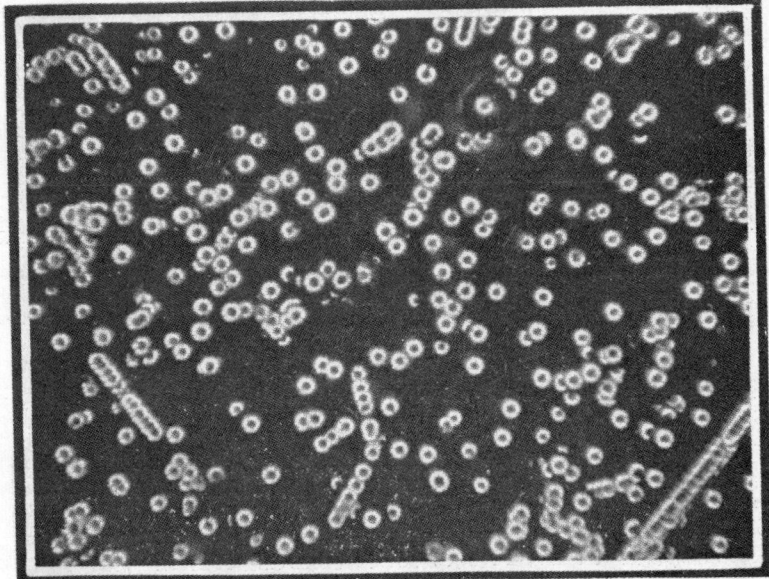

Figure 8.12 Lamina of decomposed glass with hollows forming optical images of the light source (hollow cone) by reflected light. (After ref. [8.11].)

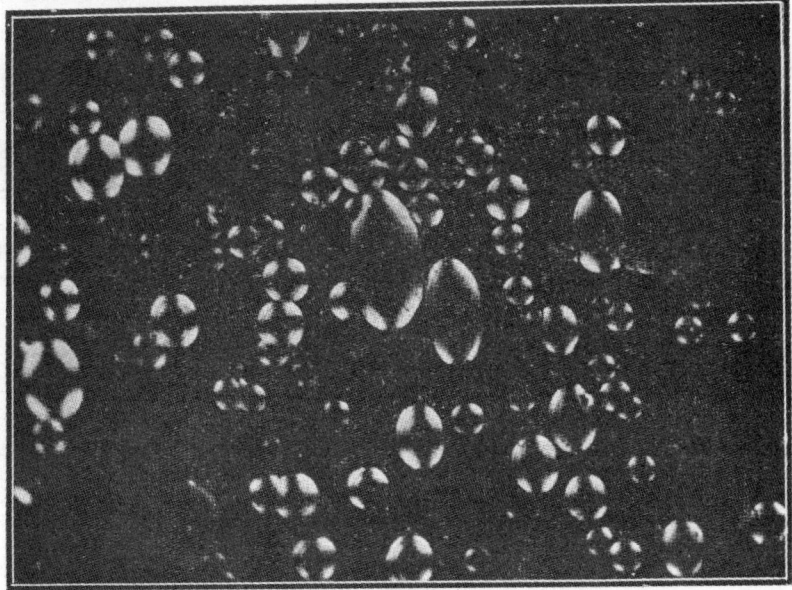

Figure 8.13 Deep ellipsoidal and spherical cavities in iridescent glass in monochromatic light under crossed nicols. (After ref. [8.11].)

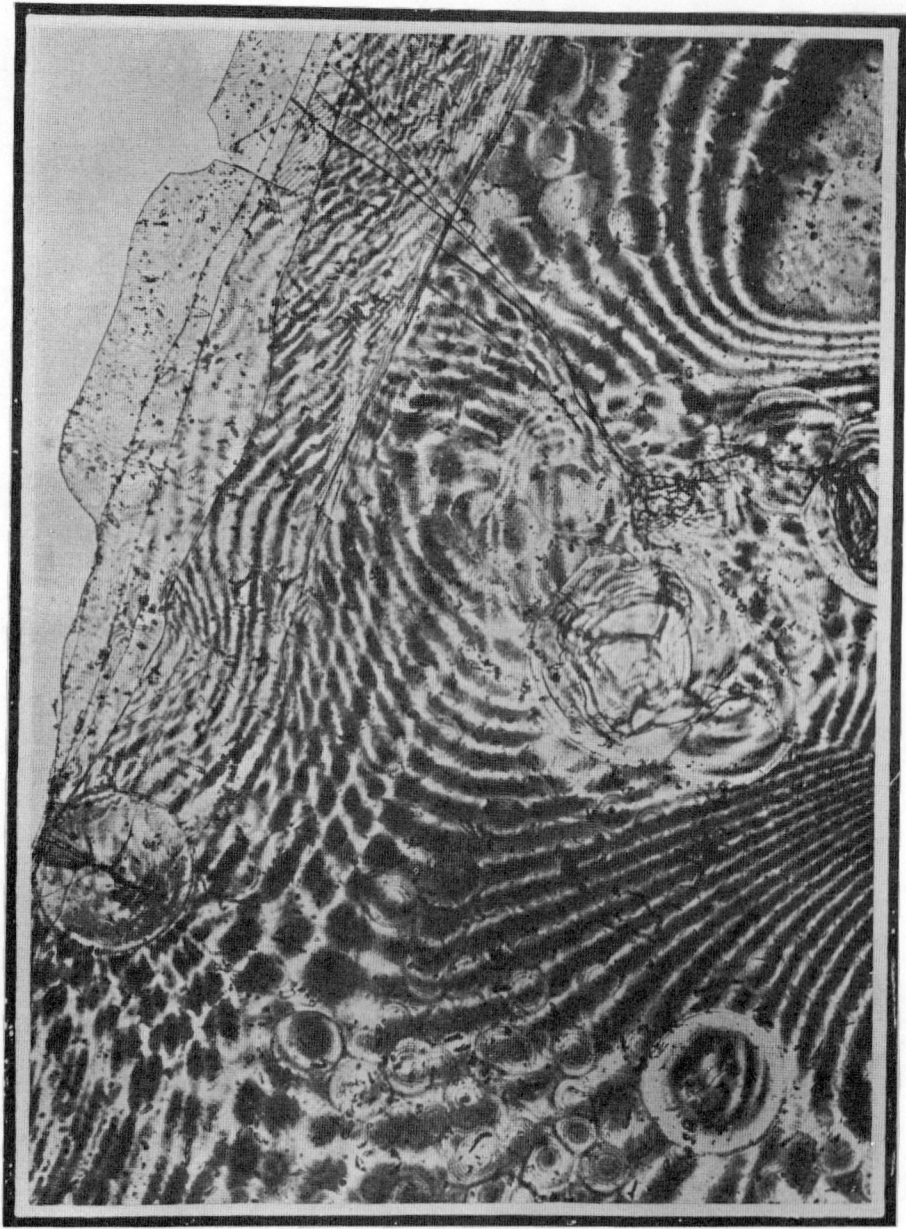

Figure 8.14 Photomicrographs of decomposed glass in transmitted monochromatic light showing laminar edges, intruding films of air, and fringes due to the air films. (After ref. [8.11].)

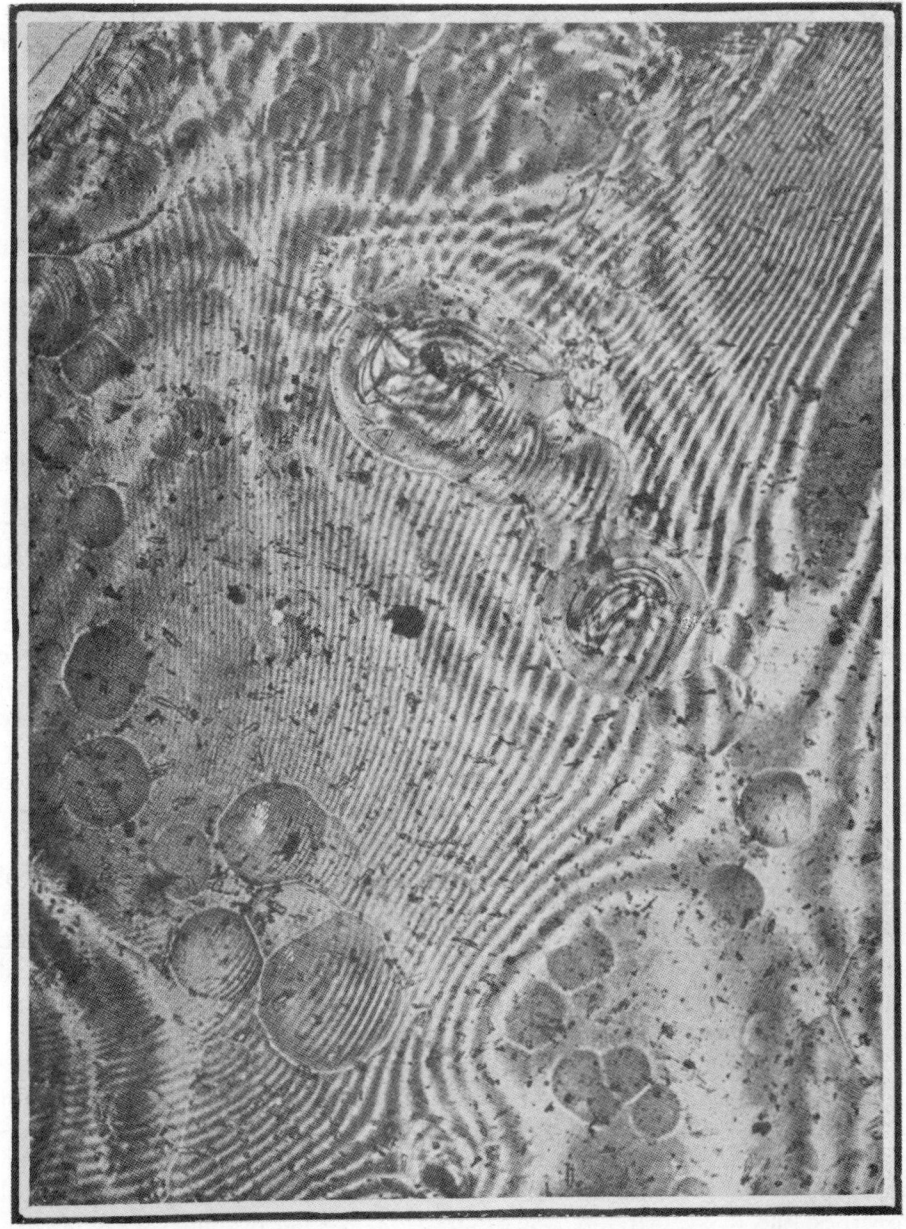

Figure 8.14 (*Contd.*)

Pretty colours are seen when the laminar structure is observed in transmission. Brewster was of the view that the colours are due to thin plates of air separating the laminae, but Raman disagrees and remarks,

> Indeed, the fact to which Brewster himself drew attention, namely that the elementary films of glass adhere with such force that it is difficult to separate them, is very strong evidence that they are in optical contact and not separated by continuous films of air.

Could this be demonstrated? Yes, by breaking the optical contacts and admitting air, whereupon colour variations are observed depending on the thickness of the air film. That the colour changes observed were due to air films was confirmed by gently pressing on the flakes and squeezing the air out, whereupon the (newly formed) coloured bands moved about in the field of view. However, the bands were restored to their original positions when the pressure was removed.

Figure 8.14 shows two examples of the structures observed in transmission with monochromatic light. To Raman's mind, the sharp bands suggest an analogy with the sharp interference pattern seen in a Fabry–Perot etalon.

The effect of immersion in liquids was also studied, and by observing the capillary flow of liquids Raman and Rajagopalan concluded that the laminations consisted essentially of a periodic distribution of cavities or pores, and were not due to an alternation of solid layers of different refractive index in the material of the glass. The structure was most likely due to the leaching out of the more soluble layers, the periodicity of the decomposition being probably analogous to the Liesegang effect. Strong mechanical pressure (applied by rolling a blunt steel point firmly on the surface) caused the cavities to collapse, removing the colours permanently.

Finally, a spectroscopic examination of the transmitted as well as the reflected light was carried out. The results were reminiscent of those obtained earlier by Ramdas, characteristic of regularly stratified media.

While Raman occupied himself with surveying the various manifestations of optical phenomena in stratified media and enquiring into the basic causes for the effects observed, he left to Ramachandran the task of developing a formal treatment. Ramachandran provided a perfect foil to Raman in this respect. Given his natural bent of mind towards topics mathematical he enjoyed furnishing the analytical back-ups to Raman's explorations, his paper on the reflection of light by a periodically stratified medium being a good example [8.12].

What happens to light when it falls on a stratified medium? Obviously it is not possible to deal with this problem in its widest generality. Mathematical convenience requires some simplifying assumptions, and accordingly Ramachandran supposes that the medium is non-absorbing and has a *periodic* variation of optical properties. Effectively, therefore, one can regard it as a pile of similar slabs, say n in number (see Fig. 8.15a). When light falls on such a stratified medium there will be multiple reflections of all sorts, resulting finally in two streams of energy, one passing upwards and the other downwards. What one would like to know is the amplitude T_{n+1} of the wave transmitted by the last stratification and that reflected out of the first, viz., R_1. In solving for these quantities, there is a chain of events which must be addressed. For example, if one considers R_s the wave amlitude reflected from the sth slab, it really consists of two parts, namely, that reflected by the slab from the part T_s falling on it, and that transmitted by the slab of the part R_{s+1} reflected by the next slab below. Thus if O_s and O_{s+1} are corresponding points in adjacent slabs and if we define

$$t = \frac{\text{amplitude of the transmitted wave at } O_{s+1}}{\text{amplitude of the transmitted wave at } O_s}$$

and

$$r = \frac{\text{amplitude of the reflected wave at } O_s}{\text{amplitude of the wave incident at } O_s},$$

then one may write the bookkeeping relations

$$R_s = rT_s + tR_{s+1}$$
$$T_s = tT_{s-1} + rR_s.$$

It will be observed that the effects produced by the sth slab are related to those produced by the slabs adjoining it on either side. In this way one has recursion relations or a chain of equations describing the passage of radiation through the medium. The problem is now tightly defined and Ramachandran is able to produce an elegant solution (for R_1 and T_{n+1}).

This line of attack was suggested by Raman, who was inspired by a much earlier work due to Darwin on the diffraction of X-rays by thick crystalline slabs. The crystal is not quite stratified in the same sense as that considered here, but in another sense it is, and Darwin was concerned with the progressive depletion of electromagnetic energy due to reflection from various crystal planes. There are thus analogies between the two problems.

Ramachandran's formulation is sufficiently general for him to recover as a special case the results obtained earlier by Lord Rayleigh for the stratified medium shown in Fig. 8.15b. His principal findings are captured in Fig. 8.16 where the reflectivity R (equal to the ratio of the intensity $|R_1|^2$ of the reflected beam to that, $|T_1|^2$, of the incident beam) is plotted as a function of a quantity ϕ which is related to path retardation between adjacent layers. Loosely translated, ϕ is related to the wavelength which is reflected. Thus the plots essentially indicate the nature of the reflected spectra for stratified media of various thicknesses. Three things stand out: firstly, as n increases, the intensity of the principal maximum increases, reaching saturation eventually; secondly, the principal maximum narrows in width as n increases but not indefinitely as was earlier believed; thirdly, subsidiary maxima develop as n increases, indicative of reflection at several select wavelengths.

Ramachandran's geometry is clearly too idealized to apply quantitatively to practical situations, such as that provided by shells, for example. On the other hand it clarifies the underlying physics, and is certainly relevant in dealing with problems like that studied by Ramdas.

A few years later, there was beautiful confirmation of the theory of Ramachandran

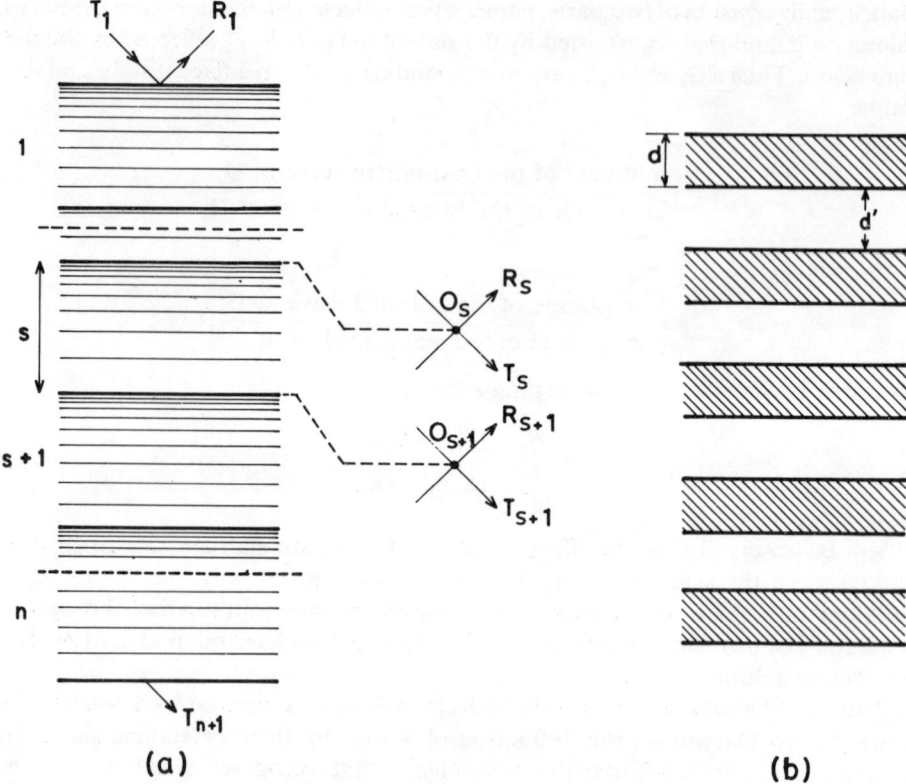

Figure 8.15 (a) Schematic representation of the periodically stratified medium considered by Ramachandran; (b) the medium considered by Rayleigh, which consists of n parallel plates each of thickness d and of refractive index μ separated by empty gaps of thickness d'.

by Raman and Krishnamurti [8.13] who obtained excellent spectra, one example of which is shown in Fig. 8.17.

In February 1941 Raman received an invitation from the Sayaji Rao Gaekwar Foundation, Baroda, to give a course of lectures on optics. This provided him with an opportunity to review and discuss the work done over the years by him and his students. Raman prepared very hard for these lectures, making many slides and organizing many demonstrations. In fact (as we shall soon see) this led to a re-examination of many old problems which had been shelved. Later he started writing out his lectures so as to reach a larger audience. Concerning this effort, he observes:

> It was the desire of the Foundation which invited the author to Baroda that the subject-matter of these lectures should be developed and written out in the form of a series of six lectures for publication. It was planned that the lectures would deal with the following topics: (i) Interference of light, (ii) Diffraction of light, (iii) Coronae, haloes and glories, (iv) Optics of heterogeneous media, (v) Light in ultrasonic fields and (iv) Molecular scattering of light....
>
> The preoccupations of the author slowed down the writing up of the volume for publication and finally brought it to a stop in the year 1943 after 160 pages had been

On to Bangalore

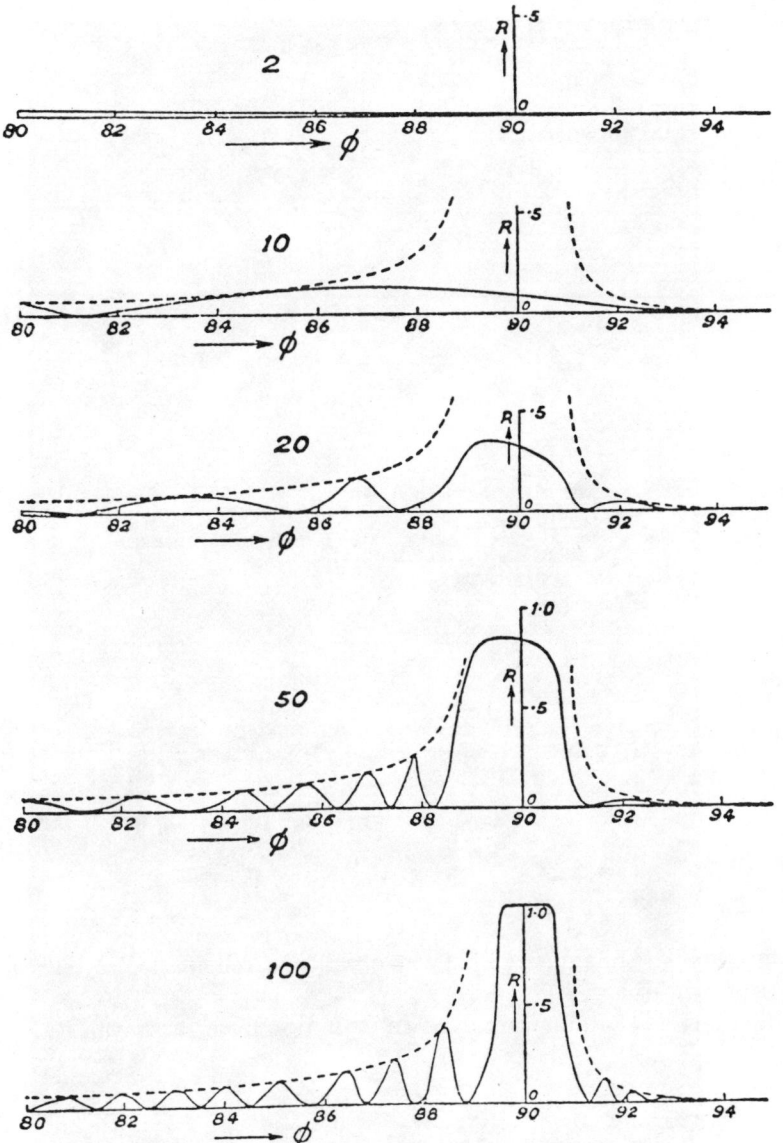

Figure 8.16 Reflectivity R of a regularly stratified medium as a function of the parameter ϕ related to path retardation. For further explanations, see text. (After ref. [8.12].)

printed off. Much labour and thought had been devoted to the work and it is believed that it contains material of enduring value and interest. Accordingly, it appeared desirable to release the part already printed as Part I of the lectures and thus make it available for perusal by those interested in optical theory and experiment. [8.14]

The year of release was 1959 although the printing itself was completed over a decade earlier! And, as usual, Part II was never published!! As Ramachandran

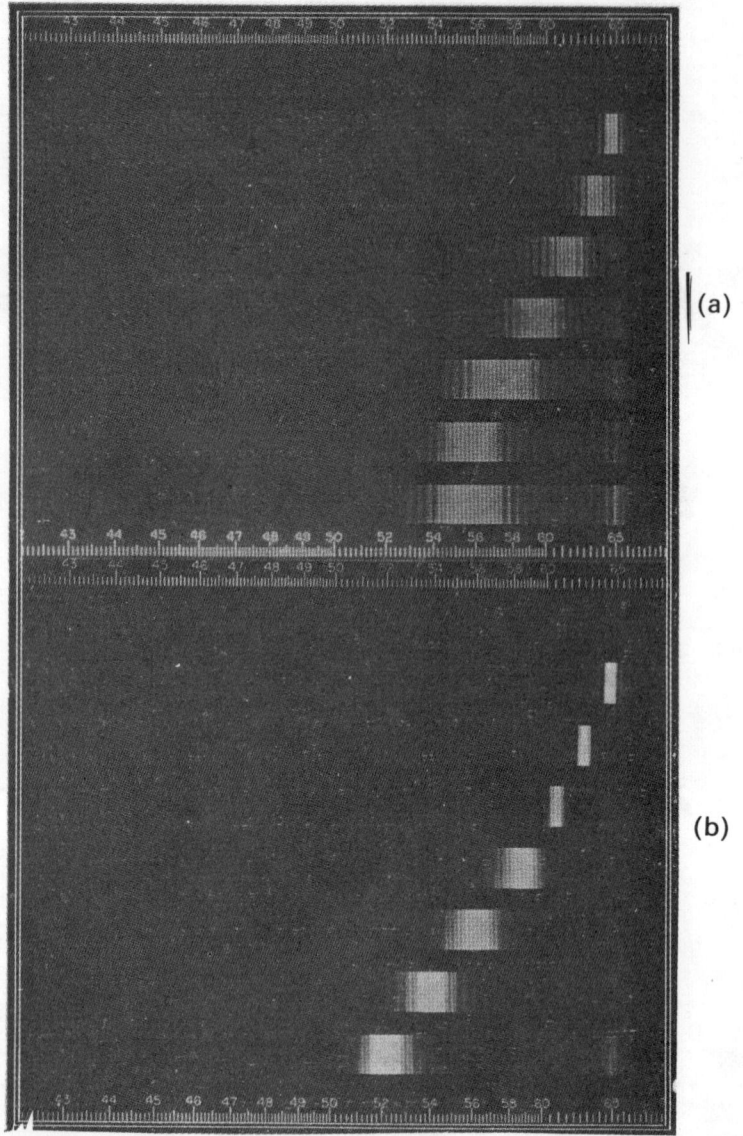

Figure 8.17 (a), (b) Reflection spectra of potassium chlorate crystals similar to those in Figs. 8.2a and 8.3 respectively. Observe the improvement in the quality of the results. (After ref. [8.13].)

remarks with reference to this incomplete effort,

> He could never sit down to write anything systematically, because he was always bubbling with ideas, and was finding it necessary to make newer and newer studies in his research programme. [8.15]

On Coronae and Speckles

As mentioned in Chapter 4, the study of coronae and related phenomena started in Calcutta with Bidhubushan Ray investigating the phenomenon of glory or brocken bow [8.16]. This refers to the effect seen when a bank of fog or cloud is viewed by an observer in a balloon or on a mountain, looking away from the Sun nearly in the direction of propagation of the Sun's rays. When favourably situated, the observer sees rings of coloured light around the shadow of the head. Rejecting all previous explanations as they were inconsistent with experiment, Ray showed how the observed effects may be understood by considering light which travels backwards towards the source from the water droplets in the cloud. Later, Mitra [8.17] photographed in monochromatic light the coronae produced by uniform clouds of droplets of various sizes (see Fig. 8.18). Meanwhile, Raman, during his studies on the radiant spectrum (see Chapter 4), had called attention to the effects observed when lycopodium powder is dusted on a glass plate and a source of light is viewed through it. Such effects had already been seen earlier by Exner, von Laue and de Haas. But as yet there was no consistent explanation of the various details, other than the vague recognition that coronae and allied effects seen both with water vapour clouds and lycopodium dust belonged to the same generic class of problems.

In 1941, when Raman was preparing for his Baroda lectures, these problems began to receive attention again, particularly from Ramachandran [8.12, 8.18, 8.19, 8.20]. The general background is well described by Raman. Both water vapour clouds and lycopodium dust consist of randomly distributed particles capable of diffracting light. There is, however, a difference in the fact that whereas in a cloud the particles execute rapid, uncorrelated movements, in a lycopodium screen they are static.

Quantitative studies on coronae in Bangalore were initiated by Balakrishnan [8.21] who studied the phenomena observed with water vapour clouds. For reasons which will become clearer later, one has, in this problem, essentially to consider the effects produced by a single water droplet, the effect of N droplets being just N times that of one.

Earlier workers had tried to explain coronae by assuming that the droplets acted like opaque spheres but from Mitra's work it was clear that the transmission of light through the droplet was an important factor. Raman has explained Balakrishnan's work in a simple and elegant way [8.14]. Consider a drop as in Fig. 8.19 and suppose it is illuminated with a plane wave as shown. If the drop is nearly transparent, then, during passage through the drop, only the *phase* of the wave is modified and not its amplitude. "The wavefront on emergence would thus exhibit a *dimple* having the same radius as the drop and a depth equal to the maximum retardation it produces." Significantly, a similar idea forms the basis of the Raman–Nath theory (as we shall see in Chapter 9).

The problem thus boils down to calculating the effects of the dimple. More accurately, "the effect of a drop may be found by *subtracting* from the optical effect of the dimple in the wavefront, the effect produced by plane waves of light passing

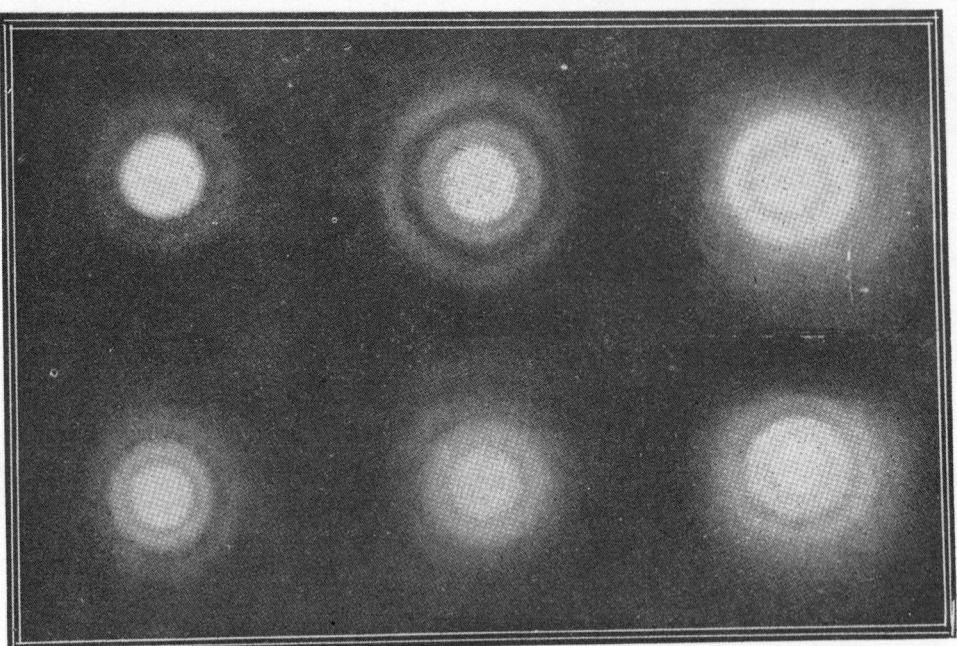

Figure 8.18 Coronae due to water droplets of different sizes. (After ref. [8.17].)

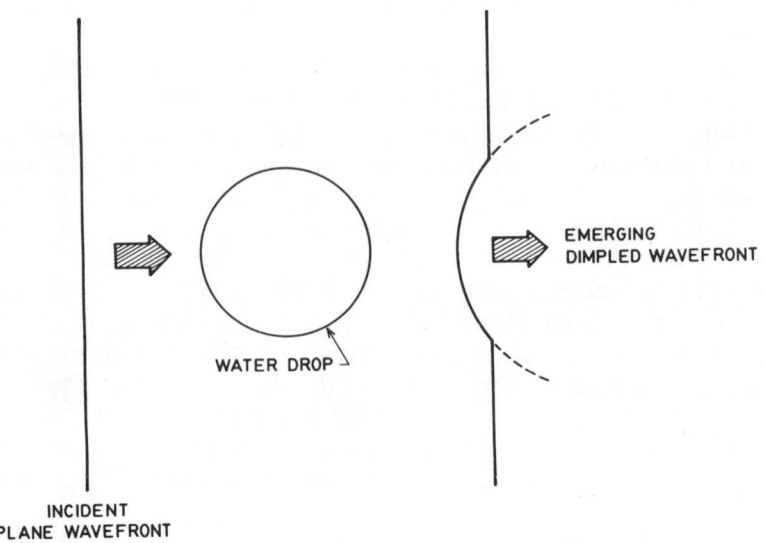

Figure 8.19 Modification of a plane wavefront on passage through a transparent water drop.

through a circular aperture of the same radius in an opaque screen". Balakrishnan restricted his attention to effects produced in the forward direction, and was able to explain the colours observed in the central part of the corona.

A more thorough treatment of this class of problems was later given by Ramachandran in a series of papers [8.18, 8.19, 8.20]. Not surprisingly, Lord Rayleigh had already touched upon some of these questions much earlier, albeit briefly; Ramachandran not only pushed the analysis further, but also added an elegant finishing touch of his own. To start with, he studied a problem not considered by Balakrishnan, i.e., light transmission through a cloud of opaque particles (like a smoke cloud for example). Clearly, the simplest approach to this problem is a geometrical one.

> Suppose that a beam of cross-sectional area A traverses a column of smoke of length l, and that N is the number of particles per unit volume of the medium. Assume further that the particles are all of the same size, and have a radius a. On a geometric basis, the effect of each particle would be to cut out a portion of the transmitted beam of area πa^2, and thus reduce the energy content by an amount corresponding to this area. In this way, each particle reduces the energy content of the beam by a certain fraction, and it would thus appear that if the number of particles n_0 be such that $n_0 \pi a^2 = A$, then the beam would be completely cut out. However, it is not so, on account of the fact that some of the particles would screen those behind them, and thus increase the chance of a portion of the direct radiation coming through. It is therefore a question of probabilities to determine what fraction of the energy is transmitted by the particles in the medium. In our case, the number of particles, n, is evidently $= NAl$.
>
> The problem at hand is identical with one in which n disks, each of radius a, are thrown at random on an area A, and it is required to find the probable area covered by the n disks. [8.18]

After calculating the probabilities, Ramachandran shows that the intensity of light transmitted through a column of length l is given by

$$I_l = I_i \exp(-\pi a^2 Nl), \tag{8.1}$$

where I_i is the incident intensity. Thus there is an exponential attenuation due to the cloud. It is of course a bit naive to assume that all particles have identical radii but a distribution in sizes is easily allowed for, leading to the modified formula

$$I_l = I_i \exp\left(-\frac{3\pi}{2} a_m^2 Nl\right), \tag{8.2}$$

where a_m is the most probable value of the radius.

The geometrical approach just discussed ignores an obvious fact, namely, the diffraction effect of the particles. Taking this into account, Ramachandran obtains (for the case of identical particles) the result

$$I_l = I_i \exp(-2\pi a^2 Nl). \tag{8.3}$$

Comparing with (8.1), one sees that the transmitted intensity is lower. As Raman puts it, diffraction effects cause "an extra loss of energy".

Ramachandran also pursued further Balakrishnan's analysis of the transparent drop. Whereas Balakrishnan had confined himself to the effects produced in the forward direction, Ramachandran was interested in the intensity observed in other directions as well. This required manipulating Balakrishnan's mathematical expressions into a form suitable for numerical calculations, a task which Ramachandran accomplished with consummate skill [8.19].

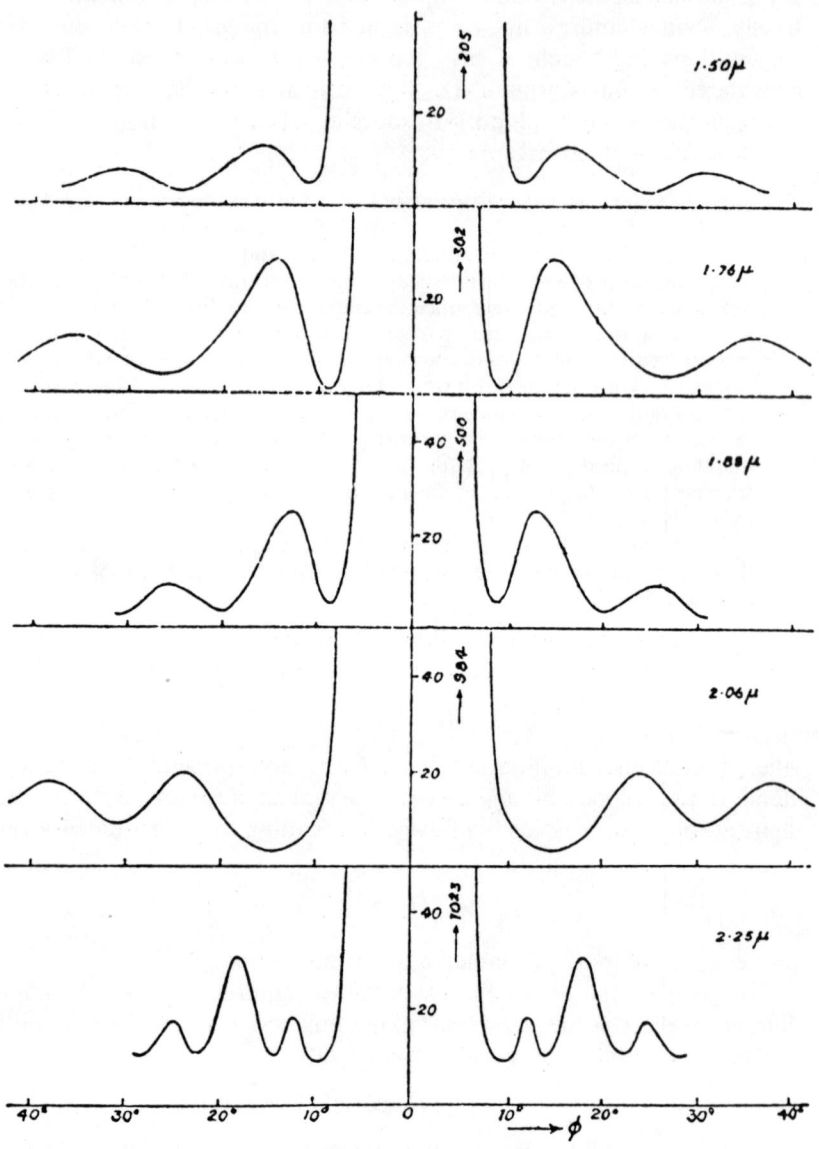

Figure 8.20 Computed intensity versus diffraction angle ϕ in the corona of clouds containing water droplets of various sizes. The peaks correspond to the different rings in the halo. (After ref. [8.19].)

Figure 8.20 shows the final outcome. The curves reveal how the nature of the ring system is altered by changes in the size of the drops. The contraction and expansion of the rings as the drop size is varied are quite evident. The theoretical results explained most of the experimental observations of the previous workers.

Perhaps the most interesting and significant of this series of papers is the one dealing with diffraction by randomly dispersed but *static* particles. Many years earlier, de Haas had attempted an explanation of the observed mottled structure of the corona (see Fig. 8.21). According to him, it was due to the cumulative effect of interference from every pair of particles. Of this theory, Ramachandran comments:

> The failure of the de Haas point of view may be brought out strikingly by the following analogy. Suppose we are considering the opposite case of a regular square diffraction grating, formed for example by a square mesh of wires. In this case, one may take each pair of meshes in the grating, and imagine it as giving rise to a set of interference fringes of the type imagined by de Haas. But such a method gives us no idea at all of what the nature of the diffraction pattern due to the complete grating would be. [8.20]

How then does one go about the problem? Raman argued that since the scattering particles are stationary in the lycopodium screen, there would be *definite* phase relationships between the waves scattered by them, and associated interference effects. Does this mean that the random distribution of particles has no effect? Actually, it does, and in a rather interesting way. To appreciate the last remark, consider Fig. 8.22 which shows the cumulative amplitude of light scattered by eight particles in different situations. All particles give rise to scattered waves with the same amplitude but the waves are different in phase, the latter depending on the direction of observation. One thus draws vectors to represent the various scattered waves in

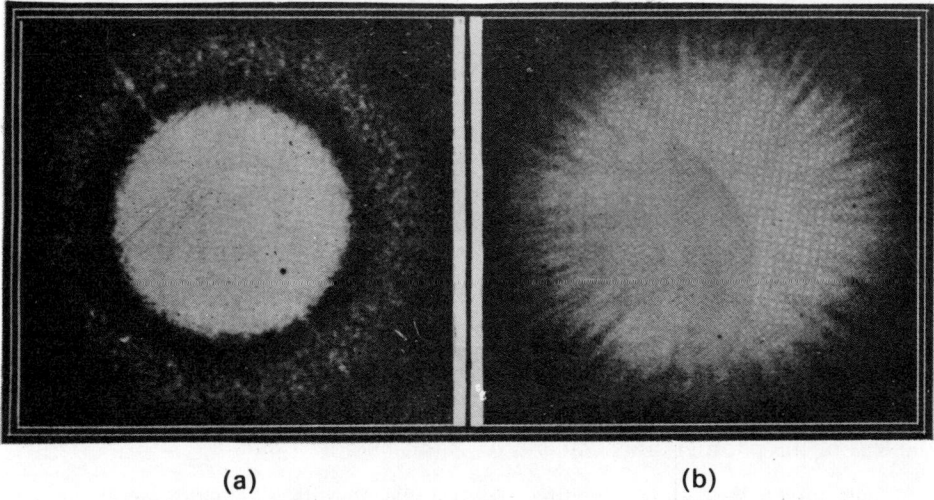

(a)   (b)

Figure 8.21 Diffraction corona due to lycopodium spores showing (a) granular structure in monochromatic light and (b) radial streaks in white light. (After ref. [8.20].)

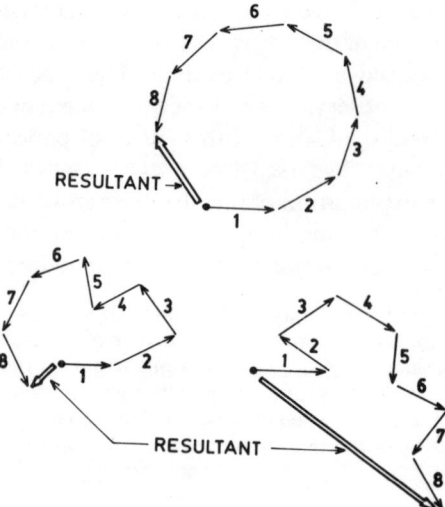

Figure 8.22 A scattered wave is characterized by its amplitude and its phase. Suppose we have eight such waves produced by eight particles whose positions are fixed in space. Depending on the position from which the constellation of particles is viewed, the eight waves can combine in various ways, some scenarios for which are shown here schematically. The resultant intensity is equal to the square of the resultant amplitude. As is evident, wide variations in the resultant intensity are possible.

amplitude and phase. As is evident, different possibilities can exist for the resultant amplitude (and therefore the resultant intensity). In the same way, with the lycopodium powder also, one has to consider the waves diffracted by all the particles together. Each particle is the source of a secondary wave, and in those directions in which the phase relations between the waves diffracted by the particles happen to be such that there is a large co-operative effect, there will be bright spots. Conversely, where there is cancellation there will be less intensity. It is easy then to understand that there will be a large number of such spots, irregularly arranged in the field of view.

Ramachandran carried out simple but elegant experiments to substantiate this argument. The source of light was a pinhole S illuminated by the 5461 Å radiation derived from a mercury lamp (see Fig. 8.23). At a certain distance from S was placed a glass plate G dusted with lycopodium powder. The resulting diffraction pattern could be brought to a focus on a plate C by a lens L. The two patterns shown in Fig. 8.24 were obtained with a small circular aperture and a triangular aperture respectively. As can be seen, each of the bright spots in the field is a focused image of the original source of light, formed by the joint action of the diffracting particles and the lens of the photographic camera. As Ramachandran puts it,

> Each point in the corona exhibiting an observable intensity is, therefore, essentially an optical image of the original source produced by the entire cloud of particles functioning as a randomly distributed secondary source of light.

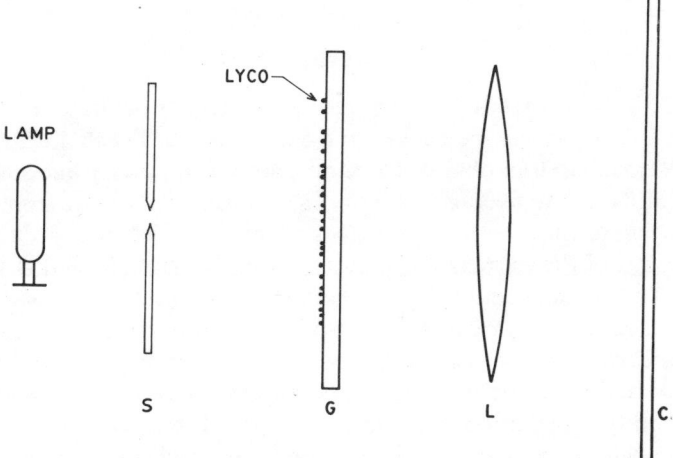

Figure 8.23 Experimental arrangement used by Ramachandran to study the corona due to a static distribution of particles.

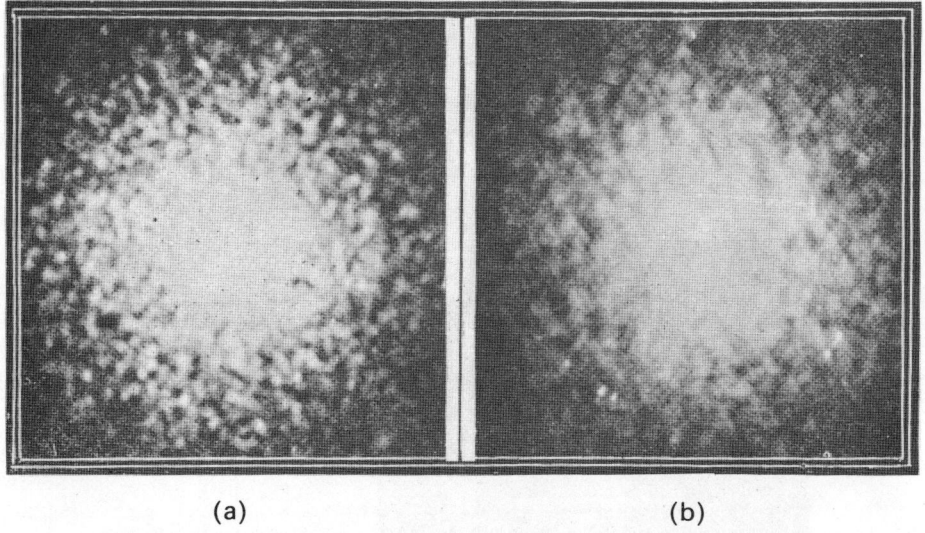

(a)   (b)

Figure 8.24 Central disc of the lycopodium corona observed in monochromatic light with (a) a circular pin hole and (b) a triangular aperture as source. (After ref. [8.20].)

All of this is of course still qualitative. To make it quantitative, Ramachandran appealed to an analysis due to Lord Rayleigh who had considered the general problem of the distribution of intensity due to N vibrations whose phases are random. Adapting from Rayleigh's work and writing $f = (I/N)$, where I is the intensity, Ramachandran showed that the probability distribution for intensity is given by

$$p(f)df = \exp(-f)df, \tag{8.4}$$

where $p(f)df$ is the probability that the resultant intensity lies between the fractions f and $f + df$. One expects the images in a corona to obey such a distribution. To verify this, Ramachandran obtained a photograph using a very fine source of light (see Fig. 8.25). Now the average intensity in a corona falls away from the centre. (Experts may note that the variation goes as $J_1^2(x)/x^2$, where $x = 2\pi a \sin(\phi)/\lambda$, a being the radius of the particles, λ the wavelength of the light and ϕ the angle of diffraction.) To take account of this fall, the part of the photograph where the spots were clearly seen was divided into five annular regions, as seen in the photograph. The spots in each annulus were then classified on an intensity scale by a suitable technique. Once the intensities of the individual spots were known, the construction of the probability distribution was straightforward. The final outcome is shown in Fig. 8.26 from which one sees that the Rayleigh statistical law of intensity variation is completely verified.

In the course of these investigations, an interesting phenomenon was observed.

> If one moves the screen containing the lycopodium powder keeping the eye fixed on the source, then the ring system is not found to undergo any change, but the fine-structure appears to move relative to the pattern of rings in the same direction as the motion of the screen. Vice versa, keeping the screen fixed, if one moves the eye, all the while looking at the source, then the fine-structure appears to move in a direction opposite to the motion of the eye.

These effects are geometrical in origin, and Ramachandran was able to explain them completely. They are in fact the forerunner of the speckles phenomenon

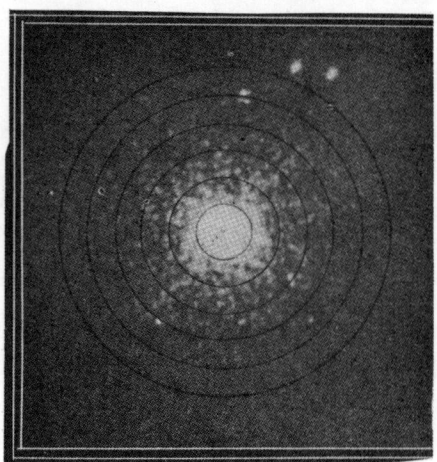

Figure 8.25 Corona with lycopodium spores. For data analysis, the spots were arranged into groups defined by the annular regions. Spots belonging to the same ring have the same geometrical factor $J_1^2(x)/x^2$ and must obey Eq. (8.4). (After ref. [8.20].)

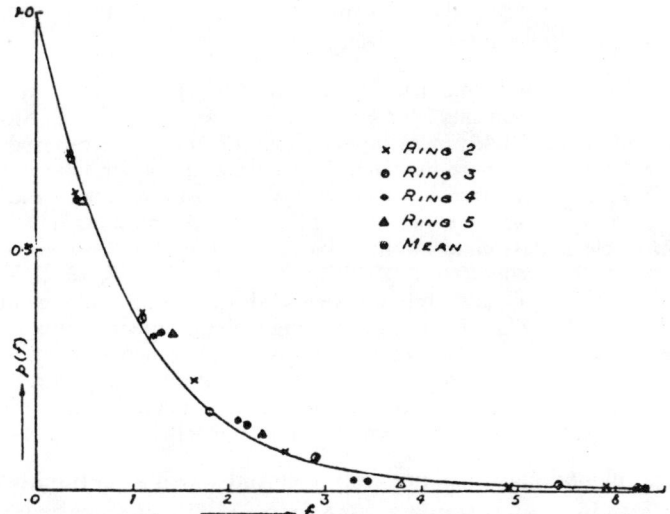

Figure 8.26 Verification of the Rayleigh law. Spots belonging to the various groups of the previous figure were analysed according to their intensities and were found to satisfy Eq. (8.4). (After ref. [8.20].)

discovered after the advent of the laser. When a smooth, say metallic, surface is illuminated by a laser beam, it will be observed to have a granular appearance. The field of view is filled with bright specks of light of varying intensity, making a fanciful display as the scatterer or the laser is moved. This is the speckles effect and arises due to random height variations in the reflecting surface of the order of a wavelength of the illuminating light. As in the lycopodium grating, every point on the rough surface produces a (scattered) wave, and the intensity variations observed by the eye are due once again to cancellations and reinforcements to varying degrees.

Going back to the corona, it is pertinent to ask why the results for the water vapour cloud and the lycopodium cloud are so different. Ramachandran explains:

> If the particles themselves are in motion, as in a gas or a cloud, then a continuous redistribution of phase takes place, and what one sees is the integrated effect of all these over a definite period. As Lord Rayleigh has shown, such a redistribution tends to make the resultant intensity (perceived as an average over a certain period) approach the average value n, this tendency being greater the larger the number of redistributions. Hence, for the light scattered by a gas, or a cloud of particles, one is quite justified in regarding the scattered intensity in inclined directions as equal to n times that due to a single particle.

An interesting case occurs when the motion is slow.

> If the motion of the particles is slow, then the alteration in the position of the spots in the fine-structure will also be slow, and can be observed. Such a slow random motion takes place in Brownian movement, so that it must be capable of detection by this technique.

Raman amplifies in his Baroda lectures under the heading "Observation of Brownian movements without a microscope".

> As is well known, the individual particles in colloidal suspensions and emulsions execute "Brownian movements", which are most lively when the particles are very small and are suspended in an inviscid fluid. For our present purpose, it is necessary to select a substance in which the particles are of fair size so that the coronal disk is of sufficient intensity and also exhibits a visible structure. Fresh milk is the most easily available material satisfying this requirement. When a little of it is flowed on to a clean glass plate and then allowed to drain away as completely as possible, a thin film remains firmly adherent to the plate. A small aperture illuminated by a mercury arc lamp and viewed through such a film exhibits an extended field of diffuse illumination surrounding it. Fixing attention over a limited area of the field, it is noticed that this exhibits a structure which is not static but is continually changing. Bright points of illumination continually appear in the field and others disappear. These changes become less rapid and ultimately stop when the film is dry; the structure of the field is then completely static[13]. [8.14]

The speckles phenomenon has not only received much attention, but has also found applications (see, for example, refs. [8.22, 8.23]). It is recognized that the net amplitude of the reflected field at every given point is the vector sum of a large number of small, independent contributions, and that the speckle effect is really an interference effect. This is precisely what Raman and Ramachandran said many years earlier, but few seem to know they did. The intensity of the bright spots in a speckle pattern was found to be distributed according to the Rayleigh law given by Eq. (8.4), a fact already verified by Ramachandran at least two decades before. And Isenor [8.24] found that when the scatterer was moved so did the specks, preserving an object–image relationship exactly as Ramachandran had observed (but alas without any knowledge of that fact)[14].

Is there a danger in making a discovery long before the time is ripe for it?

Light Scattering

After the marvellous success registered in Calcutta, it was only natural that light scattering received attention in Bangalore also. While Raman was not personally involved in most of these studies, he kept in close touch and provided the necessary inspiration.

The Bangalore studies fall into two basic categories: (i) Raman scattering experiments on crystals, and (ii) Brillouin and related scattering experiments, especially on liquids. The studies on crystals culminated in the Born–Raman controversy which will be dealt with in Chapter 10. Here we will mostly survey the work on liquids, starting with a brief review of some of the basics (the reader who has gone through Chapter 6 may find some of this material repetitive, though in diluted form).

Three things are of interest in a light scattering experiment: (i) the intensity, (ii) the polarization, and (iii) the spectrum. In the beginning only the first two aspects received attention (including in Calcutta). Following the discovery that the spectrum itself could be modified, questions relating to (i) and (ii) became sharpened. In other words, one now became interested in the intensity and polarization characteristics of each part of the scattered spectrum.

When light is scattered by a medium, frequency changes occur on account of the translational, rotational and vibrational motions of the molecules (see Fig. 8.27). The frequency change which Raman first detected is associated with molecular vibrations and is fairly large, being of the order of a few hundred cm^{-1} on the

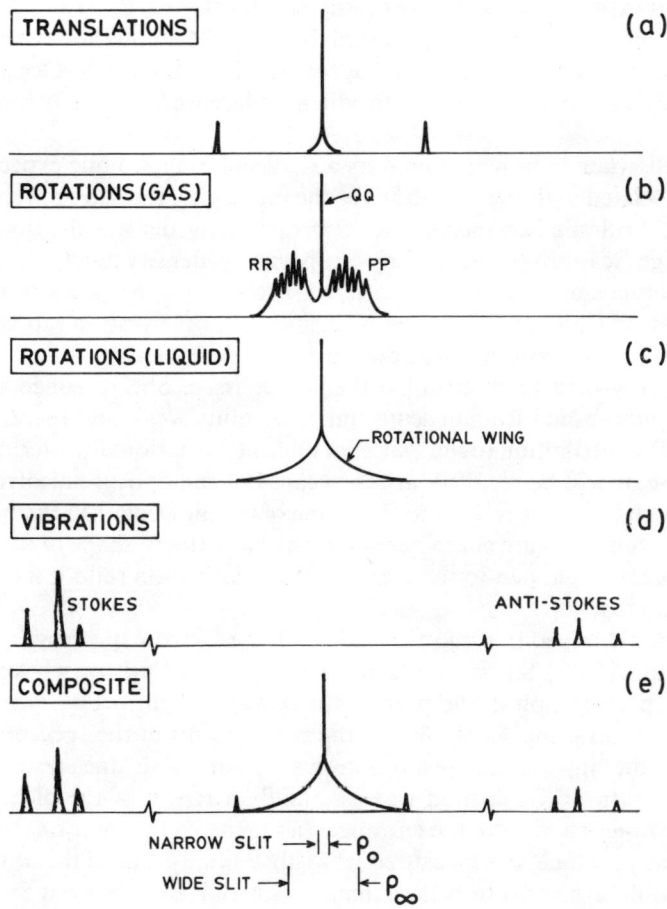

Figure 8.27 (a)–(d) Various components of the spectra of scattered light associated with translations, rotations and vibrations; (e) the composite spectrum. Bulk of the intensity of scattered light is associated with Rayleigh and rotational scattering. Below (e) are shown schematically the slit widths used at Bangalore to include or exclude the rotational wings as the case may be.

average. Soon after Raman's discovery, the rotational Raman effect was detected. The rotational spectrum consists of three branches, QQ which is unshifted in frequency, and PP and RR on either side of the central line. In gases under low pressure, the PP and RR branches reveal themselves as a collection of closely-spaced lines but at high pressures they appear as two humps adjoining the QQ branch (hydrogen is an exception). In liquids (hydrogen excepted) the rotational spectrum is once again smeared out, though extending slightly farther out. Also, there is no perceptible hump, the wings decreasing monotonically on either side from the QQ branch. On account of the smaller frequency shifts involved in rotational scattering, good spectral resolution is needed to observe it.

The translational movement of the molecules results in density fluctuations and one would thus expect it to contribute to Rayleigh scattering, i.e., scattering without frequency shift. However, this is only part of the story. Brillouin showed that density fluctuations can act as "moving mirrors" accompanied, of course, by Doppler shift in frequency. This is Brillouin scattering, to which a reference has already been made in Chapter 5.

In a nutshell, when light is scattered by a (molecular) liquid one expects Raman scattering associated with the vibrations of the molecules, Raman scattering due to the rotations, Brillouin scattering due to propagating density fluctuations, and finally Rayleigh scattering due to non-propagating density fluctuations. In the twenties the composite nature of the scattered spectrum was not known, and people (including those in Calcutta) measured intensities and polarization ratios assuming that all the observed scattering was due to Rayleigh scattering.

Sunanda Bai wished to re-examine the whole issue. She reasoned that while Brillouin and vibrational Raman scattering were quite weak and therefore would not make much contribution to the scattered intensity, rotational scattering was not. Depolarization, it will be recalled, arises because of the optical anisotropy of the molecules, which is in turn related to the shape of the molecule. On the other hand, the rotational wings also are characteristic of the anisotropic shape of the molecule. What, therefore, will happen to the measured depolarization ratio if the rotational wings are excluded?

Sunanda Bai examined this question using a large Littrow spectrograph capable of high resolution [8.25]. She scattered light exactly as the Calcutta workers did, but introduced a spectrograph in the path of the scattered light before measuring the depolarization ratio using nicols. When the entrance slit of the spectrograph was opened wide the instrument resolution was poor, with the result that the depolarization ratio ρ_∞ measured was essentially a repeat of the older measurements. On the other hand, with the entrance slit narrowed the rotational wings were clipped and the ρ_0 which was measured was substantially that of the unshifted line alone. One would expect ρ_0 to be less than ρ_∞ which is exactly what Sunanda Bai found. Typical results obtained by her are shown in Fig. 8.28. The difference $(\rho_\infty - \rho_0)$ was smaller the greater the viscosity of the liquid. With increase of temperature the difference $(\rho_\infty - \rho_0)$ was found to increase, consistent with the decrease of viscosity.

Sunanda Bai also extracted a quantity C_Q which is the ratio of the intensity of the

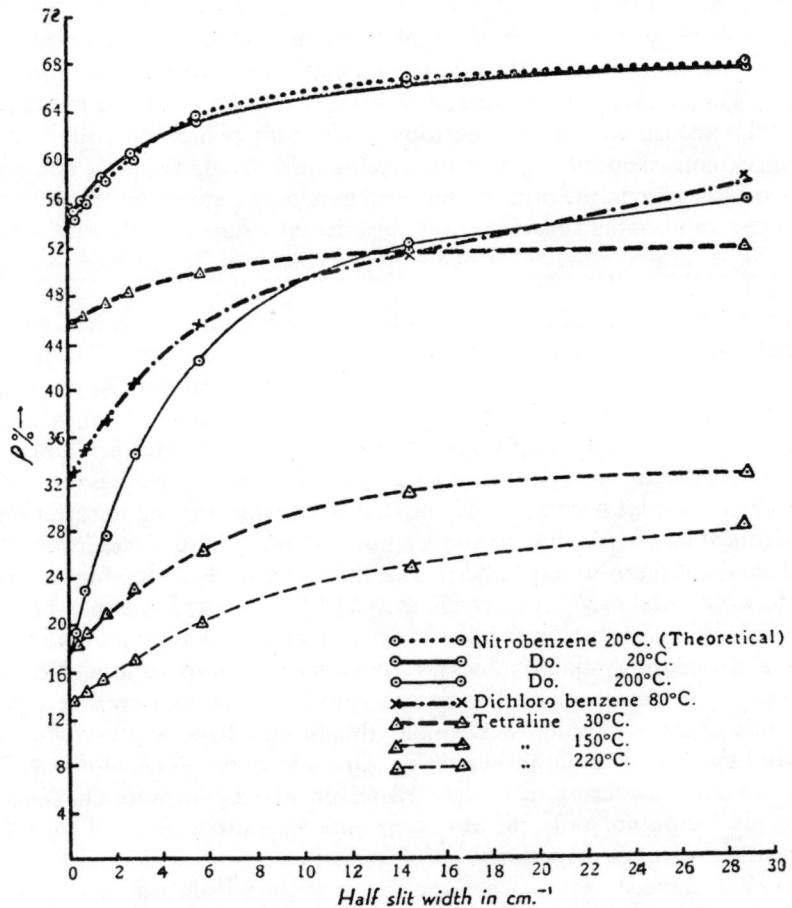

Figure 8.28 Typical results for depolarization obtained by Sunanda Bai. Zero slit width gives ρ_0 while large slit width yields ρ_∞. Sunanda Bai studied sixteen liquids, several of them at various temperatures. (After ref. [8.25].)

QQ branch to the total intensity of the rotational wing. Earlier, Placzek and Teller had derived a formula for this quantity assuming that the molecules in a liquid rotate freely. By comparing the measured values of C_Q with those computed using the Placzek–Teller formula, Sunanda Bai showed that "the assumption of complete freedom of rotation for molecules in a liquid is not valid". She also measured the shape of the rotational wings and proposed a model to explain it.

Newton had deduced that v the velocity of sound in a fluid is give by $v = \sqrt{E/\rho}$, where ρ is the density and E the elastic modulus. Newton observed that one must take the isothermal value, i.e., the value measured under conditions of constant temperature, for the elastic modulus. Laplace disagreed, and reasoned that when sound waves pass, any point fixed in space experiences, alternately, compression and dilation effects, the compression involving a generation of heat and dilation a cooling. But these opposite states follow each other too quickly for the temperature change to be transmitted from point to point and so for the temperature to be maintained constant everywhere. Hence, clearly, the isothermal modulus cannot apply. Laplace argued that one must instead use the adiabatic modulus, i.e., the modulus measured with the entropy S constant (which permitted temperature changes, consistent, of course, with S being held fixed). At first it was not obvious that one must go to the other extreme suggested by Laplace but Stokes pointed out that any appreciable departure from the adiabatic state would result in a stifling of the sound, which, however, is contrary to experience. So Laplace's viewpoint was upheld.

The question of adiabatic versus isothermal conditions is also relevant to the case of light scattering by liquids, for the scattering occurs because of fluctuations, and the question is whether they are isothermal or adiabatic. At a slightly more quantitative level, the expression for the scattering coefficient can be expressed in terms of the piezo-optic coefficient which measures the rate of variation of refractive index with pressure ($(\partial n/\partial p)$). One would like to know whether it is the isothermal or the adiabatic piezo-optic coefficient that is relevant for light scattering. Clearly, experiment had to provide the verdict, but first one needed values for the isothermal and adiabatic piezo-optic coefficients for the scattering liquids. Many measurements of the isothermal piezo-optic coefficients had been made but none of the adiabatic coefficients. This lacuna was filled by Raman and Venkataraman [8.26]. Sunanda Bai then used the available values for the isothermal piezo-optic coefficients and the adiabatic ones measured by Raman and Venkataraman to show that "for all liquids the bulk of the scattering is essentially due to an adiabatic process" [8.27].

We turn now to Brillouin scattering. Gross in the Soviet Union was the first to observe such scattering in liquids. However, his results were challenged, for he reported seeing not only the Brillouin lines but also several of their overtones. Repeating Gross's experiment, Cabannes failed to find the Gross components but claimed instead that the Rayleigh line itself was slightly frequency-shifted (towards the red). Meyer and Ramm and later Ramm confirmed the first Gross components (i.e., the Brillouin lines) but could not detect any red-shift of the Rayleigh line. It was at this stage that Raghavendra Rao entered the picture. Exploiting the high resolution capability of the Fabry–Perot interferometer, he not only confirmed beyond doubt the existence of the Brillouin lines in several liquids but also verified that the frequency shifts were indeed well predicted by Brillouin's formula (see Chapter 5) [8.28].

Brillouin scattering studies in liquids were later pursued intensely by Venkateswaran [8.29], particularly from the point of view of verifying various aspects of the Landau–Placzek theory (see Chapter 6). Venkateswaran was also interested in the influence of viscosity, and investigating glycerine he made an interesting discovery. The viscosity of glycerine at room temperature is so high that the sound waves postulated by Brillouin (i.e., the waves which are manifest as the practical level as "moving mirrors") are completely damped. No Brillouin peak is therefore expected but Venkateswaran observed a peak. Analysis however showed that this peak was due to *transverse* sound waves instead of the usual *longitudinal* sound waves. It was believed that liquids, lacking rigidity, do not sustain transverse waves in contrast to solids. But Venkateswaran's results on glycerine and castor oil showed that highly viscous liquids did exhibit some solid-like behaviour.

Brillouin scattering by crystalline solids also received attention, starting with the studies of Raman and Venkateswaran who were able to observe both the longitudinal and the transverse Brillouin peaks in gypsum [8.30]. Particular mention should be made of the research of R. S. Krishnan whose work on Brillouin scattering from diamond is a bench-mark [8.31, 8.32]. All these papers, and those on fluorescence, received notice after the revival of light scattering following the advent of the laser. Comparing the early Bangalore studies with later work done using lasers, one is struck by the quality of the former, especially when account is taken of the difficulties of experimentation in the pre-laser era.

Haidinger's Rings in Curved Plates

It is well known that interference figures exhibited by transparent plates are of two distinct kinds, typical examples being the familiar Newton's rings and the less-familiar Haidinger's rings (already introduced in Chapter 4). Whereas Newton's rings arise because of surface curvature, Haidinger's rings are observed (usually) in plane parallel plates, the rings arising because of varying inclinations of the incident light. It was generally held that plane-parallelism is a *sine qua non* for the observation of the Haidinger phenomenon, and one went even so far as to prescribe its observation as a test for plane-parallelism. Raman and Rajagopalan showed by elegant experiments that such a belief is incorrect, and that Haidinger interference patterns are observable in curved plates of uniform thickness but of arbitrary form [8.33]. Why not then look for such patterns in soap-bubbles which offer convenient specimens of curved films?

Raman and Rajagopalan did precisely that [8.34]. There is, however, a small problem, for the bubbles that one usually forms do not have uniform thickness. Soon after they are blown, there is a natural flow of liquid downwards causing the lower levels to gain in thickness at the expense of the upper. To prevent this accumulation of fluid at lower levels, Raman and Rajagopalan employed a simple trick. A bubble was first blown and allowed to sit on the circular end of a vertical glass tube. Two *very gentle* currents of air were then blown upwards from two glass tubes placed below the level of support of the bubble and displaced from it in two directions 90° apart. The effect of the air currents impinging at an angle on the surface of the bubble is to set up an upward drift of the liquid within the film. The resulting circulation soon results in the thickness of the film becoming the same everywhere. By controlling the air jets, the bubble was held steady and prevented from going into oscillations. The effectiveness of this technique may be readily judged from Fig. 8.29.

Raman and Rajagopalan gave a simple theoretical argument to establish that the observed patterns are indeed of the Haidinger variety. Interestingly, one seems to

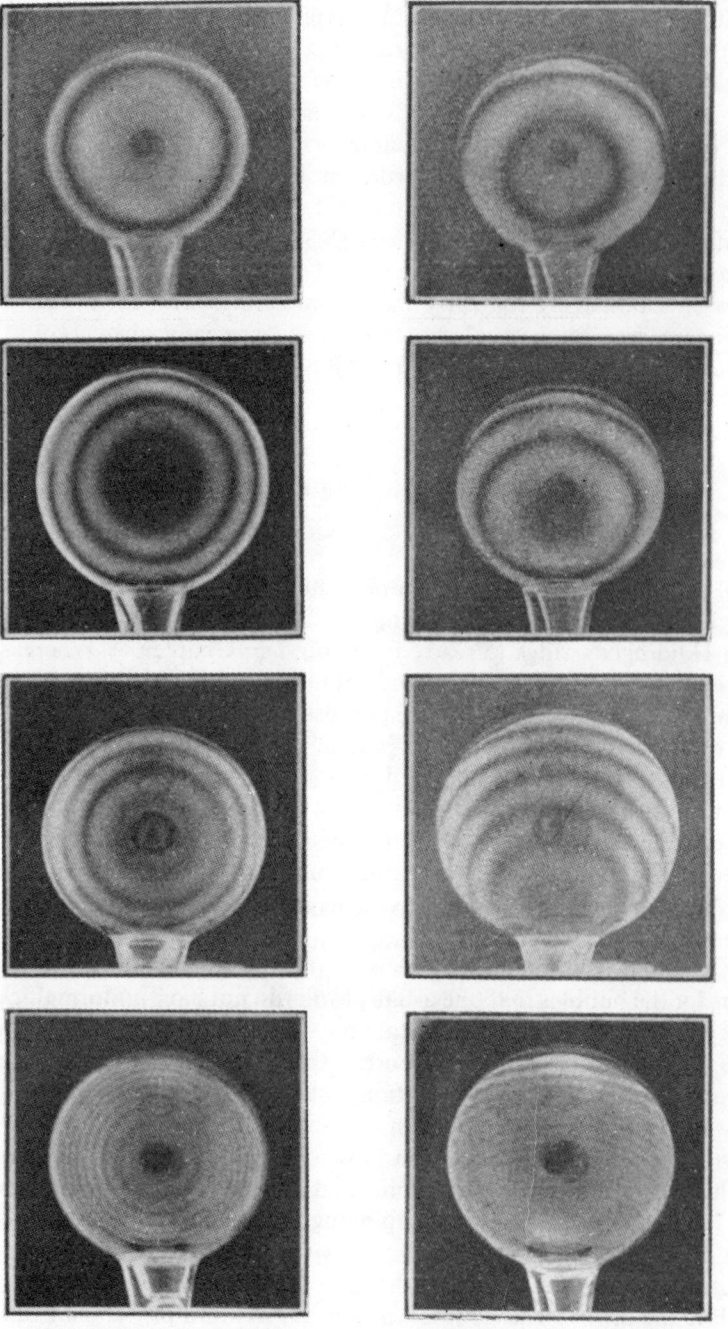

Figure 8.29 Interference fringes in soap-bubbles by reflected light. (After ref. [8.34].)

observe Haidinger's rings even in non-uniform films, prompting Raman and Rajagopalan to make a separate study of Haidinger's rings in non-uniform plates using fragments of spherical bulbs blown from soft glass [8.35].

The noted American physicist Willard Gibbs had earlier observed that the changes occurring within a soap film (formed by immersing a wire ring into a soap solution) are due not so much to the action of gravity as to the suction exercised by the ring of liquid formed along the line where the film meets its solid supports. Raman and Rajagopalan comment:

> Indeed it appears that in the case of a plane film this suction is the agent principally responsible for its thinning down. It would, therefore, seem to be important to investigate whether this is the case also for a spherical bubble when the perimeter of its support is reduced to the absolute minimum necessary. [8.34]

The technique of Haidinger's rings offers a means of studying the problem but Raman and Rajagopalan did not go beyond pointing the way.

Liesegang Rings

Around the turn of the century, Liesegang discovered that spatially periodic precipitation could occur when certain chemical reactions take place in gels. As it was of interest to the geologist and the biologist, the Liesegang pattern, as it was called, received much attention.

Liesegang rings may be quite easily demonstrated, e.g., by the precipitation of silver chloride in a gel. One starts with good quality gelatin and allows it to swell in water. The gelatin is then dissolved in hot water, treated with a small quantity of sodium chloride, filtered and poured on a glass plate. After it sets the gel is treated with a drop of concentrated silver nitrate solution, and the plate is covered to prevent photochemical action. If the plate is examined after 24 hours, one sees a precipitate of silver chloride. While to the naked eye the precipitation would appear continuous, under a microscope one would find very fine rings arranged in a regular pattern.

In 1936 Subba Ramaiah started a study of the Liesegang rings. Besides rings of silver chloride in gelatin, Subba Ramaiah also studied silver chromate rings in gelatin and lead iodide rings in agar [8.36]. Figure 8.30 shows some representative photographs of the patterns obtained by him and their fine structure. Subba Ramaiah also made microphotometric records of the ring structure (being one of the first to do so), and two examples of the records obtained by him may be seen in Fig. 8.31.

Why do such patterns occur? The answer was not precisely known but there was a speculation that "diffusion waves" were responsible for the pattern. Raman and Subba Ramaiah felt that

> the analogies between a wave and a periodic precipitate would be without physical content unless it can be shown by investigation that superposition effects can be observed in periodic precipitates analogous to interference and diffraction phenomena in acoustics and optics. Indeed, the existence of such superposition effects, if established, would enable the wave-like character of Liesegang precipitates to be regarded as an established fact besides giving it a real physical significance. [8.37]

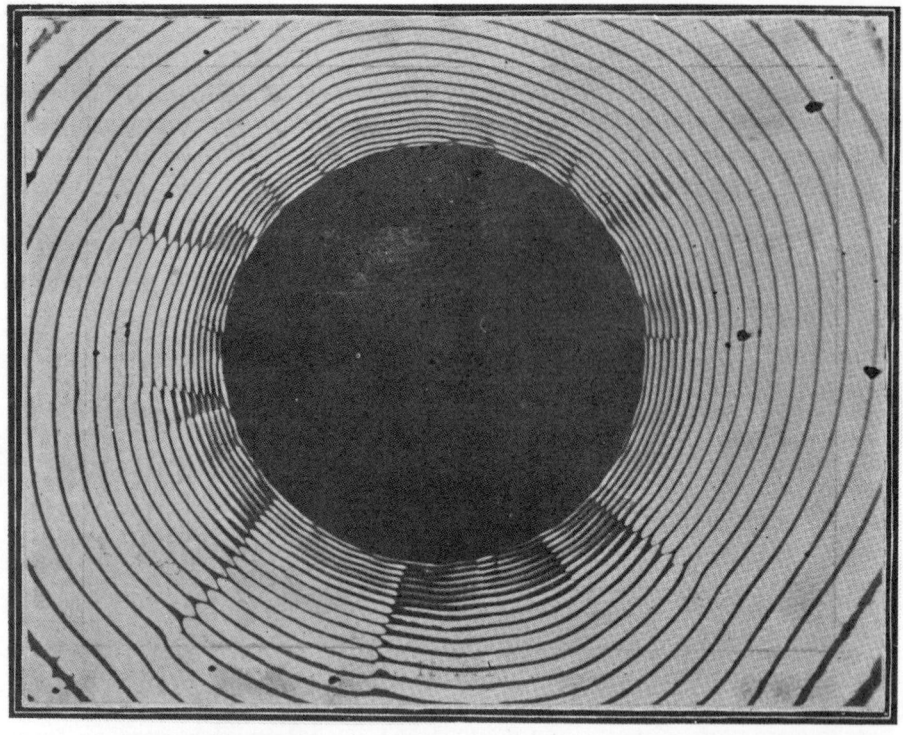

(a)

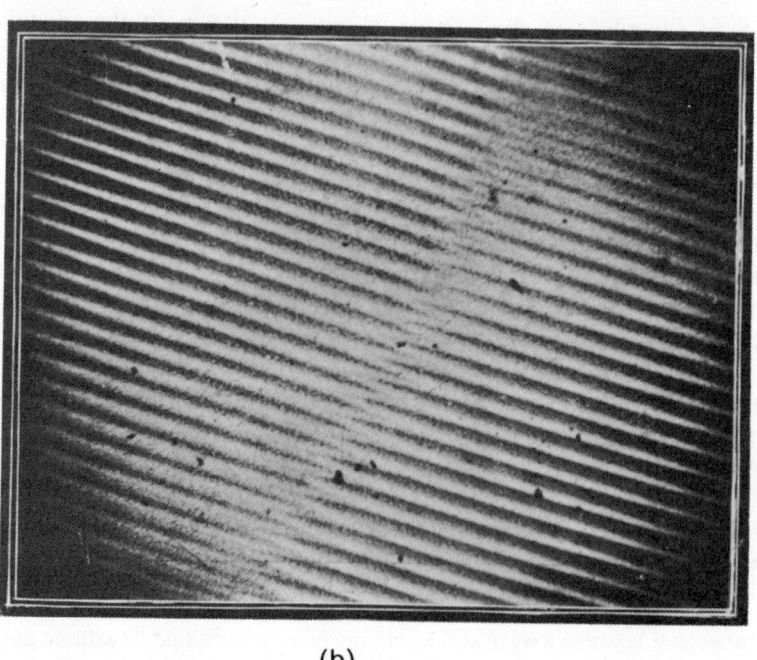

(b)

Figure 8.30 (a) A Liesegang pattern; (b)–(d) magnified portions of Liesegang patterns, showing examples of fine structure. (After ref. [8.37].)

On to Bangalore 325

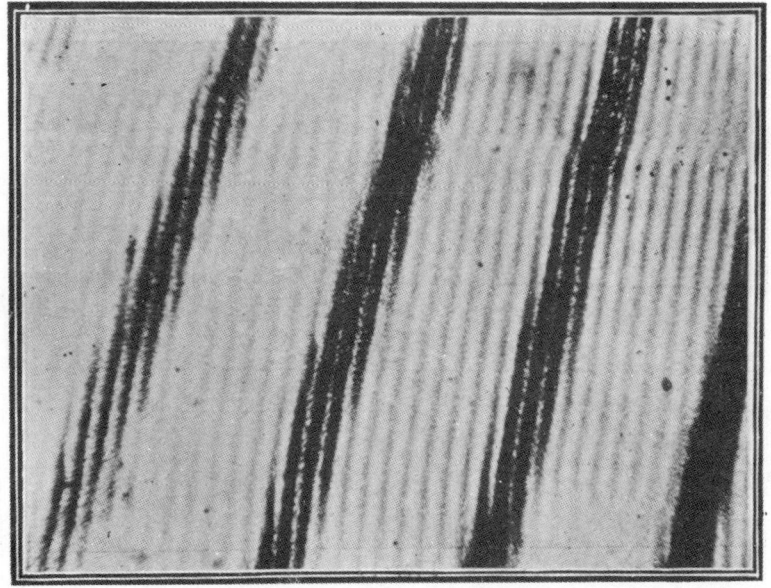

(c)

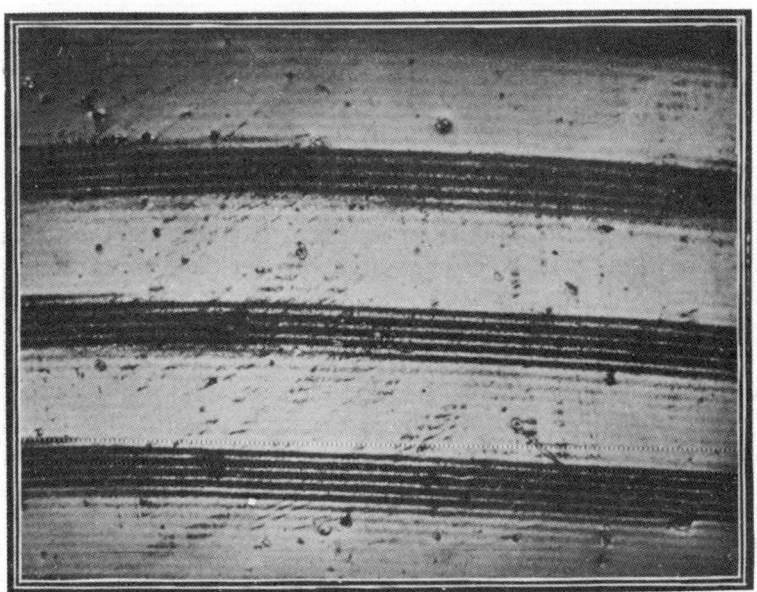

(d)

Figure 8.30 (*Contd.*)

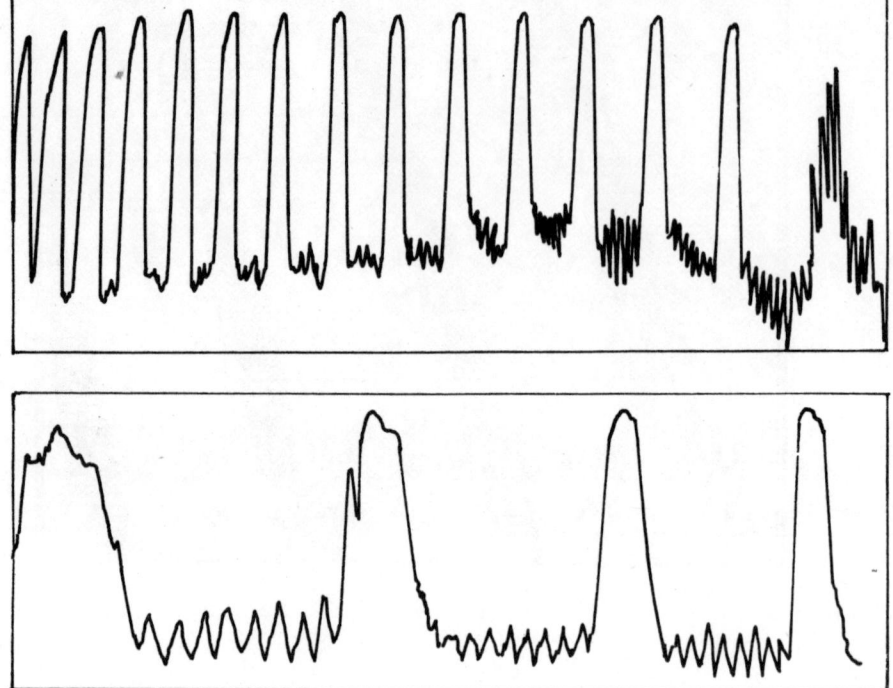

Figure 8.31 Microphotometric records of fine structure in Liesegang patterns. (After ref. [8.36].)

Raman and Subba Ramaiah then explored various scenarios of wave superposition like two intersecting wave-trains of same wavelength and amplitude, two intersecting wave-trains of different wavelengths and equal amplitudes, etc. The interference effects to be expected in situations like these are conveniently studied, for example, with capillary waves on water. Figure 8.32 is a drawing made by Raman and Subba Ramaiah of one such pattern observed many years earlier by Goverdhan Lal Datta in Calcutta. If the Liesegang pattern formation is driven by waves, one should expect similar structures and indeed this appears to be the case if one examines the two photographs in Fig. 8.33. The first shows a precipitation pattern started by two circular drops of silver nitrate running into an oval figure while the second shows a pattern due to two separate drops separated by an interference dead space. Further confirmation of the underlying wave-like character was obtained by manipulating the conditions of experiment to simulate the variation of the wavelength, the direction and the amplitude of the two participating wave-trains either individually or collectively.

The rhythmic precipitation one sees in the Liesegang pattern is one example of pattern formation that occurs under non-equilibrium conditions. In Raman's days, the origin of such spatial order in chemical reactions was a mystery. There was,

On to Bangalore

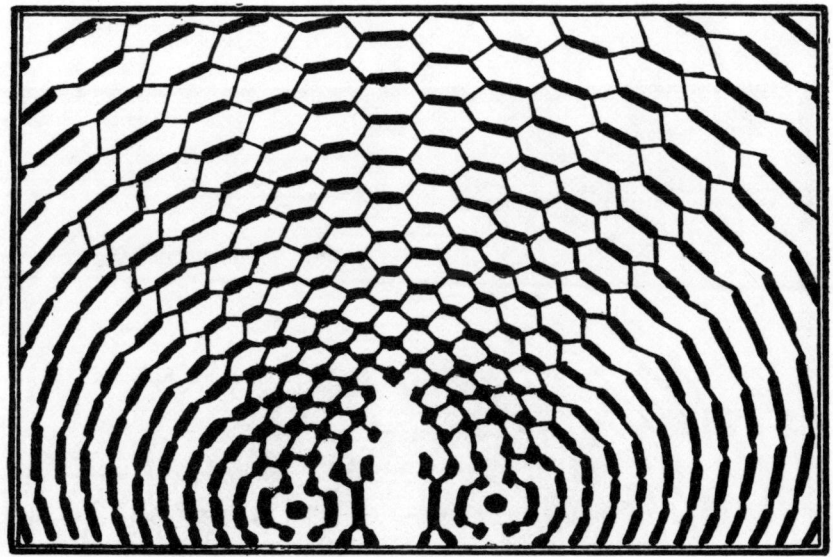

Figure 8.32 Drawing from a ripple photograph. (After ref. [8.37].)

however, a restricted but specific question to be answered, namely, whether there was an underlying wave-like character to the pattern formation. Raman and Subba Ramaiah convinced themselves that there indeed was. Today one is in a much better situation and can not only mathematically model the chemical process leading to spatial order but also simulate such patterns using a computer. Indeed, the modern theory does involve wave-like fluctuations. The tremendous progress achieved recently in this field may be discerned by glancing through the several volumes on synergetics which have appeared lately [8.38].

The Raman–Nath Theory

Perhaps the most important contribution made by Raman during his Institute period is the Raman–Nath theory. Considering the impact this theory has subsequently made on applied acousto-optics, one might well rank it next in importance only to the Raman effect.

The association of Raman with Nagendra Nath, the other personality involved in this theory, began almost by chance. Young Nath, a twenty-year-old mathematics graduate from Bihar, had just secured admission to the electrical technology course offered by the Institute, and was all set to become an electrical engineer. He was consumed with curiosity to see the Director who was by then almost a legend. To

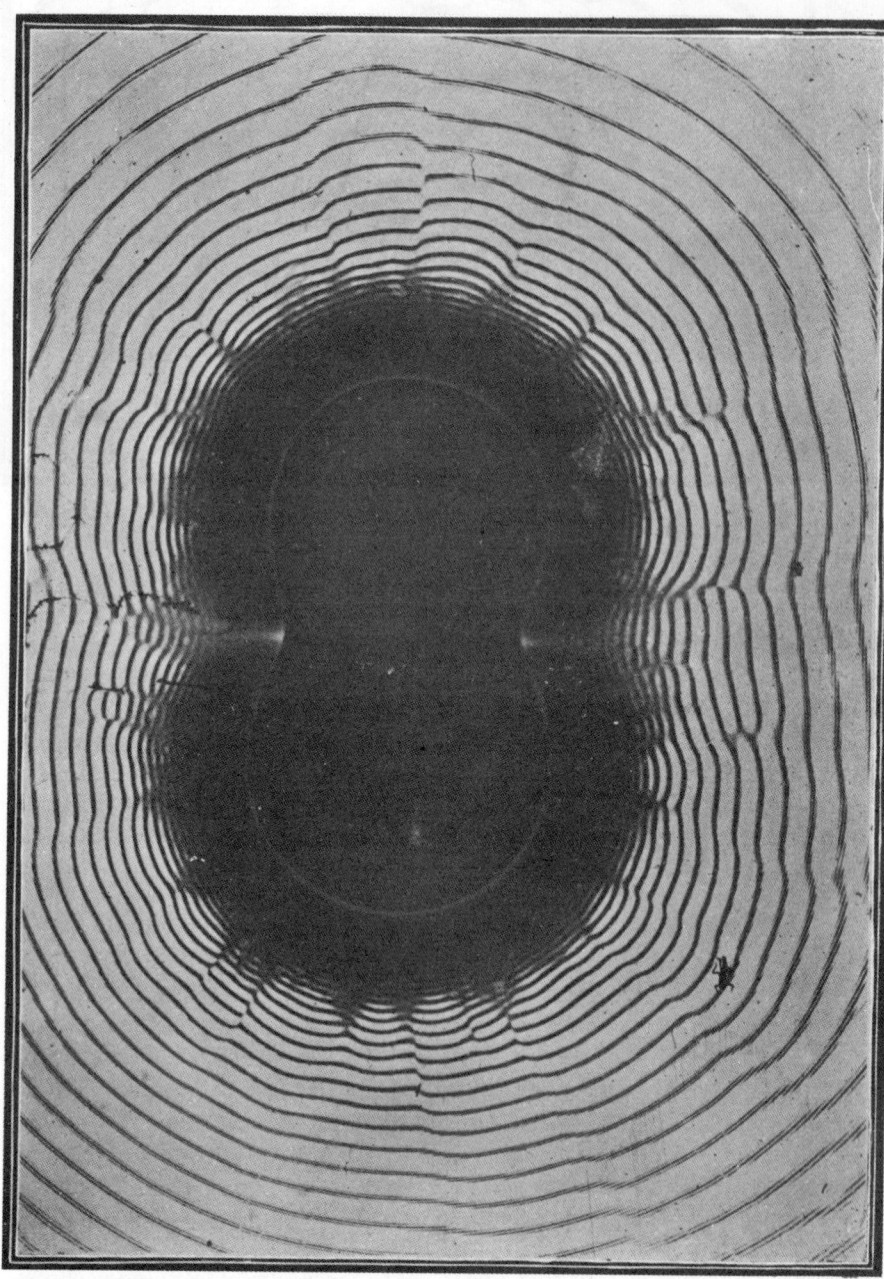

Figure 8.33a Liesegang precipitation pattern started with two merging drops. (After ref. [8.37].)

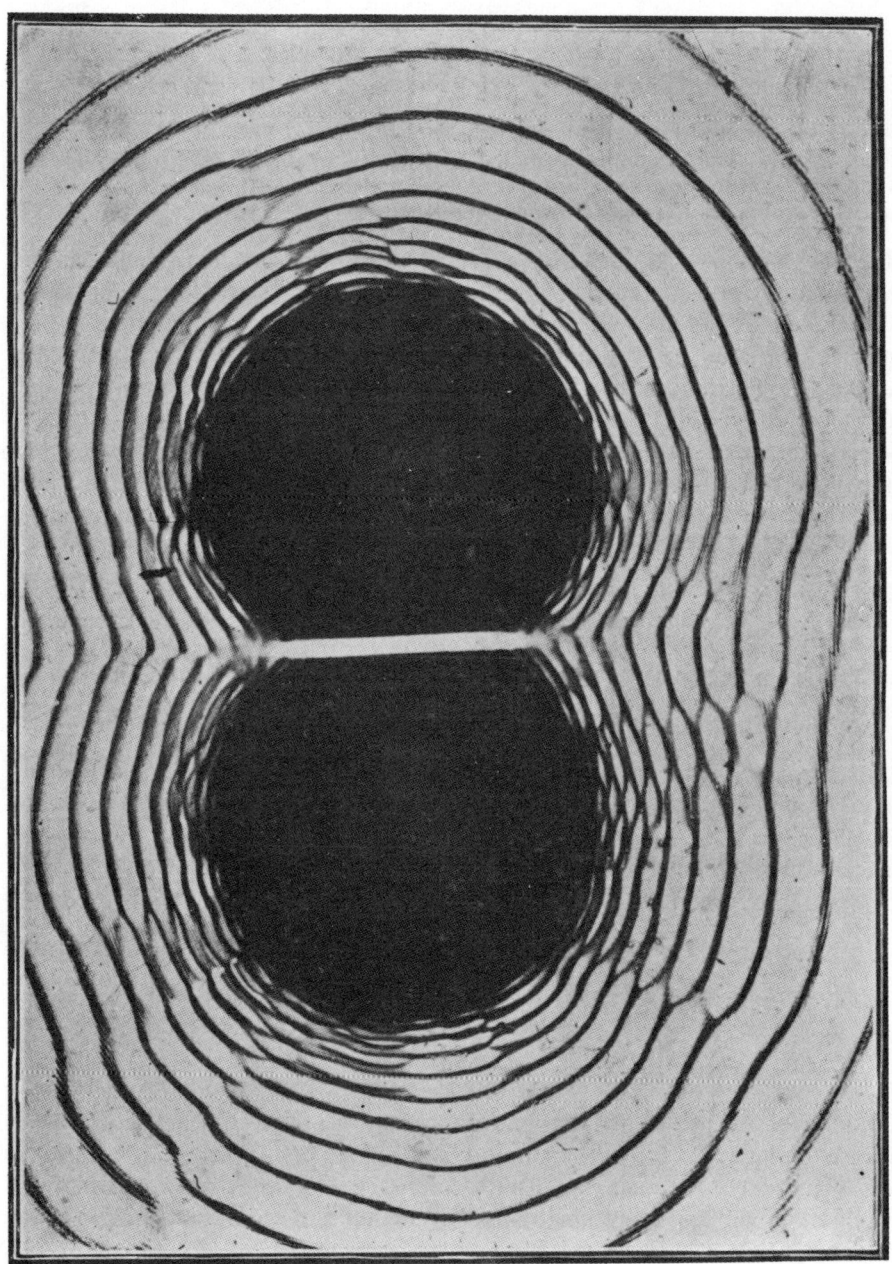

Figure 8.33b Liesegang precipitation pattern started with two separated drops. (After ref. [8.37].)

arrange a meeting with Raman, Nath concocted an excuse. As he himself described it later:

> I said to myself, why not meet the Nobel Laureate on the plea of becoming his research scholar! I had no hope that I would get admission in the Physics Department as a research scholar as I had graduated in mathematics. After a thorough interview lasting till the evening, he stood up and patted me and said that I could join the department as his research scholar. I then told him my plight that I had already secured admission in the Electrical Technology Department. He appeared to feel annoyed but then he said that I had to give up the admission in the E. T. Department. I said that I would very gladly do so.... [8.39]

Nagendra Nath quickly acquired a reputation for his talents in mathematical physics, winning praise from, among others, Born. His first significant achievement was his explanation of the so-called Raman line in diamond which delighted Raman so much that he immediately proposed Nath for election to the Academy (see Sec. 8.9).

At that time Parthasarathy was trying to measure the velocity of sound in liquids by the technique of diffraction of light by ultrasonic waves. One day he made a presentation discussing his experiments as well as his results. It became evident during the seminar that not much was known about how light is diffracted by sound waves. There was a theory due to Brillouin but it was clear that Brillouin's theory was not adequate. Writing about the seminar Nath says:

> Professor went to the board and said that the theory should be developed in a different way. A sound wave creates compressions and rarefactions. A light beam would be slowed in the region of compression and it would move faster in the region of rarefaction, and so a plane wavefront would become corrugated like a zinc sheet used for building purposes. Professor said that an analysis of this corrugated wavefront would explain the unexplained results. [8.39]

Raman was basing his thoughts on some ideas expressed by Lord Rayleigh. Nath promptly went to work and the very next day produced a mathematical formulation of Raman's ideas. This led to the first of the several papers on the subject. The scientific content of the Raman–Nath theory being of some importance, we devote a separate chapter to it later.

Soft Modes

In Chapter 7 it was mentioned that one of the important uses to which laser Raman spectroscopy has been put is the study of soft modes [8.40]. In the literature, the genesis of the soft mode concept is usually credited to Cochran and to Anderson but it is a little known fact that Raman and Nedungadi had discovered the soft mode some twenty years earlier during their spectroscopic investigations of quartz [8.41, 8.42]. Their experiment was undertaken not only as a part of a larger programme to study the Raman spectra of various crystals, but also because quartz exhibits a phase transformation at 575°C from a structure with trigonal symmetry (α-phase) to one with hexagonal symmetry (β-phase). Although the phase change itself is quite abrupt, it is preceded by many precursor signals.

> The thermal expansion coefficients, for example, gradually increase over this range of temperature [200°C–575°C], becoming practically infinite at the transition point and then suddenly dropping to small negative values. Young's moduli in the same temperature range fall to rather low values at the transition point and then rise sharply to high figures. The piezo-electric activity also undergoes notable changes. [8.41]

It was therefore of interest to investigate whether the Raman frequencies of quartz showed any changes with temperature.

The experiments were performed using a clear cylindrical piece (5 cm long and 2.5 cm diameter) supplied by a firm in Tanjore. Raman spectra were recorded at several temperatures from that of liquid air to 530°C. Of the three prominently observed Raman lines, the one corresponding to a lattice vibrational mode of frequency 220 cm^{-1} (at liquid air temperature) behaved "in an exceptional way,

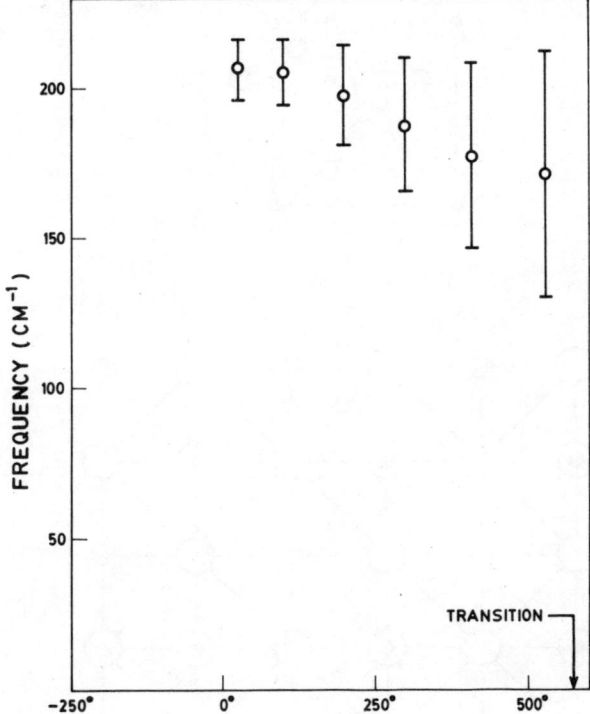

Figure 8.34 Temperature variation of the soft mode in quartz. Vertical lines show breadth of the Raman line corresponding to 207 cm^{-1} at 25°C and 220 cm^{-1} at liquid air temperature. The graph is plotted from the data presented by Nedungadi [8.42]. Nedungadi remarks that measurements were particularly difficult at the higher temperatures. He also notes that compared to the other modes, the mode giving rise to the Raman line corresponding to 207 cm^{-1} showed anomalous broadening.

spreading out greatly towards the exciting line and becoming a weak diffuse band as the transition temperature is approached". Figure 8.34 shows the temperature variation of the mode frequency. By contrast, the other Raman lines do *not* show such a behaviour. This prompts Raman and Nedungadi to observe,

> The behaviour of the 220 cm^{-1} line clearly indicates that the binding forces which determine the frequency of the corresponding mode of vibration of the crystal lattices diminish rapidly with rising temperature.

(In fact Nedungadi, in a subsequent paper, even asserts that the binding forces contributing to this mode actually *vanish* at the transition temperature.) Raman and Nedungadi conclude by saying:

> It appears therefore reasonable to infer that the increasing excitation of this particular mode of vibration with rising temperature and the deformations of the atomic arrangement resulting therefrom are in a special measure responsible for the

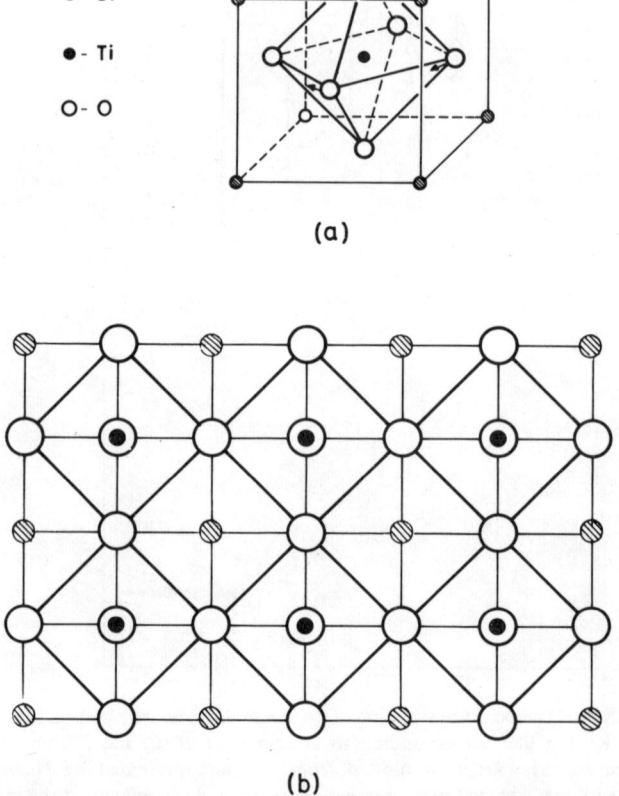

Figure 8.35 (a) Unit cell of SrTiO$_3$. The arrows show the movements of O atoms during the angular oscillations of the TiO$_6$ cage. (b) Projection of the structure on the xy plane.

On to Bangalore

remarkable changes in the properties of the crystal already mentioned, as well as for inducing the transformation from the α to the β form.

Though they did not use the specific term, there can be no question that Raman and Nedungadi not only discovered the soft mode but also understood its significance.

Phase transition was then not yet a frontier topic in physics, and for two decades the discovery of Raman and Nedungadi remained unnoticed. Towards the late fifties, Cochran and Anderson independently proposed that phase transitions leading to the onset of ferroelectricity occurred via the softening of a lattice-vibrational mode in the high-temperature phase. Neither was aware that a linkage between soft mode and structural phase transition had already been established in Bangalore much earlier. Following the publication of Cochran's paper, his student Cowley demonstrated the existence of soft modes using the technique of neutron scattering. While it was hailed as a new discovery, it was in fact a rediscovery!

Strontium titanate ($SrTiO_3$), the system Cowley studied, offers a convenient example for amplifying the concept of the soft mode. Figure 8.35 shows the (cubic) unit cell of $SrTiO_3$ at room temperature. The material has the so-called perovskite structure, and of interest is the disposition of the TiO_6 cage. When cooled to $-165°C$, $SrTiO_3$ undergoes a structural phase transition, lowering its symmetry from cubic to tetragonal. However, long before this change of structure occurs, there is an early warning in the shape of a soft mode.

$SrTiO_3$ has many modes of vibration, but the one which goes soft is associated with the angular oscillations of the TiO_6 cage (see Fig. 8.36). As the transition temperature T_c is approached from above, the oscillations become more and more sluggish (owing to the weakening of the restoring force associated with oscillations), until at the transition temperature the cages become locked into set positions. There

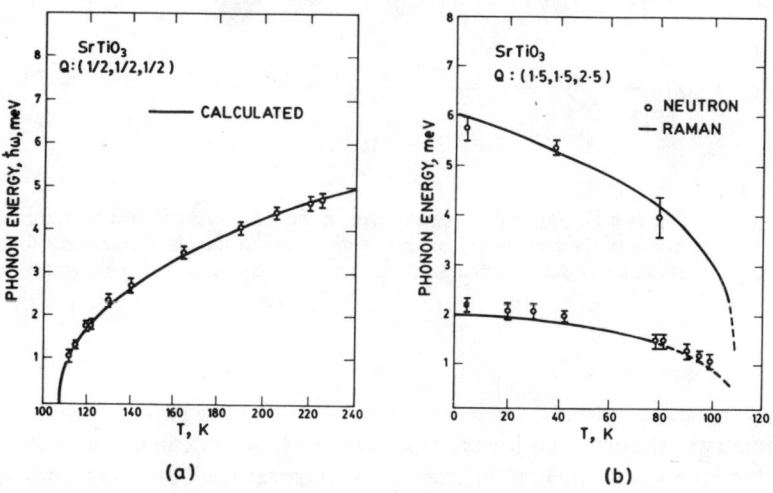

Figure 8.36 Temperature variation of the soft mode frequency (a) above T_c and (b) below T_c. The soft mode above T_c is triply degenerate; below T_c there is one non-degenerate mode and one doubly degenerate one. (After ref. [8.43].)

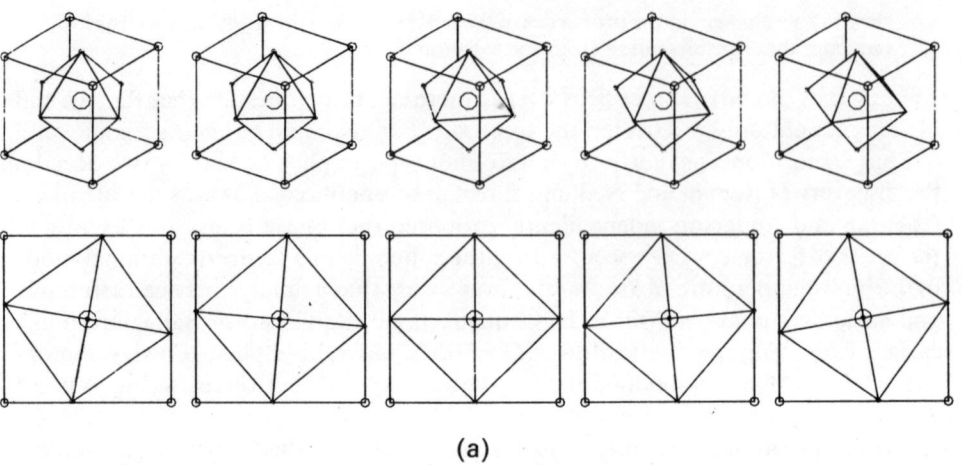

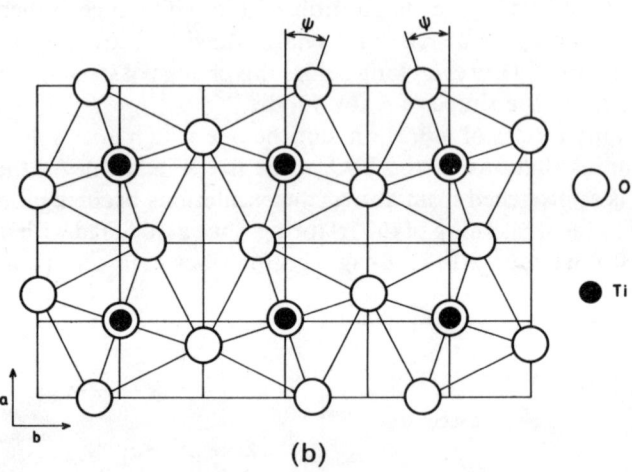

Figure 8.37 (a) Angular oscillations of the TiO$_6$ cage. Shown here are various orientations of the cage as well as their projections. At the phase transition, the cages lock in rotated positions, leading to the structure shown in (b). Compare with Fig. 8.35b.

is now a new equilibrium orientation for the TiO$_6$ cage, as a result of which the symmetry of the crystal is lowered (see Fig. 8.37). Attention is drawn to the fact that soft modes also exist below the transition temperature. This is not unexpected, for if the structure change and the lowering of symmetry are due to the "condensation" of a soft mode above T_c, then correspondingly there must be agencies below T_c which seek to restore the symmetry of the high-temperature phase as the temperature is

On to Bangalore

increased from below T_c. Thus it is that as many soft modes appear below T_c as exist above it (see Fig. 8.36). Raman and Nedungadi discovered the soft mode of quartz operative below T_c.

Apart from the general indifference to their discovery, it is a pity that even many of the papers on the structural transformation in quartz published in the seventies failed to make references to the Bangalore work. However, it is gratifying that the basic discovery at last finds mention in a book (ref. [8.44]).

☆ The soft mode concept is conveniently set in the framework of the well-known Landau theory of phase transitions. Central to this theory is the concept of an order parameter ψ (say) which is zero in the disordered (i.e., high-temperature) state but is non-vanishing in the ordered (i.e., low-temperature) state. For simplicity, let us assume ψ to be a scalar. In terms of this quantity, the free energy can be expressed as

$$F(\psi) = F_0 + \alpha\psi^2 + \beta\psi^4, \quad \beta > 0,$$

in the vicinity of the transition temperature T_c, F_0 being a reference level. The coefficient α is a smooth function of T in the neighbourhood of T_c, with a leading

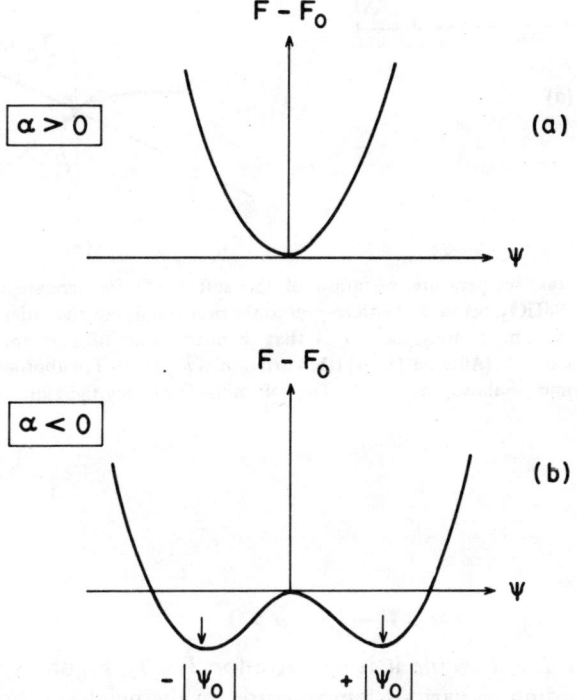

Figure 8.38 Variation of the free-energy with the order parameter.

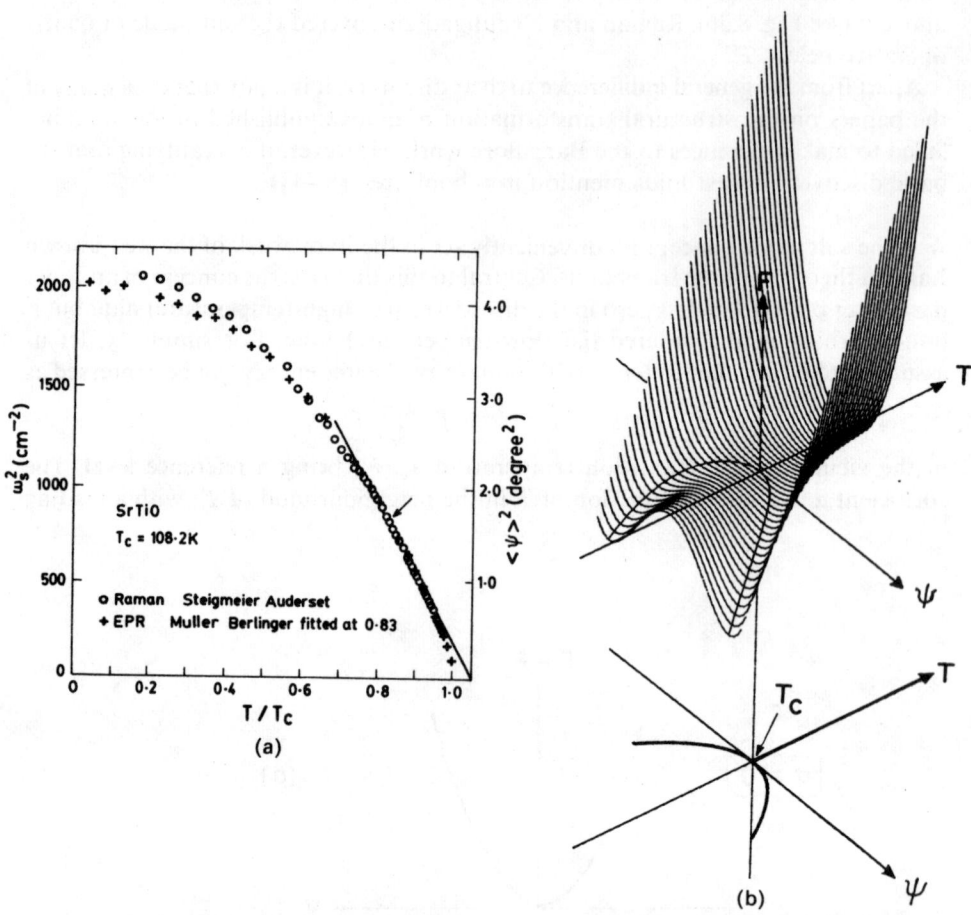

Figure 8.39 (a) Temperature variation of the soft mode frequency and the order parameter in $SrTiO_3$, below T_c. With respect to the previous figure, the order parameter is ψ while the soft mode frequency ω_s is that of fluctuations of ψ in the well in the neighbourhood of T_c. (After ref. [8.45].) (b) Variation of $F(\psi)$ with T; α approaches zero and the well becomes shallower as $T \to T_c$. The soft mode frequency therefore decreases.

behaviour

$$\alpha \approx a(T - T_c), \quad a > 0.$$

Thus α is positive for $T > T_c$ while it is negative for $T < T_c$. Figure 8.38 shows a sketch of F corresponding to various temperatures in the neighbourhood of T_c.

The thermodynamically stable state ψ_0 of the system is found by minimising F

with respect to ψ. The stability conditions are

$$(\partial F/\partial \psi)_{\psi_0} = 0, \quad (\partial^2 F/\partial \psi^2)_{\psi_0} > 0.$$

A simple calculation shows that for $T > T_c$, $\psi_0 = 0$ is the stable state whereas for $T < T_c$, stable states with non-vanishing values for the order parameter are possible, as Fig. 8.38 illustrates.

The simplified treatment sketched above no doubt explains, from free energy considerations, why a transition can occur, but does not envisage fluctuations of the order parameter. In fact ψ does fluctuate, and it is the fluctuations around ψ_0 which manifest as the soft mode in the low-temperature phase (see Fig. 8.39). Notice that instead of a lattice vibrational mode, one is now talking about fluctuations of the order parameter. This is obviously a generalization.

Schneider et al. [8.46] offer further insight into the relationship between the static and the dynamical aspects of (second-order) phase transitions. To understand their work, a few definitions are necessary. First we consider the static susceptibility $\chi_{\psi\psi}$ with respect to the order parameter, defined operationally by considering a small external field $V_{\text{ext}}(\mathbf{r}) = V_0 \exp(i\mathbf{q} \cdot \mathbf{r})$ which couples to the local order parameter $\psi(\mathbf{r})$. In the spirit of the well-known linear-response theory, the response $\delta\psi(\mathbf{r})$ may be expressed as

$$\delta\psi(\mathbf{r}) = \int \chi_{\psi\psi}(\mathbf{r}, \mathbf{r}') V_{\text{ext}}(\mathbf{r}') d\mathbf{r}',$$

where $\chi_{\psi\psi}(\mathbf{r}, \mathbf{r}')$ is the *static* order parameter response function, and may be expressed in terms of the order parameter correlation function $\langle \psi(\mathbf{r})\psi(\mathbf{r}') \rangle$. For a system that is translationally invariant,

$$\chi_{\psi\psi}(\mathbf{r}, \mathbf{r}') \to \chi_{\psi\psi}(\mathbf{r} - \mathbf{r}').$$

Near a phase transition, the Fourier component

$$\chi_{\psi\psi}(\mathbf{q}) = \int d\mathbf{r} \exp(i\mathbf{q} \cdot \mathbf{r}) \chi_{\psi\psi}(\mathbf{r})$$

diverges for some particular $\mathbf{q}$. A divergence at $\mathbf{q} = 0$ results in a state of uniform order (i.e., infinite period) while a similar divergence at a finite wavevector $\mathbf{q}_c$ ($=$ zone boundary, for example) leads to order with a period with a finite wavelength $\lambda = 2\pi/|\mathbf{q}_c|$.

Turning next to the dynamical behaviour of the system, one could consider a *weak*, time-dependent external field $V_{\text{ext}}(\mathbf{r}, t) = \lim_{\eta \to 0} V \exp(\eta t) \times \exp i(\mathbf{q} \cdot \mathbf{r} - \omega t)$ switched on adiabatically from $t = -\infty$. The response will be linear in the perturbation and given by

$$\langle \delta\psi(\mathbf{r}, t) \rangle = \lim_{\eta \to 0} \int d\mathbf{r}' \int_{-\infty}^{+\infty} dt' \chi''_{\psi\psi}(\mathbf{r} - \mathbf{r}', t - t') V_{\text{ext}}(\mathbf{r}', t') e^{\eta t'}.$$

The complex, frequency- and wavevector-dependent susceptibility is related to the above response by

$$\chi_{\psi\psi}(\mathbf{q}, z) = \mathbf{P} \int_{-\infty}^{\infty} \frac{d\omega}{\pi} \frac{\chi''_{\psi\psi}(\mathbf{q}, \omega)}{\omega - z}.$$

Schneider *et al.* point out that if the total number of degrees of freedom is large compared to the degrees of freedom associated with ψ, then the latter may be regarded as an ergodic variable. Using this assumption they show

$$\chi_{\psi\psi}(\mathbf{q}) = \lim_{\varepsilon \to 0} \chi_{\psi\psi}(\mathbf{q}, z)|_{z = i\varepsilon},$$

where $\chi(\mathbf{q})$ is the *static* response function defined earlier and $\chi(\mathbf{q}, z)$ is the *dynamic* response introduced subsequently. In other words, if the order parameter is an ergodic variable, then the isothermal susceptibility that one usually considers in the context of phase transitions can be expressed as above as an appropriate limit of the dynamic susceptibility.

The question now is whether the singular behaviour of $\chi(\mathbf{q})$ near a phase transition can be related to features in the dynamics through a general consideration of the properties of the dynamic susceptibility. The answer is obtained if one considers the

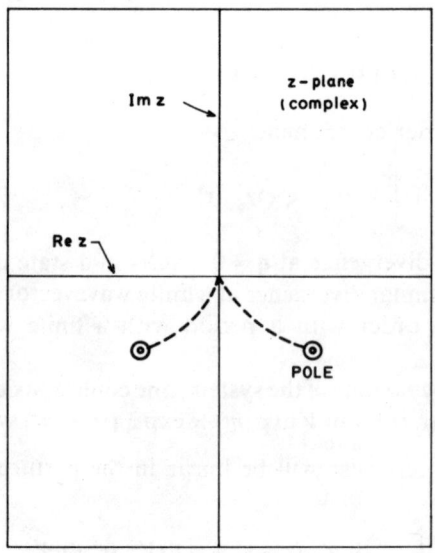

Figure 8.40 Movement of the poles of $\chi(\mathbf{q}, z)$ towards the origin as the temperature is varied. Such a behaviour for the poles implies the existence of a soft mode subject to the condition ψ is ergodic.

result

$$\chi_{\psi\psi}(\mathbf{q}) = \chi_{\psi\psi}(\mathbf{q}, z=0) = \mathbf{P} \int_{-\infty}^{\infty} \frac{d\omega}{\pi} \frac{\chi''_{\psi\psi}(\mathbf{q}, \omega)}{\omega}$$

and the expression

$$\int_{-\infty}^{\infty} d\omega\, \omega \chi''_{\psi\psi}(\mathbf{q}, \omega)$$

for the first moment of $\chi''_{\psi\psi}$. Schneider et al. remark that "in any equilibrium phase, the first moment of $\chi''(\mathbf{q}, \omega)$ exists and is finite". The only way in which the area under $\chi''(\mathbf{q}, \omega)/\omega$ (i.e., $\chi(\mathbf{q})$) can diverge while that under $\omega^2 \chi''(\mathbf{q}, \omega)/\omega$ (i.e., the first moment) remains finite is if the main contribution to the former comes from small ω, i.e., if $\chi''(\mathbf{q}, \omega)/\omega$ is peaked for small ω. In other words, for the desired behaviour, at least one of the poles of $\chi(\mathbf{q}, z)$ in the lower half of the complex plane has to move towards the origin (see Fig. 8.40), implying that a collective mode becomes soft.

Schneider et al. apply these concepts to a number of well-known phase transitions (like the spin flop and the superfluid transitions) to identify the corresponding soft modes. ☆

The Born–Raman Controversy

Max Born features in Raman's life not only because Raman tried to get him appointed in the Institute, but also on account of a scientific controversy which developed between the two. The seeds of the dispute were sown even while Born was in Bangalore, but it was only somewhat later that the dispute between the two became intense. The entire episode is discussed at length in Chapter 10 since it is important from both a scientific and a historical point of view. However, a few general comments are appropriate at this stage.

It all started with the interpretation which Raman sought to give to the second-order Raman spectra of various crystals recorded in Bangalore. To Raman the spectra appeared to consist of sharp lines to explain which he formulated a lattice dynamical theory of his own, in a sense extending an earlier version proposed by Einstein at the turn of the century to explain the observed temperature dependence of the specific heat of solids. Now Born was an old hand at the theory of crystal dynamics, having worked on the subject with von Karman (who later became famous for his studies in aerodynamics) as far back as 1913. The Born–von Karman theory was entirely classical, and needed a fresh dressing after the discovery of quantum mechanics in the twenties. Not only was this duly provided, but Born went even further towards developing a complete theory of dynamical lattices which would provide a comprehensive description of a wide range of properties of the crystalline state – elastic, thermal and optical. In Bangalore Born gave a course of lectures expounding on the emerging trends. Raman attended these lectures but did not see eye to eye with Born especially since Born's theory appeared to be incapable of explaining the observations of the Bangalore school. At that time Born's preoccupation was with the creation of a total edifice, and he therefore did not pay much attention to the question of whether or not his theory could explain the experimental observations. He turned to this question only much later.

Despite differences on scientific matters, the personal relationship between Born and Raman remained quite strong, as should be evident from the sustained efforts made by Raman to get Born a permanent appointment at the Institute. It was only after Born left Bangalore that the disagreement over the theory of lattice dynamics snowballed into a major controversy. To the scientist, the question of interest would be: who, between the two, was technically correct? As we shall see in Chapter 10, it was Born who was correct, Raman having grasped only part of the truth. But to the historian, the more interesting question is: why did Raman slip? Were there factors other than the purely scientific? And what, if any, were the consequences of such a dispute? It seems worthwhile to postpone a consideration of all such questions till the end of this chapter when one may attempt a total assessment of the Institute period of Raman.

The Diamond Story

For more reasons than one, diamond may truly be described as the prince of solids. As Raman puts it,

> It exhibits in a characteristically striking fashion many phenomena which are scarcely noticeable with other solids in ordinary circumstances. [8.47]

Combining as it does interesting physical properties with a glittering beauty of its own, it is not surprising that diamond held a deep fascination for Raman. His references to it were always enthusiastic if not ecstatic, as in the following example:

> Personal observation is, however, necessary to enable one to appreciate the remarkable beauty of these diamonds in their natural condition. With their exquisite form and their smooth lustrous faces, they look absolutely fresh from nature's crucibles although actually taken from sedimentary formation which, according to the geologists, are a thousand million years old.

Raman's interest in diamond appears to have started soon after the discovery of the Raman effect.

> I have since the year 1930, been deeply interested in physical investigation on the diamond. The difficulty of obtaining the material in a form suitable for exact studies has, however, been a serious obstacle to progress. Indeed in the early days, I was reduced to the expedient of borrowing diamond rings from wealthy friends who, though willing to oblige, were slightly apprehensive about the fate of their property.

Serious pursuit of science obviously cannot be subjected to the whims and fancies of wealthy friends, and so Raman began acquiring diamonds from various sources and building up a personal collection, naturally out of his own money. A particularly big increase in his collection was registered after participation in an auction in 1942 at the Panna diamond mines in Central India. Anna Mani, one of his students, records that in 1944, Raman had as many as 310 diamonds, including some from Hyderabad. Concerning these Raman and Ramaseshan write:

> They were picked out and purchased from the stock of unset stones in the possession of a firm of jewellers at Hyderabad (Deccan). No information was available regarding the origin of these stones beyond the statement that they had been detached for sale from some ancient jewellery. Since the city of Hyderabad is the nearest market to various places in the Deccan where diamonds are found, it is not improbable that the stones are of South Indian origin. All the eleven specimens are small, but they are of particular interest, being, with one exception, quite different from the Panna diamonds in their general features. [8.48]

There were also sixteen specimens presented by the de Beers of Kimberley, South Africa, which

> have proved very useful in enabling us to compare the South African [external] forms with the Indian ones and determine the relationships between them. Two items of particular interest in the collection may be mentioned here. One is a remarkably perfect example of the form of diamond first described by Haidinger, illustrations of which are to be found in the standard texts on mineralogy. The other is a triangular twin of flat tabular form with beautifully sculptured edges, presenting an interesting comparison with the rounded contours of the triangular twins found at Panna.

Raman took much pains to study, characterize and classify the external forms of the diamonds in his collection.

Single crystals (of various materials) of natural origin and those grown in the laboratory are often highly faceted and have flat faces. The edges are usually sharp and straight, being the intersections of various symmetry planes. The Panna diamonds, however, seemed to be an exception for they invariably exhibited curved faces. As Raman and Ramaseshan describe:

> During the senior author's two visits to the Panna State Treasury, he had the opportunity of examining several hundreds of these diamonds, including several very large and exceptionally fine specimens and never once came across a crystal showing plane faces either alone or in combination with the usual curved forms. It is very remarkable also that though the Panna diamonds are found in conglomerate beds of obviously sedimentary origin, it is exceptional to find a specimen exhibiting signs of having undergone any wear and tear during the transit from the original site of formation to its final resting place in those beds. Indeed, amongst our 43 specimens, there are only two which give any indication of having suffered in this manner. Most of our specimens, in fact, exhibit a remarkable transparency and smooth lustrous faces on which the details are seen beautifully clear and sharp. There cannot therefore be any doubt that the Panna diamonds exhibit precisely the same form as that in which they originally crystallized.

Is there a *geometrical* explanation for these forms? Yes. Let us first

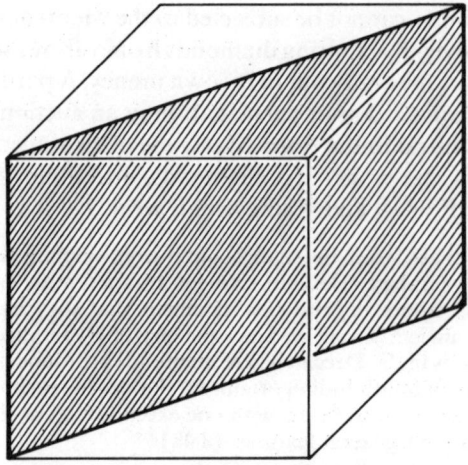

Figure 8.41 Cube showing one of the diagonal planes; there are six such planes. If a crystal possesses such symmetry planes, then it would belong to the tetrahedral class.

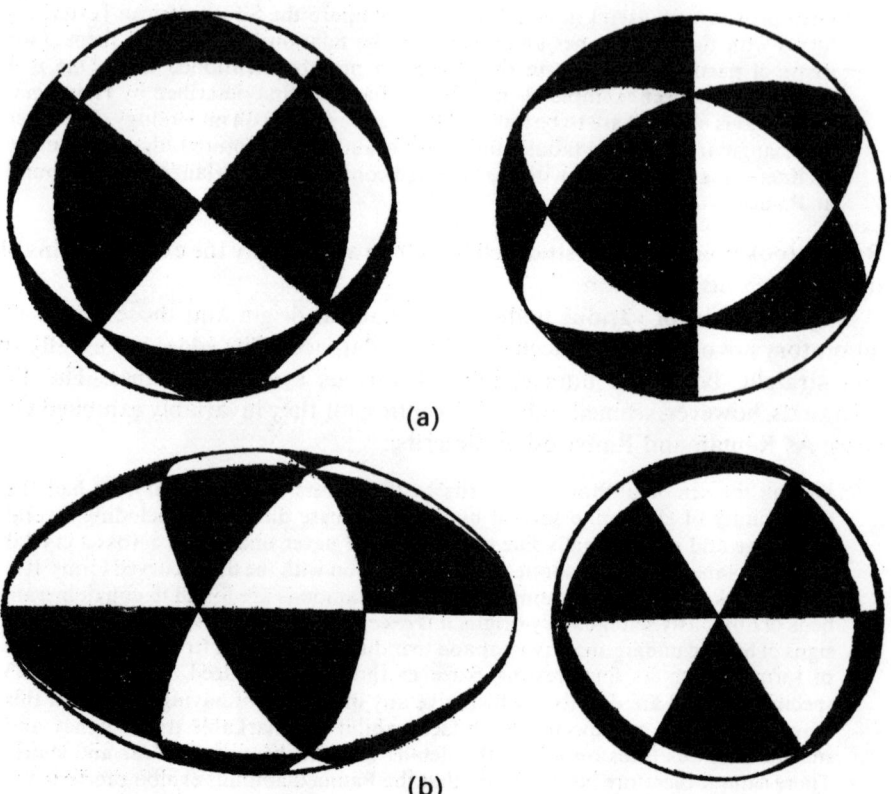

Figure 8.42 (a) Division of a spherical surface by tetrahedral symmetry planes; (b) similar division of a prolate spheroid. (After ref. [8.48].)

recall the symmetry properties of the crystal classes belonging to the cubic system. All five classes in that system have as a common feature the four axes of trigonal symmetry which are the cube body-diagonals. Taking these axes in pairs and drawing planes through them, we obtain the six diagonal planes of the cube [see Fig. 8.41]. If these are symmetry planes, the crystal would belong to the tetrahedral class. All the elements of symmetry appearing in that class are represented by drawing through the centre of a sphere the six diagonal planes. The sphere then appears divided up into 24 equal spherical triangles [see Fig. 8.42a].

When the exercise is repeated with a prolate spheroid, one obtains the patterns shown in Fig. 8.42b. Now it is known that when the general shape of a crystal departs from regularity, the pattern of edges is altered. While the *directions* of the edges which persist remain the same, their positions are shifted, with new edges appearing along the lines of intersection of the planes which did not previously meet (see Fig. 8.43). With this background, consider now Fig. 8.44 which shows sketches of diamond N. C. (New Collection) 18 which is a 57 mg piece from Udasna (Panna) "perfectly water-white in colour". By recalling the earlier figures, one can obtain some idea of the relationship of the external form to the internal symmetry.

Next come the questions: Why curved faces at all? What is their physical basis? Raman has an explanation.

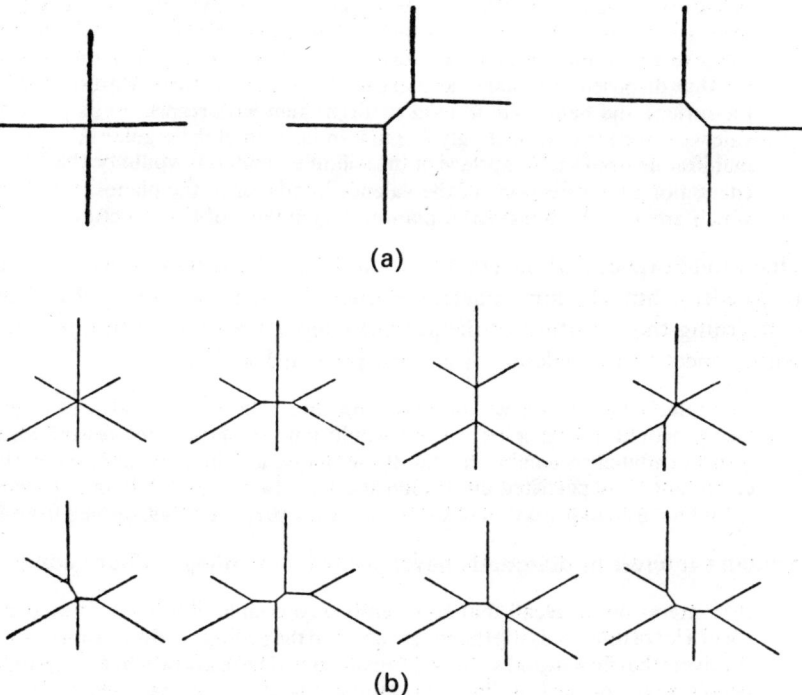

Figure 8.43 Pattern of edges in the vicinity of (a) two-fold and (b) three-fold axes of symmetry. (After ref. [8.48].)

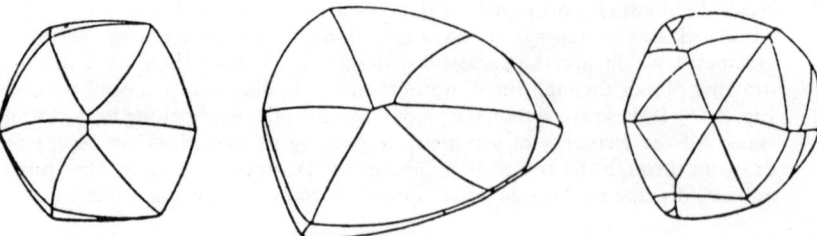

Figure 8.44 Front and end views of diamond N.C. 18. This form can be visualized by considering Fig. 8.42b and modifying some of the edges as in Fig. 8.43. (After ref. [8.48].)

It appears highly probable that diamond results from the solidification of carbon which has assumed the liquid state under conditions of high pressure and temperature. The state of the atoms in the liquid state is an important point needing consideration. The thermal agitation would certainly prevent a perfect ordering of the valence bonds within the liquid. Hence, it follows that the molten carbon would assume a rounded shape and this would be the more likely, the smaller the volume under consideration. Solidification is accompanied by a fixation of the valence bonds but not necessarily by any radical change of shape. On this basis, it is easy to understand why the crystals formed have curved faces. It may be remarked that the smaller diamonds in our Panna collection exhibit a highly marked curvature of the surface on which a pattern of edges appears.... The formation of these patterns is readily explained. At the surface of the molten carbon there would be some free valences which may attach themselves to the surrounding material. The valences not thus disposed of would link each carbon atom to its three nearest neighbours on the surface and hence would tend to align them with respect to its position in the valence directions. Accordingly, the first indication of the regular internal structure manifesting itself on the surface of the solidified material would be the formation of edges along the directions of the valence bonds, or in the planes containing them which are also the tetrahedral planes of symmetry of the structure.

One would expect that the crystal would be a spherical modification of a geometric polyhedron, but why non-spherical shapes? An investigation by Ramaseshan [8.49] concerning the variation of the surface energy with the orientation of the surface with respect to the valence directions provided a clue.

The interfacial tension would vary with direction and the surfaces of minimum energy would not be spherical but would tend to show some resemblance to the forms exhibited by a cubic crystal. If the shapes assumed by diamond in the liquid crystalline state persisted on solidification or else suffered only minor changes, we would have an explanation of the forms now observed. [8.49, quoted from Raman]

Raman's interest in diamonds never waned, and Bhagavantam comments,

It is interesting to recall that so recently as a couple of years before his death, he spent a lot of time studying the geography and the geology of the Krishna valley and of the rivers that flow in and around it because it had been known that at one time 60,000 people were engaged in diamond mining operations in that area. [8.50]

Raman's involvement with diamond grew to such an extent that at one stage, "every student working with him was engaged in studying one aspect or other of the properties of diamond". So intense was this activity that symposia devoted to diamond were held in 1944 and 1946, with participation only by Raman and his students in Bangalore! A variety of investigations were reported: X-ray diffraction, X-ray topography[15], birefringence, the Faraday effect, diamagnetic susceptibility, luminescence, phosphorescence, ultraviolet absorption, infra-red absorption, thermal expansion, Raman scattering, Brillouin scattering, etc.

Naturally there was a focal theme, and this was the structure of diamond. At first sight this might appear puzzling, for was not the structure of diamond well known? Yes, if one is talking about the *positions* of the atoms or, rather, the carbon nuclei. In fact the architecture (see Fig. 8.45) is unique enough to have earned the distinctive name, the diamond lattice. What then is the mystery?

In 1934 Robertson, Fox and Martin made the important discovery that while a large majority of diamonds exhibited strong infra-red absorption around $8\,\mu$, some did not [8.51]. Since that time, these two types have come to be referred to as Type I and Type II respectively. This finding triggered a chain of investigations, and it was

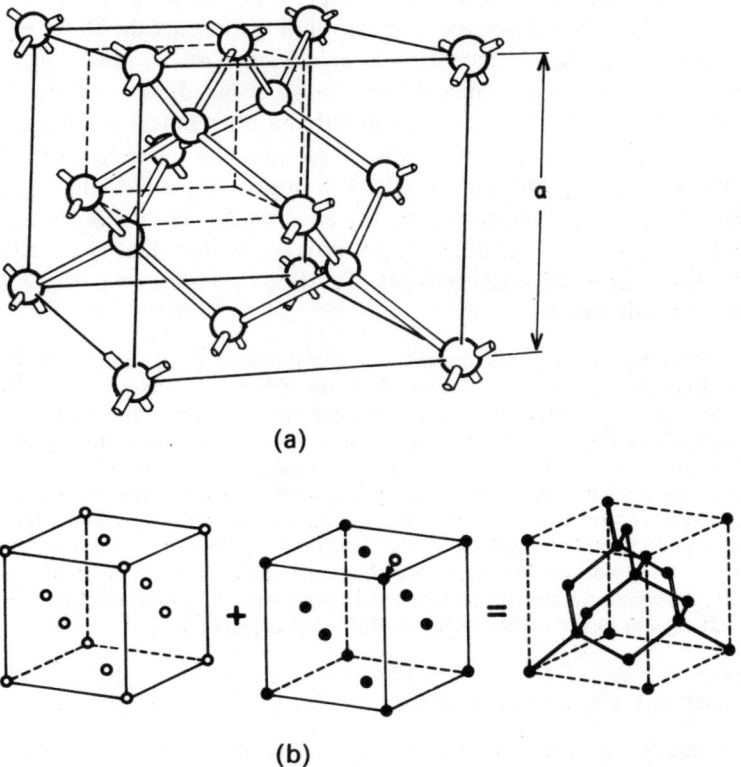

Figure 8.45 (a) Crystal structure of diamond. This possesses a centre of symmetry at the mid-point of each line connecting nearest neighbour atoms. (b) The diamond structure may be obtained by starting with a face-centred cubic lattice and moving it along a cube diagonal by one-fourth the diagonal distance.

observed that the two types of diamond differed in many other respects also. For example, the UV absorption edge in Type I was close to $0.4\,\mu$ (3000 Å) while in Type II it was shifted to $0.2\,\mu$ (2250 Å). Studies carried out in Bangalore seemed to suggest there were not two but in fact four types of diamond. The question was, why so many types? There were suggestions that peculiarities in the infra-red absorption arose, possibly owing to impurities or crystal imperfections, but the Bangalore investigations found that crystals which were near-perfect (as judged by the low mosaic spread revealed in X-ray studies) showed *more* infra-red absorption than those with larger mosaic spread. So infra-red absorption properties did not seem to be linked to crystal imperfections.

Now there are well-defined rules based on crystal symmetry called selection rules, according to which infra-red absorption in the cubic class of crystals is possible if the crystal symmetry is tetrahedral (T_d) but not if the symmetry is octahedral (O_h). It seemed (to the Bangalore school) that there were in fact four possible structures for diamond, two with tetrahedral symmetry and two others with octahedral symmetry.

The question still remains as to how one understands the different structures in relation to that depicted in Fig. 8.45. The point is that the positions shown in that figure refer only to the carbon *nuclei*. Surrounding these are the electrons, and unlike in isolated atoms, the electron clouds around the nuclei need not be spherical. Rather, they tend to be concentrated along some directions, owing to forces responsible for chemical bonding. In slightly more technical language, the electron orbitals in a diamond crystal are highly directional as a consequence of hybridization between different atomic orbitals. The optical, infra-red and other properties mentioned earlier are determined by the *electron distribution in the crystal*. It was argued by Raman that four different patterns of electron density distributions were possible, each anchored to the network of carbon nuclei shown in Fig. 8.45. Raman's thinking goes as follows:

> We shall accept the X-ray finding that the structure of diamond consists of two interpenetrating face-centred cubic lattices of carbon atoms which are displaced with respect to one another along a trigonal axis by one-fourth the length of the cube-diagonal [see Fig. 8.45]. Each carbon atom in the structure has its nucleus located at a point at which four trigonal axes intersect. Hence, we are obliged to assume that the electronic configuration of the atoms possesses tetrahedral symmetry. It must also be such that the alternate layers of carbon atoms parallel to the cubic faces have the same electron density. This is shown by the X-ray finding that the crystal spacings parallel to the cubic planes are halved. Hence, the possibility that the two sets of carbon atoms carry different total charges is excluded. In other words, diamond is not an electrically polar crystal in the ordinary sense of the term. [8.52]

What then is the situation? Raman explains:

> It is readily shown, however, that the charge distributions may satisfy both of these restrictions and yet not exhibit a centre of symmetry at the points midway between neighbouring carbon atoms. To show this, we remark that when two similar structures having tetrahedral symmetry interpenetrate, centres of symmetry would not be present at the mid-points between the representative atoms unless the tetrahedral axes of the two structures point in opposite directions.

On to Bangalore

Raman's arguments may be better understood by referring to Fig. 8.46 which is adapted from the one presented by Raman himself. Sketched here are decorations in the neighbourhood of four nearest-neighbour carbon nuclei such that each decoration has a local tetrahedral symmetry. The decorations do not purport to show the actual electron distributions but merely the geometrical symmetry. As indicated in the figure,

> we may, in fact, have *four* possible kinds of arrangement. Of these the arrangements shown in T_d I and T_d II have tetrahedral symmetry, while O_h I and O_h II would be distinct forms, both having octahedral symmetry.

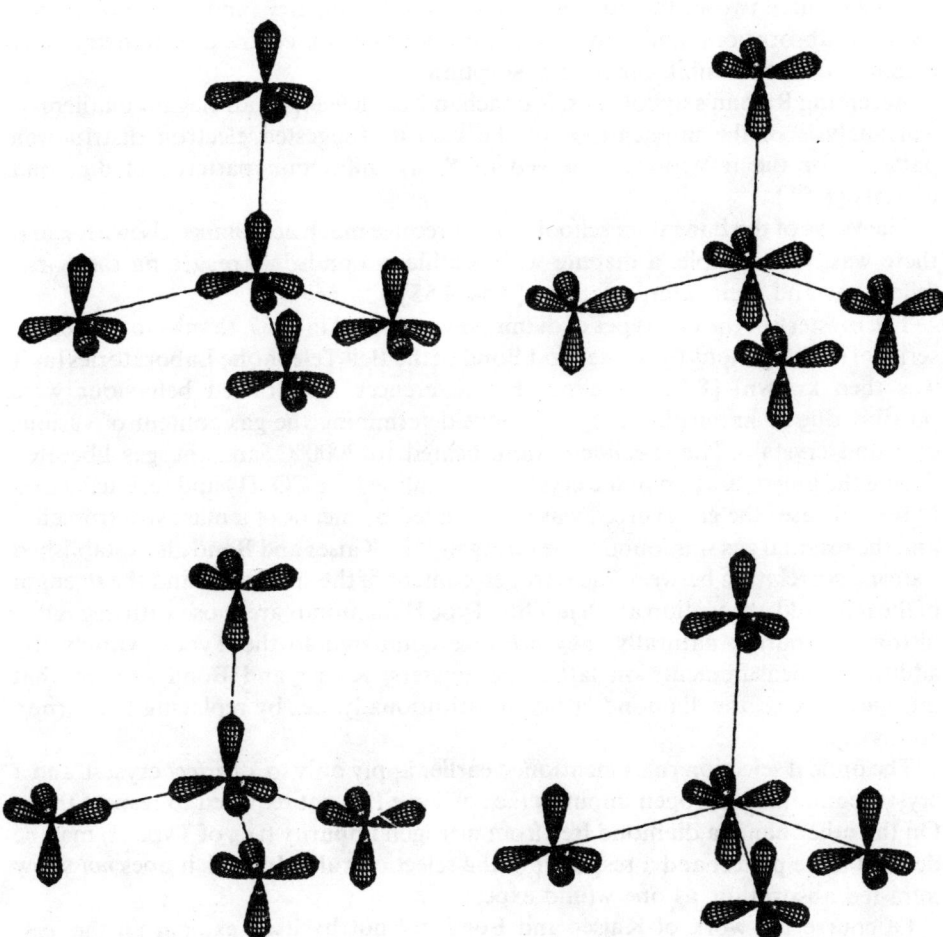

Figure 8.46 The four forms of diamond proposed by Raman. Sketched here are patterns indicative of the symmetry of the various structures. For explanations, see text.

Raman gives further arguments pointing out how the proposed structures are consistent with a very important experimental fact, namely, that diamond is diamagnetic.

At this stage let us paraphrase Raman's line of thinking. Symmetry considerations, i.e., group theory, dictates that infra-red absorption is possible if the symmetry is tetrahedral but not if it is octahedral. An essential difference between the two is that the latter has a centre of symmetry (between (000) and $(\frac{1}{4}\frac{1}{4}\frac{1}{4})$) while the former does not. In addition, it was already known from chemistry that the atomic orbitals of carbon hybridize (in molecules and in solids) to become orbitals with a local tetrahedral symmetry. Raman thus poses the following puzzle: What are the different ways in which decoration patterns having local tetrahedral symmetry can be arranged in a diamond lattice? Not surprisingly he arrives at the answer in Fig. 8.46 where two of the lattices lack centre of symmetry (and therefore permit infra-red absorption), and two of the lattices possess a centre of symmetry (and therefore do not exhibit infra-red absorption).

Accepting Raman's hypothesis, Ramachandran developed an elegant mathematical analysis of the implications of the various suggested electron distribution patterns for the reflections observed in X-ray diffraction patterns of diamond crystals [8.53].

The views of the Bangalore school did not receive much acceptance elsewhere, and there was, for example, a dispute with Kathleen Londsdale regarding the X-ray reflections and their interpretations [8.54, 8.55].

The mystery of the two types of diamond was settled in 1957, thanks to an elegant series of investigations by Kaiser and Bond of the Bell Telephone Laboratories (as it was then known) [8.56]. Feeling that differences in infra-red behaviour were possibly due to impurities, they set about determining the gas content of various diamond crystals. The specimens were heated to 2000°C, and the gas liberated during the graphitization of the crystal was analysed for CO, H_2 and residual gases. In several cases the gas evolved was investigated by means of a mass spectrometer, and the residual gas was found to be nitrogen ^{14}N. Kaiser and Bond also established a strong correlation between the nitrogen content of the specimens and the strength of the infra-red absorption at 7.8 μ. Thus Type II diamonds are those with negligible nitrogen impurity; naturally they are rare compared to the Type I variety. By additional measurements on lattice parameters, Kaiser and Bond showed that nitrogen entered the diamond lattice substitutionally, i.e., by replacing the carbon atoms.

The optical selection rules mentioned earlier apply only to a *perfect* crystal, and a crystal containing nitrogen impurity (i.e., of Type I) is not required to respect them. On the other hand, a diamond free from nitrogen impurity (i.e., of Type II) may be deemed to be perfect and a respecter of the selection rules. Indeed, it does *not* show infra-red absorption, as one would expect.

Of course the work of Kaiser and Bond did not by itself explain all the less-understood features exhibited by diamond, but it was clear that a model postulating four possible forms for diamond was not required.

Experience shows that a physical theory can be wrong for one of two reasons. It may violate already known, tested and well-accepted principles, in which case it is seldom taken seriously. On the other hand, it may be logically consistent with respect to already known facts but be unable to explain or be inconsistent with facts discovered subsequently. Theories of the latter type are quite common and the hotter the field, the more the candidates; of course, eventually most of them fall by the wayside. In fact progress is often stimulated by a vigorous competition. Raman's theory for diamond structure belongs to the second of the two categories mentioned above. As Ramaseshan comments:

> It is interesting that Raman was led on to the discovery of his [i.e., Raman] effect because of his intuitive belief dating back from 1922 that the "weak fluorescence" that was observed in light scattering in liquids was not due to impurities but was of molecular origin. It must have been this unshakable conviction that made him drive his students and collaborators into purifying and repurifying hundreds of liquids to look for specific characteristics in the scattered light which would distinguish it from the normal molecular scattering or fluorescence due to impurities. However, the same feeling seems to have played him false in the case of diamond. Unable to purify his diamonds, he studied hundreds of diamond plates, unwilling to believe that this "Prince of Solids" could have major impurities in it. Years later it was established that many of the phenomena he and his students discovered arose due to impurities (like nitrogen) in the diamond lattice. It is ironic that the symmetry changes induced by these are similar to those that Raman proposed to explain the observed phenomena and which he believed intrinsic to the carbon atom. [8.57]

8.9 The Franklin Medal

In 1941 the Franklin Institute, Philadelphia, awarded Raman the Medal of Merit, its highest honour, and elected him an honorary member. A similar distinction had earlier been conferred upon Einstein, Millikan and Compton. America had not yet entered the war, and Raman was planning to go there and receive the medal in person, taking the opportunity not only to visit the Franklin Institute (whose centenary celebrations he had participated in during his 1924 visit), but also to renew contacts with various American scientists. The plan was to go to Hong Kong via Calcutta and Rangoon, and then take a ship from there to be in time for the Franklin Day celebrations on May 21. The Council of the Indian Institute of Science offered felicitations, and agreed to place him on deputation to America for five months. But something happened and Raman was unable to go. Thereupon, the British Ambassador in Washington requested the Franklin Institute that the Medal be presented in India. Thus in July 1942 the Governor of Madras made a trip to Bangalore, and decorated Raman at a special function at the Residency. A pleasant interlude, and a moment in which to forget the troubled past.

8.10 Academy

The years 1934 and 1935 are important in the history of Indian science for the reason that they saw the establishment of our academies of science. Normally every country has only one academy of sciences but India has two, thanks to the Saha–Raman differences! While the reader has to wait till Chapter 11 for the full story it is pertinent to remark here that the passage of time has mellowed feelings all round, removing all traces of the original controversy. In his day Raman was sharply criticized for what was described as a breakaway action but in retrospect it must be conceded that Raman (i) helped strengthen scientific publication in India and (ii) actively promoted the concept of topical symposia. Earlier in Calcutta the Bulletin was the vehicle for propagating the research findings made at the IACS. Likewise, the *Proceedings of the Indian Academy of Sciences* in its earliest days provides an admirable record of the researches of the Bangalore school. Today the Academy journals have grown into a large family serving various sciences from astrophysics to genetics. All this would not have been possible but for the sustained work put in by Raman for several decades.

8.11 Visit of Gandhiji

An event of interest was the visit of Mahatma Gandhi to the Institute. Raman was a very great admirer of Gandhiji, his first personal contact with the Mahatma going back to the Calcutta days. Independently, Lady Raman, with her deep involvement with *Sarvodaya*[16], had built up her own contacts with Kasturba Gandhi as well as with the Mahatma.

In 1936 Gandhiji came to Nandi Hills near Bangalore to convalesce after a surgical operation. Raman called on Gandhiji accompanied by the noted Swiss biologist Rahm who was working on tiny organisms which could survive without food and water for many years. Gandhiji was very much interested in this work and desired to have a copy of the paper when the investigations were completed. The idea of a visit to the Institute was suggested at that time, and Gandhiji accepted. He came accompanied by Kasturba, Mahadev Desai and Sardar Patel[17]. As they were being received by Raman, Gandhiji quipped that he had come more to see his old friend Lady Raman than to see the Institute. Raman naturally took great pleasure in showing the visitors around, waxing eloquent on the scientific achievements of the Institute's scientists. If Gandhiji was mildly curious about science, the scientists at the Institute were even more curious about the Mahatma. At the conclusion of the visit there was the inevitable autograph hunting. As usual Gandhiji readily obliged – for a fee of course[18]! It is said that the autograph collection was led by Raman himself. Gandhiji borrowed Raman's Parker pen but later decided to treat it as his

During the inauguration of the Indian Academy of Sciences. To Raman's left is Max Born, and Sir Mirza Ismail is on his right.

Gandhiji at the Indian Institute of Science.

Driving with friends in Europe.

Lakeside chat in Europe. Seated with his back towards the camera is Heisenberg.

Delivering a lecture in Europe.

An informal address.

Hedi Born in Indian dress.

With Dirac.

Raman Research Institute under construction.

With Jawaharlal Nehru.

In a pensive mood.

With the Nobel citation.

Inspecting a guard of honour presented by students of a Sainik (Public) School.

With visitors at the Raman Research Institute.

Lecturing on his theory of lattice dynamics.

Joyously romping with children in the garden of the Raman Research Institute.

fee. The pen was then auctioned on the spot, Raman being invited to join in the bidding!

As the visitors were about to leave, Raman wanted to know whether they had any comments or observations to make. Gandhiji had none but Sardar Patel wondered whether anything could be done to improve the *charka*. Not perhaps the remark Raman was hoping for!

A footnote to this visit is provided by Raman himself.

> In the year 1945 I was staying at Bombay in the residence of the Sarabhai family in Nepean Sea Road and was being helped by Dr Vikram Sarabhai to collect funds for the construction of my Research Institute at Bangalore [see Chapter 12]. One evening Vikram suggested that I might call on Gandhiji who was conducting a prayer meeting on the beach sands. I waited on the outskirts of the crowd till the meeting was over and then moved forward to meet him. To my surprise, he immediately recognized me, and made enquiries about me and about Lady Raman. Then he proceeded to recall his visit to my laboratory at Bangalore several years earlier and specially mentioned the demonstrations of the harmonic modes of vibration of the Indian musical drum which I had shown him and which had evidently impressed him. [8.58]

Bhagavantam recalls that in 1937 an article appeared under the title "Gandhiji and Raman" after a visit by the author (of the article) to the International exhibition in Paris [8.58]. The author was quite disappointed to find nothing about India in the exhibition, even in the stalls of England. But curiously there were two items of Indian interest in the American pavilion. One was a bust of Mahatma Gandhi, placed facing that of Rockefeller (!), and the other was a demonstration of the Raman effect, accompanied by a write-up on Raman. One does not know how these two exhibits found a place in the American pavilion.

8.12 Fiftieth Birthday

On November 7, 1938, Raman attained the age of fifty. The event called for a celebration, and the Council of the Academy decided to bring out a *festschrift* in his honour. The response to the call for contributions was quite generous, with Kohlrausch, a noted Raman spectroscopist in Germany, acting as the European coordinator. The commemorative volume [8.59] appeared as a special issue of the Academy *Proceedings*, a unique feature being that it carried articles not only in English but also in French, German and Italian – perhaps the only time ever.

There are many familiar names amongst the contributors, starting with Brillouin who is "very much pleased to seize the opportunity of this Jubilee volume to pay my tribute of admiration for the wonderful work of your President Sir Venkata Raman, whom I am glad to count among my personal friends since a good many years" [8.59, p. 251].

The controversy with Max Born had not yet erupted, and, not surprisingly, Born is one of the prominent contributors. He writes, in an article entitled "Some remarks on reciprocity":

> As I had the privilege of collaborating with Sir C. V. Raman in his Institute for a period of six months in 1935–36, I have a great desire to show my admiration of his discoveries and my appreciation of his unceasing efforts for the advancement of Indian science by contributing to this volume of homage devoted to papers on scattering of light and Raman effect. But since the olden days when I did some work in this field, it has developed so marvellously that it has become the domain of a class of specialists with whom a 'general practitioner' of physics cannot compete. Therefore, I shall content myself with presenting a few remarks of a very general nature about the difficulties which theoretical physics encounters when dealing with the nature of light and ultimate particles. [8.59, p. 309]

In a footnote Born points out that when Heisenberg, Jordan and himself developed quantum mechanics, they applied the matrix formalism to the problem of dispersion and computed the Kramers–Heisenberg formula. As we know, the incoherent scattering of light predicted by this formula is what Raman had discovered. Born is proud that his work "had some connection with this fundamental discovery".

It is curious that whereas in this volume Born contributed a paper on what might be called the theory of elementary particles, about a decade later in a similar volume brought out on the occasion of Niels Bohr's seventieth birthday, he wrote about problems of crystal physics that had arisen out of researches of the Bangalore school.

Of special interest to the historian is a paper by James Hibben of the Carnegie Institution. Hibben is interested in a trend analysis and his result which is reproduced in Fig. 8.47 shows an interesting dip in 1933. Hibben wonders why.

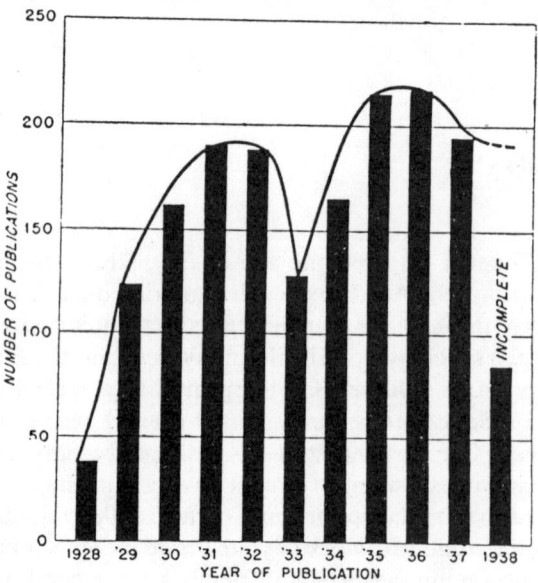

Figure 8.47 The number of publications per annum on the Raman effect. (After Hibben, ref. [8.59].)

With regard to this behaviour there are several comments which may be of interest.

The first of these is that an increase in the average length of papers might cause a diminution in the average number, but would not account for the second rise in the curve. Furthermore, a generous sampling of publication-length indicates that the average paper is about five pages in length and this has not altered greatly with time.

The possibility of the effect of the depression in the United States – which reached a maximum in 1932 – on research funds, endowments and scholarships would occur to the average American. Such an effect would be a delayed one and would ostensibly account for both the diminution and the subsequent increase in the publication rate. However, a closer examination of all the pertinent facts shows that this hypothesis is, at best, only partially tenable. The overall decrease in 1933 from the general average is 30%, but in the United States where the financial situation was more acute at this particular period than abroad, it reached a value only slightly lower, namely 33%, while in Germany, combined with Austria, the reduction was 47%. Part of the latter reduction can undoubtedly be attributed to the known diminution in fundamental research in Germany. Finally, it can be shown that the world-wide recession in all types of chemical and physical articles in 1933 amounted to but five per cent. This seems to confirm the view that the drop in the rate of publication in 1933 on the subject of Raman spectra is significant, that it was not greatly affected by economic conditions, and that the length of articles played no obvious role. [8.59, p. 295]

What then was responsible for the dip? Hibben advances the interesting thesis that the dip is associated with a change in the character of Raman scattering research. The first bloom of novelty had worn off, and physicists were satisfied that they understood the origin of the effect; naturally there was a decline of interest. If this is accepted as an explanation, then the reawakening of interest after 1933 must also be explained. This is readily done, for it is during this period that chemists began to turn to the Raman effect as an analytical tool. As Hibben remarks, the Raman effect "became the adopted son of chemistry".

Hibben also presents a graph of the distribution of the international effort which we reproduce in Fig. 8.48. The strong activity in India is quite worthy of note. By contrast, when laser Raman scattering came to the fore, the contribution from India was practically negligible.

8.13 Bhabha and Raman

It is time to introduce Bhabha whose name was briefly mentioned in Chapter 1. Homi Bhabha was truly a remarkable person who in the space of a few short years gave Indian science and technology a mighty push, the like of which has never been seen either before or after. Not surprisingly, J.R.D. Tata described him as "one of the few authentic men of genius I have met in my lifetime".

Homi Jehangir Bhabha was born in Bombay on October 30, 1909. His grandfather Dr Hormusji Bhabha had served as the Inspector-General of Education in the Princely State of Mysore, while his father Jehangir H. Bhabha was a barrister who practised in Bombay. Both of them had been in the Council of the Institute as

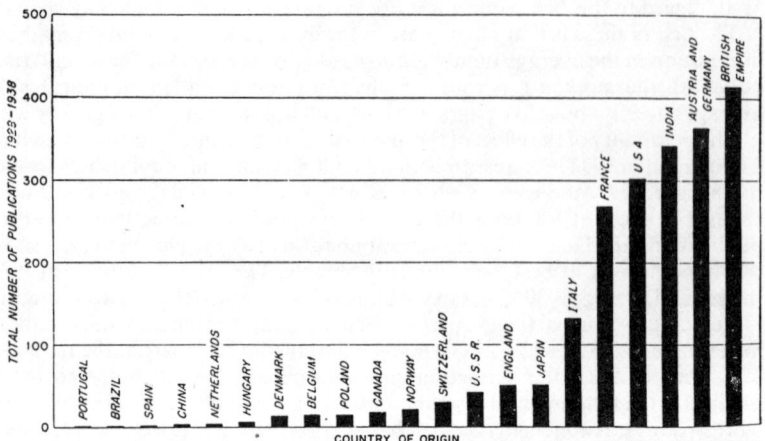

Figure 8.48 Geographical distribution of research on the Raman effect. (After Hibben, ref. [8.59].)

representatives of the Tata family and were known to Raman. Bhabha was well connected. His mother was the granddaughter of the popular philanthropist Sir Dinshaw Petit, while his paternal aunt was married to Sir Dorabji Tata.

Young Homi attended the Cathedral School in Bombay and grew up in a truly aristocratic environment. His biographer recalls:

> As a boy, he had access to his grandfather's large library which had been supplemented by his father's fine collection of books on painting and art in general, collected during his student years at Oxford and London.
>
> Thanks to his father's and aunt's collections of recorded music, Homi Bhabha, by the age of 17, was already familiar with the recorded symphonies, concertos, quartets and sonatas of Beethoven and Mozart, and of the recorded operas of Wagner and Verdi. Listening to music for Homi and his younger brother and boyhood friends was a serious matter. Lights were dimmed and listeners even sat in hushed absorption during the frequent changes of gramophone disks in the days before the long-playing records. [8.60]

It was not just art and culture, for young Homi was also exposed to the affairs of the world. He was accustomed in his boyhood to listen to

> discussions between older members of the family relating to projects for the industrial development of India, ranging from iron and steel and hydroelectric power generation to the manufacture of heavy chemicals. [8.61]

Mahatma Gandhi was a frequent visitor to his aunt's house (where Homi used to go to have his lunch), and there were also friendly ties between the Tata and the Nehru families. As a result of these connections, Bhabha got a ringside view of the freedom movement. Outwardly, Bhabha always conveyed the impression of a highly Westernized man, which indeed he was, but thanks to these early exposures, deep inside he also had strong Indian roots.

After passing the Senior Cambridge Examination at the age of sixteen, Homi did a small stint at the Elphinstone College, Bombay, before setting off to join the Gonville and Caius College in Cambridge. His uncle Sir Dorabji Tata was an alumnus of the college and had made generous endowments to it. Bhabha did not go to England to study physics.

> The intention of his father and of his uncle, Sir Dorab Tata, was that he should obtain an engineering degree with a view to joining the Tata Iron and Steel Company at Jamshedpur. During the course of his engineering studies, Bhabha wanted to change to mathematics, but his father urged that he should complete the engineering course and promised that if he got a first he would finance further studies in mathematics. Bhabha duly obtained his first class in the mechanical sciences tripos in June 1930.... [8.60]

The elder Bhabha kept his promise and Homi now switched to theoretical physics. It was the best of times, and he was in the best of places – Cavendish, where he could personally witness great discoveries, and rub shoulders with famous men.

The years 1930–1939 were heady and exciting for Bhabha. First he won the Rouse Ball Studentship in mathematics, and later was elected to the Isaac Newton Studentship. He became a member of an informal group, the Kapitza Club, which met every Tuesday to discuss physics, and was active in dramatics and in athletics. On the professional side he became deeply involved in theoretical studies on cosmic rays as well as on elementary particles. His first paper was published in 1933; others quickly followed, including one written with Heitler which became a benchmark. He now began to travel to the leading centres of physics in Europe, visiting Bohr in Copenhagen, Pauli in Zurich and Fermi in Rome. Strong ties were also made in Cambridge, particularly with Cockcroft, and with W. B. Lewis. Later many of these persons were to help Bhabha with the realization of his dreams.

In 1939 Bhabha was holidaying in India when war broke out in Europe. He could perhaps have returned if he really wanted to, but there did not seem much point in it since the European scientific scene was badly disrupted. Moreover, most of his Cavendish friends had been sucked into the war effort in one form or another. India seemed a better place to pursue science, at least for the moment. Since Bhabha already enjoyed a reputation there were a few feelers, but most of the places did not look attractive. Bangalore, however, was a different cup of tea. The city was nice, the climate good, and he had ancestral roots there. Above all, the Physics Department of the Indian Institute of Science had a dynamic head who was vigorous and active; perhaps Bangalore was not as good as Cambridge, but it was certainly far better than anything else the country had to offer. And so in 1940, under persuasion from Raman, Bhabha joined the Institute as Reader in physics. Even prior to this he had signalled his intention to stay in India by publishing some work he had done in England in the *Proceedings of the Indian Academy of Sciences*; the byline of the paper has the address, Gonville and Caius College.

The Bangalore period of Bhabha marks a turning point not only in his own life but in the history of Indian science itself, and therefore merits a close look. Bhabha was quite different from all the others in the Physics Department. Besides having received his training elsewhere, he had also built up an independent international reputation (although it had not yet reached a peak). For Raman, Bhabha was like a breath of fresh air – someone who knew about the latest developments in a frontier area, and with talents complementary to his own. Bhabha too was attracted by Raman's love of Nature and his quest for aesthetics. Equally important was Raman's perpetual, bubbly enthusiasm for science. For an "exile" like himself, there was nothing so refreshing as to have someone take keen and genuine interest in the progress of his work. Thus it was "love at first sight". While they differed in many respects, they also had much to share and enjoy besides science[19].

Reference has already been made to Raman's keen desire to enter nuclear physics. Lacking grants, he often wished that he had used his Nobel Prize money to buy some radium (instead of spending it on other things like diamonds, for example). The arrival of Bhabha rekindled Raman's hopes – maybe one could use Bhabha's connections and persuade the Tatas to chip in. Bhabha and Raman had many discussions, with the latter trying to steer Bhabha's attention in the direction of neutron stars. But Bhabha's thoughts seemed to lie elsewhere. There was a powerful mathematical streak in him which now surfaced, and he became drawn to the aesthetic beauty inherent in exact mathematical solutions. M. G. K. Menon comments that Bhabha's papers of this period are "very much more mathematical and deeply introspective in character" [8.62]. He adds, "Homi has often told me that though the work he did in India was less recognized internationally, it gave him much greater intellectual pleasure than the earlier and more celebrated work done in Europe."

Within two years of coming to Bangalore, Bhabha was elected to the Royal Society, Raman having proposed him. A year later he won the coveted Adams Prize, the first Indian to do so. Menon observes, "He received many honours and awards throughout his life but he always had particular personal pride and affection for these two." Following these triumphs there was an offer of a chair in physics from the University of Allahabad, and one from the Indian Association. Bangalore countered with its own offer of a professorship, which was taken.

Although by training as well as disposition Bhabha was a theoretical physicist, his Cambridge experience had taught him that theory cannot flourish in isolation of experiment. With much enthusiastic support from Raman, he now began to turn his attention towards experiments also. Cosmic rays seemed the best area to work in, especially since India offered not only a wide range of latitudes from the magnetic equator in the south to about 25° magnetic in the north, but also a wide range of hill stations (besides, as was later noted, some of the deepest mines in the world). Already Millikan had recognized India's potentialities and had made a trip to carry out a series of experiments. Why not therefore start a full-time cosmic ray programme in India? With seed money obtained from the Tatas, Bhabha now became an experimenter[20]. Sreekantan tells us that story:

> He designed, based on his first-hand familiarity with cascade theory, a unique Geiger counter telescope that preferentially selected the penetrating component [of cosmic radiation] without requiring too much lead absorber.... The telescopes were flown in an aircraft from Bangalore [these belonged to the US Air Force and they were stationed in Bangalore] on December 26 and 28, 1944, and had half-hour

exposures at 5000, 10000, 15000, 20000, 25000, and 30000 feet. These constituted the first measurements.... He also got constructed a 12″ diameter circular cloud chamber identical to the one operating at that time in Prof. P. M. S. Blackett's laboratory at Manchester University and initiated a systematic study of the scattering properties of the penetrating component. [8.63]

Bhabha now started playing an active role on the Indian scene. He was promptly elected a Fellow of the Indian Academy of Sciences, and regularly attended its meetings. In 1943 he was elected President of the Physics Section of the Indian Science Congress. Following that event, he stayed back on Saha's invitation to give a course of lectures. Bhabha also began to publish most of his work in Indian journals, patronizing particularly the *Proceedings of the Indian Academy of Sciences*.

Life, however, was not just science. During the preceding decade he had become deeply involved in Western art and music but now it was time to explore their eastern counterparts. He delighted in visiting places like Ajanta and Ellora, Sanchi and Elephanta, and his interests also spread to classical Indian music and dance. Quite possibly these experiences later influenced his selection of sites for the location of atomic research centres. Jawaharlal Nehru noticed the remarkable juxtaposition of the research centre at Trombay and the Elephanta caves on an island across the water: the new and the old facing one another in ageless harmony. Indira Gandhi similarly observed the juxtaposition of the nuclear power station in Kalpakkam to the famous temples of Mahabalipuram.

For Bhabha, the Bangalore period was not only a discovery of India but of himself. Earlier he had felt that the Indian experience would be but a brief interlude – after the war it would be either back to England or maybe even America. But now a deep feeling of identity between himself and his country began to develop, and he became conscious of the great cultural heritage of India. He also began to feel an urge to shock India out of her slumber and transform her with the aid of science and technology.

Bhabha now began to dream big, and the Institute seemed too small a setting for realizing his ambitions. In a letter addressed to J. R. D. Tata he strongly pleaded for support to basic sciences in India, pointing out that

> there is no genuine knowledge of the universe that is not potentially useful for man, not merely in the sense that action may one day be taken on it but also in the fact that every new knowledge necessarily affects the way in which we hold all the rest of our stock. [8.62]

Bhabha concluded by asking whether the Dorab Tata Trust would be interested in supporting an institute *fully devoted to fundamental research*. In his reply Tata observed:

> From what you say in your letter, it is evident that there is scope for rendering valuable service to the country and to the cause of scientific research in India. The advancement of science is one of the fundamental objects with which most of the Tata Trusts were founded. [8.62]

Encouraged by J. R. D. Tata's reply, Bhabha now wrote a formal letter to Sir Sorab Saklatvala, Chairman of the Sir Dorab Tata Trust, in which he said:

> There is at the moment in India no big school of research in the fundamental problems of physics, both theoretical and experimental. There are, however, scattered all over India, competent workers who are not doing as good work as they would do if brought together in one place under proper direction. It is absolutely in the interest of India to have a vigorous school of research in fundamental physics, for such a school forms the spearhead of research not only in less advanced branches of physics but also in problems of immediate practical application in industry. If much of the *applied* research done in India today is disappointing or of very inferior quality it is entirely due to the absence of a sufficient number of outstanding *pure* research workers who would set the standard of good research and act on the directing boards in an advisory capacity.... [8.60]

One cannot but recall at this point Raman's defence in the Institute Council when he described how he had helped the Railways and various Princely States in applied areas. Bhabha continues in the letter:

> I had the idea that after the war I would accept a job in a good university in Europe or America, because universities like Cambridge or Princeton provide an atmosphere which no place in India provides at the moment. But in the last two years I have come more and more to the view that *provided proper appreciation and financial support are forthcoming,* it is one's duty to stay in one's own country and build up schools comparable with those that other countries are fortunate in possessing. [8.62]

He now comes to the point, which is that the Trust must support an institute fully devoted to fundamental research, and adds,

> The scheme I am now submitting to you is but an embryo from which I hope to build up, in the course of time, a school of physics comparable with the best anywhere. [8.62]

The institute would work on fundamental problems in theoretical physics with special reference to cosmic rays and nuclear physics, and experimental research on cosmic rays. In the same letter are the truly prophetic words:

> Moreover, when nuclear energy has been successfully applied for power production, in say a couple of decades from now, India will not have to look abroad for its experts but will find them ready at hand. [8.62]

As Bhabha himself put it much later, this was

> more than a year before the explosion of the first atomic bomb on Hiroshima and before nuclear physics had become what might be called the "bandwagon" of science. [8.64]

The Trustees of the Sir Dorab Tata Trust decided to accept Bhabha's proposal, and thus was born the Tata Institute of Fundamental Research (TIFR), marking a turning point for both Bhabha and the country.

One cannot avoid the temptation of comparing Bhabha's experience with that of Raman. For supporting basic research, Raman was accused of being untrue to the spirit of Jamshedji Tata and the Institute was wrested from his control. Bhabha now finds the same Institute inadequate and is able to elicit support from the Tatas for what was earlier denied to Raman! Would Bhabha have continued in Bangalore if the Institute had been allowed to blossom as a centre for basic research as Raman had hoped? We will never know. Perhaps destiny willed that Raman's efforts be thwarted so that Bhabha's may flower!

While discussing Bhabha one must not forget Vikram Sarabhai who also passed through the Bangalore school during a crucial period of his life. Young Sarabhai had gone to England to study science but the outbreak of war compelled him to return. His father Ambalal Sarabhai, a leading magnate of Ahmedabad, then requested Raman to take the young lad in his department so that he might stay in touch with research. Thus Sarabhai began to work with Bhabha and became interested in cosmic rays.

Around that time there appeared a paper which reported that cosmic rays were capable of leaving tracks in photographic plates. Raman immediately sensed the potentialities of this discovery and strongly commended it to Sarabhai declaring in a characteristic fashion, "Take my advice, young man; this will surely lead to the Nobel Prize!" It did, not for Sarabhai who ignored the advice, but for Powell in Bristol who independently got the same idea as Raman.

After the war Sarabhai returned to England to complete his PhD and, following the Bhabha model, started his own research centre, the Physical Research Laboratory in Ahmedabad. He became a member of the Atomic Energy Commission (AEC) and when Bhabha died succeeded him as the Chairman of the AEC. If Bhabha is the father of atomic energy in India then to Sarabhai belongs the credit of being the founder of space science and technology in India. For both Bhabha and Sarabhai, Bangalore was a fruitful period. In an unobtrusive yet very significant way, Raman made their stay a pleasant experience influencing their decision to continue their careers in India rather than go abroad, which both could easily have done.

We shall pick up further pieces of the Bhabha story in a later chapter but before we wind up the present digression, it is necessary to make a few remarks on the impact Bhabha had on technology development, especially as it offers in some measure a vindication of Raman's stand on the subject.

Science is no doubt the fountainhead of technology but the manner in which the discoveries of science transform and blossom into technology is quite complex. There is a maze through which knowledge has to diffuse before reappearing as a useful process or a product. And it takes not only many individuals of diverse disposition, outlook and motivations to achieve this transformation, but equally, many organizations with correspondingly varied structures. Bhabha seems to have intuitively understood this. First he created the TIFR. From this then emerged the atomic energy programme, encompassing not only oriented research but also various types of industrial activities like the production of nuclear fuel, the production of heavy water, and, last but not least, the construction and operation of nuclear power plants. Bhabha also sowed the seeds of the space programme which Sarabhai later nursed and grew, again with its own ramifications. Bhabha was getting involved in electronics development too but he died before that activity could take strong root.

The Bhabha strategy for development has close parallels both in England and in Canada where in fact the efforts in nuclear energy were spearheaded by Bhabha's contemporaries at Cambridge. And in France also the nuclear energy industry came out of basic research programmes due to Joliot Curie, Horowitz, Vendreyes, and others. Seen in this light, the thwarting by the Institute management of Raman's efforts to build up basic research was a fundamental mistake. In the Raman plan, the Institute by itself might not have engaged in applied research but it is more than likely that many scientists emerging from it would subsequently have made a mark in applied science. Perhaps in those days the complex relationship between science and technology was not fully appreciated. On the whole, it was a setback for both Raman and the Institute.

8.14 Into Retirement

Back now to the Raman story. November 1948 – sixty years of age; it was time to retire, not from active life but from the Institute professorship. A grand farewell was arranged, and friends and students assembled from near and far to participate. At the reunion dinner the preceding night, K. S. Krishnan (now occupying the prestigious position of Director of the National Physical Laboratory) casually revealed that he would be reading a letter from Prime Minister Nehru conveying felicitations and announcing a Government grant for pursuing research even in retirement. Raman was free to use the money as he pleased, the only requirement being an annual report of expenditure and progress to be submitted to the Government. The gathering was hardly prepared for Raman's response. He flew into a rage declaring that he could very well manage on his own thank you, and that there was no need for Government funds with strings attached. So sensitized had he become to the Establishment and its power that he distrusted even normal administrative requirements. Financial pinch, though painful, seemed preferable to Government control and there did not seem to be any compelling reason to submit to it even in retirement. The much-expected announcement was therefore not made.

The Raman we see leaving the Institute is a very different person from the one we saw earlier leaving Calcutta. It is a much older man, somewhat battered by senseless vilification and with fewer scientific triumphs to boast of. But it would be erroneous to conclude that the second phase of Raman's life was a failure. In fact, a careful appraisal leads to a different conclusion.

There can be little doubt that surrendering the directorship must have been a severe below and a loss of face for Raman. The parallel might not be exact but in terms of the treatment meted out, one cannot help recalling Oppenheimer's fate. If Raman had stopped doing science after this painful experience, it would have been perfectly understandable. On the contrary, such was his spirit that he threw himself into science with even greater vigour.

Raman's work at the Institute is sometimes dismissed as not being "contemporary". In a limited and clinical sense this is perhaps true. However, in making this judgement one must remember that Raman was very eager to move with the times and pursue nuclear physics but did not have the resources to do so. Such money as was pumped in for this purpose was earmarked for Bhabha's use.

Was it merely a question of money? Given his classical background would he have been a success in this new frontier even if he had obtained the funds? In debating this question one must remember that in the thirties nuclear physics was still going through an exploratory, "qualitative" phase dominated by experimentalists like the Curies and the Cavendish group. Much of Fermi's celebrated work also falls into this category. None of these studies really called for any profound understanding of quantum mechanics, and it is reasonable to suppose that Raman might have held his own. The conclusion is particularly tempting in the light of Raman's attempts to persuade Bhabha to follow certain leads in astrophysics, and the advice he gave to young Vikram Sarabhai to look for cosmic ray tracks in photographic emulsions.

As for the work which Raman actually did during this period, if one looks carefully behind the feathers, the shells, the lycopodium powder and the broken pieces of glass, one would observe that Raman was becoming interested in the passage of waves through heterogeneous media. Even the Raman–Nath theory is nothing but an example of that. Viewed thus, it becomes apparent that Raman was taking the next major step in classical wave theory, in the process unconsciously opening up a new chapter in applied physics! One has merely to browse through various journals on applied physics (particularly those devoted to acoustics and optics) published after the sixties to be convinced of this fact.

There is also a curious sociological detail worth observing in passing. The Raman–Nath theory received immediate attention and recognition because it was also current elsewhere. On the other hand, the work on speckles and the discovery of the soft mode, though both important in their own respective ways, remained ignored since the world at large did not seem to regard them as significant just then.

While applauding Raman for not losing his touch in optics, one must at the same time recognize that in solid state physics he seemed rather out of depth. The study of the physical properties of solids goes back to the last century and perhaps even much earlier. The advent of quantum mechanics brought forth, however, an entirely new perspective, the emphasis shifting to an understanding of all the usually studied properties (thermal, optical, electrical, etc.) in *microscopic* terms. At this level classical physics was found totally wanting and quantum mechanics inescapable. The new mechanics almost came to Bangalore via Born, but alas!

Nowhere is Raman's lack of grasp of the quantum theory of solids more evident than in his papers on diffuse X-ray streaks [8.65–8.68]. The discovery was an interesting and intriguing one. The Laue diffraction patterns of many crystals – diamond, sodium nitrate, calcite, hexamethylene tetramine, etc. – showed not only the familiar diffraction spots arising from crystalline periodicity but also several other unexplained streaks, very interesting modifications to which occurred when the orientation of the crystal was altered and the temperature was changed. How does one explain all this? Raman is on the right track and suspects, as is reasonable, that the new effects might well be due to the thermal motion of the atoms in the crystal, with possible masking effects from Compton scattering on the one hand and crystal imperfections on the other. By tortuous reasoning running to several pages he arrives at two conservation equations governing the scattering from the thermal agitation, equations which are pretty standard for one schooled in solid state physics. However, for Raman this is a discovery. His arguments while laborious are fascinating nevertheless, revealing as they do his struggle to grasp quantum truths with a classical background. That he eventually succeeds is entirely due to the fact that only wave concepts are needed thus far. Having arrived at these conservation equations, i.e., for the wavevector (in nonlinear optics one would call it the equation for phase matching) and for the energy, he stops. What should have been the starting point becomes for him the end. The logical follow-up should have been a calculation of the *intensity* of scattering, its angular and temperature dependence, etc., all vitally necessary for a thorough test of his hypothesis. Unfortunately Raman is not able to proceed further for it calls not merely for a qualitative appreciation of the principles of quantum mechanics but a mastery over the technical manipulations involved.

Why did Raman shy away from the quantum theory of solids, especially when he had such a good opportunity to learn it from Max Born? One can only speculate. Maybe it was his age; many in his age group even in Europe and in America did find it difficult to adjust to the new developments. But there is also another possibility. Born's stay at the Institute coincided with the period when Raman was experiencing great mental stress. Among other things, the opposition as a part of its smear campaign was claiming that the Raman effect was really Krishnan's discovery! Perhaps Raman felt an overpowering urge to re-establish his reputation via new discoveries and findings, and in the process started to become blind to his shortcomings. When Born left there was none to offer an independent and critical opinion. His students too did not correct him when the occasion demanded, whatever be the reason.

The one-sided growth in Bangalore becomes particularly evident when one compares the work on light scattering done in Bangalore with similar work done around the same time in the USSR, where strong theoretical support was available. While recalling his experiments to me, Thosar (see Sec. 8.8) remarked that the fluorescence exhibited by the ruby necklace was most spectacular, evoking amazement in everyone, including Raman. Did the shape of the ruby pieces have anything to do with it and could they have discovered the laser principle if they had probed further, wondered Thosar. An interesting speculation, but one has one's doubts, given the absence of a guiding spirit well-versed in the theory.

It is easy to play up Raman's failings, but one must be understanding in one's criticism. Hypothetical though it may sound, the question to be asked is whether anyone else in similar circumstances fared any better[21]. While it might be argued that such considerations are irrelevant in the bar of history, a summary dismissal would, at the same time, reflect an insensitivity to human problems. Further, the difficulties of pursuing science in a backward and isolated environment cannot be minimized, and it is my submission that Raman was a creature of circumstances, even as many others of his generation were. In a somewhat different manner, both Saha and Bose also appear to have found the atmosphere debilitating. While at least

On to Bangalore

Raman somehow struggled to keep himself busy with research, the scientific output of the other two rapidly petered out.

The most notable contribution Raman made during the Bangalore period is one which has gone practically unnoticed. Scarcely is it realized that he founded again a viable school of physics from which came many leaders. Nor must one forget the tradition he built up in experimental physics despite the acute shortage of funds. This is particularly worthy of recall at the present time when experimental science is generally at a low ebb in the country. And, as many remembered with feeling years later[22], Raman never failed to infect those around him with his undying enthusiasm for science.

Perhaps the best tribute to Raman's leadership was unconsciously paid by Homi Bhabha when he resolved to continue in India after the war.

Altogether, the Institute period was quite different from Calcutta. While a first glance seems merely to show a somewhat faded Raman with but modest triumphs, a closer look reveals a man of great fighting spirit tenaciously hanging on and enthusing his students, unmindful of his own personal disappointments. It is regrettable that the leadership he provided has been largely forgotten though it has, in its own way, made a deep impact on Indian science.

9 *Son et Lumiere*

The Raman–Nath theory is unquestionably the high point of Raman's work at the Indian Institute of Science. The theory was progressively unfolded in a series of five papers by Raman and Nath together [9.1], following which Nath [9.2] published two papers independently. There were also other papers from the Institute relating to the experimental verification [9.3, 9.4].

Raman loved waves, and this problem had light waves as well as sound waves. What more could he ask for? The problem was not formulated by him, and had existed even earlier; what Raman and Nath did was to provide an original solution.

The Raman–Nath papers have a lyrical quality, and even today their charm as well as value remain undiminished.

9.1 The Problem

Consider a parallel beam of light incident normally on a rectangular cell filled with a liquid. A sound wave is now propagated through the fluid in a direction perpendicular to that of light, as illustrated in Fig. 9.1. The sound wave produces alternate layers of compression and rarefaction, and thereby a periodic modulation of refractive index. The question now is: what happens to the light beam?

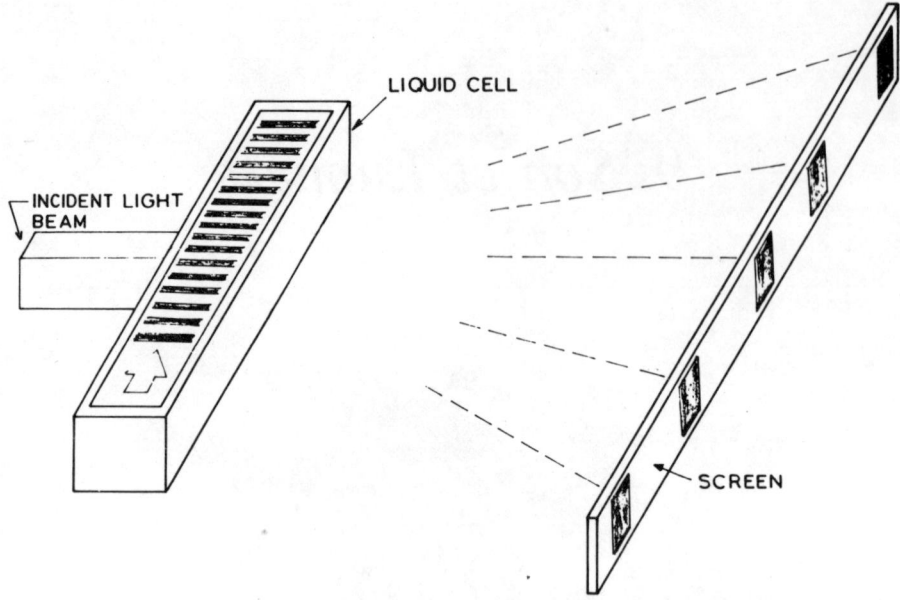

Figure 9.1 Geometry of the experiment to demonstrate the diffraction of light by density fluctuations in a liquid produced by an impressed ultrasonic wave. The fat arrow shows the direction of the sound wave.

In 1921 Brillouin [9.5], who was considering the scattering of light by liquids, argued that the thermal agitation of atoms can be regarded as equivalent to the passage of sound waves travelling in various directions. He then showed that consequent to the periodic disturbance created by a sound wave of wavelength λ^*, the light would be diffracted, the diffracted beams emerging at an angle θ given by

$$\sin \theta = (\lambda/\lambda^*),$$

a formula familiar in diffraction theory. This is Brillouin scattering, already mentioned in Chapters 5 and 8.

A few years after Brillouin made his prediction, Debye and Sears [9.6] in America and Lucas and Biquard [9.7] in Europe attempted to verify whether light would in fact be diffracted by sound waves by *deliberately* imposing a sound wave on the liquid (as shown in Fig. 9.1). They then found several diffraction bands at angles θ_n given by

$$\sin \theta_n = \pm n\lambda/\lambda^*, \tag{9.1}$$

where n is an integer. Shortly thereafter, Bär [9.8, 9.9] in Switzerland carried out more thorough investigations, and his beautiful results in fact provided much incentive to Raman and Nath.

All these workers found diffraction bands exactly as Brillouin had predicted. However, there was a problem. As the amplitude of the sound wave was varied[1], the nature of the diffraction pattern changed considerably, with some lines becoming more intense and others disappearing etc. Raman and Nath refer to this phenomenon as a wandering of the intensity between various orders. Why did this happen?

Lucas and Biquard attempted an explanation of the observed results assuming that the intensity of the bands was controlled by the refraction or bending of the light beam by the periodic disturbance in the liquid cell. In other words, they tried to reduce the problem to *geometrical optics*, but their approach was not successful.

In 1933 Brillouin [9.10] attempted a quantitative explanation of the various experiments cited above, offering what we would today describe as a perturbation theory for calculating the band intensities. Brillouin's theory is valid only when the externally imposed sound beam is faint, and predicts (in the first instance) only two first-order side bands in addition to the central, zero-order band. But experiments showed bands of many orders, besides a wandering of the intensity amongst the various orders. Thus, neither the theory of Brillouin nor that of Lucas and Biquard could provide an explanation of the diverse experimental facts.

9.2 Raman's Ideas

In the Lucas–Biquard theory, the rays of light *bend* while travelling inside the liquid. Raman and Nath argued that the rays do *not* bend. Then how does one explain the intensity of the bands on the screen? This is precisely where Raman's genius shows. Drawing inspiration from Lord Rayleigh, Raman conjectured that in the presence of the sound wave, the liquid behaved like a *phase grating*.

The effect of a phase grating can be appreciated by considering the analogy of a 100-metre race. At the start, all the competitors are together. During the race, each runner stays in his own track or lane but, on account of variations in speed, the arrival of the sprinters at the finish line is not simultaneous. Some arrive ahead while others reach later. Something similar happens when light passes through a phase grating. But how can light travel at different speeds like the runners in a race? Is not the speed of light a constant? Yes, but that is for light travelling in vacuum. When light travels in a medium, its speed is determined by the refractive index of the medium. The latter in turn depends on the density and when the density varies periodically on account of the sound wave, there are alternate slow and fast tracks for the light. The net result is that at the exit end of the liquid cell, the phase is not uniform across a plane as it is at incidence. In slightly more technical language, one has a *corrugated wavefront* (see Fig. 9.2), and such a wavefront is equivalent to a collection of light beams propagating in different directions. It is not surprising therefore that on a distant screen, one sees an array of well-separated bands as in Fig. 9.1.

Given the basic idea of a phase grating and a resulting corrugated wavefront, many things follow automatically. First of all there is a wandering of the intensity in

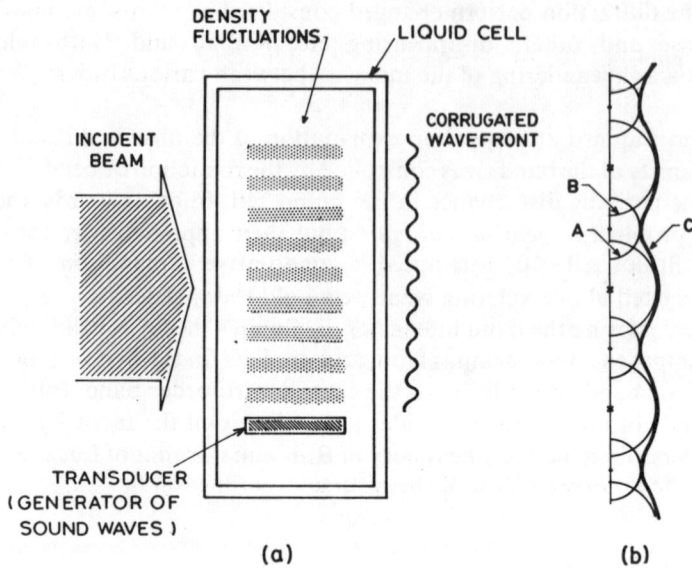

Figure 9.2 (a) Emergence of a corrugated wavefront after passage of light through a liquid tank in which sound waves are generated. (b) A and B show Huyghens secondary waves emerging from the right-hand side of the tank, from regions of compression and rarefaction respectively; C is the envelope which shows corrugations.

the various orders, this being a consequence of the variations in the corrugation which result when the amplitude of the sound wave is altered. The wanderings can be quite intricate, as Fig. 9.3, taken from the first paper of Raman and Nath, shows. When there is no sound wave there is a single emergent light beam, leading to the so-called central band. As the amplitude of the sound wave is gradually increased, additional bands start appearing but with some interesting features. For instance, after a stage, the central band starts diminishing in intensity; subsequently, "the first-order bands fall in their intensity, giving up their former exalted place to second order", and so on.

Suppose the light is incident obliquely, as in Fig. 9.4? This is no problem and similar arguments can be given. However a new feature emerges. Let ϕ be an angle as defined in Fig. 9.4.

> The diffraction spectrum will be most prominent when $\phi = 0$. The intensity of the various components wander when ϕ is increased. When ϕ increases from zero to α_1, the number of the observable orders in practice decreases and when $\phi = \alpha_1$ all the components disappear except the central one which will attain maximum intensity. This does not mean that the intensities of all the orders except the central one decrease to zero monotonically as ϕ varies from zero to α_1, but some of them may attain maxima and minima in their intensities before they attain the zero intensity when $\phi = \alpha_1$. This is obvious in virtue of the property that the intensity of the nth component depends on the square of the Bessel function J_n[2]. As ϕ increases from α_1 to β_1 the intensity of the central component falls and the other orders are reborn

Son et Lumière

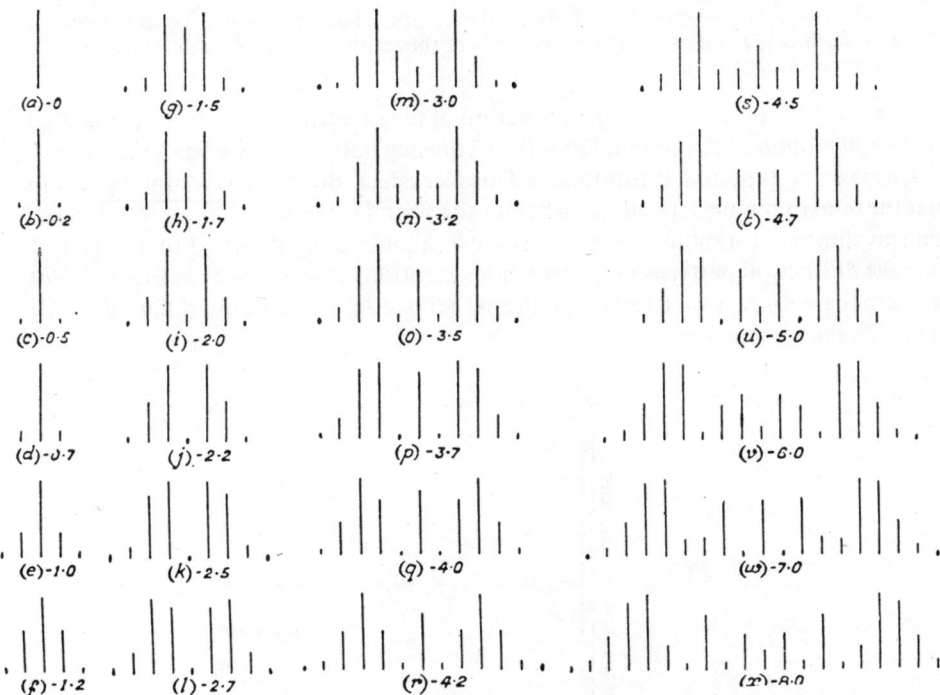

Figure 9.3 Raman–Nath diffraction patterns for various values of a parameter called v by Raman and Nath (and called x by us – see Eqn. (9.4)). Varying v is, in a sense, tantamount to varying the amplitude of the sound wave. (After ref. [9.1].)

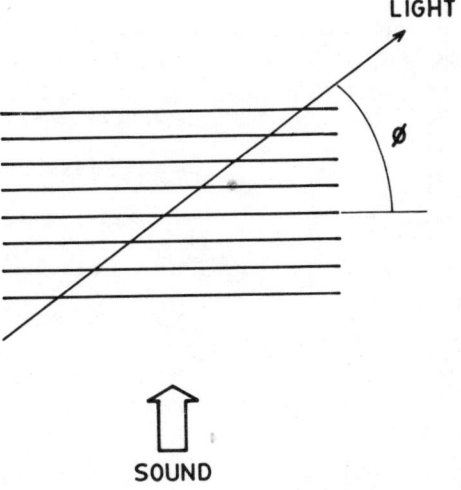

Figure 9.4 Geometry of Raman–Nath experiment for oblique incidence.

one by one. As ϕ increases from β_1 to α_2, the number of observable orders decreases and when $\phi = \alpha_2$ all the orders vanish except the central one which will attain the maximum intensity, and so on. [9.1]

Formula (9.1) predicts that the diffraction of the incident light wave is controlled by the *wavelength* of the sound. Does the frequency not have any effect? Yes indeed, and, as is to be expected, it produces a Doppler effect. But there is a subtle point in that the final outcome depends on whether the sound wave is or not progressive, i.e.. is a travelling or a standing wave. If v is the frequency of light and v^* that of sound, then the diffraction patterns in the two cases mentioned would be as in Fig. 9.5. Not only are there diffraction bands as in the earlier (static) analysis, but different bands have different frequencies.

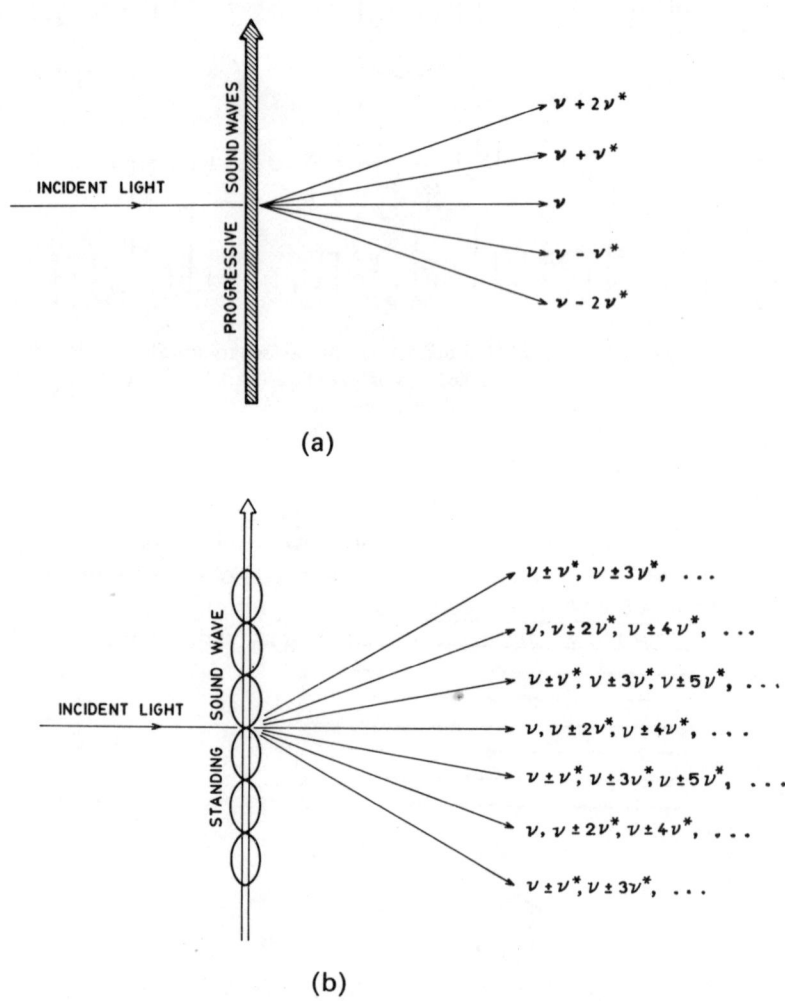

Figure 9.5 Doppler shifts produced in a Raman–Nath experiment. (After ref. [9.1].)

So far, the explanation of the observed effects has been based on the assumption that the liquid behaves like a phase grating when a sound wave is impressed on it. But it is a little too simplistic to assume that only the phase of the light wave is affected in its passage through the liquid, and not its amplitude. To take account of amplitude changes also, Raman and Nath tighten up their discussion from their third paper onwards. What emerges then is the final form of the Raman–Nath theory which is capable of dealing with all the various facets: (i) normal and oblique incidence, (ii) Doppler effect, (iii) intensity wandering, and (iv) effect of an arbitrary periodic disturbance as opposed to one of a simple sinusoidal form.

All the Raman–Nath papers were published in the *Proceedings of the Indian Academy of Sciences*[3] but they did not suffer obscurity on that account. Like the papers of Landau in the USSR and of Tomanaga in Japan, the Raman–Nath papers proved that ultimate recognition depends on the quality of the work and not on where the journal concerned is published. Ramaseshan [9.11] has observed that following the publication of the Raman–Nath papers, the circulation of the *Proceedings* went up.

Nath subsequently went to Cambridge and pursued his investigations of some of the mathematical aspects of the Raman–Nath theory. The interesting point is that he continued to publish his findings in the *Proceedings of the Indian Academy of Sciences* instead of in a journal published in England, which he could easily have done. What a contrast to present tendencies!

9.3 Experimental Verification

One of the earliest to provide detailed quantitative evidence for the correctness of the Raman–Nath theory was Sanders [9.12] working at the National Research Laboratories in Ottawa, Canada. Sanders not only observed the wandering of the intensity among the various orders as many other previous workers had done, but also verified that the wandering was in conformity with the predictions of Raman and Nath.

Sanders' results are reproduced in Figs. 9.6 and 9.7. In Fig. 9.6 are shown by vertical bars the intensities of the various orders as computed from the elementary version of the Raman–Nath theory given in their very first paper. The circles correspond to the measured intensities in the various orders. As can be seen, the agreement is excellent. Further reinforcement of this conclusion is provided by Fig. 9.7.

Before publishing his full paper [9.12], Sanders drew attention to his results in a letter to *Nature* [9.13] while commenting on work by Lucas and by Bär. In that letter, Sanders remarks, "The theory of Raman and Nath, though based on an assumption which is somewhat erroneous in the practical case, describes most accurately the nature of the diffraction effects." This criticism was unjustified based as it was on Raman and Nath's first paper where it was assumed that only the phase

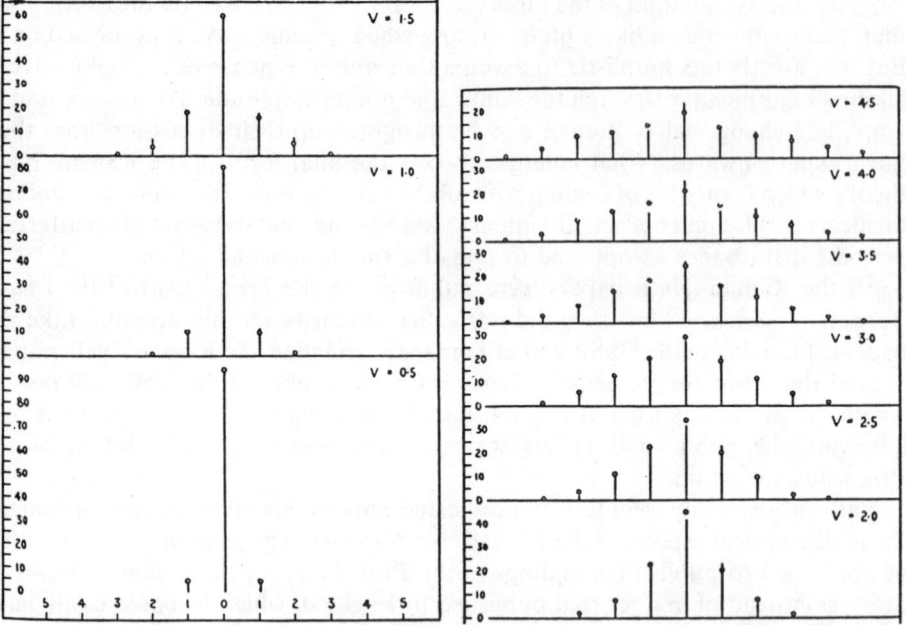

Figure 9.6 Observed and computed intensities of five orders obtained with standing waves. Each graph refers to a single value of the Raman–Nath parameter (see Eq. (9.4); our symbol for this is x whereas Sanders uses v). (After ref. [9.12].)

was affected and not the amplitude. As we know, this assumption was relaxed later but perhaps Sanders was not aware of it. So Raman and Nath replied to Sanders through the columns of *Nature*, pointing out that "in our papers IV and V, the restrictions mentioned above [i.e., that the phase alone changes] were dispensed with, and the theory of the phenomenon was developed quite rigorously" [9.14].

The full-fledged Raman–Nath theory was put to a stringent test only in 1963 by Klein and Hiedemann [9.15] in the United States. Figure 9.8 shows a sample of their results. What is displayed is the wandering of the intensity of the central order, but explored over a much wider range than in earlier experiments of Sanders and of Nomoto [9.16]. The three curves in the figure refer to the elementary Raman–Nath theory, a slight improvement computed by Mertens [9.17] using the more exotic version of the Raman–Nath theory (i.e., paper IV), and a proper solution again based on paper IV, this time computed by Berry [9.18]. Berry's results show that the Raman–Nath theory is as good as one would want it to be, provided one takes the trouble of carefully solving the equations. To obtain the full flavour of Berry's work, the reader must consult Sec. 9.5.

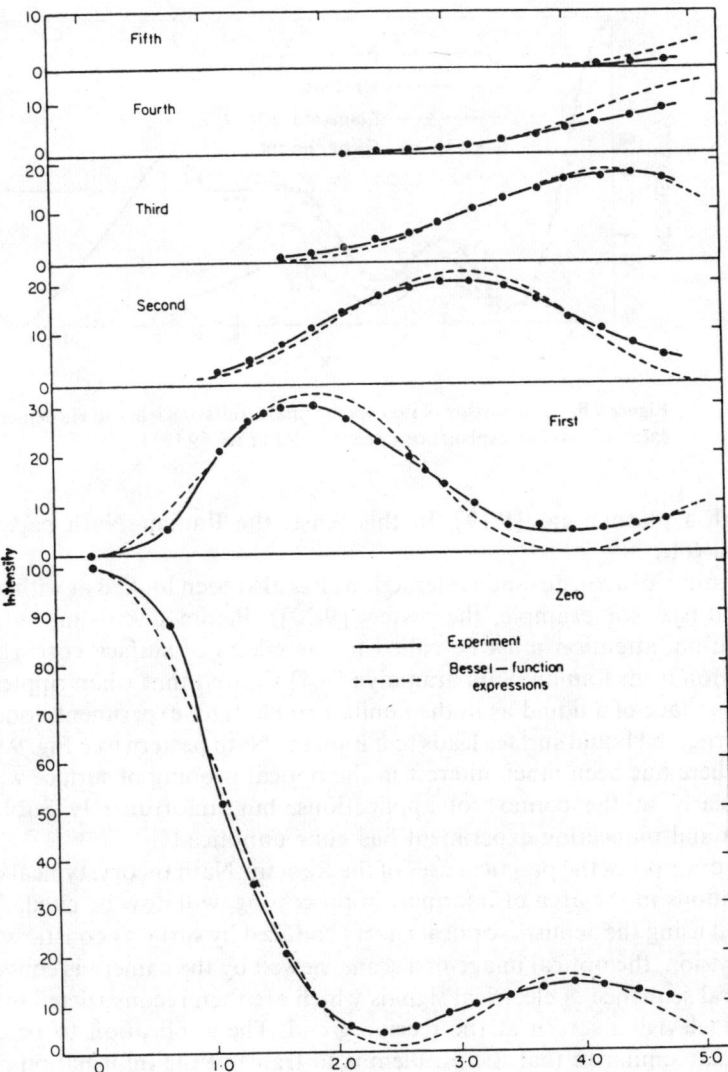

Figure 9.7 The results of Sanders presented in a different form. Shown here are the intensity variations of the different orders as a function of x. (After ref. [9.18].)

9.4 Applications

The Raman–Nath theory dealt with the archetypal problem of wave propagation in inhomogeneous media, of which there are several well-known examples such as propagation of starlight through the atmosphere, propagation of laser waves

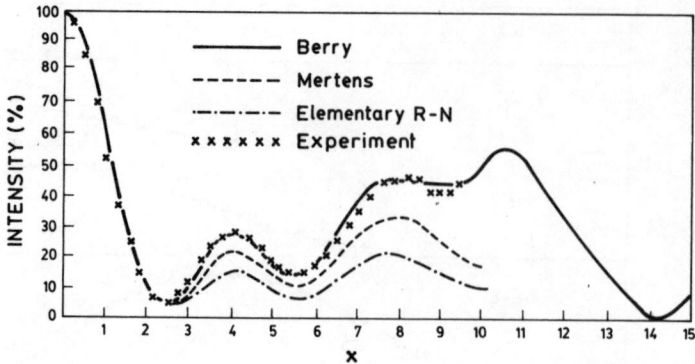

Figure 9.8 Comparison of the experimental results of Klein and Hiedemann with various calculations. For explanations, see text. (After ref. [9.18].)

through a plasma, etc. [9.19]. In this sense, the Raman–Nath papers played a seminal role.

The study of acousto-optic interactions has also been looked at with applications in mind (see, for example, the review [9.20]). Before discussing an example of application, attention must be called to the effects of surface corrugation. On a suggestion from Raman, Subbaramaiya [9.4] showed that when ripples are set up on the surface of a liquid as in the familiar ripple-tank experiment done in college, the corrugated liquid surface leads to a Raman–Nath pattern (see Fig. 9.9). In recent times there has been much interest in the optical probing of surface waves [9.20], particularly in the context of applications, but unfortunately Subbaramaiya's elegant and pioneering experiment has gone unnoticed.

One example of the practical uses of the Raman–Nath theory, typical of the many applications in the area of information processing, will now be cited. An image is scanned using the acoustic-optical effect produced by surface acoustic waves [9.21]. In television, the optical image of a scene viewed by the camera is converted into a temporal sequence of electrical signals which are then reconstructed into a picture on the television screen at the receiving end. The application to be discussed is somewhat similar in that the problem is to transmit the information content of a transparency, using light as the carrier.

The arrangement used is shown in Fig. 9.10. The incoming beam is focused by a cylindrical lens L_1 onto a drum on which is wrapped the transparency to be scanned. As the drum rotates, the transparency is scanned line by line, the emergent beam falling on a lithium niobate crystal. A surface acoustic wave is propagated on the crystal and when the light beam interacts with the surface wave, a sequence of signals is produced which replicates the spatial pattern of brightness of the image along a line. The light falling on the crystal is back-scattered, and, owing to the Raman–Nath process, is also diffracted. The intensity in the diffraction band varies in time as

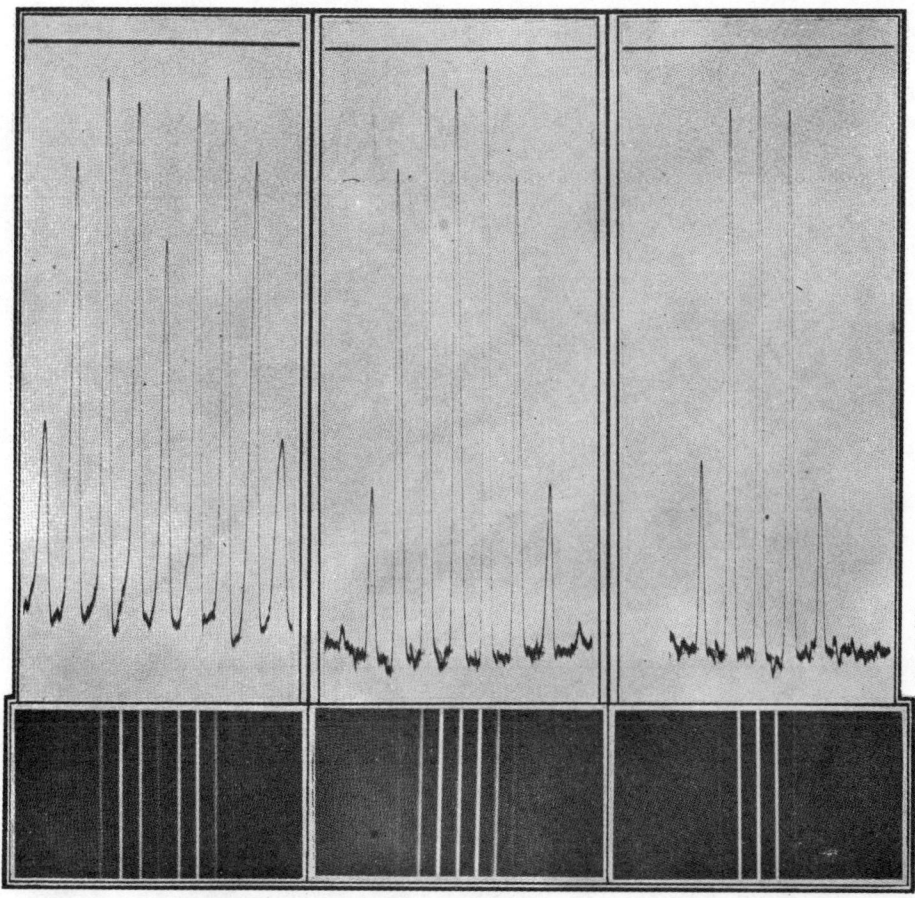

Figure 9.9 Raman–Nath patterns obtained by corrugating a liquid surface using a ripple tank. (After ref. [9.4].)

already described, and contains the information across one line of the transparency. As the drum revolves, the picture is scanned line by line. A demonstration reconstructed image is shown in Fig. 9.11. Considering that the information carrier is light, this technique has great potentiality with the advent of fibre-optic communication.

A Raman–Nath encoder is superior to a fully mechanical one employing point-by-point scanning since (in the former) the scanning is mechanical only in one direction and acoustic in the other. On the other hand, it should be obvious that a Raman–Nath encoder would be far less expensive than a TV camera (which employs full electronic scanning).

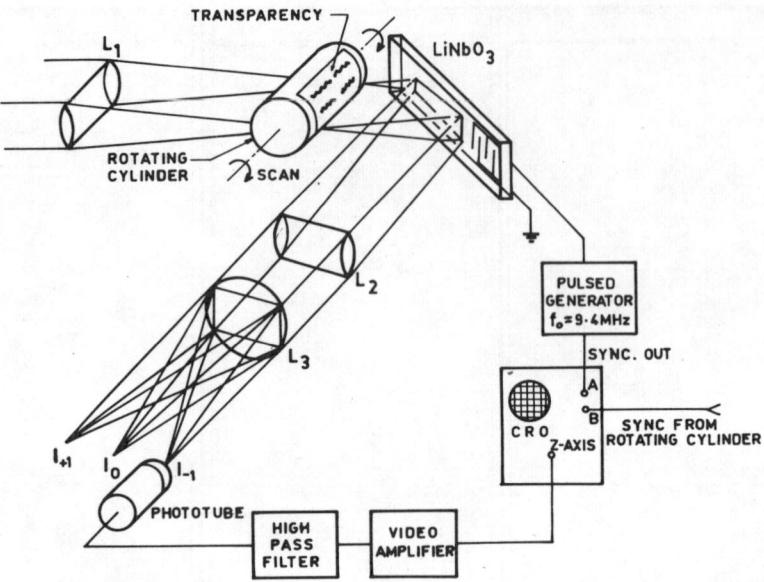

Figure 9.10 Arrangement used for information processing using the Raman–Nath effect. The scanning is mechanical in one direction and acoustical in a perpendicular direction. For further explanations see text. (After ref. [9.21].)

Figure 9.11 Photograph of image obtained after transmission using the arrangement shown in the previous figure. (After ref. [9.21].)

Son et Lumiere

☆ 9.5 The Theoretical Aspects

This section is devoted to the more technical aspects of the Raman–Nath theory. The *cognoscenti* would immediately see that I have leaned heavily on the charming little monograph of Berry [9.18]. I do this because the latter offers a convenient link with the literature that appeared subsequent to the classic papers of Raman and Nath.

Figure 9.12 provides a reference framework for describing the geometry of the experiment. Assuming that the sound wave is sinusoidal, the refractive index $\mu(\eta, t)$ can be written as

$$\mu(\eta, t) = \mu_0 + \mu_1 \cos(b\eta - v^* t), \tag{9.2}$$

where μ_0 is the value in the undisturbed medium and μ_1 the amplitude of the periodic modulation. The quantity $b = 2\pi/\lambda^*$ where λ^* is the wavelength of sound (v^* is the frequency). Since the speed of sound is less than the speed of light by a factor of 10^5, one can, to a first approximation, assume that in the short time taken for the light to traverse the liquid, the sound wave presents a stationary profile, and thus write

$$\mu(\eta) = \mu_0 + \mu_1 \cos b\eta. \tag{9.3}$$

This approximation is *not* equivalent to considering the ultrasonic beam as a standing wave but is tantamount to neglecting the Doppler effect.

Two dimensionless parameters important in the theory are:

$$\rho = b^2/\mu_0\mu_1 k^2$$

$$x = k\mu_1 D, \tag{9.4}$$

where $k = 2\pi/\lambda$, λ being the wavelength of light. In experiments, ρ and x are in the

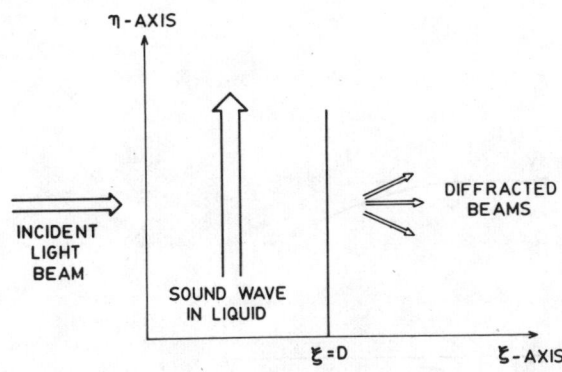

Figure 9.12 Reference framework for describing the Raman–Nath experiment. (After ref. [9.18].)

ranges

$$0.01 < \rho < 100$$
$$0 \leqslant x < 20. \tag{9.5}$$

The theory of Lucas and Biquard, which preceded the Raman–Nath theory, is based on geometrical optics, and attempts to follow the path of the rays of light in a stratified medium.

Consider a ray of light incident at $\xi = 0$ and $\eta = \eta_I$, and let ψ be the angle between the η-direction and the direction of the ray (see Fig. 9.13). According to Snell's law,

$$\mu(\eta) \sin \psi(\eta) = \text{constant}. \tag{9.6}$$

The initial condition gives

$$\mu(\eta = \eta_I) \sin(\pi/2) = \mu(\eta_I) \equiv \mu_I. \tag{9.7}$$

From Fig. 9.13,

$$d\xi/d\eta = \tan \psi. \tag{9.8}$$

Using (9.7),

$$d\xi/d\eta = 1 \Big/ \left(\frac{\mu^2}{\mu_I^2} - 1 \right)^{1/2}, \tag{9.9}$$

whence, upon integration,

$$\xi = \int_{\eta_I}^{\eta} d\eta' \Big/ \left(\frac{\mu^2}{\mu_I^2} - 1 \right)^{1/2}. \tag{9.10}$$

One can now introduce (9.3) into (9.10) and evaluate the integral numerically. This gives the path of the ray through the cell. It is now assumed that the nth order diffracted beam making an angle θ_n (defined by formula (9.1)) with the ξ-axis is

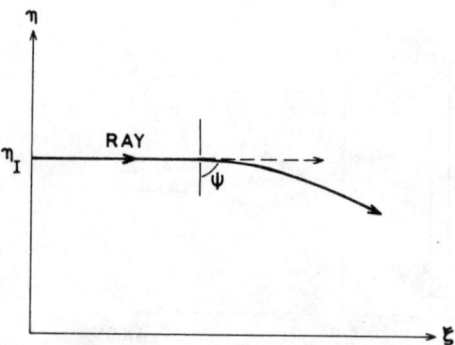

Figure 9.13 Illustration of the quantities entering a geometrical-optic description of ultrasonic diffraction. (After ref. [9.18].)

produced only by those rays emerging from the vessel at angles θ in the range

$$\theta_n - \frac{1}{2}(\theta_n - \theta_{n-1}) < \theta < \theta_n + \frac{1}{2}(\theta_{n+1} - \theta_n). \tag{9.11}$$

How good is the geometrical optics approach? Figure 9.14 provides the answer.

We next turn to the basic principle underlying the phase grating theory. Consider now a light beam represented by a scalar wavefunction $\exp(ik\xi)$ incident on an object having infinite extension along the η-axis but finite extension along the ξ-axis. Inside the medium the wavefunction may be written

$$\phi(\xi, \eta) = T(\eta) \exp(ik\xi), \tag{9.12}$$

where $T(\eta)$ is the *transmission function*. In our problem, it is periodic, i.e.,

$$T(\eta) = T(\eta + \lambda^*), \tag{9.13}$$

whence the medium behaves like a diffraction grating. If in addition $|T(\eta)| = 1$, i.e., if

$$T(\eta) = \exp[i\gamma(\eta)], \tag{9.14}$$

where $\gamma(\eta)$ is a real periodic function of η, then the object is called a phase grating. Using (9.14),

$$\phi(\xi, \eta) = \exp[i(k\xi + \gamma(\eta))]. \tag{9.15}$$

In his book on sound, Lord Rayleigh considered the effects produced when a reflecting surface has corrugations. In other words, he dealt with a *reflection* phase grating. Raman's genius lay in recognizing that the medium behaved like a *transmission* phase grating, an idea missed by earlier workers but which readily came to him because of his great familiarity with the works of Lord Rayleigh.

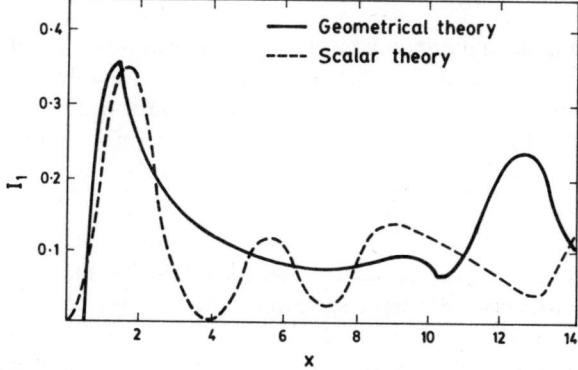

Figure 9.14 Comparison between the results of geometrical optics and the exact scalar wave theory of Berry. Plotted is the intensity variation of the first order as a function of the parameter x. The exact scalar wave theory agrees very well with experiment, as may be seen by comparison with Fig. 9.8. (After ref. [9.18].)

Since ϕ is a periodic function of η, it can be written

$$\phi(\xi,\eta) = \sum_{n=-\infty}^{\infty} \psi_n \exp[i(k\xi + nb\eta)], \quad (9.16)$$

where

$$\psi_n = (b/2\pi) \int_{-\lambda^*/2}^{+\lambda^*/2} \exp[i(\gamma(\eta) - nb\eta)]d\eta. \quad (9.17)$$

Equation (9.16) represents a set of plane waves travelling at angles

$$\theta_n = \sin^{-1}(nb/k) \quad (9.18)$$

to the ξ-axis with amplitude ψ_n. These waves constitute the diffracted beams.

In their very first paper, Raman and Nath showed

$$\psi_n \approx \text{phase factor} \times J_n(x), \quad (9.19)$$

where $x = k\mu\eta$ and J_n is the Bessel function of the first kind of order n. The intensity I_n of the nth order diffraction line is given by

$$I_n = [J_n(x)]^2. \quad (9.20)$$

This is the famous Bessel-function expression of Raman and Nath. However, its applicability is restricted to a limited range given by

$$\rho n^2 < 1$$

$$\rho x^2 < 2. \quad (9.21)$$

The first of these conditions was stated by Raman and Nath and the second by Extermann and Wannier [9.22]. Compared to the range in (9.5), condition (9.21) is rather restrictive. A more accurate estimate of the intensity I_n valid in regions beyond (9.21) can be obtained by solving the full-fledged Raman–Nath difference-differential equation.

Reference has already been made in Chapter 4 to the scalar diffraction theory of Kirchhoff. The differential equation satisfied by the (time-independent) wave amplitude in this theory is referred to as the Helmholtz equation. This is the starting point for obtaining the Raman–Nath equation, and in the present instance reads

$$\frac{\partial^2 \phi}{\partial \xi^2} + \frac{\partial^2 \phi}{\partial \eta^2} + k^2 \mu^2(\eta)\phi = 0. \quad (9.22)$$

The Raman–Nath equation just alluded to is obtained by making a Fourier expansion of ϕ. But before considering that, let us go back to the geometrical optics approach and examine it in the light of (9.22).

Geometrical optics is a good approximation if the optical character of the medium is roughly constant over a large number of wavelengths of light. This implies $k^2 \mu^2$ is large and slowly varying. Under these conditions one may write

$$\phi(\xi,\eta) = a(\xi,\eta) \exp[ik\mathscr{L}(\xi,\eta)], \quad (9.23)$$

where the phase function $\mathscr{L}$ is called the *eikonal*. Introducing (9.23) in (9.22), one gets an equation for the eikonal which is the starting point of the geometrical optics approach. The work of Lucas and Biquard is essentially in this spirit.

Next let us consider Brillouin's work. His is a perturbation theory approach. Using (9.3) one writes (9.22) as

$$\frac{\partial^2 \phi}{\partial \xi^2} + \frac{\partial^2 \phi}{\partial \eta^2} + k^2 \mu_0^2 \phi = -2k^2 \mu_1 \mu_0 \phi \cos b\eta. \tag{9.24}$$

The zeroth approximation to the wavefunction is

$$\phi \approx \exp(ik\mu_0 \xi). \tag{9.25}$$

To get the first correction to the trivial solution, the wavefunction is written

$$\phi = \exp(ik\mu_0 \xi) + \chi(\xi, \eta), \tag{9.26}$$

which, when substituted in (9.25), leads to a differential equation for χ. Solving this yields

$$\phi(\xi, \eta) = \exp(ik\mu_0 \xi) \left[1 + \frac{2i}{\rho} \exp(-i\rho x/4) \right.$$
$$\left. \times \sin(\rho x/4) \{\exp(ib\eta) + \exp(-ib\eta)\} \right], \tag{9.27}$$

which shows that in addition to the central band, there are two side bands characterized by wavevectors $\pm b$ in the direction of η. These bands constitute the zero-order and first-order diffraction beams with amplitudes

$$\psi_0 \approx 1$$
$$\psi_{\pm 1} \approx (2i/\rho) \exp(-i\rho x/4) \sin(\rho x/4), \tag{9.28}$$

from which one may readily compute I_0 and $I_{\pm 1}$.

We now turn to the Raman–Nath equation to which a reference was made earlier. This was obtained in their fourth paper, starting from the Helmholtz equation (9.22). Exploiting the periodicity of $\phi(\xi, \eta)$ one may Fourier-expand it as

$$\phi(\xi, \eta) = \sum_{n=-\infty}^{\infty} f_n(\eta) \exp(inb\xi). \tag{9.29}$$

Putting

$$f_n(\eta) = \psi_n(\eta) \exp(-ik\mu_0 \eta) \tag{9.30}$$

and introducing the dimensionless variable

$$x = k\mu_1 \xi \tag{9.31}$$

immediately leads to the famous Raman–Nath difference-differential equation

$$2\frac{\partial \psi_n}{\partial x} - \psi_{n-1} + \psi_{n+1} = i\rho n^2 \psi_n, \tag{9.32}$$

written here in Berry's notation rather than in that of the original paper. The important thing to note here is that the amplitude of the nth order is related to those of the neighbouring orders on either side. If there were no term on the r.h.s. of (9.32) then $\psi_n(x)$ would be the Bessel function $J_n(x)$, an approximate solution which Raman–Nath reported in their first paper.

In his monograph, Berry devotes as many as eight chapters to the solution of the Raman–Nath equations! A connoisseur of applied mathematics would find this part of Berry's book particularly enjoyable. Of special interest is a diagrammatic perturbation scheme, very similar to the Feynman diagram technique in quantum field theory.

We have already seen some comparisons between experiment and the full-fledged theory. For small values of ρx, the wanderings of the intensity are adequately described by the Bessel-function approximation (9.20); but for large values a proper solution of the Raman–Nath equations is necessary.

A word now about the Doppler effects. To obtain these Raman and Nath used the time-dependent wave equation in which $\phi(\xi, \eta)$ is replaced by $\phi(\xi, \eta, t)$ and $\mu(\eta)$ by $\mu(\eta, t)$.

A few additional remarks. Berry has given what he calls an exact scalar wave theory. It is an alternative approach for solving the Helmholtz equation using Mathieu functions. The solution obtained using the Raman–Nath equation does not require Mathieu functions. Berry's theory and the Raman–Nath theory are thus completely equivalent.

Earlier, Bhatia and Noble [9.23] gave another version of the exact treatment based on Maxwell's equations for the electromagnetic radiation. Their theory involves an integral equation. In physical terms, Bhatia and Noble treat the light disturbance at a point as a sum of the incident light and the light scattered to that point from matter present at other points. Allowance is made for the time of transit of the scattered wave between the scattering point and the point of observation. Berry has demonstrated that a fully scalar theory suffices to explain all the present experiments. He has also pointed out that the Raman–Nath treatment has been used by Molière [9.24] to discuss the scattering of high-energy particles by a centre of force.

It must be noted that Berry's exact scalar wave theory neglects Doppler effects, and does not treat the case of oblique incidence. However, both these effects can be allowed for.

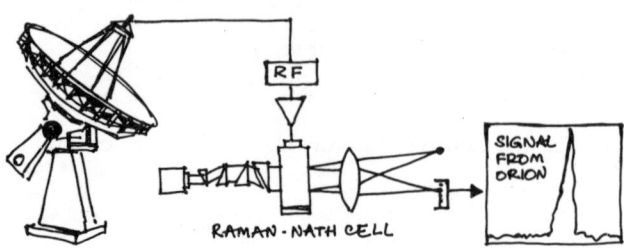

10 *The Born–Raman Controversy*

He who never made mistake, never made a discovery.
SAMUEL SMILES

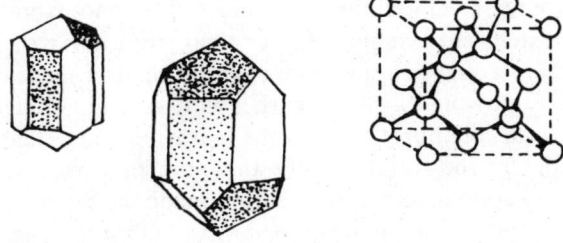

10.1 The Beginnings of Crystal Dynamics

In 1819 the Frenchmen Dulong and Petit noticed that the specific heat per gram atom appeared to have the same value, ~ 6 cal/mol degree, for all solids. This result, known as the Dulong–Petit law, was explained only in 1871. Boltzmann, who gave the theory, argued that atoms in a solid executed vibrations about their mean positions rather like a harmonic oscillator. Applying the principle of equipartition of energy, he then deduced that C_V the specific heat at constant volume is ~ 6 cal/mol degree, the Dulong–Petit value. One would have felt delighted by this explanation except that, as noted by Boltzmann himself, there were some substances, like carbon, boron and silicon, which seemed to violate the Dulong–Petit law.

Not much notice was taken of these deviations to the rule until later experiments revealed that the Dulong–Petit law held only at high temperatures, and that the specific heat of *all* solids actually decreased with temperature, approaching zero as the temperature approached absolute zero. There was no way in which this new result could be understood in classical statistical mechanics. The breakthrough came in 1907 when Einstein used Planck's law for the distribution of energy among the oscillators, instead of the classical equipartition law to which Boltzmann had appealed.

Einstein's theory, while successful in broadly describing the temperature dependence of C_V, is rather naive as far as the description of the atomic vibrations is concerned. Einstein himself recognized this shortcoming but did not bother about it. For him it was sufficient that the mystery regarding the breakdown of the Dulong–Petit law had been explained.

10.2 The Debye Theory

It was soon noticed, particularly by Nernst and Lindemann, that Einstein's theory showed certain deficiencies at very low temperatures. This led Debye to speculate in 1912 that the shortcomings of Einstein's theory lay in the frequency distribution implied by it. Debye reasoned that the atoms in a solid cannot vibrate independently of each other as Einstein had assumed. Rather, their vibrations would be coupled, as a result of which there would be a spectrum of vibrational frequencies instead of just a single frequency. Following Rayleigh's earlier work on a one-dimensional chain of atoms, Debye tried (in collaboration with the mathematician Haar) to construct a mathematical model for the vibrations of a three-dimensional crystal lattice[1]. The problem proved very complicated and so Debye made the *ansatz* that as a result of the coupling of the motion of adjacent atoms, the vibrations are propagated as waves. One knows quite generally that vibrational waves of long wavelength are the same as the acoustic waves that can be set up by coupling the crystal externally to an ultrasonic source. Debye asserted that *all* atomic vibrational waves are of this type. The crystalline lattice was thus swept away and replaced by an elastic continuum, which in turn led to a continuous frequency spectrum. The different elastic waves were regarded by Debye as independent oscillators, and the enumeration of their energy levels, their occupancy, etc. was then done *a la* Einstein. In this way, Debye was able to provide a better description of the temperature dependence of the specific heat than was possible with the Einstein theory.

10.3 The Born Theory

About the same time as Debye, Born and von Karman were also independently tackling the problem of vibrations of a crystal lattice. They started with a one-dimensional chain of atoms and later generalized the results to three-dimensional lattices. According to Born and von Karman, the normal modes of vibration of an infinite crystal can be described in terms of travelling waves. In this respect there is a certain amount of similarity with the Debye theory. However, while Debye analysed the waves associated with atomic motions by assuming the solid to be an elastic

The Born–Raman Controversy

continuum, Born and von Karman described them with due regard to the lattice structure.

10.4 Periodic Boundary Conditions

One important feature of the Born–von Karman theory is that it employs what are called the *periodic* or *cyclic boundary conditions*. It is necessary to understand what these conditions imply as they were the target of attack by Raman. Crystals are finite and have boundaries. Depending on how one grows the crystal, cuts it, polishes it, etc., the size and shape can vary widely. But theoreticians do not like to deal with such arbitrary situations. Born circumvented this problem with the periodic boundary condition, the essence of which is illustrated in Fig. 10.1.

Consider a finite crystal AB as in Fig. 10.1a. This has boundaries at A and B. Now using AB as a module, repeat it so as to obtain an infinitely (i.e., periodically or cyclically) extended crystal as in Fig. 10.1b. The segment AB is now embedded in the infinite lattice and appears as A′B′ without any boundaries. It is now claimed (and this is the essence of the periodic boundary condition idea) that the frequency spectrum of any arbitrary real crystal is practically the same as that of an equivalent crystal of the same shape *and* with cyclic or periodic boundary conditions. With respect to the example just considered, this means that the crystals AB and A′B′ have

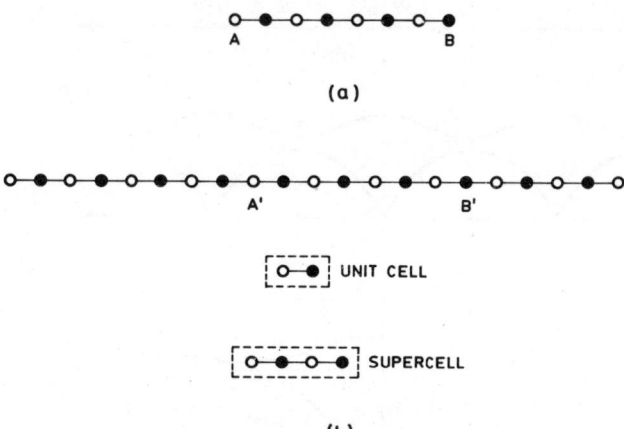

Figure 10.1 (a) A finite diatomic crystal AB. When this is periodically repeated on either side, one obtains the infinite one-dimensional chain shown in (b). The segment A′B′ is the analogue of the crystal AB. However, while AB has free boundaries, A′B′ is subject to cyclic boundary conditions. Calculation of frequency spectra is easier in the latter case. In using this simplification, it is tacitly assumed that the spectra of AB and A′B′ are very nearly the same. Later, Peierls made this argument more rigorous (see Sec. 10.9).

vibration spectra with practically the same frequency distribution. The great merit of the cyclic condition approach is that it considerably simplifies the calculation of the normal modes of the lattice. Once the normal mode frequencies are known, the calculation of the frequency spectrum is straightforward; and from the frequency spectrum to the specific heat C_V is a well-defined step.

Figure 10.2 amplifies some of the ideas stated above. With cyclic boundary conditions, the normal modes are travelling waves, each characterized by a frequency ω, a wavelength λ and a polarization (which we will ignore). Information about the normal modes is comprehensively represented by a graph of ω versus $q(=2\pi/\lambda)$, referred to usually as the dispersion relation for lattice vibrational waves. Observe the following:

1) As the size of the sampled segment A'B' increases, the number of normal mode frequencies also increases. In fact, if N is the total number of atoms in A'B', then

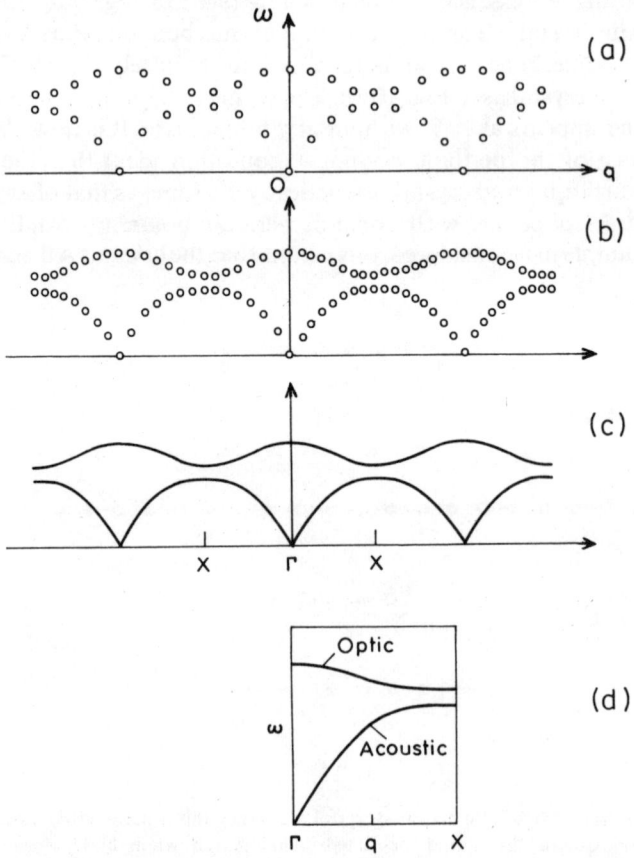

Figure 10.2 (a)–(c) Plots of ω versus q for the diatomic chain as N is increased. It should be evident that as the sample size is increased, one would eventually obtain the continuous curves shown in (c). On account of the repetitive nature, the curves are usually represented only over a limited region ΓX, as in (d). The region $X\Gamma X$ is referred to as the Brillouin zone.

The Born–Raman Controversy

the sample would have N normal modes. (For a three-dimensional lattice, this number would be $3N$.)

2) Although the sample size may be varied, the *shape* of the dispersion curves remains unaltered; but the curves are better defined, having more points on them.
3) The dispersion curves have two branches of which the lower one is called *acoustic* and the upper one *optic*. (However, if, unlike in Fig. 10.1, there was only one atom in the unit cell, there would be no optic branch.)
4) The low-frequency region of the acoustic branch is linear. Waves associated with normal modes in this region are the same as those introduced by Debye.

Figure 10.3 shows some of the modes of vibration associated with what Raman would call a supercell. In terms of the dispersion curve, these modes are somewhat special. We highlight them here as they form the hub of Raman's theory of crystal dynamics.

In the Einstein model there are no lattice vibrational waves. The Debye model accepts their existence but makes the sweeping assertion that *all* such waves can be

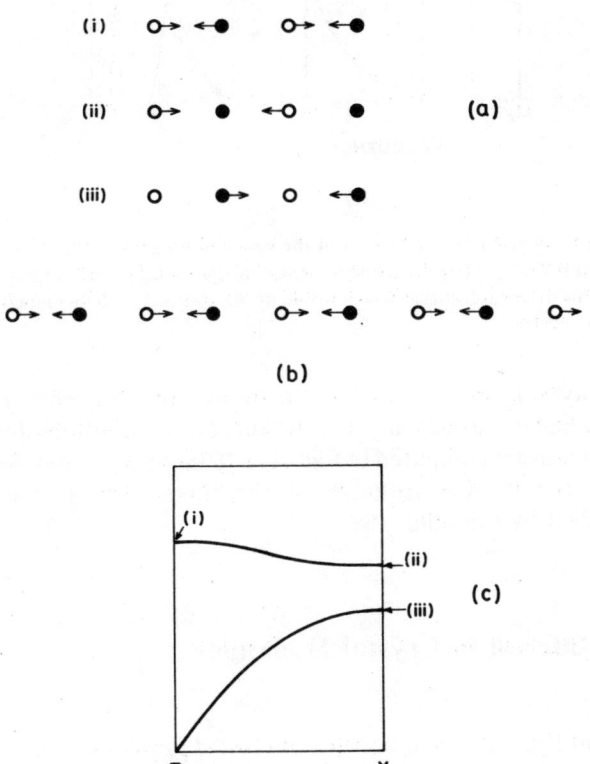

Figure 10.3 (a) Vibrations of the supercell associated with the one-dimensional chain of Fig. 10.1. According to Raman, there would be three modes, as illustrated here. From such modes one can obtain the vibration pattern in the chain by repetition, as in (b). In the Born theory, these modes correspond to those shown with arrows in (c).

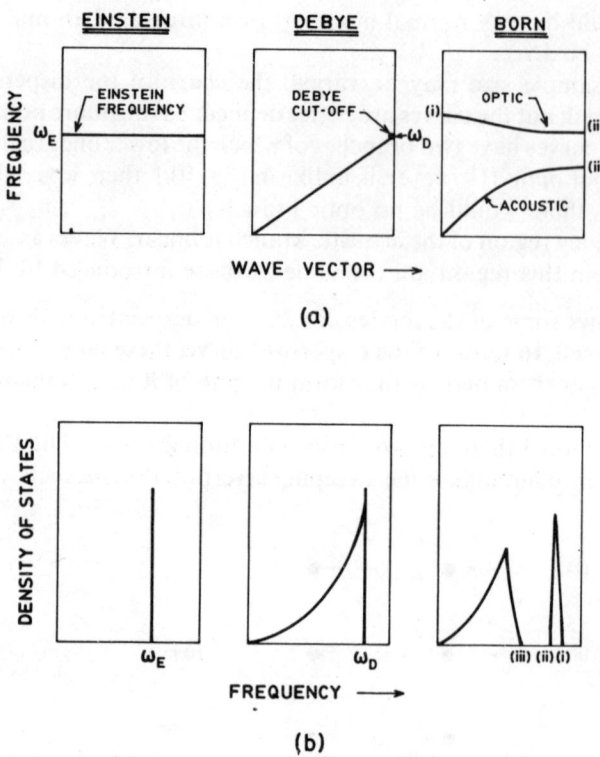

Figure 10.4 Schematic comparison of the essential features of the theories of Einstein, Debye and Born. We consider a one-dimensional diatomic lattice as in Fig. 10.1. In (a) are sketched the dispersion curves which would be appropriate to these models, and in (b) the frequency spectra.

treated as elastic waves. In other words, there are no optic branches. Further, the acoustic branch is linear *throughout*. The frequency distributions for the three models under discussion are compared in Fig. 10.4. If one accepts that the Born–von Karman model is the most reasonable of the three, then one can see the simplifications implied by the other two.

10.5 Raman's Interest in Crystal Dynamics

It was inevitable that Raman would become interested in the dynamics of atoms in crystals. Crystals always held a fascination for him, and when he acquired a new tool, it was only natural that he used it to study crystals. The Raman spectrum of diamond figures prominently in the Born–Raman controversy on lattice dynamics[2]. Bhagavantam has recounted how the experimental study of diamond started.

In the search for new materials of interest, wherein Raman effect could be studied, the manner in which Raman picked up the case of diamond and the story of how a study of its remarkable physical properties and of its structure came to be a fascinating and life-long involvement for him are both interesting and illuminating. Immediately after the discovery, Raman noticed quite casually that his younger brother was wearing a wedding ring with a diamond on it and made some provocative remarks[3]. The latter, who [had] just then graduated in physics and was a research student made a brief spectroscopic examination and found that diamond exhibits a strong and sharp Raman line corresponding to the now well known frequency shift of 1332. This result was immediately confirmed by me, repeating the experiment at Calcutta, in Raman's laboratory. It was further observed that besides this sharp line, a complex luminescence spectrum, the principal feature of which was a well defined band at 4115 A.U. was also present. With these results as the starting point, Raman built up his interest in diamonds to such an extent in a few years, that at one time, every student working with him was engaged in studying one aspect or other of the properties of diamond.

The need for diamonds, in all sizes, shapes and quality, became so much that Raman began acquiring them by all methods such as purchasing, borrowing from shops and wealthy owners, buying in auctions and from individuals and so on[4]. He soon became a connoisseur competing with professional jewellers. Quite early in this process, he borrowed one big piece (140 carats) for me from the collection of one of the then wealthy Indian Maharajahs. The loan was for a limited period and against a security bond for a fairly large sum of money which he had to sign before borrowing it. When he put this rare diamond in my hand for intensive spectroscopic studies in the following 48 hours, he cautioned me to be careful and not lose it and told me how history is replete with stories of professional thieves following great Indian diamonds to stumble on the right opportunity for whisking them away from the owners.

For two nights, I slept under the spectrograph with protecting glasses on my eyes. We got some very exciting results and my paper was immediately published. The results were abstracted in Science Abstracts under the title "Raman spectrum of diamond". Within brackets, the abstractor noted that the weight of the diamond used was 140 carats and to show his surprise, he inserted three or four exclamation marks. He could not believe that a poor Indian scientist could handle a diamond of that large size. During the process, I also learnt with some pride that India was the original home of some of the well known diamonds which found their way to other parts of the world and helped to spread the fame of this gemstone and of Indian Maharajahs. [10.1]

10.6 Raman Spectrum of Diamond

A characteristic feature of the Raman spectrum of diamond is the occurrence of sharp lines. (However, as we shall soon see, these lines are more apparent than real.) Figure 10.5 shows some examples of the Raman spectrum of diamond due to Raman and various colleagues of his. The questions which arise are: Why sharp lines, and how does one account for their number? Obviously these would lead one to the problem of normal modes of vibration of a crystal lattice. There was already

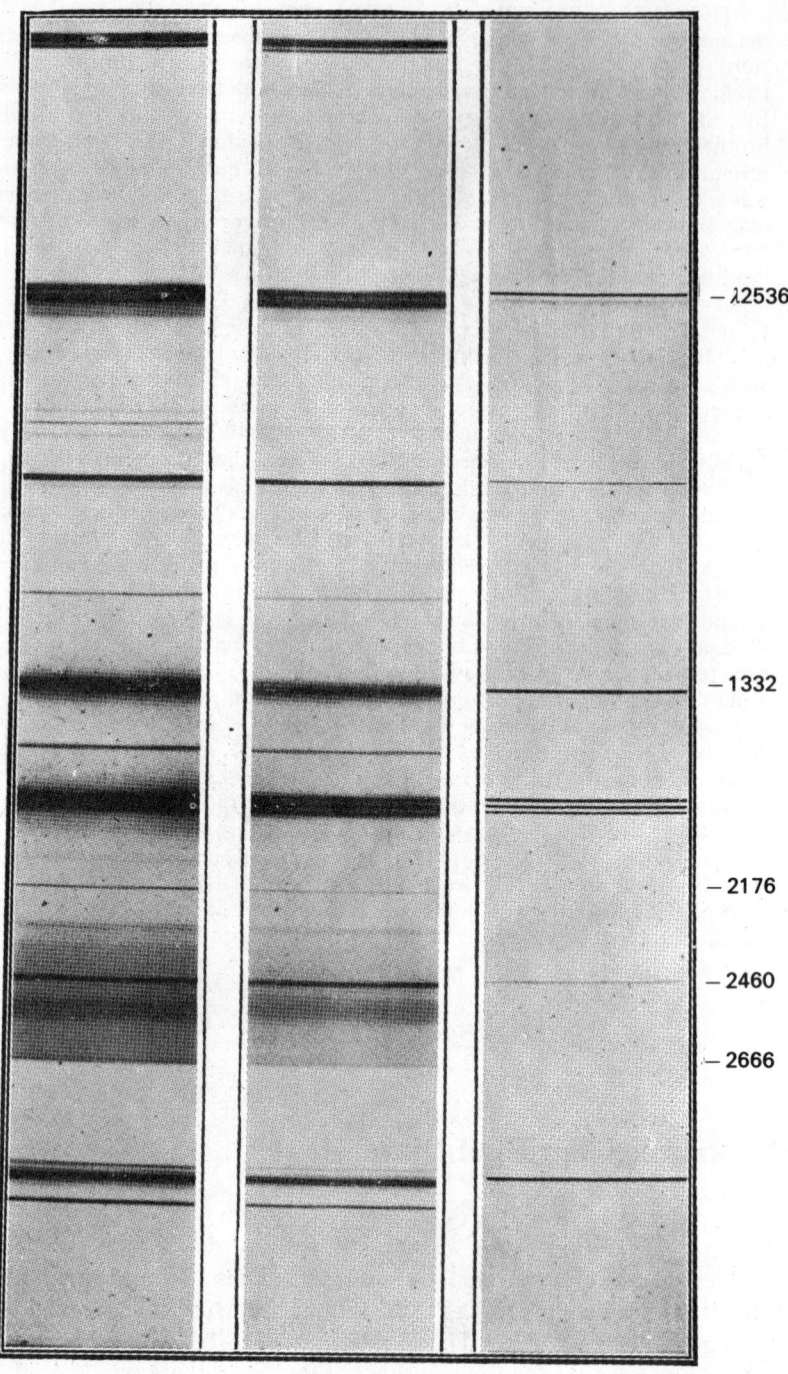

Figure 10.5 Raman spectrum of diamond observed with a spectrograph of medium resolution at increasing exposures. (After ref. [10.2].)

available a scenario for this, due to Born. But Raman finds it unsatisfactory, and decides to construct his own.

Before considering Raman's theory, let us ask what one may expect for Raman scattering in the Born picture. Raman scattering by a crystal involves exchange of energy by the light quanta with the normal modes of vibration. If only one mode is involved, the scattering is by what is called a one-phonon process. When many normal modes participate, one has multiphonon scattering.

One-phonon Raman scattering is possible only by the $q = 0$ modes of the optic branch[5]. The number of such modes is generally small. If there are p atoms in the primitive unit cell, then one can expect at most $(3p - 3)$ such frequencies (in three dimensions).

Multiphonon Raman scattering is a different story. Two normal modes with frequencies ω_1 and ω_2 can jointly participate in Raman scattering, provided their wavevectors add up (vectorially) to zero. Given this rather weak constraint, one can envisage all sorts of combinations of ω_1 and ω_2 leading to two-phonon Raman scattering. In a nutshell, the two-phonon Raman spectrum would not consist of sharp lines (as in the case of molecules) but would rather be a blur or a continuum.

10.7 Raman's Theory

Let us now turn to Raman's interpretation of the spectra obtained in his laboratory. He is struck by the appearance of sharp lines in the recorded spectra. Molecules exhibit such spectra and it is natural therefore to attempt an explanation along the lines of molecular models. Does this mean that one regards the entire crystal as a giant molecule? Obviously not, for that would result in millions of frequencies leading to a blur in the spectrum, which is contrary to observation. The molecular picture therefore needs to be introduced with some finesse.

Raman bases his strategy on some energy considerations. We now know that these arguments are rather tenuous but Raman decides that the

> energy of excitation would necessarily be localized in...a group of lattice cells whose linear dimensions are of the same order of magnitude as the range of the intermolecular forces. A quantum of vibrational energy when distributed over such a small volume would result in atomic movements of finite amplitudes and hence, as the result of optical anharmonicity, give rise to scattered radiations with overtones or summational frequency shifts. [10.2]

What about the elastic waves of Debye? Raman has this to say:

> The thermally excited vibrations of the elastic solid type would be incapable of giving any observable second-order effects. For, the energy of any such vibration having a specified frequency would be distributed over the volume of the crystal and the amplitudes of vibration could, therefore, only be infinitesimal. The interatomic displacements associated with the translatory movements would be of a still smaller order of magnitude. Hence, the local variations in optical polarizability associated

> with the elastic vibration of any particular frequency would be excessively small, and since they vary in phase from point to point within the crystal, their external effects would cancel out completely.

There is an exception, however, which

> arises when the separation between the nodal planes of the pattern is so related to the wavelength of the light traversing the crystal and the angle of incidence on the nodal planes that there is a coherent reflection of the incident light waves by the elastic wave pattern.

This of course is Brillouin scattering.

Let us paraphrase Raman's ideas in the language of Born. He claims that (i) normal modes associated with the acoustic branches can be seen via Brillouin scattering, and (ii) second-order Raman scattering is restricted to those modes in which the oscillations are confined to a local region. There is no controversy over (i); the differences are all concerning (ii). It is worth noting here that Raman's theory of crystal dynamics really had its origin in his attempt to explain prominent features in the observed Raman spectrum. We shall return to this point later.

The problem now is to describe the "dynamical behaviour of the elementary units of which it [the crystal] is composed". The way Raman chooses this elementary dynamical unit (or the unit cell for lattice dynamics, as one might call it) is interesting.

> The fundamental property of crystal structure is that it comes into coincidence with itself following a unit translation along any one of the three axes of the lattice. Hence, the normal modes of vibration characteristic of the structure of a crystal should satisfy a similar requirement. This can evidently happen in two ways, viz., the amplitudes and phases of oscillation of equivalent atoms in adjacent cells are the same; alternatively, the amplitudes are the same while the phases are all reversed following the unit translation. Since these two possibilities exist for each of the three axes of the lattice, we have $2 \times 2 \times 2$ or eight possible situations. In each of these situations, the equations of motion of the p atoms contained in the unit cell can be completely solved, yielding us $3p$ solutions. Thus in all, we have $24p$ solutions from which the three simple translations must be excluded. We are then left with $(24p - 3)$ normal modes and frequencies of vibration. Thus, the fundamental result emerges that a crystal consisting of p interpenetrating Bravais lattices of atoms has $(24p - 3)$ characteristic modes of vibration, each of which is characterized by a specific frequency. In $(3p - 3)$ of these modes, equivalent atoms have the same amplitudes and phases of oscillation in the adjacent cells, while in the $21p$ other modes the amplitudes are the same, while the phases alternate in adjacent cells along one, two or all three of the axes of the lattice.
>
> The $(24p - 3)$ normal modes of vibration indicated by the preceding argument may obviously be regarded as the modes of internal vibration of the group of $8p$ atoms comprised in a supercell of the crystal lattice whose linear dimensions are twice as large as that of the unit cell containing p atoms. The three omitted degrees of freedom would then represent the three degrees of translatory freedom of movement of the whole group of $8p$ atoms included in the supercell. Thus, it emerges that the structural unit whose dynamical behaviour is representative of the entire crystal is not the unit cell of the crystal structure but is twice as large in each direction as the latter. [10.3]

The Born–Raman Controversy

The enlarged cell introduced by Raman for elucidating his theory has been referred to by us as the supercell, and is illustrated for the diatomic chain in Fig. 10.1b. As far as the translatory motions of the supercells are concerned, Raman accepts the coupling of such motions of various supercells, leading to Debye-type elastic waves. It is only with respect to the internal vibrations of the atoms inside the supercell that Raman rejects the coupling between cells. Thus the supercell behaves and scatters like a molecule, and *a crystal is replaced by a collection of independent, non-interacting supercells* as far as the optic modes of vibration are concerned. No wonder the Raman spectrum looks like that for a molecular system in this model. Observe that since the scattering entity is the supercell, there are no complications due to surfaces, boundary conditions, etc.

10.8 The Dispute

To fix our ideas concerning the nature of the frequency spectrum in Born's theory as well as in Raman's theory, let us consider a specific example, namely, solid argon. This is a monatomic solid, with face-centred cubic structure. Thus $p=1$, and Raman's theory demands 21 characteristic frequencies. However, owing to various

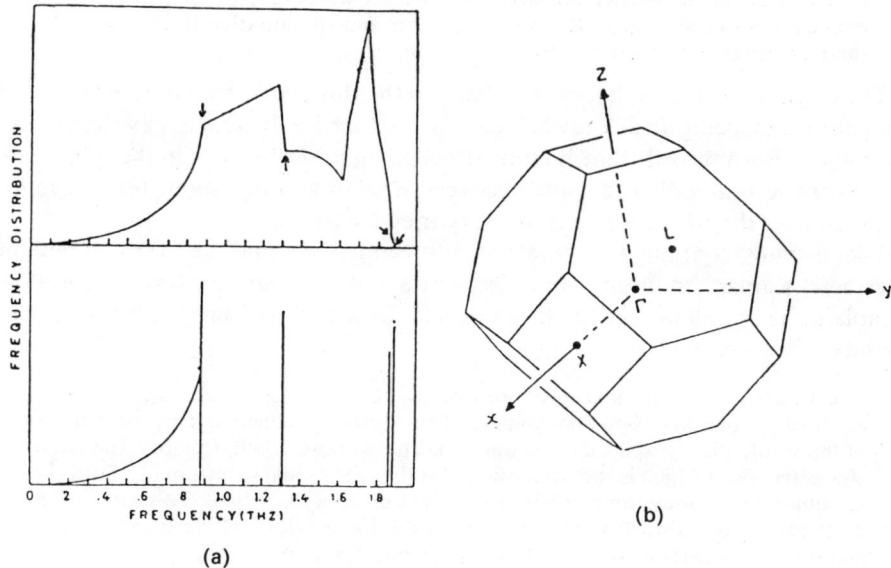

Figure 10.6 Frequency spectrum of solid argon. In the top half of (a) is shown the spectrum calculated according to Born's theory. The arrows indicate the so-called van Hove critical frequencies (see Sec. 10.9). These frequencies are associated with the points L and X of the Brillouin zone shown in (b). In the bottom half of (a) is sketched the frequency spectrum expected in Raman's theory.

degeneracies due to the symmetry of the lattice, there are only four distinct frequencies. Figure 10.6 shows the frequency spectrum of solid argon in Born's theory as well as in that of Raman. There are interesting correspondences to which we shall return shortly. But first we must hear Raman's objections to the theories of Born and Debye.

> One of the basic objections to their method of approaching the specific heat problem is that since wave motions involve progressive changes of phase along the direction of propagation and may have any frequency assigned to them, they can neither be treated as normal modes nor enumerated. The theories of Debye and Born seek to escape this difficulty by postulating that the number of wave motions is identical with the number of degrees of freedom of the system, while the choice of wavelengths is determined by still another postulate, e.g., the so-called postulate of the cyclic lattice which is claimed to represent the effect of the external boundary of the crystal. Since it is obviously impossible to formulate any boundary conditions for the atomic movements at the external surface of a crystal, the procedure is clearly artificial. [10.4]

It would seem that Raman was influenced by Lord Rayleigh's definition (made in a particular context) that all particles in a normal vibration have at any instant the same or opposite phases of vibration. No wonder he rejected the theories of Born and Debye. Equally understandable is the support he seeks to derive from Einstein.

> Einstein's approach to the problem is fundamentally correct....
> [His] view of a crystal as an assembly of immense numbers of quantized oscillators having a common set of vibration frequencies is not only the logical and correct view of the matter but also proves itself when fully developed to be an eminently successful view. It gives us a deep and quantitative insight into the thermal behaviour of solids. [10.3]

The origin of Raman's theory goes back to the thirties. At that time Born was in Bangalore and, being an acknowledged expert on lattice dynamics, gave lectures on the subject. Born records that Raman attended all his lectures. "On the other hand there were several violent disputes between Raman and me about his theoretical ideas. But on the whole we were on very friendly terms."

The dispute continued after Born left Bangalore, and as Born complains elsewhere, Raman "induced his pupils to attack me in *Nature*". Born did not just complain; he counterattacked! In an article dedicated to Bohr on his seventieth birthday, Born remarks:

> The Indian physicists have produced many new and accurate observations of electrical, optical and X-ray phenomena. These results led them to deny the validity of the whole theory of lattice dynamics and to propose another theory, the main characteristic of which is the contention that the vibrational spectrum of a lattice is not quasi-continuous, but consists of a small number of sharp lines. This contention is, of course, too absurd to be taken seriously. For in classical mechanics and in quantum mechanics as well, a vibrating system of N particles has $3N - 6$ normal modes of vibration (6 being the number of translational and rotational degrees of freedom). Raman would hardly deny that a molecule consisting of 10 atoms has 24 normal modes; perhaps he would also agree that a system of 100 atoms has 294 normal modes – but a system of 1000 atoms which may be already called a microcrystal would, according to his theory, not have 2994 normal modes but only 48 (if I

understand his somewhat vague statements). The attacks of the Indian physicists against lattice dynamics are mainly directed against the use of the cyclic boundary condition.... There is no doubt that lattice dynamics as represented in my old book is correct[6]. But the new observations of the Indian scientists are not concerned with *dynamical* but with *thermal, electrical and optical* phenomena, and I perfectly agree with the Indians that the theory as developed in my book is incapable of accounting for many of these experiments. The reason is that my book was written before the discovery of quantum mechanics.... [10.6]

Born's theory is no doubt correct; nevertheless, Born was unjustified in his criticism of Raman's counting of the number of degrees of freedom. It must be conceded that Raman did not express himself in sufficient detail in his early papers concerning this aspect. Perhaps he regarded it as self-evident, failing to see the genuine confusion that might arise in the minds of those accustomed to a different kind of bookkeeping. Actually Raman's bookkeeping is not at all at fault (as we shall see in a later section devoted to the more technical aspects). His mistake lay elsewhere. Anyway, the impression had been created in the pages of the prestigious *Reviews of Modern Physics* that Raman was confused and did not understand the basics.

Before we get back to the physics of the problem, it is interesting to note what Born has recorded of his personal relationship with Raman after the eruption of the controversy.

> We have met Raman twice since we left India. The first occasion was a conference in Bordeaux to celebrate the twenty-fifth anniversary of the discovery of the Raman effect. He received an honorary doctor's degree[7], but the same degree was also conferred upon me. I am sure that the French colleagues did this to demonstrate that in the dispute about lattice vibrations not Raman but I was right. At the first reception in Bordeaux we greeted each other very cordially and had a lively talk. Then Raman abused some theoretical physicist (I have forgotten whom) because he had done experiments which Raman regarded as poor; I replied: 'But, my dear Raman, what about the other way round, when experimentalists venture to make theories?' or something like that. Though he first remained quite friendly he later became furious and said to Hedi [Born's wife], his neighbour at the banquet, that I had given him deadly offence and that he would leave the conference. She had great trouble in appeasing him, but during the whole congress he was nervous, excitable and aggressive.
>
> The second time, we met Raman at one of the Lindau meetings of Nobel Laureates. He was sitting at the next table in the dining room of the Schachen Hotel, greeted us in a very friendly manner and talked with us in his lively manner from one table to the next. But the next day his attitude had changed. He avoided us and went out of the way when we met in the house or garden. He must have suddenly remembered that I was his 'enemy'.
>
> Actually I never was. I still admire his fascinating personality, his devotion to science and research. It makes me sad to think that by inviting me to India and trying to keep me there permanently he has brought himself into a precarious situation, and had to give up his leading position at the Institute of Science. But I cannot see that I am to blame for this misfortune. Nor can I accept a scientific theory which I regard as wrong. Hedi and I regret all this and particularly the split between us and Lady Raman whom we loved dearly. [10.7]

10.9 History's Verdict

Let us now return to the physics underlying the Born–Raman controversy, and consider the verdict given by experiment. In the forties, R. S. Krishnan[8] obtained data of very good quality on the second-order Raman spectrum of diamond [10.8]. These experiments deserve high commendation, representing as they did the limits of experimental capability at that time. We show in Fig. 10.7 the actual spectrum in juxtaposition with the microphotometer record. Krishnan analysed the sharp features and identified what he called Raman lines. Figure 10.8 is a diagrammatic

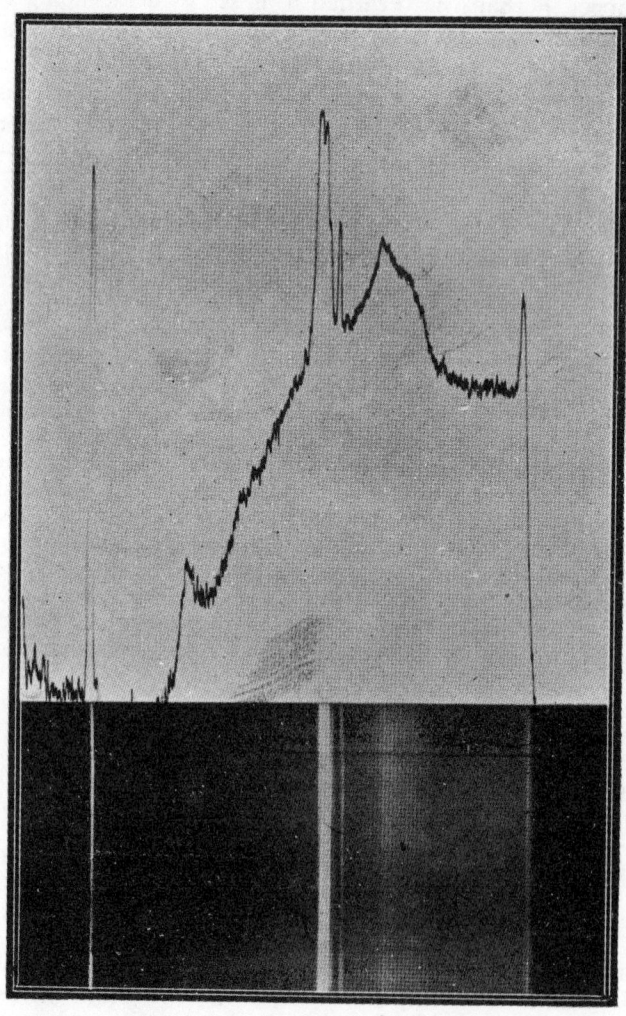

Figure 10.7 High resolution Raman spectrum of diamond. (After ref. [10.9].)

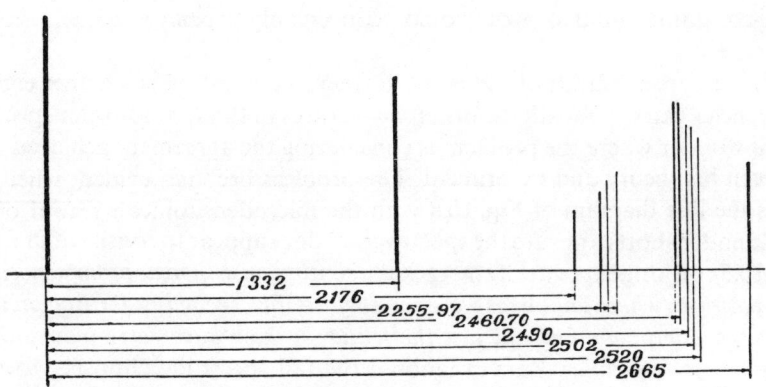

Figure 10.8 Diagrammatic representation of the second-order Raman spectrum of diamond as prepared by Raman. (After ref. [10.9].)

representation of these features as prepared by Raman [10.9]. To explain this line spectrum, Raman constructed a force-constants model for the vibrations of the diamond lattice. Essentially it was a model for the vibration spectrum of his supercell, and yielded eight distinct frequencies, which is what one expects. The numbers computed by Raman for these frequencies are given in Table 10.1. With these eight frequencies, Raman was able to explain all the observed prominences in terms of overtones and summations. One might not consider the agreement achieved by Raman as perfect but if one is prepared to invest the effort to refine the

Table 10.1 Comparison of critical frequencies (in cm^{-1}) for diamond obtained by different methods. The first column gives the Brillouin zone points (see Fig. 10.6) associated with these frequencies. The numbers in the second column were computed by Raman using his theory.

	Raman[a]	Neutron[b]	Laser Raman[c]
Γ	1332		1332 ± 0.5
X	1088	1072 ± 26	1069 ± 5
X	1239	1184 ± 21	1185 ± 5
X	740	807 ± 32	807 ± 5
L	1149	1210 ± 37	1206 + 5
L	1250	1242 ± 37	1252 ± 5
L	1008	1035 ± 32	1006 ± 5
L	621	552 ± 16	562 ± 5

[a] After ref. [10.9].
[b] After ref. [10.11].
[c] After ref. [10.14].

force-constants model, one could conceivably achieve better accord with experiment.

Perfect agreement is not the point. Rather, the question is whether eight distinct frequencies can explain *all* the observed features of the second-order spectrum. One might wonder where the problem is considering the agreement achieved by Raman between his theory and experiment. The problem becomes evident when one compares the line diagram of Fig. 10.8 with the microdensitometer record of Fig. 10.7. While under short exposure the spectrogram does appear to consist of sharp lines (see Fig. 10.5), on long exposure it emerges as *a continuum with some prominences in it, at the very positions where faint lines were seen before. The second-order Raman spectrum is the entire continuum and not just the features which gave early indications of their presence via faint lines.* Raman ignored the full spectrum, choosing to retain and explain only the prominent features, and that was his mistake[9].

Is Born's theory able to do a better job? Indeed it is. Provoked by Raman, Born made quite an effort to rebut Raman's objections. First he patiently worked out the quantitative theory of second-order Raman scattering [10.10]. This was in 1947. Shortly thereafter, his student Helen Smith[10] published a detailed calculation of the second-order spectrum of diamond. This is shown in Fig. 10.9.

When the differences between Born and Raman first erupted not many were interested in lattice dynamics, and few paid attention to this controversy. But whenever expert theoreticians heard about the clash, they had no difficulty in supporting Born. This was not on account of the persuasive result painstakingly

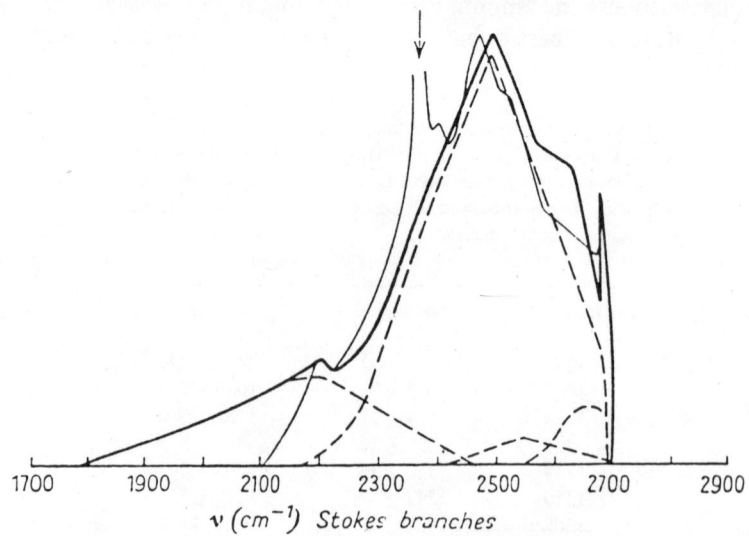

Figure 10.9 Comparison of the second-order Raman spectrum computed by Helen Smith with that measured by Krishnan. The fine line represents Krishnan's microphotometric record while the thick line represents Smith's calculations. The broken lines represent theoretical contributions to the intensity of frequency density functions from which the thick line was obtained. (After ref. [10.5].)

computed by Helen Smith, but because the cyclic boundary condition is a cornerstone of crystal physics. Without it one could not have Bloch's theorem, electron energy bands in crystals, and so on. Time and again theorists openly disagreed with Raman. It is said that when Raman visited Kapitza and gave a seminar at the Institute for Physical Problems in Moscow, Landau rushed to the blackboard to counter Raman's arguments[11]. In 1951 Peierls came to India to attend a conference on elementary particles arranged by Bhabha. After this conference Peierls went to the Science Congress at Bangalore where Raman gave a lecture on his theory. Peierls would not accept Raman's criticism of the periodic boundary conditions, having used them himself in 1936 to study the equation of state of a relativistic gas. A confrontation ensued with Raman demanding an explanation of the observed sharp lines. Peierls declined to be drawn into discussing the experiments, preferring to rest his case on purely theoretical arguments. Unable to convince Raman, Peierls finally published his reasoning in the *Proceedings of the National Institute of Sciences*. This proof is reproduced by Born in his celebrated book on lattice dynamics [10.5].

The burden of Peierls' argument is that the frequency distribution of a crystal slab of arbitrary shape is indistinguishable from that of a crystal of similar shape but subject to periodic boundary conditions, unless one probes the frequency spectrum with very high frequency resolution. For a crystal 1 cm × 1 cm × 1 cm, Peierls estimates that a frequency resolution $(\Delta\omega/\omega) \sim 0.01\%$ would be needed to detect effects due to the breakdown of cyclic conditions, and experiments are never done with such precision. Thus, according to Peierls, Raman's criticism is academic. But if one deals with *fine particles*, the story is different. Surface effects *do* play a role, as recent experiments have shown. So there is a regime where Raman's objections become meaningful.

Till the late fifties, the study of lattice vibrations was not quite in the mainstream of solid state physics, and interest in the subject was confined to spectroscopists, and those measuring specific heats. Things changed dramatically with the advent of slow neutron inelastic scattering. The neutron scattering experiment is rather similar to the Brillouin scattering one, with the important difference that one can obtain information not only about the acoustic but also about the optic branches of the dispersion curves. This technique was applied in the mid-sixties to the study of diamond [10.11], and the results obtained are summarized in Fig. 10.10 and Table 10.1. Strange as it might sound, these results, while lending support to the Born–von Karman theory, have also served to partially vindicate Raman's point of view!

To understand the last remark, we need to appreciate two facts. Firstly, it was pointed out by Raman's student Viswanathan that the $(24p-3)$ frequencies enumerated by Raman correspond, in Born's language, to normal mode frequencies for which the group velocity (given by $d\omega(q)/dq$) vanishes [10.12]. Now group velocity can vanish for select normal modes either for reasons of symmetry or on account of the nature of the interatomic forces. We shall ignore the latter possibility as the former is the more important one. We then find that Raman's $(24p-3)$ frequencies are a select, symmetry-determined subset of the innumerable number of frequencies possible in Born's theory, the corresponding normal modes being characterized by zero group velocity.

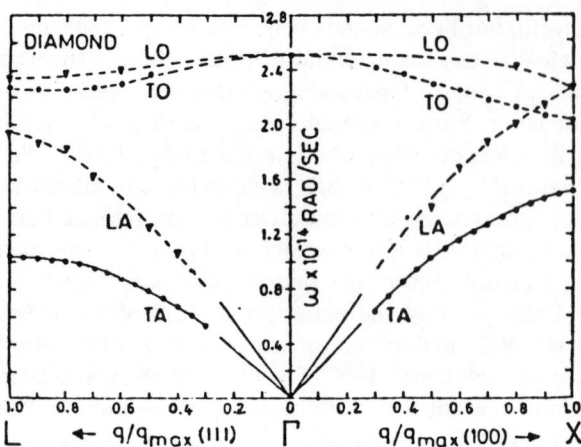

Figure 10.10 Phonon dispersion curves for diamond as obtained by neutron spectrometry. The labels Γ, X and L refer to different points of the Brillouin zone (see Fig. 10.6). (After ref. [10.11].)

What is special about such modes with zero group velocity? The answer came from an important paper due to van Hove [10.13], and this is our second point. Using topological arguments, van Hove showed quite generally that *the frequency spectrum would have singularities at those frequencies the normal modes for which have zero group velocity*. Referring back now to Fig. 10.6, which shows the frequency spectrum for argon in the Born as well as the Raman theory, we observe that the frequencies predicted by Raman correspond precisely to the van Hove singularities.

In the case of second-order Raman scattering, the van Hove singularities assume importance, for it is these that control the occurrence of prominent features in the observed spectrum. This is clearly seen in the work of Solin and Ramdas[12] [10.14], performed after the advent of lasers (see Fig. 10.11). Thanks to the laser as well as modern photon-counting techniques, recording a second-order Raman spectrum is no longer the agony it was in the old days. On comparing the spectrum obtained by these latter-day workers with the earlier spectrum, one is filled with admiration for those early experiments. All the prominences seen earlier in Bangalore are seen again, this time far more clearly, and practically at the same frequencies.

Solin and Ramdas have carefully analysed the singularities in their second-order Raman spectrum and shown that these frequencies can be explained by a set of eight critical frequencies. Guided by the values measured for these frequencies in the neutron scattering experiment (see Fig. 10.10), they have deduced their own set which is quoted in Table 10.1 along with the neutron scattering results and Raman's own values.

One can now see clearly where Raman went wrong. The irony is that he was not completely wrong but only partially so. His mistake was that *he interpreted the singular features as sharp lines*, possibly misguided by their appearance in

The Born–Raman Controversy

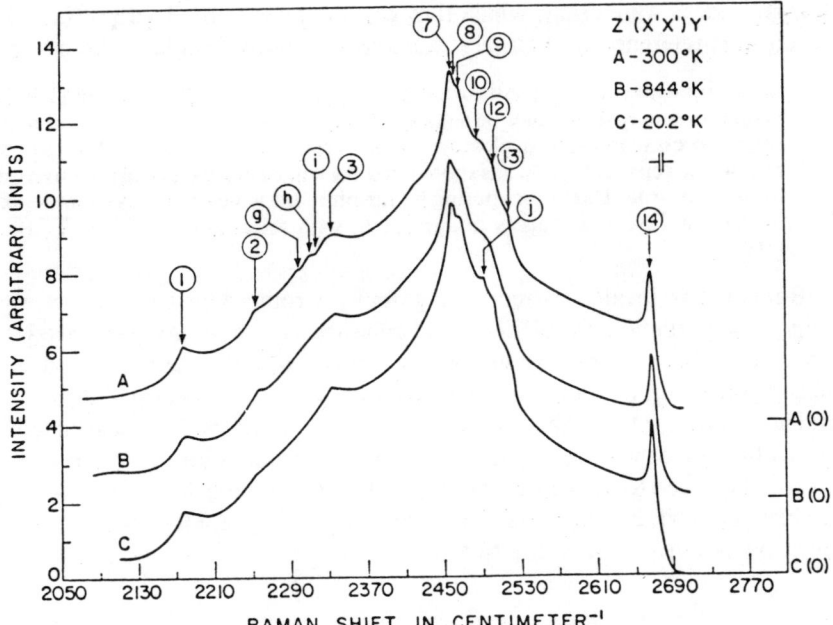

Figure 10.11 Second-order Raman spectrum of diamond obtained by laser Raman spectroscopy. The clearly recognizable sharp features are labelled with encircled Arabic numbers. (After ref. [10.14].)

photographs obtained with limited exposure. He dismissed the continuous background seen in spectra recorded with long exposure. This he should not have. Possibly he thought it was due to fogging. We do not know. But it is interesting that having decided to look at only a *part* of the entire spectrum, he produced the *correct explanation* for that part! The types of vibrations of the supercell visualized by him are precisely those predicted by Born's theory. And the part of the frequency spectrum Raman chose to look at was not an arbitrary part but a special one related to the van Hove critical points.

Looking back, one cannot but marvel at Raman's ingenuity. While it is easy to criticize him, it is remarkable that in his own intuitive way he saw the importance of the zone-boundary modes, and went after them via his supercell. It is sad that Raman and his opponents never sorted out their differences. Raman did not appreciate the nuances of theory, while the theorists did not understand the experiment (or else they might have corrected Raman even then). To compound matters, the van Hove singularities were not known then. Raman was a loner battling against a strong opposition which preferred mathematical logic to intuition, however brilliant the latter might be. By the time van Hove's critical points were discovered, neutron scattering was established, and laser Raman scattering came to the fore, Raman's theory was totally forgotten and no one thought it fit to assuage the wounded feelings of an old man. Instead the impression was created that

Raman's ideas were crazy, which they certainly were not. There is but one passing favourable reference in the literature, and that is by Loudon who observes:

> Since important critical points usually occur at Γ, X and L [for diamond], Raman's assumption may sometimes be a good one and he and his coworkers have met with some success in interpreting features of the second-order spectra.... Using Raman's approach, Venkatarayudu has shown that the rule of mutual exclusion breaks down for second-order Raman scattering by the phonons which are at symmetry point X in the usual picture. This is in agreement with selection rules given by Birman. [10.15]

Raman was locally correct but globally wrong. The part he got right is an important part and reflects his native brilliance. He would not have erred the way he did had he been trained properly in quantum physics. Raman could have saved himself much agony had he established proper communication with Born. But that did not happen. Maybe there were psychological barriers. As we asked in Chapter 8, was it his ego, or was he weighed down by his problems in the Institute and felt the over-riding need to establish his superiority? Here we reach a dark edge. All we see is a proud man, stubborn and uncompromising, being rapidly left behind and denied even the credit that was due to him.

☆ 10.10 The Theoretical Aspects

In the Born theory, the dynamics of the crystal lattice is studied as follows [10.5, 10.16]. Consider a crystal with p atoms in the primitive unit cell. We identify these p atoms with the index $k (= 1, \ldots p)$ and the unit cell itself with the index l. The classical equations of motion of the atoms in the lattice can be written as

$$M_k \ddot{u}_\alpha(lk) = - \sum_{l'k'\beta} \phi_{\alpha\beta}(lk; l'k') u_\beta(l'k'). \tag{10.1}$$

Here M_k denotes the mass of the kth atom in the cell and $\mathbf{u}(lk)$ its displacement from the equilibrium position. The Greek subscripts denote Cartesian components. The quantity $\phi_{\alpha\beta}(lk; l'k')$ is one of the spring constants coupling atoms (lk) and $(l'k')$. The harmonic approximation has obviously been made while writing (10.1).

Equation (10.1) represents a system of simultaneous linear differential equations, infinite in number. However, one can exploit the translational periodicity of the crystal to simplify the problem. Thus we introduce wave-like solutions of the form

$$u_\alpha(lk) = (1/M_k)^{1/2} w_\alpha(\mathbf{q}k) \exp[i\{\mathbf{q}\cdot\mathbf{x}(l) - \omega t\}], \tag{10.2}$$

where $\mathbf{x}(l)$ denotes the position of the lth unit cell, and is defined more formally by

$$\mathbf{x}(l) = l_1 \mathbf{a}_1 + l_2 \mathbf{a}_2 + l_3 \mathbf{a}_3,$$

l_1, l_2, l_3 being integers and $\mathbf{a}_1, \mathbf{a}_2, \mathbf{a}_3$ the basis vectors of the crystal lattice.

Substituting (10.2) in (10.1) and carrying out some simplifications, we find that the infinite number of equations reduce to

$$\omega^2 w_\alpha(\mathbf{q}k) = \sum_{\beta k'} D_{\alpha\beta}(\mathbf{q}; kk') w_\beta(\mathbf{q}k'), \qquad (10.3)$$

where

$$D_{\alpha\beta}(\mathbf{q}; kk') = (M_k M_{k'})^{-1/2} \sum_{ll'} \phi_{\alpha\beta}(lk; l'k')$$
$$\times \exp[i\mathbf{q} \cdot \{\mathbf{x}(l') - \mathbf{x}(l)\}] \qquad (10.4)$$

is called the *dynamical matrix*. Equation (10.3) is an eigenvalue problem, and the condition for its solubility is that the secular determinant must vanish, i.e.,

$$\|\omega^2 \delta_{\alpha\beta} \delta_{kk'} - D_{\alpha\beta}(\mathbf{q}; kk')\| = 0. \qquad (10.5)$$

We shall denote the $3p$ solutions of the above equation as $\omega_j^2(\mathbf{q})$ ($j = 1, \ldots 3p$), and the corresponding eigenvectors by $\mathbf{e}(\mathbf{q} j)$ with components $e_\alpha(\mathbf{q} j; k)$. The eigenvectors satisfy the following orthonormality and completeness conditions:

$$\sum_{k\alpha} e_\alpha^*(\mathbf{q} j'; k) e_\alpha(\mathbf{q} j; k) = \delta_{jj'} \qquad (10.6a)$$

$$\sum_j e_\beta^*(\mathbf{q} j; k') e_\alpha(\mathbf{q} j; k) = \delta_{\alpha\beta} \delta_{kk'}. \qquad (10.6b)$$

Until now we have considered the problem of determining the vibrational frequencies and the eigenvectors associated with wave-motion corresponding to the wavevector $\mathbf{q}$. Next comes the task of specifying the allowed values of $\mathbf{q}$ and this is where the periodic or cyclic boundary conditions come into the picture.

We imagine the infinite crystal to be subdivided into large *macrocells* each containing a very large number N of primitive cells, where N is typically of the order of the number of cells in the finite specimens used in the laboratory. For convenience, although this is not necessary, we choose the macrocell to have the same shape as the primitive cells and with edges $L\mathbf{a}_1, L\mathbf{a}_2, L\mathbf{a}_3$. It follows that $N = L^3$. The periodic boundary conditions are now applied to the macrocells. In practical terms, one demands

$$\mathbf{u}(l,k) = \mathbf{u}(l+L, k). \qquad (10.7)$$

This implies

$$\mathbf{q} = \xi_1 \mathbf{b}_1 + \xi_2 \mathbf{b}_2 + \xi_3 \mathbf{b}_3, \qquad (10.8)$$

where $\xi = n_i/L$, n_i being an integer. In (10.8), $\mathbf{b}_1, \mathbf{b}_2, \mathbf{b}_3$ are the basis vectors of the lattice reciprocal to that defined by $\mathbf{a}_1, \mathbf{a}_2, \mathbf{a}_3$. The values of $\mathbf{q}$ defined by (10.8) form a fine mesh of points in reciprocal space. Not all the $\mathbf{q}$ values defined above are physically distinct, since $\omega_j^2(\mathbf{q}) = \omega_j^2(\mathbf{q} + \mathbf{G})$, where $\mathbf{G}$ is a vector of the reciprocal lattice (i.e., the lattice defined by $\mathbf{b}_1, \mathbf{b}_2, \mathbf{b}_3$). All the distinct values of $\mathbf{q}$ are obtained by restricting attention to one primitive cell of the reciprocal lattice. The Brillouin zone in Fig. 10.6 is such a region.

The macrocell we have considered has $3pN$ normal mode frequencies. For many purposes it is necessary to consider the spectrum of these frequencies. This spectrum, also known as the frequency distribution, is defined by

$$g(\omega) = \frac{1}{3pN}\sum_j \sum_\mathbf{q} \delta\{\omega - \omega_j(\mathbf{q})\}. \tag{10.9}$$

In Raman's theory, it is given by [10.6]

$$g(\omega) = g_1(\omega) + g_2(\omega),$$

where

$$g_1(\omega) = A\omega^2, \quad 0 < \omega < \omega_D,$$

$$g_2(\omega) = B \sum_{R=1}^{24p-3} \delta\{\omega - \omega_R\},$$

A and B being suitable normalization constants, and ω_D an appropriate cut-off (see also Fig. 10.6). A reference to the paper cited earlier shows that Raman carefully chose A and B to ensure that the bookkeeping of the degrees of freedom was in order. He did not slip on that score as Born alleged.

We now briefly consider the argument of Peierls in support of the cyclic conditions. The idea underlying his proof is that there is a connection between the frequency distribution and the propagation of disturbances through the crystal. In particular, if we know how the disturbance propagates up to a certain time τ, then we can uniquely determine the frequency distribution except for its fine structure on a frequency scale $\Delta\omega \sim 1/\tau$.

Viswanathan [10.12] used the Born theory to study the normal modes with zero group velocity, and arrived at the conclusion that there must be $(24p - 3)$ of these. We now know that this does not exhaust the set of zero group velocity modes. Viswanathan also showed that any disturbance can be written asymptotically (for large time) as a superposition of Raman's vibrations but with time-dependent amplitudes. Born comments that this does not mean that eventually only these frequencies are excited. He maintains that the initial energy distribution in the frequency spectrum is unaltered [10.5].

Frequency spectra exhibit singularities. Their existence was first pointed out by van Hove [10.13]. The van Hove singularities appear as discontinuities in $g'(\omega) = [dg(\omega)/d\omega]$ or in the higher derivatives of the frequency spectrum. In general, they arise when either grad $\omega_j(\mathbf{q})$ vanishes completely or one or more components of the gradient change sign discontinuously while the remaining components vanish. Points in q-space associated with critical frequencies are sometimes referred to as *critical points*. Van Hove not only drew attention to their existence, but showed, using some powerful topological results due to Morse, that the existence of some critical points in a given branch of the dispersion curve implies the existence of others in the same branch. Further, the total numbers of the different types of critical points are related to each other by what are called *Morse relations*.

Phillips [10.17] has shown how group theory[13] may be used to locate the critical points required by symmetry. This is the Raman–Viswanathan set of $(24p - 3)$ frequencies[14]. Phillips emphasizes that the symmetry-determined set does not exhaust all the critical points. Others could exist and are conveniently hunted for using the Morse relations.

We now consider second-order Raman scattering. This involves two normal mode frequencies $\omega_j(\mathbf{q})$ and $\omega_{j'}(\mathbf{q}')$. The conservation equations are:

$$\mathbf{q} + \mathbf{q}' = 0 \qquad (10.10a)$$

$$\Omega = \omega_j(\mathbf{q}) + \omega_{j'}(\mathbf{q}') \quad \text{(sum mode)} \qquad (10.10b)$$

$$\Omega = \omega_j(\mathbf{q}) - \omega_{j'}(\mathbf{q}') \quad \text{(difference mode)}. \qquad (10.10c)$$

Here $\Omega(=2\pi\Delta v)$ is the Raman frequency shift. The constraints (10.10) are rather weak, and several modes $(\mathbf{q}j)$ and $(-\mathbf{q}j')$ can combine to give second-order Raman scattering. As a result, one expects the scattered spectrum $I(\Omega)$ to be a continuum rather than a set of delta functions. However, the Raman spectrum does show singularities similar to the van Hove singularities. To a rough approximation, these singularities can be understood by considering the *combined frequency distribution*, or the *joint density of phonon states* as it is sometimes called. This quantity is defined by

$$g_{jj'}^+(\Omega) = \frac{1}{(3p)^2 N} \sum_{\mathbf{q}} \delta\{\Omega - [\omega_j(\mathbf{q}) + \omega_{j'}(\mathbf{q})]\}, \qquad (10.11)$$

($\omega_j(\mathbf{q}) = \omega_j(-\mathbf{q})$). The superscript $+$ in (10.11) signifies a sum mode. Experiment reflects

$$\sum_{jj'} g_{jj'}^+(\Omega) + \sum_{jj'} g_{jj'}^-(\Omega), \qquad (10.12)$$

where $g_{jj'}^-$ is similarly defined.

Like $g(\Omega)$, $g_{jj'}^\pm(\Omega)$ also exhibits singularities [10.16]. The structures seen by Raman (and erroneously interpreted by him to be sharp lines) and more clearly later by Solin and Ramdas reflect singularities in the quantity in (10.12). Raman retained only these critical features out of the entire second-order spectrum, and explained them (correctly) – features which we now know can be interpreted in terms of the $(24p - 3)$ van Hove critical points.

To complete the picture it must be mentioned that strictly speaking, second-order Raman scattering is not governed by (10.12) but by a related quantity called the polarizability-weighted joint-frequency spectrum [10.16]. It is this which Helen Smith painstakingly computed in order to compare with R. S. Krishnan's results. Of course at that time the van Hove singularities were not known. (Also there were no computers to relieve the extreme tedium of calculation, which makes Helen Smith's work almost heroic.)

11 *The Academy*

One of Raman's gifts to India is the Indian Academy of Sciences, or the Academy as it is often called. With Raman involved there was, inevitably, high drama accompanying the birth of the Academy; but first let us survey the background.

Starting from the early years of this century, a multitude of bodies – services, societies, universities and research institutes – came into existence in quick succession. Some of the services like the Survey Department, the Meteorological Department and the medical services have already been mentioned. Scientific research was also carried on in several Government-supported institutes like the Imperial Institute of Veterinary Research in Mukteswar, the Imperial Agricultural Institute in Pusa, the Central Research Institute in Kasauli, the Imperial Forest Research Institute in Dehra Dun, and so on, not to mention various university departments.

Where societies were concerned, the geologists gave an early lead with the founding of the Mining and Geological Institute of India in 1906. The Indian Mathematical Society was started in 1907 in Poona. This was followed by the Institution of Engineers, India, which was founded in 1921 with its headquarters in Calcutta, and today has branches in several cities. The Botanical Society was also formed in the same year. The Indian Chemical Society with its headquarters in Calcutta came into existence in 1924, and the Geological, Mining and Metallurgical

Society of India was founded in the same year. In addition, there were the Indian Society of Soil Science, the Indian Physiological Society, the Society of Biological Chemists, the Institution of Chemists, and of course the Indian Physical Society.

With all these developments taking place, there was a growing feeling that there should be some national forum where scientists of different disciplines could meet and exchange views. The Asiatic Society (see Chapter 1) served this purpose to a certain extent but it was not enough. To counter geographic as well as specialist isolation, J. L. Simonson of the Presidency College, Madras, and P. S. MacMahon of the Canning College, Lucknow, proposed in 1911 the founding of an Indian Association for the Advancement of Science, analogous to the British Association for the Advancement of Science. As a result, the Indian Science Congress Association was born, and the inaugural meeting was held in 1914 in the rooms of the Asiatic Society under the patronage of Lord Carmichael and the presidentship of Sir Asutosh Mookerjee. (The Indian Science Congress survives to this day, having grown in strength as well as prestige.) But the Science Congress met "only once a year, and it is for one week only during 52 that scientists are afforded the opportunity for fruitful intercourse. During the rest of the year, the centres of research tend to remain in geographical isolation from one another" [11.1]. The Science Congress was not enough and something extra was needed. An Indian Academy of Sciences seemed the right answer, and its creation was urged by many.

In May 1933 an editorial was published in *Current Science*[1] echoing this urge for the creation of an Academy. Actually Raman wrote the editorial in which he said:

> The conviction that research is civilization, and determines the economic, social and political development of a nation has not yet been unreservedly accepted as part of the administrative policy of India, and we are disposed to ascribe the tardy and perhaps unwilling recognition of this fundamental fact to the absence of an all-India scientific organization whose function would be to concentrate enlightened public opinion on the doctrine that science is material and spiritual wealth.... It seems to us that the early establishment of a National Academy of Science should secure closer and better organized co-operation of activities among all research institutes in India, and exercise through its official journal a wider influence for the consolidation and promotion of the best interests of science. [11.2]

Raman also attached considerable importance to publishing, and he was concerned that "papers of outstanding merit frequently gravitate to foreign periodicals" for lack of a proper national journal. He added,

> While the foundation of the scientific reputation of a country is established by the quality of work produced in its institutions, the superstructure is reared by the national journals which proclaim their best achievements to the rest of the world.

The achievements of Indian science were national assets, and only an Academy could treasure and display them collectively.

But would not the proposed Academy clash with the Science Congress? No.

> The scope and functions of the Academy are different from those of the Indian Science Congress which offers principally the advantage of human contacts while giving opportunities to discuss the preliminary stages of work still in progress. Thus

the aims of the two institutions will be distinct, but complemental. Among other functions which the Academy will exercise should be included the protection and advancement of the professional interests of its members. It should acquire the necessary authority to advise Government, the universities and other institutions on all scientific matters and other problems referred to it for consideration and to negotiate on behalf of Indian scientific workers with similar institutions abroad.

The editorial was followed up with a questionnaire in September of the same year, and the scientists of India were asked to express their views. Those in Calcutta deliberated on the questionnaire in several meetings, and then proposed that in view of the accepted position of the Indian Science Congress Association as an all-India body, the question of forming an Academy should be discussed at a Science Congress session. An opportunity for this presented itself in January 1934 when the Science Congress met in Bombay. In his presidential address, Saha expressed approval of the idea of an Indian Academy and proposed the Royal Society of London as a model. The whole question was thereafter discussed in a special meeting of the General Committee, and it was decided to form a special Academy Committee to consider the issue further in depth.

The Academy Committee (which included Raman) was quite large, with twenty-five members! But then so many interests had to be represented – a typically Indian syndrome. As if twenty-five members were not enough, others were co-opted on various occasions. Trouble started in the very first meeting of the Academy Committee in February 1934. Raman found that instead of a quick decision being taken, the issue was being endlessly debated and dragged. There were also serious differences between Raman and several others in the Committee over the draft minutes of the first meeting.

Raman could take it no more. He resigned from the Academy Committee, and on April 27 formed an Academy on his own, registering it in Bangalore under the Societies Registration Act. Including him, there were 160 Foundation Fellows.

Calcutta was deeply perturbed and upset (naturally). The Calcutta members of the Academy Committee (they were the leading group) met in an emergency session in May when the news of the sudden developments in Bangalore reached them. Raman's various charges were duly rebutted, and an explanation was produced of the Academy Committee's stand.

The Committee continued to meet and deliberate but things did look rather confused for a while until Raman applied a healing touch. On June 16 he addressed a letter to L. L. Fermor, the representative of the Asiatic Society in the Academy Committee. In this letter Raman suggested a federation of the various existing academies in India, namely, the Asiatic Society, the United Provinces Academy of Sciences in Allahabad[2] and the Indian Academy of Sciences in Bangalore. This federation was to be an alternative to the Academy originally proposed to be created under the auspices of the Science Congress. There was some bargaining, and eventually concord was established, whereupon Raman withdrew his resignation from the Academy Committee.

According to the new arrangement, there would be a National Institute of Sciences of India. It was agreed that the Academy in Bangalore would not promote the formation of branches. Likewise, the National Institute also would refrain from forming its own branches.

The National Institute of Sciences came into existence on January 3, 1935, with 125 Foundation Fellows, Raman being one of them. Commenting on those developments, Born wrote to Rutherford:

> Saha intended to found an all-India Academy, but things went too slowly for Raman's temperament and he founded his own Academy (Indian Academy of Science) in Bangalore, with his own *Proceedings*. Now there are two Academies in India, not too many for such an enormous country, but they are bitter adversaries. All the North Indians joined Saha's party, and the South Indians that of Raman. [11.3]

While this remark of Born might superficially appear true, it is not quite so. If one carefully studies the lists of Foundation Fellows of the two organizations, one finds not only many names common to the two, but also names of people from all parts of India. There are quite a few British names too, reflecting the presence of Britishers on the Indian scientific scene. Concerning the composition of the Academy, Raman observed in his address at the first annual meeting:

> Our list of Fellows is also representative of all parts of India. Bombay heads the list with 38 Fellows, closely followed by Madras Presidency with 35, and Mysore State with 33. Other provinces are also well represented. We have 21 Fellows in the United Provinces, 13 from the Punjab, 11 from Bengal, 8 from Central Provinces; Bihar, Orissa, Hyderabad, Travancore and Burma are also represented in our list.

But there is no denying that there were two camps. However, the numbers of camp followers appear to have been relatively small. A large proportion of Foundation Fellows of both organizations seem to have been motivated by considerations of science alone, and joined simply because they were invited. It is noteworthy that while Raman agreed to be a Foundation Fellow of the National Institute, Saha and some of his close associates never joined the Academy. In fact, many years later when an eminent scientist from Bengal was elected to the Indian Academy of Sciences, he rejected the Fellowship. But Time the healer has worked, and all traces of hard feelings have now disappeared.

The inaugural meeting of the National Institute of Sciences was held in Calcutta on January 7, 1935. Lewis L. Fermor, who came to India in 1902 and served in the Geological Survey of India, was elected the first President. His inaugural address is interesting, for it comprehensively traces the growth of science in India during the British period, and also the events leading to the formation of the National Institute. Originally there was a desire for an Academy under the umbrella of the Science Congress. But events took a different turn, and the National Institute came into existence instead. So Fermor had the task of explaining away the original need and its replacement with a new one!

> During the past year the word *Academy* has been much before us.... It seems desirable, therefore, that we should first enquire what the word Academy means. It will surprise most of you to learn that the first Academy was a pleasure garden in Athens which is supposed to have belonged to an ancient Attic hero named

Academus.... In this garden the Greek philosopher Plato taught for nearly 50 years; and the Academy thus started lasted from the days of Plato to those of Cicero, that is, for over 300 years.... While Academies, if we go to the original meaning, must, therefore, function locally or regionally in the most important portion of their activities, they can also legitimately make a wider appeal.... the Asiatic Society cannot hope to cater for... the whole of India.... The United Provinces Academy of Sciences, founded at Allahabad in 1930, was, therefore, on this argument, a desirable creation to provide for the meeting of students of all branches of science in Northern India.... When in 1933, the proposal was mooted to found an Indian Academy of Sciences, some of us overlooked the fact that there were already two such Academies in existence – one called the Asiatic Society of Bengal and the other the United Provinces Academy of Sciences. The proposal, therefore, to found a third Indian Academy... logically meant either the creation of a fresh garden in another part of India, or of a body to co-ordinate the already existing gardens. Our friends in Bangalore knew all the time that they needed a Society of Academy status with its headquarters in Bangalore. Had they boldly said so at the beginning, the confusion that has arisen in scientific circles during the past year would have been avoided, because it is obviously correct that Southern India should have its own philosopher's garden. However, Bangalore did not do this.... Object as we may to the manner in which our Bangalore friends cut adrift... we... welcome the Indian Academy of Sciences founded at Bangalore.... But we still need a co-ordinating body; and that is why it is necessary to found the National Institute. [11.1]

Fermor wanted, in addition, a National Institute of Letters and also a National Institute of Arts, and the three Institutes to be linked by the Institute of India, like the Institute of France. Fermor concluded his long address with a vision of "a magnificent palace of learning in Calcutta equivalent to Burlington House in London".

As it turned out, the National Institute functioned in Calcutta till 1946, when it moved to Delhi. In 1970 its name was changed to Indian National Science Academy. In other words, it did not evolve as Fermor had envisaged. As someone remarked, "Raman's editorial in *Current Science* was so good that India got two Academies instead of just one!"

Let us now return to the Academy which Raman founded. At the time it was formed, it was decided that the headquarters would be temporarily in Bangalore, the intention being that a permanent place would be identified later. But soon the Maharajah of Mysore gifted ten acres of choice real estate in the vicinity of the Indian Institute of Science as a permanent location for the Academy. Thus the Academy came to stay in Bangalore, and its offices function there to this day.

Raman had clear ideas about the objectives of the Academy. These were to be three-fold:

(i) to hold meetings for discussing the results of research,
(ii) to hold symposia on special subjects, and
(iii) to publish the *Proceedings*.

The inaugural meeting of the Academy was held in the campus of the Indian Institute of Science in August 1934. The scientific part of the meeting consisted of the presentation of a few original papers, followed by a symposium on "Molecular spectra".

The style of the inaugural meeting set the fashion for all the subsequent annual meetings, with the Business Meeting of the Fellows and symposia becoming standard menu. Every meeting was a colourful affair, and with a characteristic Raman touch. Throughout his life, Raman was elected the President. To people in other countries this might appear strange and undemocratic, but in India nobody complained. The academic community – at least one section of it – simply accepted Raman as its patriarch, and did not regard his repeated election as a negation of democratic principles. The Indian ethos *is* different! As is only to be expected, such a practice was not without its deficiencies. Particularly in the later years when Raman went through an emotional slump (see Chapter 12), the management of the Academy appears to have been somewhat slack.

Raman's Academy, as some used to call it, was one large happy family, with Raman himself jocularly referring to it as his circus. Its mood was jovial but its purpose was ever serious – and always a low profile. Raman was particular about holding the meetings in college or university campuses so that students, even if they could not follow all the proceedings, would be exposed to the general scientific atmosphere. There were, of course, the evening lectures of a popular nature, always a part of the menu. Many were inspired and drawn to science and the academic world by such lectures. Satish Dhawan, the Academy President for the period 1977–79, once recalled how inspired he was after listening to a lecture by Bhabha.

The scientific programmes were carefully arranged, to include, wherever possible, a symposium on a topic of interest to the host institution. Despite the fact that he was a physicist, Raman took care to see that the symposia covered a wide spectrum of subjects, in keeping with the objectives of the Academy. In Raman's lifetime, over sixty symposia were held on topics such as the physics of the upper air, nitrogen transformations in nature, stellar evolution, technology and biochemistry of rice, physics of thunderstorms, natural resources of Andhra, oceanographic research in the Indian seas, chemical and biological control of pests, and earthquakes.

The annual meetings of the Academy in Raman's days were somewhat unique. Raman never missed them except just once, on account of non-availability of airline reservation. The meetings were invariably like family reunions, with personal warmth and camaraderie among the Fellows helping to relieve the strain imposed by the crowded scientific sessions. In India there is a weakness for converting meetings, at times even scientific meetings, into festivals or occasions for elaborate speech-making, garlanding, etc. With Raman science *always* came first. And what was missed by way of fanfare, flags and buntings was more than made up with humour.

Raman's capacity for humour seems to have been almost legendary. There was light-hearted banter all the time – at the sessions, during the breaks and at mealtimes. He had the gift of making people roar with laughter with his remarks, stories and mannerisms. The annual meetings thus became intense personal experiences for the participants, and not merely a scientific get-together.

On the occasion of the Golden Jubilee in 1985, many Fellows of the Academy nostalgically recalled the Raman days[3]. Thosar, for example, remembered an incident at the Nagpur meeting in 1941 [11.4]. Bhabha was to give a lecture on cosmic rays, and the venue was the Convocation Hall of the university. Naturally, there was no blackboard. Raman introduced Bhabha as a brilliant theoretician, and then, looking around, noticed the absence of a blackboard. He then said, "Well, there is no facility to draw diagrams or write equations here", and, with a twinkle in his eye, added, "It is good in a way – it stimulates the imagination."

Bhagavantam [11.5] recalls a moment of pathos at the Madurai meeting of 1966. Raman had invited him (Bhagavantam) to deliver a lecture, and while introducing the speaker referred to him as an able scientist who unfortunately frittered away his abilities as the occupant of a position of high governmental responsibility. Bhagavantam made a spirited defence of himself. He recalls that when Raman made his closing remarks, he "broke down and was in tears of affection and appreciation of all that I said in my address".

Apart from the scientific discussions, Academy business was also transacted during these meetings, an important item being the election of new Fellows[4]. Other Academy matters were also discussed. On one occasion, there were complaints that the quality of papers published in the *Proceedings* was going down. Raman patiently listened to all the criticism and replied, "Gentlemen, we publish what you produce." That gave the Fellows something to think about!

The evening lectures have already been mentioned. They were intended to be popular, the idea being that they should reach the public at large. Raman chose the speaker and the topic. Naturally the subject would be one he wanted to learn about. However, the purpose was not merely to bring Raman up to date but to inspire the audience, especially the young ones in it.

Sometimes Raman himself delivered the evening lecture. Once, in Madras, he spoke on the physics of musical instruments. On that occasion, Raman's brother Mr C. S. Iyer played on the violin suitably to illustrate the talk. The violin used belonged to Raman's father, and is now preserved in the museum of the Raman Research Institute.

Right from the beginning, Raman took great pains over the publication activities of the Academy. A science journal is no good if it does not appear regularly, and on the appointed dates. Recognizing this, Raman promised punctuality of publication, a promise which he maintained throughout. Bringing out journals on schedule is far from easy in India, especially if one has to operate on a shoe-string budget. There are also complications due to photographs, line drawings and mathematical symbols. Even today it is not easy to handle such problems since our printing industry is generally not geared to produce technical publications. How much more difficult it must have been for Raman! But it was his duty and he had the *sraddha*.

When it was first started in 1934, the *Proceedings of the Indian Academy of Sciences* consisted of a single journal whose objective was "to provide all scientific men with an opportunity of obtaining at least a general idea of what is being done in India in fields of knowledge other than their own speciality". But the amount of published matter grew so rapidly that, starting from July 1935, it was found necessary to separate the *Proceedings* into two parts – A, Physical and Mathematical Sciences and B, Biological Sciences – rather in the pattern followed by the

Proceedings of the Royal Society. This scheme remained in force till 1977 (seven years after Raman's death), when the *Proceedings* was split into six theme journals. These were named *Proceedings – Chemical Sciences, Earth and Planetary Sciences, Mathematical Sciences, Animal Sciences, Plant Sciences* and *Engineering Sciences* (recently renamed *Sadhana*). No, physics was not forgotten! How could it be? The physics part of the original *Proceedings* was reborn as *Pramana, Journal of Physics*, happily published in collaboration with the Indian National Science Academy and the Indian Physics Association – a sure sign that all old wounds have healed. Three more journals were added soon after, namely, *Bulletin of Materials Science, Journal of Biosciences* and *Journal of Astronomy and Astrophysics*.

The latest addition to the family is the *Journal of Genetics*. This journal, founded in 1910 by William Bateson in England, was brought by J. B. S. Haldane to India when he came to settle down here. The journal ceased publication in 1977. However, there was a desire in the community of geneticists that this important journal should be revived. The Academy agreed to do so, and obtained permission to resume publication of the journal.

Many changes have occurred in the publication mechanics as well. The appearance and get-up of the journals have been spruced up considerably, and computer typesetting is used for printing. Despite the increase in number of journals, publication regularity is maintained as in Raman's days. In addition to having an editor, each journal has an editorial advisory board which, in some cases, is international in character. In overall charge of all the journals is an Editor-in-Chief. In Raman's days papers submitted for publication had to be communicated by a Fellow of the Academy, in the tradition of the Royal Society[5]. This procedure is no longer followed and authors may now submit papers directly. There is also a refereeing system. Statistics reveal that there are many rejections as well as revisions.

There have been other innovations also. Some journals like the *Bulletin of Materials Science* often publish the proceedings of topical meetings. Occasionally, special issues are brought out either to felicitate a distinguished scientist on the occasion of an important anniversary or to honour an active Fellow who has passed away.

Right from the beginning, the pages of the Academy's journals have been illuminated, as Mukunda puts it, by important contributions. In the old days, there were, for example, papers by Bhabha and by Harish Chandra, a "towering figure of modern mathematics". This tradition has continued and there always has been a sprinkling of excellent papers. One even sees papers from other countries in the Academy's journals. But when the total score is taken, one cannot help feeling that the best Indian contributions *still* continue to gravitate outside the country. Why is this so? The issue is complicated but worth pondering over[6].

Given Raman's dominating personality and the fact that he personally carried the burden of managing the Academy's affairs, one would have thought that the Academy would have had problems after his death. This was not the case, and therein lies the strength of the Academy. Although Raman facetiously referred to it as a circus, the Fellowship is a well-knit community, as much committed to the ideas of the Academy as its founder was. Thus the Academy not only survived

Raman but actually grew in strength. In fact, thanks to the collective dynamism of the office-bearers who succeeded Raman, the Academy has risen to even greater heights. It has instituted a prestigious Raman Professorship. Another innovation is the Young Associateship which gives an opportunity for outstanding scientists in their late twenties and early thirties to be associated with the Academy for a few years. All the improvements concerning the journals are also due to the new guard.

Sponsorship of symposia and discussion meetings continues as before with the added feature that the proceedings of many of these are brought out as special publications of the Academy. Much work is being done to improve the visibility of the journals abroad and thereby reassure those who fear that publishing in India carries the penalty of obscurity.

The vigour of the Academy was amply demonstrated on the occasion of the Golden Jubilee. In the framework of Indian practices, it was an occasion tailor-made for VIP appearances and speech-making. But the Academy's traditions were different. In keeping with the spirit of the founder's vision, the celebrations got off to a start with a thought-provoking lecture entitled "The pursuit of science: its motivations" by S. Chandrasekhar, the Nobel Prize winning astrophysicist and a Foundation Fellow. It was a nostalgic occasion for the old guard, especially the surviving Foundation Fellows. The presence of Riazuddin Siddiqui, a Foundation Fellow, and the then President of the Pakistan Academy of Sciences, was particularly touching.

When Raman wrote the *Current Science* editorial, he envisaged many functions for the proposed Academy. As it turned out, the Academy he founded could discharge only some of these functions. Many of the others, for example, liaison with foreign Academies, have been taken up by the Indian National Science Academy. Thus the two Academies are today playing complementary roles. A pleasing feature is that, invariably, there are many Academicians serving simultaneously on the Councils of both bodies. This has not only led to better understanding but also co-ordination.

Increased activity has also increased the expenditure. In Raman's days, the Academy was a rather low-budget affair, unavoidable since Government subsidy was neither sought nor received. Even now, the budget of the Academy is a pittance compared to the expenditure of Academies in other countries. But, in relation to the revenue from subscriptions and other sources, the expenditure *is* high and external support becomes necessary. Whether one likes it or not the Government is the only source and the Academy has to turn to it, much as Raman would have disapproved. Unfortunately, the Government subsidy received is not adequate. A recent report presented by the treasurer showed how some activities had to be curtailed for want of funds. Actually, the amount the Government is spending on the Academy is negligible. Considering the outstanding record of the Academy, especially in the field of scientific publication, it is beyond comprehension that the Government grant is so meagre. Much greater sums are wasted by way of improper project management, unwanted tours, and so on. One wonders whether considerations of parity are swaying the minds of those who administer funds. "If we give to the Academy, then would it not be setting a precedent and encourage other claimants?" Are there really so many organizations in the country with a record of sustained service to the cause of science, especially in crucial areas like journal publication? In spite of Raman and fifty years of the Academy, and an equal amount of service by a few other dedicated organizations, the culture of recognizing and rewarding excellence (in organizations) has not penetrated our bureaucratic echelons.

Raman once declared that "an Academy of science is not an ornament but an indispensable institution". The Academy founded by him has fully risen to this expectation.

12 The Final Years

> "My own garden is my own garden" said the Giant. "Anyone can understand that, and I will allow nobody to play in it but myself." So he built a high wall all round it, and put up a noticeboard: TRESSPASSERS WILL BE PROSECUTED.
>
> OSCAR WILDE,
> in **The Selfish Giant**

For Raman, leaving the Indian Institute of Science was a liberating experience. Not that he was at any time subservient to authority; nevertheless, it was a welcome relief for him to be on his own during the evening of his life.

Three distinct phases may be discerned in this, the final period. In the beginning Raman was full of exuberance, busy creating and consolidating his own institute. Then followed a long period of disillusionment and discontent during which he became a recluse. But eventually he regained his cheer, transformed into his old self once again, and began to mix freely. Though his moods went through many changes he never took his mind off research; and more than ever before, Raman's research became a quest for aesthetics.

12.1 The Raman Research Institute

It was obvious to Raman (as well as to everyone else) that retirement did not mean a stoppage of research but just a change of venue. Thus the creation of a new research institute began to engage Raman's attention even before his departure from the IISc. As he once described it:

> You know, I was in the Indian Institute of Science and I was due to retire at 60. So two years before my retirement, I started building this Institute so that on the day I retired I took my bag and walked right into this Institute. I cannot remain idle for a single day. [12.1]

A typical Raman statement, but it mildly glosses over some historical details. We have already seen that way back in 1934 the Maharajah of Mysore made a gift of a ten-acre plot of land to the Academy. One of the stipulations accompanying the grant was that a building for the Academy would be put up within a reasonable period of time. But years passed and nothing happened until in late 1941 the Government of Mysore made it known that the land would be resumed if it continued to remain unutilized. Putting up a building now became an urgent necessity, and Raman made a representation praying for an extension of six months to start construction.

Now the plot was gifted not merely to house the Academy's offices but also for use in connection with all the activities of the Academy, consistent, of course, with its stated aims. Since the promotion of science was the principal objective of the Academy, Raman suggested that the establishment of a research institute in the premises be accorded priority. He even volunteered to raise funds for the proposed institute, subject to the condition that it would be an independent entity and not a part of the Academy as such. Retirement, which was now slowly drawing near, must undoubtedly have influenced Raman's thinking. The proposal was discussed at an extraordinary meeting of the Academy in February 1943, and shortly thereafter a formal agreement was executed between Raman and the Academy concerning the founding of the institute. For many practical purposes Raman was no doubt indistinguishable from the Academy, but at the same time, keeping the long-term future in view, he sought right from the beginning to give distinct identities to the Academy and to the Institute.

Fund raising now became urgent, and Raman went up and down the country appealing to philanthropists, princes and wealthy industrialists for donations. He once described his mission as begging but was not ashamed of it. "Our greatest men were beggars – the Buddha, Sankara and even Gandhi" he declared.

A small sum was soon collected. It was not enough but adequate to start construction, and by 1948 there was a building of sorts. The new centre was named the Raman Research Institute, and in a characteristic fashion Raman moved in and started using it even though the facilities were far from complete. Reminiscing about it, Raman's former student Jayaraman writes:

> For the first year at the Raman Research Institute there was no electricity, but that did not deter Raman from carrying out several beautiful optical experiments with sunlight, a few lenses and a pair of polaroids. He considered a beam of sunlight as the best source and in Bangalore there was no shortage of blue sky and bright sun. A manually operated heliostat, kept in order by voice communication, produced astonishing results. [12.2]

In 1948 Raman was appointed National Professor, a post which carried an honorarium sufficient for his personal needs. The Institute finances, however, were not in a comfortable state. The laboratories had to be equipped, the staff had to be paid, the building had to be maintained and the surrounding area cleared and landscaped. Determined not to approach the Government, Raman now started a few chemical industries (in association with a former student of his), the earnings from which he ploughed into the Institute. Indeed, the Institute meant so much to him that later he gifted away most of his property to the Academy for the benefit of the Institute, as also the Lenin Peace Prize money.

Visitors to the Institute never fail to be impressed by its beautiful garden. This is largely the creation of Raman, who spent much time in planning it and supervising its layout. Jayaraman recalls:

> Raman loved trees, flowers and, above all, his rose garden. All the best roses that Bangalore nurseries could supply were bought and planted in his rose garden under his supervision, and he admired them like a child would admire a new toy. He knew the botanical name of every tree in the campus, and had them planted carefully to maximize the effect of their floral display. Everyday he would go round the garden twice, to enjoy, to relax, and to think.

Raman also loved to collect crystals, gems, minerals, rock specimens, shells, stuffed birds, butterflies – anything that displayed colour. Commenting on Raman's passion for the collection of these objects, Jayaraman observes:

> Raman had an exquisite collection of quartz crystals and quartz family minerals.... He was never tired of acquiring quartz crystals, although he would say he would not go in for any more of quartz. There was a man from Coimbatore who used to work for the Telegraphic Department. He would bring quartz crystals, small and large, to sell, and Raman's resolution would break down at the sight of the material!

Raman's collections grew so rapidly that soon a museum became necessary. About the latter Jayaraman says:

> The specimens were neatly arranged in glass shelves and were lighted appropriately for the best possible viewing. One little room had luminescent minerals and these came to life with brilliant colours when the ultraviolet (UV) lights were turned on. This was one of the thrilling experiences for a visitor to the Raman Institute. When this section of the museum was put into operation, Raman must have turned the UV lights on and off a hundred times, each time enjoying the sight like a child.

Raman loved displaying his collection to visitors. The late Felix Bloch told me that when he visited the Institute, Raman took him on a tour and when they arrived at the museum Raman said: "Professor Bloch, this museum contains my collection of crystals, gems, minerals, etc. But then you are a scientist and you know all about these things. So we shall not spend more than a few minutes here." Bloch added with a twinkle in his eye: "You know Raman. He got so carried away that a few minutes became a couple of hours!" However, a museum tour was never a dull experience as the visitor would invariably be entertained to a fund of colourful stories and anecdotes.

Raman also started the construction of a small observatory. He was always fascinated by the vast canopy of stars that stretches above, and once he declared that if he had his life all over again, he would devote it entirely to astronomy – not to make great discoveries and gain fame but simply to lose himself in wonder. However, the pursuit of astronomy was not to be regarded merely as a matter for personal enjoyment. It was also serious business requiring careful attention and planning. Through the pages of *Current Science* he had earlier repeatedly made strong pleas for astronomical research in India, some of which are interesting to recall. He first reminds us about our past tradition.

> Astronomy is the oldest of the natural sciences, its beginnings being traceable to the remotest periods of recorded human history. There is ample indication in ancient Sanskrit literature of the interest with which the subject was studied in India from the earliest times, while the later writings of Aryabhatta, Varahamihira, Brahmagupta and of Bhaskaracharya, which have come down to us, show that astronomy was actively studied in India at a time when the lamp of learning lighted by the ancient Greeks had burnt out, and Europe was passing through the dark ages. The vicissitudes of Indian history in the later centuries of the present millennium were not favourable to the development and expansion of cultural interests. Some indication that active interest in astronomy nevertheless did not altogether disappear in India is furnished by the astronomical instruments of an earlier era which have been preserved to us, and by the curious structures known as Jaisingh's observatories which are still to be seen at Delhi, Benares and Jaipur. [12.3]

What left India behind was the telescope, and from the days of Galileo there was a ceaseless drive to build bigger and better telescopes. Such instruments are possible only through a skilful combination of high quality optics and precision mechanical engineering. Besides, big telescopes cost money and therefore questions may be raised about whether at all India needs them. Raman is ready for such questions.

> It might be urged that a poor country in which the vast majority of the people live at or below the marginal level of human existence, should not trouble itself about astronomy – a non-utilitarian pursuit, as some might be disposed to regard it. To convert those who hold this view to a different state of mind, it might be useful to point out clearly the enormously important part that astronomical studies have played and are playing in the development of both scientific knowledge and general culture. [12.3]

Raman then points to many examples, mentioning that even subjects like chemistry and energy production have a relationship to astronomy.

> Chemistry is a subject of vast practical importance, and to the uninitiated, it might seem that it could have nothing in common with the science of astronomy which lives with its head up amongst the stars! It is useful to dispel such an illusion if it

> exists in the minds of any. The vital link between chemistry and astronomy is to be found in the problems of the origin of the elements, of their abundance, and of their associations and segregations, all of which are of the utmost importance not only to the chemist, but also to the geologist, the mining engineer and the metallurgist. The spectroscope reveals that all or nearly all the elements present in the earth are also present in the stars. Even the mysterious nebulium proved to be nothing more mysterious than oxygen and nitrogen under somewhat unfamiliar conditions. The problem of the origin of the elements is, therefore, not so much a terrestrial problem as an astronomical one. The transmutations of the chemical elements successfully effected, though on a very minute scale, in the atom-smashing laboratories of the world suggest that such or other analogous transmutations are in progress in the cosmic crucibles which we call the stars. Indeed, the suggestion has been made (and is probably well-founded) that such transmutations are the origin of the tremendous outpouring of energy continually going on from the Sun and the stars. Such extra-terrestrial knowledge cannot but prove ultimately of the highest value and importance to terrestrially-minded capitalists and *entrepreneurs* of industry! [12.3]

It should be appreciated that this remark was made long before research to harness fusion power was started anywhere in the world and long before Homi Bhabha made the famous forecast about fusion power in his presidential address to the first Atoms for Peace Conference in Geneva in 1955.

Raman tries yet another interesting argument calculated specially to appeal to Indian pride.

> ...the work of S. Chandrasekhar, now Professor of Astrophysics at Chicago University,...[is] an indication of what could be accomplished in this country under favourable conditions. It would require an entire number of *Current Science* and not a paragraph or two to sketch the many fields of astronomical and astrophysical research traversed by Chandrasekhar and the results obtained by him during the last fifteen years. The *Monthly Notices* of the Royal Astronomical Society during the years Chandrasekhar was at Cambridge, and the last ten volumes of the *Astrophysical Journal* since he went to the United States bear witness to his energy, the strength and range of his scientific interests and his powers of investigation and exposition. His two treatises on "Stellar Structure" and "Dynamics of Stellar Systems" published by the Chicago University Press make his work in the respective fields conveniently accessible to specialist and non-specialist alike. A memoir on "Stochastic Problems in Physics and Astronomy" which appeared as the January 1943 issue of the *Reviews of Modern Physics* establishes links between the problems of stellar astronomy and those arising in colloid chemistry, and is a very remarkable effort in scientific synthesis. [12.3]

After appeals of various kinds, finally it is his turn to ask questions!

> Our politicians and philosophers are constantly reminding us of India's great spiritual heritage. Should they not raise their voice also to remind us of India's intellectual heritage as reflected in our age-old interest in astronomy, and help to build up a renewed and active interest in its study? [12.3]

Raman does not appear to have completed the observatory he started[1], but it is gratifying that in due course his campaign for astronomical research bore fruit. Thus we have today research programmes not only in the traditional area of optical astronomy but also in the newer ones like radio astronomy, infra-red astronomy,

X-ray astronomy and gamma-ray astronomy. However, contrary to Raman's expectations, most of these activities are being carried on in specialized institutes, universities being hardly in the picture. But then this is part of the general neglect of our universities.

The Raman Research Institute started off essentially as a one-man show, which was no new experience to its founder. Though he was now somewhat older, Raman did not mind the work-load. Rather he enjoyed being active, and there was enough for him to keep busy about, including the various affairs of the Institute and those of the Academy, besides of course the regular publication of the Academy journals. Above all there was the research, to which we now turn.

12.2 Science at the Raman Institute

As during the other periods, optics continued to dominate Raman's research. Naturally there was a continuity with the past, his studies on the Christiansen effect being a good example. Let us start with Raman's own description of the effect:

> In the well-known experiment due to Christiansen (1884), an optically isotropic solid, e.g., glass, is powdered and put inside a flat-sided cell which is then filled with liquid and the refractive index of the latter is adjusted suitably by varying its composition or altering its temperature. Beautiful chromatic effects are observed when the refractive index of the liquid is thus brought into coincidence with that of the powder for some chosen wavelength in the spectrum. The cell becomes transparent for a restricted region of the spectrum in the vicinity of that wavelength, while the rest of the incident light passing through the cell is diffused out in various directions and appears as a halo surrounding the light source. [12.4]

There were no serious attempts to explain the observations of Christiansen, barring a passing suggestion by Lord Rayleigh. At Raman's initiative, N. K. Sethi in 1920 performed some experiments using chromatic emulsions prepared by shaking glycerine and turpentine. However, these emulsions are transient as the two liquids quickly separate, and the experiments are rather difficult to perform. A few years later Sogani improved upon Sethi's work by using better emulsions and clarified many aspects. In 1949 Raman came back to the subject, offering a comprehensive explanation of the observed facts.

> The theory as now developed gives us a clear account of the phenomena and yields results in satisfactory accord with the facts of observation. Its publication has appeared desirable in view of the fact that of recent years, the importance of the Christiansen effect has been more widely appreciated. Many papers have been published and many references to it have appeared in textbooks, concerning themselves chiefly with its practical application in optical filters capable of isolating narrow regions in the spectrum with the minimum loss of light. Strangely enough, however, one does not find in the literature of the subject any recognition of the fact that the performance of such a filter is determined by the principles of wave optics.

The Final Years

Not surprisingly, the theory which Raman advances leans heavily on the concept of wavefront corrugation, and is similar in many respects to the earlier work by Ramachandran on clouds (discussed in Chapter 8), except that instead of air and water, one has here two media of refractive indices μ_1 and μ_2. Raman is able to explain not only the intensity of the transmitted spectrum but also aspects of the diffraction halo. Further, the theory has relevance to practical applications.

> In the practical use of a Christiansen filter, it is necessary, among other things, by suitable methods to separate the light regularly transmitted by the filter from that diffused or scattered by it.... If the particles are very large, the diffracted light would appear in directions very close to that of the regularly transmitted light, and its separation from the latter would obviously become difficult. If, on the other hand, the particles are very small, the spectral width of the transmission would itself be greatly increased. There is thus an optimum size for the particles if the filter is to function most usefully.

Experiments to demonstrate the Christiansen effect are usually performed with irregular fragments of glass. With spherical particles one expects even more spectacular effects, and Raman and Ramaseshan carried out an elegant demonstration of these. They describe the genesis of these experiments:

> During a visit by the senior author to the works of Messrs Chance Brothers at Birmingham in May 1948, he noticed in their showroom an exhibit of glass in the form of tiny spherules about a millimetre in diameter. A sample of the material very kindly presented by the firm was brought back to India, and the studies now reported were made with it. [12.5]

Although the particles were spherical in shape, they were not all of the same size. The best effects are achieved with uniformly sized particles, and the first item of business therefore was to sift out and prepare a sample of spheres having nearly the same diameter. The experiment itself is simple to perform, it being merely necessary to pack a transparent cell with the spherules and then add liquid to it.

> Acetone and carbon disulphide, the two liquids used, have refractive indices respectively lower and higher than the glass for the whole range of the visible spectrum. Hence, when a cell one centimetre thick containing the spherules is filled up with either one or the other liquid by itself, the medium is incapable of regularly transmitting an incident light beam. Nevertheless, a good deal of light does find its way through the cell in these circumstances. There is, however, a remarkable difference in its appearance in the two cases. When the spherules are surrounded by a liquid of lower index, the cell presents a brilliant sparkling appearance, due evidently to the emergence of light beams of considerable intensity from localized areas on its surface. The addition of a little carbon disulphide, though insufficient to render the cell transparent to any part of the spectrum, enhances the sparkling effect and makes it more attractive by reason of a play of colours similar to the "fire" of a diamond. On the other hand, when the liquid surrounding the spherules has a higher refractive index, these effects are not observed. The emergent light is then faint and diffuse, and the cell presents a dull appearance. The striking difference between the two cases, as well as the intermediate stages in which the cell is transparent to particular regions of the spectrum, can all be simultaneously

observed by pouring enough carbon disulphide to fill the lower half of the cell, and then adding acetone to fill the upper half. The acetone, being lighter, floats above the carbon disulphide and inter-diffusion takes place only slowly. The region of mixing appears as a bright band of transmitted colours, violet above and red below, while the upper and lower parts of the cell exhibit the effects arising from a penetration of the light through the cell without regular transmission.

Some of the photographs obtained by Raman and Ramaseshan can be seen in Fig. 12.1. Undoubtedly the pictures would have been more spectacular had they been in colour but unfortunately colour photography was then rather beyond reach in India. The authors note that the walls of the cell influence the arrangement of the spheres within it, and "at least four layers of particles running parallel to the walls of the cell may be clearly seen".

Further elucidation of the diffraction effects was provided by Raman and Ramaseshan by studying what happens when a single spherule immersed in a liquid is illuminated by a distant, monochromatic point source.

> ...the nature of these [i.e., the diffraction patterns] depends notably on the shape of the particle, being widely different for the two most interesting cases in which it is respectively a sphere and a spheroid of revolution. The configuration of the patterns changes progressively as the plane of observation is shifted away from the spherule; the difference in the refractive indices of the sphere and of the liquid in which it is immersed determines how rapid this change is. [12.6]

As Raman himself has noted, a full explanation of the observed facts is available only via wave theory. On the other hand, a qualitative understanding is readily obtained using geometric theory.

> The result of the passage of the light through the sphere and the enclosing liquid is most readily visualized by considering the cylindrical bundle of rays incident on the sphere to be divided up to a great number of concentric hollow cylindrical beams. Each such cylinder of rays would converge to a focus on the axis and subsequently diverge, but the focal points would all be different, being farthest from the sphere for the axial pencils and nearest to it for the marginal ones. In other words, instead of all the rays passing through the sphere converging to a single focus on the axis, we would have a continuous line of foci or concentration of intensity along the axial ray. Further, since each hollow cylinder of rays emerges from the cell as a hollow cone of rays and the convergence of these is different, it follows that the successive cones would intersect each other and form a caustic surface, the cross-section of which by any plane normal to the axis would be a circle. The bundle of rays emerging from the sphere would therefore exhibit a concentration of intensity along its circular periphery. The area enclosed by this would be largest when the beam emerges from the cell and would contract as we recede therefrom, finally collapsing to a point when we reach the focus of the axially incident rays.

The foregoing remarks are illustrated in Fig. 12.2. As is evident, rays which have traversed the sphere by different paths intersect each other after emergence from it. It follows then that one should also be able to observe interference effects. The experimentally significant phenomena of interference in this case are the result of diffraction effects, which indeed are quite evident in the photographs in Fig. 12.3. Even more beautiful patterns are obtained with spheroidal particles, as Fig. 12.4 shows.

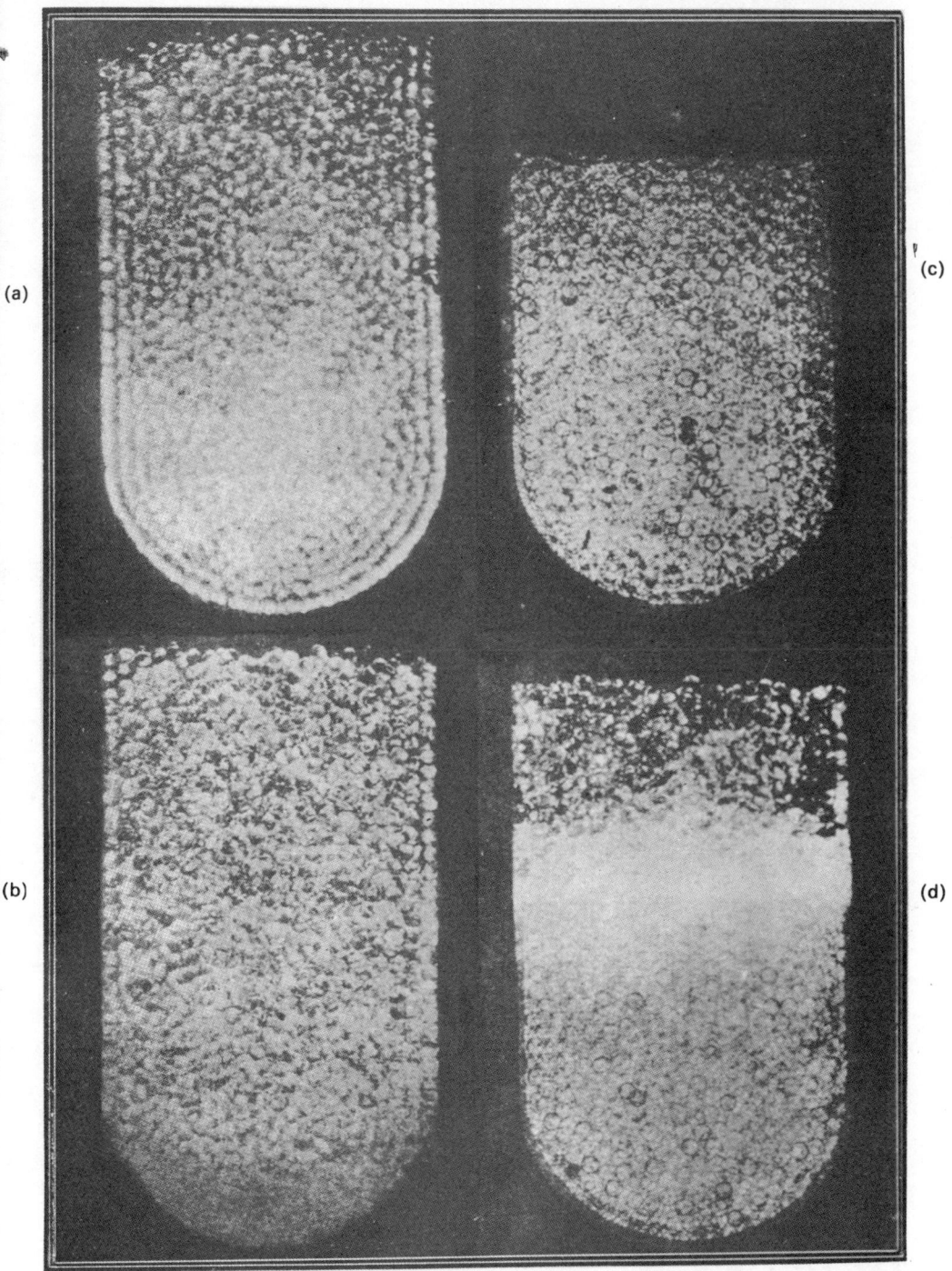

Figure 12.1 The Christiansen effect. Appearance of the cell in transmitted light in different circumstances: (a) without any liquid, (b) when filled with acetone, (c) when filled with carbon disulphide, and (d) when filled half and half with the two liquids, carbon disulphide below and acetone above. (After ref. [12.5].)

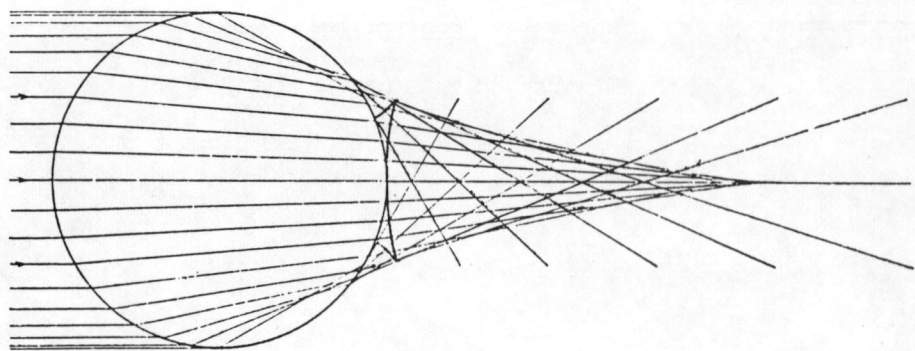

Figure 12.2 Paths of parallel rays incident on a glass sphere of refractive index 1.54 immersed in water (refractive index 1.33) and emerging therefrom. (After ref. [12.6].)

A few years after these studies, Raman and Bhat observed the Christiansen phenomenon using powdered quartz, barium sulphate, calcium sulphate, lithium carbonate, and magnesium fluoride suspended in suitable liquids [12.7]. In contrast to glass, all these materials are birefringent. It is interesting that Christiansen himself tried to detect the effect with birefringent materials but failed to do so. Raman and Bhat observed not only brilliant colours but also many interesting polarization effects.

The Christiansen phenomenon exemplifies the propagation of light through heterogeneous media, of which, actually, there are many examples. For instance, most minerals occur in Nature as polycrystalline aggregates consisting of anisotropic crystallites variously oriented and firmly adhering to each other to form a solid. It is clear that the optical properties of the individual crystals, e.g., their birefringence and absorption, would influence the optical characteristics of the aggregate. For example, the greater the birefringence, the greater would be the reflection at the boundaries of the polycrystalline grains. In fact the brilliant whiteness of pure marble is sometimes explained as being due to strong reflection arising from the strong birefringence of the calcite crystallites in it. One can be certain that such a geometric theory would not appeal to Raman. Indeed it did not, for if one took it seriously then one should expect that the more fine-ground the material is, the more it should reflect, which, however, is contrary to experience. Raman and Viswanathan therefore developed an alternative scenario [12.8] based on the wave theory. In their model, the polycrystal is replaced by a space-filling aggregate of cubes (see Fig. 12.5) each of which is really a crystalline piece of the birefringent material. It is further assumed that the edges of the cube are parallel to the three axes of an index ellipsoid (for a discussion of the index ellipsoid, see Sec. 12.3). However, the index ellipsoids associated with the various elementary cubes are not similarly oriented, the orientation fluctuating randomly between the various possibilities. Raman and Viswanathan now ask what happens if plane-polarized light is incident on such an aggregate with its plane of polarization parallel to a cube edge.

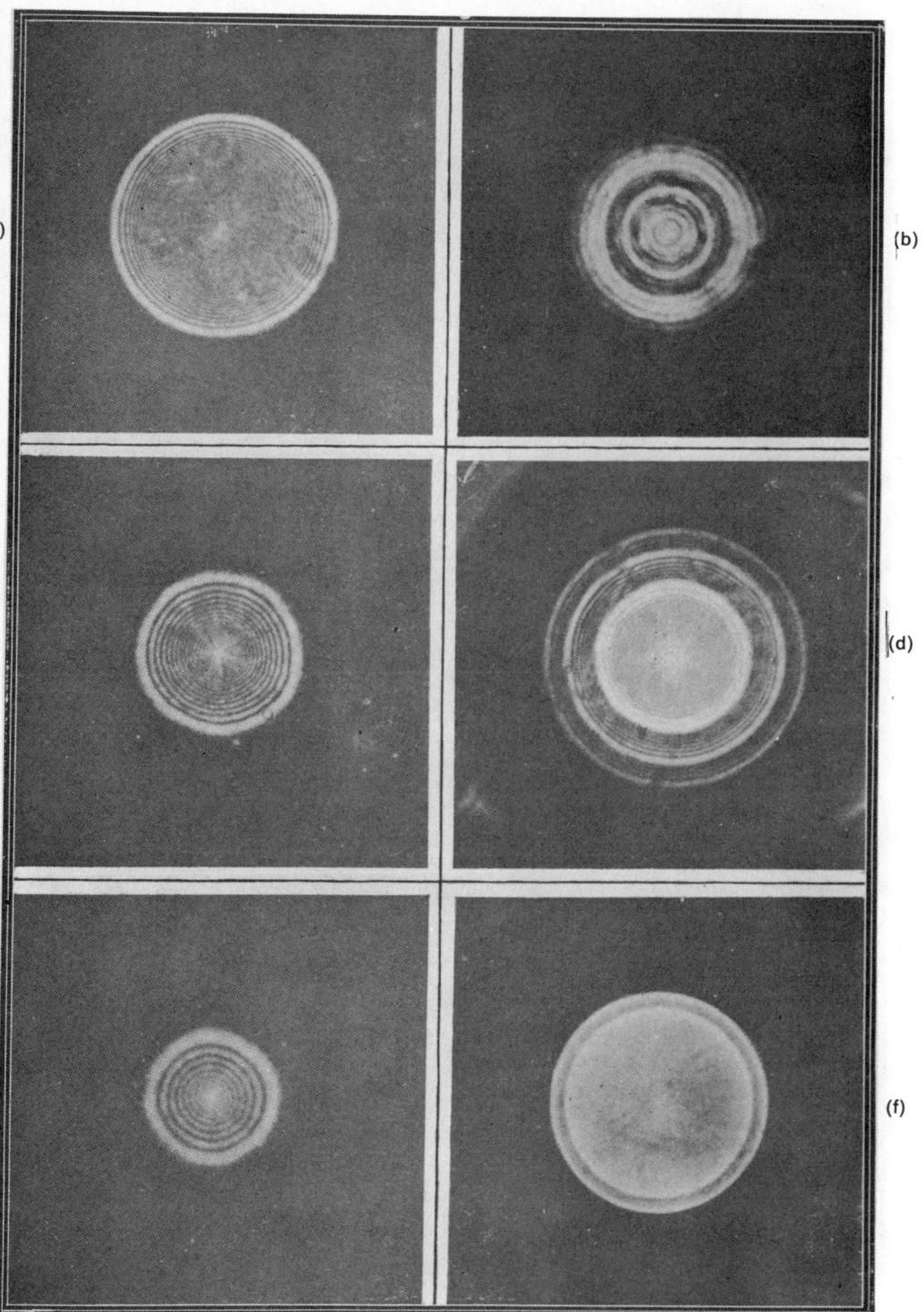

Figure 12.3 Diffraction patterns of (a)–(e) a 1 mm spherical glass ball, and (f) a 5 cm quartz sphere, immersed in various liquids. The plane of observation was different in the different cases. (After ref. [12.6].)

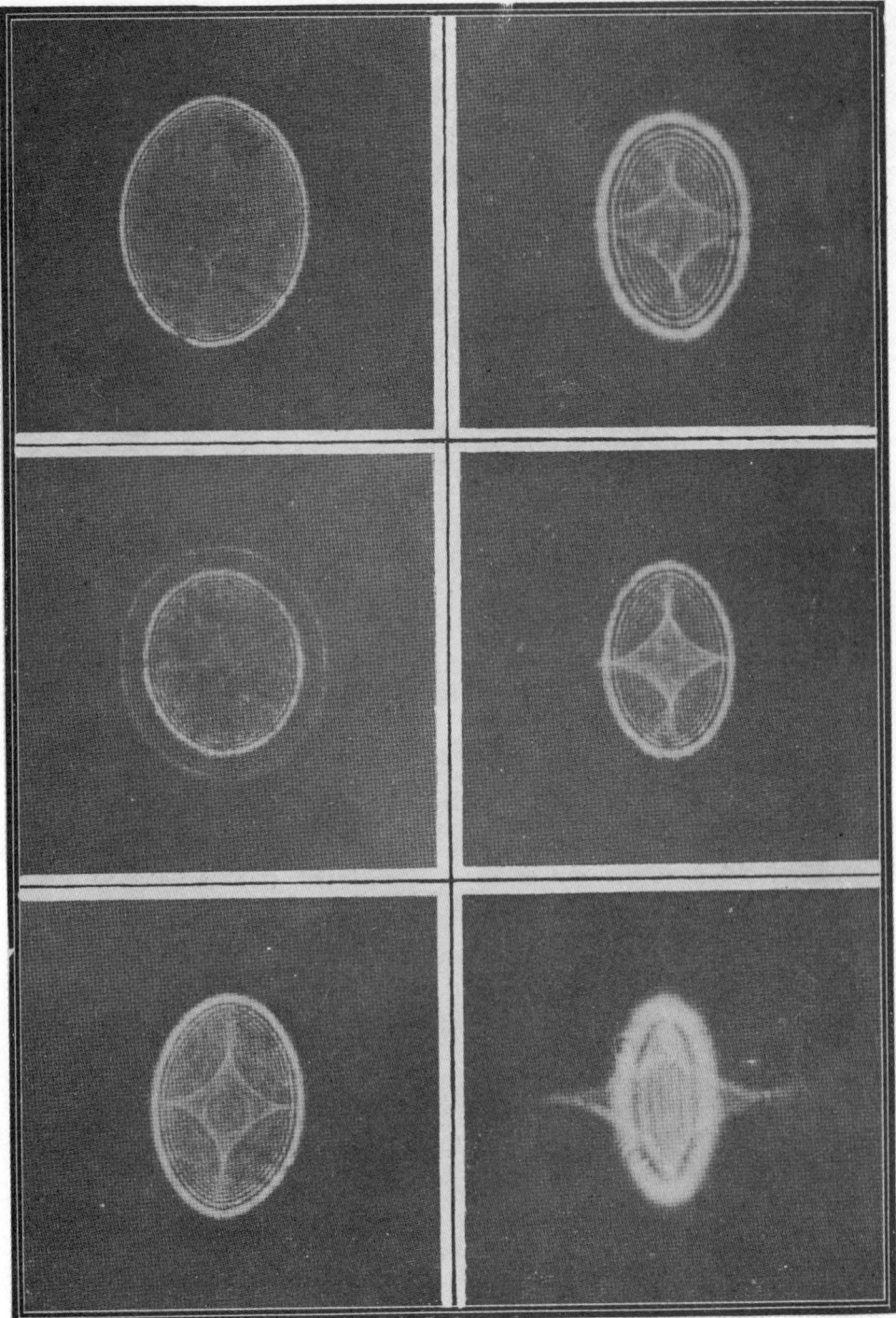

Figure 12.4 Diffraction patterns of a spheroidal ball. (After ref. [12.6].)

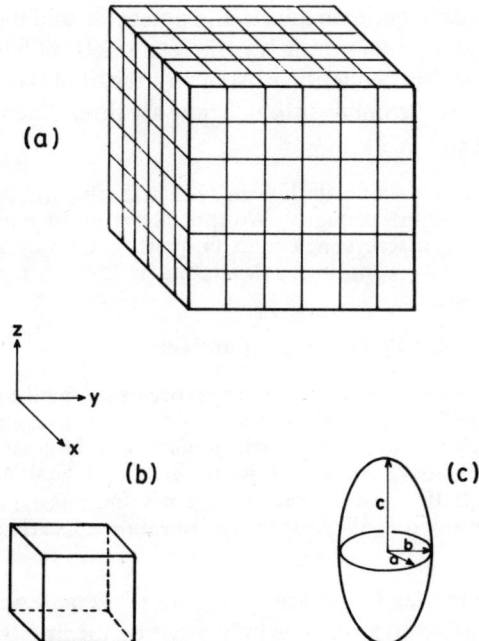

Figure 12.5 A polycrystal visualized as an aggregate of cubes (a). Each cube has its edges parallel to x, y, z axes, as in (b). The optical properties of the crystalline material are described by an index ellipsoid (c). However, the axes of the ellipsoids associated with the various elementary cubes could adopt different orientations. Thus, the x, y and z axes may be respectively parallel to a, b, c; or c, a, b; or c, b, a; etc. In other words, the orientation fluctuates randomly from cube to cube. This is the model of ref. [12.8].

Undoubtedly their model implies several strong assumptions, but despite this they were able to explain many of the observed facts.

Can one use the same model to explain the Christiansen effect? This was indeed done by Raman and Bhat by assuming that some cubes in the aggregate are randomly filled with liquid instead of with the crystalline material. Many deductions of this theory were verified by Raman and Bhat.

Raman's work on amethyst [12.9, 12.10], which we consider next, is typical of his studies on gems and minerals carried out during this period.

By reason of its very beautiful colour, amethyst has been used as a gemstone since ancient times. It is a close cousin of quartz, which, in contrast, is colourless. Raman is interested in this difference and launches "a fact-finding study of those properties of amethyst which are calculated to throw light on its physico-chemical make-up". In association with Jayaraman he (i) makes a comparative study of the densities of quartz and amethyst, (ii) compares the Rayleigh scattering by the two materials, (iii) examines their diamagnetic susceptibilities, (iv) studies the radiographs of various amethyst specimens, and finally (v) makes a topographic examination of the gemstone. Many interesting conclusions emerge, the most important of which relates to the structural relationship between quartz and amethyst, and the genesis of the latter in Nature. Quartz has trigonal symmetry whereas amethyst has the lower monoclinic symmetry, the symmetry reduction arising as a result of the silicon and oxygen atoms moving to slightly different positions. There are three different ways in which this can occur, leading to

> three possible species of amethyst which are however different only in respect of their orientation within the colourless quartz. We thus arrive at an immediate explanation of the fact that the amethystine colour in quartz usually appears in three sectors, the orientation of its monoclinic axis altering by 120° when we pass from one sector to the next. [12.10]

How does this transformation occur? There is an answer:

> ... the transformation from colourless quartz to amethyst occurs during the growth of the crystal [in Nature], and the change of structure from trigonal to monoclinic symmetry is brought about by the presence of ferric impurities in the crystallizing material and during their progressive expulsion from the growing crystal. At some stage during this process, the ferric oxide separates completely, forming aggregates, and thereafter ceases to play any role, the further crystallization appearing in the form of colourless quartz.

In other words, from the same molten liquid are born both these materials, but in a certain sequence: amethyst, the first to appear, is influenced by the impurities, while quartz, which appears later, is not, since by that time the impurities have become segregated. One is not entirely convinced that it is structure alone that is responsible for the beautiful colour of amethyst and not the impurities, but Raman does make the point that anisotropic *intrinsic* absorption influences the colours displayed in polarized light.

Comparing with the work done during the Calcutta period, one cannot help noticing a certain change in Raman's attitude. Whereas earlier he took pains to convince the world at large with his findings, he now appears to be looking more inwards. Many questions which should be raised are not, while others are probed inadequately. He has seen, he has understood, and he is happy – that is all that seems to matter. The further probing which he himself would once have undertaken is now left to his protégé Pancharatnam.

12.3 Raman and Pancharatnam

Unlike in the past, Raman did not have many students during the last phase of his life; but of the few that he had, most acquitted themselves very well, making distinctive contributions of their own later. Unique among them was Pancharatnam, not only because he carried on Raman's tradition in optics, but more so because he was in some respects a reflection of Raman himself.

Pancharatnam was the son of one of Raman's sisters. Born in 1934, he studied right up to college entirely in Nagpur. Earlier his father was employed there, but

The Final Years

even after he moved out Pancharatnam continued to remain in that city. His interests lay in physics and, after passing the M Sc examination (securing a first class and standing first in rank besides), he was on the look-out for a position and a place to pursue research. This was in the early fifties. Curiously enough there had been no meeting between him and his uncle after his childhood days. To him Raman was practically a stranger, living in retirement and pursuing his own special interests. The thought of joining Raman as a student therefore did not occur to him. After the examination he came down to Bangalore for a visit, and a desire to meet his brother Chandrasekhar (then a research scholar and now a professsor at the Raman Institute, and well known for his work on liquid crystals) brought him to the Raman Institute. Raman, who was pottering around in the garden at that time, spotted the young lad, but did not recognize him as his nephew! Nor did the latter care to reveal his identity immediately. It would seem that there was not much opportunity for such exchanges, for, upon learning that the visitor had passed the M Sc examination in physics, Raman plied him with questions in the subject. Impressed with the answers and learning that Pancharatnam wanted to pursue research, Raman asked him whether he would like to enrol as a student in the Institute. To Panch the offer was quite unexpected for he had not come there looking for one. For his part, Raman was utterly surprised when the young visitor did not jump at the offer; instead he wanted time to think it over! Raman agreed, and the following day Pancharatnam came back to the Institute to convey his acceptance. Recalling the incident later, Raman remarked that it was a unique experience for him to have a prospective student ask for time to make up his mind. "He made me humble, you know", he observed.

Raman suggested to Pancharatnam to take up the study of lattice dynamics but the latter's preference was for crystal optics. This is a subject one is likely to dismiss as lacking in sufficient scientific interest, having been explored for nearly a century. Such a conclusion would, however, be totally in error. In a series of beautiful but deceptively simple-looking papers on the subject, Pancharatnam reached the very fringe of optics, leading to crucial questions relating to coherence. In the fifties high-class physics generally meant the anti-proton, meson resonances, parity non-conservation, nuclear magnetic resonance, slow neutron scattering and semi-conductors. These were indeed the exciting frontiers, all except one having been in the limelight of the coveted Nobel Prize. However, a great new revolution in optics was also in the offing. The maser was already there and the laser just around the corner. Coherence and quantum optics were soon to become as fashionable as any other topic, and it is in this perspective that one must view Pancharatnam's work and the influence Raman had on it. But first we must digress somewhat, and supplement the background on optics furnished in Chapter 4.

We start by recalling the concept of linearly polarized light (illustrated in Fig. 4.43). Consider now two linearly polarized light waves of the same frequency but with different amplitudes and a phase difference between them (see Fig. 12.6). The resultant vector will both rotate and change in magnitude so that, effectively, the tip of the vector traces an ellipse. This is elliptic polarization. Indeed, as in the case of the Lissajous figures (recall Fig. 4.5), the ellipse may assume various configurations,

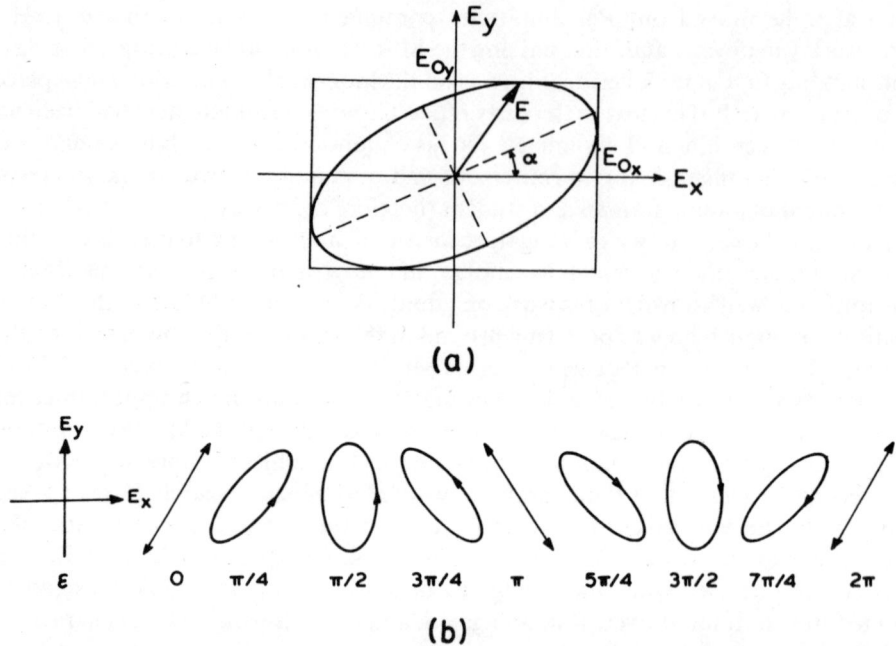

Figure 12.6 Two vibrations $E_x = E_{0x} \cos(kz - \omega t)$ and $E_y = E_{0y} \cos(kz - \omega t + \varepsilon)$ combine into elliptically polarized light, as illustrated in (**a**). For a given ratio of E_{0x}/E_{0y}, the configuration of the polarization can vary with ε, as in (**b**).

as shown in Fig. 12.6b. Clearly, circular polarization (see also Fig. 12.7) is a special case of elliptic polarization, where the two amplitudes are equal and the phase difference ε equals either $\pi/2$ or $3\pi/2$. Depending on the direction of rotation of the resultant vector, one speaks of either right- (R) or left-circularly (L) polarized light. In passing one also notes that (see Fig. 12.8)

(i) two linearly polarized waves travelling in the same direction and with the same frequency combine into an elliptically polarized wave even if their planes of vibration are not mutually perpendicular,
(ii) a right-circularly polarized wave and a left-circularly polarized wave (of the same frequency and travelling in the same direction) can combine, in general, into an elliptically polarized wave, and
(iii) two elliptically polarized waves of the same frequency and travelling in the same direction combine, in general, into a single elliptic polarization wave.

Implicit in the discussion presented thus far is the assumption that the light wave is monochromatic, i.e., has a single frequency component, say ω_0. By definition, a monochromatic wave is an *infinite* wave train, unfortunately never realizable in practice. Actual light sources are always polychromatic, although in sources like the laser, the spread in frequencies is rather small.

The Final Years

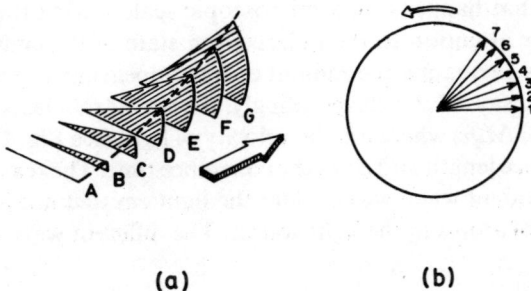

Figure 12.7 Concept of circularly polarized light. Suppose light is propagating in the direction shown by the fat arrow in (a). At any given instant, the orientation of the electric field vector at different points A, B, etc., will be as shown, with the tips tracing a helix. At a given point such as A, the orientation of the electric field vector at successive instants will be as shown in (b). The tip of the vector will appear to rotate in an anti-clockwise direction. This is left-circularly polarized light.

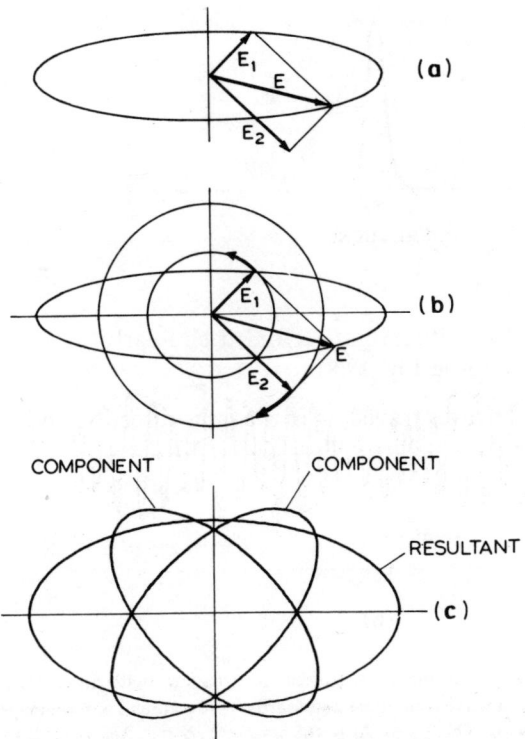

Figure 12.8 Elliptic waves synthesized by combination of (a) two linearly polarized waves, (b) two circularly polarized waves, and (c) two elliptically polarized waves.

Let us now examine what happens on a microscopic scale during the emission process, paying particular attention to the polarization state of the emitted wave. Light is emitted by atoms as quanta, the radiant energy appearing as a wave train with a finite spatial extent, say ΔL. Correspondingly, the wave train lasts for only a duration Δt of the order of $\Delta L/c$, where c is the velocity of light (see Fig. 12.9a). One refers to ΔL as the coherence length and Δt as the coherence time. The reason for this nomenclature becomes evident when we consider the light emitted not just by one atom but by a collection of atoms in the light source. The different wave trains will

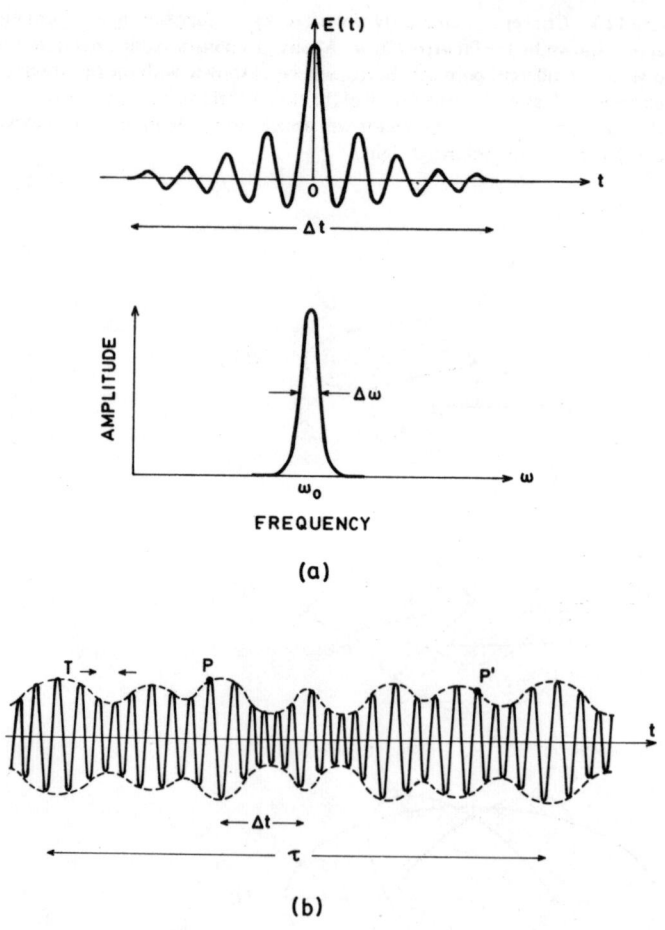

Figure 12.9 A wave train essentially consists of just a few oscillations as in (a). Because of its finite lifetime Δt, a wave train must necessarily have a spread of frequencies $\Delta \omega$ about a nominal frequency ω_0. The longer Δt is, the smaller becomes $\Delta \omega$. In (b) is shown a quasi-monochromatic wave train. There are basically three time scales: (i) microscopic, of the order of T the period, (ii) mesoscopic, of the order of Δt the coherence time, and (iii) macroscopic, of the order of $\tau \gg \Delta t$. On scale (iii), the amplitude and phase fluctuate randomly.

The Final Years

then combine as in Fig. 12.9b to produce a quasi-monochromatic wave train. Two points P and P' on this wave train will have a definite phase relationship only if their time separation is less than Δt introduced earlier. In other words, while the phase of a quasi-monochromatic wave will remain sensibly constant on a time scale of Δt or less, from a macroscopic standpoint it fluctuates rapidly (see Fig. 12.9). Correspondingly, the polarization will remain steady only during intervals Δt or less, fluctuating on a larger time scale, i.e., the light would be unpolarized. Thus, the concepts of polarization and coherence are related in a fundamental way. White light is a superposition of polychromatic trains, with ω_0 spread over a wide range. Note, however, that white light can be polarized by subjecting it to a selection procedure (recall Fig. 4.46).

At this stage one might wonder, and legitimately too, as to how at all Newton and Young could observe interference effects with their simple experiments (recall, for example, Fig. 4.41). The answer becomes clear if one invokes the concepts just described.

Consider a point source placed at S, and two points S_1 and S_2 far away from it (see Fig. 12.10). Our usual description of interference effects is based on the Huygens wavelets emitted at S_1 and S_2 which will arrive at P with a path difference ΔX. A little reflection should show that if ΔX is larger than the coherence length, interference effects will not be visible. In other words, the observation of interference effects depends on a careful choice of h and d as defined in Fig. 12.10. Michelson, who understood these subtleties very well, exploited the knowledge to design beautiful experiments to measure the diameters of stars.

We turn next to the optical properties of crystals. The propagation of light through a crystal is controlled not only by the species of atoms in it but also by their geometrical arrangement in space, in other words by the crystal structure. However, at a macroscopic level one is not concerned with the details of the atomic architecture but merely with how light propagation varies with direction.

Reference has already been made to the phenomenon of birefringence. This, as well as the other aspects we are about to discuss, flow essentially from the symmetry

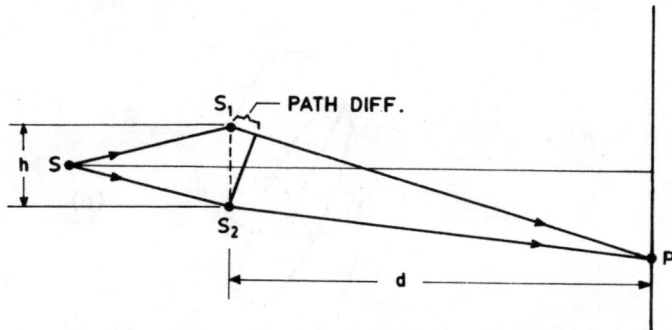

Figure 12.10 Illustration of the relevance of coherence length to the formation of interference fringes. For explanation, see text. Also recall Fig. 4.41.

of the crystal. From the standpoint of the optical properties, one may classify crystals as isotropic, uniaxial and biaxial. A crystal is said to be isotropic if it exhibits the same refractive index in all directions. Crystals with cubic symmetry (e.g., diamond and sodium chloride) belong to this category. Uniaxial and biaxial crystals, on the other hand, exhibit anisotropy.

A convenient representation of the directional variation of the refractive index is provided by the so-called index ellipsoid (see Fig. 12.11). For a given direction **s**, there are in general two values of the refractive index, n_1 and n_2, obtained by taking a section of the ellipsoid perpendicular to **s**, as illustrated in Fig. 12.11b. Further explanations will be provided shortly. For cubic crystals, $n_x = n_y = n_z$, and the index ellipsoid degenerates into a sphere. In uniaxial crystals, $n_x = n_y$, and the ellipsoid becomes an ellipsoid of revolution, i.e., the surface obtained by rotating the ellipse about the z axis. For biaxial crystals, the ellipsoid is a general figure with $n_x \neq x_y \neq n_z$ (Fig. 12.11a).

A point source of light placed in vacuum emits spherical waves; but when it is placed in a crystal the situation is quite different, on account of the refractive index having (i) two values for a given direction, and (ii) angular variations. Figure 12.12 illustrates what happens in crystals with optical anisotropy. Observe that although there are two sheets to the wavefront, the sheets touch along certain directions known as the *optic axes*; along these directions, there is no double refraction. Uniaxial crystals have only one optic axis whereas biaxial crystals have two.

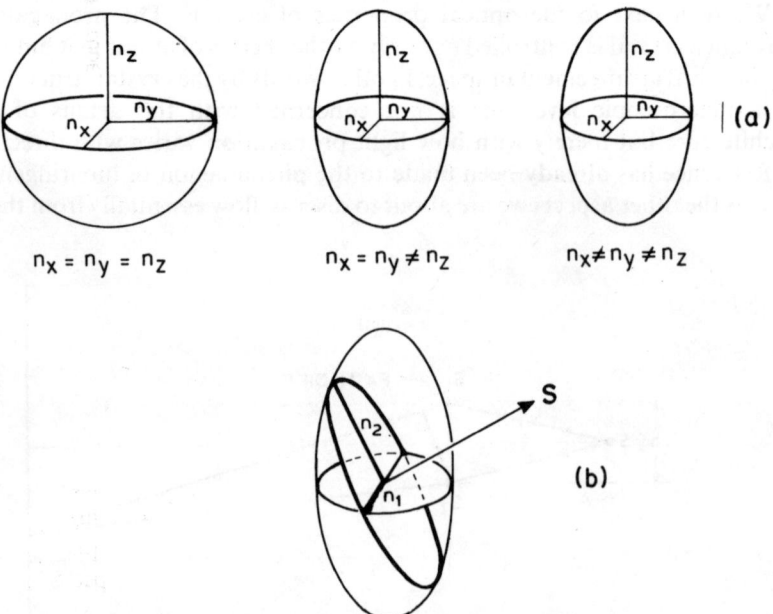

Figure 12.11 (a) The index ellipsoids; (b) illustrates how the two refractive indices n_1 and n_2 associated with light propagating along a particular direction **s** may be obtained.

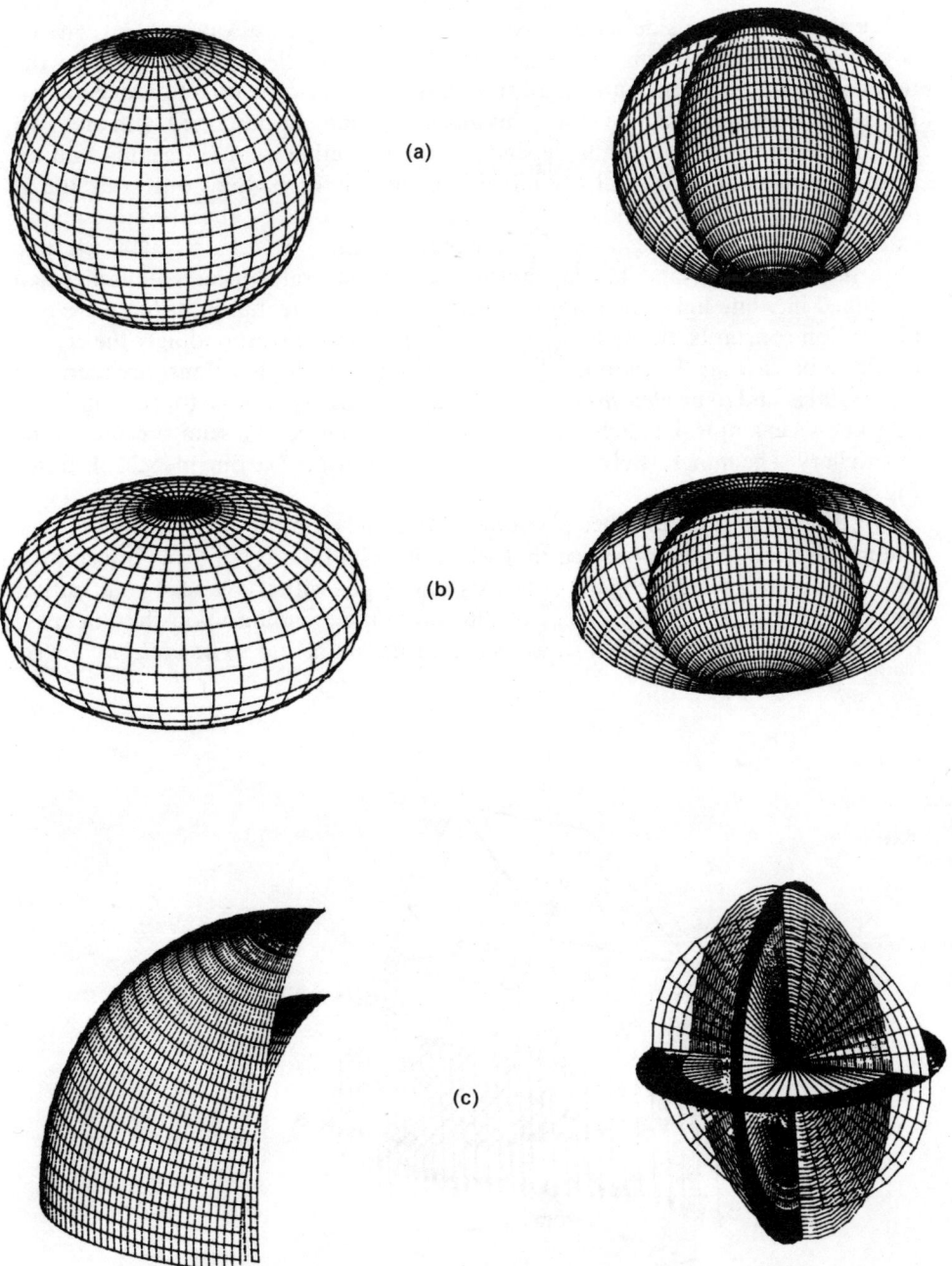

Figure 12.12 (a), (b) Wavefronts in a uniaxial crystal, i.e., a crystal with just one optic axis. (c) The situation in a biaxial crystal, i.e., one with two optic axes; on the left are shown portions of the two surfaces at a particular instant of time, and on the right, the edges in three mutually perpendicular planes.

We have not yet considered (intrinsic) absorption, which is always present; when it is strong and depends on frequency, the diversity of effects increases. Like the refractive index, absorption of light in a crystal can also be direction-dependent. Indeed, as with the refractive index, this anisotropy may be described with one, two or three absorption constants, depending upon whether we are dealing with an isotropic, uniaxial or biaxial crystal. When there is more than one constant, it implies that the ordinary and the extraordinary rays would be absorbed to different extents, as Fig. 12.13 illustrates for a particular case. Thanks to such selective absorption, uniaxial and biaxial crystals can in general appear coloured when examined in white light, the colour depending on the orientation. If there are two absorption constants, two colours become evident and correspondingly the crystal is said to be *dichroic*. With more than two constants multiple colours are seen and the crystal is said to be *pleochroic*. The naturally occurring mineral tourmaline is the best-known example of a dichroic crystal and is often used as a semi-precious stone in jewellery. The minerals vivianite and cordierite[2] display strong pleochroism (see Fig. 12.14).

We turn now to yet another phenomenon, namely, optical activity. In 1811 the French physicist Arago observed that when plane-polarized light was transmitted through a quartz plate along the optic axis, it emerged with its plane of polarization rotated (see Fig. 12.15), the extent of the rotation depending upon the thickness traversed. Any material which causes such a rotation is said to be *optically active*.

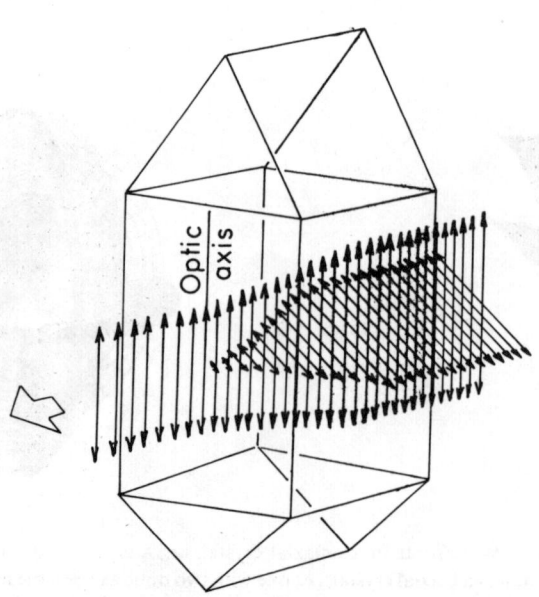

Figure 12.13 Selective absorption of one of the two components.

The Final Years

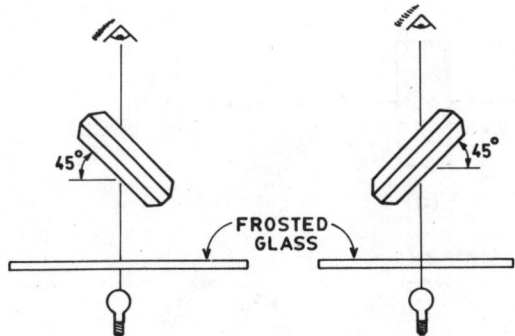

Figure 12.14 Demonstration of pleochroism in vivianite.

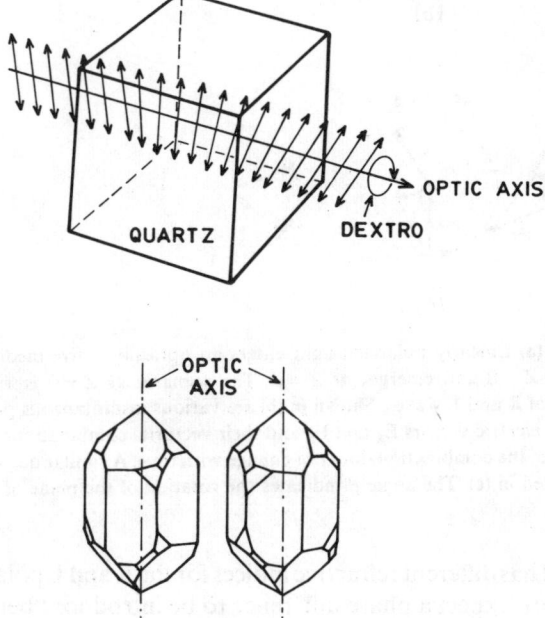

Figure 12.15 Illustration of optical activity in quartz. As discussed in the text, one can have right-rotating as well as left-rotating quartz crystals.

In 1822 the English astronomer Herschel recognized that there actually existed two forms of quartz, one right-rotating and the other left-rotating; they are referred to as *dextrorotatory* and *levorotatory* respectively. It is now known, in terms of atomic structure, that one form is the mirror image of the other; crystallographers refer to the two forms as enantiomorphs. A simple explanation of optical activity is provided in Fig. 12.16. One supposes that the incident linearly polarized light is a superposition of right-circularly (R) and left-circularly (L) polarized waves. It is further assumed that the

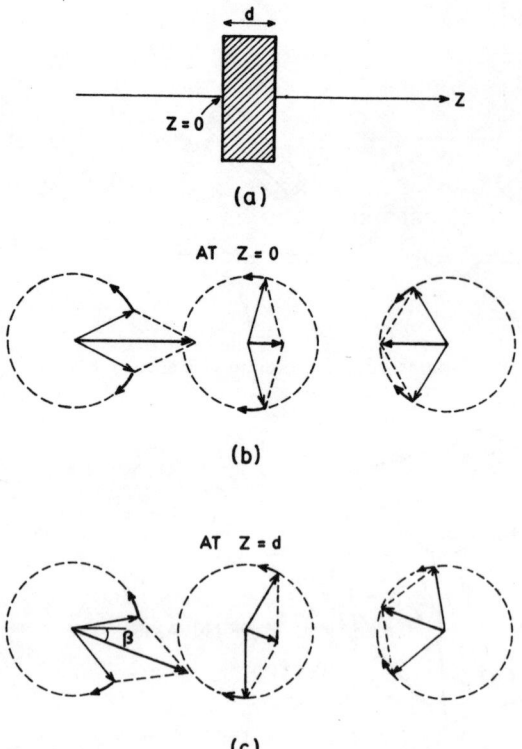

Figure 12.16 (a) Linearly polarized light enters an optically active medium along the optics axis at $Z=0$ and emerges at $Z=d$. The scenario at $Z=0$ is visualized as a superposition of R and L waves. Shown in (b) are various instantaneous positions of the corresponding electric vectors E_R and E_L and their vectorial combinations. Observe that the alignment of the combination does not change with time. A similar decomposition for $Z=d$ is depicted in (c). The angle β indicates the rotation of the plane of polarization.

optically active material has different refractive indices for the R and L polarized light, whereupon it is natural to expect a phase difference to be introduced between them owing to traversal through the medium. Because of this extra difference imposed by the crystal, when the R and L waves recombine at the exit end to form a linearly polarized wave, the direction of polarization appears rotated with respect to that at entry. What now if the R and L waves are absorbed to different extents? Indeed this is possible, leading to what is known as *circular dichroism*.

In the fifties Pancharatnam was almost totally preoccupied with the study of polarized light for the representation of which he made repeated and skilful use of an elegant concept known as the Poincaré representation introduced at the turn of the century. More popular with the pundits, however, was the alternative representation based on the so-called Stokes parameters which had the merit of linkage to Maxwell's electromagnetic theory. In retrospect, Pancharatnam's approach is to be applauded not only for the reason that it provided a simple means of understanding complex phenomena, but also because of deeper implications that have only recently been discovered.

Conceptually the Poincaré representation is quite simple but, thanks to the popularity of the Stokes parameters, remained neglected for a long time until its great utility was pointed out [12.11, 12.12][3]. Very simply stated, there is a one-to-one correspondence between all the points on the surface of a sphere (of unit radius and called the Poincaré sphere) and all possible forms of elliptic vibrations that can be conceived, circular and linear vibrations being special cases of elliptic vibrations. More explicitly, if the major axis of the ellipse makes an angle l with a reference direction (see Fig. 12.17) and if ω denotes the ellipticity, then this particular state of elliptic vibration can be represented as a point P on the sphere having a longitude $2l$ and a latitude 2ω. Elliptic vibration with the same characteristics but the opposite sense of rotation will be represented by a point P′ diametrically opposite. Clearly, the north and the south poles represent the two possible states of circular polarization while points along the equator denote all possible states of linear polarization. In passing one notes that all surface points represent states of *complete* polarization. One could extend the concept and represent partially polarized states by points *inside* the Poincaré sphere such that their distance from the centre is a measure of the degree of polarization. In this picture, an unpolarized beam would be

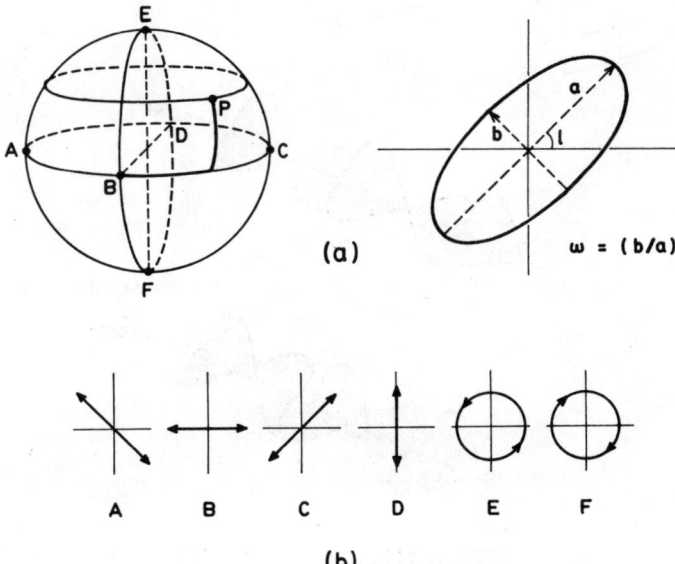

Figure 12.17 (a) The Poincaré sphere. A point P of longitude $2l$ and latitude 2ω represents an elliptic vibration of azimuth l and ellipticity ω, as shown in (b). The poles represent circular vibrations and the equator represents various states of linear polarization (c).

represented by the centre of the sphere. Our immediate concern would be with completely polarized waves.

Given the Poincaré sphere, the change in the state of polarization of light as it passes through any medium can be visualized as a trajectory on the surface of the sphere. Let us now see how Pancharatnam exploits this idea to design an achromatic quarter-wave plate [12.13].

The quarter-wave plate is an optical element which introduces a relative phase difference $\varepsilon = \pi/2$ between the two constituent components (see Fig. 12.18). It should be immediately evident that when plane-polarized light passes through such an element, it will, in general, emerge as elliptically polarized light. The action we have described is for light of a particular wavelength. If the element is to act as a quarter-wave plate for all wavelengths, then the refractive index of the material of which the plate is made must not vary with wavelength. In general this is not true, and what acts as quarter-wave plate for the deep red of the spectrum might well act as a half-wave plate for the blue region. The problem now is to design an achromatic device,

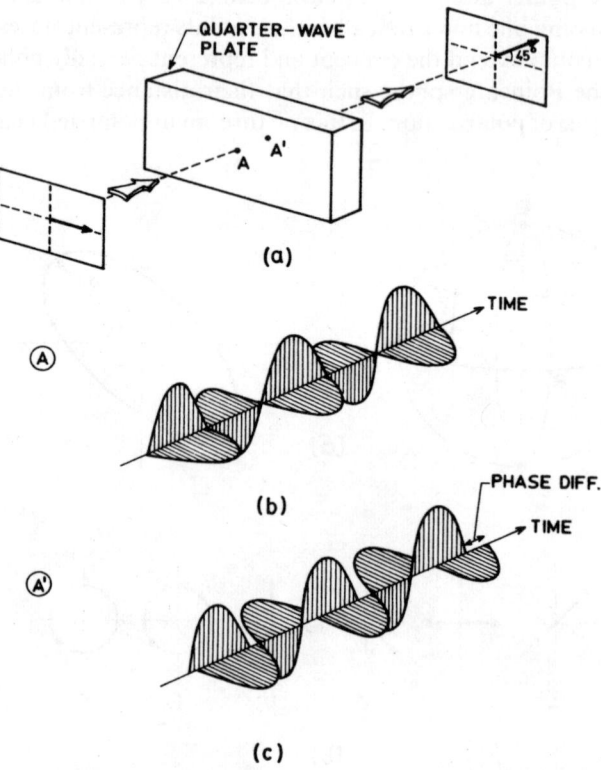

Figure 12.18 A quarter-wave plate is a device such that linearly polarized light passing through it emerges rotated by 45°, as shown in (a). In (b) and (c) are shown the electric vibrations associated with the ordinary and the extraordinary waves at entry and exit respectively.

The Final Years

i.e., one which will function as a quarter-wave plate over a range of wavelengths instead of at only one. Clearly a homogeneous plate would not suffice, but a composite, if properly designed, could be made to do the job.

Pancharatnam's composite consists of three birefringent plates of which the first and the last produce a phase difference or retardation of say $2\delta_1$ and the central one a retardation of $2\delta_2$. How do these plates combine? The Poincaré sphere provides the answer, if one remembers the following simple rule of spherical geometry: if ABC is a triangle described on a sphere whose centre is O, then a rotation about AO through twice the internal angle at A followed by a rotation about BO through twice the internal angle at B is identical to a rotation about CO through twice the external angle at C (see Fig. 12.19a). Referring now to the Poincaré sphere in Fig. 12.19b, the action of the first two plates in the composite is equivalent to rotation about AO by $2\delta_1$ followed by another about BO by $2\delta_2$. But according to the theorem just stated, this is equivalent to a rotation about CO by 2ϕ. As for the effect of the third plate, it is represented again by a rotation $2\delta_1$ about AO. The action of the composite as a whole may thus be represented by appropriate rotations about CO and AO, which, according to Fig. 12.19b and our theorem, are also equivalent to a single rotation through 2δ about DO. The design problem is now reduced to the following: given δ, which is half the final rotation required (which here must equal $\pi/4$), what should be δ_1, δ_2 and c? Pancharatnam simplifies matters by fixing δ_2 at $\pi/2$, which leaves only δ_1 and c to be determined. Actually the problem is slightly more complicated in that δ_1 and c must be so determined that the composite element functions as a quarter-wave plate over a range of wavelength. But this did not pose any serious problem to Pancharatnam, and based on his design he actually built an element which performed satisfactorily over the wavelength range 4100 Å to 6800 Å. A German company is now marketing such a device, and it is gratifying that it bears Pancharatnam's name.

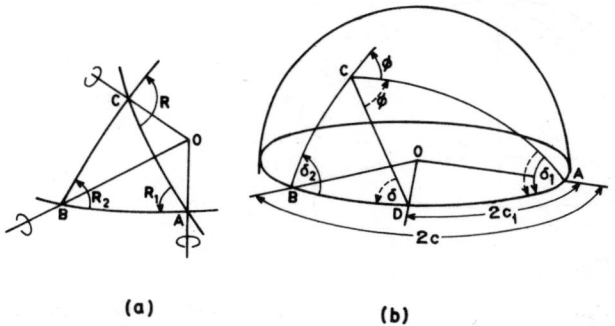

(a) (b)

Figure 12.19 (a) Construction for the composition of two rotations. A rotation about AO followed by another about BO is equivalent to a single rotation about CO according to the composition rule $2R_1 + 2R_2 = 2R$. (b) Illustration of the composition involved in the design of an achromatic quarter-wave plate. For explanations, see text. (After ref. [12.13].)

Another exercise of a similar nature was the design of an achromatic circular polarizer. The latter is a device which transforms linearly polarized light into circularly polarized light. It is usually stated that a quarter-wave plate would do the job, and in terms of the Poincaré sphere it implies the transformation of a point such as R on the equator to the pole (see Fig. 12.20). However, this applied to that wavelength λ for which the plate produces a phase shift of $\pi/2$. Wavelengths λ_1 and λ_2 on either side of λ will experience different shifts, say $\pi/2 - \varepsilon$ and $\pi/2 + \varepsilon$, and therefore will not be circularly polarized. Nevertheless, a combination can be made to do the trick if, as Pancharatnam showed, the ellipticities of the incident beam are dispersed in a proper fashion along the arc $R_1 R_2$ in Fig. 12.20. It turns out that the composite carries all points on the arc $R_1 R_2$ to S. Pancharatnam built an achromatic circular polarizer and showed that it performed satisfactorily over the range 4400 Å to 7400 Å.

Pancharatnam liked to present his findings in carefully structured papers spread over several parts, being, in this respect, different from Raman, who often tended to miss the follow-up. One such series is devoted to light propagation through absorbing biaxial crystals, and was inspired by the very interesting properties of the mineral iolite found in South India. Because of its pleochroism, certain specimens of this material appear practically colourless and transparent in certain orientations but show colour in others. In general, absorbing biaxial crystals display a variety of remarkable phenomena which arise owing to a combination of birefringence and dichroism. Pancharatnam showed that their combined effect could be obtained by recourse to what he called the superposition principle, an idea of which he made repeated use.

Pancharatnam's analysis of light propagation along the so-called singular axes is illustrative of the new findings he made using the superposition principle. It is known that close to the optic axis (in a biaxial crystal) and on either side of it, there exist two other directions along each of which only *one* state of polarization can be

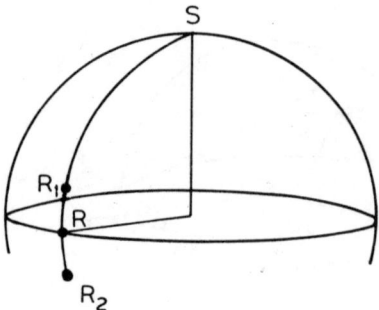

Figure 12.20 A circular polarizer converts linearly polarized light into circularly polarized light, i.e., takes a state such as R and on the Poincaré sphere to the state S. Pancharatnam produced an achromatic combination consisting of two half-wave plates and one quarter-wave plate. Effectively, if the states of polarization of the incident light corresponding to different wavelengths were dispersed along the arc $R_1 R_2$, the device collapsed all these states to the state S. (After ref. [12.13].)

propagated, i.e., only a right-circularly (R) polarized wave can be propagated unchanged along one of these axes and a left-circularly (L) polarized wave similarly along the other. These directions are referred to as singular axes, and since there are two optic axes there will be four singular axes. The question now arises as to what precisely happens if an R wave is incident in the direction of a singular axis along which an L wave is transmitted unchanged. It was earlier supposed that the R wave would be totally reflected. Using the principle of superposition, Pancharatnam came to the surprising conclusion that the R wave is transmitted with greater intensity than the L wave itself! Actually, what happens is that the R wave gets changed into an L wave. These deductions were verified experimentally.

Of particular importance is the four-part series, "Generalized theory of interference and its applications". These investigations were inspired by the beautiful interference phenomena one sees with crystal plates in convergent light (see Fig. 12.21 for an explanation of this geometry). Figures of the most diverse description are observed, some examples of which may be seen in Fig. 12.22.

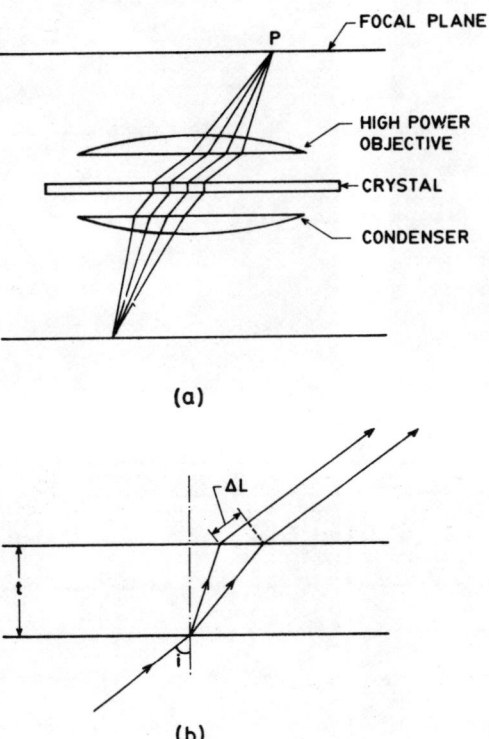

Figure 12.21 (a) The so-called convergent light arrangement for observing interference effects produced by crystals. (b) Illustration of the origin of the interference effects. The incident ray upon entering the birefringent crystal splits into two rays (also recall Fig. 4.47) which, upon emergence, have a path difference ΔL. It is this which leads to the interference effects.

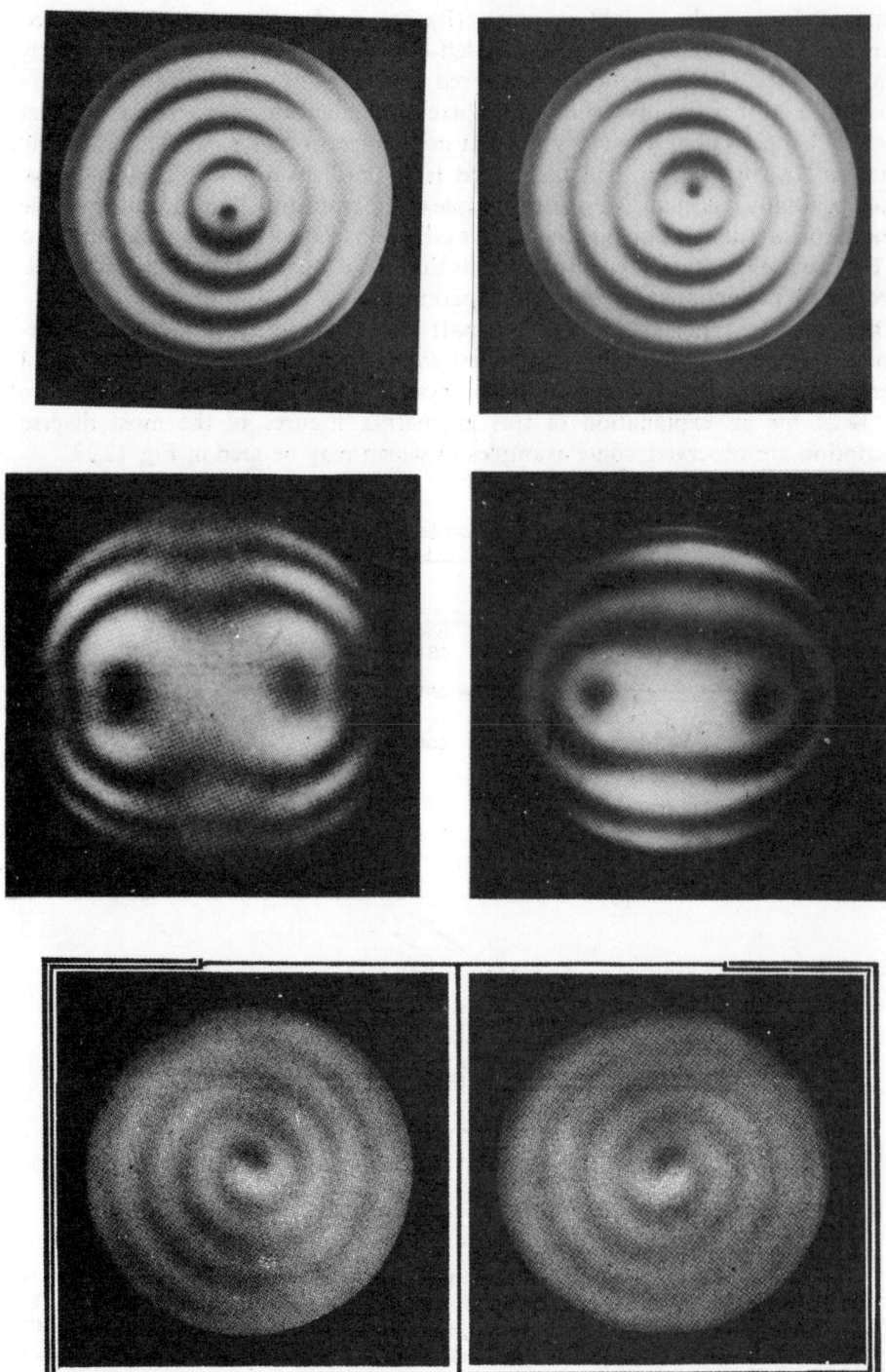

Figure 12.22 Some examples of interference patterns observed by Pancharatnam. (After ref. [12.13], and G. N. Ramachandran and S. Ramaseshan, in *Handbuch der Physik* (Springer, Berlin, 1961), vol. 25, part 1.)

The Final Years

We pursue now the question of coherence since it is in the study of partial coherence that Pancharatnam arrived at results of contemporary importance via the unlikely medium of crystal optics. Two light beams are said to be completely incoherent when their intensities merely add and no interference effects are exhibited. For interference patterns to occur, the two beams must be coherent or at least partially so. To probe further, consider first a monochromatic beam with elliptic polarization, which means that we disregard amplitude and phase fluctuations such as were illustrated in Fig. 12.9. One knows that any elliptic vibration can be decomposed into two orthogonal linear vibrations (the states of which would be represented by diametrically opposite points on the equator of the Poincaré sphere, as shown in Fig. 12.23a). In general, any elliptic vibration G can be decomposed into two orthogonal elliptic vibrations A and A', as in Fig. 12.23b. From the properties of

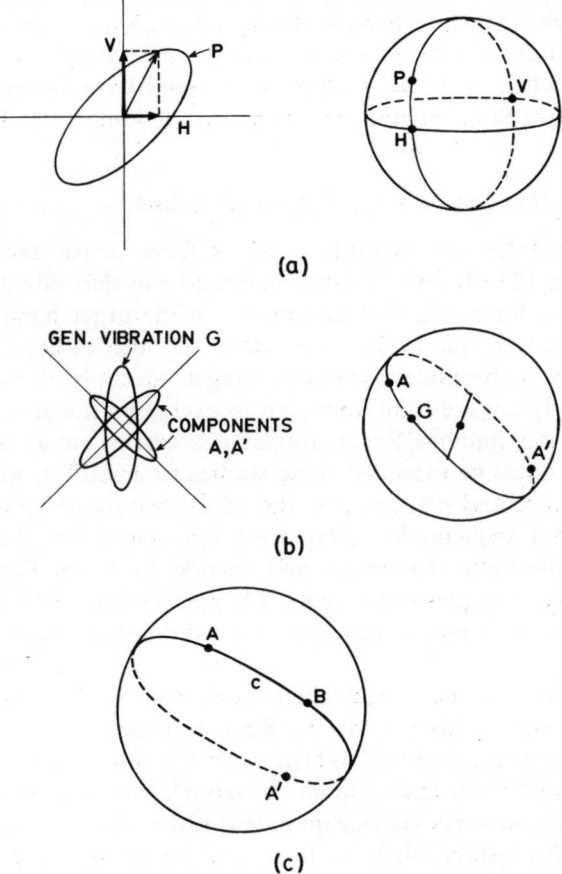

Figure 12.23 (a) Elliptically polarized light resolved into two orthogonal linearly polarized waves and represented by points H and V on the Poincaré sphere. (b) Elliptically polarized light decomposed into two orthogonal elliptic waves A and A'. (c) Interference of elliptically polarized beams in states A and B; further comments in text.

the Poincaré sphere, Pancharatnam is able to show that beams such as A and A' which are necessarily oppositely polarized cannot ever interfere. Then follows a result relating to the interference of two elliptically polarized beams in arbitrary states A and B, as in Fig. 12.23c. If I_1 and I_2 denote the respective intensities of the two beams, the total intensity I is given by

$$I = I_1 + I_2 + 2\sqrt{I_2 I_2} \cos(c/2) \cos \delta, \qquad (12.1)$$

where c is the angular separation of the states A and B on the Poincaré sphere (shown in Fig. 12.23c) and δ is the phase difference between the two beams. (More technically δ is the phase difference between the beam 1 and the component of beam 2 when resolved in the state A.) The interference effects are described by the last term on the right-hand side of Eq. (12.1). When A and B become antipodes, this term automatically vanishes. If now one considers instead two *partially* polarized beams, a new element enters into the picture, namely, the *degree of coherence* γ, which can take values between 0 and 1. If $\gamma = 0$, the two beams are said to be completely incoherent. At the other extreme, when $\gamma = 1$, the two beams are completely coherent. When the effects due to the degree of coherence are taken into account, Eq. (12.1) becomes modified to

$$I = I_1 + I_2 + 2\gamma \sqrt{I_1 I_2} \cos(c/2) \cos \delta, \qquad (12.2)$$

Physically, the need for the quantity γ arises from phase and amplitude fluctuations (recall Fig. 12.9). Result (12.2) was independently derived by Wolf, using the idea of correlation functions. Pancharatnam, on the other hand, leaned on earlier work on diffraction due to Zernike (where the idea of correlations was implicit but not explicit in the modern sense), proving incidentally the point that the pioneers had intuitively got the right approach to averaging fluctuations.

To illustrate his various findings, Pancharatnam also carried out a few simple but elegant experiments. Later he extended these studies to amethyst, which has, in addition to birefringence and pleochroism, the added feature of optical activity. Though not numbered sequentially, there were four papers in this series on amethyst, dealing with both experiment and theory. In them, Pancharatnam answers for the first time the question as to what happens when light is propagated along the singular axis of an absorbing biaxial crystal that has optical activity as well.

Many interesting aspects come to light when one examines as a whole the papers Pancharatnam wrote during his stay at the Raman Institute. He is not shy of experiments, but logical deductions suit his taste more. For some, theoretical physics necessarily means formal mathematics in all its rigour[4]. Pancharatnam is no less accurate, but gets to the answers via elegant means rather than painstaking ones. On one occasion, after establishing an important result by crisp arguments demanding not more than a couple of paragraphs, he remarks, "One may compare the simplicity of the above proof [essentially a geometric one] with that used in the usual treatment of Stokes parameters (Chandrasekhar, Rayleigh, ...)." At the same time, Pancharatnam did not shy away from the traditional route, and, whenever

The Final Years

needed, corroborated his results by those obtained by more popular methods. During the course of this work he also became aware that Clark Jones in America was developing an alternative method (now known as the Jones calculus), and occasionally he made references to it.

It is interesting and in a sense fortuitous that Pancharatnam's preoccupation with questions of coherence came about just around the time when there was a global focus on the topic, thanks to the innovative and far-reaching experiments of Hanbury Brown[5]. With the arrival of the maser and the laser shortly thereafter, coherence in optics assumed even greater significance. Indeed, even the long-forgotten crystal optics (earlier left mainly to the mercy of mineralogists) suddenly became important in the context of devices for nonlinear optics. Of course, when Pancharatnam started on his research career in the early fifties he could not have foreseen all this, but it is reassuring that the pursuit of apparently simple problems can also be quite rewarding, provided one asks deep questions and looks for the unusual.

A comparative study of Raman and Pancharatnam is quite fascinating. There are many similarities, like a predilection for optics, for example. Indeed, even in Pancharatnam's style of writing, one finds touches of Raman. But there are differences too – significant differences. While Raman loved to leap over mathematics and soar on conceptualization, Pancharatnam prefers mathematical logic. But it is not a turgid grind; algebra is an aid and not a mask to obscure the intrinsic beauty of the phenomenon. Nowhere is the contrast between the two more evident than in the only joint paper they ever wrote, the subject being mirages.

The phenomenon is well known and refers to the appearance (particularly in deserts) of the reflection of an object in what looks like a pool of water (which of course does not exist). The usual explanation is that when the surface of the earth is strongly heated by the Sun's radiation, the layers of air immediately close to the earth become warmer than those somewhat higher. As a result, the density of air increases instead of decreasing with height, at least over a certain distance in the vertical direction. Under these conditions, light from an object some distance above the surface of the earth may reach an observer not only along the direct path but also along a curved path (see Fig. 12.24a), creating the illusion of a reflection.

Raman and Pancharatnam are of the opinion that the conventional explanation is both "inadequate and indeed unsatisfactory", and decide on further explorations from both the theoretical and experimental angles. What is their criticism of conventional wisdom? Essentially that it relies on ray optics instead of on wave optics as it should. As they remark,

> It is usually stated that the ray continues to curve round, but it must be noted that on the basis of geometrical optics the ray should really continue parallel to the stratifications along the limiting layer [see Fig. 12.24b]. [12.13, p. 213]

What follows is quite striking. In a few crisp sentences, the explanation of the phenomenon is given in terms of wave theory. Geometrical optics is inappropriate, for it implies that the amplitude of the light wave varies quite slowly in space, and this is contrary to reality. As the authors argue:

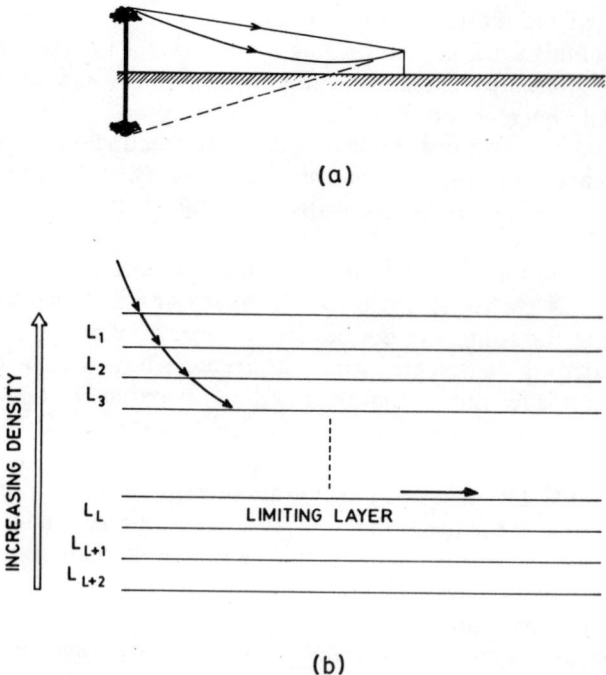

Figure 12.24 (a) Schematic illustration of the origin of the mirage. (b) The bending of a light ray due to decreasing density of layers of air. If ray optics is taken seriously, then the path of the ray through the various layers $L_1, L_2, \ldots$ will be as shown. In the limiting layer L_L, the ray will become parallel to the layer and curving up again is inconceivable. This paradox is eliminated in the wave theory, but it implies some previously unknown nuances.

This condition is clearly violated in the present case, for ... the disturbance would be infinitely great at the limiting layer ... and zero immediately below it. It follows that while the actual facts may bear some resemblance in their general features to this elementary picture, the real situation would nevertheless be rather different; firstly, a reflected wave must emerge from the limiting layer; secondly, there should exist a finite intensity below the limiting layer, while above that layer, the intensity though large would exhibit successive maxima and minima arising from the interference of the incident and reflected waves and progressively diminish as we move from the limiting layer toward the first medium. [12.13, p. 213]

No mathematics, no formulae – just words clothing an idea. There is no mistaking that this part of the paper is entirely due to Raman. He is completely convinced about his view of things. If the world at large wants decisive proof, what better than experiments?

The arrangement first tried for producing the mirage under laboratory conditions was similar to that described by R. W. Wood. A long steel plate, approximately 1.5 metres long, 10 cm in width and 2 cm in thickness was supported horizontally and heated from below by four gas burners, the upper surface being covered with soot.

> The arrangement did not prove very satisfactory since the rising of the hot air from the heated surface prevented a sufficiently sharp temperature gradient from being established, while at the same time such phenomena as could be observed were of an exceedingly labile nature owing to the turbulent convection of air near the hot surface. However, the conditions were greatly improved when an electric fan was used to blow away the hot air from above the surface. When this was done, it was possible to see clearly the reflected image of the "mountain peaks" formed by holding the serrated edge of a cardboard against an illumined ground glass screen at the far end of the plate. The action of the fan in sharpening the temperature gradient was picturesquely revealed by suitably lowering the eye, when a luminous thin cushion of air was seen to form over the heated surface. In order to study the phenomena critically, the object viewed was replaced by an illuminated slit kept parallel to the heated surface and the light diverging from it was rendered parallel by a collimating lens and allowed to fall obliquely on the hot plate. The beam was allowed to cover the whole length of the plate, the angle of incidence being adjusted merely by moving the slit. Further, the plate was heated electrically to ensure a fairly uniform temperature. [12.13, p. 214]

There was, however, still some difficulty in observing the interference fringes. This was removed by turning the plate edgewise so that the hot surface became vertical although the length remained horizontal. The hot air now flowed up in streamlines parallel to the surface of the plate, and a fan was no longer necessary. The first picture in Fig. 12.25 shows the pattern observed with the slit held vertical and illuminated by bright sunlight. The fringes are clearly seen, vindicating the wave theory approach.

Interference fringes are fine, but out in the desert one deals with real-life objects like trees and houses. What are the implications of the wave theory for such a situation? Raman and Pancharatnam provide an answer.

> In order to make the nature of the image evident, the serrated edge of a hacksaw blade has been used to form one of the edges of the slit. An aperture was kept in front of the lens and the succession of photographs exhibit the alteration in the phenomena as the aperture is gradually moved to the left [see Fig. 12.25]. [12.13, p. 215]

A remarkable feature of the sequence of photographs of the serrated edge is the occurrence of a third erect image close to the reflected image in the fourth and the fifth photographs of the sequence. The figure also shows photographs taken using a small model of a bird as the object. The appearance of a third erect image in addition to the usual reflected image may be discerned in the last two photographs of the sequence.

Pancharatnam obviously accepts Raman's ideas, but his way of convincing the world is by direct mathematical analysis. The problem itself is simple and straightforward, leading to the result shown in Fig. 12.26. Three things are evident from it: (i) immediately above the limiting layer there is a large concentration of intensity; (ii) slightly beyond, there is a series of interferences which progressively diminish in spacing and intensity; and (iii) below the main maximum, the intensity rapidly falls to zero.

In 1960 Pancharatnam received the degree of Doctor of Science from Nagpur University. That was also the year he went abroad for the first time, travelling to

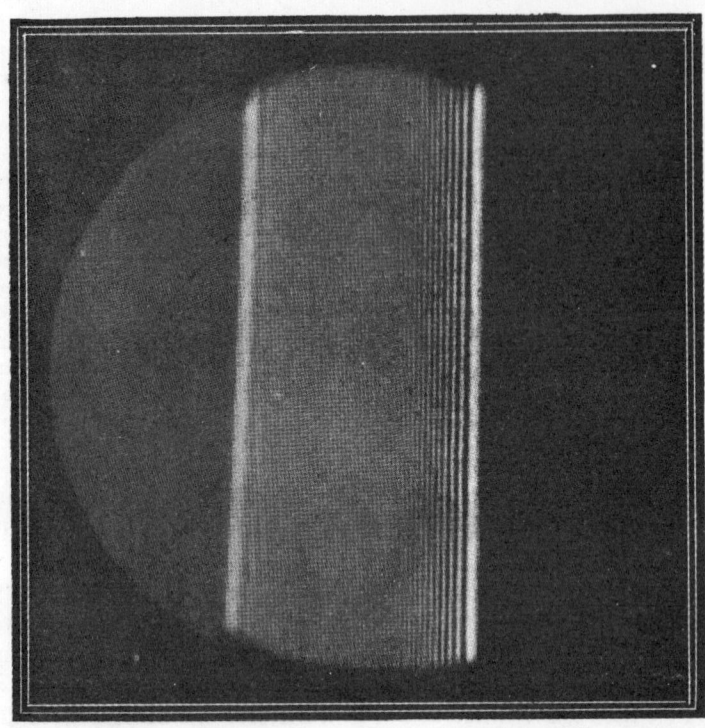
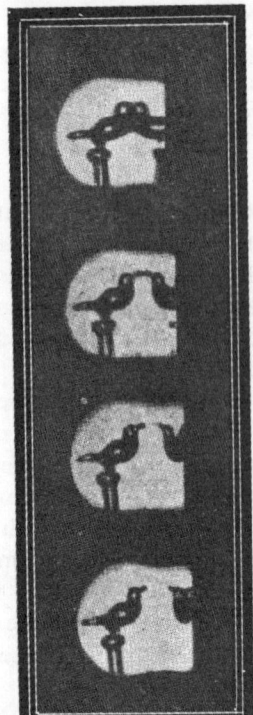
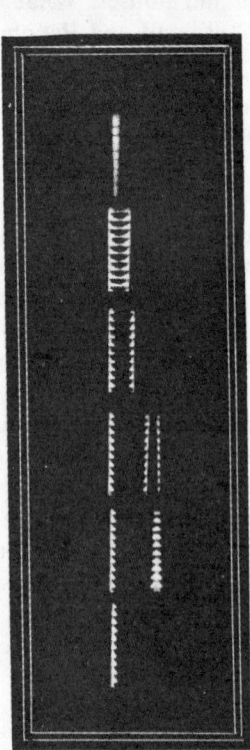

Figure 12.25 Photographs obtained by Raman and Pancharatnam to illustrate the subtle features implied by the wave theory of mirages. (After ref. [12.13].)

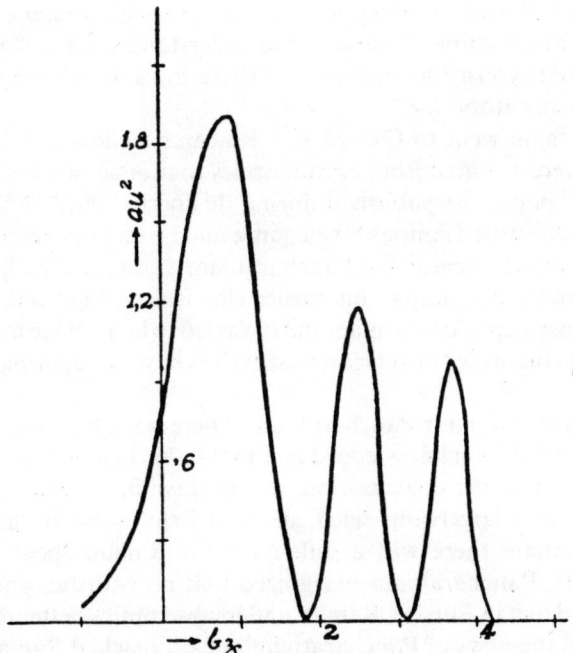

Figure 12.26 Variations in intensity near the limiting layer. A quantity proportional to the "reflected" intensity is plotted as a function of another proportional to the height z above the limiting layer. (After ref. [12.13].)

Rochester to attend the first ever international conference on "Coherence and quantum optics". On his return to India he joined the Mysore University as a reader[6].

The exposure abroad obviously broadened Pancharatnam's mind as is evident from two articles he contributed to *Current Science*. Continuing his studies on partial coherence, he now finds that the usual Stokes parameters are inadequate to deal with the state of partial polarization of a polychromatic beam, and introduces instead the spectral representation of the Stokes parameters. Once again there is evidence of the slightly different style of analysing problems.

Shy by nature, Pancharatnam was quite retiring and did not indulge in socializing. But he had deep concern and compassion for humanity, and took active part in the *Sarvodaya* movement. Writing about this aspect George Series observes: "... the feeling that he himself was overprivileged grew to the point where he felt compelled to go out and serve the underprivileged, the out-castes and the undernourished. He gave not only money, but time and personal service, and all this privately, as if he were afraid that public knowledge would vitiate its value." [12.14]

Towards late 1963 Pancharatnam began to think of leaving Mysore for personal reasons. Raman who at this stage, with all his students gone, was practically alone, invited Pancharatnam back to the Institute, especially as he saw in the latter the ideal successor. With utmost courtesy and respect, Pancharatnam declined the invitation adding at the same time, "I am sure you understand that this does not mean that I am not being true to you in the larger sense of the term, i.e., true to science, which is what you have basically stood for."

In 1964 Pancharatnam went to Oxford as a Research Fellow at St Catherine's College where his interest shifted from crystal optics to atomic physics and optical pumping. One of the papers he published during this period elicited appreciation from Victor Weisskopf, himself famous for elegance and beauty in scientific analysis. The papers of this period also reveal that Pancharatnam is completely adjusted to the quantum theory of radiation, despite his earlier classical background. This is the transition Raman himself should have made but it was left to his protégé to complete it. In May 1969 Pancharatnam died of a chest illness even as he was preparing to return to India.

Raman had a deep affection for Pancharatnam. There was also a unique sense of mutual respect. It is a widely acknowledged fact that in his meetings with students, Raman usually dominated the conversation. As Bhagavantam remarks, "Conversation with him used to be largely one-sided, and so it went on till the last day of his life." With Pancharatnam there was a difference, for Raman spent much time listening. For his part, Pancharatnam recognized that notwithstanding the many leading experts he had met in Europe, Raman understood optics better than anyone he had known. When the news of Pancharatnam's death reached Raman he broke down. It is said that Raman never wept for any man; clearly Pancharatnam was an exception.

☆ 12.4 Pancharatnam and Berry's Phase Angle

In a recent paper [12.15] which has since excited much interest, Berry has drawn attention to the existence of a previously ignored phase factor which appears when a quantum-mechanical system is subjected to adiabatic changes. Subsequently it has been noted that suggestions of such a phase angle occur in many previous papers, though it was Berry who first drew pointed attention to it, besides discussing its significance. Ramaseshan and Nityananda [12.16] observe that Berry's phase angle is implicit also in the work of Pancharatnam, a remark which is significant considering the fact that the first demonstrations of the effects predicted by Berry have been in optics [12.17, 12.18].

First let us see what is meant by Berry's phase, as it is often known today. Let $\hat{H}(R)$ be the Hamiltonian of a system, dependent on the parameter set $R = (X, Y, \ldots)$. One now varies the parameters $X, Y, \ldots$ slowly in time, effectively making R sweep a trajectory in parameter space. Suppose that the excursion of the system occurs adiabatically between the times $t = 0$ and $t = T$, where T is such that $R(T) = R(0)$. In other words, R makes a closed circuit, say C.

The Final Years

We now turn the focus on the eigenstates. At any instant t, the eigenstates satisfy ($R = R(t)$)

$$\hat{H}(R)|n(R)\rangle = E_n(R)|n(R)\rangle.$$

Adiabatically, a system prepared in the state $|n(R(0))\rangle$ will evolve with $\hat{H}$ and be in the state $|n(R(T))\rangle$ at time T. Quantum mechanics also tells us that $|n(R(T))\rangle$ is just a phase factor times $|n(R(0))\rangle$. The question is: what phase factor? Surprisingly, the traditional answer

$$|n(R(T))\rangle = \exp\left\{-\frac{i}{\hbar}\int_0^T E_n(t')dt'\right\}|n(R(0))\rangle$$

does not tell the full story. Berry showed that one must instead have

$$|n(R(T))\rangle = \exp\left\{-\frac{i}{\hbar}\int_0^T E_n(t')dt'\right\}\exp(i\gamma(C))|n(R(0))\rangle,$$

where $\gamma(C)$ is an extra phase angle now known as Berry's phase.

Berry's phase angle has deep and wide implications (some pointed out by Berry himself), which is why there is widespread interest in it. Can the existence of $\gamma(C)$ be demonstrated? Yes, and an elegant way would be by means of interference effects. Berry himself outlined such an experiment.

A polarized monochromatic neutron beam is split into two beams which are then passed through regions where magnetic fields exist. Along the path of one of the beams the field **B** is kept constant while along the other the magnitude $|\mathbf{B}|$ is kept constant but its direction is slowly varied so that the circuit C in parameter space subtends a solid angle Ω. The two beams are then combined and the count rate measured as a function of Ω, whereupon interference effects should be seen.

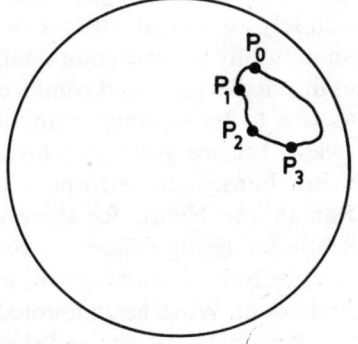

Figure 12.27 Polarization of light and Berry's phase angle. Consider polarized light, represented by the state P_0 on the Poincaré sphere, entering an optical medium. Let its passage result in polarization changes described by the closed trajectory $P_0 P_1 \ldots P_0$. On emergence, the light beam will have suffered a phase change equal to half the solid angle subtended by the loop at the origin.

In his paper Berry noted that although his discussion was based on the behaviour of a quantum-mechanical system, the phase which he discovered had "implications throughout wave physics". More recently, Hanney [12.19] has drawn attention to the classical analogue of the Berry problem. In their paper, Ramaseshan and Nityananda recall how some of Berry's ideas are latent in Pancharatnam's work, and then proceed to make the following remark:

> Consider any closed curve on the Poincaré sphere, starting and ending at a point P_0 [see Fig. 12.27]. The state of polarization P_0 can be analysed along a sequence of states P_1, P_2, P_3 along the curve, with the last step being the analysis of P_n along P_0. In the limit $n \to \infty$ with the separations $P_0P_1, P_1P_2, \ldots$ all tending to zero, the final state P_0 differs from the initial by a phase equal to half the solid angle subtended by the curve $P_0P_1 \ldots P_n$ at the centre of the sphere.

As one can see by recalling the experiment suggested by Berry, the correspondence is quite striking[7].

12.5 The Recluse

Raman in retirement should have had an ideal and peaceful existence. He was his own master, there was enough to keep busy, and there were beautiful problems to study. But it did not turn out that way at all. Somewhere along the line, something snapped, and suddenly he became a tormented man. On reflection, such a transformation was perhaps to be expected.

Raman's entry into retirement coincided with the dawn of India's freedom. A new feeling swept the country, and under Nehru's dynamic leadership India struggled hard to break loose from centuries of backwardness. Science was the magic wand that was supposed to transform society, and the intelligentsia joined in the chorus, "Progress through science". Having been closely associated with the promotion of science in pre-Independence India, Raman naturally became emotionally involved with these new developments; and given his distinctive views and painful experiences, he also found much to disagree with. As was to be expected, Raman expressed himself strongly, and not surprisingly his views became good copy for the press as they were regarded as a criticism of Nehru himself. In retrospect, one cannot interpret it simply as a clash between Raman and Nehru, for there was a deep commitment to science on both sides. Before analysing differences such as there appeared to be, one must first digress to describe Nehru's own concern with science.

Jawaharlal Nehru was a man of many dimensions. While he undoubtedly played a prominent role in the freedom struggle, he will probably be applauded more for the contribution he made in setting this country on the course towards modernity. Amongst all the leaders of that period, Nehru was perhaps the only one to understand the historical role of scientific and technological development in modern society. As Saha often sarcastically remarked, most of the politicians (of that period) were wedded to the "philosophy of the bullock-cart", and a "back-to-the-*Vedas*"

The Final Years

policy. But Nehru was a unique exception, and it was a matter of supreme luck for India that he was in the saddle during the first few critical years.

Nehru was not a scientist but he was drawn to science in a way few politicians have been. His first contact with science was during his student days, writing about which he observes, "I haunted the laboratories of that home of science, Cambridge." He adds, "Though circumstances made me part company with science, my thoughts turned to it with longing." To a scientist science can be an end in itself but to Nehru it was also a vital force which has changed our world. As he remarks in a message sent to the Silver Jubilee session of the Indian Science Congress held in 1938 in Calcutta (over which Lord Rutherford was to have presided but did not – he passed away before this event):

> In later years, I arrived again at science when I realized that science was not only a pleasant diversion and abstraction, but was of the very texture of life, without which our modern world would vanish away. Politics led me to economics, and this led inevitably to the scientific approach to all our problems and to science itself. It was science alone that could solve the problems of hunger and poverty, of insanitation and illiteracy, of superstition and deadening custom....[8] [12.20]

Familiarity with the European scene, and in particular a tour of the Soviet Union during the thirties, were responsible for this transformation. Nehru concludes his message with the words,

> And so I hope, with Lord Rutherford, "that in the days to come India will again become the home of science, not only as a form of intellectual activity, but also as a means of furthering the progress of her people".

Again and again he returned to this theme of science with a social purpose. Lest scientists become preoccupied with their discoveries, he cautions them:

> Surely, science is not merely an individual's search for truth. It is something infinitely more than that if it worked for the community. It must have a social objective before it. For a hungry man or a hungry woman, truth has little meaning. He wants food. For a hungry man, God has no meaning. He wants food. And India is a hungry, starving country.... So science must think in terms of the 400 million persons in India.

This was at the Science Congress of 1947, a few months before Partition. Some years later, speaking at the 38th Science Congress in Bangalore (where Homi Bhabha who was President of the session had just delivered an address on elementary particles), Nehru observed:

> You have just been listening to Dr Homi Bhabha's address, and, no doubt, all of you found it interesting as I did. All these are very fascinating subjects and yet I was trying to correlate what he said and the subject to the kinds of problems we have to face.... Inevitably, a person like me who is concerned with day-to-day problems of great importance has always to think a little less of pure research but more of the application of research to the problems of human society....

Interestingly, Saha too held somewhat similar views. As he comments in one place:

> Even "pure" science, it is generally admitted now, subserves directly or indirectly human and social needs and the expression "science for science's sake" like the sister

adage "art for art's sake" is fast passing out of the vocabulary of those who have looked into the genesis, history and future of both science and art. Not that scientific research cannot or should not be carried out with the interest centred chiefly in itself, but it is fallacious to think that, for this reason, it is objectively dislodged from the social framework in which the work is proceeding. [12.21]

It is clear from the various pronouncements that he made that science for Nehru was sometimes synonymous with technology. For a man in his position and with his responsibilities, such an interpretation is understandable. At the same time, it would be wrong to conclude that Nehru's approach to science was either narrow or mercenary because he also saw science as a triumph of the human intellect, besides admiring the discipline it imposed and the objectivity it fostered. Thus, scientific temper too became an obsession with him. As he declared in an address to the Ceylon Association for the Advancement of Science:

> The scientific method or attitude has to start in the minds of men. It is not merely a matter of some students going and pottering about in the laboratory and calling themselves scientists. They have to develop that attitude of mind which is a search for truth, a rather ruthless search.... It is extraordinary how we find scientists, sometimes very noted scientists, being very good in their own domain, and coming out of it in some other department of life, being hopelessly at sea and most unscientific.... [12.20]

There can be no doubt that Nehru understood perfectly the ethos of modern science. His concern with it was both complex and emotional, and it is against this backdrop that we must view his various actions as well as pronouncements.

Nehru did not confine himself to going up and down the country dispensing advice. He also acted, drawing heavy assistance from Bhatnagar (whom we have already met in Chapter 8) and Bhabha. Their styles of functioning were poles apart but Nehru accepted them with equal warmth and extended all possible facilities to both.

To Bhatnagar goes the credit of building up the CSIR (Council of Scientific and Industrial Research) organization. Contrary to popular belief the CSIR existed even before Independence, having been created in 1941 as a part of the Department of Supplies and Industries with the objective of aiding the war effort. In 1944 Sir A. V. Hill was invited to India to advise the Government on post-war planning, development and reconstruction. It was on his suggestion that a Department of Planning was set up, and the CSIR was shifted to that department. After Independence Nehru himself became the president of CSIR, and with his blessings Bhatnagar vastly expanded the CSIR network, dotting the country with a chain of National Laboratories[9].

There was no reference to atomic energy in the profile of development proposed by Hill – understandable since at that time atomic energy was a closely guarded secret; but when atomic energy became public, it was in the most dramatic fashion possible. While Hiroshima and Nagasaki aroused mortal fear, there was also the hope that, handled properly, atomic energy could be a great boon to mankind. And with his Cambridge background and international reputation, Bhabha was the ideal prophet. To Nehru, the exploitation of atomic energy was particularly attractive as

The Final Years

it would be the best display of his Government's determination to harness science for progress.

In 1948 the Atomic Energy Commission was established in the Ministry of Natural Resources with Bhabha as the Chairman and Bhatnagar and K. S. Krishnan as the other members. The Commision used the existing institutions, the foremost among them being the TIFR, to do preliminary scientific studies and train scientific personnel. Likewise, in a small chemistry laboratory set up in Peddar Road, Bombay, was produced the first piece of metallic uranium. It is from such modest beginnings that the Indian atomic energy programme was born. If Bhatnagar believed in the chain model, Bhabha relied on the percolation model.

With Bhabha and Bhatnagar in full swing, science and technology got a boost of a kind Mahendra Lal Sircar and J. N. Tata could scarcely have dreamt of. Spectacular results were achieved on the agricultural front and India leap-frogged from seemingly perpetual dependence on PL–480 food imports from the US to self-sufficiency. Within a few fleeting years, science became a vast nationally organized activity and there were people working on all sorts of subjects ranging from astrophysics to antibiotics, from computers to catalysts, from mesons to the monsoon, from number theory to nuclear physics, from pulsars to polymers, from quarks to earthquakes, from reactors to remote sensing, from semiconductors to sewage treatment, from turbulence to tuberculosis,... One cannot help being amazed at this phenomenal progress which would do any country proud, more so a developing country like India. Indeed, it is almost a unique example in the annals of history. Where then was there room for Raman to complain? This is a delicate question and requires careful examination.

Raman had no quarrel with Nehru's devout espousal of science. In fact, on one occasion he himself declared in a characteristic fashion, "There is only one solution to India's economic problems and that is science, more science and still more science." What seems to have upset him deeply was the strategy used. One is still too close to the events to say categorically whether Raman was right or wrong. Indeed, notwithstanding his many critical observations, one is inclined to feel that much good did flow from the momentum generated by Bhatnagar and Bhabha. At the same time, it is inappropriate to brush aside Raman's criticisms, especially as they were born out of deep anguish. It is worthwhile therefore to pause and to ponder over some of his remarks, looking a little deeper than he himself cared to reveal.

Of the several critical remarks that Raman made, we select a representative few for comment. Rapid expansion of scientific activity necessarily called for a certain amount of import of scientific equipment. Compared to the levels one sees today, imports then were on a very much smaller scale, especially on account of the prevailing acute foreign exchange shortage. But even that low volume jarred Raman's sensitivity.

Amongst all the sciences, experimental physics in particular has always thrived on innovative instrumentation by scientists. In Calcutta, for example, Raman's students devised many clever methods for studying magnetic properties, managing to remain competitive with the West. Recognizing the constraints of the Indian situation, Raman always set much store by ingenuity. The story is told that once he saw one of his students in a crest-fallen mood. Upon enquiry he learnt that (spectroscopic) experiments similar to those being performed by his student were also in progress in England at the same time and the student's worry was that whereas he had merely a 1 kW lamp his competitor abroad had a 10 kW lamp. "Don't worry," Raman told the student, "put a 10 kW brain on the problem." No doubt an exercise in levity to cheer up a depressed student, but clearly there was also a message.

The new mood seemed quite contrary to Raman's philosophy and naturally he was rather upset by the flurry of imports that occurred. To him it was brains that produced good science rather than costly instruments, and he never tired of citing the example of the discovery of the Raman effect which was made with very simple equipment. Further, by importing instruments instead of building them, "we are paying for our ignorance", he declared. Predictably, the response was that things had changed very much since Raman's days and that he was oversimplifying matters somewhat. Quite true, and one cannot really expect an analytical chemist, for example, to function without the tools needed for analysis. Elsewhere everyone used commercial, ready-made equipment and so why should not we? On the other hand, was Raman's point totally devoid of reason?

There can be no question that after World War II, scientific instruments did start becoming increasingly sophisticated. So specialized did instrument design become that in the West, while some concentrated on research others began to devote full time to instrument development. However, the two species usually co-operated, and often industry too joined in the developmental effort. Thanks to this symbiosis, there emerged a viable instrumentation industry which, incidentally, also afforded a career to many.

The story in India was altogether different. Since instrument development fetched little reward or recognition it was largely ignored, and scientists (physicists included) generally waited and pined for imported, ready-made equipment. For the lucky few who could get it the era of push-button research had arrived. But import did not exactly make our research competitive. Foreign exchange being scarce, not enough instruments could be imported, besides which there were problems of break-downs, spares procurement, after-sales service, and so on.

It would be wrong to assume that everyone uniformly fell a prey to the import disease and that indigenous attempts in instrumentation were totally lacking. Far from it, and indeed there were stunning exceptions. Bhabha in particular deserves much credit for his bold and unfailing support to local effort. However, the fact remains that India failed to copy what it really should have, namely, the balanced development of research, the tools for research, and an industry to furnish these tools. Lest the impression be created that instruments are something needed only by the scientists, it must be pointed out that modern technology is inconceivable without instrumentation, whether it concerns steel production, oil exploration or even dairying. Thanks to the absence of a viable instrument industry in the country, our import bill is now soaring to disturbing proportions. Are we not paying for our ignorance?

Another of Raman's critical observations relates to the Science Congress. For many years Raman regularly attended the annual sessions of that body, and in fact during the Allahabad Science Congress of 1949 (over which K. S. Krishnan presided), Nehru offered special felicitations to Raman who had just crossed sixty. Later the Science Congress soured for him, even as many other things did. He certainly was not the first to be critical of that body, but the manner in which he did made news. Asked why he had of late stopped attending the annual meetings, Raman shot back saying he wanted to have nothing to do with a Science Congress which was repeatedly inaugurated by a politician.

The reference here was to the annual inauguration by Nehru – later it was Indira Gandhi. One must record that such a criticism was not fair. Nehru's association with science and the Indian Science Congress was not only long-standing but also quite personal. Year after year he came rather like on a pilgrimage, invariably beginning his speech with a reference (sometimes humorous) to this annual ritual[10]. His intentions were however always quite serious – to listen, to learn, to reflect, and to share his thoughts. As he once declared:

> It has become the custom of the Science Congress to invite me year after year to the annual session and for me to come and utter, if I may say so, some platitudes. I come here realizing I don't throw any particular light on situations that you might have to consider. Nevertheless, I come here, partly because it satisfies me and I am interested in the development of science in India. I also wish to convey to you the sympathy of the Government, its message of encouragement and its faith in the future of science in India.

He would always begin on a low key like this but soon drift to matters of importance, like ethics, the use of science for constructive rather than destructive purposes, and so on. Nehru always spoke extempore at these meetings, and this allowed him to speak straight from the heart.

Given this background it is difficult to understand Raman's outburst. But in a *larger* and in a *generic* sense, Raman's criticism had (and continues to have, even more so today) a valid point, especially considering our penchant for glamourizing serious functions with the presence of VVIPs, a practice which seems to know no bounds. Nowhere else in the world is this done but we are yet to break away from our feudal past. Perhaps Raman's was the last powerful voice to be raised against such vulgarization.

The "ivory tower" issue was another that attracted much attention. Writing about it Bhagavantam observes:

> It is on record that when the late Jawaharlal Nehru, then Prime Minister of India, admonished India's scientists and asked them to come out of the ivory towers in which they had confined themselves, Raman reacted in a typically sharp manner and said: "The men who matter are those who sit in ivory towers. They are the salt of the earth and it is to them that humanity owes its existence and progress." [12.1]

One presumes that Raman's comment was made as a response to a query, probably from the press. Once again it is an impulsive remark, not valid perhaps in the limited context in which it was made, but very true in a broader perspective. To appreciate this, one must first go back to Nehru's specific references to the ivory tower.

Nehru brought up the question of ivory tower several times, but, surprisingly, not in the sense in which one would assume it to have been. One of the first references to it may be found in his address to the Lucknow Science Congress of 1953. Nehru is very much worried about the cold war (then at its height), and is disturbed that though the human mind was able to create extraordinary things like nuclear weapons, the mind itself was lagging very much behind its own creations. He adds:

> Parts of the world resent even the existence of the other parts. Parts of the world want to destroy other parts of the world. Surely this is not the kind of prelude to one world.

International co-operation was the only sensible answer, but unfortunately the record of the politicians in promoting it was rather poor.

> And so how are we to approach this problem? If the scientists go on functioning in an ivory tower way, they will no doubt do some good. If they come out of the ivory tower and help in solving the problems of the age, they will do a great deal of good....

The message is clear. Speaking in a somewhat similar vein at the Agra Science Congress he says:

> Something very essential has grown out of science. What we do in our daily life is closely connected with science and its application. Science today dominates us and is likely to dominate us in the future.... The time has come when it is difficult for anyone to live in an ivory tower and isolate himself from his surroundings. We have to get out of narrow shells and compartments and take an integrated view of life. Scientists who have an important part to play in moulding the destinies of the people must not lose sight of this fact.

And so, getting out of the ivory tower meant, essentially, becoming conscious of society and its problems, and in particular of the danger of misuse of science. However, on one occasion Nehru did make a reference to ivory tower in the sense in which one usually tends to understand it[11]. Speaking at the Baroda Science Congress of 1955 he said:

> Some eminent foreign people have told me that while Indian scientists are doing excellent work and the National Laboratories have a great and fine staff, there is an element of ivory-tower attitude among the scientists. I do not think this criticism is correct. Still it is a fact that scientific research work and its practical application have not been properly co-ordinated with big plans of development.

Viewing them in the proper perspective, one cannot take exception to any of Nehru's observations. Indeed, his point is well taken. Raman's reaction cannot therefore be regarded as a retort against Nehru, for the latter, despite his keen anxiety for material progress, never really demanded his pound of flesh from the scientists. And in the sense of being concerned about society, Raman was always well outside the so-called ivory tower. Why then was Raman so incensed? It was partly because the ivory-tower syndrome was invoked by all and sundry to admonish scientists about their supposed duties to the masses. (Indeed, this flood of advice has not abated.) In a nutshell, it was once again cudgels against basic research even

One of the abalone shells studied by Raman. (See Chapter 8.)

Some stones from Raman's collection. *Clockwise from top left*, cameo opal, labradorite and amethystine quartz.

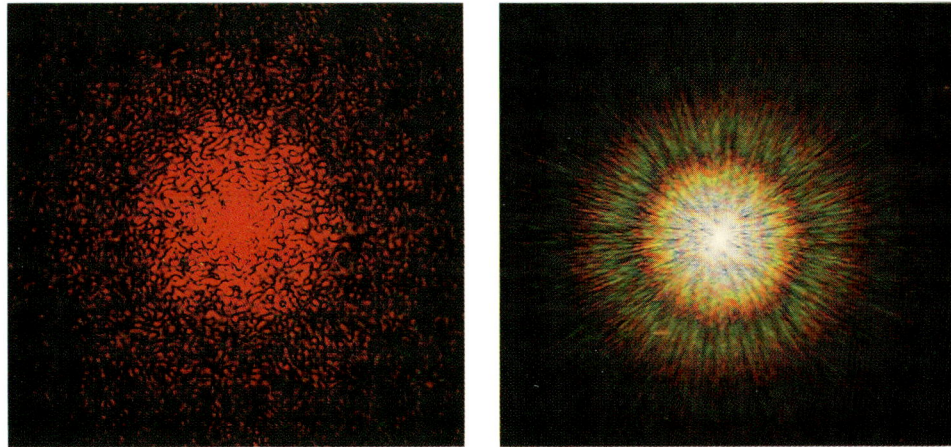

Speckles in monochromatic (*left*) and white (*right*) light. (Compare with Fig. 8.21.)

Quetelet fringes in colour.

A tree-lined avenue in the Raman Institute.

The tree marking the spot where Raman was cremated.

though there was still so little of it. It is this that Raman was implicitly protesting against, and quite rightly too.

It should not be assumed that Raman was merely making a series of arbitrary and disconnected pronouncements. On the contrary, they were symptoms of a deep concern he had begun to feel about the way science was being promoted. It seemed to him that in the rush for development, scientific excellence and the objectives of science had begun to take a back seat. Sycophancy was on the rise, and ill-equipped people were being propelled into seats they were not ready to occupy. Everyone paid lip sympathy to the universities, but when it came to funding them, they were generally forgotten. What was worse, mediocrity was slowly allowed to become institutionalized. In retrospect, Raman's utterances, though harsh, implicitly carried a warning that was unfortunately not heeded. And despite all the pious hopes of that period, the linkages between science and technology in India continue to be quite tenuous. If they had grown and become strong would we be witnessing the screwdriver technology phenomenon that has become so rampant today?

It is interesting that while Raman's caustic comments attracted attention, he was not the only one to make critical remarks. Bhabha, for example, had his own criticisms. Discussing the establishment of various National Laboratories, he observes:

> All these laboratories were brought into existence in the same way[12]. A Planning Officer was appointed for planning the work and building of each laboratory. The plan was usually drawn up on the basis of the work of similar laboratories abroad, divided into divisions and sections, and an estimate of the staff required made on this basis. An attempt to fill the posts was then made on the basis of advertisement....
>
> The standard method of planning laboratories and filling posts is often forced on many by the administrative and financial requirements of Government. As Professor Blackett has said, "we must endow ability wherever it is found, and we must guard against subsidising mediocrity". The standard method is certainly not conducive to achieving this aim. Government is spending large sums now on supporting scientific research and technological development, and it is in Government's interest to study and devise *de novo* the best administrative and financial procedures for scientific institutions and for getting the maximum return on the money spent. To apply existing administrative and financial procedures, devised for an entirely different purpose, to scientific institutions is largely to defeat the purpose which the Government has in view by letting the tail wag the dog[13].
> [12.23]

Bhabha was also critical of the CSIR for surrendering its autonomy and for its omnibus adoption of Government procedures.

Surprisingly, Nehru too was critical of the Governmental set-up and machinery. He was particularly disturbed that bodies created to function in a non-Governmental fashion soon succumbed to the Governmental style, either willingly or unwillingly. Speaking about the Planning Commission he says:

> When we started it, I definitely thought it should not function as a part of the Government. But now it is just like any other part of Government.... the same hierarchy of Secretaries, Under-Secretaries and what not – it is frightening. What

was thought of as a close body of people who think and advise Government has now grown into a huge organization with all the Departments of Government almost duplicated there, and each one sending papers to the other which is the normal habit of Government[14].

Nehru is also worried that scientific establishments are going the same way.

> I had hoped that science at least would escape that numbing influence of the Governmental way of working. I do not know how far we have succeeded. I rather doubt it. I am inclined to think some of our laboratories are gradually succumbing to the Governmental way.

And he is concerned also about the bullying of the scientist by the administrator.

> There might be an Einstein amongst them [i.e., the scientists] but he had to be kept in his place. The administrator was the boss.

So Raman was not alone in his criticism of the establishment. Others criticized but stayed within the system as they believed they had responsibilities of their own to discharge. Raman, on the other hand, would make no compromises and preferred to stay outside it[15]. With his reputation and prestige he could have enjoyed power if he wanted to but he rejected it. He was even sounded out for the Vice-Presidentship. "What will I do with that ship?" he is reported to have asked. He was pained whenever one of his former students or colleagues took up a Government assignment involving administrative duties, for to him that implied a desertion of science. While others couched their comments in arguments that sounded reasonable, Raman was terse and did not care if people thought he sounded unreasonable. Most people misunderstood him and few supported him. Later Bhagavantam wrote,

> Men like him are not thrown up everyday and, if the rugged contours and the sharp corners of this giant did not compromise with the soft-spoken ways of the successful world, we can only describe the phenomenon by stating that "it is no reproach on Everest that one cannot play golf on it". [12.1]

History will probably see him as a conscience-keeper for science, which in a sense he was, even as Rajaji played a similar role *vis-a-vis* politics around the same time. Not surprisingly, both of them were not exactly popular with the Establishment[16].

In Raman's case there was also a sly counter-attack by certain unscrupulous elements. Once again questions were raised about his scientific abilities, and about his great discovery. No wonder he became like a wounded tiger. Already he was rather lonely, all his students having gone away. Deeply upset by the lies being spread against him, he withdrew further to become a recluse, adding even more to his loneliness. He put up a board asking visitors to keep off and not disturb him. He was pained, bitter and angry, his anguish heightened by memories of the past[17]. In one of his spells of depression, he resigned from the Royal Society[18]. It is said that when the letter of resignation was placed before the Council of the Society, there was a hush around the table. Resignation from the Society was most uncommon, the previous known instance being that of Newton!

Ramaseshan provides a poignant description of the Raman of this period.

> To those who knew him, he was a soul in agony. He was veritably like Timon of Athens – bitter and cynical....
>
> To him... science was a personal endeavour, an aesthetic pursuit and above all a joyous experience....
>
> He saw the expenditure of large sums of money – in the belief that science, and therefore technology, will automatically be created. He saw the replacement of quality by quantity. He saw the choice of research topics – dictated mostly by foreign fashions and this hurt him most....
>
> "My life" he once cried, "has been an utter failure. I thought I would try to build true science in this country. But all we have is a legion of camp followers of the West."
>
> To my mind the agony of Raman can only be compared to that of Gandhi in Noakhali when he too found that all his life's work was as nought, the apostle of non-violence witnessing his countrymen beating each other to death. [12.24]

Raman's mood also reminds one of a touching story entitled "The Selfish Giant" by Oscar Wilde in which the Giant, as indicated at the beginning of this chapter, shuts himself off from the world. Fortunately, like the character in that story, Raman eventually got over his despondency through the company of the young.

Raman said all the right things but perhaps the "wrong" way. That his comments were neither misdirected nor irrelevant is quite evident from the several deficiencies we witness today in our educational and science and technology scenes. Of course, one still does not know whether it is better to grow slowly but carefully, or in an explosive manner allowing weeds also to come up, leaving their eradication to later times. Can weeds, once they are allowed to spread, ever be successfully eliminated? Only time can tell!

12.6 The End

In the story by Oscar Wilde just alluded to, the Giant, lying awake in bed one morning, hears some lovely music. It was just a little linnet but so long had it been since he had heard a bird sing that it seemed to him to be the most beautiful music in the world. Looking out through the window he saw a wonderful sight. Children had crept into his garden and were everywhere, playing. And in celebration, the trees had covered themselves with blossoms and were waving gaily in the breeze.

Raman came out of his despondent mood in an essentially similar manner, by discovering the happy world of children. In them he saw the future of India, the India that he dreamt of. "This is our real strength" he declared. "If they [i.e., the young] are enthused and if they are instilled with a spirit of adventure, the sleeping giant will wake up and we can conquer the world." Ramaseshan describes the new mood:

> Almost every day he was seen taking school children and college students round his Institute, regaling them with stories and showing them experiments. He began accepting lecture engagements. His charisma was still potent with the younger generation. He attracted enormous audiences, and university halls began overflowing again. [12.24]

There is a touching photograph in the Institute showing the septuagenarian Raman romping in the gardens along with a group of young students. Raman also instituted, in the memory of Mahatma Gandhi, a Gandhi Memorial Lecture which he himself delivered every October 2.

In the midst of all such activities, his preoccupation with science continued unabated. But it was now science with a difference. Fascinated by the beautiful flowers in his garden, he began to ask himself how flowers got their wonderful colours. He examined them *in vivo* and *in vitro* and wrote papers describing his studies. In fact his paper entitled "The red oleander and the purple *Petrea*" was published barely a few months before his death. I have heard some criticism of these papers but it would appear that the critics have missed the point. This was not modern physics but recreational science, and one presumes there is still a place in the present busy world for such pursuits[19]. In fact Raman did not publish all that he studied. For example, he wondered about the way shells are built up layer by layer, day and night through the seasons; perhaps this autobiography of the unknown living creature in the depths of the sea could tell something about the temperature variations in its environment through the abundance ratio of the oxygen isotopes; and so on. A good part of Raman's explorations were of this nature, and were communicated as comments, speculations and questions to willing listeners. Like the heroes of the detective books he enjoyed so much, Raman loved puzzles, and many problems of natural phenomena are indeed puzzles challenging the keenest intellect. Physics of course is unique in that many of the puzzles can now be reduced to a mathematical level[20].

Reference must be made here to Raman's work on vision which kept him busy during his last days, culminating finally in a book entitled *The Physiology of Vision*. As already mentioned, Helmholtz wrote a book bearing the same title almost a hundred years earlier. Raman is interested in the nature of vision and visual processes, and is dissatisfied not only with various theories of the past but also with the newly emerging view based on photochemical processes. His investigations appear to be extensive but are clearly without any linkage to current activities in the field. In fact the book does not carry any references as scientific volumes do!

Raman's views did not find acceptance, which is not surprising. While Raman's theory as such may not be of much interest, it is, however, worth recalling a simple but ingenious method he devised for exhibiting the functioning of the retina.

> The technique employed is the use of a colour filter which freely transmits light over the entire range of the visible spectrum except over a limited and well-defined region which it completely absorbs. It is possible by the use of suitable dye-stuffs in appropriate concentrations to prepare colour filters of gelatine films on glass exhibiting the spectroscopic behaviour described. Holding such a colour filter before his eye, the observer views a brilliantly illuminated screen for a brief interval of time and then suddenly removes the filter while continuing to view the screen with his attention fixed at a particular point on it. He then observes on the screen a picture in colours which is the chromatic response of the retina to the light of the colour previously absorbed by the filter and which impinges on it when the filter is removed. Actually,... what the observer sees is a highly enlarged view of his own retina projected on the screen and displaying the response of the retina in its different areas produced by the incidence of the light of the selected wavelengths....

> Why the phenomenon described above manifests itself is not difficult to understand. A colour filter completely absorbing a selected part of the spectrum when placed before the eye of the observer protects the retina from the incidence of light from that part of the spectrum, and if such protection continues for a sufficient period of time, it has the result of sensitizing the retina for the reception of light of those wavelengths when the filter is removed. *Per contra*, light of wavelengths not absorbed by the filter being incident on the retina both when the filter is in position and after its removal, the visual sensation which it excites becomes enfeebled by the continued exposure. Accordingly, when the filter is removed, the visual response of the retina to light of the wavelengths for which its sensitivity has been enhanced is far stronger than the continuing response to the other wavelengths and manifests itself vividly to perception. [12.25]

In short, one sees an image of portions of the retina sensitive to the wavelengths absorbed by the filter. By using filters that absorbed different wavelengths of light, one could explore the retina in detail. True, the image seen is transient, but one can repeat the exercise any number of times by holding the filter before the eye for a while and then removing it.

In 1968 Raman attained the age of eighty, another milestone. The annual Academy meeting in Ahmedabad was an occasion not only for celebration but also for nostalgia. As expected, Raman's erstwhile colleagues and students gathered in strength to honour him. Replying to the felicitations, Raman, in a pensive and humble mood, pointed out that his own achievements were small compared to those of the giants. He also commented on the singular absence of reference to the one quality he proudly cherished, remarking, "I wish someone had said that I had the heart of a lion."

Vikram Sarabhai had just completed setting up the Community Science Centre in Ahmedabad where the young could amuse themselves and experience at first hand the thrills of exploring science. What better than asking Raman to dedicate it? Raman thoroughly enjoyed himself talking to the young audience. His theme was "Why is the sky blue?", and he gave a glimpse of how unbridled curiosity can lead one to fundamental problems in science.

> As I look up and see, the sky is blue... Just look up, look at the sky... You learn science by keeping your eyes and ears open... In this world, we see all kinds of miracles happening in Nature. To me, everything I see is incredible, something absolutely incredible. We take it all for granted. But I think the essence of the scientific spirit is to look beyond and to realize what a wonderful world it is we live in.... The moment you ask "Why is the sky blue?" you go deeper and deeper into the problems of physics....

And slowly he led his youthful audience through Rayleigh scattering, the nature of atoms, of molecules, of light itself.

The Films Division of the Government brought out a documentary entitled "C. V. Raman – The Man of Science", and for the first time the public at large got a glimpse of Raman in his Institute setting. He was seen in his laboratory, in his office, and strolling in the garden. Sitting on a bench and gazing at the trees he had planted long ago and which had now grown tall and majestic, he reminisced about his career, adding that it was always better to work under slightly hostile conditions as it brought out one's best!

The year, 1970; Raman's health was showing signs of strain, and so the annual meeting of the Academy usually held in November was now scheduled for September. The meeting was in Bangalore and lasted a whole week. Young scientists presented reviews of progress in different subjects – nuclear sciences, radio astronomy, meteorology, seismology, crystallography, genetics, agricultural science and neurophysiology. In spite of failing health Raman attended all the lectures, taking keen interest and asking many questions.

October 2 saw Raman deliver as usual the Gandhi Memorial Lecture. In a reference to this particular lecture[21], M. G. K. Menon says,

> ... when I saw him a few weeks ago, when he was getting on to almost 82 years in age, he rushed me to the first floor of the Raman Research Institute in Bangalore to show me an experiment consisting of tubes of the same diameter but of different lengths, each closed at one end, from the other ends of which he blew with the pleasure of a schoolboy; this was to illustrate his views on the theory of hearing at the Gandhi Memorial Lecture.... [12.22]

As exuberant and as enthusiastic as ever.

Shortly before his birthday, Raman became ill and was removed to a private nursing home. He told his doctors, "I do not want to survive my illness if it means anything less than a hundred per cent active and productive life." Since he seemed to be slightly improving, he was, in accordance with his wishes, moved back to his residence in the Institute campus. Though once again in his favourite surroundings, he could no longer go out and enjoy his garden. In typical fashion he conveyed his disappointment, saying, "If I had known I was going to die here, I would have arranged for the windows to be lower." Puzzled at first, those around him soon realized that Raman was complaining that he could not even look out and catch a glimpse of the trees and the flowers! The bed was then raised so that the garden came into view.

Raman knew that the end was approaching, and there were still many loose ends to be tied up. He had been managing both the Institute and the Academy single-handed but the time had come to bring about a formal separation between the two bodies, define their functions, and legalize their respective holdings. He conducted the necessary meetings from his bed and those about to inherit the various responsibilities realized what a load he had been carrying so long all by himself. He also dictated what might be called his Last Testament, mainly referring to the Raman Research Institute and its future. In it he said:

> The Raman Research Institute was created by me in 1948 to provide a place in which I could continue my studies in an atmosphere more conducive to pure research than that found in most scientific institutions. To me the pursuit of science has been an aesthetic and joyous experience. The Institute has been to me a haven where I could carry on my highly personal research work.
>
> This personal character of the Raman Research Institute should obviously change after me. It must blossom into a great centre of learning embracing many branches of science. Scientists from different parts of India and from all over the world must be attracted to it.

It was not known who the successor would be but whosoever it was, he would have a "perfect nucleus for the growth of a centre of higher learning".

There was the question of funds. Raman had functioned in a particular style but it might not be either convenient or possible for his successor to do likewise. And so Raman said:

> I have always felt that science can only flower out when there is an internal urge. It cannot thrive under external pressures. I strongly believe that fundamental science cannot be driven by instructional, industrial, governmental or military pressures. This was the reason why I decided, as far as possible, not to accept money from Government.
>
> I am a very practical man and I am practical enough to see that it would not be possible for others to run or grow a good institution without funds. I have bequeathed all my property to the Institute. Unfortunately, this may not be sufficient for the growth of this centre of learning. I, therefore, will not put it as a condition that no Governmental funds should be accepted by the Institute. I would, however, strongly urge taking only funds that have no strings attached.

Money alone would not do, and there were other things to worry about besides, like maintaining quality. Emphasizing this Raman declared,

> Nothing is so detrimental to the growth of science in an institution than the existence of dead wood floating aimlessly which cannot participate in the scientific growth of an institute.

Clearly, he had not forgotten his earlier experiences. Raman's thoughts were of course on his own Institute but one realizes that his advice has universal applicability.

The end came in the early hours of Saturday, November 21, 1970. By a special arrangement, his mortal remains were cremated in the grounds of the Institute, amidst surroundings he loved and enjoyed. According to his wishes, there were no religious ceremonies. Thousands of people – school children, students, colleagues, friends and others – thronged to pay homage. Raman had passed into history. No plaque or monument marks the spot – just a solitary tree, precisely as he himself would have wished. Recently, the tree bloomed for the first time.

And so as we bid goodbye, let us briefly reflect on Raman's personality. In a certain sense he was rather a simple person, easily delineated and predictable in his responses. Those who knew him well always spoke of his child-like simplicity[22]. Indeed, this is reflected in the joy he experienced whenever he beheld something beautiful and also in the anger he felt when hurt. Even his famous ego was direct and unsophisticated – almost naive, one might say – and that, as Born wrote to Rutherford, was part of Raman's problem[23]. His exuberance while speaking, his sullen withdrawal, and the happiness he finally found in the company of the young, all testify to his transparent simplicity. And yet in some respects he is also an enigma. Consider, for example, his adherence in later life to his pet theories. Was it just child-like stubbornness or was it something more complicated, a result of the tribulations he had to pass through?

Raman knew no fear and, totally unmindful of the consequences, never hesitated to call a spade a spade. Youthful daring, one might say, but he displayed this trait right till the very end of his long life. He preferred to fight openly and stand by the things he believed in, instead of scheming or saying things behind people's backs. In the clouded atmosphere in which he lived, this quality of Raman stands out.

There was also a curious mixture of characteristics. Born spoke of Raman's European intensity, and in this as well as in many other things related to science Raman was quite un-Indian. But in his response to adulation it seemed the other way about. As the only Nobel Prize winner in science in a society that lacked science heroes, he received effusive praise all his life, and one is not sure that he escaped its numbing influence. Perhaps the erosion of scientific objectivity and the staunch adherence to his pet theories in later life are manifestations of this numbing effect.

Raman was very critical of things he did not like but he could also rise above differences. Nehru, for example, was a welcome guest at the Institute[24], and Raman had good personal links with Bhabha although he was critical of the latter about many things. Whenever Bhabha took a vacation in Bangalore (which he sometimes did to coincide with the TIFR Summer Schools[25]), he made it a point to call on "Sir C. V.". Bhabha also corresponded with Raman and once even suggested that he take up the measurement of the spins and the magnetic moments of nuclei, adding, "You would be the best person in India to take up some of this work and train young men." When Saha passed away there was an obituary in *Current Science*, and when Bhatnagar retired Raman wrote to Nehru suggesting that Bhatnagar in retirement be remembered and extended facilities such as he might require.

Scientists reared in a liberal atmosphere generally tend to be international in outlook but Raman grew up in a colony. Like that of his contemporaries, his outlook was therefore quite nationalistic. He did not involve himself in politics, for he believed that politics and science were poles apart, and there were other ways of serving the cause of the nation. He disapproved of the Raj but, interestingly, seemed to relish the knighthood conferred on him; his name always appeared in his scientific papers as Sir C. V. Raman.

Raman had very little interest either in religion or in religious practices, something very uncharacteristic of people of his generation, especially those belonging to his community. As Bhagavantam says:

> There were very few occasions when to him, Divinity and God-head meant anything very different from all that is manifest to man in the wonderful world around him. He would not generally let himself be drawn into conversation about God and if anyone tried to do so, his reluctant reaction was that while there is so much to learn about man, in fact much more than what one can chew, why worry about God. [12.1]

There is but one known reference to God in Raman's writings. It occurs in a statement he made about the institute he established.

> It is my earnest desire to bring into existence a centre of scientific research worthy of our ancient country where the keenest intellects of our land can probe into the mysteries of the Universe and by so doing help us to appreciate the transcendent

Power that guides its activities. This aim can only be achieved if by His Divine Grace, all lovers of our country see their way to help the cause. [12.22]

Ethics came naturally to Raman, as it does to all great scientists, and he always set much store by integrity[26]. It is curious that Raman had a small tuft, traditionally a symbol of religious orthodoxy. However, in Raman's case it would seem that the tuft was merely a carry-over from childhood, springing from habit rather than a religious belief[27].

Science dominated his life. As he once wrote to a correspondent in the US:

My first scientific paper was published ... when I was just 18 years of age. I am now over 75 years old and do not recollect any time during this long period when I took my mind off from my scientific interests. [12.1]

It is interesting that a substantial number of his publications were single-author papers. There is but one paper I am aware of with three authors, Raman one of them. In all papers with a co-author, Raman was always the first author. However, from the style of such papers it is abundantly clear that Raman was not only personally involved in the work, but also that he actually wrote the manuscript himself. But whenever a student did the work entirely on his or her own, it was published in the student's name – so it was right from S. K. Banerji and Chinmayanandam, down to Pancharatnam towards the end. And never did his students complain that Raman hogged credit. On the contrary, Bhagavantam remarks,

He was always more than generous in acknowledging the share of his co-workers and encouraged every deserving student that came to him.... [12.1]

For Raman, science was a window to Nature's majesty but there were moments when she had to be admired directly. Bhagavantam tells the story of how once in Darjeeling Raman decided to watch sunrise over the Kanchenjunga from the Tiger Hill and took Bhagavantam along with him. As they both set forth towards the vantage point, it seemed that they were going to be a bit late. Unwilling to miss the grand sight Raman began to run, literally an uphill task. But they made it in time and were treated to a glorious spectacle. Raman then turned to Bhagavantam and asked, "Wasn't it worth running?"

Ramaseshan recalls a similar incident that occurred several decades later.

A few months before he died, I remember, while walking with him one evening amongst the eucalyptus groves that he loved, he stopped me in his characteristic manner and pointing towards the sky, said, "Have you seen anything so beautiful?"

Above, one saw little wisps of multi-coloured clouds passing close to the Moon, which gleamed over the shimmering leaves of the trees – trees which appeared like Nature's own cathedral.

"This is happiness" he said. "That we should be alive, and that we should be endowed by Nature the faculty to perceive this fleeting vision of beauty – this is happiness indeed." [12.24]

13 Sharing the Pleasures

> *Who was enroll'd 'mongst wonders, and when we,*
> *Almost with ravish'd list'ning, could not find*
> *His hour of speech a minute...*
>
> **Henry VIII** (Act I, sc. II)

At the Golden Jubilee meeting of the Indian Academy of Sciences[1], the noted agricultural scientist M. S. Swaminathan narrated the following anecdote. When Raman won the Nobel Prize and was very much in the news, an aged (and illiterate) aunt of his asked him what the much-discussed Raman effect was all about. When Raman explained it to her in simple terms she exclaimed, "Is it for this they are making so much fuss?"

About Raman's science there may be many different views but about his lecturing ability there seems to be only one. People who have heard him – be it in Calcutta in the twenties or in Bangalore in the sixties or anywhere else at any other time – are unanimous that Raman was a lecturer *par excellence*. He had all the desirable qualities – a stentorian voice, a powerful command of the language (English, of course), a great capacity for imagery, a keen sense of humour, and last but not least, an ability to demonstrate experiments and let the audience see for themselves what he was talking about.

While Raman has lectured to diverse audiences, it would appear that he was at his best when he gave popular lectures, for then he could set aside murky technicalities, reduce everything to the simplest of terms, and invite the listener to share with him the thrills and the excitement of science. No biography of Raman would therefore be complete without a portrait of Raman the lecturer. At the same time, it does not make much sense to *write* about his lectures, for that can hardly capture the pleasure his audiences must have experienced over the years. Obviously the best way of dealing with the subject would be to place before the reader a suitable selection from Raman's lectures themselves. While perhaps there is a wide field to choose from in this respect, I intend to confine myself to a beautiful set of radio broadcasts which Raman made over the years. Raman seemed to like the radio and even persuaded Max Born to make a broadcast during the latter's brief stay in India[2].

The radio is a powerful medium, for it can reach far and wide. At the same time it has its limitations. The speaker not only does not see his audience, but, more important, there can be no spontaneous questions and answers. There is the further restriction that the speaker must be constantly seated in front of a microphone, being thereby denied the pleasure of strutting around on the stage (which Raman appeared to like). Gesticulations (of which Raman was fond) carry no meaning, and the lecturer cannot write on a board or draw explanatory diagrams. Finally, he cannot demonstrate experiments (as Raman often enjoyed doing). He has only one tool at his disposal, namely, words, which of course Raman could use most effectively. Indeed, it is my personal view that Raman was more eloquent in his radio broadcasts than anywhere else.

The entire set of broadcasts has been gathered together in a charming (but unfortunately hardly known) little book entitled *The New Physics: Talks on Aspects of Science*, published by the Philosophical Library of New York [13.1]. In an introduction to this volume, the distinguished theoretical physicist Francis Low (then at the Institute for Advanced Studies, Princeton) says:

> Physics by its very nature requires extreme specialization on the part of its students. Its conclusions, which must eventually predict numbers for the results of actual measurements, are best expressed in mathematical formulae. This has the disadvantage of making the subject well-nigh unintelligible to the layman. There are unfortunately few teachers who are able to surmount this handicap. Professor Raman has written a book which avoids this pitfall and thus should give the lay reader an opportunity of penetrating at least part of the way into the mysteries of this interesting and important science.

The book contains the texts of nineteen broadcasts on diverse topics ranging from the microscopic world of atoms to the vast universe itself. A purist, while readily conceding that Raman does touch upon various aspects of science, might nevertheless question the applicability of the title The New Physics to all the lectures[3]. Be that as it may, Raman does talk about The New Physics in his first lecture, and explains what he means by it. One must of course begin by indicating when this so-called new physics was born. As Raman puts it:

An Irishman, if there be one amongst my listeners tonight, might feel inclined to ask me the question, how old is this New Physics about which you are talking to us? My answer would be – exactly forty-three years old and still going strong.

Raman treats Röntgen's epoch-making discovery of X-rays as marking the beginning of the new physics. As one knows, following this came many other equally important discoveries, like that of radioactivity, for example. Raman comments,

During the past four decades, the spate of new phenomena has flowed into physics with undiminished vigour, so much so that it is becoming increasingly difficult even for a man of science – excepting, of course, the discoverer himself, to feel thrilled by a new physical phenomenon.

Raman would not, however, like his listeners to think that it was all the work of experimenters.

Indeed the amazing progress of the new physics has been due to no small extent to the courageous leadership and constant guidance given to experimenters by the theoretical physicists or natural philosophers, who in turn build on the foundations firmly laid by experimental discovery. I do not believe there is a single listener to my talk who has not heard of Einstein and his relativistic philosophy which forms an integral part of the framework of thought in the new physics. Not all my listeners, however, will have heard of Professor Niels Bohr of Copenhagen, whose picture hangs facing that of Lord Rutherford at the head of the staircase in my house at Bangalore. Professor Bohr, as he often reminds his listeners, comes from one of the smallest countries in Europe, namely Denmark. But, in the view of many including myself, he is the greatest natural philosopher of the day. The work of Niels Bohr in building up a theory of atomic structure which has inspired a host of experimenters in their work is one of the greatest triumphs of the human mind. In the still unsolved problems presented by the facts of atomic disintegration and transmutation, he as the foremost thinker of the day may yet lead us to a completer understanding of the experimental results[4].

One may well ask, what has the new physics achieved? Raman has the answer:

One has only to look back to the physics of my college days at Madras thirty years ago and to look at the physics of today to appreciate the difference. The old physics was successful chiefly in giving what might be called a macroscopic or large-scale description of natural phenomena, that is to say, a statement of observed facts regarding the properties of matter, heat, light, sound, electricity and magnetism. On the other hand, its attempts to interpret the observed facts in terms of atomistic and molecular concepts was definitely a failure, except in a severely restricted field. The reason for this failure was that the old physics had practically no foundation on which to build.

This shortcoming had since then been eliminated, and in turn, the new developments in physics had even contributed to an improved understanding of chemistry. How did this come about? Raman explains:

One of the essential facts of chemistry is that the strength of chemical combination and the energy required for or released by such combination is very different in different cases, and it is only in physical theories that it is possible to find any real understanding of these facts, and indeed also of the real nature of chemical combination. The success of the new science known as "chemical physics" has transcended all expectations. Indeed, it is not unreasonable to hope that before many years pass, theoretical chemistry will come to be regarded as a branch of mathematics.

What has been the secret of all these amazing successes?

> Simply stated, it is the elimination of the Newtonian mechanical laws from the field of atomic and molecular physics and their substitution by other and new laws governing sub-atomic processes.

The reference here, of course, is to quantum mechanics. At that time quantum mechanics was still rather new and quite mystifying. Raman correctly remarks:

> The present generation has not yet had time to fully understand and absorb the new theoretical outlook; but the latter has abundantly justified itself by its success in handling problems of the most varied nature. To the next generation, the new modes of thinking now required in our science will doubtless become quite habitual.

How true indeed[5]!

The spectacular successes of the new physics are many, most notable of these being the achievement of the age-old dream of alchemy, i.e., the conversion of one chemical species into another. This was possible only with ingenious devices which we familiarly refer to as particle accelerators. Of them Raman says,

> In their boldness and novelty of conception, and in the purposes for which these are used, these contrivances fittingly represent the spirit of the new physics.

This was way back in the forties. One wonders how he would describe today's breathtakingly giant machines like the Large Electron Positron (LEP) collider rapidly nearing completion in Geneva, and the even bigger Superconducting Super Collider (SSC) recently authorized for construction in the US.

The public at large would no doubt be mesmerized by these mind-boggling achievements (even as scientists themselves are) but, when they snap out of their reverie, they (i.e., the public) are well prone to ask: Of what practical use is all this? Raman has a word for listeners of this type.

> The vast body of new knowledge which the new physics has created naturally represents a greatly increased power to use the forces of Nature for good and evil. In a hundred different ways, physics has during this period of advance influenced human life and activity. But I would not have you forget that the greatest leaders of our science have always been those whose aim has been the promotion of knowledge for its own sake.

A gentle defence of the ivory tower!

What is the purpose of science? Why do physicists pursue whatever it is that they are pursuing so intensely? Raman tells us:

> The purpose of scientific study and research is to obtain an ever deeper understanding of the workings of Nature. To the physicist falls the task of discovering the ultimate units or entities that constitute the material universe and of ascertaining the principles which govern their behaviour.

This is an endless quest, and, as Raman notes, it "takes the physicist farther and farther every year". But there are rewards.

> The wealth of knowledge gathered on the way has however been immense. It is not hoarded in a treasure chest like a miser's money but given freely away to all who will take it.

Many scientists would of course complain that Nature often guards her secrets well. However, the history of science also shows that under relentless pursuit, Nature does reveal herself. There are other benefits as well.

> The discoveries and inventions of the physicist become the working tools of the engineer, the chemist and the biologist, and in due course of time add immeasurably to the world's wealth and welfare.

It is amazing that society stubbornly refuses to learn this lesson despite the ever increasing numbers of spectacular examples. Not too infrequently, the basic scientist is shoved aside in matters of funding on grounds of social priorities. Nevertheless, the basic researcher stoically perseveres, and the physicist in particular tries to

> delve deep into the nature of things, and in doing so he has to concern himself with matters which to the uninitiated might seem very remote indeed from the affairs of the workaday world.

As an illustration, Raman chooses the example of cosmic rays, an intelligent choice, in view of a recent visit then by an American team of researchers, duly reported in the newspapers. He says:

> My listeners may remember that we had recently staying with us at Bangalore Prof. R. A. Millikan[6] who is a leading pioneer in the study of these rays. Prof. Millikan came out to India with his assistants specially to study the cosmic ray effects in our latitudes in the hope of getting some light on the origin of the cosmic radiation. This appears to be very mysterious yet, but the study of the effects produced by cosmic rays has already been most fruitful for physics in many ways. The total energy received by the earth as cosmic radiation is about the same as that which comes to us every night as starlight and may not therefore seem at first sight particularly significant. But actually, it is most significant because of the form which it takes, namely in discrete units expressible in many millions or even billions of electron volts, a quantity which is immensely larger than anything that can be produced in our laboratories even with the most modern equipment. It is this enormous energy of the individual particles or units that invests the study of the cosmic radiations with extreme interest, opening out as it does the possibility of observing phenomena which we could never hope to reproduce in our laboratories.

Raman then describes various ingenious methods employed for the study of these cosmic rays. Later, some of these were to be employed by Bhabha in Bangalore. Raman again touches upon the question of practical utility:

> The new knowledge that is reaching us through the study of the cosmic rays is of the profoundest interest to the physicist. That this knowledge will in due course influence the whole scientific outlook of mankind and ultimately contribute to human welfare goes without saying. But the onlookers who do not see most of the game should have patience.

Up to the present time, cosmic rays have not found *direct* practical utility in the same sense that nuclear fission has, for example. However, one could argue that they have in a sense contributed to advancement as Raman predicted. The discovery in the late forties of various species of elementary particles in cosmic rays created such an appetite for the organized study of interactions at high energies that a strong campaign was made for high-energy accelerators. And in the mood that prevailed in the post-war years, funding for accelerators was quite easy. From those days, the accelerator community has maintained steady progress, going from one big machine to another. Not only did this provide physicists with more and more powerful tools, but there were also numerous spin-offs in various branches of technology like cryogenics, ultra-high vacuum, magnet technology, radio frequency engineering, radiation detection, real-time computing, and so on. Recently, letting their imagination have a free run, some scientists even proposed a floating accelerator on the oceans to explore for oil by neutrino tomography!

For us in India, the study of cosmic rays has more than a passing significance in another sense. Experimental physics at TIFR revolved largely around cosmic rays in the early days. In turn this spurred interest in the design of radiation counters, electronic counting equipment, and so forth, all of which made no small contribution in shaping the beginnings of our atomic energy programme. Indeed, the giant Electronics Corporation of India Limited (ECIL) can be directly traced back to the small electronics production unit set up at TIFR in its early days when it was functioning out of World War II barracks in Colaba, Bombay.

Parenthetically, it is interesting to add that in a public lecture in 1953, Saha sharply criticized Bhabha's atomic energy programme, observing, "Cosmic ray work [at TIFR] has no doubt great importance for knowledge of fundamental particles, but I challenge the Indian Atomic Energy Commission to prove that it has so far played any role in atomic energy developments in any country." As it turned out, Saha's criticism was somewhat premature. He did not live long enough to see how the cosmic ray experience not only provided valuable seed but also ramified in various ways (as just illustrated with an example). Further, the French model of atomic energy development which Saha strongly commended in that address was in fact emulated to quite an extent though somewhat slowly. In the same lecture, Saha also criticized some statements made by K. S. Krishnan in his capacity as a member of the Atomic Energy Commission.

Cosmic rays are fine but they are also somewhat exotic. Is there not a simpler way of bringing science to the listener, say in terms of something he is familiar with? Indeed there is, as Raman shows by talking about shells.

Most physicists would be stumped if asked whether there was any connection between science and shells. Not so Raman. He first introduces the subject of shells with characteristic flavour.

> Gathering shells on the seashore would probably be regarded today by many as an amusement fit only for young children. A century ago, however, shell collecting was a very fashionable hobby. Vast sums of money were expended by amateur conchologists who paid high prices for rare and beautiful shells. Many of the collections made during this period have passed by gift to the various national museums where they can now be examined at leisure by students of the subject. I am

Sharing the Pleasures

a collector myself, though not in a large way, and would be happy to be able to infect some of my listeners with the enthusiastic admiration which I feel for Nature's handiwork in this field.

People collect shells and admire their various forms but seldom wonder where a shell comes from. Raman would like his listeners to

> please remember that it was once occupied by a living pulsating creature, and was built up, little by little, by this creature around itself as a home and as a mantle of defence in its struggle for existence – a home which grew up with its occupant as the latter increased in size and reached maturity. The study of shells, fascinating in itself, becomes doubly so when we regard it, in its correct perspective, as the study of one of the most ancient forms of life on this planet of ours, a lowly form of life, no doubt, but nevertheless of the deepest significance and interest. The forms, the sizes, the colours, and the architectural characters of shells are manifold in their variety and charm. But the mystery and the interest deepen when we ask ourselves why and how the humble mollusc builds for itself these forms of beauty.

Man not only admires shells, but also makes use of them! Raman skilfully introduces this fact to gently lead up to some interesting aspects of the physics and chemistry of shells.

> Many of you have no doubt seen sea-shells collected into large heaps and then burnt in order to make lime. This will remind you that...the substance of a shell is common chalk or calcium carbonate; this when burnt is converted into quicklime or calcium oxide. The chalk which forms the material of the shell is deposited as a secretion by the animal from the edges of its living substance and gradually builds up the substance of the shell. Along with the chalky substance, there is a small quantity of organic matter or horny substance which helps to bind the inorganic material chalk firmly together and give it mechanical strength. The shape of the shell is also in many cases evidently calculated to give the structure special rigidity and capacity to withstand impact without fracture. Apparently, in some cases, even if the shell is accidentally damaged, the animal is capable of repairing the break by depositing fresh material.

The chalky material of the shell presents different appearances in different cases. It is hard and white in shank, translucent in window-pane oyster, and beautifully lustrous and iridescent in mother-of-pearl. Raman traces these differences to the *atomic structure* of these materials.

> Chalk is, in its natural state, a crystalline substance.... The natural tendency of chalk to form large crystals is resisted during its deposition on the edge of the shell by the presence of the organic horny matter mixed with it. Nevertheless, it does crystallize, and the larger the size of the crystals, and the more regularly they are laid down, the more nearly transparent the substance of the shell tends to become. This is actually the case with the window-pane oyster....
>
> It is also necessary to remark that there are two distinct crystalline forms of chalk known, which are chemically the same but physically different, namely calcite and aragonite. Calcite is by far the more common in Nature, and indeed this is also the form in which it appears in the substance of all molluscan shells....Mother-of-pearl...consists of chalk in the rarer form of aragonite, but as exceedingly numerous crystals. These are imbedded in a horny substance forming layers approximately parallel to the surface of the shell.... there may be twenty thousand or thirty thousand such layers per inch of thickness of the mother-of-pearl.

Even pearls are nothing but

> common chalk with a little horny substance thrown in. But this does not detract from their beauty and value any more than the fact that diamond is just carbon affects the prestige of that gem.

Why does Nature manifest itself in so many diverse forms? Is there some Grand Design? We do not quite know.

> The geometrical forms of shells and the carving of their external surfaces are often exquisitely beautiful. What part these forms play in the life of the animal, or whether this beauty is just part of the exuberance of Nature in creating forms of life and endowing them with grace is more than I can venture to say, not being a professional zoologist.... There is little doubt, however, that the external form of the shell and the internal architecture of its substance are intimately related to each other....

While one may not yet know whether the beautiful patterns on shell surfaces have a functional role, one is at the present time at least able to explain how such patterns could arise. As in the case of Liesegang precipitation (see Chapter 8), it appears that certain nonlinearities called into play during shell growth are the controlling factor. In fact, Meinhart and Klingler [13.2] of Tubingen have recently succeeded in simulating many of the observed shell patterns using a suitable nonlinear reaction–diffusion model.

Raman's deep interest in geometry has already been referred to (see Chapter 2). Geometry finds expression in Nature not only in the limited manner expounded in Euclid's book but in fact in a somewhat more generalized fashion. Recognizing this, the celebrated mathematician Felix Klein started a programme to reduce natural phenomena to a (generalized) geometric foundation. In a deeper sense, modern gauge theories, so popular with high energy physicists (and also with a certain section of condensed matter physicists), are also geometrical in character in so far as they reflect various subtle internal symmetries. Raman's appreciation of geometry in Nature was of a different type, his interest being in the symmetry of the external form.

> The concept of beauty defies abstract analysis. On the purely physical plane, however, we can recognize certain elements of beauty, some of which, such as symmetry and proportion, are geometrical in character. When we survey the forms of living beings, it becomes evident that such geometrical characters form a large part of natural beauty. An essential element of its external aspect, this geometrical character, namely the balancing of right against left, is noticeable in most living forms. In the botanical world, we find other and more highly developed patterns of symmetry which give physical beauty to the foliage and the flowers of plants. Under the microscope, the minuter forms of life, both plant and animal, exhibit an infinite variety of symmetric structures. The study of the humbler types of life, as for instance of the mollusca found in fresh water or on the seashore, reveals to the observer a wonderful wealth of geometrical forms. Nature's artistry, however, does not exhaust itself in producing patterns of symmetry. The beauty of a forest, for example, does not consist merely in the beauty of the foliage and flowers of the trees, however ornamental these may be. The tall straight trunks, the spreading crowns and the interlacing branches of the trees of the forest are other elements of geometrical form which hold and delight the eye of the observer.

It is interesting, to note in passing, that in recent years, *random* geometry has found not only quantification in some measure, but also application, e.g., in describing the atomic structure of amorphous materials.

In the case of shells, Raman had speculated whether there was any *purpose* in the multifarious form. He now offers some thoughts in a similar vein, *vis-a-vis* plants and animals.

> In seeking to understand and interpret these creative efforts of Nature, we can approach them from two different points of view. In the first place we can consider the biological aspect, which is that of the successful functioning of life. The form and structure of the animal or plant determine its activities, and there can be little doubt that the geometrical characters of the structure serve a biological purpose. The balancing of right and left, for example, is almost universal amongst the living things which move on the earth, sail in the air or swim under water, and we recognize that it serves the purpose of facilitating locomotion. We do not find it, for example, in the mollusca which are content to attach themselves to solid supports, and are indeed provided with a special adhesive mechanism for the purpose. We can safely generalize and say that nearly every detail of geometrical form has some purpose to serve in the functioning of life.

Life has a wide variety of building materials to choose from, and the manner in which the choice is made is quite interesting. Inorganic substances like, for instance, the carbonates and phosphates of calcium, are preferred for solid structures like bone, proteins being quite unsuitable for this task. On the other hand, proteins are favoured in certain other circumstances. Interestingly, the macroscopic geometry is influenced by the microscopic symmetry of molecules. Raman explains:

> The nature and properties of these inorganic substances and the manner in which they are associated with the protein substance is one of the essential factors which has determined the geometrical forms of living animals. The protein substance is however adapted for building certain special types of quickly growing structure, as for instance, the horns and antlers of the ruminants and the hair and the wool of mammals. The beautiful silky hair which is the crowning ornament of a woman's head owes its geometrical character of a long continuous fibre to the structure of a certain type of protein molecule. The glory of human hair thus stands revealed as a special effort in protein chemistry made by Nature.

What is it that influences the geometry of plant life?

> The geometry of plant life is largely determined by the properties of that wonderful substance which is called cellulose. This material is a compound of carbon, oxygen, and hydrogen akin to sugar in some respects but with very different properties. Whereas sugar is soluble in water, cellulose is not, and its molecules are so built that they readily form groups, or micelles as they are called, which can join together and build elongated fibres. Cellulose fibres, cemented together by an amorphous substance called lignin, form the woody structure of trees and plants. When therefore we admire the tall trunks of the giant trees and the wonderful geometric tracery of the branches and leaves in a forest, it is not inappropriate to remember that this wealth of beauty is made possible by the geometry of the cellulose molecule.

Let us pause here and absorb how Raman has been developing the theme of geometry in Nature. He started by talking about the external geometry of the living form and then gradually led on to the structure of living matter and the influence of molecular structure on the external form. What about crystals, which clearly are the most dramatic manifestation in Nature of geometric regularity? Raman reserves them for the last.

> When you visit the geological museum, you will see exhibited beautiful examples of natural crystals, as for instance octahedral crystals of diamond, hexagonal columns of quartz, cubes of rock-salt, rhombohedra of calcspar, dodecahedrons of garnet, and huge prismatic plates of mica, to mention only a few items. Indeed, nearly all solids known to Nature or made by man are essentially crystalline in structure, though not always in external form. Modern research has revealed the internal architecture of crystals to be an array in space of atoms or molecules, piled up, row after row, column by column and layer upon layer in perfect order and with equal spacing. Geometrical theory enables us to discuss the possibilities of such regular arrangement of the ultimate particles of matter and to classify crystals according to the types of external or objective symmetry which they exhibit, and also according to the types of internal arrangement or grouping of the particles of matter within them. It is astonishing but very satisfactory that the 32 different crystal classes and the 230 different types of internal grouping of particles indicated to be possible by pure geometrical reasoning are precisely the numbers respectively of crystal classes and of atomic groupings which have been found by experimental study to exist in solids.

This is an example of what Eugene Wigner refers to as the unreasonable power of mathematics!

Besides geometrical form, colour and light in Nature are constant sources of pleasure to the sensitive soul. Remember Wordsworth? We have seen how much preoccupied Raman himself was with these subjects. We should not be surprised therefore that Raman expresses himself eloquently on this subject.

> The face of Nature as presented to us is infinitely varied, but to those who love her it is ever beautiful and interesting. The blue of the sky, the glories of sunrise and sunset, the ever-shifting panorama of the clouds, the varied colours of forest and field, and the star-sprinkled sky at night – these and many other scenes pass before our eyes on the never-ending drama of light and colour which Nature presents for our benefit. The man of science observes Nature with the eye of understanding, but her beauties are not lost on him for that reason. More truly it can be said that understanding refines our vision and heightens our appreciation of what is striking or beautiful. Many a time also has it been the case that the study of natural phenomena has pointed the way to a far-reaching advance in knowledge.

Raman cites a few examples like the connection between electricity and thunderstorms, but, curiously enough, does not make a reference to his own studies on the colour of the sea which led to a major discovery.

The scientist is not merely captivated by the beauty of natural phenomena but also asks questions. For example:

> What is meant by the colour of an object? Is it the colour of the light reflected by the surface of the object, or is it the colour of the light which has passed through it, or the colour of the light diffused within its interior, and thence emerging? It may seem surprising to raise so many questions about what appears at first sight to be a simple matter. Actually, however, the colour of a substance as defined in these three ways may be entirely different. A typical example is the colour of water. The light reflected

by the surface of water is evidently of the same colour as the light falling on it. If, for instance, sunlight falls on the surface of water, the reflected light will be of the same colour as sunlight. The colour of a beam of white light which has passed through a column of water will on the other hand be influenced by any specific absorption which water may possess for the different parts of the solar spectrum. Actually, even the purest water exercises a sensible absorption for the red and yellow rays of the spectrum. Hence, sunlight which has passed through a long column of water exhibits a distinct greenish tinge. Then again, the passage of light through water is attended by a diffusion of the light, firstly by any suspended particles and secondly by the molecules of the water themselves. If the suspended particles are sufficiently small in number and therefore of negligible importance, the diffusion of light within the water will be due principally to the molecules of the water, and the colour of the diffused light will be a sky-blue colour.

It thus becomes evident that the colour of water as seen in any particular circumstances depends on the extent to which the reflection of light at its surface, the specific absorption of the red and yellow rays of the spectrum in the passage of the light through the water, and finally, the diffusion of light within the interior of the liquid, determine the observed effects. It is not surprising therefore that the apparent colour of even the clearest water varies with the circumstances of observation. If the surface reflection is eliminated, as for instance, by looking vertically downwards, and the water is sufficiently clear and sufficiently deep, then the colour is determined by the joint effects of absorption and diffusion by the molecules and is of a dark blue colour, much deeper than the blue of even the clearest sky. If on the other hand, only relatively small depths come into play, as for instance, when the water is churned up and is full of air bubbles or is contained in a relatively shallow basin, the diffusion effect becomes negligible and the water appears green or greenish blue.

While Nature offers many dazzling displays of light and colour, man too has sought to generate both light and colour for many reasons of his own. How did it all begin? Raman tells us:

> To rise from bed at dawn, work the livelong day, and go home to roost like the birds at sunset may be a splendid way of ordering human affairs. But I doubt if even the stoutest-hearted dictator would venture to enforce this as a rule of life. It might conceivably serve for those who live in the tropics with twelve clear hours of daylight, but would be utterly impracticable in northern latitudes, to say nothing of the Arctic Zone where the sun is not seen for six whole months and only the auroral lights save the world from perpetual darkness.

And so, man devised various sources of illumination with which to extend his hours of activity. His first source was a fire or a blazing torch which, in colder climates, had the merit of also supplying heat.

> A good deal could be said about the development in successive eras of man-made sources of light. We in India are familiar with our castor oil lamps picturesquely made in bronze, but now laid out of sight in odd corners of our homes. The old order has passed away and given place to new....

Yes indeed. We have gone through a progression of various lamps like the kerosene lamp, gas mantles, carbon filament lamps, metal filament lamps, neon lamps, mercury lamps, fluorescent lamps, sodium lamps, halogen lamps, and so on.

As already noted, the primitive methods of illumination combine the production of heat and light. This, however, is wasteful.

The ideal light is one in which the whole of the energy sent out by the source is concentrated within the rays of the visible spectrum. There would then be no waste of energy as unwanted heat.

Interestingly, Nature seems to have already solved this problem. There is, of course, the familiar glow-worm, and

> in the waters of our seas and oceans live many denizens of the deep endowed with what has been happily termed living light.

Today, the term more commonly used to describe this phenomenon is bioluminescence. Man has not yet produced commercially viable "chemically manufactured light", as Raman calls it, but he has been able to exploit luminescence in his own way. It all started with a cobbler of Bologna who discovered that gypsum could be converted by calcination into a substance that had the strange property of emitting light in the dark.

> This phenomenon was long regarded as a scientific curiosity and excited intense interest, with the result that it was extensively studied. These studies have taught us a great deal about the phenomenon of luminescence and the methods of creating it. A body is said to be luminescent when it gives more light in a given range of the spectrum than can be obtained by merely heating it up. Luminescence may arise in various ways, the most familiar being the emission of light under ultraviolet illumination. A striking example of such emission is the behaviour of the so-called blue diamonds which when placed in invisible ultraviolet light shine visibly. I have in my possession a diamond of this kind which emits enough light in a dark room under ultraviolet irradiation to enable a newspaper held close to it to be read.

Obviously, this is not a practical way of generating cold light, for even Maharajahs cannot afford it! However, from a scientific point of view, the study of such phenomena, or of the luminescence of substances like zinc sulphide when bombarded with rays from radioactive substances, is quite interesting. Studies of light emission have revealed that luminescence is of two kinds.

> The first is the light emission observed during the excitation. This is called fluorescence. The other kind is phosphorescence, or the emission which continues when the excitation is removed. The latter property enables the body to continue shining in the dark, slowly giving up the energy which it has stored up during excitation.

Both types of luminescence have found many practical applications. For example, fluorescent lamps (which at the time Raman made this particular broadcast were just beginning to find use)

> are essentially electric discharge tubes with a thin transparent coating of a luminescent material on the interior of the glass walls. The invisible ultraviolet radiation generated by the electric discharge is converted by the coating into visible light.

There are then the luminescence excited by electron bombardment which has made possible the television screen, and the less known luminescent paints. The latter are, however, becoming increasingly popular.

Sometimes, luminescence arises owing to the presence of "almost incredibly small traces of impurities".

> A beautiful example is the intense red luminescence of ruby which is due to the presence of chromic oxide as an impurity in the crystalline alumina which is the main constituent of ruby. By varying the nature and the extent of the impurity or impurities present, as well as by suitable physical treatment of the basic material, the colour and the intensity of its luminescence as also its duration after the removal of excitation can be controlled in a very striking way.

Raman does not lose the opportunity to emphasize how it was basic research that made possible all the diverse applications.

> The phenomenon of luminescence furnishes many fine examples of how investigations undertaken solely in view of their scientific importance have led to important practical developments.

Raman made three broadcasts on what he called the "Physics of the countryside" which are truly remarkable for a variety of reasons. Firstly, the title is a stroke of pure genius; it is doubtful if one could have come up with a more compelling one. Secondly, the broadcasts illustrate to the average person how science is both relevant and important in day-to-day living. Even more significant, they illustrate that Raman, though deeply immersed in his own research, was fully alive to the practical aspects of science, especially in relation to *rural life*. A lesser man anxious to please politicians might well have entitled these lectures "Science and rural development"!

Raman deals with three topics (under the omnibus title alluded to), namely, the soil, water (ground water particularly) and weather.

The soil of our earth is the basis of all agriculture and therefore also of human civilization. It is

> both physically and chemically of complex and variable nature. This is not surprising when we consider its origin and the vicissitudes which it has undergone both naturally and by human agency. The bulk of an ordinary arable soil consists of a heterogeneous collection of mineral particles of all shades and sizes, ranging from large stones to dimensions below the limit of visibility even in a powerful microscope.

As one knows, depending on the sizes of these particles, one can have soils of great variety and with different properties. All these are important. The mechanical properties of the soil, for example, determine the labour involved in cultivation.

> The impedance which the soil offers to the movement of a plough varies very greatly with the circumstances. It depends, of course, on the nature of the soil, and particularly also on its condition at the time. A soil which has hardened under the influence of a long drought is difficult to plough....

Understandably, therefore, much ingenuity has been exercised in devising ploughs suited to soils of varying nature, for different depths, and adjusted to the power of draught available. Agriculturists are equally concerned about moisture in the soil.

> A soil which can hold no moisture is obviously of no use. Hence, the study of the moisture content of a soil, and of the distribution and movements of the fluid within its volume, is of prime importance. It was formerly thought that the behaviour of water in a soil was closely analogous to the familiar rise of water in a capillary tube.... It was believed that this simple physical principle afforded an easy explanation of the well-known observed difference between drought-prone, coarse gravel or sandy soil, and drought-resisting, fine gravel or clayey ones. Careful quantitative studies, however, have shown that the capillary theory in its simple form is incapable of explaining the observed behaviour of soil.

How, then, does one explain it?

> In reality, the pore space in soil is of a cellular nature, consisting of relatively large empty spaces communicating with each other through relatively narrow necks. The water distributes itself within these cells and necks in such manner as to reduce its free surface to a minimum.

In passing we note that percolation through a random network of pores is a much-studied subject lately, with deep implications also for oil exploration.

The wide variation in cross-section of the gaps between adjacent soil particles has most important consequences for the flow of water through the soil. It turns out that

> the water in soil tends to resist changes, whether these are in the direction of increasing or decreasing moisture content. Instead of moving through the pores from the regions of high moisture content to low, it will adapt itself to the suction gradient by an alteration of the internal distribution of the water inside the cavities.

Naturally one would expect all this to have a relevance to the manner in which plants obtain water from the ground. Indeed it has, and Raman amplifies:

> The capillary theory taught that the drying due to water imbibition by the roots was met by the movements of water from the moister regions, in other words that the water was brought to the plant roots. The correct view is however, just the opposite; the plant roots have to ramify extensively through the soil in search of moisture. This, of course, we know to be actually the case, and the tremendous range of root development of a vigorously growing plant must be seen to be believed.

The porosity of soil has other implications as well.

> It is obvious that the soil being a porous material will contain gas in its interstices. The composition of the enclosed air will be influenced by various factors, including especially the absorption of oxygen by plant roots and micro-organisms and the evolution of carbon dioxide. Hence, it depends upon the rapidity with which carbon dioxide can escape to the atmosphere and be replaced by oxygen. The expansion and contraction of air owing to temperature changes of the soil, the changes of the pore space in the soil owing to rain, irrigation and evaporation of soil moisture, the action of wind and of barometric changes in the external air have all to be considered in this connection.

From soil, it is a natural step to water – a substance that is essential for life, vital for agriculture, captivating to the poet in the various forms in which it appears in Nature, and fascinating to the scientist on account of its properties and structure.

> ...this common substance which we take for granted in our everday life is the most potent and the most wonderful thing on the face of our earth. It has played a role of

vast significance in shaping the course of the earth's history and continues to play
the leading role in the drama of life on the surface of our planet.

There is nothing which adds so much to the beauty of the countryside as water, be
it just a little stream trickling over the rocks or a little pond by the wayside where the
cattle quench their thirst of an evening. The rainfed tanks that are so common in
South India – alas often so sadly neglected in their maintenance – are a cheering
sight when they are full.... Some of these tanks are surprisingly large and it is a
beautiful sight to see the sun rise or set over one of them. Water in a landscape may
be compared to the eyes in a human face. It reflects the mood of the hour, being
bright and gay when the sun shines, turning to dark and gloomy when the sky is
overcast.

The importance of water is of course readily appreciated but it is not so often that
one thinks of the flow of water and its consequences.

One of the most remarkable facts about water is its power to carry silt or finely
divided soil in suspension.... Such particles [of soil] are, of course, extremely small,
but their number is also great, and incredibly large amounts of solid matter can be
transported in this way. When silt-laden water mixes with the salt water of the sea,
there is a rapid precipitation of the suspended matter. This can be readily seen when
one travels by steamer down a great river to the deep sea. The colour of the water
changes successively from the muddy red or brown of silt through varying shades of
yellow and green finally to the blue of the deep sea.

One cannot, of course, talk about the flow of water in the countryside without calling
to one's mind the problem of soil erosion.

The problem of soil erosion is one of serious import in various countries and
especially in many parts of India. The conditions under which it occurs and the
measures by which it can be checked are deserving of the closest study. Soil erosion
occurs in successive steps, the earliest of which may easily pass unnoticed.... The
menace which soil erosion presents to the continuance of successful agriculture is an
alarming one in many parts of India, calling urgently for attention and preventive
action. The terracing of the land, the construction of bunds to check the flow of
water, the practice of contour cultivation and the planting of appropriate types of
vegetation are amongst the measures that have been suggested. It is obvious that the
aim should be to check the flow of water at the earliest possible stage before it has
acquired any appreciable momentum and correspondingly large destructive power.

One regrets to say that soil erosion not only remains neglected but is threatening
to assume menacing proportions, thanks to the rapid denudation of our forests.
Raman points out that conservation of water and control of water movement not
only check erosion but can serve other purposes as well.

By far the cheapest form of internal transport in a country is by boats and barges
through canals and rivers. We hear much about programmes of rail and road
construction, but far too little about the development of internal waterways in
India.

Indeed, proposals have been made from time to time to have a network of canals
across the length and breadth of the country to remedy the frequently witnessed
spectacle of severe drought in several parts of the country and devastating floods in
several others at the same time. But nothing has come out of these. Raman concludes
with the remark:

> In one sense, water is the commonest of liquids. In another sense, it is the most uncommon of liquids with amazing properties which are responsible for its unique power of maintaining animal and plant life. The investigation of the nature and properties of water is, therefore, of the highest scientific interest and is far from an exhausted field of research.

Dependent so much as we are on rain for our supply of water, Raman turns next to the subject of weather.

> To the dweller in the towns, the weather is nothing more than a minor inconvenience which can be minimized by a little forethought in the matter of taking an umbrella instead of a walking stick when going out of the house. I will go so far as to say that the average city dweller is scarcely conscious of the weather except when he is reminded of it in some particularly unpleasant fashion....
>
> The weather, on the other hand, plays a vital part in the life of the countryside. Those who dwell in the wide spaces of the earth for ever watch the skies and grow weather-wise, for their work and prosperity, nay their very lives, depend on what the skies may bring forth. The cycle of the seasons so beautifully described in Kalidasa's *Ritusamhara* is also the cycle of the life of the countryside in India. If one overslept like Rip Van Winkle in Washington Irving's story and woke up unconscious of the lapse of time, a glance at the agricultural scene in any familiar area would enable the date to be fixed within a week or two. Vast tracts in our country still depend exclusively on rainfall for the possibility of any kind of agriculture. The opening and shutting of the sluice-gates in the sky are therefore the most important events in the calendar of the man who tills the earth in these areas.

One is not only aware of the dependence of our agriculture on rainfall but also of the impact a good monsoon has on the economy. Indeed, it is not only our present Finance Ministers who anxiously look to the skies.

> A former Finance Member of the Government of India is reported to have said that the budgets he had to prepare and present every year were a "gamble in rain". This expression puts in a neat and forcible way the existing preponderance of agriculture in the economy of India, and the controlling influence of the weather on the same. This relation between the weather and public finance appears to have been the principal reason for the establishment by Government of a Department of Meteorology during the last century....
>
> The enormous importance of meteorology to an agricultural country like India will however bear being repeatedly emphasized. It is remarked that so far no Indian university has thought it worthwhile to provide for instruction and research in this subject. This perhaps is only one more illustration of the existing lack of co-ordination between India's real needs and her educational activities.

Mercifully the situation is slightly better at present in that about a dozen of our universities now offer master's level courses. However, the quality of instruction leaves much to be desired! Parenthetically it may also be remarked that many of Raman's students, starting with S. K. Banerji and K. R. Ramanathan, found a berth in the India Meteorological Department, and played a leading role in strengthening its research activities. About the scientific potentiality of meteorology, Raman says:

> The widest possible diffusion of meteorological knowledge, and the promotion of an active interest in study and research in Indian meteorology are I think of vital

> importance.... I think the impression has prevailed in the past that meteorology is a subject in which nothing need be done or could be done except under official auspices. I think that such an impression is without real justification and that much work of great value could be accomplished by academically-minded scientists working under unofficial or university auspices.

Unfortunately, the importance of this message remains lost even today. Our monsoons continue to be a major scientific puzzle to the point of attracting considerable research attention abroad, but we here can hardly boast of strong schools of meteorology in our campuses.

Getting back to the scientific aspects of weather, what is it that a meteorologist is interested in?

> The pressure, temperature and humidity of the earth's atmosphere are the factors which control its behaviour and interest the meteorologist. Any distribution of pressure over the earth's surface which departs from the condition in which equilibrium is possible necessarily gives rise to horizontal movements of the atmosphere or winds as we call them. It is evident that if such winds were to blow continuously from one part of the earth's surface to another, the air would pile up over the latter, leaving the former empty. Since this is not possible, it follows that there must be a reverse movement elsewhere, and the most natural place to look for such a movement is the upper levels of the earth's atmosphere. It is this simple consideration that invests the study of the condition of the upper levels of the earth's atmosphere and the movements which occur in them with very great importance in meteorology. It is necessary, in fact, in order to understand what is happening near the earth's surface, to know what is happening far above it, and to correlate the two sets of facts. It is for this reason that it is now a regular practice in meteorology to investigate the upper air by observation of the movements of free balloons and also by sending up balloons containing instruments which automatically record the condition of the atmosphere at the higher levels or send radio signals to an observer below.

Today a new dimension has been added to the study of meteorology by pressing into service such sophisticated tools as satellites and computers. Raman's prediction that "active study and investigation in the subject will prove fruitful" is certainly coming true. And there is certainly a growing awareness that meteorology is also essential "from the standpoint of the promotion of India's most vitally important industry, namely agriculture". However, it must be added that our concern for integrated water management is not as intense as it should be. Several years ago, the late K. R. Ramanathan suggested to the Government that we must have a national water budget (even as we have a financial budget), so that every drop of water is properly accounted for and utilized. It is said that Israel has made a great success of water budgeting. Nothing much has been heard of what happened to Ramanathan's proposal, and one would not be surprised if, like many other proposals, it got lost in the miles of filing cabinets that line the Secretariat corridors.

Not infrequently, rain occurs as part of a thunderstorm. Raman devoted a separate lecture to the topic of atmospheric electricity because, again, the topic afforded a convenient means of discussing aspects of science by talking about a familiar phenomenon. As is well known, it all started with Benjamin Franklin, who

> demonstrated the idea put forward earlier by him that the lightning in clouds is an electric discharge of the same general nature as can be obtained on a much smaller

scale in the laboratory. Ever since, the study of atmospheric electricity has been continued by generations of physicists with unabated vigour and interest. In spite of the lapse of nearly two centuries, however, it cannot be said that the subject has reached the stage of exhaustion.

Of course, the physicist is interested in the phenomenon purely from a scientific point of view. However, as always, practical benefits do ensue from such purely basic studies. Indeed, the idea of protecting tall structures by the use of lightning-conductors is a good example, and goes back to Franklin himself.

The spark produced in lightning is qualitatively similar to the spark produced in the laboratory by electrical discharge but is of course very much more intense.

> We may well wonder how the astonishingly large electrical forces necessary are generated inside the thunder-cloud.
>
> Observations of the clouds in which lightning discharges occur make it evident that these clouds are involved in a rapid upward movement of the air. Such a movement is characteristic of thunder-clouds and evidently furnishes the mechanical power necessary for a separation of the electricity to occur as between the different parts of the clouds. It is natural to assume that for some reason or other the moving upward current of air carries with it electrical charge of one sign in excess, leaving behind in the part of the cloud below it an excess of the electrical charge of the opposite sign. This separation and transport of electricity from one part of the cloud to the other naturally results in the two parts becoming charged up in opposite ways. The electric tension then increases until the resistance of the air breaks down and the lightning discharge occurs within the cloud.

The transport of electricity can also occur by the movements of water drops carrying electric charges. But how exactly do the water drops acquire electric charges? Raman mentions that there are two rival theories, one due to Sir George Simpson and the other due to C. T. R. Wilson (of cloud chamber fame). He avoids commenting on their relative merits claiming that the matter is *sub judice* (!) but nevertheless describes some experiments which had been done to clarify some of the issues involved. This was in the forties. Apparently, one is no wiser about the origin of atmospheric electricity even at the present time. As Feynman [13.3], who gives a delightful account of lightning in his famous book, remarks: "The fundamental origin of lightning is really not thoroughly understood. We only know it comes from thunderstorms."

As he often does, Raman raises the question of practical use, especially since in this case "the electrical energy dissipated in an average thunderstorm is quite a formidable amount". Raman then advises the listener to abandon such notions since "trying to make practical use of thunderstorm energy is like trying to hitch a tiger to a tonga[7]". He adds:

> I am reminded of the fate of a bright young physicist who investigated the electrical power of thunderstorms in an Alpine valley. The investigation was successful, but the published paper describing its results included an obituary notice of the author.
> ...[!]

The listener must accept that the outlook for practical use of thunderstorm energy is not very hopeful. But one should not be dismayed; the phenomenon itself is beautiful and the mystery is still there.

A rather unusual broadcast is the one where Raman introduces a scientific theme via a book review. He has come into possession of a substantial monograph (900 pages!) entitled *The Properties of Glass* by G. W. Morey of the Carnegie Institute, and would like to use it to illustrate the interrelation between science and industry.

> The glass industry, as Mr Morey tells us in the first chapter of his book, goes right back to the dawn of human civilization. Glass is found in Nature in the form of the mineral obsidian, and primitive man learnt by experience that this substance is easily broken into sharp, often elongated pieces which lend themselves readily to the fashioning of arrowheads, spearheads and knives; its use for such purposes by people of Stone-Age culture was widespread. The beginnings of the artificial manufacture of glass also go back to a remote period of human history. It would appear that the early civilizations of China, Mesopotamia and Egypt share the honour of having discovered how to make glass. The manufacture of glass utensils was a well-established industry in the days of the early Roman empire, and glass had indeed then become a common material in household use. During the Middle Ages, Venice became a great centre of the glass industry and built up a reputation for skilled craftsmanship and beauty of design which has survived to the present day.

In the nineteenth century, glass began to be used in scientific instruments, and this created a demand for a special type of glass.

> Optical glass must be free from unmelted particles, and from air bubbles, and must be completely homogeneous or uniform. It should also be free from colour and have a refractive index and dispersion for light which have been specified in advance. The production of glass meeting these very exacting requirements has contributed enormously to our knowledge of glass technology, and this in its turn has reacted towards the improvement of all types of glass.

Why does one find glass to be so useful?

> The practical utility of glass for a great variety of purposes is evidently the result of the fact that this material possesses several valuable physical properties. Amongst these are the possibility of fashioning glass into any desired shape at a moderately high temperature and its capacity for retaining the form given to it on cooling down, its impermeability to gases and liquids alike, its chemical durability against the action of wind, water and even of corrosive chemicals, its elastic and mechanical strength, its capacity to resist large changes of temperature, its hardness, its transparency to a considerable part of the spectrum, the wide range of colours which can be given to it, and its electrical resistivity. Scientific investigation has shown that every one of these properties is determined by the chemical composition of the glass and can be controlled and is capable of being varied over a wide range by adjustment of the chemical composition. Indeed, great improvements have already been effected in glass by such studies.

Such advances have led to the familiar pyrex glass on the one hand, and high quality mirrors for giant telescopes on the other. After commenting on various aspects of Morey's book, Raman comes to the last chapter which deals with the constitution of glass. This chapter

> is the most interesting and at the same time one of the shortest in the whole book. This is not surprising because in spite of the intensive research in the last two decades on the fundamental question of the nature of glass, it remains as yet more or less unanswered. The general nature of the problem may be illustrated by comparing the two naturally-occurring substances, obsidian and granite. Chemically,

the two substances have nearly the same composition. Nevertheless, they are wholly different in their structure and properties. Granite, as everyone knows, is a coarse crystalline rock in which the constituent particles of quartz, felspar and mica may be readily observed with the naked eye. Obsidian, on the other hand, is a structureless amorphous solid. In crystals such as those constituting felspar, mica or quartz, the ultimate atomic particles are arranged in regular geometrical order, as is beautifully shown by X-ray investigations. On the other hand, in an amorphous solid or glass the atomic particles are distributed in such a way that the X-ray diagram of the substance is essentially similar to that of a liquid. The latest view of the structure of glass is that it consists of a network of silicon atoms joined to each other through oxygen atoms and forming an irregular network somewhat like a badly-made mosquito curtain in three dimensions. The metallic atoms of sodium, calcium, etc. find places for themselves in the various holes in the network. There is no regular repetition in the pattern and hence the structure is non-crystalline. The idea conveyed by this picture, namely that the structure of glass is irregular though on an intelligible plan, indicates that glass need not have a completely defined chemical composition. Indeed, as is well known, the chemical composition of glass can be varied within wide limits. The best glass of all is pure silicon dioxide, that is, fused quartz or vitreous silica.

Concerning the structure of glass (i.e., the arrangement of the atoms in it), one is tempted to quote also from a famous paper by Zachariasen in the thirties. Zachariasen wrote, "It must be frankly admitted that we know practically nothing about atomic arrangement in glasses." Fifty years later, the structure of glass continues to be an enigma (!) though the situation is perhaps not so bad.

Raman concludes his review of Morey's book by saying that it is most scholarly, which is not surprising considering that it took him (Morey) fourteen years of hard work to prepare it. Raman also makes the following comments:

It is not an exciting book, and indeed this could scarcely be hoped for in a technical publication of this kind. The lesson which it brings to the reader is that the true path to success in industry is through patient labour guided and sustained by the spirit of scientific research, and not through haphazard and hasty efforts.

Raman made many other broadcasts dealing with such topics as crystal symmetry, solid state physics and the stellar universe. Perhaps the best is the one entitled "The scientific outlook", appearing as the concluding chapter of the anthology of broadcasts. In it Raman asks: "What is meant by a scientific discovery? How is it made?"

These are questions of perennial interest which are often asked and to which the most varied answers have been returned. A discovery may obviously be either of a new fact or of a new idea. It is clear however than an unexplained observation is of no particular significance to science. An idea unsubstantiated by facts is equally devoid of importance. Hence to possess real significance a scientific discovery must have both an experimental and a theoretical basis. Which of these aspects is the more important depends on the particular circumstances of the case, and a rough distinction thereby becomes possible between experimental and theoretical discoveries....

The word discovery suggests a dramatic and exciting event, like finding a fifty-carat diamond in a ploughed field, for example. The history of science is indeed full of such dramatic discoveries, the drama and the excitement being particularly manifested in the personal behaviour of the scientist immediately following the event.... The classic story is that of Archimedes who rushed into the street straight from his bath with nothing on, crying "Eureka eureka", when his famous principle of hydrostatics flashed into his mind. The point of the story is the intense emotion aroused by a sense of the overwhelming importance of the new idea[8]. The joy and exaltation felt at such a moment are indescribable.... They are the greatest reward of a lifetime spent in the pursuit of knowledge for its own sake.

Raman reminds his listeners that the reception given at first to even major discoveries "is not always one of respectful admiration". Perhaps he is remembering some of his own experiences!

One of the commonest ways in which the achievement is sought to be minimized by the unthinking or the envious is by attributing it to accident or a stroke of luck akin to the winning of a lottery ticket. Such comments are of course deplorable and indeed quite meaningless. The idea that a scientific discovery can be made by accident is ruled out by the fact that the accident, if it is one, never occurs except to the right man.... Rarely indeed are any scientific discoveries made except as the result of a carefully thought-out programme of work.

At what stage during a scientific career is a person most likely to make a great discovery? According to popular opinion, it is in one's youth. Raman confirms this opinion:

If there is one fact more than any other which stands out in the history of science, it is the remarkable extent to which great discoveries and youthful genius stand associated together. Scores of instances can be quoted in support of this proposition. Indeed, if one were to attempt to write a treatise on any branch of science in which all discoveries made by youthful workers were left out, there would be very little left to write about[9]. The fact of the matter appears to be that, other things being the same, the principal requisite for success in scientific research is not the maturity of knowledge associated with age and experience, but the freshness of outlook which is the natural attribute of youth. The conservatism which develops with increasing age is thus revealed as a factor which militates against great achievement in science.

Finally, what is it that motivates scientists (at least a select few) to dedicate their entire lives to the pursuit of the unknown? As Chandrasekhar [13.4] observes in a thought-provoking lecture, this is a difficult subject if one is to avoid the common and the banal. There is so much variation amongst scientists in taste, attitudes and temperament, that discerning a common denominator might even seem a hopeless task. For Chandra himself, "a certain modesty towards understanding Nature is a pre-condition to the continued pursuit of science". Raman approaches the subject by asking:

... what is it that drives men to devote themselves to any type of idealistic activity? I think it will be readily conceded that the pursuit of science derives its motive power from what is essentially a creative urge. The painter, the sculptor, the architect and the poet, each in his own way, derives his inspiration from Nature and seeks to represent her through his chosen medium, be it paint, or marble, or stone, or just well-chosen words strung together like pearls on a necklace. The man of science is just a student of Nature and equally derives his inspiration from her[10]. He builds or paints pictures of her in his mind, through the intangible medium of his thoughts. He seeks to resolve her infinite complexities into a few simple principles or elements of action which he calls the laws of Nature. In doing this, the man of science, like the exponents of other forms of art, subjects himself to a rigorous discipline, the rules of which... he calls logic. The pictures of Nature which science paints for us have to obey these rules, in other words have to be self-consistent. Intellectual beauty is indeed the highest kind of beauty. Science, in other words, is a fusion of man's aesthetic and intellectual functions devoted to the representation of Nature. It is therefore the highest form of creative art.

> *To understand and so become aware*
> *And, thus, mine beauty from the crystalled air*
>
> JOHN KEATS

14 Looking Back

I have, in this rough work, shap'd out a man....
Timon of Athens (Act I, sc. I)

The task of writing the biography of a scientist is normally well cut out. One describes the academic background, the scientific work, the discoveries and their significance, and then throws in for good measure the various personal details plus of course the inevitable anecdotes. The job is not so straightforward in the case of Raman, for his life is intertwined with the history of science in modern India. Also, Raman was not a simple personality as his one-track mind and clear-cut likes and dislikes might tend to suggest. Beneath a simple exterior, he was a highly troubled personality. Therefore, as one looks back, one is tempted to ask to what extent Raman was the person he was on account of the situation he found himself in. It is my submission that Raman was, to a certain extent (as were, perhaps, several of his contemporaries), a creature of circumstances. Naturally, the reader would draw his or her own conclusions.

14.1 Raman the Scientist

Like all great scientists Raman was passionately committed to science and scientific research, right till the very end. Neither money nor power (and this is significant in the Indian context) could draw him away.

In terms of style, one could characterize him as a linear combination of Faraday and Helmholtz, with a dash of Lord Rayleigh added. Perhaps the Faraday component was the strongest. But unfortunately he lived about thirty or forty years too late, for, in the latter half of his life, the style he was used to had completely disappeared from vogue.

His approach and spirit were those of a rugged pioneer. Pioneers explore and do not settle down to build skyscrapers. Later in his life he was expected to – that is, if he wanted to belong. And when he did not oblige, he was ignored. But one cannot take away the fact that in his day, he was king.

His forte was waves – both of sound and of light – and when he played with them he attained heights comparable to those attained by masters like Helmholtz and Rayleigh, both of whom he admired. On occasions, he even outdid Lord Rayleigh (recall the examples cited in Chapter 5). Where optics was concerned, he never lost his touch, as is evident from the Raman–Nath theory and from the work he inspired Pancharatnam to do. If only the latter had been with Raman in the thirties, and if only Raman had the requisite peace of mind at that time! Who knows what they might not have accomplished with their combined genius and complementary talents? Optics bloomed again via Fourier optics, speckle interferometry, holography, phase conjugation and what not[1]. But they all came too late for Raman, and Pancharatnam was anyway already gone.

Scientists approach natural phenomena in diverse ways reflecting their individual tastes. Einstein, Dirac, Chandrasekhar, and the likes of them are in quest of the mathematical basis of the Grand Design. Bohr, on the other hand, is intrigued by the philosophical implications. Raman belongs to yet another category. He approaches Nature with almost child-like curiosity; he must touch, feel and play. If he can get away with it, he does not hesitate to use Nature itself as a laboratory. Remember the cruise on the Mediterranean and the climb up the Dodabetta? G. I. Taylor, another great scientist of that period, liked to describe himself as an amateur scientist who worked mainly for pleasure. That description fits Raman aptly. In those days, physics had not yet become highly formalized and still retained what Einstein once nostalgically referred to as "musicality in the sphere of thought". Alas, those days are gone, and so are natural philosophers like Rayleigh and Raman.

An irrepressible curiosity about the molecular basis of optics led Raman to a mighty discovery, but thereafter he wandered into solid state physics, a somewhat unfamiliar terrain. The advent of quantum mechanics completely changed the character of physics (including solid state physics), but Raman and his colleagues (with the possible exception of K. S. Krishnan) missed that sharp turn and their thinking continued to remain classical. By and large, geography was the culprit. While a strong new wind was blowing across Europe, there was hardly a flutter in India – it was so far away. Also, there were no native couriers like America's Oppenheimer, van Vleck and Slater. True, there was Riazuddin Siddiqui[2], who had studied with Heisenberg. But Siddiqui was a mathematical type and did not actively spread the message amongst physicists, who in fact needed it most. To make matters worse, Raman developed a mental block and refused to absorb the subtleties of the new developments and their potentialities for application from Born during their brief association. Quite possibly, Raman's attitude coloured those of his colleagues as well.

Raman's obstinacy is puzzling considering the efforts he had made earlier to bring Sommerfeld to Calcutta to lecture on quantum mechanics. Did he begin to feel out of depth with modern developments? Or were his personal problems beginning to wear him down? (Remember, the Indian Institute of Science controversy was at its peak.)

Missing the bus of quantum mechanics had some painful consequences. Excellent experiments were being done in Raman's laboratory, and as Born put it, "the skill of Indian physicists produced a host of interesting results". But there was no matching theoretical effort – this in spite of Born's early presence. If only the Institute had retained Born and blessed Raman's effort to build up a theory group, if only Raman had succeeded in his efforts to get Schrödinger and a few others, if only... What a tragic miss for Indian physics! To complete the story of quantum mechanics in India, it was only towards the late forties that some repair of the damage was finally effected. Bhabha, in particular, deserves credit for inviting Dirac to spend a few months at TIFR. A new generation now had the chance to make up for past omissions, and learn the subject from a great master.

14.2 Raman the Man

The personality of Raman is a jumble of patterns, some understandable and others puzzling. Those in the former category essentially flow from his cultural background. The Indian household, particularly that of the South Indian Brahmin, was highly patriarchical in those days. Raman extended the role of the patriarch to the laboratory as well, something which did not seem unnatural either to him or to his immediate colleagues.

It was alleged both at the IACS in Calcutta and at the IISc in Bangalore that Raman was a poor administrator. This is a strange charge, for Raman had earned such high praise for his administrative abilities from the British during his civil service days. How could such skill and talent desert him overnight? His problem was not lack of administrative skill but that he ran the laboratory like a strong boss. There is nothing unusual in that, then or now, in India or indeed even in parts of Europe, but somehow Raman had to pay a price, and a heavy one at that.

He was accused of having embezzled funds at IISc. We have already seen how hollow that charge was. All he did was to reallocate funds amongst the various departments. Years later, Bhabha handled a budget *several hundred times* that which Raman controlled at Bangalore, and he certainly allocated funds according to the priorities he set. Nobody asked any questions, there were no review committees, and Bhabha was not obliged to resign.

Raman had the supreme ill luck of facing rebellion twice in quick succession. This is most unusual, and I cannot think of a parallel either in India or anywhere else. Shortage of funds was certainly a contributory factor. But resource crunch cannot explain it all, for financial hardship has always been part of the Indian scene. Bossism too is not a convincing explanation, for that too is not unusual. There have been tyrants who have got away, and Raman certainly was no tyrant. Did he, with his ego, wound his adversaries and incite them to blind fury?

Raman's ego is well known, and even his best admirers concede that. But ego is not such a rare commodity, especially in those who have some achievement to speak of. Most well-known scientists have a strong streak of arrogance, although some manage to conceal it beneath a mask of urbanity. Rare indeed are individuals like Peter Ewald or Dorothy Hodgkin, who are totally self-effacing. Bhabha was haughty, Saha acerbic, and Raman caustic. Why then did Raman alone pay so heavily for something not so unusual? I have hazarded an explanation in Chapter 8. It may or may not be convincing.

Sustaining achievement is always a problem, especially for those who have reached the pinnacle. A rare few manage to produce more successes and continue to stay at the top; most choose the option of drifting to less-demanding activities, often organizational in nature. There remain some who continue to strive but cannot maintain their original stature. Raman belonged to the last category.

Remaining in science was a natural choice for Raman but he had to face several handicaps. The first was that of playing a new game where, thanks to quantum mechanics, the rules were different. Second, he lacked the requisite peace of mind. For decades he was the unquestioned leader and yet, overnight, he had been unceremoniously rejected. Only a yogi could have come out unscathed from such a shattering experience. No wonder Raman lost his old magic touch.

Raman, after the Calcutta and Bangalore episodes, was no longer the same. Something had changed, for ever. No doubt the warmth was still there and also a bit of the old sparkle; but Raman also carried deep scars which occasionally produced inexplicable and abnormal behaviour. Such aberrations were not long-lasting but their consequences were.

The older Raman was never intolerant of viewpoints different from his own. He recognized when he had erred and quietly accepted it (as in the case of photon spin and negative scattering; see Chapter 6). He knew the role of facts and of logic in modern science. He himself was a master of clear and logical thinking. And yet his logic seems to have failed him at crucial times in the later years; for how else is one to explain his obstinacy over some of his pet ideas? He began to shut himself off from concepts other than those he believed in. One cannot attribute it to senility. Disappointed, hurt, angry, bitter – yes; but senile? No, never. Till the very end, Raman was absolutely clear-headed. No question about that. And yet...

Nowhere is this occasional aberration more painfully evident than in some of Raman's references to Krishnan in later years. Like all scientists, Raman was proud of his achievements. He received much credit for it, all of it fully deserved. Krishnan never contested this credit. Not only was Krishnan a perfect gentleman but more than anything else, in the classic Indian tradition, he revered Raman as his *guru*, even as many others did. But there were mischief-mongers who revelled in claiming that Raman was an ignoramus who became famous by hanging on to the coat-tails of Krishnan, and that Krishnan had actually discovered the effect bearing Raman's name, while he (i.e., Raman) did not understand what that discovery was all about. An altogether contemptible and ridiculous fabrication, but the canard was projected with some force, particularly in the later years when Raman began to suffer a slow eclipse. This was indeed the last straw. First he was denied his position and now he was even being denied credit for his achievements. It must have truly been a nightmarish experience. I think Raman must have lost his self-control as a result, for otherwise how could he have said those unpleasant things about Krishnan, who would not hurt even a fly?

These unfortunate lapses, though few in number, tarnished Raman's image. But take those away and you see the Raman of old. Rather egoistic as usual, a bit dominating, but warm-hearted and *without a trace of malice in his heart*. Even his worst enemies never accused him of being a schemer or of stabbing one in the back. He had no need to. He feared no man, and he spoke out what he felt. In fact there was a child-like simplicity about him which endeared him to his friends and colleagues, even if at times he gently bullied them. And he always had deep compassion. Ramaseshan [14.1] recalls, "Some of us still remember how he wept like a child during one of the Academy lectures when pictures were shown of children of our land suffering from nutritional ailments."

One cannot help feeling that events took a heavy toll of Raman's personality, producing distortions which he never had in his earlier years.

14.3 A Comparison

In Chapter 1, I introduced Bhabha, Raman and Saha as distinguished physicists who play an important role in this book. This is a good juncture to compare them before they bow out.

Let us start with Raman. To him basic science and research were everything. He was gregarious and believed in schools of research. A steady stream of distinguished students came from his school, almost till the very end. He understood moreover that the pursuit of science requires support that transcends equipment, buildings and money, and for this reason founded an Academy which offered a forum for publication as well as for discussion.

Unlike Raman, Saha accorded greater priority to teaching than to research. He did not have a grip over experimentation and perhaps this prevented him from providing the necessary leadership when the cyclotron project he initiated ran into technical problems. But he saw the importance of nuclear physics quite early and tried his best to see that it took root in India. What he valiantly sought to achieve was left for Bhabha to accomplish.

After Independence, Saha managed to realize his dream of an institute of nuclear physics but the institute was beset with problems. Possibly they were compounded by Saha sharing his time between academic and political activities. An introvert, he lacked the ebullient dynamism of Raman and the go-getting spirit of Bhabha. He also seems to have been circumscribed by the limitations and tribulations of his early career, and the prejudices shaped by them. But he did excellent science which has left an indelible mark, particularly on astrophysics. One must also not forget that he gave theoretical physics a start in India. Saha's tradition was ably carried on by his student D. S. Kothari in Delhi (in association with R. C. Mazumdar and others). For years Delhi University churned out with unfailing regularity bright students trained well in the basics of mathematical physics.

Bhabha was in a different class altogether, and his impact too was of a different nature. He was a great institution builder. To start with he created the TIFR where, despite his own training as a theoretical physicist, he encouraged a variety of disciplines ranging from radio astronomy to molecular biology. And he also helped to found an excellent school of mathematics even though (and this was exceptional) he had to concede full autonomy in its management.

Bhabha always did things in style, style which Raman would never dream of nor Saha approve. But it was not all show. Like the others, he was deeply committed to excellence. He managed to establish rapport and understanding with Nehru which neither Raman nor Saha could manage. Actually, Saha, with his socialistic leanings and pre-war connections with Nehru, should have been where Bhabha was. But it did not happen that way. Even if Saha had been where Bhabha was, it is doubtful if he would have achieved as much. His vision was rather constrained, undoubtedly a consequence of his austere youth. Bhabha was just the opposite. He was an aristocrat, and never felt cramped by the proverbial Indian poverty. He dreamed big and also managed to translate some of his dreams into reality. More than anything else, he taught India to throw off the shackles of limited thinking, and take both big science and modern technology in her stride.

They were altogether different personalities. Their styles differed, their opinions differed, and so did their contributions. Raman galvanized research, Saha (especially through his students) gave theoretical physics a much-needed foothold, and Bhabha launched a leap-frog programme. They were of course also distinguished physicists in their own right, and brought credit to the country by their professional achievements. They came into contact with each other but, unfortunately, these associations were not entirely fruitful; only the Raman-Bhabha combination worked for a while[3]. The base that Indian physics enjoys today is in a large measure due to these three. If only they had got along! Ah, one more of those wistful wishes!!

14.4 Hazards in the Indian Scene

While evaluating the achievements of Raman, one must keep in mind the difficulties researchers faced in those days. Indeed, many of those handicaps still persist, which is why Indian science as an entity is not yet at the forefront, although outstanding individuals continue to be produced.

Perhaps the most important drawback is the educational system. In Raman's days, it was good where arts and law were concerned but definitely inadequate for engineering as well as for science. Except in civil engineering, design was hardly taught since very little was manufactured in India. Engineering graduates were needed only for plant and equipment maintenance. Likewise, in physics, the emphasis was on creating a broad awareness of the subject rather than in inculcating a problem-solving ability. How else can one explain weird questions like: write an essay on wave mechanics, or short notes on the Heisenberg uncertainty principle, Lorentz transformations, etc.[4]? Raman, Bose, Saha, Krishnan and their tribe are therefore worthy of our highest esteem and admiration, for they not only overcame this first obstacle but also raised themselves to international level by their own effort, *staying in the country*.

In more recent times the educational system has no doubt changed somewhat. But it has also come under enormous pressure, both social and political. The net outcome is that input manpower for scientific research continues to be under-prepared and ill-trained.

Next in the list of handicaps is the lack of opportunities. In Raman's days, opportunities for basic research were practically non-existent in India. But Raman *created* opportunities for himself, even when he was posted in cities like Rangoon and Nagpur where there was nothing like the IACS. And for the first time in the history of the country he built up a thriving school of research, making Mahendra Lal Sircar's dream a reality.

Funds. As we have noted earlier, money for supporting basic research was always a problem. The British spent millions in maintaining their army in India but for science there was hardly any money. One does not accuse the British of deliberately suppressing Indian science, for such a charge would be untrue. But it is true that British administration in India was indifferent to the fate of Indian research. It was generous enough to tolerate and even applaud it, so long as it (i.e., Indian scientific research) could raise funds from elsewhere. But Crown funds? Well, there were other priorities, like protecting the Empire, for example. Maybe an occasional contribution like to a charitable cause, but not much more.

In spite of an acute paucity of funds, Raman's enthusiasm remained undiminished. In fact he managed to whip up almost a mania for research. Remember Venkateswaran with his job in the Test House blowing every moment of his spare time at the Bow Bazar laboratories? And Ramanathan who dashed to Calcutta from Rangoon on every conceivable occasion, paying his own way? And remember Bhagavantam, a young lad of eighteen, who wanted to pursue research instead of eternal security in a Government job? What about Raman himself supporting impoverished research scholars with money from his pocket, besides looking after their meals at times? He must have been simply crazy about research!

One cannot however minimize the problem of scarcity of funds. The famous split between Raman and Saha was, reduced to its essentials, just a Darwinian struggle for survival. Bangalore was no different. With all the accolades and honours, Raman could not manage a paltry few thousand rupees to keep his research going. Remember Thosar's story about the fused bulb? What a stunning contrast to the huge amount Stalin spent in buying up Kapitza's Cavendish equipment to get him started again in Moscow!

One wonders if Raman might have had a more even career, like Bragg and Rutherford, had he emigrated while young. Perhaps yes, perhaps not. He certainly had the drive and dynamism one saw more of in European circles than on the Indian scene. On the other hand, he might not have adjusted enough socially to become a permanent member of Western society, given his strong South Indian cultural roots. But all that is speculative. The more pertinent fact is that unlike Bragg or Rutherford he did not leave the land of his birth, even in his darkest hour.

14.5 An Introspection

If Raman's life epitomizes anything, it is the difficulty of pursuing pure science in a country that is not only geographically distant from countries doing mainstream science, but is poor and backward as well. Bose lost himself first in teaching and then in culture, while Saha drifted to social uplift and politics. Bhabha was more successful, but in institution building rather than in pure science itself. Raman was too fond of Nature and of science to cut loose in this manner. And instead of support, he received worthless praise from an ignorant public, brickbats from jealous colleagues, and indifference from a disdainful bureaucracy[5].

Raman's tragedy is by no means unique, and there must be many more one has not heard of who have succumbed to an unfavourable environment. It is true that nowadays money for scientific research is more easy to come by than in Raman's days. But money alone does not produce good science.

Our public often wonders what is wrong with our science, and why there is no parade of heroes like Ramanujan or Raman. It has not yet woken up to the fact that unless there are a proper environment, organized support, and nourishment, occurrence of excellence on the Indian scene would continue to be infrequent. The example of sports may be used to convey the point, since everyone is familiar with sports.

Time and again we send large teams to international competitions, amidst fanfare and high expectations. Pretty soon reality stares us hard in the face. There are few medals and our dreams crash. Everyone is furious – the press and the public alike. The odd hero or heroine is singled out, duly paraded, and lavished with an embarrassing shower of praises, but soon everything is forgotten. And the story repeats all over again! Meanwhile, the cry of professional sportsmen like Sunil Gavaskar about where things are wrong goes unheeded.

Public attitude to excellence must change. Excellence (at least in science) is not a spectacle to be paraded and treated just with civic receptions, Republic Day honours, and the like. Rather, it must be nursed, given necessary support, and finally left alone. Also, there must be a certain amount of boldness in keeping out the charlatans and the mediocre. We tend to be too legalistic and confuse discrimination in merit with social injustice. Excellence has its own aristocracy, and considerations of parity which are valid elsewhere are not meaningful in this domain. This is accepted even in the USSR. One yearns for the day when we would have learnt all the various lessons that are to be learnt, and when our country would have evolved to the point of retaining people of the calibre of Chandrasekhar and Harish Chandra.

It is a moment for introspection. More than the public, our scientific community has many issues to ponder over. Raman, Saha, Krishnan – they all faced odds, but they remained here, to struggle and to build up science in this country. Today one sees young talent itching to flee to greener pastures, even though conditions *have* improved in many respects. There is hardly any commitment to our national development. On the other hand, one sees the strange spectacle of many of these *emigrés* becoming deeply concerned about the land of their adoption! Even among those who have chosen to stay, many have hardly any roots here. As someone remarked, it is as if their bodies are here but their souls are elsewhere! Should we live in dreamland and let our institutions wither?

To many dedicated scientists in this country, things often look desperate and hopeless, but where there is no hope, there is no life. Such a defeatist attitude – unfortunately widely prevalent amongst young people – is not warranted. In fact it is dangerous. Raman recognized this and administered a sound piece of advice while once addressing young graduates. He observed:

> I would like to tell the young men and women before me not to lose hope and courage. Success can only come to you by courageous devotion to the task lying in front of you and *there is nothing worth in this world that can come without the sweat of*

our brow [italics mine]. I can assert without fear of contradiction that the quality of the Indian mind is equal to the quality of any Teutonic, Nordic or Anglo-Saxon mind. What we lack is perhaps courage, what we lack is perhaps driving force which takes one anywhere. We have, I think, developed an inferiority complex. I think what is needed in India today is the destruction of that defeatist spirit. We need a spirit of victory, a spirit that will carry us to our rightful place under the sun, a spirit which will recognize that we, as inheritors of a proud civilization, are entitled to a rightful place on this planet. If that indomitable spirit were to arise, nothing can hold us from achieving our rightful destiny.

One might disagree with many of the things that Raman said, but can anyone disagree with this?

Notes

Chapter 1

1. Some of the material that follows has been derived from ref. [1.1].
2. A short biographical sketch of Sir J. C. Bose may be found in Chapter 3.
3. British India was divided into Provinces (which were under direct British control) and Princely States each ruled by a Nawab or a Maharajah. However, all Princes had to toe the British line, and a watchful eye was kept by the British Resident in the state.
4. A biographical sketch of Sir P. C. Ray may be found in Chapter 3.
5. The Crown representative in India was the Viceroy who reported to the Secretary of State for India in the British Cabinet. The Viceroy was assisted by a Council whose members were mostly Europeans. This Council was the equivalent of a ministerial Cabinet.
6. The well-known conductor Zubin Mehta is a Parsi.
7. Of Vivekananda, Nehru writes: "He was a powerful orator in Bengali and English and a graceful writer of Bengali prose and poetry. He was a fine figure of a man, imposing, full of poise and dignity, and sure of himself and his mission, and at the same time full of dynamic and fiery energy and a passion to push India forward.... Wherever he went, he created a minor sensation not only by his presence but by what he said and how he said it.... In America he was called the 'cyclonic Hindu'."
8. Bankim Chandra Chatterjee was a great novelist whose inspiring works have been translated into many Indian languages. His inspiring song *Vande Mataram* was the anthem during the freedom struggle.
9. The great poet also recalls the mood of those times. First there was the inspiration provided by the liberal ideas and then the disillusionment of how they were disowned! "As a boy in England, I had the opportunity of listening to the speeches of John Bright, both in and outside Parliament. The large-hearted radical liberalism of those speeches, overflowing all narrow national bounds made a deep impression on my mind.... Then came the parting of ways, accompanied by a painful feeling of disillusion, when I began increasingly to discover how easily those who accepted the highest truths of civilization disowned them with impunity whenever questions of national self-interest were involved."
10. Aurobindo's career is remarkable. His father was a confirmed Anglophile, and though short of resources, insisted on sending his children to England for their education. Young Aurobindo excelled in the classics and won many prizes for proficiency in Greek and Latin verse while in King's College, Cambridge. He passed with distinction the Indian Civil Service (ICS) examination but failed in the riding test and was therefore not selected for appointment. On return to India he worked as an English lecturer for a while, then turned a revolutionary and finally drifted to philosophy. He founded the internationally famous Aurobindo Ashram in (the then French territory of) Pondicherry. The habitat Auroville in Pondicherry is named after him.

Chapter 2

1. The names given here are Anglicized versions prevalent during the British days, while those in parentheses are the names given after Independence (being the ones by which these places were known much earlier, especially to the local population).
2. Marquis of Wellesley is the brother of Arthur Wellesley who later became famous as the Duke of Wellington. Marquis of Wellesley conquered more territory in India than Napoleon did in Europe.
3. The two principal forms of classical music in vogue are the Carnatic music (practised mainly in the South), and the Hindustani music (popular mainly in the North).
4. Indeed, a Carnatic music vocal recital concert today without a violin accompaniment is unthinkable.
5. In Andhra Pradesh, initials numbering three and more are not uncommon. I do not claim to understand the system prevalent there!
6. Names like Subramanian are often written in English in several ways like Subramaniam, Subramaniyan, Subrahmanian, etc.! Similarly, there are minor variants of Chandrasekaran like Chandrasekar, Chandrasekhar and so on. There is however no ambiguity in the original Sanskrit names.
7. Besides Raman and Chandrasekhar, several others directly descended from the old patriarch Mr Chandrasekara Iyer have made a mark in science. I cannot think of any comparable example of a family of talented scientists other than the celebrated Curie family.
8. Incidentally, people from the South of India now settled abroad have solved the problem of family name by adopting their own given names as family names! Thus, Mr Sampath's son would be known as "something" Sampath, "something" being a given name now playing the role of the first name.
9. Along with the Elphinstone College in Bombay and the Presidency College in Calcutta, the Presidency College, Madras, was one of the three important Government colleges established soon after the introduction of Western education into India.
10. Porter of the University College, London, figures again in Chapters 3, 4 and 5. In the early twenties, Raman provided an ingenious explanation for some intriguing observations made by Porter which defied even the great Lord Rayleigh. Later Porter played host to Raman when the latter made his first visit to England in 1921, and offered facilities for work besides arranging for experiments to be conducted at the Whispering Gallery of St Paul's Cathedral in London.
11. Edwin Arnold spent some time as a lecturer in the Deccan College, Poona (now Pune). Attracted to Hindu philosophy, he once observed: "I would wish to see cultivated those fields of Eastern philosophical thought which I have feebly and hastily traversed, as affording a sweet sovereign medicine against the fever of a too busy national life." Arnold commanded such respect that, later, the great patriot Lokamanya Tilak arranged for Arnold's portrait to be mounted in the College, notwithstanding the fact that Arnold was an Englishman.
12. S. Chandrasekhar told me that even in 1929 a scientific career in India was far from attractive, and that he was under strong pressure to enter Government service. Fortunately, he was able to overcome that pressure and have his own way!
13. The *veena*, and Raman's studies on the instrument, are described in Chapter 4.
14. Bhagavantam is one of Raman's distinguished students. Studying under Raman in Calcutta, he actively worked on the Raman effect for many years before branching off into other areas. He was also a pioneer in the use of group theory in problems of molecular spectroscopy. We shall be quoting him on several occasions.
15. The rupee in those days was a very stable currency, directly convertible into the English Pound at the (constant) exchange rate of twelve rupees to the pound. The rupee was divided into sixteen annas, each anna into four pice, and each pice into three pies! The present decimal system was introduced soon after Independence. The purchasing power of the rupee was incredibly high by today's standard. Hearing old-timers talk makes one feel that one is listening to fiction!

Chapter 3

1. Calcutta was the capital of India till 1922, when New Delhi was made the capital.
2. The material for this section has been drawn largely from *A Century*, published by the IACS in 1975 on the occasion of its centenary.
3. In those days automobiles were scarce, and taxi here refers to a hansom cab.
4. Burma was a part of British India till 1935.
5. Concerning publishing, Ramdas has this to say:
    The writer recalls that hardly a week passed in the Association at Calcutta without a detailed paper, and often shorter notes, being despatched to foreign journals or the Calcutta University Press which published the *Proceedings of the Indian Association for the Cultivation of Science*. He [Raman] was so very critical in composing and editing scientific papers, whether his own or of his pupils, that often it may get too late for posting the paper to the publishers in the ordinary manner. But he would hail a taxi and rush to the General Post Office, pay late fee and get a paper despatched in the nick of time. After such an adventure, he very often would share his joy in sending away an important paper for publication by treating the anxiously waiting group of his pupils to a solid feast of 'sandesh' and 'rasgolla' [Bengali sweets] that he would send for from the famous 'Bhim Nag' of College Street. [ref. 3.6]
6. This paper appeared in the *Proceedings of the Royal Society* (**95**, 533 (1919)) and carried a foreword as well as an appendix by Raman.
7. A lakh is one hundred thousand while a crore is ten million. In India these terms are more common than million and billion.
8. Gilbert Walker, FRS, was the Director-General of Observatories. He communicated many of Raman's early papers to the Royal Society. To Walker also belongs the distinction of having helped the mathematical genius Srinivasa Ramanujan at a crucial stage. The story goes that once when Walker visited the Madras harbour to inspect the meteorological instruments kept there, the Chairman of the Port Trust asked Walker to take a look at the mathematical scribblings of an apparently eccentric clerk. Walker quickly recognized that the dreamy clerk had most unusual talent and helped him to get a scholarship in the Madras University. Later Ramanujan went to England; the rest of the story is well known.
9. Later S. K. Banerji joined the India Meteorological Department and rose to become the first Indian to occupy the post of the Director-General of Observatories while S. K. Mitra achieved fame in radio physics. Observe that Raman misspells the names of Saha, Bose and Banerji. See Sec. 3.5 for the biographical sketches of Saha and Bose.
10. Mahalanobis, FRS, founded the Indian Statistical Institute in Calcutta and later headed the Planning Commission.
11. To appreciate the intensity of Saha's commitment to India's development, see the *Collected Works of Meghnad Saha*, edited by S. Chatterjee (Orient Longman, Calcutta, 1987). Saha's scientific approach and professional touch are particularly evident in his studies on our major rivers and the havoc they periodically cause via floods. Some of these papers were written almost fifty years ago but they continue to carry a message, and are worth reading even *now*.
12. J. C. Ghosh and other friends from Dacca wrote to Bose asking him to apply for the post of Professor of Physics, adding that since Bose did not have a doctorate, would he please get a testimonial from Einstein? At first Bose hesitated but later he talked to Einstein, who was shocked. "Are not the papers you have published sufficient?" he asked, incredulous. Poor Einstein, he did not know about this subcontinent! Incidentally, J. C. Ghosh succeeded Raman as the Director of the Indian Institute of Science (see Chapter 8).
13. This late recognition is illustrative of the eclipse Bose suffered after his early rise to fame. I have heard that when Dirac came to India in the mid-fifties, he met Bose in Calcutta and was surprised to learn that he was not a Fellow of the Royal Society. It is believed that Dirac was instrumental in getting Bose elected FRS.
14. Chandrasekhar told me an anecdote related to this incident. During a visit to Madras after this European tour, someone (at C. S. Iyer's residence where Raman was staying) asked Raman whether

he did not feel embarrassed wearing a turban in England. Raman replied: "Young man, when I was in London I attended a lecture by Lord Rutherford. I was sitting in the last row, when Rutherford looked up, and seeing me said: 'Prof. Raman, why are you sitting there in the back all alone? Come up here in front.' After the lecture I asked Rutherford how he recognized me since we had not met before. He said: 'Well, I know you of course by your papers. And when I saw a person with a Madrasi turban in the audience I knew it must be you.' So, young man, what is wrong with a Madrasi turban?"

15. The reference here is to the famous incident alleged to have occurred during the clash in 1756 between Siraj-ud-daulah, the Nawab of Bengal, and the British forces. Siraj-ud-daulah stormed Fort William and the garrison there under Holwell surrendered. British historians claim that the Nawab locked up a large number of British soldiers into one small room as a result of which many died. However, Indian historians describe this as a trumped-up charge. The *Oxford History of India* records, "The emphasis upon the incident grew so great that the Black Hole of Calcutta became, along with Plassey and the Mutiny, one of the things every schoolboy knew about India."

16. Concerning these events, Chandrasekhar has the following to say:

    I have an equally vivid recollection of a day in early March in 1928, when Professor Raman visited our home in Madras on his way to Bangalore where on the 16th of March he was to give the address announcing his discovery of what was soon called the Raman Effect. I remember well his showing slides of the first Raman spectra ever taken and of the state of euphoria he was in. On that occasion someone drew attention to the discovery of the Compton Effect a few years earlier, and Raman responded with 'Ah, but my effect will play a great role for chemistry and molecular structure!' That statement was indeed prophetic. Later during the summer of 1928, I spent two months at the Indian Association for the Cultivation of Science at Raman's laboratory where at that time there were many young men who together with Raman were pursuing the new discovery. Among them were several who were later to become leaders of Indian Science.... You can imagine what a marvellous experience it must have been for a young man [Chandrasekhar was only seventeen then] to have witnessed at such close quarters a group of enthusiastic scientists caught in the wake of a great discovery.
    [Quoted by S. Ramaseshan in *Current Science* **57**, 163 (1988)]

17. A thousand rupees in those days was really a substantial sum of money. One wonders whether the Palit Professor can today make such an instant offer without going through tortuous "proper channels"!

18. The notes of these lectures were prepared by Krishnan, and published by Calcutta University. Krishnan's biographical memoir, published by the Royal Society, says, "He developed five of the lectures in an independent way and was commended by the visiting Professor for his originality and scholarship in supplying elegant mathematical proofs."

19. These events have been reconstructed based on conversations with persons who had first-hand knowledge. Saha's publication record for that period shows that he was quite active although perhaps not as prolific in output as Raman was. One therefore wonders about Raman's contention. Was it their strong personal differences peeping through?

20. Dr L. L. Fermor to whom this comment is due was a distinguished geologist, and a member of the Geological Survey of India. He appears again in Chapter 11.

Chapter 4

1. *Ekalavya* is a character in the Hindu epic *Mahabharata* who, secretly watching from the side while the great *Dronacharya* was instructing his pupils, picked up great skills in archery, outdoing even the best student *Arjuna*.
2. With reference to the papers published by Raman in the *Journal of the Indian Mathematical Club*, it is interesting to recall that the mathematician Dr Ganesh Prasad in his Presidential Address to the Physico-Mathematics Section of the IACS said: "Professor Raman is well known to you as an

experimentalist of worldwide reputation. But some of you will feel surprised to learn that by his recent mathematical researches, he has firmly established his claim to be considered a sound mathematician. The two papers which he read before the Calcutta Mathematical Society during the current year [1913] are very valuable, and I trust he will continue his mathematical researches."
3. Raman employed another method also for ascertaining when the motor "bites" but that was not capable of demonstration to a large audience. In common with many others of that period, Raman delighted in devising and describing methods suitable for exhibition on the stage.
4. Some reservations have been expressed concerning Raman's explanation of the wolf note. See reference [4.18].
5. Following Raman many others have employed violin players of their own, but Raman's is the only one in which the violin moves. For further details, see reference [4.18].
6. For some unknown reason, Raman refers to Stradivarius as Straduarius.
7. For example, F. A. Saunders in ref. [4.18] compares the Raman curves for six violins which include a cheap one!
8. Ghosh later published a book entitled *Sound and Vibration* (Indian Press Ltd, Allahabad, 1951). This is probably the first textbook on the subject to be published in India.
9. Ghosh also published (after Raman left Calcutta) a paper entitled "The elastic impact of a pianoforte hammer" (*J. Acoust. Soc. Am.* **7**, 254 (1936)). Shortly thereafter W. E. Koch, then at the Indian Institute of Science, published a paper entitled "The vibrating string considered as an elastic transmission line" (*J. Acoust. Soc. Am.* **8**, 227 (1937)). Raman too was at the Institute at that time, and Koch acknowledges Raman for suggesting the problem and for guidance. This paper is reproduced in a benchmark collection. Later Koch became the Director of the Acoustics Division at the Bell Laboratories.
10. The computer graphics were kindly prepared by Mr Sunder Kingsley.
11. Jagjit Singh narrates the following story. "[Raman] posed the same problem [of the *mridangam*] as a brain-teaser in one of the question papers he set for a post-graduate examination of Allahabad University. Of course, he never expected anyone to tackle such an off-beat problem in an examination hall. But a talented one among the candidates, Harish Chandra, took up the challenge. Although for want of time this was the only question Harish tackled, Raman was so fascinated with the answer that he personally congratulated Harish for his marvellous performance." Later Harish Chandra briefly worked with Bhabha, before settling down at Princeton where he emerged as a "towering figure in mathematics".
12. The Catgut Acoustical Society started off as an informal group of violin buffs, the founders including F. A. Saunders, C. M. Hutchins, J. C. Schelling and R. E. Fryxell. Raman had much correspondence with Saunders in the thirties, and was made an honorary member of the Society.
13. I am much indebted to Mr M. C. Valsakumar for an extensive discussion of the works of Raman and of Lord Rayleigh on vibrations, which greatly helped to clarify my thoughts.
14. The preparation of this section has benefited much from a reading of *Optics* by E. Hecht and A. Zajac (Addison–Wesley, Reading, Mass., USA, 1974).
15. Strictly speaking, Fourier optics may be said to have commenced with experiments by Abbe (1893) and by Porter (1906) who intentionally manipulated the spectrum of an image. The aim of these experiments was to verify Abbe's theory of the microscope.
16. It is interesting to note that Banerji's work is cited in Lord Rayleigh's collected works as a footnote.
17. Concerning Raman's interest in coronae, Ramaseshan (*Curr. Sci.* **57**, 163) has this to say:
    In 1910, when he was an Assistant Accountant-General in Nagpur, his clerks noticed him at lunch-time studying the solar coronae reflected in a pool of water in front of his office. Later, in Calcutta, he was seen often making observations on the lunar coronae when taking his evening stroll in the *maidan*.... He was often up very early in the morning at Bangalore observing the coronae formed around the planet Venus.... I myself have seen him measuring the polarization of the coronae in 1967 in Bangalore when he was 79!
    In passing, attention is also called to the beautiful book by Robert Greenler, *Rainbows, Halos and Glories* (Cambridge University Press, Cambridge, 1980).
18. Not many may know that there is a charming volume in the *Romance of Science* series by C. V. Boys,

entitled *Soap Bubbles: Their Colours and the Forces which Mould Them* (Macmillan, New York, 1920, for the Society for the Promotion of Christian Knowledge). Raman had a copy of this book and seems to have used it heavily.

19. The only other person I can think of who often preferred words to mathematics is John C. Slater. Slater wrote many books which, it is said, were largely dictated by him!

Chapter 5

1. A purist might regard such a classification as naive and an oversimplification! Maybe so, but we need not get involved in a semantic debate here. For a historical review on Rayleigh scattering as well as a discussion on terminologies, see Young [5.1].
2. Later, this interest in the Doppler effect was to assume importance in the context of the Raman–Nath theory. See Chapter 9.
3. This quantity is formally defined in Chapter 6. Essentially it is a measure of the intensity of the scattered light.
4. For further details, see Chapter 6.
5. For an interesting historical account of some of the determinations of the Avogadro number, see Pais [5.13].
6. Quoting B. N. Sreenivasaiah, Ramdas [5.19] tells the following story: "On a December evening in 1927 when I was at Calcutta after an examination and interview at Delhi, I visited the Association at Calcutta to pay my respects to Prof. Raman. His elder brother Sri C. Subrahmanya Iyer was also with Prof. Raman when I entered his office. Soon after I went in, K. S. Krishnan rushed in and excitedly informed the Professor of the announcement in the evening papers that Prof. A. H. Compton had been awarded the Nobel Prize.... On hearing this news, Prof. Raman beamed with delight and burst out in his characteristic fashion: 'Excellent news ... very nice indeed. But look here Krishnan. If this is true of X-rays, it must be true of light too. I have always thought so. There must be an optical analogue to the Compton effect. We must pursue it and we are on the right line.'"
7. Ramaseshan told the author that years later he met Ashu Babu during one of his visits to Calcutta. Ashu Babu clearly recalled those momentous days and Raman giving instructions in his booming voice on how exactly to look for the Kramers–Heisenberg effect.
8. This is essentially a pocket spectroscope. Till the end Raman carried one just as a doctor carries a stethoscope. It is an inexpensive instrument and Raman was always proud of the fact that he made a major discovery spending very little money on equipment.
9. Ramdas [5.19] states that the term Raman effect was first coined by him in a paper he communicated to *Nature* on May 29 and which appeared in print on July 14. However, there is an even earlier paper due to Cabannes published on June 18 where the discovery is termed the Raman effect (*l'effect Raman*).
10. Exceptions to the Stokes law are known [5.24].
11. Though third in terms of preparation, this paper appeared in print at the same time as the first note to *Nature* despatched on February 16, 1928.
12. It is curious that both the paper communicated on May 7 to the *Indian Journal of Physics* and the one sent to the *Proceedings of the Royal Society* are marked Part I. Part II never seems to have been published!
13. In 1925 Raman and Ramdas studied this problem experimentally.
14. See Chapter 10.
15. We shall have further occasion to consider these branches in Chapter 10.
16. Born and Huang [5.30] remark, and correctly, that from a strictly quantum-mechanical point of view, there is no difference between Brillouin scattering and Raman scattering in crystals.
17. This is our reference [5.26].
18. This is our reference [5.25]. Fabelinskii does not reproduce this letter, but he quotes from it.
19. This is our reference [5.33], minus the figure!

Notes

20. Several units are used in spectroscopy, depending on the circumstances and convenience. As mentioned in Chapter 2, wavelength of light in the visible region is usually expressed in Angstrom units ($Å = 10^{-8}$ cm). More common for wavelength in the infra-red region is micron, symbol μ, which is equal to 10^{-4} cm (thus $1\mu = 10,000$ Å). The unit of frequency is hertz, symbol Hz, for cycles per second. However, in spectroscopy one more often expresses frequency as inverse centimetres (cm^{-1}), or wavenumber. Conversion to hertz is effected using the rule Freq. in Hz = $c \times$ Freq. in cm^{-1}, where $c = 3 \times 10^{10}$ cm/sec is the velocity of light.

    Consider now the experiment of Mandel'shtam and Landsberg on quartz in which the incident wavelength λ was 2536 Å. The wavelength change observed by them was 30 Å, giving the value 2566 Å for the scattered wavelength λ'. The corresponding frequency is given by $v' = c/\lambda'$. The frequency change $\Delta v = c(\frac{1}{\lambda} - \frac{1}{\lambda'})$ is thus ~ 461 cm^{-1}. This is close to an infra-red frequency of quartz, and is quoted in units of wavelength ($=c/\Delta v$) by Mandel'shtam and Landsberg.

21. In fact even in his Nobel lecture Fermi repeats this thesis but is careful enough to add the following footnote: "The discovery by Hahn and Strassmann of barium among the disintegration products of bombarded uranium, as a consequence of a process in which uranium splits into two approximately equal parts, makes it necessary to re-examine all the problems of transuranic elements, as many of them might be found to be products of a splitting of uranium."

22. In 1923 Ross of Stanford University looked for Compton-like shift by scattering radiation of wavelength 5461 Å off paraffin. The expected shift was 0.708 Å but none was observed (*Phys. Rev.* 9, 246 (1923)).

23. In the early days of slow neutron inelastic scattering, I heard many experts introduce the concept to an unfamiliar audience by describing it as the neutronic analogue of the Raman effect! Unlike in Raman scattering, there is no polarizability involved. But then, as the dictionary meaning implies, use of the word analogy does *not* demand correspondence in *all* respects.

24. When Mt Everest was first conquered, there was much speculation in the press as to who was the first to set foot on the summit, i.e., whether it was Tenzing Norgay or Edmund Hillary. An irrelevant question. The important fact was that at last Everest had been vanquished. And the man who perhaps deserved the maximum credit was Sir John Hunt who masterminded the expedition but never set foot on the peak!

25. It is learnt that in 1929 a letter was received by the Nobel Committee from the Calcutta University in response to the usual invitation circulated to make nominations. The nominee was not Raman but someone else hardly heard of!

26. Present practice seems to be to make the announcements in October.

27. It is said that Raman while accepting the Prize told the gathering that he wished he was receiving the honour as a citizen of free India rather than one of a colony. I have not been able to confirm this story, but one need not be surprised if the story is true. Such remarks would be typical of Raman.

28. A popular story is that many of the dinner guests tried to persuade Raman to drink saying that while they knew about Raman effect in alcohol, they would now like to study the effect of alcohol on Raman!

29. Observe that Lady Raman is having the problem discussed in Chapter 2. Once it is Sir Chandrasekhara and next time it is Sir Raman!

Chapter 6

1. Kastler later won the Nobel Prize for his work on optical pumping. On the occasion of his sixtieth birthday Bloembergen dedicated an article entitled "Conservation of angular momentum for optical processes in crystals", see *Topics in Nonlinear Optics* (Indian Academy of Sciences, Bangalore, 1982).

Chapter 7

1. The concept of a coherent light source is discussed in Chapter 12.
2. Bloembergen, to whom a reference has already been made in Chapter 6, came to India in 1979 as Raman Professor of the Indian Academy of Sciences (see Chapter 11). During this period he delivered, among others, the Gandhi Memorial Lecture founded by Raman (see Chapter 12). A collection of his reprints was brought out by the Academy in 1982 under the title *Topics in Nonlinear Optics*.

Chapter 8

1. For the preparation of this section, I have relied heavily on the book by F. R. Harris: *Jamsetji Nusserwanji Tata* (Oxford University Press, London, 1925).
2. It is interesting that decades later Bhabha established his research centre for atomic energy (now appropriately renamed the Bhabha Atomic Research Centre) in Trombay, close to the Tata estate. The Tata Institute for Social Sciences is also located in the neighbourhood.
3. Further details of this connection are provided in Sec. 8.12.
4. One must note that high salaries have been used to lure away eminent scientists, for example by the universities in Texas.
5. Saha's stand is curious, considering that his own teacher J. C. Bose protested against differential treatment (see Sec. 3.5)
6. When Sir Martin Forster was due to retire, the Tatas informally approached the Royal Society for some suggestions concerning a successor, indicating that if possible, they would prefer an Indian. Rutherford, who was then President of the Royal Society, immediately pointed out that Raman was an automatic choice.
7. It is said that when the complaint was made to Raman that he was giving preference to the students of Madras University he retorted, "Yes, that is the only university which does not teach the students anything, and when they come to me they do not have to unlearn nonsense as students from other universities have to!"
8. Apparently Guha had a large number of children, and Raman was told that at least on this count he should be sympathetic. Raman is reported to have retorted, "I am not responsible for that!"
9. I have personally participated in one such seminar held in Delhi. There was another to which I was invited but did not go as I did not see much point in attending it.
10. For the benefit of non-physicists, one should perhaps say a few words about Pyotr Kapitza. A legendary figure of Soviet science, Kapitza made his early reputation at Cavendish in the late twenties and early thirties. During this period he decided to go home on vacation but when he arrived in Russia, Stalin refused to let him go back to England to resume his research despite strong pressure from the British. When told that Cavendish very much wanted Kapitza to return, Stalin wryly remarked, "We too would like to have Rutherford." Kapitza had no choice but to settle down in the Soviet Union, and to help him get started, Stalin arranged to buy the equipment Kapitza had at Cavendish. Later Kapitza did excellent work, which eventually won for him the Nobel Prize. Perhaps even more significant than his research is the crucial role he played in saving the brilliant theoretical physicist Landau from the concentration camp. In later years Kapitza did much to popularize science amongst the young, even as Raman did.
11. The reference here is to the tour of India which Lord Rutherford was shortly to make, during which he was expected to address the Science Congress. Unfortunately he died before that, and his address was read by Sir James Jeans.
12. Professor A. Jayaraman (now at AT & T Bell Labs) tells the following story about two stuffed pheasants Raman once acquired. These were for the museum which included a collection of various colourful objects like rocks, minerals, butterflies, etc. Jayaraman writes:

> There were two beautiful stuffed Himalayan pheasants, which he had bought from a taxidermist in Calcutta. We left the pheasants in an open almirah and forgot to lock the door

of the room....As ill luck would have it, two mongrels got wind of the stuffed birds. They gained entry into the room in the night and destroyed one of the pheasants beyond recognition. The next morning when we [i.e., Jayaraman and his colleague Padmanabhan] came in and saw the spectacle, it was such a shock. We knew that Professor would get very upset with us for not locking the room and something had to be done to cover up the catastrophe. We quickly ordered the servants to clean up the place and bury the remnants of the pheasant in a far-off place. Prof. Raman came, and went straight into the museum room. Not finding one of the pheasants he wanted to know what happened. It was part of our decision to maintain that only one pheasant was there. Professor repeatedly said he definitely bought two. Since we strongly maintained that there was only one pheasant, he convinced himself that although he paid for two, only one specimen was brought to Bangalore.

At the time the incident took place, Jayaraman was Raman's student. Jayaraman concludes by observing that after being converted by their story Raman went out, only to discover feathers scattered by the mongrel while tearing the bird to pieces! Jayaraman adds that Raman must surely have wondered but did not reopen the case of the missing pheasant!

13. My colleague Ajay Sood informs me that he has observed similar phenomena in his experiments on colloidal systems. In fact, by noting whether the speckle pattern is completely stationary or slowly moving, he is able to ascertain whether the system is in a glassy or in a liquid phase, a conclusion confirmed by supplementary measurements of relaxation as well as of the structure factor.

14. The pioneering work of Raman and Ramachandran was recalled by Hariharan at a special commemoration meeting held soon after Raman's death (see *Curr. Sci.* **41**, 376 (1972)).

15. X-ray topography was independently discovered in Bangalore by Ramachandran. While obtaining Laue photographs of diamond, Hariharan noted that some of the Bragg spots had dark streaks across them. Raman immediately recognized that these were signatures of defects, and suggested to Ramachandran an X-ray diffraction experiment which would highlight the surface features. Ramachandran worked hard all night, and by the time Raman walked into the lab next morning, he had the desired picture ready. Raman was delighted, for, exactly as he had anticipated, the texture of the crystal stood revealed. Warmly complimenting Ramachandran, he said, "We shall call this technique topography." Later it became known that the technique had already been employed earlier and reported upon in the literature.

16. The *Sarvodaya* is a social service movement launched by Mahatma Gandhi. I recognize that this bland description does hardly any justice to the spirit and the noble objectives of that movement! Kasturba Gandhi was very active in it.

17. Mahadev Desai was a close associate, and will perhaps be best remembered for his long introduction to the English translation of Gandhiji's work on the *Bhagavad Gita*. Patel (later known as the Iron Man of India) became the Deputy Prime Minister in Nehru's Cabinet in Independent India. To Patel goes the credit of bringing around five hundred princely states into the Indian Union.

18. People loved to collect Gandhiji's autograph, and he skilfully exploited this craze. There was always a minimum fee, all the collection going to the Harijan Welfare Fund.

19. Bhabha excelled in pencil portraits and there is a nice one of Raman by him. Raman described it as a picture of one scientist by another. "No," replied Bhabha, "it is a picture of one artist by another." Air India brought out a calendar of Bhabha's portrait sketches which included pictures of the British physicist P.M.S. Blackett, artist Hussain, and a mountain villager.

20. The Tatas also made a lump sum and a recurring (for a few years) grant to Saha to help him with his cyclotron project.

21. Abraham Pais, whose book on Einstein has already been cited, has this to say concerning the violent attacks on Einstein during the anti-Jewish wave in Germany: "After that, the creative period ceases abruptly, though scientific efforts continue unremittingly for another thirty years. Who can gauge the extent to which the restlessness of Einstein's life in the 1920s was the cause or the effect of a lessening of creative powers?"

22. In May 1971 there was a special gathering of several of Raman's students where there were many nostalgic recollections of Raman the teacher.

Chapter 9

1. Strictly speaking it is not the amplitude but a quantity related to it. See Sec. 9.5.
2. For further details, see Sec. 9.5.
3. This journal (to which a reference has once been made earlier in Chapter 8) was founded by Raman. See Chapter 11.

Chapter 10

1. For a summary of the early history of lattice dynamics, see G. Venkataraman, in *Current Trends in Lattice Dynamics*, edited by K. R. Rao (Indian Physics Association, Bombay, 1979), p. 7. See also references therein to papers by Born and by Debye at the Copenhagen Conference of 1964.
2. I discussed this controversy during the first Academy meeting held after Raman's death, i.e., in 1971. I still recall the trepidation I felt while entering the portals of the Raman Institute (for the first time), and describing how Raman had erred! An account of this lecture may be found in *Current Science* **41**, 349 (1972).
3. Raman's brother C. Ramaswamy confirmed this to me. Ramaswamy was at that time a Demonstrator of Physics in the Presidency College, Madras, and when he came for a visit, Raman told his brother, "I say, why don't you put that thing on your finger to some use?" The reference was to a diamond ring that Ramaswamy was wearing! The advice was taken, and Ramaswamy was able to get a letter published in *Nature*. By that time Raman was back in Calcutta from where he sent a telegram of congratulations. As Ramaswamy recalled, that was "the best use I made of my father-in-law's gift!"
4. Concerning diamonds. *The Hindu* of August 30, 1938, has the following interesting report:
   All India Radio, Madras, broadcast a talk on "Three diamonds" by Mr Vummidi Pandurangaiah. In the course of his talk, he said:

   The number of cut diamonds which exceed a hundred carats in weight which exist in the world today is very limited. Almost all of them came from India. Of such large stones, I propose to talk to you of only three which have influenced great empires of the world.

   Take the Koh-i-noor. The name means 'Mountain of Light'. In its original cutting, it weighed 186 carats. In romantic history, it stands foremost. Legend says that it was found in the Godavari river, thousands of years ago and that it was the Syamantaka-Mani coveted by Krishna and acquired by him. In 1304 the Moghal emperors took it from Malwa. In 1739 Nadir Shah, the Persian conqueror, seized it. He it was that gave the gem the name of Koh-i-noor.

   An amusing story is told as to how this famous gem once changed hands. Shah Mahomed had the habit of keeping this stone within his turban which he never took off. Shah Mahomed gave a banquet to Nadir Shah. At the banquet, Nadir Shah took off his crown adorned with beautiful pearls and placed it on Shah Mahomed's head as a mark of eternal faith and friendship, removing at the same time Mahomed's turban which he put on his own head. Whether the wily Nadir was aware of the contents of Mahomed's turban or not, the rules of etiquette prevented Mahomed from making any objection to a seemingly spontaneous act of good-will.

   After an eventful career of adventure, Koh-i-noor became the spoil of Ranjit Singh. His successors kept it till 1850 and upon the fall of Sikh power, it passed to the East India Company. Lord Dalhousie, on behalf of the Company, presented it to Queen Victoria. The Queen had it recut, reducing its weight to 106 carats, and had the gem mounted in the Imperial Crown. The wisdom of recutting the gem has been doubted.

   A peculiar feature of the history of Koh-i-noor is that it never changed hands through purchase, as far as we know. The Conqueror always got it.

From Krishna came another diamond which influenced the destinies of France. It is known as the Pitt or Regent Diamond. Somewhere about 1701, about 150 miles from Golconda, a labourer found a stone weighing 410 carats. The stone was too big to conceal on the person of the man who found it. He deliberately cut his leg and concealed the stone in the folds of his bandage. He wanted to go to distant lands and sell this unique find abroad. But the English skipper who took him on board found the secret out, threw the man to the sharks and sold the diamond to a Parsee merchant. Sir Thomas Pitt, Governor of Fort St George, bought the diamond for about three lakhs of rupees. In 1717 he sold the stone to the Duke of Orleans, the Regent of France, for Rs. 18 lakhs. Till 1792, the fateful year of the French Revolution, the stone remained among the Crown Jewels of France. During the days of Revolution, it was stolen, but the thieves were unable to find a buyer and threw it into a ditch where it was found later. The Republican Government of France pledged the stone to Holland and raised money to meet the Napoleonic Wars. Napoleon redeemed the jewel and mounted it on his State Sword. Today it rests, the pride of a great nation, in the Apollo Gallery at Paris.

A third stone whose history was mixed up with that of the Russian Empire came also from South India. It is called the Orloff.

5. More technically, this possibility is subject to selection rules.
6. The reference here is to the book *Dynamik der Kristallgitter* published in 1915. There were a few updates later by way of review articles. Finally, in 1954, Born, along with Huang, published the classic *Dynamical Theory of Crystal Lattices* [10.5].
7. The following is the text of a message signed by thirty leading scientists and issued from Bordeaux:
    The scientists who have met together in Bordeaux to celebrate the Twentieth Anniversary of the Discovery of the Raman Effect, and who have great pleasure in having Sir C. V. Raman in person among them, express to their colleagues in the great country of India their cordial greetings. They have great admiration for the great work of Sir C. V. Raman and the team of workers in India which he has created, and which he has inspired with his faith.
8. As already mentioned in Chapter 8, R. S. Krishnan worked with Raman in Bangalore. He is not to be confused with K. S. Krishnan referred to in Chapters 3, 4 and 5.
9. Krishnan remarks that "superposed over the Raman lines there is a feeble continuum which extends from the Raman line 2253 cm^{-1} to the line 2666 cm^{-1}".
10. Helen Smith is a remarkable lady. Deeply upset by the atomic bomb, she left physics altogether and turned to another profession!
11. Landau's propensity for objecting is well known. It is said that a cartoon exists of Landau sitting in Niels Bohr's seminar with a gag in his mouth and tied to the chair – so that the lecturer could get a word in edgeways!
12. A. K. Ramdas involved in this experiment is the son of L. A. Ramdas who worked with Raman in Calcutta. In fact A. K. Ramdas also briefly worked with Raman at the Raman Research Institute. He is now in Purdue University.
13. Bhagavantam and Venkatarayudu wrote a book on group theory, with particular emphasis on applications to molecular spectroscopy. For many years, this was the only authoritative text on the subject, and a whole generation of spectroscopists, both in India and abroad, have been reared on it. I still vividly recall the letter I received from Prof. Joseph Birman of New York enquiring where he could obtain a copy of the book.
14. V. C. Sahni informs me that it is not a rigorous theorem that the symmetry-determined set always contains $(24p - 3)$ frequencies. However, this is true in the case of the diamond lattice.

Chapter 11

1. The journal *Current Science* was itself founded as a result of resolutions passed at a special meeting held during the session of the Indian Science Congress in Bangalore in 1931.

2. This Academy was formed in 1930. It has now been renamed National Academy of Sciences, Allahabad.
3. With rare exceptions, the Academy meetings almost always started on November 7, Raman's birthday. The Golden Jubilee event was likewise scheduled to start on November 7, 1984, but had to be postponed as the country was then in mourning following Indira Gandhi's assassination. The meeting was then held in February 1985.
4. Owing to a large increase in the number of aspirants for the Fellowship, the election procedure followed at present is different and somewhat more elaborate, involving specialist Committees etc.
5. Bhabha not only published many of his own papers in the *Proceedings*, but encouraged others in TIFR to do likewise. I have personally seen Bhabha scrutinize manuscripts with care before forwarding to the Academy, bestowing the same attention he would have given were he communicating to the Royal Society.
6. Lately the issue of publishing in Indian journals has aroused much concern, leading to several editorial comments. While papers continue to "gravitate" abroad, the discovery of high-temperature superconductors has produced a new trend as far as *Pramana* is concerned. Previously the advice used to be: publish a short letter abroad and the full paper in an Indian journal (as Raman did in the late twenties). Anxious to establish priority, workers in high-temperature superconductivity seem to prefer to publish their letters in *Pramana* and their full papers abroad!

Chapter 12

1. Instead of the optical telescope that Raman hoped to instal, there is now a small radio telescope operating in the GHz region in the observatory dome. In addition, there is also a large millimetre-wave telescope similar to the one built by Leighton at Caltech. Radio astronomy is one of the subjects actively pursued now at the Raman Research Institute.
2. Cordierite occurs in large chunks. When light is polarized along one axis it is almost completely absorbed and the crystal looks dark. When it is polarized along either of the other two axes it is transmitted and the crystal looks transparent.

   It is said that the Vikings used cordierite (which they found in Denmark) as an aid in navigation when the Sun was obscured by the clouds. They would scan the sky, looking through a crystal of cordierite. When they looked in the direction of the Sun, the light was unpolarized and could not be extinguished by the crystal no matter what its orientation. However, in other directions the situation was different, and in particular there was a direction for which the light could be completely cut off by suitably orienting the crystal. We know the reason, but the Vikings thought that in the forward direction the Sun was so powerful it could not be extinguished! Anyway, knowing the position of the Sun, they could navigate.
3. An exhaustive article by G. N. Ramachandran and S. Ramaseshan on crystal optics in *Handbuch der Physik*, vol. 25, part 1 (1961) makes extensive use of the Poincaré sphere. George Series [12.14] remarks that Pancharatnam made an important and substantial contribution to this article but was unwilling to have his name included as a co-author.
4. In the preface to the old version of his book *Statistical Physics*, Landau says, "We are talking here about theoretical physics, and therefore of course mathematical rigour is irrelevant and impossible." Commenting on this, P. W. Anderson observes, "This is not quite true, but it is very close to it."
5. Hanbury Brown was born in India. In 1974 he visited the country as Raman Professor. In 1985 the Indian Academy of Sciences brought out a special publication entitled *Photons, Galaxies and Stars* containing reprints of his important papers.
6. The Mysore University asked Raman to recommend suitable names for appointment in the Department of Physics. Raman recommended the name of Pancharatnam, among others, and the latter was duly appointed. Some disgruntled elements then went to court alleging nepotism on the part of Raman! In a perceptive verdict the judge ruled that experts must necessarily be biased – in favour of excellence! The case was dismissed.

7. Bhandari and Samuel of the Raman Research Institute carry this discussion farther, not only emphasizing the subtle difference between Berry's phase and Pancharatnam's phase, but also report on an experimental study of the latter using laser interferometry (*Phys. Rev. Lett.* **60**, 1211 (1988); see also Samuel and Bhandari, *ibid.* **60**, 2339 (1988)).
8. It is worth drawing attention here to the close relationship that existed between Saha and Nehru in the pre-Independence days. Towards the late thirties, Subhas Chandra Bose (later to escape from house arrest and form the Indian National Army) became the President of the Indian National Congress. For Saha, who knew Bose from his student days, this was a matter of great joy because Bose advocated in unequivocal terms the large-scale industrialization of the country. Bose further proposed the establishment of a National Planning Committee (NPC) to draw up detailed schemes for the promotion of industries when India became free. Saha strongly endorsed Bose, and he once wrote:

> There are those who want to preach asceticism for everyone, but there is only one Mahatma amongst 400 millions and his way of living, however noble, cannot satisfy the large mass of ordinary mortals which make up our continent and the world. Further, as far as economics and national reconstruction are concerned, the Mahatma has unfortunately allowed himself to be surrounded by a number of ill-informed fanatics of the dubious Gospel of the Spinning Wheel and the Bullock-cart who do not allow the Truth to get anywhere near him. Besides the fanatics of the Charka and Bullock-cart cult, we must also mention the large number of hypocrites, mostly big businessmen, who find it convenient to advertise the utility of the Charka, and the village oil-press. Fanatics can after all be understood, but the support of the money-bags is a subtle kind of hypocrisy which is not so apparent to the unsophisticated mind.

There was a move to have Sir M. Visveswarayya (famed for his enlightened development programmes implemented in the State of Mysore) head the proposed NPC. However, Saha was of the view that the recommendations of the NPC would not be effective unless a national leader like Nehru headed the Committee. He worked hard behind the scenes to persuade Visveswarayya not to accept the post, and Nehru to accept it instead. It is believed that Nehru's message to the Science Congress of 1938 was at Saha's initiative.

Saha's deep commitment to India's development becomes evident from his stirring speeches and articles. It is a pity he fell apart with Nehru after Independence. As he said in Parliament, "Fate has ordained that I shall be in opposition, but I hope my friendship with Panditji will stand the strain."
9. Speaking at the Baroda Science Congress of 1955, Nehru paid a tribute to Bhatnagar who had just passed away. He said: "I have always associated with many prominent figures closely connected with the Science Congress and among them the chief was Dr S. S. Bhatnagar. It is not necessary to say anything formal about him. You all knew him. But I would like to pay a tribute to Bhatnagar... who, I think, has done, and I say this with due respect to others, more than anyone else for scientific development in India."
10. Though not a scientist, Nehru had been invited to preside over the 30th session of the Indian Science Congress in Calcutta. However, he could not do so as he was in prison. He was invited again in 1947 when he was Vice-President in the Interim Government.
11. About ivory tower, Saha has this to say:

> Scientists are often accused of living in the ivory tower and not troubling their minds with realities. Apart from my association with political movements in my juvenile years, I had lived in the ivory tower up to 1930. But science and technology are very important for administration nowadays, at least as much as law and order. I have gradually glided into politics because I wanted to be of some use to the country in my own humble way. [12.21]

12. Raman's comment about the National Laboratories was more brusque: "Shah Jehan built the Taj Mahal to bury one of his favourite women. The National Laboratories were built to bury scientific instruments." [12.22]
13. Things were much better during Bhabha's time. These days one sometimes wonders if the dog itself is about to disappear!
14. Speaking to scientists and technologists in Delhi in November 1970, Indira Gandhi too said something similar. "It was rather disturbing to find" she said, "that the leaders of the scientific

community, who should guide the Government in identifying the imbalances to be corrected and initiatives to be taken, themselves looked towards the Governmental bodies most of the time." [12.22]

15. In this context, it is interesting to recall Feynman's observation, "I have a principle of not going anywhere near Washington or having anything to do with Government." (*Phys. Today*, February 1988)

16. Paraphrasing various quotes and remarks, Sarvepalli Gopal, in his biography of Nehru (*Jawaharlal Nehru—A Biography* (Oxford University Press, New Delhi,) vol. 3, p. 203) observes:

> The way of the prophet, wedded to basic principles whatever happens, cannot always be the way of the leader of men, who has to consider all the time how far he can take with him those whom he leads. It is the lot of the prophet to be stoned; but a leader has to strike a compromise between truth and men's receptivity to truth. The leader always has a problem as to how far he should compromise with his principles. If he compromises too much he loses his principles; if he does not compromise enough he loses his leadership. To compromise is to embark on a slippery slope, yet refusal to compromise can sometimes mean isolation.

Gopal obviously has Gandhi and Nehru in mind. These remarks could easily be translated to the post-Independence scientific scene, casting Raman on one side and Bhabha and Bhatnagar on the other.

17. Speaking once about his experiences in Bangalore, Raman said:

> Bangalore is a nice place and I thought there were people here who were interested in science. I did not come here to make money. I wanted to make Bangalore the centre of science in Asia. Then at the Indian Institute of Science they gave me hell. You may be a scientist, but you are no administrator, they said. They cut the grants and they even said I had embezzled money.
>
> What I suffered was indescribable. You must have political support to succeed, you see, and I had none. Then I quit and set up my own Institute. I am my own master and I am happy about it. This Institute of mine is a fitting reply to all the torture heaped on me. I work here quietly, silently. The work of science is done quietly and silently. Look at out National Laboratories. They have become vested interests for some people.

Raman never forgot his IISc experience. Years later he would often remark that he grew tall trees in the campus of the Raman Research Institute so that they would block out the view of the other Institute!

18. It was magnanimous of the Royal Society to have published a biographical memoir on Raman even though he resigned from the Fellowship. The memoir was written by Bhagavantam, and occasionally I have quoted from it.

19. Recreational science is not without utility (if one should be narrow-minded enough to have such expectations). The most outstanding example which occurs to me is the discovery of the so-called Penrose tiling, which shattered the central dogma of crystallography and in a sense paved the way for the discovery of quasicrystals. For further details, see (i) M. Gardner: *Sci. Amer.* **236**, 110 (January 1977), and (ii) P. Gratias: *Contemp. Phys.* **28**, 219 (1987).

20. As another example of puzzles of this nature, the reader's attention is directed to the article by Luis Alvarez on mass extinctions which appeared in the July 1987 issue of *Physics Today*. It must however be added that the complexion of science is quietly undergoing a change and even a branch like biology which was once believed to be way beyond the reach of mathematics is no longer so. Recently, I saw an advertisement from a cancer research centre asking not for a molecular biologist (as one would have expected), but a mathematician! No wonder Eugene Wigner spoke of the "unreasonable power" of mathematics.

21. The Gandhi Memorial Lectures delivered by Raman always dealt with science in one form or the other. Among the topics he lectured on were: Light, colour and vision; Musical instruments; Gems and gemmology; The colour of flowers; Green leaves; The lure of astronomy; Voice, speech and language. The Gandhi Memorial Lectures still continue, now delivered by men of eminence in various fields. Naturally, the topics are no longer restricted to science.

22. Perhaps the following incidents give a glimpse into this aspect of Raman's personality.

In the mid-fifties, Bulganin and Khruschev made an extended tour of India, the first ever such visit by Soviet leaders outside the so-called Iron Curtain. In Bombay the Soviet visitors were taken to the

TIFR, and in the same spirit a visit to the Raman Institute was included as part of the Bangalore programme. Raman generally distrusted politicians and more so communists, but Bulganin and Khruschev were honoured guests in the country. So he not only agreed to receive them but enthusiastically put up banners and buntings. Everyone was keenly awaiting the arrival of the guests but there had been some delays in the earlier engagements and the local organizers decided at the last minute to cancel the visit to the Raman Institute. Raman naturally became angry and tore off the decorations, saying, "See, I told you one cannot trust these politicians!"

On another occasion, Raman decided to see, in the company of some of his students and associates, the film "Anna and the King of Siam" starring Yul Brynner, then showing in town. Soon after the show started Raman became angry as he thought Asians were being portrayed in poor light in the film. "I am not going to pay money and watch us brown people being humiliated in our own environment." So saying he walked out; and those with him had to do likewise although they were all keen to watch the rest of the film!

23. Bhagavantam's analysis of Raman's ego is interesting. He narrates the incident of how, provoked by a reporter, Raman once said: "This Institute is a monument to my egotism. I am an egotist and just as Egyptian kings used to build pyramids before their death, so is this Institute my pyramid." Bhagavantam adds: "Although this statement taken out of context appears to give prominence to the ego in Raman, it is the opinion of this writer that it is no more than the expression of an uncommon degree of self-confidence which Raman always displayed." In support of this argument Bhagavantam mentions that Raman himself later pointed out to the reporter the main reason for establishing the Institute. This part of Raman's statement has already been quoted in Sec. 12.1.

24. Nehru's visit to the Raman Institute is interesting not only for the manner in which it happened but also because of the distorted account of the event which subsequently appears to have gained currency. In those days Indira Gandhi used to accompany Nehru, and during one of the visits of the latter to Bangalore, a visit to the Raman Research Institute was suggested to Indira Gandhi as a way of keeping her busy while Nehru was attending to political matters. Raman played the genial host to Indira and after giving her a guided tour, took her to the verandah of the first floor from where one could command a magnificent view of the surrounding landscape, including the distant Nandi Hills. In characteristic fashion, Raman then remarked, "Why don't you ask your papa to come and enjoy this grand view instead of messing with politics?" Mrs Gandhi laughed but it would appear that she did speak about this to Nehru, for he came to the Institute the following day. Raman was absolutely delighted. Once again the tour, but this time, after explaining the activities in progress, Raman requested Nehru to institute a professorship so that work might be carried on even after he [i.e., Raman] was gone. Nehru smiled and assured Raman that the Institute would always be well taken care of. Raman was not so easily convinced. He said, "Mr Prime Minister, how can I be sure? Politicians often forget what they promise." Nehru had a hearty laugh.

I have heard this account from one who was actually present during the visits of both Mrs Gandhi and Nehru. However, I have also heard another version according to which Raman made some caustic remark, and Nehru, taking exception, walked out of the Institute in a huff. Unfortunately, such was Raman's image that the latter version was widely believed as it sounded very credible though it was incorrect.

25. One particular Summer School, in which Murray GellMann was the principal lecturer, is memorable, for the audience got the first ever glimpse of the famous Eight-Fold Way, later to earn for its author the Nobel Prize. For Bhabha, now increasingly absorbed in scientific administration, these Schools were occasions to charge his "academic batteries".

26. The story is told of a person who came to the Institute seeking a job. Found unsuitable for appointment, he was sent away after being paid the travel and other expenses due to him. Some time later, Raman observed that the man was still lingering near the clerk's office, and sharply remarked: "I told you we cannot take you. Why are you still hanging around here?" The applicant answered, "I know that Sir, but I came back to return the excess cash paid to me by mistake by your office." Impressed, Raman immediately offered him the job, saying: "It matters not if your physics is inadequate, I can teach you that. You are a man of character!" [12.22]

27. Although personally Raman was neither religious nor given to the observance of religious practices, he respected others' sentiments in the matter. For example, during the Academy meeting at the Annamalai University, he was taken to the famous Nataraja temple in Chidambaram where the authorities received him with due temple honours. As was expected of him, Raman went without his headgear and in a *dhoti* – a rare sight in public.

During the Madurai meeting, there was no such visit to the Meenakshi temple in the city. But many Fellows sneaked out saying they were going "for a walk". Noticing this, Raman asked: "Why this mass going-for-a-walk business? I have never heard of Fellows being so concerned about exercise." One presumes he understood!

Chapter 13

1. See Chapter 11.
2. Born describes his experience in his autobiography (*My Life: Recollections of a Nobel Laureate*, Taylor and Francis, London, 1978). He had gone to Delhi as part of an extended North Indian tour. Raman was also visiting Delhi then on some administrative business. He (i.e., Born) then received an invitation to make a radio broadcast. "There was only about an hour in which to prepare a script for a thirty-minute talk, and I had written only half of it when I had to begin talking. So I had to improvise the other half, a most exciting and fatiguing experience. Later I learned that the radio system had just been inaugurated a few days before, so there were not many listeners, and it did not matter what I said."!
3. Actually the broadcasts, made over a period of time, were not all intended to deal with the new physics. It would seem that the title has been appended to the anthology by the publisher.
4. The model of the atom which shot him to fame is probably the contribution by which Bohr is best remembered. However, Bohr's contemporaries probably valued even more his profound contribution to the interpretation of quantum mechanics. Bohr also made many important contributions in nuclear physics, and his model for nuclear fission still has utility.
5. Recently I heard a high energy physicist of yester-year remark that he found the currently popular string theory somewhat beyond him. At the same time, he foresaw that the younger generation would be quite comfortable with it.
6. Recall that Millikan played host to Raman in Caltech in 1924 (see Chapter 3). Millikan's visit to Bangalore is also referred to in Chapter 8.
7. The *tonga* is a horse-drawn cab. It may still be seen in the older sections of many Indian towns.
8. Einstein remarks: "The emotional state which enables such achievements is similar to that of the religious person or a person in love; the daily pursuit does not originate from a design or a programme but from a direct need." (Quoted by Abraham Pais in '*Subtle is the Lord...' The Science and the Life of Albert Einstein* (Clarendon Press, Oxford, 1982)).
9. Medical science seems, however, to be somewhat of an exception. I have heard many eminent doctors assert that it takes many cases and therefore years to arrive at important findings.
10. Quoting Einstein again: "Above stands the marble smile of implacable Nature which has endowed us with more longing than with intellectual capacity." (From '*Subtle is the Lord...*' cited above).

Chapter 14

1. *Physics Today*, in its issue of June 1986, gives statistics about physics PhD's produced in the US for the period 1969–1984. While condensed matter physics registered a decline from about 400 to 258, particle physics from 220 to 138, nuclear physics from 188 to 72, and atomic physics from 127 to 77, in optics the output rose steadily from 16 to 53.

Notes

2. Siddiqui's name has been mentioned earlier in Chapter 11.
3. For further comparisons between Bhabha and Saha, see R. S. Anderson: *Building Scientific Institutions in India* (Centre for Developing Area Studies, McGill University, Montreal, 1975).
4. The mania for essay questions and descriptive answers does not seem to be unique to India. Richard Feynman, in *Surely You Must be Joking, Mr Feynman!* (W. W. Norton, New York, 1985), describes how a similar outlook prevailed in Brazil in the fifties. Quite possibly, Brazil has since then evolved!
5. In 1953, when the Silver Jubilee of the discovery of the Raman effect was being celebrated, Raman observed: "The more I look back on my career, the more I feel that it has been a long history of frustration, disappointment, struggle and every kind of tribulation. But there have been a few gleams of success. It was poverty and the poor laboratories that gave me the determination to do the very best I could."

References

Chapter 1

1.1 B. V. Subbarayappa: *Western science in India*, in: *A Concise History of Science in India*, edited by D. M. Bose, S. N. Sen and B. V. Subbarayappa (Indian National Science Academy, New Delhi, 1971)

Chapter 2

2.1 S. Ramaseshan: *C. V. Raman Memorial Lecture 1978* (Indian Institute of Science, Bangalore, 1978)
2.2 C. V. Raman: *Books that have Influenced Me* (G. A. Natesan & Co., Madras, 1947) p. 21
2.3 P. R. Pisharoty: *C. V. Raman* (Publications Division, Govt. of India, New Delhi, 1982)
2.4 C. V. Raman: *Philos. Mag.* **12**, 494 (1906)
2.5 C. V. Raman: *Philos. Mag.* **14**, 591 (1907)
2.6 C. V. Raman: *Phys. Rev.* **32**, 307 (1911)
2.7 H. O. Peitgen and P. H. Richter: *The Beauty of Fractals* (Springer-Verlag, Berlin, 1986)
2.8 S. Bhagavantam: in *Biographical Memoirs of the Royal Society* (Royal Society, London, 1971) vol. 17, p. 565

Chapter 3

3.1 C. V. Raman: *Nature (London)* **128**, 362 (1931)
3.2 C. V. Raman: in the Annual Report of the IACS, 1927–1928
3.3 C. V. Raman: *The Calcutta School of Physics* (Calcutta University Press, Calcutta, 1917)
3.4 S. Ramaseshan: *C. V. Raman Memorial Lecture 1978* (Indian Institute of Science, Bangalore, 1978)
3.5 S. Bhagavantam: in *Biographical Memoirs of the Royal Society* (Royal Society, London, 1971) vol. 17, p. 565
3.6 L. A. Ramdas: *Indian J. Phys. Edn.*
3.7 Annual Report of the IACS, 1926–1927
3.8 C. V. Raman: *Proc. Phys. Soc.* **42**, 309 (1930)
3.9 S. Ramaseshan: *Science Age*, February 1987, p. 33
3.10 Annual Report of the IACS, 1926–1927
3.11 G. Torkar: *J. Raman Spectrosc.* **17**, 13 (1986)
3.12 S. Chandrasekhar: *Am. J. Phys.* **37**, 577 (1969)
3.13 L. L. Fermor: Presidential Address, 20th Science Congress, Patna

Chapter 4

4.1 C. V. Raman: *Nature (London)* **82**, 9 (1909)
4.2 C. V. Raman: *Phys. Rev.* **32**, 309 (1911)
4.3 C.V. Raman: *J. Indian Math. Club* October 1909, p. 170; February 1910, p. 14

4.4 Lord Rayleigh: *Scientific Papers* (Cambridge University Press, Cambridge, 1900) vol. II, p. 188
4.5 C. V. Raman: *Philos. Mag.* **24**, 513 (1912)
4.6 C. V. Raman: *Nature (London)* **82**, 156 (1909)
4.7 C. V. Raman: *Phys. Rev.* **4**, 12 (1914)
4.8 C. V. Raman: *Phys. Rev.* **35**, 449 (1912)
4.9 C. V. Raman: *Phys. Rev.* **5**, 1 (1915)
4.10 C. V. Raman and Asutosh Dey: *Philos. Mag.* **34**, 129 (1917)
4.11 C. V. Raman: *Philos. Mag.* **29**, 15 (1915)
4.12 R. P. Feynman: *Lectures on Physics* (Addison–Wesley, Reading, Mass., USA, 1964) vol. I, chapter 50
4.13 Hermann Helmholtz: *On the Sensations of Tone*, translated by A. Ellis (Dover, New York, 1954)
4.14 C. V. Raman and S. Appaswamaiyar: *Philos. Mag.* **31**, 47 (1916)
4.15 C. V. Raman: *Bull. Indian Assoc. Cultiv. Sci.* No. 15 (1918)
4.16 B. S. Madhavarao: *Curr. Sci.* **40**, 232 (1971)
4.17 C. V. Raman: *Philos. Mag.* **32**, 391 (1916); see also *Philos. Mag.* **35**, 493 (1918)
4.18 *Musical Acoustics, Part I*, edited by Carleen M. Hutchins (Dowden, Hutchinson and Ross, Pennsylvania, 1975)
4.19 C. V. Raman: *Proc. Indian Assoc. Cultiv. Sci.* **6**, 19 (1920); see also *Philos. Mag.* **39**, 535 (1920)
4.20 R. N. Ghosh: *Indian J. Phys.* **1**, 141 (1926)
4.21 C. V. Raman and B. Banerji: *Proc. R. Soc. London* **A97**, 99 (1920)
4.22 R. N. Ghosh and J. N. Dey: *Proc. Indian Assoc. Cultiv. Sci.* **9**, 193 (1925)
4.23 C. V. Raman: *Sir Asutosh Mookerji Silver Jubilee Volume* (Calcutta University Press, Calcutta, 1922) Vol. 2, p. 179
4.24 C. V. Raman: *Proc. Indian Acad. Sci.* **A1**, 179 (1934); see also C. V. Raman and S. Kumar, *Nature (London)* **104**, 500 (1920)
4.25 Lord Rayleigh: *The Theory of Sound* (Dover, New York, 1945)
4.26 C. V. Raman: *Proc. Indian Assoc. Cultiv. Sci.* **7**, 29 (1921)
4.27 K. C. Kar: *Phys. Rev.* **21**, 695 (1923)
4.28 See reference [4.18], p. 146
4.29 C. V. Raman: in *Handbuch der Physik* (Springer, Berlin, 1927) vol. VIII, p. 354
4.30 R. N. Ghosh: *Proc. Indian Assoc. Cultiv. Sci.* **9**, 143 (1925)
4.31 M. Panchanon Das: *Proc. Indian Assoc. Cultiv. Sci.* **9**, 297 (1925)
4.32 R. N. Ghosh: *Phys. Rev.* **20**, 526 (1922)
4.33 K. N. Rao: *Proc. Indian Acad. Sci.* **A7**, 75 (1938)
4.34 B. S. Ramakrishna and M. M. Sondhi: *J. Acoust. Soc. Am.* **26**, 523 (1954)
4.35 B. S. Ramakrishna: *J. Acoust. Soc. Am.* **29**, 234 (1957)
4.36 Lord Rayleigh: *Scientific Papers* (Cambridge University Press, Cambridge, 1912) vol. V, p. 617
4.37 C. V. Raman and G. A. Sutherland: *Proc. R. Soc. London* **A100**, 424 (1921); see also *Proc. Indian Assoc. Cultiv. Sci.* **7**, 159 (1922). 4.37a C. V. Raman: *Proc. Indian Assoc. Cultiv. Sci.* **7**, 159 (1922)
4.38 G. N. Ramachandran: *Curr. Sci.* **40**, 212 (1971)
4.39 C. V. Raman: *Philos. Mag.* **17**, 204 (1909)
4.40 C. V. Raman: *Philos. Mag.* **21**, 618 (1911)
4.41 S. K. Mitra: *Philos. Mag.* **35**, 112 (1918)
4.42 S. K. Mitra: *Philos. Mag.* **37**, 50 (1919)
4.43 Lord Rayleigh: *Scientific Papers* (Cambridge University Press, Cambridge, 1920) vol. VI, p. 455
4.44 S. K. Banerji: *Astrophys. J.* **48**, 50 (1918)
4.45 S. K. Banerji: *Philos. Mag.* **37**, 112 (1919)
4.46 C. V. Raman: *Phys. Rev.* **13**, 259 (1919)
4.47 S. K. Mitra: *Philos. Mag.* **38**, 289 (1919)
4.48 N. Basu: *Philos. Mag.* **35**, 79 (1918)
4.49 T. Chinmayanandam: *Philos. Mag.* **37**, 9 (1919)
4.50 N. K. Sethi: *Philos. Mag.* **42**, 669 (1921)
4.51 C. V. Raman and K. S. Krishnan: *Proc. Phys. Soc.* **38**, 350 (1926)
4.52 C. V. Raman and K. S. Krishnan: *Proc. R. Soc. London* **A116**, 254 (1927)

4.53 C. V. Raman: *Nature (London)* **109**, 105 (1922)
4.54 B. N. Chuckerbutti: *Proc. Indian Assoc. Cultiv. Sci.* **7**, 73 (1922)
4.55 M. N. Mitra: *Indian J. Phys.* **3**, 175 (1928)
4.56 I. Ramakrishna Rao: *Indian J. Phys.* **2**, 167 (1928)
4.57 C. V. Raman and B. Banerji: *Philos. Mag.* **41**, 338 (1921)
4.58 C. V. Raman and P. N. Ghosh: *Nature (London)* **102**, 205 (1918)
4.59 C. V. Raman and I. Ramakrishna Rao: *Proc. Phys. Soc.* **39**, 453 (1927)
4.60 C. V. Raman: *Philos. Mag.* **38**, 568 (1919)
4.61 C. V. Raman: *Nature (London)* **108**, 12 (1921)
4.62 C. V. Raman: *Philos. Mag.* **43**, 357 (1922)
4.63 T. Chinmayanandam: *Proc. R. Soc. London* **A95**, 176 (1918)
4.64 C. V. Raman and G. L. Datta: *Philos. Mag.* **42**, 826 (1921)
4.65 N. K. Sethi and C. M. Sogani: *Proc. Indian Assoc. Cultiv. Sci.* **7**, 61 (1922)
4.66 C. V. Raman: *Lectures on Physical Optics* (Indian Academy of Sciences, Bangalore, 1959)
4.67 K. S. Krishnan: *Indian J. Phys.* **4**, 385 (1929)
4.68 C. V. Raman: *Proc. Phys. Soc.* **42**, 309 (1930)
4.69 S. Bhagavantam: *Proc. R. Soc. London* **A214**, 545 (1929)
4.70 C. V. Raman and K. S. Krishnan: *Philos. Mag.* **5**, 769 (1928)
4.71 C. V. Raman: *Nature (London)* **109**, 477 (1922)
4.72 C. V. Raman: *Astrophys. J.* **56**, 29 (1922)
4.73 C. V. Raman: *Phys. Rev.* **12**, 442 (1918)
4.74 A. Venkatasubbaraman: *Proc. Indian Assoc. Cultiv. Sci.* **6**, 109 (1920)
4.75 K. Seshagiri Rao: *Proc. Indian Assoc. Cultiv. Sci.* **6**, 165 (1921)
4.76 C. V. Raman: *J. Opt. Soc. Am.* **12**, 387 (1926)
4.77 S. Smith: *Nature (London)* **127**, 855 (1931)
4.78 G. Venkataraman: *Bull. Mater. Sci.* **4**, 175 (1982)
4.79 C. V. Raman and K. R. Ramanathan: *Proc. Indian Assoc. Cultiv. Sci.* **8**, 127 (1923)
4.80 F. Zernicke and J. Prins: *Z. Phys.* **41**, 184 (1927)
4.81 S. Ramaseshan: *C. V. Raman Memorial Lecture 1978* (Indian Institute of Science, Bangalore, 1978)

Chapter 5

5.1   A. T. Young: *Phys. Today* January 1982, p. 42
5.2   C. V. Raman: *Nature (London)* **103**, 165 (1919)
5.3   C. V. Raman and Bidhubushan Ray: *Proc. R. Soc. London* **A100**, 102 (1921)
5.4   C. V. Raman: *Nature (London)* **108**, 242 (1921)
5.5   C. V. Raman: *Nature (London)* **108**, 367 (1921)
5.6   C. V. Raman: *Nature (London)* **114**, 49 (1924)
5.7   C. V. Raman: *Nature (London)* **108**, 402 (1921)
5.8   C. V. Raman: *Proc. R. Soc. London* **A101**, 64 (1922)
5.9   C. V. Raman: *Nature (London)* **109**, 42 (1922)
5.10 C. V. Raman: *Nature (London)* **109**, 75 (1922)
5.11 S. Ramaseshan: *C. V. Raman Memorial Lecture 1978* (Indian Institute of Science, Bangalore, 1978)
5.12 C. V. Raman: *Molecular Diffraction of Light* (Calcutta University Press, Calcutta, 1922)
5.13 A. Pais: *'Subtle is the Lord...' The Science and the Life of Albert Einstein* (Clarendon Press, Oxford, 1982)
5.14 C. V. Raman and K. Seshagiri Rao: *Philos. Mag.* **45**, 625 (1923)
5.15 C. V. Raman: *Nature (London)* **111**, 13 (1923)
5.16 C. V. Raman: Presidential Address to the Indian Science Congress, 1929
5.17 K. R. Ramanathan: *Proc. Indian Assoc. Cultiv. Sci.* **8**, 181 (1923)
5.18 K. S. Krishnan: *Philos. Mag.* **50**, 697 (1925)
5.19 L. A. Ramdas: *Indian J. Phys. Edn.*
5.20 C. V. Raman: *The Molecular Scattering of Light*, Nobel Lecture, 1930

5.21 The diary of K. S. Krishnan is quoted by P. R. Pisharoty in *C. V. Raman* (Publications Division, New Delhi, 1982) and by R. S. Krishnan in *K. S. Krishnan Memorial Lecture 1978*.
5.22 A. Smekal: *Naturwissenschaften* **11**, 875 (1923)
5.23 H. A. Kramers and W. Heisenberg: *Z. Phys.* **31**, 681 (1925)
5.24 R. W. Wood: *Physical Optics*, 3rd edition (Macmillan, New York, 1934)
5.25 C. V. Raman: *Nature (London)* **121**, 619 (1928)
5.26 C. V. Raman and K. S. Krishnan: *Nature (London)* **121**, 501 (1928)
5.27 C. V. Raman: *Indian J. Phys.* **2**, 387 (1928)
5.28 C. V. Raman and K. S. Krishnan: *Indian J. Phys.* **2**, 399 (1928)
5.29 C. V. Raman and K. S. Krishnan: *Proc. R. Soc. London* **A122**, 23 (1929)
5.30 M. Born and K. Huang: *Dynamical Theory of Crystal Lattices* (Clarendon Press, Oxford, 1954)
5.31 I. L. Fabelinskii: *Sov. Phys. Usp.* **21**, 780 (1978) (English translation of *Usp. Fiz. Nauk* **126**, 124 (1978))
5.32 L. D. Landau and E. M. Lifshitz: *Electrodynamics of Continuous Media* (Pergamon, London, 1963)
5.33 G. Landsberg and S. Mandel'shtam: *Naturwissenschaften* **16**, 557 (1928)
5.34 E. Amaldi, O. D'Agostino, E. Fermi, B. Pontecorvo, F. Rasetti and E. Segre: *Proc. R. Soc. London* **A149**, 522 (1935)
5.35 A. S. Ganesan: *Indian J. Phys.* **4**, 281 (1929)
5.36 Y. Rocard: *Comptes Rendus* **186**, 1107 (1928)
5.37 J. Cabannes: *Comptes Rendus* **186**, 1201 (1928)
5.38 I. Tamm: *Z. Phys.* **60**, 345 (1930)
5.39 G. Placzek: in *Handbuch der Radiologie* (Academische Verlag, Leipzig, 1934) vol. VI, part II, p. 205
5.40 C. V. Raman: *Nature (London)* **123**, 50 (1929)
5.41 E. Rutherford: *Proc. R. Soc. London* **A126**, 184 (1930)
5.42 S. Bhagavantam: In *Proceedings of the Sixth International Conference on Raman Spectroscopy*, edited by E. D. Schmid *et al.* (Heyden, London, 1978) vol. 1, p. 3
5.43 Quoted by Ramdas in ref. [5.19]

Chapter 6

6.1 C. V. Raman and K. S. Krishnan: *Philos. Mag.* **5**, 498 (1928)
6.2 H. A. Kramers and W. Heisenberg: *Z. Phys.* **31**, 681 (1925); this paper is reprinted in B. L. Van der Waerden: *Sources of Quantum Mechanics* (North–Holland, Amsterdam, 1967), where historical implications with reference to the development of quantum mechanics are discussed.
6.3 M. Born and K. Huang: *Dynamical Theory of Crystal Lattices* (Clarendon Press, Oxford, 1954)
6.4 G. Placzek: in *Handbuch der Radiologie* (Academische Verlag, Leipzig, 1934) vol. VI, part II, p. 205
6.5 L. Van Hove: *Phys. Rev.* **95**, 249 (1954)
6.6 R. Gordon: *J. Chem. Phys.* **42**, 3658 (1965)
6.7 L. D. Landau and G. Placzek: *Phys. Z. Sowjet.* **5**, 172 (1934)
6.8 C. S. Venkateswaran: Presidential Address, Physics Section, Indian Science Congress, 1951
6.9 C. V. Raman and S. Bhagavantam: *Indian J. Phys.* **6**, 353 (1931)
6.10 R. Frisch: *Z. Phys.* **61**, 626 (1931)
6.11 A. Kastler: *J. de Physique*, 159 (1931)
6.12 C. V. Raman and S. Bhagavantam: *Nature (London)* **128**, 114 (1931)
6.13 R. Bär: *Naturwissenschaften* **19**, 375, 463 (1931)
6.14 M. N. Saha and Y. Bhargava: *Nature (London)* **128**, 817 (1931)
6.15 C. V. Raman and K. S. Krishnan: *Nature (London)* **122**, 12 (1928)
6.16 M. N. Saha, D. S. Kothari and G. R. Toshniwal: *Nature (London)* **122**, 398 (1928)

Chapter 7

7.1 D. A. Long: *Raman Spectroscopy* (McGraw-Hill, New York, 1977)

7.2 N. Bloembergen: *Proceedings of the Sixth International Conference on Raman Spectroscopy*, edited by E. D. Schmid et al. (Heyden, London, 1978), vol. 1, p. 335
7.3 E. J. Woodbury and W. K. Ng: *Proc. IRE* **50**, 2367 (1962)
7.4 N. Bloembergen, H. Lotem and R. T. Lynch: *Indian J. Pure and Appl. Phys.* **16**, 151 (1978)
7.5 M. Delhaye and P. Dhamelincourt: *J. Raman Spectrosc.* **3**, 33 (1975)

Chapter 8

8.1 M. Born: *My Life – Recollections of a Nobel Laureate* (Taylor and Francis, London, 1978)
8.2 S. Ramaseshan: *C. V. Raman Memorial Lecture 1978* (Indian Institute of Science, Bangalore, 1978)
8.3 P. R. Pisharoty: *Curr. Sci.* **40**, 222 (1971)
8.4 R. Ananthakrishnan: *Curr. Sci.* **40**, 221 (1971)
8.5 C. V. Raman: *Proc. Indian Acad. Sci.* **A1**, 1 (1934)
8.6 Lord Rayleigh: *Proc. R. Soc. London* **A103**, 233 (1923)
8.7 L. A. Ramdas: *Proc. Indian Assoc. Cultiv. Sci.* **8**, 231 (1923)
8.8 C. V. Raman: *Proc. Indian Acad. Sci.* **A1**, 567 (1934)
8.9 C. V. Raman: *Proc. Indian Acad. Sci.* **A1**, 574 (1934)
8.10 C. V. Raman and V. S. Rajagopalan: *Proc. Indian Acad. Sci.* **A9**, 371 (1934)
8.11 C. V. Raman and V. S. Rajagopalan: *Proc. Indian Acad. Sci.* **A10**, 469 (1940)
8.12 G. N. Ramachandran: *Proc. Indian Acad. Sci.* **A16**, 336 (1942)
8.13 C. V. Raman and D. Krishnamurti: *Proc. Indian Acad. Sci.* **A36**, 315 (1952)
8.14 C. V. Raman: *Lectures on Physical Optics* (Indian Academy of Sciences, Bangalore, 1959)
8.15 G. N. Ramachandran: *Curr. Sci.* **40**, 212 (1971)
8.16 Bidhubushan Ray: *Proc. Indian Assoc. Cultiv. Sci.* **8**, 23 (1923)
8.17 M. N. Mitra: *Indian J. Phys.* **3**, 175 (1928)
8.18 G. N. Ramachandran: *Proc. Indian Acad. Sci.* **A17**, 171 (1943)
8.19 G. N. Ramachandran: *Proc. Indian Acad. Sci.* **A17**, 202 (1943)
8.20 G. N. Ramachandran: *Proc. Indian Acad. Sci.* **A18**, 190 (1943)
8.21 T. A. S. Balakrishnan: *Proc. Indian Acad. Sci.* **A13**, 188 (1941)
8.22 B. Ya. Zel'dovich, V. V. Shuknov, T. V. Yakovleva: *Usp. Fiz. Nauk* **149**, 511 (1986) [English translation: *Sov. Phys. Usp.* **29**, 678 (1987)]
8.23 *Laser Speckles and Related Phenomena*, edited by J. C. Dainty (Springer–Verlag, Berlin, 1975); see also *Seeing stars with speckle interferometry* by Harold A. McAlister, *Am. Sci.* March/April 1988
8.24 N. R. Isenor: *Appl. Opt.* **6**, 163 (1967)
8.25 K. Sunanda Bai: *Proc. Indian Acad. Sci.* **A13**, 439 (1941)
8.26 C. V. Raman and K. S. Venkataraman: *Proc. R. Soc. London* **A171**, 137 (1939)
8.27 K. Sunanda Bai: *Proc. Indian Acad. Sci.* **A15**, 338 (1941)
8.28 B. V. Raghavendra Rao: *Proc. Indian Acad. Sci.* **A1**, 261, 473 (1934)
8.29 C. S. Venkateswaran: Presidential Address, Physics Section, Indian Science Congress, 1951
8.30 C. V. Raman and C. S. Venkateswaran: *Nature (London)* **142**, 791 (1938); **143**, 798 (1939)
8.31 R. S. Krishnan: *Proc. Indian Acad. Sci.* **A26**, 399, 450 (1947)
8.32 R. S. Krishnan and V. Chandrasekharan: *Proc. Indian Acad. Sci.* **A31**, 427 (1950)
8.33 C. V. Raman and V. S. Rajagopalan: *J. Opt. Soc. Am.* **29**, 413 (1939)
8.34 C. V. Raman and V. S. Rajagopalan: *Proc. Indian Acad. Sci.* **A10**, 317 (1939)
8.35 C. V. Raman and V. S. Rajagopalan: *Philos. Mag.* **29**, 508 (1940)
8.36 K. Subba Ramaiah: *Proc. Indian Acad. Sci.* **A9**, 467 (1939)
8.37 C. V. Raman and K. Subba Ramaiah: *Proc. Indian Acad. Sci.* **A9**, 455 (1939)
8.38 See H. Haken: *Synergetics – An Introduction* (Springer–Verlag, Berlin, 1978), and various other volumes in the series on synergetics published by Springer.
8.39 N. S. Nagendra Nath: *Curr. Sci.* **40**, 234 (1971)
8.40 G. Venkataraman: *Bull. Mater. Sci.* **1**, 129 (1979)
8.41 C. V. Raman and T. M. K. Nedungadi: *Nature (London)* **145**, 147 (1940)
8.42 T. M. K. Nedungadi: *Proc. Indian Acad. Sci.* **A11**, 86 (1940)

8.43 G. Shirane and Y. Yamada: *Phys. Rev.* **177**, 858 (1969)
8.44 R. Blinc and B. Zeks: *Soft Modes in Ferroelectrics and Antiferroelectrics* (North-Holland, Amsterdam, 1974)
8.45 E. F. Steigmeier, H. Auderset and Harbeke: in *Anharmonic Lattices, Structural Transitions and Melting*, edited by T. Riste (Noordhoff, Leiden, 1974) p. 153
8.46 T. Schneider, G. Srinivasan and C. P. Enz: *Phys. Rev.* **A5**, 1528 (1972)
8.47 Quoted by S. Ramaseshan: *Curr. Sci.* **47**, 181 (1978)
8.48 C. V. Raman and S. Ramaseshan: *Proc. Indian Acad. Sci.* **A24**, 1 (1946)
8.49 S. Ramaseshan: *Proc. Indian Acad. Sci.* **A24**, 122 (1946)
8.50 S. Bhagavantam: *Proceedings of the Sixth International Conference on Raman Spectroscopy*, edited by E.D. Schmid et al. (Heyden, London, 1978) vol. 1, p. 3
8.51 R. Robertson, J. J. Fox and A. E. Martin: *Philos. Trans. R. Soc. London* **A232**, 482 (1934)
8.52 C. V. Raman: *Proc. Indian Acad. Sci.* **A19**, 189 (1944)
8.53 G. N. Ramachandran: *Proc. Indian Acad. Sci.* **A24**, 58 (1946)
8.54 K. Lonsdale: *Nature (London)* **155**, 144 (1945)
8.55 G. N. Ramachandran: *Nature (London)* **156**, 83 (1945)
8.56 W. Kaiser and W. L. Bond: *Phys. Rev.* **115**, 857 (1959)
8.57 S. Ramaseshan: *Curr. Sci.* **47**, 181 (1978)
8.58 Quoted by S. Bhagavantam: *Professor Chandrasekhara Venkata Raman* (Andhra Pradesh Academy of Sciences, Hyderabad, 1972)
8.59 Raman Golden Jubilee Commemoration Volume, also *Proc. Indian Acad. Sci.*, No. 5 (November), 1938
8.60 Lord Penny: in *Biographical Memoirs of the Royal Society* (Royal Society, London, 1967) vol. 13, p. 35
8.61 Sir John Cockcroft: *Commemoration Lecture* (The Royal Institution, London, 1967) p. 5
8.62 M. G. K. Menon: *Commemoration Lecture* (The Royal Institution, London, 1967) p. 17
8.63 B. V. Sreekantan: in *Collected Scientific Papers of Homi Jehangir Bhabha*, edited by B. V. Sreekantan, Virendra Singh and B. M. Udgaonkar (Tata Institute of Fundamental Research, Bombay, 1985)
8.64 H. J. Bhabha: Lecture delivered before the International Council of Scientific Unions in Bombay in January 1966.
8.65 C. V. Raman and P. Nilakantan: *Proc. Indian Acad. Sci.* **A11**, 379, 389, 398 (1940)
8.66 C. V. Raman and N. S. Nagendra Nath: *Proc. Indian Acad. Sci.* **A12**, 83, 141 (1940)
8.67 C. V. Raman and P. Nilakantan: *Curr. Sci.* **9**, 165 (1940)
8.68 C. V. Raman: *Curr. Sci.* **17**, 65 (1948)

Chapter 9

9.1 C. V. Raman and N. S. Nagendra Nath: *Proc. Indian Acad. Sci.* **A2**, 406 (1935); **A2**, 413 (1935); **A3**, 75 (1936); **A3**, 119 (1936); **A3**, 459 (1936)
9.2 N. S. Nagendra Nath: *Proc. Indian Acad. Sci.* **A4**, 222 (1936); **A8**, 499 (1938)
9.3 S. Parthasarathy: *Proc. Indian Acad. Sci.* **A3**, 442 (1936)
9.4 D. S. Subbaramaiya: *Proc. Indian Acad. Sci.* **A6**, 333 (1937)
9.5 L. Brillouin: *Ann. Physique (Paris)* **17**, 88 (1922)
9.6 P. Debye and F. W. Sears: *Proc. Natl. Acad. Sci. USA* **18**, 410 (1932)
9.7 R. Lucas and P. Biquard: *J. Phys. Rad.* **3**, 464 (1932)
9.8 R. Bär: *Helv. Phys. Acta* **8**, 591 (1935)
9.9 R. Bär: *Helv. Phys. Acta* **9**, 265 (1936)
9.10 L. Brillouin: *La Diffraction de la Lumiere par des Ultrasons* (Paris, 1933)
9.11 S. Ramaseshan: *C. V. Raman Memorial Lecture 1978* (Indian Institute of Science, Bangalore, 1978)
9.12 F. H. Sanders: *Can. J. Res.* **A14**, 158 (1936)
9.13 F. H. Sanders: *Nature (London)* **138**, 285 (1936)
9.14 C. V. Raman and N. S. Nagendra Nath: *Nature (London)* **138**, 616 (1936)

9.15 W. R. Klein and E. A. Hiedemann: *Physica* **29**, 981 (1963)
9.16 O. Nomoto: *Bull. Kobayashi Inst. Phys. Res.* **1**, 42 (1951)
9.17 R. Mertens: *Meded. K. Vlaam. Acad.* **12**, 1 (1950)
9.18 M. V. Berry: *The Diffraction of Light by Ultrasound* (Academic Press, New York, 1966)
9.19 This topic was elegantly reviewed by R. Nityananda at the Annual Meeting of the Indian Academy of Sciences in 1986. Unfortunately, the full text of this lecture is not available in print.
9.20 G. I. Stegeman: *IEEE Trans. Sonics and Ultrasonics* **SU23**, 33 (1976)
9.21 A. Alippi, A. Palma, L. Palmieri and G. Socino: *Appl. Phys. Lett.* **26**, 357 (1975)
9.22 R. C. Extermann and G. Wannier: *Helv. Phys. Acta* **9**, 520 (1936)
9.23 A. B. Bhatia and W. J. Noble: *Proc. R. Soc. London* **A220**, 356 (1953)
9.24 G. Molière: *Z. Naturforsch.* **29**, 133 (1947)

Chapter 10

10.1 S. Bhagavantam: *Proceedings of the Sixth International Conference on Raman Spectroscopy*, edited by E. D. Schmid *et al.* (Heyden, London, 1978) vol. 1, p. 3
10.2 C. V. Raman: *Proc. Indian Acad. Sci.* **A34**, 61 (1951)
10.3 C. V. Raman: *Proc. Indian Acad. Sci.* **A43**, 327 (1956)
10.4 C. V. Raman: *Proc. Indian Acad. Sci.* **A42**, 163 (1955)
10.5 M. Born and K. Huang: *Dynamical Theory of Crystal Lattices* (Clarendon Press, Oxford, 1954)
10.6 M. Born: *Rev. Mod. Phys.* **17**, 245 (1945)
10.7 M. Born: *My Life – Recollections of a Nobel Laureate* (Taylor and Francis, London, 1978)
10.8 R. S. Krishnan: *Proc. Indian Acad. Sci.* **A24**, 25 (1946)
10.9 C. V. Raman: *Proc. Indian Acad. Sci.* **A44**, 99 (1956)
10.10 M. Born and Mary Bradburn: *Proc. R. Soc. London* **A188**, 161 (1947)
10.11 J. L. Warren, J. L. Yarrell, G. Dolling and R. A. Cowley: *Phys. Rev.* **158**, 805 (1966)
10.12 K. S. Viswanathan: *Proc. Indian Acad. Sci.* **A37**, 424, 435 (1953)
10.13 L. Van Hove: *Phys. Rev.* **98**, 1189 (1955)
10.14 S. A. Solin and A. K. Ramdas: *Phys. Rev.* **B1**, 1687 (1970)
10.15 R. A. Loudon: *Adv. Phys.* **13**, 423 (1964)
10.16 G. Venkataraman, L. A. Feldkamp and V. C. Sahni: *Dynamics of Perfect Crystals* (MIT Press, Boston, 1974)
10.17 J. C. Phillips: *Phys. Rev.* **104**, 1263 (1956)

Chapter 11

11.1 L. L. Fermor: Inaugural Address to the National Institute of Sciences of India, January 7, 1935
11.2 *Current Science*, Editorial, May 1933
11.3 M. Born: Letter to Lord Rutherford dated October 22, 1936
11.4 *Indian Academy of Sciences–the First Fifty Years* (Indian Academy of Sciences, Bangalore, 1984) p. 111
11.5 S. Bhagavantam: *Proceedings of the Sixth International Conference on Raman Spectroscopy*, edited by E. D. Schmid *et al.* (Heyden, London, 1978) vol. 1, p. 12

Chapter 12

12.1 S. Bhagavantam: *Professor Chandrasekhara Venkata Raman* (Andhra Pradesh Academy of Sciences, Hyderabad, 1972)
12.2 A. Jayaraman: Private communication
12.3 C. V. Raman: *Curr. Sci.* **12**, 197, 289, 313 (1943)
12.4 C. V. Raman: *Proc. Indian Acad. Sci.* **A29**, 381 (1949)
12.5 C. V. Raman and S. Ramaseshan: *Proc. Indian Acad. Sci.* **A30**, 211 (1949)

12.6 C. V. Raman and S. Ramaseshan: *Proc. Indian Acad. Sci.* **A30**, 277 (1949)
12.7 C. V. Raman and M. R. Bhat: *Proc. Indian Acad. Sci.* **A41**, 61 (1955)
12.8 C. V. Raman and K. S. Viswanathan: *Proc. Indian Acad. Sci.* **A39**, 55 (1955)
12.9 C. V. Raman and A. Jayaraman: *Proc. Indian Acad. Sci.* **A40**, 189 (1954)
12.10 C. V. Raman and A. Jayaraman: *Proc. Indian Acad. Sci.* **A40**, 221 (1954)
12.11 H. G. Jerrard: *J. Opt. Soc. Am.* **39**, 859 (1949)
12.12 G. N. Ramachandran and S. Ramaseshan: *J. Opt. Soc. Am.* **42**, 49 (1952)
12.13 *Collected Works of S. Pancharatnam* (Oxford University Press, for the Raman Research Institute, Bangalore, 1975)
12.14 Foreword by G. W. Series in ref. [12.13]
12.15 M. V. Berry: *Proc. R. Soc. London* **A392**, 45 (1984)
12.16 S. Ramaseshan and R. Nityananda: *Curr. Sci.* **55**, 1225 (1986)
12.17 A. Tomita and R. Y. Chiad: *Phys. Rev. Lett.* **57**, 937 (1986)
12.18 G. Delacretaz, E. R. Grant, R. L. Whetten, L. Woste and J. W. Zwanziger: *Phys. Rev. Lett.* **56**, 2598 (1986)
12.19 J. H. Hannay: *J. Phys.* **A18**, 221 (1985)
12.20 Baldev Singh: *Jawaharlal Nehru on Science–Speeches delivered at the Annual Sessions of the Indian Science Congress* (Nehru Memorial Museum and Library, New Delhi, 1986). All the quotations from Nehru are taken from this book
12.21 *Collected Works of Meghnad Saha*, edited by S. Chatterjee (Orient Longman, Bombay, 1987) vol. 2
12.22 The Bharatiya Vidya Bhavan, Bombay, brought out a special issue of the *Bhavan's Journal* in honour of Raman in December 1970. Many quotations are extracted from this
12.23 H. J. Bhabha: Lecture delivered before the International Council of Scientific Unions in Bombay in January 1966
12.24 S. Ramaseshan: *C. V. Raman Memorial Lecture 1978* (Indian Institute of Science, Bangalore, 1978)
12.25 C. V. Raman: *The Physiology of Vision* (The Indian Academy of Sciences, Bangalore, 1968)
12.26 S. Ramaseshan: *Curr. Sci.* **40**, 248 (1971)

Chapter 13

13.1 C. V. Raman: *The New Physics: Talks on Aspects of Science* (Philosophical Library, New York, 1951)
13.2 H. Meinhart and M. Klingler: *J. Theor. Biol.* **126**, 63 (1987)
13.3 R. P. Feynman: *Lectures in Physics* (Addison–Wesley, Reading, Mass., USA, 1964) vol. 2, chapter 9
13.4 S. Chandrasekhar: *Curr. Sci.* **54**, 161 (1985)

Chapter 14

14.1 S. Ramaseshan: *C. V. Raman Memorial Lecture 1978* (Indian Institute of Science, Bangalore, 1978)

Appendix 1 HONOURS BESTOWED ON RAMAN

1912   Curzon Research Prize
1913   Woodburn Research Medal
1924   Elected Fellow of the Royal Society, London
1928   Matteucci Medal – Societa Instaliana Della Scienza, Rome
1929   Knighted by the British Government in India
1930   Hughes Medal – Royal Society, London
1930   Nobel Prize
1935   Rajasabhabhushana – Decoration by the Maharajah of Mysore
1941   Franklin Medal – Franklin Institute, Philadelphia
1948   Appointed National Professor
1954   Bharata Ratna – Decoration by the President of India
1957   Lenin Prize, USSR

Honorary Doctorates from the Universities of:

   Allahabad, Benaras, Bombay, Calcutta. Dacca, Delhi, Freiburg, Glasgow, Kanpur, Lucknow, Madras, Mysore, Paris and Patna, Osmania University, Hyderabad, and Sri Venkateswara University, Tirupati

Honorary Member:

   Deutsche Akademie of Munich
   Hungarian Academy of Sciences
   Indian Science Congress Association and several other Indian science organisations
   Royal Irish Academy
   Royal Philosophical Society, Glasgow
   Zurich Physical Society

Honorary Fellow:

   Optical Society of America
   Mineralogical Society of America

Foreign Associate:

    Academy of Sciences, Paris

Foreign Member:

    Academy of Sciences, USSR

Honorary Member:

    Academy of the Socialist Republic of Romania
    Catgut Acoustical Society

General President:

    Indian Science Congress, 1929

President:

    Indian Academy of Sciences, 1934–1970

Appendix 2 BIBLIOGRAPHY

(Consolidated list of C V Raman's scientific papers published in six volumes by the Indian Academy of Sciences, 1988)

Volume I. Scattering of Light

1. THE DOPPLER EFFECT IN THE MOLECULAR SCATTERING OF RADIATION [1919 *Nature (London)* **103** 165]
2. ON THE TRANSMISSION COLOURS OF SULPHUR SUSPENSIONS [1921 *Proc. R. Soc. London* **A100** 102; with B B Ray]
3. A METHOD OF IMPROVING VISIBILITY OF DISTANT OBJECTS [1921 *Nature (London)* **108** 242]
4. THE COLOUR OF THE SEA [1921 *Nature (London)* **108** 367]
5. THE MOLECULAR SCATTERING OF LIGHT IN LIQUIDS AND SOLIDS [1921 *Nature (London)* **108** 402]
6. ON THE MOLECULAR SCATTERING OF LIGHT IN WATER AND THE COLOUR OF THE SEA [1922 *Proc. R. Soc. London* **A101** 64]
7. OPTICAL OBSERVATIONS OF THE THERMAL AGITATION OF THE ATOMS IN CRYSTALS [1922 *Nature (London)* **109** 42]
8. ANISOTROPY OF MOLECULES [1922 *Nature (London)* **109** 75]
9. MOLECULAR STRUCTURE OF AMORPHOUS SOLIDS [1922 *Nature (London)* **109** 138]
10. MOLECULAR DIFFRACTION OF LIGHT [1922 The Calcutta University Press 103 pages]
11. DIFFRACTION BY MOLECULAR CLUSTERS AND THE QUANTUM STRUCTURE OF LIGHT [1922 *Nature (London)* **109** 444]
12. MOLECULAR AELOTROPY IN LIQUIDS [1922 *Nature (London)* **110** 11]
13. OPALESCENCE PHENOMENA IN LIQUID MIXTURES [1922 *Nature (London)* **110** 77]
14. TRANSPARENCY OF LIQUIDS AND COLOUR OF THE SEA [1922 *Nature (London)* **110** 280]
15. THERMAL OPALESCENCE IN CRYSTALS AND THE COLOUR OF ICE IN GLACIERS [1923 *Nature (London)* **111** 13]
16. ON THE MOLECULAR SCATTERING OF LIGHT IN DENSE VAPOURS AND GASES [1923 *Philos. Mag.* **45** 113; with K R Ramanathan]
17. ON THE MOLECULAR SCATTERING AND EXTINCTION OF LIGHT IN LIQUIDS AND THE DETERMINATION OF THE AVOGADRO CONSTANT [1923 *Philos. Mag.* **45** 625; with K Seshagiri Rao]
18. THE MOLECULAR SCATTERING OF LIGHT IN LIQUID MIXTURES [1923 *Philos. Mag.* **45** 213; with K R Ramanathan]
19. ON THE POLARIZATION OF THE LIGHT SCATTERED BY GASES AND VAPOURS [1923 *Philos. Mag.* **46** 426; with K Seshagiri Rao]
20. THE MOLECULAR SCATTERING OF LIGHT IN CARBON DIOXIDE AT HIGH PRESSURES [1923 *Proc. R. Soc. London* **A104** 357; with K R Ramanathan]
21. THE SCATTERING OF LIGHT BY ANISOTROPIC MOLECULES [1923 *Nature (London)* **112** 165]
22. THE STRUCTURE OF MOLECULES IN RELATION TO THEIR OPTICAL ANISOTROPY [1924 *Nature (London)* **114** 49]
23. THE SCATTERING OF LIGHT BY LIQUID AND SOLID SURFACES [1923 *Nature (London)* **112** 281]
24. RELATION OF TYNDALL EFFECT TO OSMOTIC PRESSURE IN COLLOIDAL SOLUTIONS [1927 *Indian J. Phys.* **2** 1]
25. THE SCATTERING OF X-RAYS IN LIQUIDS [1923 *Nature (London)* **111** 185]
26. THE NATURE OF THE LIQUID STATE [1923 *Nature (London)* **111** 428]

27. The diffraction of X-rays in liquids, liquid mixtures, solutions, fluid crystals and amorphous solids [1923 *Proc. Indian Assoc. Cultiv. Sci.* **8** 127; with K R Ramanathan]
28. On the mean distance between neighbouring molecules in a fluid [1924 *Philos. Mag.* **47** 671]
29. The scattering of light by liquid boundaries and its relation to surface tension—Part I [1925 *Proc. R. Soc. London* **A108** 561; with L A Ramdas]
30. The scattering of light by liquid boundaries and its relation to surface tension—Part II [1925 *Proc. R. Soc. London* **A109** 150; with L A Ramdas]
31. The scattering of light by liquid boundaries and its relation to surface tension—Part III [1925 *Proc. R. Soc. London* **A109** 272; with L A Ramdas]
32. Die Zerstreuung des Lichtes durch dielektrische Kügeln (*German*) [1925 *Z. Phys* **33** 870]
33. On the thickness of the optical transition layer in liquid surfaces [1927 *Philos. Mag.* **3** 220; with L A Ramdas]
34. The birefringence of crystalline carbonates, nitrates and sulphates [1926 *Nature (London)* **118** 264]
35. The electrical polarity of molecules [1926 *Nature (London)* **118** 302; with K S Krishnan]
36. Magnetic double-refraction in liquids, Part I: Benzene and its derivatives [1927 *Proc. R. Soc. London* **A113** 511; with K S Krishnan]
37. Electric double-refraction in relation to the polarity and optical anisotropy of molecules, Part I: Gases and vapours [1927 *Philos. Mag.* **3** 713; with K S Krishnan]
38. Electric double-refraction in relation to the polarity and optical anisotropy of molecules, Part II: Liquids [1927 *Philos. Mag.* **3** 724; with K S Krishnan]
39. Disappearance and reversal of the Kerr effect [1928 *Nature (London)* **121** 794; with S C Sirkar]
40. Optique—La constante de birefringence magnetique du benzene (*French*) [1927 *C. R. Acad. Sci. Paris* **184** 449; with K S Krishnan]
41. Magnetic double refraction [1927 *Nature (London)* **119** 528; with I Ramakrishna Rao]
42. The magnetic anisotropy of crystalline nitrates and carbonates [1927 *Proc. R. Soc. London* **A115** 549; with K S Krishnan]
43. A theory of electric and magnetic birefringence in liquids [1927 *Proc. R. Soc. London* **A117** 1; with K S Krishnan]
44. A theory of the optical and electrical properties of liquids [1928 *Proc. R. Soc. London* **A117** 589; with K S Krishnan]
45. The Maxwell effect in liquids [1927 *Nature (London)* **120** 726; with K S Krishnan]
46. A theory of the birefringence induced by flow in liquids [1928 *Philos. Mag.* **5** 769; with K S Krishnan]
47. The scattering of light in amorphous solids [1927 *J. Opt. Soc. Am.* **15** 185]
48. The molecular scattering of light in a binary liquid mixture [1927 *Philos. Mag.* **4** 447]
49. A theory of light-scattering in liquids [1929 *Philos. Mag.* **5** 498; with K S Krishnan]
50. The theory of light-scattering in liquids [1929 *Philos. Mag.* **7** 160]
51. Optical behaviour of protein solutions [1927 *Nature (London)* **120** 158]
52. X-ray diffraction in liquids [1927 *Nature (London)* **119** 601; with C M Sogani]
53. X-ray diffraction in liquids [1927 *Nature (London)* **120** 514; with C M Sogani]
54. Thermal degeneration of the X-ray haloes in liquids [1927 *Nature (London)* **120** 770]
55. A critical-absorption photometer for the study of the Compton effect [1928 *Proc. R. Soc. London* **A119** 526; with C M Sogani]

56. THERMODYNAMICS, WAVE-THEORY AND THE COMPTON EFFECT [1927 *Nature (London)* **120** 950]
57. A CLASSICAL DERIVATION OF THE COMPTON EFFECT [1928 *Indian J. Phys.* **3** 357]
58. A NEW TYPE OF SECONDARY RADIATION [1928 *Nature (London)* **121** 501; with K S Krishnan]
59. A CHANGE OF WAVELENGTH IN LIGHT-SCATTERING [1928 *Nature (London)* **121** 619]
60. A NEW RADIATION [1928 *Indian J. Phys.* **2** 387]
61. THE OPTICAL ANALOGUE OF THE COMPTON EFFECT [1928 *Nature (London)* **121** 711; with K S Krishnan]
62. A NEW CLASS OF SPECTRA DUE TO SECONDARY RADIATION, PART I [1928 *Indian J. Phys.* **2** 399; with K S Krishnan]
63. THE NEGATIVE ABSORPTION OF RADIATION [1928 *Nature (London)* **122** 12; with K S Krishnan]
64. POLARIZATION OF SCATTERED LIGHT-QUANTA [1928 *Nature (London)* **122** 169; with K S Krishnan]
65. MOLECULAR SPECTRA IN THE EXTREME INFRARED [1928 *Nature (London)* **122** 278; with K S Krishnan]
66. THE PRODUCTION OF NEW RADIATIONS BY LIGHT SCATTERING—PART I [1929 *Proc. R. Soc. London* **A122** 23; with K S Krishnan]
67. ROTATION OF MOLECULES INDUCED BY LIGHT [1928 *Nature (London)* **122** 882; with K S Krishnan]
68. INVESTIGATIONS OF THE SCATTERING OF LIGHT [1929 *Nature (London)* **123** 50]
69. THE RAMAN EFFECT: INVESTIGATION OF MOLECULAR STRUCTURE BY LIGHT SCATTERING [1929 *Trans. Faraday Soc.* **25** 781]
70. THE MOLECULAR SCATTERING OF LIGHT, NOBEL LECTURE DELIVERED AT STOCKHOLM, 11th December 1930
71. COLOUR AND OPTICAL ANISOTROPY OF ORGANIC COMPOUNDS [1929 *Nature (London)* **123** 494]
72. MAGNETIC BEHAVIOUR OF ORGANIC CRYSTALS [1929 *Nature (London)* **123** 605]
73. THE RELATION BETWEEN COLOUR AND MOLECULAR STRUCTURE IN ORGANIC COMPOUNDS [1929 *Indian J. Phys.* **4** 57; with S Bhagavantam]
74. DIAMAGNETISM AND CRYSTAL STRUCTURE [1929 *Nature (London)* **123** 945]
75. A NEW X-RAY EFFECT [1929 *Nature (London)* **124** 53; with P Krishnamurti]
76. ANOMALOUS DIAMAGNETISM [1929 *Nature (London)* **124** 412]
77. DIAMAGNETISM AND MOLECULAR STRUCTURE [1930 *Proc. Phys. Soc.* **42** 309]
78. A NEW TYPE OF MAGNETIC BIREFRINGENCE [1931 *Nature (London)* **128** 758; with S W Chinchalkar]
79. ATOMS AND MOLECULES AS FITZGERALD OSCILLATORS [1931 *Nature (London)* **128** 795]
80. EVIDENCE FOR THE SPIN OF THE PHOTON FROM LIGHT-SCATTERING [1931 *Nature (London)* **128** 114; with S Bhagavantam]
81. THE ANGULAR MOMENTUM OF LIGHT [1931 *Nature (London)* **128** 545]
82. EXPERIMENTAL PROOF OF THE SPIN OF THE PHOTON [1931 *Indian J. Phys.* **6** 353; with S Bhagavantam]
83. EXPERIMENTAL PROOF OF THE SPIN OF THE PHOTON [1932 *Nature (London)* **129** 22; with S Bhagavantam]
84. DOPPLER EFFECT IN LIGHT-SCATTERING [1931 *Nature (London)* **128** 636]
85. NATURE OF THE THERMAL AGITATION IN LIQUIDS [1935 *Nature (London)* **135** 761; with B V Raghavendra Rao]
86. ACOUSTIC SPECTRUM OF LIQUIDS [1937 *Nature (London)* **139** 584; with B V Raghavendra Rao]
87. LIGHT SCATTERING AND FLUID VISCOSITY [1938 *Nature (London)* **141** 242; with B V Raghavendra Rao]
88. NEW METHODS IN THE STUDY OF LIGHT SCATTERING, PART I: BASIC IDEAS [1941 *Proc. Indian Acad. Sci.* **A14** 228]

89. DETERMINATION OF THE ADIABATIC PIEZO-OPTIC COEFFICIENT OF LIQUIDS [1939 *Proc. R. Soc. London* **A171** 137; with K S Venkataraman]
90. SPECTROSCOPIC INVESTIGATION OF THE SOLID AND LIQUID STATES [1942 *Curr. Sci.* **11** 225]
91. THE NATURE OF THE LIQUID STATE [1942 *Curr. Sci.* **11** 303]
92. THE $\alpha - \beta$ TRANSFORMATION OF QUARTZ [1940 *Nature (London)* **145** 147; with T M K Nedangadi]
93. LATTICE OSCILLATIONS IN CRYSTALS [1939 *Nature (London)* **143** 679; with T M K Nedungadi]
94. SCATTERING OF LIGHT IN CRYSTALS [1945 *Nature (London)* **155** 396]

Volume II. Acoustics

1. Vibrations and Wave Motions

95. THE SMALL MOTION AT THE NODES OF A VIBRATING STRING [1909 *Nature (London)* **82** 9]
96. THE MAINTENANCE OF FORCED OSCILLATIONS OF A NEW TYPE (1909 *Nature (London)* **82** 156]
97. THE MAINTENANCE OF FORCED OSCILLATIONS [1910 *Nature (London)* **82** 428]
98. PHOTOGRAPHS OF VIBRATION CURVES [1911 *Philos. Mag.* **21** 615]
99. REMARKS ON A PAPER BY J S STOKES ON "SOME CURIOUS PHENOMENA OBSERVED IN CONNECTION WITH MELDE'S EXPERIMENT" [1911 *Phys. Rev.* **32** 307]
100. THE SMALL MOTION AT THE NODES OF A VIBRATING STRING [1911 *Phys. Rev.* **32** 309]
101. THE MAINTENANCE OF FORCED OSCILLATIONS OF A NEW TYPE [1912 *Philos. Mag.* **24** 513]
102. SOME REMARKABLE CASES OF RESONANCE [1912 *Phys. Rev.* **35** 449]
103. EXPERIMENTAL INVESTIGATIONS ON THE MAINTENANCE OF VIBRATIONS [1912 *Bull. Indian Assoc. Cultiv. Sci.* **6**, 1]
104. SOME ACOUSTICAL OBSERVATIONS [1913 *Bull. Indian Assoc. Cultiv. Sci.* **8** 17]
105. THE MAINTENANCE OF VIBRATIONS [1914 *Phys. Rev.* **4** 12]
106. ON MOTION IN A PERIODIC FIELD OF FORCE [1914 *Bull. Indian Assoc. Cultiv. Sci.* **11** 25]
107. ON MOTION IN A PERIODIC FIELD OF FORCE [1915 *Philos. Mag.* **29** 15]
108. THE MAINTENANCE OF VIBRATIONS BY A PERIODIC FIELD OF FORCE [1917 *Philos. Mag.* **34** 129; with A Dey]
109. ON THE MAINTENANCE OF COMBINATIONAL VIBRATIONS BY TWO SIMPLE HARMONIC FORCES [1915 *Phys. Rev.* **5** 1]
110. ON DISCONTINUOUS WAVE-MOTION, Part I [1916 *Philos. Mag.* **31** 47; with S Appaswamaiyar]
111. ON DISCONTINUOUS WAVE-MOTION, Part II [1917 *Philos. Mag.* **33** 203; with A Dey]
112. ON DISCONTINUOUS WAVE-MOTION, Part III [1917 *Philos. Mag.* **33** 352; with A Dey]
113. AN EXPERIMENTAL METHOD FOR THE PRODUCTION OF VIBRATIONS [1919 *Phys. Rev.* **14** 446]
114. A NEW METHOD FOR THE ABSOLUTE DETERMINATION OF FREQUENCY [1919 *Proc. R. Soc. London* **A95** 533; with A Dey]
115. WHISPERING GALLERY PHENOMENA AT ST. PAUL'S CATHEDRAL [1921 *Nature (London)* **108** 42; with G A Sutherland]

116. On the whispering gallery phenomenon [1922 *Proc. R. Soc. London* **A100** 424; with G A Sutherland]
117. On whispering galleries [1922 *Bull. Indian Assoc. Cultiv. Sci.* **7** 159]
118. On the sounds of splashes [1920 *Philos. Mag.* **39** 145; with A Dey]
119. The nature of vowel sounds [1921 *Nature (London)* **107** 332]

2. Musical Instruments—The Violin and the Pianoforte

120. The dynamical theory of the motion of bowed strings [1914 *Bull. Indian Assoc. Cultiv. Sci.* **11** 43]
121. On the "wolf-note" of the violin and 'cello [1916 *Nature (London)* **97** 362]
122. On the "wolf note" in the bowed stringed instruments [1916 *Philos. Mag.* **32** 391]
123. On the alterations of tone produced by a violin "mute" [1917 *Nature (London)* **100** 84]
124. On the "wolf-note" in bowed stringed instruments [1918 *Philos. Mag.* **35** 493]
125. The "wolf-note" in pizzicato playing [1918 *Nature (London)* **101** 264]
126. On the mechanical theory of the vibrations of bowed strings and of musical instruments of the violin family, with experimental verification of the results—Part I [1918 *Bull. Indian Assoc. Cultiv. Sci.* **15** 1]
127. On the partial tones of bowed stringed instruments [1919 *Philos. Mag.* **38** 573]
128. The kinematics of bowed strings [1919 *J. Dept. of Sci.*, Univ. Calcutta **1** 15]
129. On a mechanical violin-player for acoustical experiments [1920 *Philos. Mag.* **39** 535]
130. Experiments with mechanically-played violins [1920 *Proc. Indian Assoc. Cultiv. Sci.* **6** 19]
131. The theory of the cyclical vibrations of a bowed string [1918 *Bull. Indian Assoc. Cultiv. Sci.* **5** 1]
132. The subjective analysis of musical tones [1926 *Nature (London)* **117** 450]
133. On Kaufmann's theory of the impact of the pianoforte hammer [1920 *Proc. R. Soc. London* **A97** 99; with B Banerji]

3. Musical Instruments of India

134. 'The Ectara', [1909 *J. Indian Math. Club* 170]
135. Oscillations of the stretched strings [1910 *J. Indian Math. Club* 14]
136. Musical drums with harmonic overtones [1920 *Nature (London)*, **104** 500; with S Kumar]
137. The Indian musical drums [1935 *Proc. Indian Acad. Sci.* **A1** 179]
138. On some Indian stringed instruments [1921 *Proc. Indian Assoc. Cultiv. Sci.* **7** 29]
139. The acoustical knowledge of the ancient Hindus, Asutosh Mookerjee Silver Jubilee Volume, (Calcutta: University Press)[1922 **2** 179]

4. Monograph

140. Musical instruments and their tones [1927 *Handb. Phys.* **8** 354]

5. Ultrasonics

141. THE DIFFRACTION OF LIGHT BY HIGH FREQUENCY SOUND WAVES: Part I [1936 *Proc. Indian Acad. Sci.* **A2** 406; with N S Nagendra Nath]
142. THE DIFFRACTION OF LIGHT BY SOUND WAVES OF HIGH FREQUENCY: PART II [1936 *Proc. Indian Acad. Sci.* **A2** 413; with N S Nagendra Nath]
143. THE DIFFRACTION OF LIGHT BY HIGH FREQUENCY SOUND WAVES: PART III, DOPPLER EFFECT AND COHERENCE PHENOMENA [1936 *Proc. Indian Acad. Sci.* **A3** 75; with N S Nagendra Nath]
144. THE DIFFRACTION OF LIGHT BY HIGH FREQUENCY SOUND WAVES: PART IV, GENERALISED THEORY [1936 *Proc. Indian Acad. Sci.* **A3** 119; with N S Nagendra Nath]
145. THE DIFFRACTION OF LIGHT BY HIGH FREQUENCY SOUND WAVES: PART V. GENERAL CONSIDERATIONS—OBLIQUE INCIDENCE AND AMPLITUDE CHANGES [1936 *Proc. Indian Acad. Sci.* **A3** 459; with N S Nagendra Nath]
146. DIFFRACTION OF LIGHT BY ULTRASONIC WAVES [1936 *Nature (London)* **138** 616; with N S Nagendra Nath]
147. NATURE OF THE THERMAL AGITATION IN LIQUIDS [1935 *Nature (London)* **135** 761; with B V Raghavendra Rao]
148. ACOUSTIC SPECTRUM OF LIQUIDS [1937 *Nature (London)* **139** 584; with B V Raghavendra Rao]
149. LIGHT SCATTERING AND FLUID VISCOSITY [1938 *Nature (London)* **141** 242; with B V Raghavendra Rao]

Volume III. Optics

150. UNSYMMETRICAL DIFFRACTION-BANDS DUE TO A RECTANGULAR APERTURE [1906 *Philos. Mag.* **12** 494]
151. NEWTON'S RINGS IN POLARISED LIGHT [1907 *Nature (London)* **76** 637]
152. SECONDARY WAVES OF LIGHT [1908 *Nature (London)* **78** 55]
153. HISTORICAL NOTE ON THE DISCOVERY OF THE ULTRAMICROSCOPIC METHOD [1909 *Philos. Mag.* **17** 495]
154. THE EXPERIMENTAL STUDY OF HUYGENS'S SECONDARY WAVES [1909 *Philos. Mag.* **17** 204]
155. THE PHOTOMETRIC MEASUREMENT OF THE OBLIQUITY FACTOR OF DIFFRACTION [1909 *Nature (London)* **82** 69]
156. THE PHOTOMETRIC MEASUREMENT OF THE OBLIQUITY FACTOR OF DIFFRACTION [1911 *Philos. Mag.* **21** 618]
157. ON INTERMITTENT VISION [1915 *Philos. Mag.* **30** 701]
158. THE COLOURS OF THE STRIAE IN MICA [1918 *Nature (London)* **102** 205; with P N Ghosh]
159. ON THE DIFFRACTION FIGURES DUE TO AN ELLIPTIC APERTURE [1919 *Phys. Rev.* **13** 259]
160. THE SCATTERING OF LIGHT IN THE REFRACTIVE MEDIA OF THE EYE [1919 *Philos. Mag.* **38** 568]
161. THE "RADIANT" SPECTRUM [1921 *Nature (London)* **108** 12]
162. THE "RADIANT" SPECTRUM [1922 *Nature (London)* **109** 175]
163. ON THE PHENOMENON OF THE "RADIANT SPECTRUM" OBSERVED BY SIR DAVID BREWSTER [1922 *Philos. Mag.* **43** 357]
164. ON THE COLOURS OF MIXED PLATES—PART I [1921 *Philos. Mag.* **41** 338; with B Banerji]
165. ON THE COLOURS OF MIXED PLATES—PART II [1921 *Philos. Mag.* **41** 860; with B Banerji]

166. On the colours of mixed plates—Part III [1921 *Philos. Mag.* **42** 679; with K Seshagiri Rao]
167. On Quetelet's rings and other allied phenomena [1921 *Philos. Mag.* **42** 826; with G L Datta]
168. The colours of breathed-on plates [1921 *Nature (London)* **107** 714]
169. The colours of tempered steel [1922 *Nature (London)* **109** 105]
170. A method of improving visibility of distant objects [1921 *Nature (London)* **108** 242]
171. The spectrum of neutral helium [1922 *Nature (London)* **110** 700]
172. On the spectrum of neutral helium—Part I [1923 *Atrophys. J.* **57** 243; with A S Ganesan]
173. On the spectrum of neutral helium—Part II [1924 *Astrophys. J.* **59** 61; with A S Ganesan]
174. Anomalous dispersion and multiplet lines in spectra [1925 *Nature (London)* **115** 946; with S K Datta]
175. On Einstein's aberration experiment [1922 *Astrophys. J.* **56** 29]
176. Einstein's aberration experiment [1922 *Nature (London)* **109** 477]
177. On the convection of light (Fizeau effect) in moving gases [1922 *Philos. Mag.* **43** 447; with N K Sethi]
178. Conical refraction in biaxial crystals [1921 *Nature (London)* **107** 747]
179. On a new optical property of biaxial crystals [1922 *Philos. Mag.* **43** 510; with V S Tamma]
180. The effect of dispersion on the interference figures of crystals [1924 *Nature (London)* **113** 127]
181. The optical properties of amethyst quartz [1925 *Trans. Opt. Soc. London* **26** 289; with K Banerji]
182. On Brewster's bands—Part I [1925 *Trans. Opt. Soc. Am* **26** 51; with S K Datta]
183. On the diffraction of light by spherical obstacles [1926 *Proc. Phys. Soc. London* **38** 350; with K S Krishnan]
184. On the nature of the disturbance in the second medium in total reflection [1925 *Philos. Mag.* **50** 812]
185. On the total reflection of light [1926 *Proc. Indian Assoc. Cultiv. Sci.* **9** 271]
186. Huygens's principle and the phenomena of total reflection [1927 *Trans. Opt. Soc. London* **28** 149]
187. The diffraction of light by metallic screens [1927 *Proc. R. Soc. London* **A116** 254; with K S Krishnan]
188. Diffraction of light by a transparent lamina [1927 *Proc. Phys. Soc. London* **39** 453; with I Ramakrishna Rao]
189. The diffraction of light by high frequency sound waves: Part I [1936 *Proc. Indian Acad. Sci.* **A2** 406; with N S Nagendra Nath]
190. The diffraction of light by sound waves of high frequency: Part II [1936 *Proc. Indian Acad. Sci.* **A2** 413; with N S Nagendra Nath]
191. The diffraction of light by high frequency sound waves: Part III, Doppler effect and coherence phenomena [1936 *Proc. Indian Acad. Sci.* **A3** 75; with N S Nagendra Nath]
192. The diffraction of light by high frequency sound waves: Part IV, Generalised theory [1936 *Proc. Indian Acad. Sci.* **A3** 119; with N S Nagendra Nath]
193. The diffraction of light by high frequency sound waves: Part V, General considerations—Oblique incidence and amplitude changes [1936 *Proc. Indian Acad. Sci.* **A3** 459; with N S Nagendra Nath]
194. Diffraction of light by ultrasonic waves [1936 *Nature (London)* **138** 616; with N S Nagendra Nath]

195. On the wave-like character of periodic precipitates [1939 *Proc. Indian Acad. Sci.* **A9** 455; with K Subba Ramaiah]
196. Interference patterns with Liesegang rings [1938 *Nature (London)* **142** 355; with K Subba Ramiah]
197. Haidinger's rings in curved plates [1939 *J. Opt. Soc. Am.* **29** 413; with V S Rajagopalan]
198. Haidinger's rings in soap bubbles [1939 *Proc. Indian Acad. Sci.* **A10** 317; with V S Rajagopalan]
199. Conical refraction in naphthalene crystals [1941 *Proc. Indian Acad. Sci.* **A14** 221; with V S Rajagopalan and T M K Nedungadi]
200. Conical refraction in naphthalene crystals [1941 *Nature (London)* **147** 268; with V S Rajagopalan and T M K Nedungadi]
201. The phenomena of conical refraction [1942 *Curr. Sci.* **11** 44]
202. The theory of the Christiansen experiment [1949 *Proc. Indian Acad. Sci.* **A29** 381]
203. The Christiansen experiment with spherical particles [1949 *Proc. Indian Acad. Sci.* **A30** 211; with S Ramaseshan]
204. Diffraction of light by transparent spheres and spheroids: The Fresnel patterns [1949 *Proc. Indian Acad. Sci.* **A30** 277; with S Ramaseshan]
205. The Christiansen experiment [1953 *Curr. Sci.* **22** 31; with M R Bhat]
206. The structure and optical behaviour of some natural and synthetic fibres [1954 *Proc. Indian Acad. Sci.* **A39** 109; with M R Bhat]
207. The theory of the propagation of light in polycrystalline media [1955 *Proc. Indian Acad. Sci.* **A41** 37; with K S Viswanathan]
208. A generalized theory of the Christiansen experiment [1955 *Proc. Indian Acad. Sci.* **A41** 55; with K S Viswanathan]
209. The Christiansen experiment with birefringent powders [1955 *Proc. Indian Acad. Sci.* **A41** 61; with M R Bhat]
210. The optical behaviour of polycrystalline solids [1957 *J. Madras Univ.* **B27** 1]
211. Christiaan Huygens's and the wave theory of light [1959 *Proc. Indian Acad. Sci.* **A49** 185]
212. The principle of Huygens's and diffraction of light [1959 *Curr. Sci.* **28** 267]
213. The optics of mirages [1959 *Proc. Indian Acad. Sci.* **A49** 251; with S Pancharatnam]
214. The scintillation of the stars [1964 *Curr. Sci.* **33** 355]
215. Lectures on Physical Optics, Part I [1959 Indian Academy of Sciences, Bangalore]

Volume IV

1. Miscellaneous Papers

216. The curvature method of determining the surface tension of liquids [1907 *Philos. Mag.* **14** 591]
217. Some new methods in kinematical theory [1912–13 *Bull. Calcutta Math. Soc.* **4** 1]
218. On the summation of certain Fourier series involving discontinuities [1913–14 *Bull. Calcutta. Math. Soc.* **5** 5]
219. The viscosity of liquids [1923 *Nature (London)* **111** 600]

220. A THEORY OF THE VISCOSITY OF LIQUIDS [1923 *Nature (London)* **111** 532]
221. THE PHOTOGRAPHIC STUDY OF IMPACT AT MINIMAL VELOCITIES [1918 *Phys. Rev.* **12** 442]
222. PERCUSSION FIGURES IN ISOTROPIC SOLIDS [1919 *Nature (London)* **104** 113]
223. ON SOME APPLICATIONS OF HERTZ'S THEORY OF IMPACT [1920 *Phys. Rev.* **15** 277]
224. THE OPTICAL STUDY OF PERCUSSION FIGURES [1926 *J. Opt. Soc. Am.* **12** 387]
225. PERCUSSION FIGURES IN CRYSTALS [1958 *Proc. Indian Acad. Sci.* **A48** 307]
226. PERCUSSION FIGURES IN CRYSTALS [1959 *Curr. Sci.* **28** 1]
227. INDIA'S DEBT TO FARADAY [1931 *Nature (London)* **128** 362]
228. NEWTON AND THE HISTORY OF OPTICS [1942 *Curr. Sci.* **11** 453]
*229. ASTRONOMICAL RESEARCH IN INDIA: I [1943 *Curr. Sci.* **12** 197]
230. ASTRONOMICAL RESEARCH IN INDIA: II [1943 *Curr. Sci.* **12** 289]
231. ASTRONOMICAL RESEARCH IN INDIA: III [1943 *Curr. Sci.* **12** 313]
232. CENTENARY OF THE FARADAY EFFECT [1945 *Curr. Sci.* **14** 281]
233. SCIENCE IN EASTERN EUROPE: I [1958 *Curr. Sci.* **27** 371]
234. SCIENCE IN EASTERN EUROPE: II [1958 *Curr. Sci.* **27** 421]
235. ZONAL WINDS AND JET STREAMS IN THE ATMOSPHERE [1967 *Curr. Sci.* **36** 593]
236. THE ATMOSPHERE OF THE EARTH [1968 *Curr. Sci.* **37** 151]

2. Colour

237. THE ORIGIN OF THE COLOURS IN THE PLUMAGE OF THE BIRDS [1934 *Proc. Indian Acad. Sci.* **A1** 1]
238. ON IRIDESCENT SHELLS, PART I. INTRODUCTORY [1934 *Proc. Indian Acad. Sci.* **A1** 567]
239. ON IRIDESCENT SHELLS—PART II. COLOURS OF LAMINAR DIFFRACTION [1934 *Proc. Indian Acad. Sci.* **A1** 574]
240. ON IRIDESCENT SHELLS—PART III. BODY-COLOURS AND DIFFUSION-HALOES [1934 *Proc. Indian Acad. Sci.* **Al** 859]
241. THE STRUCTURE AND OPTICAL BEHAVIOUR OF IRIDESCENT SHELLS [1954 *Proc. Indian Acad. Sci.* **A39** 1; with D Krishnamurti]
242. THE STRUCTURE AND OPTICAL CHARACTERS OF IRIDESCENT GLASS [1939 *Proc. Indian Acad. Sci.* **A9** 371; with V S Rajagopalan]
243. COLOURS OF STRATIFIED MEDIA I: ANCIENT DECOMPOSED GLASS [1940 *Proc. Indian Acad. Sci.* **A11** 469; with V S Rajagopalan]
244. THE IRIDESCENT FELDSPARS [1950 *Curr. Sci.* **A19** 301]
245. THE STRUCTURE OF LABRADORITE AND THE ORIGIN OF ITS IRIDESCENCE [1950 *Proc. Indian Acad. Sci.* **A32** 1; with A Jayaraman]
246. THE STRUCTURE AND THE OPTICAL BEHAVIOUR OF THE CEYLON MOONSTONES [1950 *Proc. Indian Acad. Sci.* **A32** 123; with A Jayaraman and T K Srinivasan]
247. THE DIFFUSION HALOES OF THE IRIDESCENT FELDSPARS [1953 *Proc. Indian Acad. Sci.* **A37** 1; with A Jayaraman]
248. ON THE IRIDESCENCE OF POTASSIUM CHLORATE CRYSTALS—PART I. ITS SPECTRAL CHARACTERS [1952 *Proc. Indian Acad. Sci.* **A36** 315; with D Krishnamurti]
249. ON THE IRIDESCENCE OF POTASSIUM CHLORATE CRYSTALS—PART II. POLARISATION EFFECTS [1952 *Proc. Indian Acad. Sci.* **A36** 321; with D Krishnamurti]
250. ON THE IRIDESCENCE OF POTASSIUM CHLORATE CRYSTALS—PART III. SOME GENERAL OBSERVATIONS [1952 *Proc. Indian Acad. Sci.* **A36** 330; with D Krishnamurti]

251. On the polarisation and spectral character of the iridescence of potassium chlorate crystals [1952 *Proc. Indian Acad. Sci.* **A36** 419; with D Krishnamurti]
252. The structure and optical behaviour of iridescent crystals of potassium chlorate [1953 *Proc. Indian Acad. Sci.* **A38** 261; with D Krishnamurti]
253. The structure of opal and the origin of its iridescence [1953 *Proc. Indian Acad. Sci.* **A38** 101; with A Jayaraman]
254. The structure of optical behaviour of iridescent opal [1953 *Proc. Indian Acad. Sci.* **A38** 343; with A Jayaraman]
255. The structure and optical behaviour of pearls [1954 *Proc. Indian Acad. Sci.* **A39** 215; with D Krishnamurti]
256. Optics of the pearl [1954 *Curr. Sci.* **23** 173; with D Krishnamurti]
257. On the chromatic diffusion halo and other optical effects exhibited by pearls [1954 *Proc. Indian Acad. Sci.* **A39** 265; with D Krishnamurti]
258. The structure and optical behaviour of iridescent agate [1953 *Proc. Indian Acad. Sci.* **A38** 199; with A Jayaraman]
259. The structure and optical behaviour of iridescent calcite [1954 *Proc. Indian Acad. Sci.* **A40** 1; with A K Ramdas]
260. The structure and optical behaviour of jadeite [1955 *Proc. Indian Acad. Sci.* **A41** 117; with A Jayaraman]
261. Crystals of quartz with iridescent faces [1950 *Proc. Indian Acad. Sci.* **A31** 275]

3. Optics of Minerals

262. The optical anisotropy and heterogeneity of vitreous silica [1950 *Proc. Indian Acad. Sci.* **A31** 141]
263. Structural birefringence in amorphous solids [1950 *Proc. Indian Acad. Sci.* **A31** 207]
264. The lamellar structure and birefringence of plate glass [1950 *Proc. Indian Acad. Sci.* **A31** 359]
265. The Smoky quartz [1921 *Nature (London)* **108** 81]
266. The structure of amethyst quartz and the origin of its pleochroism [1954 *Proc. Indian Acad. Sci.* **A40** 189; with A Jayaraman]
267. The birefringence patterns of crystal spheres [1956 *Proc. Indian Acad. Sci.* **A43** 1]
268. Amethyst—its nature and origin [1954 *Curr. Sci.* **23** 379]
269. On the structure of amethyst and its genesis in nature [1954 *Proc. Indian Acad. Sci.* **A40** 221; with A Jayaraman]
270. On the optical behaviour of crypto-crystalline quartz [1954 *Proc. Indian Acad. Sci.* **A41** 1; with A Jayaraman]
271. X-ray study of fibrous quartz [1954 *Proc. Indian Acad. Sci.* **A40** 107; with A Jayaraman]
272. On the polycrystalline forms of gypsum and their optical behaviour [1954 *Proc. Indian Acad. Sci.* **A39** 153; with A K Ramdas]
273. X-ray studies on polycrystalline gypsum [1954 *Proc. Indian Acad. Sci.* **A40** 57; with A Jayaraman]
274. The luminescence of fluorspar [1962 *Curr. Sci.* **31** 361]
275. The two species of fluorite [1962 *Curr. Sci.* **31** 445]

4. Diamond

276. THE PHYSICS OF THE DIAMOND [1942 *Curr. Sci.* **11** 261]
277. THE STRUCTURE AND PROPERTIES OF DIAMOND [1943 *Curr. Sci.* **12** 33]
278. THE FOUR FORMS OF DIAMOND [1944 *Curr. Sci.* **13** 145]
279. THE CRYSTAL SYMMETRY AND STRUCTURE OF DIAMOND [1944 *Proc. Indian Acad. Sci.* **A19** 189]
280. THE NATURE AND ORIGIN OF THE LUMINESCENCE OF DIAMOND [1944 *Proc. Indian Acad. Sci.* **A19** 199]
281. BIREFRINGENCE PATTERNS IN DIAMONDS [1944 *Proc. Indian Acad. Sci.* **A19** 265; with G R Rendall]
282. THE CRYSTAL FORMS OF DIAMOND AND THEIR SIGNIFICANCE [1946 *Proc. Indian Acad. Sci.* **A24** 1; with S Ramaseshan]
283. THE DIAMOND AND ITS TEACHINGS [1946 *Curr. Sci.* **15** 205]
284. NEW CONCEPTS OF CRYSTAL STRUCTURE [1946 *Curr. Sci.* **15** 329]
285. THE LUMINESCENCE OF DIAMOND AND ITS RELATION TO CRYSTAL STRUCTURE [1950 *Proc. Indian Acad. Sci.* **A32** 65; with A Jayaraman]
286. THE LUMINESCENCE OF DIAMOND—I [1950 *Curr. Sci.* **19** 357]
287. THE LUMINESCENCE OF DIAMOND—II [1951 *Curr. Sci.* **20** 1]
288. THE LUMINESCENCE OF DIAMOND—III [1951 *Curr. Sci.* **20** 27]
289. THE LUMINESCENCE OF DIAMOND—IV [1951 *Curr. Sci.* **20** 55]
290. THE DIAMOND [1956 *Proc. Indian Acad. Sci.* **A44** 99]
291. THE TETRAHEDRAL CARBON ATOM AND THE STRUCTURE OF DIAMOND [1957 *Proc. Indian Acad. Sci.* **A46** 391]

Volume V. Crystal Physics

1. Diffuse X-ray Reflections

292. A NEW X-RAY EFFECT [1940 *Curr. Sci.* **9** 165; with P Nilakantan]
293. REFLECTION OF X-RAYS WITH CHANGE OF FREQUENCY PART I. THEORETICAL DISCUSSION [1940 *Proc. Indian Acad. Sci.* **A11** 379; with P Nilakantan]
294. REFLECTION OF X-RAYS WITH CHANGE OF FREQUENCY PART II. THE CASE OF DIAMOND [1940 *Proc. Indian Acad. Sci.* **A11** 389; with P Nilakantan]
295. REFLECTION OF X-RAYS WITH CHANGE OF FREQUENCY PART III. THE CASE OF SODIUM NITRATE [1940 *Proc. Indian Acad. Sci.* **A11** 398; with P Nilakantan]
296. QUANTUM THEORY OF X-RAY REFLECTION AND SCATTERING PART I. GEOMETRIC RELATIONS [1940 *Proc. Indian Acad. Sci.* **A12** 83; with N S Nagendra Nath]
297. REFLECTION OF X-RAYS WITH CHANGE OF FREQUENCY PART IV. ROCK SALT [1940 *Proc. Indian Acad. Sci.* **A12** 141; with P Nilakantan]
298. THE TWO TYPES OF X-RAY REFLECTION IN CRYSTALS [1940 *Proc. Indian Acad. Sci.* **A12** 427]
299. CRYSTALS AND PHOTONS [1941 *Proc. Indian Acad. Sci.* **A13** 1]
300. THE QUANTUM THEORY OF X-RAY REFLECTION: BASIC IDEAS [1941 *Proc. Indian Acad. Sci.* **A14** 317]
301. QUANTUM THEORY OF X-RAY REFLECTION: MATHEMATICAL FORMULATION [1941 *Proc. Indian Acad. Sci.* **A14** 332]

302. Quantum theory of X-ray reflection: Experimental confirmation [1941 *Proc. Indian Acad. Sci.* **A14** 356; with P Nilakantan]
303. Dynamic X-ray reflections in crystals [1948 *Curr. Sci.* **17** 65]
304. X-rays and crystals [1955 *Curr. Sci.* **24** 395]
305. New concepts of the solid state [1942 *Proc. Indian Acad. Sci.* **A15** 65]

2. Dynamics of Crystal Lattices

306. The vibration spectrum of a crystal lattice [1943 *Proc. Indian Acad. Sci.* **A18** 237]
307. New paths in crystal physics [1947 *Curr. Sci.* **16** 67]
308. The vibration spectra of crystals Part I. Basic theory [1947 *Proc. Indian Acad. Sci.* **A26** 339]
309. The vibration spectra of crystals Part II. The case of diamond [1947 *Proc. Indian Acad. Sci.* **A26** 356]
310. The vibration spectra of crystals Part III. Rock salt [1947 *Proc. Indian Acad. Sci.* **A26** 370]
311. The vibration spectra of crystals Part IV. Magnesium oxide [1947 *Proc. Indian Acad. Sci.* **A26** 383]
312. The vibration spectra of crystals Part V. Lithium and sodium fluorides [1947 *Proc. Indian Acad. Sci.* **A26** 391]
313. The vibration spectra of crystals Part VI. Sylvine [1947 *Proc. Indian Acad. Sci.* **A26** 396]
314. The infra-red spectrum [1947 *Curr. Sci.* **16** 359]
315. The eigenvibrations of crystal structures [1948 *Curr. Sci.* **17** 1]
316. The scattering of light in crystals and the nature of their vibration spectra [1951 *Proc. Indian Acad. Sci.* **A34** 61]
317. The vibration spectra of crystals and the theory of their specific heats [1951 *Proc. Indian Acad. Sci.* **A34** 141]

3. Elasticity of Crystals

318. The elasticity of crystals [1955 *Curr. Sci.* **24** 325]
319. The elastic behaviour of isotropic solids [1955 *Proc. Indian Acad. Sci.* **A42** 1; with K S Viswanathan]
320. On the theory of the elasticity of crystals [1955 *Proc. Indian Acad. Sci.* **A42** 51; with K S Viswanathan]
321. Evaluation of the four elastic constants of some cubic crystals [1955 *Proc. Indian Acad. Sci.* **A42** 111; with D Krishnamurti]

4. Vibrational and Thermal Energy of Crystals

322. The nature of the thermal agitation in crystals [1955 *Proc. Indian Acad. Sci.* **A42** 163]
323. The thermal energy of crystals [1955 *Curr. Sci.* **24** 357]
324. Quantum theory and crystal physics [1956 *Curr. Sci.* **25** 377]

325. THE PHYSICS OF CRYSTALS [1956 *Proc. Indian Acad. Sci.* **A43** 327]
326. THE SPECIFIC HEATS OF CRYSTALS: PART I. GENERAL THEORY [1956 *Proc. Indian Acad. Sci.* **A44** 153]
327. THE SPECIFIC HEATS OF CRYSTALS: PART II. THE CASE OF DIAMOND [1956 *Proc. Indian Acad. Sci.* **A44** 160]
328. THE SPECIFIC HEATS OF CRYSTALS: PART III. ANALYSIS OF THE EXPERIMENTAL DATA [1956 *Proc. Indian Acad. Sci.* **A44** 367]
329. THE HEAT CAPACITY OF DIAMOND BETWEEN 0–1000° K [1957 *Proc. Indian Acad. Sci.* **A46** 323]
330. THE DIFFRACTION OF X-RAYS BY DIAMOND: PART I [1958 *Proc. Indian Acad. Sci.* **A47** 263]
331. THE DIFFRACTION OF X-RAYS BY DIAMOND: PART II [1958 *Proc. Indian Acad. Sci.* **A47** 335]
332. THE DIFFRACTION OF X-RAYS BY DIAMOND: PART III [1958 *Proc. Indian Acad. Sci.* **A48** 1]
333. THE INFRA-RED ABSORPTION BY DIAMOND AND ITS SIGNIFICANCE PART I. MATERIALS AND METHODS [1962 *Proc. Indian Acad. Sci.* **A55** 1]
334. THE INFRA-RED ABSORPTION BY DIAMOND AND ITS SIGNIFICANCE PART II. A GENERAL SURVEY OF THE RESULTS [1962 *Proc. Indian Acad. Sci.* **A55** 5]
335. THE INFRA-RED ABSORPTION BY DIAMOND AND ITS SIGNIFICANCE PART III. THE PERFECT DIAMONDS AND THEIR SPECTRAL BEHAVIOUR [1962 *Proc. Indian Acad. Sci.* **A55** 10]
336. THE INFRA-RED ABSORPTION BY DIAMOND AND ITS SIGNIFICANCE PART IV. THE NON-LUMINESCENT DIAMONDS [1962 *Proc. Indian Acad. Sci.* **A55** 14]
337. THE INFRA-RED ABSORPTION BY DIAMOND AND ITS SIGNIFICANCE PART V. THE COMPOSITE DIAMONDS [1962 *Proc. Indian Acad. Sci.* **A55** 20]
338. THE INFRA-RED ABSORPTION BY DIAMOND AND ITS SIGNIFICANCE PART VI. THE FREE VIBRATIONS OF THE STRUCTURE [1962 *Proc. Indian Acad. Sci.* **A55** 24]
339. THE INFRA-RED ABSORPTION BY DIAMOND AND ITS SIGNIFICANCE PART VII. THE CHARACTERISTIC FREQUENCIES [1962 *Proc. Indian Acad. Sci.* **A55** 30]
340. THE INFRA-RED ABSORPTION BY DIAMOND AND ITS SIGNIFICANCE PART VIII. DYNAMICAL THEORY [1962 *Proc. Indian Acad. Sci.* **A55** 36]
341. THE INFRA-RED ABSORPTION BY DIAMOND AND ITS SIGNIFICANCE PART IX. THE ACTIVITY OF THE NORMAL MODES [1962 *Proc. Indian Acad. Sci.* **A55** 42]
342. THE INFRA-RED ABSORPTION BY DIAMOND AND ITS SIGNIFICANCE PART X. EVALUATION OF THE SPECIFIC HEAT [1962 *Proc. Indian Acad. Sci.* **A55** 49]
343. THE INFRA-RED BEHAVIOUR OF DIAMOND [1962 *Curr. Sci.* **31** 403]
344. THE DIAMOND: ITS STRUCTURE AND PROPERTIES [1968 *Proc. Indian Acad. Sci.* **A67** 231]
345. QUANTUM THEORY AND CRYSTAL PHYSICS [1956 *Proc. Indian Acad. Sci.* **A44** 361]
346. THE SPECIFIC HEATS OF CRYSTALLINE SOLIDS: PART I. [1957 *Curr. Sci.* **26** 195]
347. THE SPECIFIC HEATS OF CRYSTALLINE SOLIDS: PART II. [1957 *Curr. Sci.* **26** 231]
348. THE SPECIFIC HEATS OF SOME METALLIC ELEMENTS PART I. ANALYSIS OF THE EXPERIMENTAL DATA [1957 *Proc. Indian Acad. Sci.* **A45** 1]
349. THE SPECIFIC HEATS OF SOME METALLIC ELEMENTS PART II. APPROXIMATE THEORETICAL EVALUATION [1957 *Proc. Indian Acad. Sci.* **A45** 7]
350. THE SPECIFIC HEATS OF SOME METALLIC ELEMENTS PART III. THE CHARACTERISTIC FREQUENCIES [1957 *Proc. Indian Acad. Sci.* **A45** 59]
351. THE SPECIFIC HEATS OF SOME METALLIC ELEMENTS PART IV. THE RESIDUAL SPECTRUM [1957 *Proc. Indian Acad. Sci.* **A45** 139]
352. THE SPECIFIC HEATS OF CRYSTALS AND THE FALLACY OF THE THEORIES OF DEBYE AND BORN [1957 *Proc. Indian Acad. Sci.* **A45** 273]

THE VIBRATIONS OF THE MgO CRYSTAL STRUCTURE AND ITS INFRARED ABSORPTION SPECTRUM [1961 *Proc. Indian Acad. Sci.* **A54**]

353. PART I. THE RESULTS OF EXPERIMENTAL STUDY 205
354. PART III. DYNAMICAL THEORY 223
355. PART III. COMPARISON OF THEORY AND EXPERIMENT 233
356. PART IV. EVALUATION OF ITS SPECIFIC HEAT 244

THE SPECTROSCOPIC BEHAVIOUR OF ROCK-SALT AND THE EVALUATION OF ITS SPECIFIC HEAT [1961 *Proc. Indian Acad. Sci.* **A54**]

357. PART I. THE STRUCTURE AND ITS FREE VIBRATIONS 253
358. PART II. ITS INFRA-RED ACTIVITY 266
359. PART III. THE SPECTRUM OF LIGHT SCATTERING 281
360. PART IV. SPECIFIC HEAT AND SPECTRAL FREQUENCIES 294
361. THE VIBRATION SPECTRUM OF LITHIUM FLUORIDE AND THE EVALUATION OF ITS SPECIFIC HEAT [1962 *Proc. Indian Acad. Sci.* **A55** 131]

THE SPECIFIC HEATS OF THE ALKALI HALIDES AND THEIR SPECTROSCOPIC BEHAVIOUR [1962 *Proc. Indian Acad. Sci.* **A56**]

362. PART I. INTRODUCTION 1
363. PART II. THE FREE MODES OF ATOMIC VIBRATION 6
364. PART III. THE INTERATOMIC FORCES 11
365. PART IV. THE EQUATIONS OF MOTION 15
366. PART V. THE EVALUATION OF THE FREQUENCIES 20
367. PART VI. THE ATOMIC VIBRATION SPECTRA 25
368. PART VII. EVALUATION OF THE SPECIFIC HEATS 30
369. PART VIII. THEIR INFRA-RED ACTIVITY 34
370. PART IX. SPECTRAL SHIFTS IN LIGHT SCATTERING 40
371. PART X. THE LITHIUM SALTS 45
372. PART XI. THE SODIUM SALTS 52
373. PART XII. THE POTASSIUM AND RUBIDIUM SALTS 60
374. THE INFRA-RED BEHAVIOUR OF SODIUM FLUORIDE [1962 *Proc. Indian Acad. Sci.* **A56** 223]
375. THE INFRARED BEHAVIOUR OF THE ALKALI HALIDES [1963 *Curr. Sci.* **32** 1]

THE DYNAMICS OF THE FLUORITE STRUCTURE AND ITS INFRA-RED BEHAVIOUR [1962 *Proc. Indian Acad. Sci.* **A56**]

376. PART I. INTRODUCTION 291
377. PART II. THE FREE MODES OF VIBRATION 294
378. PART III. ACTIVITY OF THE NORMAL MODES 301
379. PART IV. THE SPECTROPHOTOMETER RECORDS 304
380. SPECTROSCOPIC EVALUATION OF THE SPECIFIC HEATS OF POTASSIUM BROMIDE [1963 *Proc. Indian Acad. Sci.* **A57** 1]

Volume VI. Colour and its Perception

1. Light, Colour and Vision

381. LIGHT, COLOUR AND VISION [1959 *Curr. Sci.* **28** 429]
382. ON THE SENSATIONS OF COLOUR AND THE NATURE OF THE VISUAL MECHANISM [1960 *Curr. Sci.* **29** 1]
383. THE PERCEPTION OF LIGHT AND COLOUR AND THE PHYSIOLOGY OF VISION, PART I: THE MECHANISM OF PERCEPTION [1960 *Proc. Indian Acad. Sci.* **A52** 255]

384. THE PERCEPTION OF LIGHT AND COLOUR AND THE PHYSIOLOGY OF VISION, PART II. THE VISUAL PIGMENTS [1960 *Proc. Indian Acad. Sci.* **A52** 267]
385. THE PERCEPTION OF LIGHT AND COLOUR AND THE PHYSIOLOGY OF VISION, PART III. THE CAROTENOID PIGMENT [1960 *Proc. Indian Acad. Sci.* **A52** 281]
386. THE PERCEPTION OF LIGHT AND COLOUR AND THE PHYSIOLOGY OF VISION, PART IV. FERROHEME AND FERRIHEME [1960 *Proc. Indian Acad. Sci.* **A52** 292]
387. THE PERCEPTION OF LIGHT AND COLOUR AND THE PHYSIOLOGY OF VISION, PART V. THE COLOUR OF TRIANGLE [1960 *Proc. Indian Acad. Sci.* **A52** 305]
388. THE PERCEPTION OF LIGHT AND COLOUR AND THE PHYSIOLOGY OF VISION, PART VI. DEFECTIVE COLOUR VISION [1960 *Proc. Indian Acad. Sci.* **A52** 314]
389. THE PERCEPTION OF LIGHT AND COLOUR AND THE PHYSIOLOGY OF VISION, PART VII. GENERAL SUMMARY [1960 *Proc. Indian Acad. Sci.* **A52** 324]
390. THE ROLE OF THE RETINA IN VISION [1962 *Curr. Sci.* **31** 315]
391. LIGHT, COLOUR AND VISION [1962 *Curr. Sci.* **31** 489]
392. FLORAL COLOURS AND THEIR SPECTRAL COMPOSITION [1963 *Curr. Sci.* **32** 147]
393. THE VISUAL PIGMENTS AND THEIR LOCATION IN THE RETINA [1963 *Curr. Sci.* **32** 389]
394. FLORAL COLOURS AND THE PHYSIOLOGY OF VISION [1963 *Curr. Sci.* **32** 293]
395. THE TRICHROMATIC HYPOTHESIS [1963 *Curr. Sci.* **32** 245]
396. VISUAL ACUITY AND ITS VARIATIONS [1963 *Curr. Sci.* **32** 531]
397. THE COLOURS OF GEMSTONES [1963 *Curr. Sci.* **32** 437]
398. THE GREEN COLOUR OF VEGETATION [1963 *Curr. Sci.* **32** 341]

2. Visual Perception of Colour

FLORAL COLOURS AND THE PHYSIOLOGY OF VISION [1963 *Proc. Indian Acad. Sci.* **A58**]
399. PART I. INTRODUCTORY 57
400. PART II. THE GREEN COLOUR OF LEAVES 62
401. PART III. THE SPECTRUM OF THE MORNING GLORY 67
402. PART IV. THE QUEEN OF FLOWERS 70
403. PART V. THE BLUE OF THE JACARANDA 73
404. PART VI. COMPARATIVE STUDY OF THREE CASES 76
405. PART VII. THE ASTER AND ITS VARIED COLOURS 81
406. PART VIII. THE SPECTRA OF ROSES 84
407. PART IX. HIBISCUS AND BOUGAINVILLEA 87
408. PART X. FLOWERS EXHIBITING BAND SPECTRA 92
409. PART XI. A REVIEW OF THE RESULTS 96
410. PART XII. SOME CONCLUDING REMARKS 106
411. THE VISUAL SYNTHESIS OF COLOUR [1964 *Curr. Sci.* **33** 97]
412. FLUCTUATIONS OF LUMINOSITY IN VISUAL FIELDS [1964 *Curr. Sci.* **33** 65]
413. STARS, NEBULAE AND THE PHYSIOLOGY OF VISION [1964 *Curr. Sci.* **33** 293]
THE NEW PHYSIOLOGY OF VISION [1964 *Proc. Indian Acad. Sci.*]
414. PART I. INTRODUCTORY [**60** 139]
415. PART II. VISUAL SENSATIONS AND THE NATURE OF LIGHT [**60** 143]
416. PART III. CORPUSCLES OF LIGHT AND THE PERCEPTION OF LUMINOSITY [**60** 211]

417. PART IV. CORPUSCLES OF LIGHT AND THE PERCEPTION OF FORM [**60** 287]
418. PART V. CORPUSCLES OF LIGHT AND THE PERCEPTION OF COLOUR [**60** 292]
419. PART VI. VISION IN DIM LIGHT [**60** 369]
420. PART VII. THE PERCEPTION OF COLOUR IN DIM LIGHT [**60** 375]
421. PART VIII. THE PERCEPTION OF POLARISED LIGHT [1965 **61** 1]
422. PART IX. THE STRUCTURE OF THE FOVEA [**61** 7]
423. PART X. THE MAJOR VISUAL PIGMENTS [**61** 57]
424. PART XI. THE CAROTENOID PIGMENTS [**61** 65]
425. PART XII. CHROMATIC SENSATIONS AT HIGH LUMINOSITIES [**61** 129]
426. PART XIII. BLUE, INDIGO AND VIOLET IN THE SPECTRUM [**63** 133]
427. PART XIV. THE RED END OF THE SPECTRUM [**61** 187]
428. PART XV. THE CHROMATIC RESPONSE OF THE RETINA [**61** 193]
429. PART XVI. FURTHER STUDIES OF THE RETINAL RESPONSES [**61** 267]
430. PART XVII. LOCATION OF VISUAL PIGMENTS IN THE RETINA [**61** 335]
431. PART XVIII. THE VISUAL SYNTHESIS OF COLOUR [*Proc. Indian Acad. Sci.* **62** 1]
432. PART XIX. PERCEPTION OF COLOUR AND THE TRICHROMATIC HYPOTHESIS [**62** 10]
433. PART XX. SUPERPOSITION AND MASKING OF COLOURS [**62** 67]
434. PART XXI. THE GREEN COLOUR OF VEGETATION [**62** 73]
435. PART XXII. THE COLOURS OF FLOWERS [**62** 125]
436. PART XXIII. THE COLOURS OF THE ROSES [**62** 133]
437. PART XXIV. FLORAL PIGMENTS AND THE PERCEPTION OF COLOUR [**62** 177]
438. PART XXV. THE COLOURS OF NATURAL AND SYNTHETIC GEMSTONES [**62** 183]
439. PART XXVI. STRUCTURAL COLOURS [**62** 237]
440. PART XXVII. THE COLOURS OF INTERFERENCE [**62** 243]
441. PART XXVIII. OBSERVATION WITH A NEODYMIUM FILTER [**62** 307]
442. PART XXIX. THE REPRODUCTION OF COLOUR [**62** 310]
443. PART XXX. THE PHOTOMECHANICAL REPRODUCTION OF COLOUR [1966 **63** 1]
444. PART XXXI. THE INTEGRATION OF COLOUR BY THE RETINA [**63** 5]
445. PART XXXII. DEFECTS IN COLOUR VISION [**63** 65]
446. PART XXXIII. THE TESTING OF COLOUR VISION [**63** 71]
447. PART XXXIV. THE NATURE AND ORIGIN OF DEFECTS IN COLOUR VISION [**63** 133]
448. PART XXXV. THE FAINTEST OBSERVABLE SPECTRUM [**63** 138]
449. PART XXXVI. THE POSTULATED DUALITY OF THE RETINA [**63** 207]
450. PART XXXVII. THE SPECTRUM OF THE NIGHT-SKY [**63** 213]
451. PART XXXVIII. THE ADAPTATION OF VISION TO DIM LIGHT [**63** 263]
452. PART XXXIX. DALTONIAN COLOUR VISION [**63** 267]
453. PART XL. THE COLOURS OF IOLITE [**63** 321]
454. PART XLI. PHOTOGRAPHY IN COLOUR [**63** 325]
455. PART XLII. FURTHER OBSERVATIONS WITH THE NEODYMIUM FILTER [**63** 329]
456. PART XLIII. THE COLOURS OF FLUORSPAR [**63** 333]

3. Floral colours

457. FLORAL COLOURS AND THEIR ORIGIN [1969 *Curr. Sci.* **38** 179]
458. THE FLORACHROMES: THEIR CONSTITUTION AND OPTICAL BEHAVIOUR [1969 *Curr. Sci.* **38** 451]
459. THE COLOURS OF ROSES [1969 *Curr. Sci.* **38** 503]
460. SPECTROPHOTOMETRY OF FLORAL EXTRACTS [1969 *Curr. Sci.* **38** 527]
461. BLUE DELPHINIUMS AND THE PURPLE BIGNONIA [1969 *Curr. Sci.* **38** 553]
462. THE VARIED COLOURS OF VERBENA [1969 *Curr. Sci.* **38** 579]
463. THE PELARGONIUMS [1970 *Curr. Sci.* **39** 1]
464. THE RED OLEANDER AND THE PURPLE PETRIA [1970 *Curr. Sci.* **39** 25]

4. Monograph—Physiology of Vision

465. THE PHYSIOLOGY OF VISION, Indian Academy of Sciences, Bangalore [1968 p. 1–164]

NAME INDEX

Abbe, 509
Academus, 411
Agharkar, S.P., 37
Alexander, 2
Alippi, A., 529
Alvarez, Luis, 518
Amaldi, E., 526
Ammal, Parvati, 11, 12
Ammal, Sitalakshmi, 11
Ananthakrishnan, R., 280, 527
Anderson, P.W., 333, 516
Anderson, R.S., 521
Apparao, V., 23
Appaswamaiyar, S., 82, 524
Arago, 39
Archimedes, 25, 493
Aristotle, 47
Arjuna, 508
Armstrong, E.F., 264
Arnold, Sir Edwin, 24, 506
Aryabhatta, 420
Aston, 264, 273, 277
Auderset, H., 336, 528
Aurobindo, 505

Balakrishnan, T.A.S., 307, 309, 310, 527
Bancroft, W.D., 188
Banerjee, Kedareshwar, 169, 180, 181
Banerji, B., 524, 525
Banerji, S.K., 41, 102, 111, 135, 136, 137, 265, 471, 488, 507, 509, 524
Bär, R., 236, 238, 366, 371, 526, 528
Barton, E.H., 41
Basu, N., 145, 146, 147, 524
Bateson, William, 414
Beethoven, 354
Berlinger, 336
Berry, M.V., 372, 374, 377, 379, 382, 454, 455, 529, 530
Bhabha, Homi Jehangir, 7, 48, 279, 353–357, 359, 360, 361, 363, 399, 412, 413, 414, 421, 457, 458, 459, 460, 463, 470, 477, 497, 498, 499, 500, 501, 502, 512, 513, 516, 517, 519, 521, 528, 530
Bhabha, Hormusji, 260, 353, 355

Bhabha, Jehangir H., 353
Bhagavantam, S., 27, 56, 167, 169, 215, 234, 236, 237, 238, 239, 262, 344, 351, 388, 413, 454, 461, 464, 470, 471, 502, 506, 515, 518, 519, 523, 525, 526, 528, 529
Bhandari, Rajendra, 517
Bhargava, Y., 236, 526
Bhaskaracharya, 420
Bhat, M.R., 426, 429, 530
Bhatia, A.B., 382, 529
Bhatnagar, Sir S.S., 48, 265, 458, 459, 517
Bilderbeck, 13, 14
Biquard, P., 366, 367, 378, 381, 528
Birla, G.D., 209
Birman, J., 254, 402, 515
Blackett, P.M.S., 357, 463, 513
Blinc, R., 528
Bloch, Felix, 419, 420
Bloembergen, N., 246, 251, 252, 512, 527
Bohr, Niels, 193, 246, 352, 355, 394, 475, 496, 515, 520
Boltzmann, 2, 383
Bond, W.L., 348, 528
Born, Hedi. 263, 264, 395
Born, Max, 183, 184, 192, 205, 213, 229, 263, 264, 265, 267, 269, 271, 272, 273, 274, 278, 281, 285, 339, 340, 361, 384, 385, 388, 391, 392, 393, 394, 395, 398, 399, 402, 404, 410, 469, 474, 497, 510, 514, 515, 520, 526, 527, 529
Bose, D.M., 37, 523
Bose, Sir J.C., 4, 33, 40, 42–43, 44, 55, 505, 512
Bose, S.N. 37–38, 41, 44, 45–46, 54, 234, 501, 502, 507
Bose, Subhas Chandra, 517
Boys, C.V., 509
Bradburn, Mary, 529
Bragg, Sir Lawrence, 51, 167
Bragg, Sir William, 12, 48, 167, 168, 169, 262, 264, 502
Brahmagupta, 420
Brewster, Sir David, 201, 289, 293
Bright, John, 505
Brillouin, Leon, 180, 205, 213, 254, 330, 351, 366, 367, 528

Bromwich, 149
Brown, Hanbury, 449, 516
Brush, 145
Brynner, Yul, 519
Buchanan, J.Y., 188, 190
Buddha, 418
Bukhari, Ahmed Shah 110
Bulganin, 518, 519

Cabannes, J., 212, 213, 214, 223, 510, 526
Carmichael, Lord, 408
Chandrasekhar, S., 12, 14, 23, 54, 55, 56, 415, 421, 448, 493, 496, 503, 506, 507, 523, 530
Chandrasekhar, S. (of RRI), 431
Chandrasekharan, V., 527
Chatterjee, Bankim Chandra, 7, 505
Chatterjee, S., 507, 530
Chiad, R.Y., 530
Chinmayanandam, T., 147, 158, 159, 160, 162 471, 524, 525
Chowdhuri, Sajan Kumar, 209
Christiansen, 422
Christopher, Sir Samuel, 262
Chuckerbutti, B.N., 151, 152, 525
Cicero, 411
Clibborn, Col., 258
Cochran, W., 333
Cockcroft, Sir John, 355, 528
Compton, A.H., 52, 206, 349, 510
Coolidge, 52
Cotton, A., 169, 212, 297
Cowley, R.A., 333, 529
Curie, Joliot, 360
Curie, Mme, 45
Curzon, Lord, 257

D'Agostino, O., 526
Dainty, J.C., 527
Dalhousie, Lord, 514
Darwin, C.G., 213, 214, 303
Das, Panchanon M., 121, 524
Datta, G.L., 160, 326, 525
Daure, P., 212, 214
Davy, Sir Humphrey, 33
Debierne, 178
Debye, P., 178, 205, 213, 366, 384, 387, 388, 391, 394, 514, 528
de Haas, 307, 311
Delacretaz, G., 530
Delhaye, M., 527
Desai, Mahadev, 350, 513
Dey, Asutosh (Ashu Babu), 34, 36, 39, 53, 57, 77, 510, 524
Dey, J.N., 102, 524

Dey, Kanai Lal, 32
Dhamelincourt, P., 527
Dhawan, Satish, 412
Dickens, Charles, 7
Dikshitar, Muthuswamy, 10
Dirac, P.A.M., 193, 213, 234, 265, 496, 497, 507
Dolling, G., 529
Donkin, 65, 67, 114
Drakenstein, Hendrik van Reede tot, 3, 4
Dronacharya, 508
Dulong, 383

Ehrenfest, 180, 181
Einstein, Albert, 25, 44, 45, 171, 172, 173, 188, 189, 193, 204, 220, 224, 246, 281, 339, 349, 383, 384, 387, 388, 394, 464, 475, 496, 507, 513, 520
Ekalavya, 62, 508
Elliot, E.H., 13
Enz, C.P., 528
Euclid, 24
Ewald, P., 265, 498
Exner, 307
Extermann, R.C., 380, 529

Fabelinskii, 205, 206, 210, 211, 212, 213, 215, 228, 510, 526
Faraday, Michael, 2, 30, 33, 39, 70, 129, 496
Feldkamp, L.A., 529
Fermi, E., 210, 355, 361, 511, 526
Fermor, Sir L.L., 182, 264, 409, 410, 508, 523, 529
Feynman, R.P., 79, 490, 518, 521, 524, 530
Fielding, Lionel, 36
Forster, Sir Martin, 57, 262, 268, 272, 512
Foucault, 139
Fox, 182
Fox, J.J., 345, 528
Franklin, Benjamin, 39, 489, 490
Fresnel, Jean, 128, 131, 132
Frisch, R., 236, 526
Fryxell, R.E., 509

Galileo, 2, 420
Gandhi, Indira, 357, 461, 516, 517
Gandhi, Kasturba, 350, 513
Gandhiji (Mahatma Gandhi), 39, 350, 351, 354, 418, 466, 513, 517, 518
Ganesan, A.S., 215
Gardner, M., 518
Gauss, 39
Gavaskar, Sunil, 503
GellMann, M., 519
Gerlach, 192, 236

Name Index

Ghosh, J.C., 264, 507
Ghosh, P.N., 156, 525
Ghosh, R.N., 98, 99, 124, 509, 524
Ghosh, Sir Rashbehari, 37
Gibbs, Willard, 2, 172, 323
Goldschmidt, 52, 274
Gopal, S., 518
Gordon, R., 230, 526
Gouy, 149
Grant, E.R., 530
Gratias, P., 518
Greenler, Robert, 509
Grimaldi, Francesco, 130
Guha, 273, 277, 512
Gross, 320
Guillot, Marcel, 293

Haar, 384
Hahn, 511
Haken, H., 527
Haldane, J.B.S., 414
Hall, 21
Hanle, 236
Hannay, J.H., 530
Harbeke, 528
Hariharan, 513
Harish Chandra, 414, 503, 509
Harnock, 81
Harris, F.R., 512
Hartridge, 157
Hayden, 158
Hecht, E., 509
Heisenberg, W., 183, 200, 211, 227, 228, 239, 240, 242, 352, 526
Helmholtz, Hermann, 2, 25, 61, 62, 79, 80, 84, 85, 99, 121, 122, 157, 194, 466, 496, 524
Hemmer, 52
Herschel, 201, 439
Hertz, 39, 173
Hevesy, George, 263
Hewlett, 180, 181
Hibben, James, 352, 353
Hiedemann, E.A., 372, 374, 529
Hill, Sir A.V., 70, 458
Hillary, Sir Edmund, 511
Hitler, Adolf, 263, 271
Hodgkin, D., 498
Holwell, 508
Horowitz, 360
Howard, Maj., 259
Huang, K., 510, 515, 526, 529
Hunt, Sir John, 511
Hussain, M.F., 513
Hutchins, C.M., 509, 524

Huygens, Christiaan, 129, 131

Irvine, Sir James, 265, 266, 269, 273, 275
Irving, Washington, 488
Isenor, N.R., 316, 527
Iyer, C.S., 12, 413, 507, 510
Iyer, Chandrasekara (Raman's father), 11, 12

Jatkar, 273
Jayaraman, A., 418, 419, 429, 512, 529. 530
Jeans, Sir James, 512
Jeeves, 266
Jerrard, H.G., 530
Jones, Clark, 449
Jones, R.L., 14, 15, 18
Jones, Sir William, 4
Jordan, 352

Kaiser, W., 348, 528
Kalidasa, 488
Kapitza, P., 275, 502, 512
Kar, K.C., 110, 524
Karman, Th. von, 205, 339, 384, 385
Kastler, A., 236, 239, 511, 526
Kayser, 188
Keen, 186
Kelvin, Lord, 2, 19, 39, 43, 62
Khruschev, N., 518, 519
Kingsley, Sunder, 509
Kirchhoff, 131, 132, 133, 380
Klein, Felix, 480
Klein, W.R., 372, 374, 529
Klinkert, 77
Klingler, 480, 530
Koch, W.E., 509
Kohlrausch, 351
Kothari, D.S., 239, 500, 526
Kottler, 133
Kramers, H.A., 183, 200, 211, 227, 228, 239, 240, 242, 526
Krigar-Menzel, 84
Krishnamurti, D., 304, 527
Krishnamurthy, P., 262, 284
Krishnan, Sir K.S., 45, 47–48, 53, 54, 58, 147, 149, 150, 156, 165, 169, 170, 194, 196, 203, 204, 207, 210, 211, 212, 214, 225, 234, 360, 362, 459, 478, 496, 498, 499, 501, 503, 508, 510, 515, 524, 525, 526
Krishnan, R.S., 278, 321, 396, 398, 405, 515, 526, 527, 529
Kuhn, 265
Kumar, S., 524

Lafont, Rev. Fr, 30, 32

Lady Raman (Lokasundari), 26, 27, 35, 53, 216, 263–264, 284, 350, 351, 395, 511
Landau, L.D., 206, 230, 231, 399, 512, 515, 516, 526
Landsberg, G., 191, 204, 205, 206, 209, 210, 211, 213, 214, 220, 526
Langmuir, 52
Laplace, 319, 320
Laue, Max von, 178, 307
Leighton, 516
Lewis, W.B., 355
Liesegang, 323
Lifshitz, 206, 526
Lindemann, 384
Lindsay, Bruce, 62
Lissajous, 65, 68
Londsdale, Kathleen, 348, 528
Long, D.A., 526
Lorentz, 2, 51
Lotem, H., 527
Loudon, R.A., 402, 529
Low, Francis, 474
Lowry, T.M., 265
Lucas, R., 366, 367, 371, 378, 381, 528
Lynch, R.T., 527

McAlister, H.A., 527
Macaulay, Lord, 5
Macdonald, 149
Mackenzie, A.H., 265
MacMahon, P.S., 408
Maddox, John, 208
Madhavarao, B.S., 87, 524
Mahadevan, 182
Mahajani, G.S., 265
Mahalanobis, P.C., 44, 507
Mallick, D.N., 41
Mallock, 151
Mandel'shtam, 191, 204, 205, 206, 209, 211, 213, 220, 526
Mani, Anna, 341
Manson, David, 258
Martin, A.E., 345, 528
Marx, Karl, 7
Mason, C.W., 286, 287
Maxwell, J.C., 2, 39, 62, 129, 132, 169
Mazumdar, R.C., 500
Mehta, G.L., 51
Mehta, Zubin, 505
Meinhart, H., 480, 530
Melde, 70
Menon, M.G.K., 356, 468, 528
Mertens, R., 372, 529
Metcalf, Mrs, 263, 264

Meyer, 320
Michelson, A., 2, 172, 435
Mie, 152
Millikan, R.A., 51, 349, 477, 520
Milton, John, 7
Minto, Lord, 4, 5, 258
Mitra, M.N., 152, 307, 525, 527
Mitra, P.C., 37
Mitra, Rajendra Lal, 32
Mitra, S.K., 38, 41, 42, 138, 139, 143, 144, 507, 524
Molière, G., 382, 529
Mookerjee, Sir Asutosh, 33, 37, 38, 40, 41, 44, 45, 51, 57, 58, 192, 408
Mookerjee, Shyama Prasad, 58, 273, 276, 277
Morey, G.W., 491
Morse, 405
Mouton, 169
Mowdawalla, 263, 264, 268, 272
Mozart, 354
Mueller, 336
Mukerji, Sir M.N., 264
Mukunda, N., 414

Nadir Shah, 514
Nagendra Nath, N.S., 183, 278, 327, 330, 365, 366, 367, 368, 369, 371, 372, 377, 380, 382, 527, 528
Napoleon, 506, 515
Narayan, R.K., 10–11
Nedungadi, T.M.K., 330, 331, 332, 333, 335, 527
Nehru, Jawaharlal, 44, 265, 275, 357, 360, 456, 457, 458, 459, 461, 462, 463, 464, 470, 500, 505, 513, 517, 518, 519
Nernst, 260, 384
Newton, Sir Isaac, 2, 25, 128, 173, 319, 435, 464
Ng, W.K., 527
Nicol, William, 134
Nilakantan, P., 528
Nityananda, R, 454, 456, 529, 530
Noble, W.J., 382, 529
Nomoto, O., 372, 529
Norgay, Tenzing, 511

Oersted, 39
Oppenheimer, J.R., 360, 497
Owen, 35

Padmanabhan, 513
Padshah, Burjorji, 257,
Pais, Abraham, 193, 510, 513, 520, 525
Palit, Sir Taraknath, 37, 38
Palma, A., 529

Name Index

Palmieri, L., 529
Pancharatnam, S., 430, 431, 440, 441, 442, 443, 444, 445, 447, 448, 449, 451, 452, 453, 454, 456, 471, 496, 516, 517, 530
Pandurangaiah, Vummidi, 514
Parkinson, J.E., 280
Parsons, Dave, 206
Parthasarathy, S., 278, 330, 528
Patel, Sardar, 350, 351, 513
Pauli, W., 355
Pauling, Linus, 269
Pedlar, Sir Andrew, 43
Peierls, Sir Rudolph, 184, 265, 399, 404
Peitgen, H.O., 523
Penny, Lord, 528
Perrin, 162
Petit, 383,
Petit, Sir Dinshaw, 354
Phillips, J.C., 404, 529
Pisharoty, P.R., 283, 523, 526, 527
Pitt, Sir Thomas, 515
Placzek, G., 228, 229, 230, 231, 526
Planck, Max, 193
Plato, 47, 411
Pleijel, 216
Plowden, Lt. Col., 275, 277, 280
Poincarè, 149
Pontecorvo, B., 526
Pope, Sir William, 259, 262
Porter, A.W., 23, 51, 127, 186, 506, 509
Prasad, Ganesh, 508
Prashad, Baini, 287
Prins, J., 182, 525
Pupil, M., 297
Pythagoras, 79

Queen Victoria, 514

Rahm, 350
Rajagopalachari, Rao Bahadur S.P., 276
Rajagopalan, V.S., 294, 297, 302, 321, 527
Rajaji, 38, 464
Ramachandran, G.N., 152, 302, 303, 304, 305, 307, 309, 310, 311, 312, 313, 314, 315, 316, 513, 516, 524, 527, 528, 530
Ramakrishna, B.S., 124, 126, 524
Raman, Sir C.V., *passim*
Ramanathan (Raman's grandfather), 11
Ramanathan, K.R., 46–47, 53, 56, 57, 178, 179, 180, 181, 196, 214, 488, 489, 502, 525
Ramanujan, S., 14, 503, 507
Ramaseshan, S., 11, 26, 35, 36, 37, 38, 182, 196, 208, 215, 341, 344, 349, 371, 423, 424, 454, 456, 464, 465, 471, 499, 508, 509, 510, 516, 523, 525, 527, 528, 529, 530

Ramaswamy, C., 514
Ramdas, A.K., 400, 405, 515, 529
Ramdas, L.A., 39, 40, 49, 52, 53, 287, 303, 507, 510, 515, 523, 525, 526, 527
Ramm, 320
Ramsay, Sir William, 256, 258, 260
Ranjit Singh, 514
Rao, K.N., 124, 524
Rao, K.R., 514
Rao, Raghavendra, B.V., 320, 527
Rao, Ramakrishna, I., 156, 525
Raps, 84
Rasetti, F., 526
Ray, Bidhubushan, 127, 186, 307, 525, 527
Ray, Sir P.C., 5, 7, 33, 37, 43, 44, 45, 262, 273, 505
Rayleigh, Lord, 1, 2, 21, 22, 25, 39, 42, 47, 61, 62, 65, 67, 68, 69, 70, 71, 72, 80, 105, 111, 113, 114, 115, 118, 119, 124, 126, 127, 139, 140, 141, 165, 186, 187, 188, 189, 190, 191, 223, 287, 289, 303, 309, 313, 379, 394, 422, 448, 496, 506, 509, 524, 527
Reay, Lord, 256
Richter, P.H., 523
Ripon, Lord, 32
Riste, T., 528
Robertson, R., 345, 528
Robertson, Sir Robert, 262
Rocard, Y., 212, 213, 526
Rockefeller, 351
Röntgen, W., 475
Ross, 511
Rousseau, 7
Routh, 62
Roy, Raja Ram Mohan, 5
Rozhdestvenskii, 52
Rubens, Nichols, 210
Runge, 188
Rutherford, Lord, 48, 51, 215, 216, 263, 265, 271, 272, 274, 275, 280, 410, 457, 475, 502, 508, 512, 526

Saha, Meghnad 7, 38, 41, 43–45, 54, 55, 58, 59, 236, 239, 240, 259, 260, 262, 357, 409, 410, 456, 457, 478, 498, 499, 500, 501, 502, 503, 507, 508, 512, 513, 517, 521, 526, 530
Sahni, V.C., 515, 529
Saklatvala, Sir Sorab, 358
Samuel, Joseph, 517
Sanders, F.H., 371, 372, 373, 528
Sankara, 418
Sarabhai, Ambalal, 359
Sarabhai, Vikram, 47, 351, 359, 361, 467

Sarabhoji, 9, 10
Sastri, Saptarshi, 11
Sastri, Syama, 10
Saunders, F.A., 97, 509
Savart, Felix, 80
Savornin, 150
Schelling, J.C., 509
Scherrer, 178
Schmid, E.D., 526, 527, 529
Schmidt, 293
Schneider, T., 337, 528
Schrödinger, E., 184, 265, 497
Schuster, A., 41
Sears, F.W., 366, 528
Segre, E., 526
Sen, S.N., 523
Series, G.W., 453, 516, 530
Seshadri Iyer, Sir, 257
Seshagiri Rao, K., 56, 176, 525
Sethi, N.K., 162, 422, 524, 525
Sewell, Lt. Col., 259
Shah Jehan, 517
Shah Mahomed, 514
Shakespeare, 7
Shirane, G., 528
Shivnandan, 39
Shuknov, V.V., 527
Siddiqui, Riazuddin, 415, 497, 521
Simonson, J.L., 408
Singh, Baldev, 530
Singh, Bawa Kartar, 276
Singh, Jagjit, 509
Singh, Virendra, 528
Siraj-ud-daulah, 508
Sircar, Amrita Lal, 34
Sircar, Mahendra Lal, 29, 30, 31, 32, 33, 34, 37,
   38, 48, 58, 271, 275, 283, 459
Sivaji, 9
Sivan, Ramaswamy, 26
Slater, J.C., 497, 510
Smekal, A., 199, 200, 211, 227, 526
Smith, Helen, 398, 399, 405, 515
Smith, S., 177, 525
Smoluchowski, 188, 189, 220, 224
Socino, G., 529
Sogani, C.M., 162, 181, 422, 525
Solin, S.A., 400, 405, 529
Sommerfeld, Arnold, 2, 54, 55, 130, 133, 138,
   141, 149, 150, 184
Sondhi, M.M., 124, 126, 524
Sood, Ajay, 513
Sreekantan, B.V., 356, 528
Sreenivasaiah, B.N., 510
Srinivasan, G., 528

Stalin, J., 275, 502, 512
Stark, J., 171
Stegeman, G.I., 529
Steigmeier, E.F., 336, 528
Stephenson, 46
Stephenson, Andrew, 70, 71, 84
Stokes, Sir George, 62, 160, 161, 169, 170, 201,
   320
Stokes, J.S., 22, 23
Strassman, 511
Stumpf, 112
Subba Ramaiah, K., 323, 326, 327, 527
Subbaramaiya, D.S., 374, 528
Subbarayappa, B.V., 3, 523
Subramanyam, 277
Sunanda Bai, K., 318, 319, 527
Sutherland, G.A., 127, 524
Swaminathan, M.S., 473

Tagore, Rabindranath, 7, 43, 55
Tamm, I., 526
Tata J.N., 257, 258, 266, 275, 358, 459, 512
Tata, J.R.D., 353, 357, 358
Tata, Sir Dorab, 260, 354
Taylor, G.I., 496
Temple, Sir Richard, 30–31, 32
Terrell, Sir Courtney, 264
Thomson, J.J., 2, 39, 51, 152
Thosar, B.V., 284, 362, 502
Tilak, Lokamanya, 506
Tomanaga, 371
Tomita, A., 530
Torkar, G., 523
Toshniwal, G.R., 239, 526
Travers, Morris, 258
Tulaja, 9
Tyagaraja, 10, 26

Udgaonkar, B.M., 528

Vaidyanathan, 52
Valsakumar, M.C., 509
van der Waerden, 526
van Hove, L., 230, 400, 401, 404, 526, 529
van Vleck, J., 47, 497
Varahamira, 420
Vendreyes, G., 360
Venkataraman, G., 514, 525, 527, 529
Venkataraman, K.S., 320, 527
Venkatarayudu, T., 402, 515
Venkatasubbaraman, A., 176, 525
Venkatesachar, B., 276
Venkateswaran, C.S., 320, 321, 526, 527
Venkateswaran, S., 54, 57, 197, 502

Name Index

Verdi, 354
Vijayaraghavachariar, Sir T., 262
Vincent, J.H., 41
Visveswarayya, Sir M., 262, 517
Viswanathan, K.S., 399, 404, 426, 529, 530
Vivekananda, Swami, 7, 505
Vizianagaram, Maharajah of, 33, 34
Voltaire, 7
Vorländer, 169

Wagner, 354
Walker, 3
Walker, Sir Gilbert, 507
Walter, 169
Wannier, G., 380, 529
Warren, J.L., 529
Watson, 263, 268, 272
Watson, G.N., 149
Weisskopf, Victor, 454
Wellesley, Arthur, 506
Wellesley, Marquis of, 10, 506
Whetten, R.L., 530
White, G.W., 89

Whitehead, A.N., 47
Wigner, E., 518
Wilde, Oscar, 465
Wilson, C.T.R., 490
Wilson, E.B., 269
Wood, Charles (Lord Halifax), 5
Wood, R.W., 19, 208, 450, 526
Woodbury, E.J., 527
Wordsworth, W., 213, 482
Woste, L., 530

Yakovleva, T.V., 527
Yamada, Y., 528
Yarrell, J.L., 529
Young, A.T., 510, 525
Young, Thomas, 33, 128, 133, 435

Zachariasen, 492
Zajac, A., 509
Zeks, B., 528
Zel'dovich, B.Ya., 527
Zernike, F., 182, 448, 525
Zwanziger, J.W., 530

SUBJECT INDEX

A Century (IACS), 507
Acharya, 43
achromatic circular polarizer, 444
acoustics, 63, 90, 127
acousto-optics, 327, 365–382
Acta Mathematica, 149
Aden, 187
Agraharam, 10
All India Radio, 36, 514
American Institute of Physics, 206
analyser, 134–135
ancient glass, 293, 295, 297–302
anisotropic molecules, 191
anna, 32, 506
 coin, diffraction caustic, 143–144
Arabian Nights, 264
Aradhana, 10
amethyst, 429–430, 448
 Raman's work on, 429–430
Asiatic Society of Bengal, 4, 33, 408, 409, 411
Asiatick Society, 4
Association of Russian Physicists, 213
Astrophysical Journal, 46, 172, 421
astrophysics, 44, 361
Atomic Energy Commission, 48, 359, 459
Atoms for Peace Conference, 421
Auroville, 505
Avogadro number, 194, 510

Bangalore, 43, 58, 152, 184, 198, 204, 212, 231, 255, 257, 258, 261, 263, 265, 266, 268, 271, 272, 278, 280, 283, 284, 285, 307, 316, 321, 333, 335, 339, 340, 345, 346, 348, 349, 350, 351, 355, 356, 357, 359, 361, 362, 363, 394, 399, 409, 411, 419, 431, 457, 468, 470, 473, 475, 477, 497, 498, 508, 509, 513, 515, 518, 519, 520
Baroda lectures, 304, 307
Bell Labs, 348, 509, 512
Bengal, fervent atmosphere in, 7
Berlin Academy, 171
Berry's phase angle, 454–456, 517
 relation to Pancharatnam's work, 456, 517
Bhabha, Homi Jehangir
 Adams Prize, 356

 the artist, 513
 and atomic energy, 458–459
 cosmic ray work, 356–357
 election as Fellow of Academy, 357
 election to Royal Society, 356
 letter to J.R.D. Tata, 357
 letter to Tata Trust, 358
 and the *Proceedings*, 516
 on science and administration, 463
Bhabha Atomic Research Centre, 357, 512
Bhabha family, 257
Bhagavad Gita, 513
Bhagavantam
 on conversation with Raman, 454
 on credit sharing by Raman, 471
 on the ivory tower issue, 461
 on Lady Raman, 27
 Madurai Academy session episode, 413
 on Raman's ego, 519
 on Raman's interest in diamonds, 344, 388–389
 on Raman's Nobel Prize, 215–216
 on Raman's uncompromising attitude, 464
 on Raman's views on God, 470
birefringence, *see* double refraction
black-body radiation, 193
'Black Hole of Calcutta', 54, 508
Bohr–Sommerfeld model of atom, 192
Bombay, 4, 5, 6, 187, 257, 351, 353, 354, 409, 459, 478, 518
Bombay Harbour, 187
Born, Max
 in Bangalore, 263–265, 340, 394
 book on lattice dynamics, 399
 criticism of Raman, 394–395
 on later meetings with Raman, 395
 letter to Rutherford, 271–274, 410
 radio broadcast, 520
Born–Oppenheimer (adiabatic) approximation, 228
Born–Raman controversy, 232, 316, 339–340, 351, 383–405
Bose, Sir J.C., life sketch, 42–43
Bose, S.N., life sketch, 45–46
Bow Bazar Street, 32, 34, 44, 194, 207, 208, 214, 502

Brahmins, 6, 10, 35, 497
Brillouin
  message to New York Conference, 253
  theory of diffraction by ultrasonic waves, 367, 381
Brillouin scattering, 205, 206, 230, 231, 232, 253, 316, 318, 320, 366, 392, 399, 510
  diamond, 321
  gypsum, 321
British Ambassador in Washington, 349
British Association for the Advancement of Science, 30, 51, 408
Brownian motion, 315–316
Building Scientific Institutions in India (R.S. Anderson), 521
Bulletin of the IACS, 36, 90, 350

Calcutta, 4, 5, 7, 18, 27, 29, 42, 44, 45, 46, 48, 50, 51, 52, 53, 55, 56, 57, 58, 59, 61, 127, 135, 136, 152, 183, 184, 185, 187, 191, 192, 194, 198, 204, 207, 208, 216, 217, 255, 258, 261, 271, 282, 283, 285, 287, 307, 316, 317, 318, 326, 349, 350, 360, 363, 389, 407, 409, 410, 411, 430, 457, 459, 473, 497, 498, 502, 506, 507, 509, 510, 512, 515, 517
Calcutta Journal of Medicine, 30
Calcutta Mathematical Society, 509
Calcutta school of physics, 39–42
Caltech, 51, 275, 280, 282, 516, 520
Cambridge, 268, 272, 281, 282, 284, 355, 356, 358, 360, 421, 457, 458
Carnegie Institution, 352, 491
Catgut Acoustical Society, 110, 509
Cavendish, 355, 361, 502, 512
cello, 88
Ceylon Association for Advancement of Science, 458
charka, 351, 517
Challenger expedition, 188
Christiansen effect, 422–429
  models for, 426, 429
Christiansen experiment, 147
circular/elliptic polarization, 431–433, 441, 444, 447
clarinet, 111
Clausius–Mossotti formula, 224, 225
coefficient of restitution, 175, 176
coherence, 242, 249, 431, 435, 512
coherence length, 434
coherence time, 434
Collected Papers (Kelvin), 39
Collected Works of Meghnad Saha (S. Chatterjee), 507
College
  AVN (Vizianagaram), 12, 13
  Calcutta Medical, 30
  Canning (Lucknow), 408
  Christ (Cambridge), 42
  Deccan (Poona) 506
  Elphinstone (Bombay), 355, 506
  of Engineering (Roorkee), 258
  Gonville and Caius (Cambridge), 355
  Government (Lahore), 110
  King's (Cambridge) 505
  Madras Christian, 11, 47
  Maharajah's (Trivandrum), 46
  Metropolitan (Calcutta), 43
  Presidency (Calcutta), 5, 40, 41, 42, 43, 44, 506
  Presidency (Madras), 13, 14, 18, 23, 26, 35, 46, 81, 408, 506, 514
  of Science (Calcutta), 35, 37, 38, 39, 40, 41, 43, 45, 48, 58
  of Science (Nagpur), 35
  SPG (Trichinopoly), 11
  St Xavier's (Calcutta), 30
  University (London), 51, 127, 506
  Victoria (Palghat), 46
college spectrometer, 16, 17
colours of heated metals, 150–152
  Raman on, 151
Community Science Centre (Ahmedabad), 467
complementary filters, 195, 198, 204, 206, 207, 208
compressibility limit, 179
Comptes Rendus, 211, 212, 214
Compton effect, 54, 210, 234, 362, 510
  optical analogue of, 195, 211, 510
  Raman's comments on, 196
Conduction of Electricity (Thomson), 39
Connemara Public Library, 18
conservation of angular momentum, 234–236, 238
Coracias indica, 285, 286, 287
Cordierite, 516
coronae and halos, 152–156, 307–315
  Balakrishnan's theory, 309–310
  Ramachandran's theory, 309–310, 311–312
correlations
  intermolecular, 224, 225
  rotational (orientational), 230
  dielectric fluctuations, 230
  order parameter, 337
cosmic rays, 356, 361, 363, 477, 478, 479, 480
Cotton–Mouton effect, 167–168
Council of Scientific and Industrial Research (CSIR), 458, 463
critical opalescence, 187, 215, 219, 220, 224

Subject Index

critical temperature, 187
crore, 507
Current Science, 408, 411, 415, 420, 421, 453, 508, 509, 513, 514, 515
Current Trends in Lattice Dynamics (ed. K.R. Rao), 514

Dacca muslin, 2
Debye's theory of specific heats, 191, 209, 384, 387, 394
depolarization, 166, 223, 225, 238, 239, 318, 319
depolarization ratio, 165, 166, 223, 225, 238, 318, 319
diamagnetism, 49, 165, 169
diamond
   controversy with Londsdale, 348
   dispersion curves, 400
   external morphology of, 341, 343–344
   investigations by Raman's students, 345
   nitrogen impurity in, 348
   Ramachandran's analysis, 348
   Raman's four forms, 346–349
   symmetry of, 342–343
   types I and II, 345, 346
diamonds
   famous Indian stones, 514
   from Deccan, 341
   from Panna, 341
   Raman's collection, 341, 389
   from South Africa, 341
   stories about, 514–515
diffraction, 130–133, 183, 366, 368, 370, 380
   caustic, 142–144
   from a cylinder's edge, 145–147
   Fraunhofer, 131, 140, 142
   Fresnel, 131, 142
   Kirchhoff's theory, 132–133, 380
   oblique (unsymmetrical), 15–18, 19, 136–139
   Raman's studies on, 15–19, 136–138, 142, 147–150
   semi-infinite screen, 149–150
   Sommerfeld's theory, 133, 138, 141, 149–150, 156
Dodabetta, 191, 192, 224, 496
Doppler effect, 186, 370, 510
double refraction (birefringence), 134, 136, 165, 435
Dulong–Petit's law, 383
Dynamical Theory of Crystal Lattices (Born and Huang), 515
Dynamics of Stellar Systems (Chandrasekhar), 421
Dynamik der Kristalgitter (Born) 395, 515

early scientific establishments, 407
East India Company, 2, 3, 5, 10
Ectara, 69, 70
Einstein–Smoluchowski (E–S) formula/theory, 179, 187, 188–189, 190, 195, 225
Einstein's theory of specific heats, 383–384, 387–388, 394, 396
elastic modulus
   adiabatic, 320
   isothermal, 319, 320
electric polarization, 165, 166, 167
Electricity and Magnetism (Maxwell), 39
electron spin, 198
Electronics Corporation of India Limited (ECIL), 478
The Elements of Euclid, 24
emission
   spontaneous, 247
   stimulated, 247
Everest, Mt, 511
evolute, 142
Experimental Researches (Faraday), 39

Fabry–Perot, 14, 302, 320
feeble fluorescence, 46, 53, 54, 194–196, 198, 203, 349
Fermor, L.L.
   address to National Institute of Sciences, 410–411
   on Raman's exit from Calcutta, 58
ferroelectricity, 243, 333
Feynman diagram, 233, 234, 382
Films Division, 467
Financial Civil Service (FCS), 26, 43, 44
flow birefringence, 194
   Raman's work on, 169, 170
fluctuations, 188–189, 191, 195, 205, 207, 224, 318, 368
fluorescence, 201, 202–203, 207, 284, 285, 321, 484
   polarization of, 197, 198
   Stokes law, 201, 510
flute, 111
Fort William, 53, 508
Foucault's test, 139
   Banerji's studies, 139–141
Fourier optics, 139, 141, 496, 509
Franklin Institute, 50, 349
French Academy, 281

Gandhi, Indira
   on scientists and Government, 517–518
Gandhi Memorial Lectures, 466, 468, 512, 518

Gandhiji, visit to IISc, 350–351
Geological Survey of India, 158, 182, 410, 508
Gol Gumbaz, 127
Gopijantra, 69
Göttingen, 281

Haidinger's rings
  in mica, 158–160, 162
  in curved plates, 321–323
  in soap bubbles, 321–323
Handbuch der Physik, 446, 516
  Raman's review in, 110–112, 124, 149
Harijan Welfare Fund, 513
Helmholtz
  comments of Kelvin on, 62
  equation, 380, 381
  remarks on piano, 99
  theory of bowed strings, 121–122
Hertz's experiments (on radio waves), 193
Himalayan pheasant, 285, 512–513
Hindoo Patriot, 30
The Hindu, 257, 514
Hindu College, 5, 29
Hong Kong, 349
horn, 111
Huygens principle/waves, 16, 128–131, 133, 172, 186, 435

ice, colour of, 194
impact, 173–178
  adhesion from impact experiments, 177–178
  experiments in spacecraft, 178
  Hertz's theory, 173, 176
  percussion interferogram, 176–177
  Raman's studies on, 173, 174–177, 178
(Imperial) Academy of Sciences Leningrad, 3, 50, 52
Imperial University of India, 257
index ellipsoid, 426, 429, 436
India Meteorological Department, 3, 46, 407, 488, 507
Indian Academy of Sciences, 14, 350, 351, 407–416, 418, 467, 468, 473, 499, 516
  Ahmedabad session (1968), 467
  Annamalainagar session, 520
  annual meetings, 412–413
  Bangalore session (1970), 468
  business meeting, 412
  evening lectures, 412, 413
  founding of, 409
  Golden Jubilee, 412, 415, 473, 516
  inaugural meeting, 411
  journals, 413–414, 516
  Madurai session (1966), 413
  meeting to discuss starting of an institute, 418
  Nagpur session (1941), 412
  objectives, 411
  Raman Professor, 47, 415, 512, 516,
  symposia, 411, 412
  Young Associates, 415
Indian Administrative Service (IAS), 57
Indian Association for the Cultivation of Science (IACS), 32, 34, 38, 44, 48, 136, 145, 184 198, 209, 255, 263, 271, 356, 497, 501, 507, 508, 510
  acquisition of present name, 32
  lecture courses at, 33
  Lecture Theatre, 32
  origin of, 30–32
  Raman's entry into, 34
  Sircar's last message, 33
  starting of the *Bulletin*, 36
  visit of S. Chandrasekhar, 54, 56
  visit of Sommerfeld, 54–55
  Vizianagaram Laboratory, 33, 34
Indian Audit and Accounts Service (IAAS), 26
Indian Civil Service (ICS), 25, 256, 505
Indian Foreign Service (IFS), 57
Indian Institute of Science (IISc), 50, 57, 58, 59, 255, 258, 261, 262, 264, 266, 267, 268, 269, 270, 271, 272, 273, 274, 275, 276, 277, 278, 279, 281, 282, 283, 327, 339, 340, 350, 358, 359, 360, 362, 363, 365, 395, 402, 411, 417, 418, 497, 509, 518
  Aston memorandum, 277
  Biochemistry Department, 258, 277
  Council, 58, 259, 260, 261, 262, 263, 264, 268, 269, 273, 274, 275, 276, 277, 278, 280, 282, 349, 353, 358
  Department of Mathematical Physics, 269
  Director, 258, 260, 261, 262, 267, 268, 272, 273, 276, 277, 278, 279, 280, 281, 283, 284, 327
  ET Department, 258, 261, 263, 330
  Fowler Sub-Committee, 277
  General Chemistry Department, 258, 270
  Gymkhana, 277
  Irvine Committee report, 265–271
  objectives of, 266
  Organic Chemistry Department, 258
  Physical Chemistry Section, 263
  Physics Department, 262, 263, 271, 278, 279, 284, 330, 355
  Plowden Sub-Committee, 277
  Prof. of Chemistry, 263, 272
  Prof. of Electrical Engineering, 263, 264, 272, 273
  Prof. of Mathematical Physics, 264, 269

Subject Index

Prof. of Physics, 280
Registrar, 260, 262, 276
Review Committee, 259, 264
Review Committee (Irvine), 265, 266, 268, 269, 270, 271, 272, 273, 275, 276, 279
Review Committee (Pope), 259, 261
Review Committee (Sewell), 259, 260, 261, 262, 269
Senate, 259, 264, 276
structure of, 259
student population in, 270, 271
workshop, 263, 270, 275, 278–279
Indian J. Phys., 200, 204, 207, 208, 210, 211, 212, 214, 510
Indian Mathematical Club, 4
Indian music
  accompanying instruments, 10, 506
  Carnatic, 10, 36, 506
  Hindustani, 36, 506
  violin in, 10, 506
Indian musical instruments, 102–110
  jalatarang, 111
  mridangam, 36, 102–108, 123–126, 351, 509
  Raman's studies on, 102–104, 106–110
  sand patterns on *mridangam*, 107
  tabla, 36, 102–106
  tambura, 36, 102, 108–110, 111
  veena, 26, 102, 108–110, 111, 506
Indian National Army, 517
Indian National Congress, 44, 517
Indian National Science Academy, 47, 411, 415
Indian Science Congress, 30, 54, 259, 357, 399, 408, 409, 410, 457, 460, 461, 462, 512, 515, 517
Indian Statistical Institute, 507
Industrial Revolution, 2
Institute for Advanced Studies (Princeton), 474
Institute for Theoretical Physics (Moscow), 209
intermediate state, 201, 228, 240, 244
interference, 130, 183, 435
iridescence
  of glass, 293–302
  of potassium chlorate, 287, 288, 289, 306
  of shells, 287, 289–293
irradiance, 242
Isaac Newton Studentship, 355
isogyre, 163

Jamsetji Nusserwanji Tata (F.R. Harris), 512
Jesuit missionaries, 3
J. Acoust. Soc. Am., 509
J. Indian Math. Club, 508
Journal of the Asiatic Society, 4, 42
J. Opt. Soc. Am., 214

Jones calculus, 449

Kanalstrahlen (canal rays), 171, 172
Kanchenjunga, Mt, 471
Kavalur, telescope at, 52
SS *Kaisar-i-Hind*, 187
Kerr constant, 167
Kerr effect, 167, 168
Khaira Professor, 44, 46
Knowledge and Scientific News, 23
Kodaikanal observatory, 4
Koh-i-noor diamond, 514
Kramers–Heisenberg
  effect, 197, 200, 510
  formula, 227, 352
  negative absorption, 239
  paper, 200, 201, 211, 213, 227, 228, 239, 240, 244
  theory of dispersion, 211, 213, 229
Krishnan, Sir K.S., life sketch, 47–48
Krishnan's diary, 196–198, 211

lakh, 37, 507
laser, 241, 242, 251, 252, 254
  continuous wave, 242
  helium–neon, 244, 245
  Raman, 247, 250, 251
  tunable, 244, 245
laser fusion, 242, 251
lattice dynamics, 232, 264, 339, 340, 398, 399, 431, 514
  beginnings of, 383–384
  Born–von Karman theory, 384–385, 388, 391, 398, 399, 402–405
  comparison of theories, 388
  critical points (van Hove singularities), 400, 401, 402, 404, 405
  dispersion curves/relations, 386, 387, 399
  frequency spectrum, 384, 388, 393, 399, 404
  normal modes, 386–387, 392, 399
  periodic boundary conditions, 385–388, 394, 395, 399, 403
  Raman's theory, 387, 391–393, 394
Liesegang rings/phenomenon, 293, 302, 323–327, 328–329, 480
The Light of Asia (Arnold), 24
light quantum, *see* photon
light scattering, *see also* Raman effect
  alcohols, 194, 196
  anthracene vapour, 197
  benzene, 194
  castor oil, 321
  chloroform, 194
  due to density fluctuations, 318

detection of photon spin, 234–239
dichlorobenzene, 319
dipole approximation, 221
ether, 194
ether vapour, 190
experiments at Bangalore, 316–321
glycerine, 196, 321
iceland spar, 209
due to inhomogeneity, 219–220
Landau–Placzek theory, 230, 231, 320
liquids, 224–226
liquid carbon dioxide, 208
naphthalene, 208
nitrobenzene, 319
pentane, 208
pentane vapour, 198
polarization, 189, 197
quartz, 205
Raman's comments on work of colleagues, 196
Raman's experiments on water, 189–190
tetraline, 319
water, 190, 194, 196
Lissajous technique/figures, 65, 66, 68, 72, 75, 431
luminescence, 484, 485

Madras, 4, 5, 6, 13, 14, 47, 54, 57, 280, 413, 475, 507, 508
Madras Harbour, 507
Madras turban, 39, 508
Madras University Library, 18
Mahabharata, 508
Maharajah of Mysore, 257, 258, 411, 418
Maharajah of Vizianagaram, 33, 34
Maharajahs, Indian, 389, 505
Marseilles, 187
maser, 431, 449
Massachussetts Institute of Technology (MIT), 275
Maxwell constant, 170
Maxwell's equations/theory, 132, 193, 221, 222, 238, 440–441
Mediterranean sea, 51, 187, 496
Melde's apparatus/experiment, 22, 36, 64, 65, 68, 69, 70, 75, 77, 78
Raman's mathematical analysis of, 115
membrane, normal modes of vibration of, 105–106, 125
microprobe
electron, 252, 253
Raman, 252, 253
mirages, 449–451
molecular anisotropy, 165, 167, 168, 169, 170, 191, 194, 223, 224, 225, 318

molecular diffraction of light, 185, 188
Molecular Diffraction of Light (Raman), 192
Monthly Notices of the Royal Astronomical Society, 421
Mookerjee, Sir Asutosh
lectures at IACS, 33
on Raman's appointment to Palit Chair, 38
Moore Market, 23
Morton D. Hull Distinguished Service Professor, 12
mother-of-pearl, 289, 293, 479
My Life (Born), 520
Mysore
Dewan of, 276
Govt. of, 273, 274, 279, 418
Princely State of, 353, 517

Nagpur, 29, 35, 412, 430, 501, 509
naming system in S. India, 11–12, 506
SS *Narkunda*, 186, 187
National Academy of Sciences USA, 47
National Institute of Arts, 411
National Institute of Letters, 411
National Institute of Sciences, 410, 411
National Laboratories, 458, 462, 463, 517
National Physical Laboratory, 48, 360
National Planning Committee (NPC), 517
Nature, 12, 14, 19, 37, 157, 172, 183, 186, 187, 191, 198, 204, 208, 210, 211, 212, 213, 214, 371, 372, 394, 510, 514
Naturwissenschaften, 206, 214
navya nyaya, 11
Nehru, Jawaharlal
attraction to science, 457
felicitations to Raman, 461
on ivory tower, 462
on the Planning Commission, 463–464
relationship with Raman, 456, 459, 461, 462, 470, 519
remarks on Bhatnagar, 517
on science and its administration, 464
on science and the problems of society, 457
on scientific temper, 458
on support for science, 461
support to Bhabha and Bhatnagar, 458
visit to Raman Institute, 517
neutron scattering, 333, 397, 400, 401, 511
The New Physics (Raman), 474
Newton's rings, 160, 321
Newtonian mechanics, 193
nicol, 134, 162, 163, 164, 238, 298, 299
nineteenth-century physics, 2
Nobel Prize, 51, 54, 55, 56, 178, 193, 213, 215–217, 262, 273, 356, 359, 415, 431, 473, 511, 512, 519

Subject Index

Nobel Prize ceremony, Lady Raman's description of, 216
nodes, 64
  movement of, 64–65
  phase changes at, 65–67, 113–114
  Raman's investigations, 65–67
nonlinear optics, 245, 250, 251, 449
nuclear physics, 284, 285, 520

obliquity factor, 133
oboe, 111
old quantum theory, 193
optical anisotropy, 165, 167, 169, 170, 194, 223, 224, 318
  Raman's studies on, 169, 194
optical properties of crystals, 435–440
  circular dichroism, 440
  dichroism, 438
  index ellipsoid, 436
  isotropic, uniaxial, biaxial crystals, 436, 437
  optic axes, 436
  optical activity, 438–440, 448
  pleochroism, 438, 439, 444, 448
  singular axis, 444–445, 448
Optics (Hecht and Zajac), 509
optics
  of heterogeneous media, 361
  of stratified media, 285, 287
  Ramachandran on Raman's interest in, 127–128
  Raman's work in, 61
order parameter, 335
  fluctuations of, 337
Orloff diamond, 515

Pakistan Academy of Sciences, 415
Palit Chair, 37–38, 43, 56
Pancharatnam
  appreciation by Weisskopf, 454
  on coherence, 447–448, 449
  death of, 454
  design of achromatic circular polarizer, 444
  design of achromatic quarter-wave plate, 442–443
  on interference, 445–446, 448
  on light propagating along singular axes, 444–445
  on mirages (with Raman), 449–453
  papers of, 444, 445, 448, 449
  and *Sarvodaya*, 453
  use of Poincaré sphere, 440–444, 447–448
paper of Landsberg and Mandel'shtam, 204, 206, 209–210, 211, 213, 214
Paris Expo, 297, 351
Parsis, 6, 257, 505

Penrose tiling, 518
phase transformation/transitions, 337, 338
  Landau's theory of, 335
  in quartz, 330–333, 335
  in strontium titanate, 333–334
Philosophical Magazine, 18, 19, 157, 214
phonons
  acoustic, 232, 387
  optic, 205, 387, 399
phosphorescence, 484
photoelectric effect, 171, 193
photometer, 136–137
photon, 171, 184, 193, 434
  Einstein's suggested experiment, 171
  Raman's critique of Einstein, 171–173
  spin of, 184, 234–239
Photons, Galaxies and Stars (Hanbury Brown), 516
Physical Research Laboratory, 47, 359
Physical Review, 22, 46
Physical Society London, 165
physics in India
  beginning of, 5
  role of Bhabha, Raman and Saha, 7
The Physiology of Vision (Helmholtz), 25
The Physiology of Vision (Raman), 466
piano, 99–100, 101
  partial tones of, 99
  Raman's investigations on, 100–102
  role of hammer in, 99–102
piezo-optic coefficient
  adiabatic, 320
  isothermal, 320
Pitt (Regent) diamond, 515
Placzek–Teller formula, 319
Planck's law, 192, 234, 383
Planning Commission, 463, 507
Poincaré sphere, 440, 441, 442, 443, 444, 445, 446, 455–456
polarization of light, 134, 431–435, 440–442
  circular, 234–239
  Raman's experiments on sky-light, 191–192, 224
  sky-light, 191, 192
polarizer, 134, 135
Popular Lectures and Addresses (Lord Kelvin), 19
Proc. Indian Acad. Sci., 50, 285, 350, 351, 355, 357, 371, 413–414, 516
Proc. Indian Assoc. Cultiv. Sci., 214, 507
Proc. Natl. Inst. Sci., 399
Proc. R. Soc. London, 36, 46, 50, 186, 204, 414, 507, 510
The Properties of Glass (Morey), 491

quantum mechanics, 200, 269, 339, 361
quantum optics, 453
quantum statistics, 234
quarter-wave plate, 442–443
quartz, 205, 209, 210, 330, 331, 419, 429, 430
   infra-red frequencies of, 209, 210, 511
   Raman frequencies of, 331
Quetelet's rings, 160–162
   Raman's analysis of, 160–162
   Stokes' theory, 162

radiant spectrum, 156–157, 183
   Raman's investigations on, 153
Rainbows, Halos and Glories (Greenler), 509
Red sea, 187
Raman, Sir C.V.
   address to Physical Society London, 165
   address to Royal Society Edinburgh, 157
   advice to youth, 503–504
   and aesthetics, 23, 77, 183, 285, 417
   on amethyst, 430
   and annual reports of IACS, 48–50
   appointment to Palit Chair, 38
   on astronomy and astrophysics, 420–421
   and Bangalore school of physics, 278
   Baroda lectures, 306, 309, 316
   on basic research and utility, 477, 485
   on beauty and geometric form, 480–482
   Bengali associates of, 57
   birth, 11
   books that influenced, 23–25
   in Calcutta, 34–59
   on Chandrasekhar's work, 421
   on Christiansen effect, 422–426
   clash with Saha, 55, 58, 59
   college years, 13–23, 25
   on colour, 482–483
   compared with Bhabha and Saha, 499–501
   on composition of the Academy, 410
   controversy with Born, *see* Born–Raman controversy
   correspondence with Lord Rayleigh, 22
   on cosmic rays, 477
   Curzon Research Prize, 36
   death, 469
   on Divinity, 470–471
   documentary on eightieth birthday, 467
   early days at Raman Institute, 419–420
   editorial on founding an academy, 408–409
   election as Academy President, 412
   election as Hony Secy of IACS, 48
   election to Royal Society, 51, 183
   on establishing the Raman Institute, 418
   evening lecture in Madras, 413
   exit from Calcutta, 55–59
   father's death, 35
   felicitations at Bordeaux, 395, 515
   felicitations by IISc Council, 262
   felicitations by Nehru, 461
   fiftieth birthday, 351
   first American tour, 51
   first meeting with Nath, 330
   first papers, 15–23
   first visit to IACS, 34
   founding of the Academy, 409
   founding of the *Proceedings*, 413
   Franklin Medal, 349
   fund raising for Raman Institute, 418
   on geometry of form and its function, 481
   on glass, 491–492
   Government service, 27, 35, 36, 37, 38
   and growth of Indian science, 7
   on Helmholtz, 25
   honorary degree, Freiburg, 215
   honorary degree, Glasgow, 216
   honorary membership, Physical Society Switzerland, 215
   Hughes Medal, 216
   illness and end, 468–469
   on improvization, 49–50, 94
   on India's debt to Faraday, 30
   on IISc experience, 518
   IISc period, 261–363
   industrial consultancy, 279
   on instrumentation, 460
   interest in nuclear physics, 356, 361
   on ivory tower, 461
   knighthood, 215, 262
   last Gandhi Memorial Lecture, 468
   last paper, 466
   Last Testament, 468–469
   lecturing ability, 37, 473–474
   Lenin Peace Prize, 419
   on lightning, 490
   love of trees, 419, 468
   the man, 497–499
   marriage, 26–27
   and mathematics, 87, 183, 449
   Matteucci Medal, 215
   meeting with Pancharatnam, 431
   message to New York conference, 254
   and his museum, 419–420
   name, variations of, 12
   on National Laboratories, 517
   National Professor, 419
   on Niels Bohr, 475
   Nobel Prize, 215–216
   and Pancharatnam, 430–454

Subject Index

and Pancharatnam, a comparison, 449
parentage, 11
on physics of the countryside, 485–487
posting in Nagpur, 35
posting in Rangoon, 35
praise from British officers, 37
on his preoccupation with science, 471
presidentship of Science Congress, 54
and quantum mechanics, 184, 361, 362, 402
and racial prejudice, 52
radio broadcasts, 474–493
Raman Research Institute period, 417–469
reaction to Pancharatnam's death, 454
the recluse, 419, 456–465
recollections of college days, 13–14
and recreational science, 466
references to Krishnan, 499
relations with Born, 340, 395
resignation from Academy Committee, 409
resignation from Govt. service, 38
resignation from IISc directorship, 275–281
resignation from Royal Society, 464, 518
on the role of physicists, 476–477
school years, 12–13
on Science Congress, 460–461
on scientific outlook, 492–494
the scientist, 496–497
on shells, 287–293
and solid state physics, 361–362
on sources of light, 483–484
style of writing, 183, 449
suggestion of cosmic rays, 359, 361
suggestion of neutron stars, 356
talk on colour of sky, 467
theory of diffuse X-ray spots, 361–362
tour of Europe, 216
turban incident, 508
understanding of work of Kramers and Heisenberg, 183
and vision, 466–467
visit to Chidambaram Nataraja temple, 520
visit to Kapitza's institute, 399
visit to Mount Wilson observatory, 52
visit to Royal Institution, 51
visit to USSR, 52
Visiting Professor at Caltech, 51
on weather, 488–490
Woodburn Research Medal, 36
Raman effect, 54, 183, 185, 191, 198, 199, 202, 206, 215, 216, 219–234, 241, 285, 316, 510
in ammonium chloride, 247
anti-Stokes line, 202, 226, 239, 240, 246, 250
Bangalore address, 204, 210, 211, 212, 214
benzene, 204, 249
Bordeaux meeting, 515
Bose's comments, 54
carbon dioxide vapour, 228
coherence in, Raman's observation, 208
CARS, 250, 251, 252
combinational scattering, 206, 211
cross-section formula, 230
crystals, 205, 232, 515
in diamond, 330, 388, 389–390, 396–398, 401
discovery of, 54, 198
early papers, bibliography, 212
Feynman diagrams for, 233, 234
first letter to *Nature*, 204, 206–207, 510
golden jubilee, 252
hyper-, 246
incoherence of, 228
intensity of scattering, 229
Landau's nomenclature, 206
laser-Raman spectroscopy, 242, 244, 245, 246, 249, 397, 400, 401
mechanism, 199–203
modified scattering, 197, 198, 207, 208
molecular rotations, 228–229, 230, 237–239, 317, 318, 319
molecular vibrations, 228–229, 230, 253, 317, 318
multiphonon, 391
one-phonon, 391
paper to *Indian J. Phys.*, 204, 211, 212, 214
paper to *Proc. R. Soc. London*, 204
perturbation theory, 227–228
in potassium chromate, 244, 245
in potassium selenate, 243
and quantum electrodynamics, 232
quartz, 330, 331–332
Raman's analysis of diamond spectrum, 397
Raman's letter to Mandel'shtam, 214
Raman's reply to Darwin's note, 213–214
relation to infra-red frequencies, 204
remarks by Bloembergen, 251
remarks of Rutherford, 215
resonance, 242, 244, 245
second letter to *Nature*, 204, 207–208
second order, 339, 391, 392, 396–398, 400, 401, 402, 405
secondary radiation, 198, 202
Silver Jubilee, 521
statistics of publications, 352–353, 354
stimulated, 239, 246–250
Stokes line, 202, 226, 246, 248, 250
study of soft mode using, 243
theory of, 213, 219–234
Raman–Nath theory, 285, 307, 327, 330, 365, 367, 371, 372, 373, 374, 377, 378, 382, 510

applications, 373–376
Berry's monograph, 377, 382
corrugated wavefront, 367–368
difference equation, 380–381, 382
Doppler effect in, 370, 371, 377, 382
experimental verification, 371–373, 374
impact of papers, 371
phase grating, 367, 371, 379
the papers, 365, 371
wandering of intensity, 367, 369, 371–373
Raman Professor, 47, 415, 512, 516
Raman Research Institute, 418–420, 422, 431, 465, 466, 468, 470–471, 514, 517, 519
founding, 418
garden, 419
museum, 413, 419
separation from Academy, 468
Ramanathan, K.R., life sketch, 46–47
Ramaseshan
on Chandrasekara Iyer's family, 11
on Raman and German scientists, 265
on Raman as a lecturer, 37
on Raman's daily routine, 35
on Raman's discovery, 208
on Raman's early work, 36
on Raman's four forms of diamond, 349
on Raman's love of Nature, 471
on Raman's marriage, 26–27
on Raman's new mood, 465
on Raman's Nobel Prize, 215
on Raman's sensitivity, 499
on Raman's withdrawal, 464–465
on Raman's X-ray work, 182
Ramdas on Raman, 39–40, 52–53
random-phase approximation, 188, 224
Rangoon, 29, 35, 46, 57, 349, 501, 502
Renaissance, 2
and Indian science, 2
Ray, Sir P.C., life sketch, 43
Rayleigh, Lord
life sketch, 62
on Foucault's test, 139, 140, 141
studies on membranes, 105
studies on vibrations, 67–69
theory of Melde's experiment, 114–115
theory of whispering gallery, 127
Rayleigh scattering, 185, 188, 200, 202, 203, 205, 220–224, 225, 226, 229, 230, 235, 237, 243, 317, 318, 320, 429, 510
Residency, 273, 280, 349
response function
dynamic, 337, 338
static, 337, 338

Reviews of Modern Physics, 395, 421
Rouse Ball Studentship, 355
Royal Institution, 30, 31, 33, 43
Royal Society, 3, 43, 51, 183, 187, 216, 258, 409, 464, 507, 508, 512, 516, 518
ruby, 284, 362, 485
Rutherford, Lord
address to Royal Society, 215
letter to Raman, 280
letters to Born, 271–272, 274
remarks at award of Hughes medal to Raman, 216

Saha, Meghnad
criticism of negative absorption, 239–240
criticism of politicians, 456
on ivory tower, 457–458, 517
life sketch, 43–45
relations with Nehru, 500, 517
on the spin of the photon, 231
Saha Institute of Nuclear Physics, 44, 500
St Paul's Cathedral, 51, 126–127, 506
Sangitasaramrita, 9
Saraswati Mahal, 9
Sarvodaya, 350, 453, 513
Sayaji Rao Gaekwar Foundation Baroda, 304
scattering coefficient, 189, 223, 225
Science Abstracts, 42, 389
Science and Culture, 45
Science Association, 32
scientific institutions in British period, 3, 4, 407–408
Scientific Papers (Lord Rayleigh), 39, 47, 183, 509
Scots Lane, 34, 35
sea, colour of, 187–188, 189, 190
selection rules, 234, 348, 402
Sensations of Tone (Helmholtz), 25
shell
abalone, 290–291
aragonite in, 287, 293, 479
calcium carbonate in, 287, 479
conchin in, 287
Turbo, 292
sky, colour of, 187, 467
Soap Bubbles (Boys), 510
soap films, 162–165
Krishnan's analysis of, 165
Raman's studies on, 162–165
soft mode, 242, 285, 330–339
in quartz, 331
in strontium titanate, 333, 336
Sommerfeld, Arnold
lectures on wave mechanics, 55, 184, 497

Subject Index

visit to India, 54–55
Sommerfeld–Rubinowicz theory, 236
sonometer, 69, 101, 109
Sound and Vibration (Ghosh), 509
Soviet Physics Uspekhi, 206
speckles, 157, 307, 311–316, 496, 513
    Ramachandran's experiment, 312–315
    Rayleigh's law for intensity, 313, 314
spectrograph, 389
    Littrow, 318
    quartz, 195, 206, 209, 238
spectroscope, direct vision, 198, 208, 510
sruti, 36
Star Wars, 251
Statistical Physics (Landau), 516
Stellar Structure (Chandrasekhar), 421
Stern–Gerlach experiment, 236
Stockholm, 215, 216
Stokes parameters, 440, 441, 448, 453
Stradivarius, 95, 509
stratified medium
    Ramachandran's theory of, 302–304
    Rayleigh's theory, 303
striae in mica, Raman's studies on, 156
stroboscopic technique, Raman's use of, 72–76
Stroh violin, 56, 98–99
structure
    anthracene, 168
    naphthalene, 168
    Raman's comment on anthracene and naphthalene, 169
    strontium titanate, 332, 334
'*Subtle is the Lord...*' (Pais), 513, 520
supercell, 385, 387, 393
    Raman's description of, 392
Surely You Must be Joking, Mr. Feynman! (Feynman), 521
surface tension, 19–21
synchronous motor
    Raman's experiment with, 78
    rotational resonances, 78

Taj Mahal, 517
tala, 10
A Tale of Two Cities (Dickens), 7
Tanjore, 9, 10, 11
Tata Institute (IISc), 256
Tata Institute of Fundamental Research (TIFR), 358, 359, 459, 470, 478, 497, 500, 519
Tata Institute of Social Sciences, 512
Tata University, 258
Tatas, 57, 260, 272, 273, 354, 356, 512, 513
    commercial concerns, 256

Dorab Tata Trust, 357, 358
The Theory of Sound (Lord Rayleigh), 58
Times of India, 277
tone
    partial, 79
    quality of, 79
tonga, 490, 520
Topics in Nonlinear Optics (Bloembergen), 511, 512
Trombay, 257, 512
Tyndall effect, 286, 287

Ultra-opak microscope, 286, 294, 298
ultrasonic diffraction
    early experiments, 366
    the phenomenon, 365–366
    due to surface waves, 374–375
    theory of Lucas and Biquard, 367, 378
United Provinces Academy of Sciences, 409, 411
University
    Allahabad, 44, 356, 509
    Andhra (Visakhapatnam), 279
    Annamalai, 520
    Bombay, 5, 256, 257
    Calcutta, 5, 37, 40, 41, 42, 44, 45, 51, 54, 110, 192, 216, 275, 511
    Cambridge, 358
    Chicago, 12
    Dacca, 45
    Edinburgh, 43
    Glasgow, 216
    Johns Hopkins (Baltimore), 19, 208, 257, 275
    London, 5, 42, 256
    Madras, 5, 46, 57, 507, 512
    Manchester, 357
    Mysore, 263, 453, 516
    Nagpur, 451
    Osmania (Hyderabad), 265
    Princeton, 358
    Punjab (Lahore), 265
    Purdue, 515
    St Andrews, 265, 275
    Stanford, 511
Upanishads, 265

Vande Mataram, 505
vibrating string
    bowed, 81, 82, 83, 85, 86, 88, 121–123
    displacement curve, 85–87
    plucked, 117–118
    Raman's analysis of, 119–120, 122–123
    Raman's studies of resonances, 71–74
    Raman's study of velocity waves, 82–84
    resonances of, 71–74

struck, 118–121
theory of, 116–123
velocity profile, 85–88
velocity/Raman waves, 82, 83, 101
vibrations
  amplitude, definition of, 63
  dynamical theory, 80
  frequency, definition of, 63
  instabilities, Raman's studies of, 77
  kinematical theory, 80, 85, 88, 90
  maintenance of, 64, 67–87
  period, definition of, 63
  phase, definition of, 63
  Raman's studies of, 61
Viceroy, 32, 257, 258, 259, 262, 264, 265, 271, 275, 280, 505
Viceroy's Council, 5, 37, 505
Vikings, 516
violin, 80–99
  bowing pressure, 95, 96
  parts of, 81
  photographs of vibrations, 90–93
  physics of, Raman's contribution, 81, 84–98
  pitch of, 97
  principal mode of, 80, 84, 85
  Raman curve (frequency response), 81, 96–98
  Raman's monograph on, 36–37, 90
  Raman's player, 36, 94, 509
  third mode of, 88
  wolf-note, 88–90

wavefront, 128, 129, 132, 150, 330
  corrugated, 330, 367, 368
Western education in India, 3, 4, 5, 6
Western science in India, 3
  monographs and journals, 3, 4
whispering gallery, 51, 126–127, 506
  Raman's experiments, 127
Wood's Despatch, 5

X-ray diffraction, 178–182, 345
  in diamond, 348
  diffuse spots, 361–362
  experiments at IACS, 181–182
  Raman–Ramanathan theory, 178–180, 182
  Zernike–Prins theory, 182
X-ray topography, 345, 513

Zeeman effect, 236
Zeitschrift für Physik, 45
Zoological Survey of India, 259, 287